CHEMISTRY OF CLAYS
AND CLAY MINERALS

MINERALOGICAL SOCIETY
MONOGRAPH NO. 6

CHEMISTRY OF CLAYS AND CLAY MINERALS

EDITED BY
A. C. D. NEWMAN

Mineralogical Society
1987

Longman Scientific & Technical,
Longman Group UK Limited,
Longman House, Burnt Mill, Harlow,
Essex CM20 2JE, England
and Associated companies throughout the world.

Mineralogical Society
41 Queen's Gate
London SW7 5HR
England

*Published in the United States of America
by Wiley-Interscience, a Division of John Wiley & Sons, Inc., New York*

© Mineralogical Society 1987

All rights reserved; no part of this publication may be reproduced, stored in a retrieval system, or transmitted in any form or by any means, electronic, mechanical, photocopying, recording, or otherwise, without the prior written permission of the Publishers, or a license permitting restricting copying in the United Kingdom issued by the Copyright Licensing Agency Ltd, 33-34 Alfred Place, London, WC1E 7DP.

Published in collaboration with the Mineralogical Society

*First published 1987
Reprinted 1990*

British Library Cataloguing in Publication Data
Chemistry of clays and clay materials. —
  (Mineralogical Society monograph, ISSN 0144-1485;
  no. 6)
  1. Clay minerals   2. Clay   3. Mineralogical
  chemistry
  I. Newman, A. C. D.   II. Mineralogical Society
  III. Series
  549'.67   QE389.625

ISBN 0-582-30114-9

Set in Monophoto Times
Produced by Longman Singapore Publishers Pte Ltd
Printed in Singapore

Preface

The monograph series published by the Mineralogical Society has until now been concerned mainly with the application of techniques like X-ray diffraction, thermal analysis, infrared spectroscopy and electron optical methods to the identification and study of clay minerals. Inevitably much information about the chemical constitution was contained within these volumes and was essential to them. However, it seemed that there was also a need for a monograph that contained all this information in one volume and also covered some wider aspects of clay chemistry such as their colloid behaviour and surface chemistry, their reactions with organic substances and to heating, and the chemical conditions necessary for their formation.

These were the views of the Committee of the Clay Minerals Group of the Mineralogical Society that initiated over fifteen years ago the production of a monograph on the chemistry of clays. It is perhaps unfortunate that owing to a succession of unforeseen circumstances, its publication has been delayed until now. Sadly two of the original contributors have died in the intervening period: W. T. Granquist in 1974, and the death of G. W. Brindley, for long associated with clay minerals and with the monograph series, in 1983, is still fresh in our memories. Despite these regrets about the long gestation of this monograph, a bonus is that many advances in knowledge from recent studies have been included now that were missing from earlier drafts, and it is hoped that in its present form, the monograph will be a valid source of information about clays for many years to come.

Acknowledgements

First, I most heartily thank the contributors for their cooperation and support in what has at times been a slow and frustrating venture and for their forebearance during the times when little progress was evident. Next, I thank Sir Leslie Fowden, Director of Rothamsted Experimental Station, and Dr P. B. Tinker, Head of Soils and Plant Nutrition Department, for permission to work on the monograph during my time at Rothamsted. I am most grateful to Dr A. W. Flegmann of Bath University for his editing work on three of the chapters which relieved my burden during the earlier stages of the preparation. Particular tribute should be paid to Mrs Barbara Moseley for very accurately transferring manuscripts and typescripts of the first five chapters to the word processor. Many other people have contributed to the preparation of the monograph and I thank them for their help. Finally I should like to record the thanks of the officers of the Clay Minerals Group and of the Mineralogical Society for the ready cooperation at all times of the Longman staff.

A. C. D. NEWMAN

We are indebted to the following for permission to reproduce copyright material:

Academic Press Inc. and the Authors, K. Norrish, Dr G. A. Mills for Fig. 6.4 (A. G. Oblad et al. 1951), Table 2.2a (after K. Norrish, 1975); Allen Press Inc. and the Author, S. P. Quirk for Table 8.2 (D. J. Greenland and J. P. Quirk, 1964); American Association for the Advancement of Science for Fig. 7.3 (G. W. Brindley and G. L. Millhollen, 1966); American Association for the Advancement of Science and the Author, Shirley Turner for Fig. 2.12 (S. Turner and P. R. Buseck, 1981); The American Ceramic Society Inc. for Figs. 7.1a (G. W. Brindley et al. 1967a), 7.6-7 (after G. W. Brindley and M. Nakahira, 1959); American Chemical Society for Tables 6.1-2 (H. A. Benesi, 1956); California Division of Mines and Geology for Table 6.5 (T. H. Milliken et al. 1955); the Author, Dra. Blanca Casal for Fig. 8.6 (B. Casal, 1983); The Clay Minerals Society (USA) for Fig. 7.5 (H. Negata et al. 1974), Table 7.2 (after G. W. Brindley and G. Ertem, 1971); The Clay Minerals Society (USA) and the Author, Dr A. Tsunashima for Fig. 8.1 (G. W. Brindley and A. Tsunashima, 1972); Elsevier Science Publishers BV for Figs. 8.15-16 (J. C. Serna and G. E. Van Scoyoc, 1979); Keter Publishing House for Figs. 8.8-9 (Weiss et al. 1969); the Author, Dr D. G. Lewis for Fig. 8.2 from Fig. 1 (E. E. MacIntosh et al. 1971); Mineralogical Society of America for Figs. 7.1b, 7.2a and 7.2b (after G. W. Brindley et al. 1967a), 7.2c and 7.2d (after G. W. Brindley et al. 1967b); Mineralogical Society of America and the Author, Dr J. A. Rausell-Colom for Figs. 8.11-12 from Figs. 4-5 (J. A. Rausell-Colom and V. Fornes, 1974); Sociedade Geologica de Portugal and the Author, Dr C. de S. Figueriedo Gomes for Fig. 8.10 from Fig. 5 (C. de S. Figueriedo Gomes, 1982); the Editor, *Proceedings of the International Clay Conference 1972* for Table 8.5 from Table 2 (*Proc. Int. Clay Conf. 1972, Madrid*); Societe Francaise Mineralogie for Fig. 7.8 (after J. Lemaitre and P. Gerard, 1981b); Societe Francaise de Microscopie Electronique for Fig. 8.14 from Fig. 6 (M. Rautureau, 1974); Soil Science Society of America for Figs. 2.2, 2.3 (R. G. Gast, 1977); the Author, W. Schellmann for extracts and Table 2.2b (W. Schellmann, 1959, 1976); Verlag Chemie GmbH for Figs. 6.9 (A. Weiss, 1981), 8.3 (A. Weiss, 1963), 8.4 (G. Lagaly, 1976); the Author, S. A. Wentworth for Fig. 7.4 (S. A. Wentworth, 1967); John Wiley and Sons Inc. for Figs. 3.1a,b,c, 3.2, 3.3, 3.4a,b, 3.5a,b, 3.6 (H. Van Olphen, 1977), Table 6.4 adapted (D. H. Solomon and M. J. Rossner, 1965).

Contents

Chapter		Page
1.	The Chemical Constitution of Clays A. C. D. NEWMAN and G. BROWN	1
2.	Non-Silicate Oxides and hydroxides R. M. TAYLOR	129
3.	Dispersion and Flocculation H. VAN OLPHEN	203
4.	Cation Exchange Equilibria in Clays H. LAUDELOUT	225
5.	The Interaction of Water with Clay Mineral Surfaces A. C. D. NEWMAN	237
6.	Catalytic Properties of Clay Minerals J. P. RUPERT, W. T. GRANQUIST and T. J. PINNAVAIA	275
7.	Thermal, Oxidation and Reduction Reactions of Clay Minerals G. W. BRINDLEY and J. LEMAITRE	319
8.	Reactions of Clays with Organic Substances J. A. RAUSELL-COLOM and J. M. SERRATOSA	371
9.	Petrologic Phase Equilibria in Natural Clay Systems B. VELDE and A. MEUNIER	423
Subject Index		459
Author Index		469

Contributors

G. W. BRINDLEY, deceased.

G. BROWN, Soils and Plant Nutrition Department, Rothamsted Experimental Station, Harpenden, Herts. AL5 2JQ, UK.

W. F. GRANQUIST, deceased.

H. LAUDELOUT, Department des Sciences du Sol, Faculté des Sciences Agronomiques, Place Croix du Sud, 2, 1348 Louvain-la-Neuve, Belgium.

J. LEMAITRE, Université Catholique de Louvain, Groupe de Physico-Chimie Minerale et de Catalyse, Place Croix du Sud, 1, 1348 Louvain-la-Neuve, Belgium.

A. MEUNIER, Laboratoire de Pétrologie des Altérations Hydrothermales, Faculté des Sciences, 40 Avenue du Recteur Pineau, 86022 Poitiers Cedex, France.

A. C. D. NEWMAN, Modney Cottage, Hilgay, Downham Market, Norfolk, UK, formerly at Rothamsted Experimental Station, Harpenden, Herts. AL5 2JQ, UK.

T. J. PINNAVAIA, Department of Chemistry, Michigan State University, East Lansing, Michigan 48824, USA.

J. A. RAUSELL-COLOM, Instituto de Físico-Química Mineral, Consejo Superior de Investigaciones Cientificas, Serrano, 115-dpdo, Madrid 28006, Spain.

J. P. RUPERT, Director, International Operations, NL Treating Chemicals, NL Industries ILC, PO Box 60020, Houston, TX 77205, USA.

J. M. SERRATOSA, Instituto de Físico-Química Mineral, Consejo Superior de Investigaciones Cientificas, Serrano, 115-dpdo, Madrid 28006, Spain.

R. M. TAYLOR, CSIRO Division of Soils, Private Bag No. 2, Glen Osmond, South Australia 5064.

H. VAN OLPHEN, Dalweg 114, 6865 CW Doorwerth, The Netherlands.

B. VELDE, Ecole Normale Superieure, Laboratoire de Géologie, 46 rue d'Ulm, 75230 Paris, Cedex 05, France.

Chapter 1

The Chemical Constitution of Clays

A. C. D. NEWMAN and G. BROWN

		Page
1.1	INTRODUCTION	2
1.2	ELEMENTAL ANALYSIS OF CLAYS	3
	1.2.1 Chemical analysis schemes	3
	1.2.1.1 Determination of Fe^{2+}	4
	1.2.1.2 Determination of hydrogen	4
	1.2.1.3 Determination of fluorine	5
	1.2.2 Instrumental methods of analysis	6
	1.2.2.1 Arc-spark spectroscopy	6
	1.2.2.2 Flame spectrophometry	6
	1.2.2.3 Non-flame atom cells	7
	1.2.2.4 Plasma excitation	7
	1.2.2.5 X-ray emission spectrometry	8
	1.2.3 Matrix preparation	8
	1.2.4 Purity of samples	9
	1.2.4.1 Size fractionation	9
	1.2.4.2 Magnetic separations	9
	1.2.4.3 Density separations	10
	1.2.4.4 Selective dissolution treatments	10
1.3	STRUCTURAL FORMULA CALCULATIONS	10
	1.3.1 The major clay mineral groups and subgroups	11
	1.3.2 Calculation of unit cell contents and structural formulae	12
	1.3.2.1 Calculation when unit cell dimensions, density and chemical composition are determined	12
	1.3.2.2 Calculation when full chemical composition is known, but unit cell and density are unknown	15
	1.3.2.3 Calculation when chemical composition is known except for structural water content	17
	1.3.2.4 Other methods of calculating content of structural unit	17
	1.3.2.5 The structural formula	19
	1.3.3 Anomalous structural formulae	21
1.4	1:1 GROUP MINERALS	22
	1.4.1 The kaolin group	22
	1.4.1.1 Halloysites	26
	1.4.2 Minerals of the serpentine subgroup	26
	1.4.3 Serpentine subgroup minerals with planar layers	29
	1.4.3.1 Lizardites	29
	1.4.3.2 Nepouites and Ni-lizardites	29
	1.4.3.3 Berthierine	31
	1.4.3.4 Amesite	31
	1.4.3.5 Cronstedtite	31
	1.4.4 Serpentine subgroup minerals with non-planar layers	32
	1.4.4.1 Chrysotile	32
	1.4.4.2 Antigorite	33
	1.4.4.3 Greenalite	34
	1.4.4.4 Caryopilite	37
	1.4.4.5 Tosalite	38
	1.4.5 Relations between serpentine-group and chlorite-group minerals	39
	1.4.6 Chemical differences among serpentine-group minerals	40
1.5	THE TALC–PYROPHYLLITE GROUP	40
	1.5.1 Pyrophyllite	41
	1.5.1.1 Ferripyrophyllite	42
	1.5.2 Talc subgroup	44
	1.5.2.1 Kerolite–pimelite series	47
	1.5.2.2 The 10 Å talc hydrate of Bauer and Sclar (1981)	48
1.6	2:1 GROUP WITH INTERLAYER CATIONS, $0 < X < 1.9$	48
	1.6.1 Dioctahedral smectites	49
	1.6.1.1 Aluminian dioctahedral smectites	49
	1.6.1.2 Iron-rich smectites and nontronites	53
	1.6.2 Trioctahedral smectites	54
	1.6.3 Smectites containing uncommon elements	55
	1.6.3.1 Volkhonskoite	55
	1.6.3.2 Medmontite	55
	1.6.3.3 Sauconite	55
	1.6.3.4 Nickel-smectites	55
	1.6.3.5 Vanadium-smectite	55
	1.6.4 Vermiculites	56
	1.6.5 Charge heterogeneity in swelling 2:1 layer silicates	59
	1.6.6 Chemical composition and interlamellar swelling	61
1.7	2:1 GROUP, $X > 1.6$: THE MICA MINERALS	63
	1.7.1 Dioctahedral micas $X_2Y_4Z_8O_{20}(OH)_4$: General considerations	64
	1.7.1.1 Muscovite–phengite micas	67

			Page				Page
		1.7.1.2 Paragonite	69			1.9.2.1 Aliettite	98
		1.7.1.3 Margarite	69		1.9.3	Interstratification of smectite and mica units	99
	1.7.2	Clay micas	69			1.9.3.1 Rectorite	99
		1.7.2.1 Hydromuscovites	70			1.9.3.2 "Tarasovite"	100
		1.7.2.2 Illites	71			1.9.3.3 Partially-ordered and randomly-interstratified mica-smectites	100
		1.7.2.3 Celadonite–glauconite micas	72				
	1.7.3	Trioctahedral micas $X_2Y_{5-6}Z_8O_{20}(OH,F)_4$	75		1.9.4	Interstratification of smectite or vermiculite with chlorite	101
		1.7.3.1 Trioctahedral sodium micas	79			1.9.4.1 Tosudite	104
						1.9.4.2 Corrensite	104
	1.7.4	Lithium micas	81			1.9.4.3 Less-regular interstratification of chlorite with smectite or vermiculite	104
	1.7.5	Micas with unusual compositions	81				
1.8	CHLORITES		84				
	1.8.1	Trioctahedral chlorites	84		1.9.5	Interstratification of smectite with kaolinite	105
	1.8.2	Chlorites containing dioctahedral sheets	85		1.9.6	Interstratification of mica with vermiculite	106
	1.8.3	Chlorites with imperfect or incomplete interlayer hydroxide sheets	89	1.10	PALYGORSKITE AND SEPIOLITE		107
					1.10.1	Palygorskite	109
1.9	INTERSTRATIFIED CLAY MINERALS		92		1.10.2	Sepiolite	112
				1.11	ALLOPHANE AND IMOGOLITE		114
	1.9.1	Minerals with non-expanding layers	98		1.11.1	Disordered allophanes	114
		1.9.1.1 Kulkeite	98		1.11.2	Imogolite	115
	1.9.2	Interstratification of smectite with talc-type units	98	REFERENCES			116

1.1 INTRODUCTION

Chemical analysis was an early and essential step in establishing the nature of minerals but the chemical composition of silicates is so complex that a satisfactory systematic classification was not formulated until the spacial arrangement of the atoms was revealed by X-ray diffraction studies. This application of X-ray diffraction led to an enormous increase in our knowledge of the nature of clay minerals and has helped us to understand how important chemical composition is in determining their structure and properties. As a consequence, establishing the chemical constitution of a clay mineral remains as important today as ever it was in its mineralogical description. We now know that relatively small differences in the chemical composition of clays can greatly influence their chemical and physical properties and measuring these differences may be quite demanding on analytical techniques.

In this chapter, as an introduction to the monograph, we survey the chemical constitution of the various types of layer silicates and clay minerals in particular. The non-silicate clays, the oxides of Fe, Al, Mn and Ti are reviewed in Chapter 2.

Clay minerals are essentially hydrous aluminosilicates but magnesium and iron substitute aluminium in varying degrees, and alkali and alkaline earth elements can also be essential constituents. Hydrogen is usually present as hydroxyl in the structure and as water both within the structure and sorbed on the surface. These substitutions cause wide diversity in chemical composition within the broad general class of phyllosilicates or layer silicates to which the clay minerals belong.

Within this class, phyllosilicate minerals are subdivided, by international agreement (Bailey 1980a), into a number of groups and subgroups according to their structure and chemical constitution. Structural details of the phyllosilicates are described by Bailey (1980b), but here we are concerned with expressing the chemical composition of the clay minerals on a structural basis, and the simplest way to do this is in terms of their structural formulae. Therefore, after a brief summary of methods used to analyse clays and to calculate structural formulae from the analysis, we describe the variations in the chemical constitution of the clay minerals by their groups and subgroups. However, it is becoming increasingly evident that a number of clay minerals do not have exactly the ideal anion framework listed, and several instances of this (e.g. antigorite, Section 1.14.2) will be noted.

1.2 ELEMENTAL ANALYSIS OF CLAYS

In general the techniques used in rock and silicate analysis may be used with only minor modifications for clays; indeed, because of the small particle size, the preparation of a matrix for analysis (solution, glass, pressed pellet, etc.) may be easier with clays than with coarser grained samples. However when the aim of the chemical analysis is to calculate a structural formula and to determine the distribution of cations in several structural sites, good accuracy is required and only certain analytical methods are capable of the standard necessary.

The chemical elements in the matrix can be determined by chemical analysis, by analysis based on the physical properties of the elements, or by judicious combination of both. Gravimetric and volumetric chemical determinations were the basis of the "classical" schemes (Washington 1918, Hillebrand 1919) of analysis, but with the development of commercial instruments capable of precise measurements, physical methods now often replace the traditional chemical ones. Although it is probably misleading to quote "best" methods, each element has a chemical and physical individuality that can be used to advantage in analysis. For example, the older chemical methods for determining the alkali metals are now superseded by flame emission or absorption methods, but only the heavier metals in the group, K, Rb and Cs, are advantageously determined by X-ray spectroscopy. The final choice of the method to be used depends upon the concentration of element in the sample, the accuracy required, the instruments and time available, and to some extent the individual ability of the analyst, although to obtain accurate analyses, much skill and care is necessary no matter which methods are used.

It is not possible within the space of the present chapter to give detailed analytical procedures; moreover the techniques of classical chemical analysis are described in many texts, and recent publications are available for the modern instrumental methods. The book by Maxwell (1968) is comprehensive; that by Jeffery (1970) has a bias towards colorimetric methods and describes procedures for many minor constituents as well as the major elements, whereas Volborth (1969) emphasizes the value of traditional methods when used with modern methods of separation, and also includes a useful introduction to neutron activation analysis, currently the most satisfactory method of determining oxygen, the neglected element in silicate analysis.

1.2.1 Chemical analysis schemes

In most traditional silicate analysis, the aim was to account for the total composition of the sample so that when each elemental determination was expressed as the percentage of oxide present, the sum totalled close to 100%. It was generally found that to achieve this total it was necessary at the very least to make the following determinations: Si, Al, Fe^{3+}, Fe^{2+}, Ti, Mn, Mg, Ca, Na, K, P and water evolved below 105 °C (H_2O^-) and between 105 and 1000 °C (H_2O^+).

For some minerals, additional determinations (e.g. of F, Li and less common elements) were needed to complete the total satisfactorily.

In the *classical analysis scheme*, most of the determinations were done on solutions prepared after fusing the sample with Na_2CO_3, using predominantly gravimetric and volumetric procedures. Further samples were required for determining F, H_2O, alkali metals and Fe^{2+}. The methods were accurate when done by experienced analysts but were slow, so that the introduction of *rapid analysis schemes*, based mainly on colorimetric determinations, was widely adopted in the 1960s. At the same time, however, instrumental methods of analysis were developing rapidly and a survey of clay mineral literature for the period 1982–1984 revealed that few analyses are now done by chemical methods. The determination of F, H_2O^+ and H_2O^-, and Fe^{2+}, however, is still done chemically, unless the instrumental methods for them are available. The following sections outline the main features of these determinations.

1.2.1.1 Determination of Fe^{2+}
Schafer (1966) and Hey (1982) have reviewed the many methods proposed for determining Fe^{2+} in silicates. They are grouped according to the initial decomposition stage: (a) acid decomposition in a sealed tube; (b) acid decomposition in an inert atmosphere; (c) acid decomposition in the presence of an oxidant; (d) fusion in a sealed tube. Method (d) is useful for resistant minerals; method (c) is attractive in principle but in practice some additional decomposition of the oxidant usually seems to occur during the mineral dissolution, causing large blanks. The method most widely used is decomposition with acid, usually hydrofluoric acid, air being excluded by boiling the reactants in a closed vessel or by displacement with nitrogen or carbon dioxide. Aerial oxidation of Fe^{2+} is accelerated in the presence of fluoride, so that it is usual to complex fluoride immediately after the decomposition by adding boric acid. Dichromate solution is then added and the excess reagent titrated with standard Fe^{2+} solution using diphenylamine sulphonate indicator. References for this and other methods of determining Fe^{2+} in the digest solutions are given in the articles cited above.

With the increasing availability of the instrument, Fe^{2+}/Fe^{3+} ratios are sometimes determined by Mossbauer spectrometry (Goodman 1981), and this technique also offers the possibility of determining whether Fe is present in octahedral or tetrahedral sites.

1.2.1.2 Determination of hydrogen
Hydrogen is present in clays as water and as structural hydroxyl groups, and also combined with organic compounds and ammonium cations if these are present. Molecular hydrogen may be evolved when minerals are heated, particularly in a non-oxidizing environment, but this is formed by a reaction between ferrous iron and structural hydroxyl; the percentage of hydrogen gas in clay minerals is normally very small. The major problem in determining hydrogen in clay minerals is in distinguishing between the several categories of hydrogen that make up the total.

Total hydrogen is determined by heating the mineral in an oxidizing atmosphere and weighing the water evolved. Modifications of the Penfield method (Shapiro and Brannock 1955; Wilson 1962) are quite often used and are quick, but the relatively low temperatures attainable in a gas flame may be insufficient to ensure complete dehydroxylation; the results from these methods can at best be regarded as approximate. More accurate is the closed-circuit circulation system devised by Jeffery and Wilson (1960) in which water collection is more efficient and closer temperature control is possible. Temperatures above 900 °C are required to drive off all structural hydrogen from magnesium chlorites, serpentines and vermiculites; phlogopites may be heated at 1000 °C for several hours with very little decomposition. At these high temperatures other constituents such as fluorine compounds are volatile and retained in the absorption tube (El-Attar et al. 1972), so that in the determination of total water in refractory minerals there can be a balance between positive and negative errors. Sodium tungstate is often recommended as a flux to facilitate decomposition but as the evolution of non-aqueous volatiles is also accelerated, "retainers" must be included as well: Groves (1937) suggests covering the sample with freshly ignited quicklime to absorb fluorine. Because of the relative volatility of lead fluoride, the use of litharge in the flux seems undesirable.

If the hydrogen determination is required for recalculating the chemical composition of a mineral as a structural formula, the surface-adsorbed and hydration water must be determined separately so that the structural hydrogen may be found by difference from the total hydrogen. In the traditional analysis scheme, the hygroscopic moisture is frequently determined by oven drying in a weighing bottle at 105 or 110 °C; the weight loss after cooling is termed $H_2O - 105$ or H_2O^-, and the difference between this weight and the total water, H_2O^+.

With clay minerals this procedure is unsatisfactory for two reasons. First, many clays are exceedingly hygroscopic after drying and resorb water so quickly that an accurate dried weight cannot be obtained in this way; one of the gas circulation methods proposed by Jeffery and Wilson (1960) should be used instead. A very convenient method of determining hygroscopic and structural water on the same sample, termed "thermohygrometry" by Creer et al. (1971), measures evolved water continuously as a function of temperature.

The second more fundamental objection to any procedure is that the evolution of hydration water from many clay minerals is far from complete at 110 °C. For example, Fripiat et al. (1960) showed that water was still present in montmorillonites and vermiculites at 300 °C, and a vermiculite derived from a phlogopite containing 6% F retained molecular water up to nearly 600 °C (Newman 1967). In these investigations, the infrared adsorption in the region 1650 cm^{-1}, attributed to the angular deformation vibration of molecular water, was used as an indicator of water present; structural hydrogen in the form of hydroxyl does not absorb in this spectral region. Therefore infrared absorption can be used to determine the minimum temperature necessary to release all molecular water, and water evolved in heating beyond this temperature, measured by adsorption in a desiccant tube, is structural hydrogen. With some minerals, however, the dehydration and dehydroxylation temperature ranges overlap, and for these a satisfactory separation of hydration water and structural hydrogen cannot at present be achieved by thermal analysis.

1.3.1.3 Determination of fluorine

Most methods of determining fluorine are subject to interference from elements common in silicates, so that after decomposition by fusion, a separation of fluorine from these constituents is necessary. Often this was done by distillation from concentrated acid at 140–160 °C (Willard and Winter 1933); accurate control of the conditions is essential (Ingamells 1962) but recovery of fluorine is sometimes incomplete (Peck and Smith 1964; Evans and Sergeant 1967) and the analysis is time consuming. Hydrolysis at 600–800 °C by moist carrier gas in the presence of a catalytic flux ("pyrohydrolysis"; Nardozzi and Lewis 1961) is an attractive alternative because preliminary fusion is unnecessary and the separation is complete in 15–30 minutes (Newman 1968; Jeffery 1970; Clements et al. 1971). For best decomposition of the silicate and evolution of HF, a mixed flux of vanadium pentoxide, bismuth trioxide and tungstic oxide has been recommended. The hydrolysate is collected in dilute alkali and fluoride determined by titration or by a colorimetric method; these are described in the references quoted.

Methods have been described for determining fluorine without distillation (Huang and Johns 1967; Ingram 1970); as these require a preliminary fusion they appear to offer little advantage over pyrohydrolysis. Ingram determined fluorine electrometrically with the fluoride-selective Eu^{2+}-lanthanum fluoride electrode (Lingane 1967); the excellence of this electrode suggests that it could be used to best advantage in determining F in the condensate from pyrohydrolysis.

1.2.2 Instrumental methods of analysis

Measurement of the electromagnetic radiation absorbed or emitted when atoms or molecules undergo a transition from one energy state to another forms the basis of spectroscopic methods of elemental analysis. Transitions in the valence shell electrons involve light in the visible and ultraviolet regions of the electromagnetic spectrum, whereas radiation of X-ray frequencies is absorbed or emitted during transitions between energy states of the inner shell electrons. Spectroscopic methods are grouped according to the spectral region studied and the method of excitation, and also by whether the absorption or emission of radiation is measured.

1.2.2.1 Arc-spark spectroscopy
In arc excitation the elements in a solid matrix are vaporized and excited by the heating effect of the discharge; the lines emitted arise principally from neutral atoms. By contrast, an a.c. spark excites predominantly ionic spectra; because of its lesser heating effect the direct analysis of solutions is possible. Complex spectra are excited by both methods so that a monochromator with medium to large dispersion is required to isolate spectral lines for measurement. Modern instruments have multichannel direct-reading spectrometers, which measure the intensity of lines electronically, using an internal standard to compensate for source fluctuations.

1.2.2.2 Flame spectrophotometry
The sample, in solution form, is nebulized into the oxidant gas stream of a flame and because of the relatively low temperatures, the atomic and molecular spectra excited are comparatively simple so that less sophisticated dispersive devices are required to isolate the spectra lines for measurement. The methods are versatile and modern instruments have excellent precision. Nitrous-oxide-supported flames greatly extend the range of analyses possible and most elements of the periodic table can now be determined by flame methods; there remain practical difficulties in the determination of H, C, N, O and F, sensitivity for a few elements is relatively poor (e.g. Ce, Th, U), and although flame methods are available for Ce, Br, I, P and S, these have not been applied in silicate analysis. There remain sixty or more elements for which methods are readily available and comprehensive analysis schemes have been devised to take advantage of this versatility by combining a single decomposition of the sample with flame spectrophotometric analysis of the solution produced (Medlin *et al.* 1969; Boar and Ingram 1970).

Two main types of measurement are made in flame spectrophotometry. In *atomic emission* methods, the analyte atoms are excited by collision with flame gas molecules and a portion of the radiation emitted when excited atoms are deactivated is measured; the intensity of radiation is proportional to the concentration of excited atoms, the number of which is a small fraction (usually much less than one-hundredth) of the total number of atoms in the flame. Band emission from molecular species in the flame may also be used analytically, for instance, in the methods for the non-metals mentioned above.

In *atomic absorption*, the analyte atoms are excited by an external light source and the fraction of radiation absorbed is measured. In practice this measurement is best made with radiation from a sharp line source containing the element being determined; the incident radiation is modulated and the measurement system ideally detects only modulated radiation, rejecting unmodulated direct emission from the flame. In this way, the inherently narrow line width of atomic lines is used to enhance the effective resolution of the spectrophotometer and to eliminate spectral interferences (for example, the measurement of small amounts of Mg at 2852.1 Å in the presence of Na, emitting at 2852.8 Å). Under ideal conditions, the absorbance ($=\log I_0/I$) is proportional to the concentration of atoms in the flame; in practice, calibration lines are sometimes curved. Its chief disadvantage is that normally only one element is measured at a time.

A full discussion of the relative merits of the methods is beyond the scope of this chapter. Winefordner *et al.* (1970) have critically reviewed the analytical possibility of flame techniques and list detection limits for 68 elements. Dean (1960) discusses the general principles of flame emission methods, though some of the applications he lists are now superseded. Texts by Slavin (1968), Ramirez-Munoz (1968) and Rubeska and Moldan (1969) deal specifically with atomic absorption; Dean and Rains (1969) consider fundamental topics of general importance in all three methods and the text edited by Mavrodineanu (1970) contains articles of particular value for the practical use of all these techniques.

1.2.2.3 Non-flame atom cells

Atomization by injection of aerosol into a flame is so inefficient (1–15% for the premix burners necessary for precise measurements by atomic absorption) that very large improvements in sensitivity are possible with more efficient atomization. Vapour phase analysis by atomic absorption will determine 10^{-8} to 10^{-9} g Hg; this technique is vital in pollution studies but unlikely to be needed in silicate analysis. More generally useful to the silicate analyst is the production of atoms by heating solid or dried solutions in a graphite tube or on a graphite rod to temperatures in the range 2000–3000 °C. Advantages are the very small sample required and limits of detection 10^{-3} lower than are possible in flame cells. Disadvantages are that many interferences have been reported and good precision is difficult to achieve.

1.2.2.4 Plasma excitation

Plasma sources maintained by inductively coupled radio frequency heating are excellent media for exciting atomic spectra (Greenfield *et al.* 1964). In modern equipment, an aerosol of the solution to be analysed is fed into the centre of an argon plasma and the atomic emission focused into a multichannel spectrometer. This records the integrated intensity of 25 to 40 spectral lines simultaneously; for best accuracy, an internal standard may be included in the analyte solution (Thompson and Walsh 1983). As the plasma generates temperatures in the region 6000–10 000 K, interelement compounds are decomposed and there are no interelement interferences. Spectral lines overlap can however occur owing to the large number of lines excited, and these are allowed for in the computation using a dedicated microcomputer. The system is being used successfully in silicate analysis (Walsh 1980; Walsh and Howie 1980); clearly the multielement analysis facility has much advantage in rapidity, and good precision and accuracy are claimed. In a comparative study, Hannaker *et al.* (1984) found that there was close agreement between ICP and XRF for many elements in a wide concentration range.

1.2.2.5 X-ray emission spectrometry

In X-ray emission spectrometry the X-ray spectra of the elements in the sample to be analysed are excited either by electrons, as in the electron microprobe, or by X-rays in X-ray fluorescence spectrometry. Analysis is possible for elements from fluorine upwards in the periodic table using fluorescent X-rays. The spectra are analysed either by diffraction from large single crystals or in non-dispersive systems by solid state detectors in combination with multichannel analysers.

The electron microprobe is used to determine the elemental composition of very small volumes, a few cubic micrometres, in the surface of sections of rocks and minerals. Long (1977) has given an excellent review of the method and its application to mineralogical problems, and Sweatman and Long (1969) have carefully investigated factors affecting the accuracy of analyses. Weaver (1968) used the electron microprobe to study the distribution of impurities in kaolins. Although the method cannot be used to study compositional variations across small clay mineral flakes, it provides useful information about the chemical composition of impurities.

Analytical electron microscopy (Duncomb 1962; Cooke and Duncomb 1968) provides an even more powerful tool for studies of clay minerals and clay materials. This technique enables the correlation of features observed in the transmission electron microscope having a point to point resolution of less than 10 Å, electron diffraction data for areas less than 1 μm in diameter and chemical analysis from areas 1000 to 2000 Å in diameter. The method requires thin specimens, typically 1000 Å or less for silicates and requires analysed specimens for calibration (Lorimer *et al.* 1973; Cliff and Lorimer 1975; Lorimer and Cliff 1976). In this application the energy-dispersive technique has advantage over wavelength-dispersive systems because intensities of all the elements are measured simultaneously and are therefore independent of probe current.

The method has been used to compare the composition of different kaolinite flakes as small as 0.2 to 0.3 μm in diameter by Jepson and Rowse (1975) (i.e. particles with mass of the order of 10^{-13} g) with a coefficient of variation of about 3% for silicon and aluminium (see Section 1.4.1). Hayashi *et al.* (1978) made semi-quantitative analyses of asbestos fibres and clay minerals using analytical electron microscopy and Tazaki (1982) was able to show that spherical and crinkly forms of halloysite contained more iron than tubular forms. The same techniques enabled Veblen (1983a) to establish the compositional differences between the finely lamellar components of wonesite that had been detected by high resolution electron microscopy.

The most common analysis method using X-ray spectra of the elements is X-ray fluorescence spectrometry. Bulk samples may be analysed in the form of solids, pressed powders or liquids. When standards and unknowns differ little in composition direct comparison of intensities often gives acceptable results. Because matrix composition affects the intensity of the fluorescent X-rays, intensities are not directly proportional to amount when the composition of the standards is not close to that of the unknowns. For accurate analysis of samples with a range of compositions, corrections have to be made that take account of the effect of differences in composition on intensities. Methods for rocks and minerals for both major and minor elements have been reviewed by Norrish and Chappell (1977).

1.2.3 Matrix preparation

In the classical analysis scheme the mineral was fused with Na_2CO_3 and the glass dissolved in water. This obviously prevented the determination of sodium and a separate sample was required for the Lawrence Smith procedure. In the rapid schemes using flame photometric determination of the alkali metals, the silica was volatilized by digestion with HF and a less volatile acid such as $HClO_4$ or HNO_3, and the residue dissolved in dilute acid. Both schemes thus needed two digestions, and furthermore, some minerals resist decomposition by $HF/HClO_4$. To overcome these limitations, new decomposition methods were developed, based either on high-pressure HF decomposition or fusion with lithium metaborate.

Langmyhr and Sveen (1965) studied the decomposition of silicates by HF and concluded that a very wide range of minerals was decomposed if the vessel was pressurized so that the HF could be heated above its boiling point at atmospheric pressure. Decomposition vessels made from PTFE are now commercially available (Bernas 1968; Paus 1971) and analysis schemes have been based on the solutions obtained (Bernas 1968; Langmyhr and Paus 1968, 1969, 1970). Si is retained within the solution if the bomb is cooled before opening, and after addition of boric acid to complex fluoride, the solution may be analysed by colorimetric or flame emission and atomic absorption methods.

A more widely applied decomposition method is fusion with lithium metaborate ($LiBO_2$) followed by acid dissolution. Although first developed as a flux for colorimetric and spectrographic determinations (Ingamells 1966; Suhr and Ingamells 1966), it has been found to be an excellent technique for bringing silicates into solution (Ingamells 1970) prior to atomic absorption analysis (Medlin *et al.* 1969; Yule and Swanson 1969; Van Loon and Parissis 1969; Boar and Ingram 1970) or plasma emission methods (Hannaker *et al.* 1984).

1.2.4 Purity of samples

Clay minerals are seldom, if ever, monomineralic and before attempting to analyse a sample or use it for research, it is advisable to examine it by X-ray diffraction for the presence of other phases. If these are found, or the presence of X-ray amorphous substances is suspected, the clay can be purified by one or a combination of the following techniques.

1.2.4.1 Size fractionation
The aim is to prepare a stable dispersion of the clay sample that can be separated in different fractions by centrifugation. Soluble salts, e.g. gypsum, must first be dissolved by washing with water and the recommended treatment for dissolving $CaCO_3$ is heating with acetic acid buffered to pH 5 by the addition of sodium acetate. This reagent has the added advantage that when Na replaces Ca, Al and other flocculating cations on the clay surface, a stable dispersion is formed when the buffer is removed by washing. It is also recommended that organic matter removed by peroxidation with H_2O_2 is done in the same buffer, otherwise the excessive acidification that sometimes results may alter or even destroy the clay mineral present. Some manufacturers stabilize hydrogen peroxide with phosphoric acid and such reagents should *not* be used to peroxidize clays. It may also be necessary to remove free iron oxides by selective dissolution (see below, Section 1.2.4.4) as these can have a cementing or flocculating action on the clay. The principles and practice of fractional sedimentation are described in detail by Jackson (1956).

1.2.4.2 Magnetic separations
Clays frequently contain relatively small quantities of iron oxides which may adversely affect properties like colour and also make it impossible to calculate a structural formula for the layer silicate mineral fraction. The separation of iron oxides from aluminosilicates in soil clays is possible in a high field gradient magnetic separator (Schulze and Dixon 1979; Hughes 1982) and

such systems are used commercially on a large scale to remove coloured materials from kaolinitic clays (Ianicelli 1976). It is reported (Berry and Jorgensen 1969) that chlorite, which is more paramagnetic than illite, can be separated from other clays electromagnetically.

1.2.4.3 Density separations
The isolation of mineral species by flotation, using density differences, is well established in coarse grained mineral analysis but has only rarely been applied to clays (Halma 1969; Francis et al. 1972). There are several reasons for this: clays are readily flocculated by the concentrated solutions or heavy liquid mixtures needed; surface active agents added to keep clay dispersed are strongly sorbed and probably impossible to remove after treatment (Francis 1973); when it has proved possible to separate clay into bands in density gradient columns, different bands have contained the same minerals, that is, little or no separation was achieved (Francis and Tamura 1972). Because the origin of these effects is still not understood, little has been done to refine density separations further.

1.2.4.4 Selective dissolution treatments
Although not exclusively selective in their action, dissolution methods are very convenient and so are often used to purify clay minerals. Because they may attack the clay mineral itself as well as the impurity component, they must be used with care, but in general provide information not obtained in other ways. Schwertmann (1979) has reviewed the methods and their application to some standard clay samples of American and European origin and this article should be read for details that cannot be given here.

Briefly, soluble salts are dissolved by washing with water, but if in consequence clay disperses in water, and this is to be avoided, a dilute solution of a flocculating cation, e.g. Ca^{2+}, must be used instead. Carbonates are best dissolved in acetic acid–acetate buffer with $pH > 5$; this solution may also dissolve manganese compounds. "Free" iron oxides, that is iron oxides not present in the structure of clay minerals, are dissolved by treatment with Na-dithionite with the addition of Na-citrate as a complexing agent and $NaHCO_3$ to buffer the solution. The extracting solution usually also dissolves some Si and Al, which may originate from the iron oxides but also from limited attack on the clays. Acid ammonium oxalate solutions in the dark dissolve non-crystalline iron oxides well. Mildly alkaline reagents like 5% Na_2CO_3 will dissolve free Si and Al compounds like aluminium hydroxides and allophane but leave layer silicate clays relatively unaffected.

In general, it is important that some estimate is made of the impurities present in clays before any attempt is made to calculate structural formulae. It will be shown later that the presence of impurities in a sample being analysed can make important changes to the apparent structural formula calculated.

1.3 STRUCTURAL FORMULA CALCULATIONS

There are several ways of recalculating the chemical analysis of a mineral so that its chemical constitution is expressed as the numbers of atoms within the crystallographic unit cell or hypothetical unit cell where this may not be known directly. It is also possible in many circumstances to allocate these atoms to particular sites within the unit cell and draw up a structural formula. For minerals available as crystals large enough to allow unit cell dimensions and density to be determined directly, and a complete chemical analysis can be done, the calculation is straightforward and uses a minimum of assumptions. For clays, however, this amount of information is seldom accessible and the calculations are based by analogy on the structures of layer silicates for which the full details have been obtained.

Even if these assumptions are valid, the calculations implicitly assume that the analysis refers to a single homogeneous phase. For clays, these implied assumptions may require careful justification. Not only do clays frequently contain minor impurity components, perhaps sorbed on their surfaces, but they may also be structurally inhomogeneous, for example, an interstratified clay mineral containing two or more differing layer silicate units. Even when the clay is apparently homogeneous to the usual X-ray examination tests, a more detailed study may reveal that there are inhomogeneities of charge density within the bulk material, and this will usually imply an inhomogeneity of chemical composition as well. As experimentation becomes more sophisticated (for example, lattice imaging techniques in the electron microscope), more and more examples of structural inhomogeneity are being discovered in clay minerals. For all of these reasons, it may be advisable to err on the side of caution when interpreting the results of structural formulae calculations.

Nevertheless, when the possible limitations of such calculations are allowed for, the structural formula of a clay is a vital piece of information about it and enables it to be placed in a quite detailed classification scheme for the layer silicates, which is summarized in the next section. The sections following then describe the various ways that structural formulae can be calculated, starting with *ab initio* calculations for quartz and a brittle mica, and then showing how particular hypotheses about the clay composition and structure can also be used in the calculations.

1.3.1 The major clay mineral groups and subgroups

All phyllosilicates contain articulated silicate or aluminosilicate layers in which sheets of tetrahedrally coordinated cations, Z, of composition Z_2O_5, are linked through shared oxygens to sheets of cations, Y, octahedrally coordinated to oxygens and hydroxyls. When one octahedral sheet is linked to one tetrahedral sheet a 1:1 layer is formed as in kaolinite; when one octahedral sheet is linked to two tetrahedral sheets, one on each side, a 2:1 layer is produced as, for example, in talc and pyrophyllite.

Structural units that may be found between aluminosilicate layers are sheets of cations, A, octahedrally coordinated with hydroxyls, as in chlorites, and individual cations, X, which may or may not be hydrated, as in micas, vermiculites and smectites.

In most clay minerals octahedrally coordinated sheets with cations of type Y are found, in which all or almost all the available cations sites are filled mainly by divalent cations, such as Mg and Fe^{2+}, or in which about two-thirds of the available cations sites are occupied mainly by trivalent cations, such as Al and Fe^{3+}. The former form the trioctahedral series of minerals and the latter the dioctahedral series. Each major group is subdivided in this way and Table 1.1 presents the classification and general structural formulae for crystalline clay minerals in terms of the units discussed above.

There are six main groups of minerals distinguished according to structural type. These are:

1. The kaolinite–serpentine group characterized by uncharged 1:1 layers and 7 Å repeat units when they do not contain interlayer water or approximately 10 Å units in varieties with sorbed interlayer water.
2. The pyrophyllite–talc group with uncharged 2:1 layers.

TABLE 1.1. Classification and generalized structural formulae of phyllosilicates

Mineral group	Nature of octahedral sheet(s)	Negative charge per silicate layer[†]	Within silicate layers Octahedral cations
1. Kaolinite	dioctahedral	0	Y_4
Serpentine	trioctahedral	0	Y_6
2. Pyrophyllite	dioctahedral	0	Y_4
Talc	trioctahedral	0	Y_6
3. Micas	dioctahedral	2	Y_4
	trioctahedral	2	Y_6
Brittle micas	dioctahedral	4	Y_4
	trioctahedral	4	Y_6
4. Chlorite	dioctahedral	variable	Y_4
	di,trioctahedral	variable	Y_4
	trioctahedral	variable	Y_6
5. Smectite	dioctahedral	0.5–1.2	Y_4
	trioctahedral	0.5–1.2	Y_6
Vermiculite	dioctahedral	1.2–1.9	Y_4
	trioctahedral	1.2–1.9	Y_6
6. Palygorskite		?	Y_4
Sepiolite		?	Y_8

[†] Negative charge per formula unit is twice that given by Bailey et al. (1980a) because the formula unit used here applies to the contents of a volume defined by the a b unit cell base area that is one layer thick. For example for chlorites this volume is approximately $5 \times 9 \times 14$ Å^3, for palygorskite $18 \times 5 \times 6.5$ Å^3.

3. The micas with 10 Å units.
4. Chlorites with 14 Å units.
5. The smectites and vermiculites with units of variable thickness.

Micas, chlorites, smectites and vermiculites all have 2:1 type layers with interlayer materials that balance a negative charge on the layers; micas have unhydrated cations between the layers, chlorites have interlayer positively charged hydroxide sheets, and smectites and vermiculites have cations whose hydration varies with humidity.

The sixth major group consists of the fibrous palygorskites and sepiolites which consist of narrow ribbons of 2:1 type layers linked by their edges to produce a network of alternating channels occupied by water molecules bounded by aluminosilicate material. For further discussion of the crystal structures of the phyllosilicate minerals, the reader is referred to Bailey (1980a) and Brindley (1980).

1.3.2 Calculation of unit cell contents and structural formulae

1.3.2.1 Calculation when unit cell dimensions, density and chemical composition are determined
If V is the volume of the crystallographic unit cell in Å^3 and D is the density of the mineral in kg dm^{-3} (g cm^{-3}), the total mass U of the unit cell is given by

$$U = NDV \times 10^{-24} \text{ a.m.u.} \qquad (1)$$

where N, Avogadro's constant, is the number of atoms of unit mass (atomic mass units or a.m.u.) in 1 g. Taking $N = 6.022 \times 10^{23}$ mol^{-1} (IUPAC 1979)

Within silicate layers	Anions	Between silicate layers		
Tetrahedral cations		Cations[‡]	Hydroxide sheet cations	OH or H$_2$O
Z_4	$O_{10}(OH)_8$			
Z_4	$O_{10}(OH)_8$			nH$_2$O[§]
Z_8	$O_{20}(OH)_4$			
Z_8	$O_{20}(OH)_4$			
Z_8	$O_{20}(OH)_4$	X_2		
Z_8	$O_{20}(OH)_4$	X_2		
Z_8	$O_{20}(OH)_4$	X_2		
Z_8	$O_{20}(OH)_4$	X_2		
Z_8	$O_{20}(OH)_4$		A_4	$(OH)_{12}$
Z_8	$O_{20}(OH)_4$		A_6	$(OH)_{12}$
Z_8	$O_{20}(OH)_4$		A_6	$(OH)_{12}$
Z_8	$O_{20}(OH)_4$	$X_{0.5-1.2}$		nH$_2$O
Z_8	$O_{20}(OH)_4$	$X_{0.5-1.2}$		nH$_2$O
Z_8	$O_{20}(OH)_4$	$X_{1.2-1.9}$		nH$_2$O
Z_8	$O_{20}(OH)_4$	$X_{1.2-1.9}$		nH$_2$O
Z_8	$O_{20}(OH)_2(OH_2)_4$[‖]	$X_?$		4H$_2$O
Z_{12}	$O_{30}(OH)_4(OH_2)_4$	$X_?$		8H$_2$O

[‡] X represents a monovalent cation except in brittle micas where X is a divalent cation.
[§] For kaolinites $n=0$; for the halloysite (7 Å and 10 Å) minerals n ranges from about 0.6 to about 4.
[‖] Following Bradley (1940); according to Gard and Follett (1968) the anions are $O_{20}(OH)_3(OH_2)_3$.

$$U = 0.6022DV \text{ a.m.u.} \quad (2)$$

The chemical composition is expressed, usually, as $q\%$ oxide (say, QO) for every constituent, that is every 100 a.m.u. of mineral contains q a.m.u. of QO, or q/M atoms of Q, M being the formula weight of QO. Therefore, by simple proportion, the number of Q atoms in the unit cell of mass U is $(0.6022DV \times q)/100M$.

For example, the density of alpha-quartz is 2.65, the unit cell volume is 113 Å^3 so that the unit cell mass $U = 0.6022 \times 113 \times 2.65 = 180.3$ a.m.u. The determined chemical composition might be 46.6% Si for clear colourless quartz, with 0.2% other cations; oxygen, by difference, is therefore 53.2%. The number of Si atoms per unit cell is

$$\frac{180.3 \times 46.6}{100 \times 28.086} = 2.992$$

and the number of oxygen atoms is

$$\frac{180.3 \times 53.2}{100 \times 15.9994} = 5.996$$

The ratio of oxygen to silicon atoms is 5.996/2.992 or 2.004, that is, the formula is very close to SiO$_2$. If, by contrast, the chemical composition had been given as 99.8% SiO$_2$, the stoichiometry of Si:O would have been already assumed and the proportion of Si:O could not have been calculated. In principle, therefore, it is more informative to calculate the number of cations and anions independently, but in practice, unless a direct oxygen analysis is available, the calculation of the oxygen content by difference for a mineral with a complex composition is usually inaccurate. Even though it is elemental composition rather than oxide composition that

14   THE CHEMICAL CONSTITUTION OF CLAYS

TABLE 1.2.   Unit cell contents for the brittle mica, clintonite* (Forman et al. 1967)

1.	2.	3.	4.	5.
		Type I: Formula using oxygen figure by difference		
	$\dfrac{\text{Element weight \%}}{\text{Atomic weight}}$	Atoms (A) (%)	Atoms per unit cell $\dfrac{(A \times U)}{100}$	Charge
Si	9.059/28.086[†]	= 0.322 54	2.711	+ 10.844
Ti	0.348/47.90	= 0.007 26	0.061	+ 0.244
Al	21.006/26.9815	= 0.778 53	6.543	+ 19.629
Fe^{3+}	0.245/55.847	= 0.004 39	0.037	+ 0.111
Fe^{2+}	1.150/55.847	= 0.020 59	0.173	+ 0.346
Ca	9.091/40.08	= 0.226 82	1.906	+ 3.812
Mg	12.659/24.312	= 0.520 69	4.376	+ 8.752
Mn	0.0077/54.938	= 0.000 14	0.001	+ 0.002
Sr	0.118/87.62	= 0.001 35	0.011	+ 0.022
H	0.3402/1.007 97	= 0.337 51	2.836	+ 2.836
F	1.91/18.9984	= 0.100 54	0.845	− 0.845
Cl	0.07/35.453	= 0.001 97	0.017	− 0.017
Total	56.008			
O (by diff)	43.992/15.9994	= 2.749 60	23.109	− 46.218
			Total anions 23.971	Net total −0.482

* Crystal system: monoclinic, space group $C2/m$.
  Cell dimensions: $a = 5.14(1)$, $b = 9.01(0)$, $c = 9.86(4)$ Å, $\beta = 100.04°$, $\sin \beta = 0.984\,605$.
[†] Atomic weights based on carbon = 12.

is determined in silicate analysis (except where the oxides are weighed in gravimetric determinations), it is normal to express the results in terms of oxide composition, assuming the normal stoichiometry of each oxide.

This is illustrated in the following calculations for the brittle mica, clintonite, using data given by Forman et al. (1967), and calculating back the analysis published as oxide composition to elemental composition (Table 1.2). In the Type I calculation, the oxygen content was calculated by difference from the elemental analysis and the numbers of all atoms including oxygen were calculated independently. The Type II calculation assumes stoichiometry for each oxide, so that, for example, the presence of 2.71 Si atoms in the unit cell implies the presence of 2×2.71 oxygen atoms as well.

In *Type I* calculation, the percentage of each element is divided by its atomic weight to give the number of atoms per 100 a.m.u. (Atoms %, column 3), then divided by 100 and multiplied by the unit cell mass, U, to calculate the numbers of each type of atom in the unit cell (column 4). As a check, the charge from each element assuming its normal valency is calculated in column 5, and it can be seen from the sum at the bottom of column 5 that there is an apparent net negative charge on the mineral structure. This is an unreasonable conclusion, and the net charge on mineral structures when *all* the atoms are considered should be zero. This imbalance should almost certainly be attributed to finding the oxygen content by difference, for the discrepancy ($\sim 0.5\%$) is related to the shortfall in the sum of oxides determined in the analysis (column 7). In *Type II* calculation, it is normal to assume stoichiometry of oxides in calculating the oxygen contents of unit cell; allowance must, however, be made for non-oxide anions present (F and Cl in this example, see columns 7 and 9). Each oxide per cent is multiplied by the number of atoms in the oxide formula and divided by the formula weight to give atoms per cent (column 8 for non-

	6.	7.		8.	9.	10.	11.
				Type II: Formula assuming stoichiometry of oxides			
		Oxide weight % / Formula weight		Element atoms (A) (%)	Oxygen atoms (%)	Element atoms per unit cell	Charge
SiO_2		19.38/60.0848[†]	=	0.322 54	0.645 09	2.711	+ 10.844
TiO_2		0.58/79.8988	=	0.007 26	0.014 52	0.061	+ 0.244
Al_2O_3		39.69 × 2/101.9612	=	0.778 53	1.167 78	6.543	+ 19.629
Fe_2O_3		0.35 × 2/159.6922	=	0.004 38	0.006 57	0.037	+ 0.111
FeO		1.48/71.8464	=	0.020 60	0.020 60	0.173	+ 0.346
CaO		12.72/56.0794	=	0.226 82	0.226 82	1.906	+ 3.812
MgO		20.99/40.3114	=	0.520 70	0.520 70	4.376	+ 8.752
MnO		0.01/70.9374	=	0.000 14	0.000 14	0.001	+ 0.002
SrO		0.14/103.6194	=	0.001 35	0.001 35	0.011	+ 0.022
H_2O		3.04 × 2/18.015 34	=	0.337 49	0.168 75	2.836	+ 2.836
F		1.91/18.9984	=	0.100 54		0.845	− 0.845
Cl		0.07/35.453	=	0.001 97		0.017	− 0.017
	Total 100.36				2.772 34		
				less (F + Cl) ≡ O	0.051 26		
less O ≡ F	0.80			Total O	2.721 08	22.869	− 45.738
	99.56					Total anions 23.731	Net − 0.002

* Specific gravity = 3.102 (24.03/4 °C), density = 3.102 g cm^{-3}.
  Unit cell volume, $V = abc \sin \beta = 449.87$ Å^3. Unit cell mass, $U = 840.44$ a.m.u.
‡ This total differs from 0.000 only because of round-off errors.

oxygen atoms and column 9 for oxygen), and atoms per unit cell are found by multiplying by $U/100$ as before (column 10).

Column 11 checks the balance of cation charge against the anion charge; owing to the way the calculation is done from the oxide composition, making allowance for F and Cl substituting for O, the algebraic sum should be exactly zero, any departure being from round-off errors only.

For the worked example in Table 1.2, the net charge balance is almost exactly zero, but for many published structural formulae, this is not so. It must be emphasized that in any calculation of unit cell contents that uses or implies an anion–cation balance, the net charge should be zero, and if it is not, there must be an error in the calculation or in transcribing the results. It is advisable therefore in any calculation of unit cell contents to calculate also the anion–cation charge balance as a check. In the present chapter, whenever possible, the structural formulae have been recalculated by a computer program that includes such a check so that any remaining errors should be attributable only to transcription mistakes.

1.3.2.2 Calculation when full chemical composition is known, but unit cell and density are unknown
As discussed earlier, the density of clay mineral particles is subject to uncertainties that are not totally understood so that it is seldom possible to calculate unit cell contents *ab initio*. Determination of the unit cell volume may also be difficult because of structural imperfections and the small particle size. The ratios of atoms to each other are known from the chemical analysis, as before, but the scaling factor $U/100$ to relate these ratios to the unit cell is unknown because the density cannot be determined.

Clay structural formulae therefore are calculated by making further assumptions that fix the scaling factor. One of the most widely used is to assume anion–cation balance and to balance

16   THE CHEMICAL CONSTITUTION OF CLAYS

the total cation charge against the ideal anion charge for the structural group to which the clay belongs. Phyllosilicate clays adopt only a limited number of structures, and these are represented by formulae in Table 1.1 that include the ideal anion contents and number of cation sites of different kinds in each group. The main structural features are set by the arrangement of the anions, and in the absence of other information, it is usual to assume that the anion framework is complete. Provided the structural type is known, the number of anions per structural unit given in Table 1.1 can be assumed and the scaling factor, and hence the number of cations of different kinds in the structural unit, can be calculated.

Using the example given previously, clintonite is a mica and ideally would contain 24 anions per structural unit (the calculations in Table 1.2 show that the determined anion content is within 1% of this number), which in this mineral are O, F and Cl; the anion content can therefore be written formally as $O_{24-x-y}F_xCl_y$. As before it is assumed that the structure is electrically neutral (Type II calculation), so that the sum of the charges on the cations including hydrogen (number of atoms multiplied by valency) must equal the anion charge. Representing the total cationic charge by M, we have

$$M = \Sigma(f \cdot V_1 \cdot A_1 + f \cdot V_2 \cdot A_2 + \cdots) \quad (3)$$

where f is a scaling factor, $V_1, V_2, \ldots$ are the valencies of the elements and $A_1, A_2, \ldots$ are the numbers of the element atoms per cent (column 3, Table 1.3). The scaling factor f is to be determined by setting the cationic charge equal to the anionic charge, which is equivalent to the condition

$$M = 2(24 - x - y) + x + y$$
$$= 48 - x - y \quad (4)$$

Combining equations (3) and (4) and dividing by the scaling factor,

$$48/f = \Sigma(V_1 A_1 + V_2 A_2 + \cdots) + x/f + y/f \quad (5)$$

where x/f and y/f are the atoms per cent for F and Cl (column 3, Table 1.3).

TABLE 1.3.   Atoms per structural unit for clintonite

	Oxide weight % / Formula weight		Atoms (%)	Equivalents (%)	Atoms per structural unit[†]			
					A	Charge	B	Charge
SiO_2	19.38/60.0848	=	0.322 54	1.290 18	2.742	+ 10.966	2.725	+ 10.902
TiO_2	0.58/79.8988	=	0.007 26	0.029 04	0.062	+ 0.247	0.061	+ 0.245
Al_2O_3	39.69 × 2/101.9612	=	0.778 53	2.335 59	6.617	+ 19.852	6.578	+ 19.735
Fe_2O_3	0.35 × 2/159.6922	=	0.004 38	0.013 15	0.037	+ 0.112	0.037	+ 0.111
FeO	1.48/71.8464	=	0.020 60	0.041 20	0.175	+ 0.350	0.174	+ 0.348
CaO	12.72/56.0794	=	0.226 82	0.453 64	1.928	+ 3.856	1.917	+ 3.833
MgO	20.99/40.3114	=	0.520 70	1.041 39	4.426	+ 8.852	4.400	+ 8.800
MnO	0.01/70.9374	=	0.000 14	0.000 28	0.001	+ 0.002	0.001	+ 0.002
SrO	0.14/103.6194	=	0.001 35	0.002 70	0.011	+ 0.023	0.011	+ 0.023
				Partial sum 5.207 17		+ 44.260		+ 43.999
H_2O	3.04 × 2/18.015 34	=	0.337 49	0.337 49	2.869	− 2.869	2.852	− 2.852
F	1.91/18.9984	=	0.100 54	0.100 54	0.855	− 0.855	0.850	− 0.850
Cl	0.07/35.453	=	0.001 97	0.001 97	0.017	− 0.017	0.017	− 0.017
O					20.259	− 40.518	20.141	− 40.282
				Total sum 5.647 17	24.000	− 44.259	23.86	− 44.001

[†] A obtained using scaling factor $f = 48/5.647\,17 = 8.499\,833$ for 24(O, OH, F, Cl).
  B obtained using scaling factor $f' = 44/5.207\,17 = 8.449\,887$ for 44 equivalents.

If we call the product VA "equivalents per cent", then the right-hand side of equation (5) is the sum of the equivalents per cent for all the elements including H, F and Cl but excluding O, and this sum must be scaled to 48 by the scaling factor f, so that

$$f = 48/\Sigma(V_1 A_1 + V_2 A_2 + \cdots + x/f + y/f) \qquad (6)$$

The calculation takes this simple form when monovalent anions replace oxygen; this appears to be always true for phyllosilicates.

Having calculated f, the unit cell contents are found by multiplying atoms per cent by f. The steps of this calculation are shown for the example of clintonite in Table 1.3. The unit cell contents (column 4) are slightly larger than those obtained by Type II calculation (Table 1.2) because in the latter, the total anion content totalled 23.731, less than the ideal 24. As before a check was made of the anion-cation balance: this is shown in the next column.

1.3.2.3 Calculation when chemical composition is known except for structural water content

The determination of structural water in layer silicate minerals presents several problems of both practice and principle that are discussed earlier (Section 1.2.1.2). To avoid these difficulties clays may be ignited before analysis, or else the chemical composition is determined as usual but the water contents are excluded from further calculations. If structural water (and F, Cl, etc.) is not available for calculating the scaling factor, f, a different method must be used to find it. The commonest method is to assume that the anionic charge is that of the ideal anionic composition, e.g. for clintonite with ideal anion composition $O_{20}(OH,F,Cl)_4$, the anionic charge is -44. Note that the anionic composition in the previous calculation to 24(O,OH,F,Cl) worked out to $O_{20.259}OH_{2.869}F_{0.855}Cl_{0.017}$ whereas calculation to -44 implies that the anion composition is assumed to be $O_{20}OH_{(4-x-y)}F_x Cl_y$. Thus the total cation charge exluding hydrogen must be equal to $+44$ so that the structure is neutral overall. This assumption is equivalent to a "water free" anionic content of 22 oxygen atoms, for if $(4-x-y)OH$ dehydroxylates to $(4-x-y)/2$ H_2O, and $(4-x-y)/2$ oxygen atoms remain, the total "water-free" anion content is $(20 + (4-x-y)/2)$ oxygen, x fluorine and y chlorine atoms, with an anion charge of 44 equivalents or 22 "oxygen" atoms. The scaling factor on this basis, f', is therefore given by

$$44/f' = \Sigma(A_1 V_1 + A_2 V_2 + \cdots)$$

The right-hand side of this equation is the sum of the equivalents per cent of the cations, but excluding H, and this total is divided into 44 to give the scaling factor f'; the calculation for clintonite is given in Table 1.3.

1.3.2.4 Other methods of calculating content of structural unit

An assumed anion total is not the only basis for calculation. We may assume a number for the total of any cation or group of cations in the structural unit and calculate a scaling factor. For example, in calculating the structural formula of oxidized chamosite in which the anion framework was changed by oxidation, Brindley and Youell (1953) assumed that the tetrahedral cation content remained as in the unoxidized chamosite, and used this to calculate the scaling factor. Newman and Brown (1966) used a similar method to compare mica and the vermiculite derived from the mica by cation exchange.

This method can also be used to calculate the composition of a product derived from a clay mineral if there is good reason to believe that the aluminosilicate layers have not been altered by the reaction. For example, in their study of the reaction between Al solutions and

TABLE 1.4. Calculation of atoms per structural unit by setting the number of Si atoms: example of Ba/Al Wyoming montmorillonite

Ba-Wyoming montmorillonite			
	Oxide weight % / Formula weight	Atoms (%)	Equivalents (%)
SiO$_2$	55.2/60.085	0.918 70	3.674 80
Al$_2$O$_3$	19.0/50.981	0.372 69	1.118 07
TiO$_2$	0.0/79.899	0	0
Fe$_2$O$_3$	3.52/79.846	0.044 08	0.132 25
MgO	1.94/40.311	0.048 13	0.096 25
CaO	0.07/56.079	0.001 25	0.002 50
BaO	6.16/153.339	0.040 17	0.080 34
Na$_2$O	0.03/30.989	0.000 97	0.000 97
K$_2$O	0.06/47.102	0.001 27	0.001 27
			5.106 45

Ba/Al-Wyoming montmorillonite			
	Oxide weight % / Formula weight	Atoms (%)	Atoms for $Si = 7.9160$
SiO$_2$	48.5/60.085	0.807 19	7.9160
Al$_2$O$_3$	22.4/50.981	0.439 38	4.3089
TiO$_2$	0.0/79.899	0	0
Fe$_2$O$_3$	3.15/79.846	0.039 45	0.3869
MgO	1.77/40.311	0.043 91	0.4306
CaO	0.0/56.079	0	0
BaO	1.79/153.339	0.011 67	0.1145
Na$_2$O	0.0/30.989	0	0
K$_2$O	0.04/47.102	0.000 85	0.0083
			$f_{Si} = 7.9160/0.807\ 19 = 9.806\ 86$

montmorillonite, Brown and Newman (1973) analysed the original clay and the reaction products after saturating with Ba, with the aim of determining the charge on the interlayer AlOH complex. The structural unit contents of the original Ba-montmorillonite were calculated to balance an anion charge of -44 and the Si content so determined was used to calculate the unit contents for Ba/Al-montmorillonite (Table 1.4: lower section, left-hand side). This showed that although the Fe and Mg contents were very similar to those in the original clay, the Al content had increased due to the reaction and the Ba content decreased. There was a net increase in the total cationic charge from 44 to 46.85 which must have been balanced by additional OH. As the Al content had increased by $(4.309 - 3.211)$Al, the ratio of OH to Al was $(46.85 - 44)/(4.309 - 3.211) = 2.596$. The average formula of the interlayer complex was therefore Al(OH)$_{2.596}^{+0.404}$, which agreed with estimates made by analysing the solutions.

A calculation of the product analysis to a cationic charge of 44 is also given in Table 1.4 (lower part, right-hand side) and a quite different structural unit content is found. The contrast

Atoms per unit	Charge
7.9160	31.6640
3.2113	9.6339
0	0
0.3798	1.1394
0.4147	0.8294
0.0108	0.0216
0.3461	0.6922
0.0083	0.0083
0.0109	0.0109
	43.9997

$f' = 44/5.10645 = 8.61653$

Charge	Equivalents (%)	Atoms per "unit"	Charge
31.6640	3.22877	7.4344	29.7377
12.9267	1.31815	4.0468	12.1404
0	0	0	0
1.1607	0.11835	0.3633	1.0900
0.8612	0.08782	0.4044	0.8088
0	0	0	0
0.2290	0.02335	0.1075	0.2150
0	0	0	0
0.0083	0.00085	0.0078	0.0078
46.8499	4.77729		43.9997

$f' = 44/4.77729 = 9.21024$

between the results of these two calculations has extension to the effect of impurities on structural unit contents, and is discussed further in the next section.

Other methods of obtaining scaling factors could be devised: for example, if it were known that the mineral was in a group with an invariant site occupancy, this might be a more appropriate way of calculating the scaling factor. The principle is, however, the same as that given above.

1.3.2.5 The structural formula

Having calculated the number of atoms of different kinds in the structural unit by one of the methods given in the previous paragraphs, the structural formula is formed by assigning the cations to appropriate sites available in the structure; in pyllosilicates the sites occupied by the cations are:

1. Tetrahedral sites within layers: Si, Al and Fe^{3+}.

TABLE 1.5. Structural formulae of clintonite using unit cell contents calculated by three different methods

		Unit cell, density (Type II)	24 anions (O, OH, F)	44 equivalents (22 oxygen)
Tetrahedral sites	Si	2.711	2.742	2.725
	Al	5.289	5.258	5.275
	Total	8.000	8.000	8.000
Octahedral sites	Al	1.254	1.359	1.303
	Ti	0.061	0.062	0.061
	Fe^{3+}	0.037	0.037	0.037
	Fe^{2+}	0.173	0.175	0.174
	Mg	4.376	4.426	4.400
	Mn	0.001	0.001	0.001
	Total	5.902	6.060	5.977
Interlayer sites	Ca	1.906	1.928	1.917
	Sr	0.011	0.011	0.011
	Total	1.917	1.938	1.928
Anions	O	20.033	20.259	20.141
	OH	2.836	2.869	2.852
	F	0.845	0.855	0.850
	Cl	0.017	0.017	0.017
	Total	23.731	24.000	23.860

2. Octahedral sites within layers and in the interlayer hydroxidic sheets of chlorites: Al, Fe^{3+}, Cr^{3+}, Mg, Fe^{2+}, Mn^{2+}, Ni^{2+}, Li, Ti.
3. Sites external to aluminosilicate layer: Na, K, Ca and other large cations; in phyllosilicates these sites are usually between the layers.

It is customary in phyllosilicates to put all Si into tetrahedral sites and to allocate sufficient Al and Fe^{3+} to fill remaining tetrahedral sites. The rest of the Al and Fe^{3+}, and Cr^{3+}, Mg, Fe^{2+}, Mn^{2+}, Ni^{2+}, Ti and Li are then allocated to octahedral sites. The structural formulae for clintonite obtained from the contents of the structural unit obtained by the three methods of calculation outlined above are given in Table 1.5.

For the analyses of montmorillonite with Al-hydroxyl interlayers (Table 1.4), setting out the unit contents as structural formulae is an informative way of comparing the two methods of calculation (Table 1.6). It can be seen that the calculation based on Si content set equal to that in the original non-interlayered montmorillonite gives a more reasonable result than that based on a total of 44 cation charges. Anomalies that result from the latter calculation are:

1. A major redistribution of charge between tetrahedral and octahedral sites:
2. An octahedral site occupancy significantly greater than the normal 4 in dioctahedral minerals, which causes a net *positive* charge from the octahedral sheet.
3. A large decrease in the net layer charge.

In this example, the calculation based on 44 charges is obviously unsatisfactory but this would not be so obvious if there was no analysis of a clay without interlayer for a reference. Many natural clays are partly chloritized and calculation of a structural formula for such clays on a 44-charge basis would not be correct. However, the calculation may alert the investigator

TABLE 1.6. Structural formulae for Al-interlayered montmorillonite

Ba-montmorillonite		Ba/Al-montmorillonite			
Calculation basis: 44 charge		Constant Si		44 charges	
Si	7.916	Si	7.916		7.434
Al	0.084	Al	0.084		0.566
Total	8.000		8.000		8.00
Al	3.127	Al	3.127		3.481
Fe	0.380	Fe	0.387		0.363
Mg	0.415	Mg	0.431		0.404
Total	3.922		3.945		4.248
Ca	0.011	Ca	0.0		0.0
Ba	0.346	Ba	0.115		0.108
Na	0.008	K	0.008		0.008
K	0.011	$Al(OH)_{2.596}$	1.098		0.0
Total	0.376		1.221		0.116
Tetrahedral charge	−0.084		−0.084		−0.566
Octahedral charge	−0.649		−0.596		+0.340
Total	−0.733		−0.680		−0.226
Interlayer charge	+0.733		+0.681		+0.224

to the presence of hydroxy interlayers, because typically, the octahedral occupancy is significantly greater than normal and the net charge on the octahedral sheet may appear to be positive. Two such examples are noted later in the hydrous micas in Section 1.7.2.1 (hydromuscovite) and in Section 1.7.2.3 (celadonite and glauconite).

1.3.3 Anomalous structural formulae

The previous sections have shown how structural formulae may be calculated systematically from chemical analyses of a clay. There are, however, a number of situations where the methods described do not give a crystallo-chemically satisfactory result.

One such instance is the montmorillonite with Al-hydroxy interlayer described above and this type of example may be extended to any sample where there is impurity present. For example, amorphous or surface-sorbed silica may lead to a structural formula in which there is a greater number of Si atoms than sites in which to place them. An example of this is anauxite, which consists of kaolinite layers partly interleaved with sheets of silica (Langston and Pask 1968).

Another type of error occurs when there is a significant amount of an element present that is not included in the calculation, possibly because it was overlooked in the analysis, or when there is a serious underestimate of a determined element. Neither of these situations should occur if the oxides in the analysis totalled $100 \pm 0.5\%$, but with many modern analyses being done by electron microprobe a check on the oxide total may not be possible. Using the analysis of Ba-montmorillonite as an example (Table 1.4), if the iron content had been estimated as 2.52% Fe_2O_3 instead of 3.52%, each of the other constituents in the structural formula would have been increased by about 1%, e.g. Si would increase from 7.92 to 7.98.

If the mineral is structurally inhomogeneous, e.g. an interstratified mineral like mica–smectite, it is likely that the substitution in the aluminosilicate layers is also inhomogeneous. In this example, because it may be assumed that all the aluminosilicate layers have a common anionic structure, a calculation to 44 cation charges can validly be made and the main problem is how best to distribute the cations through the structure. But with an interstratified kaolinite–smectite or chlorite–smectite, a structural formula cannot be calculated without further structural information. If it had been determined by X-ray diffraction study that the mineral was 40:60 kaolinite–smectite, a formula could be calculated on the basis that the anion framework consisted of $0.4(O_{10}(OH)_8) + 0.6(O_{20}(OH)_4)$ or an average of $O_{16}(OH)_{5.6}$, so that the cations should balance an anion charge of 37.6. Then, in the allocation to sites, 1.6 Si and 1.6 Al should be allocated to the kaolinite component and the remainder to the smectite. This type of calculation has not yet been tested systematically, and relies on having a quantitative structural model for the interstratified mineral.

Other instances of anomalies are found when the structure departs from the regular layer silicate model. Some of the serpentine minerals provide a good example of this (see Section 1.4.2). Owing to tetrahedral–octahedral misfit, the sheets are curved and, in antigorite, the layers are periodically inverted at Si—O bonds, giving a wave-like structure. As pointed out by Bailey (1980b), the composition that is the consequence of this ideally modulated structure is $Mg_{5.647}Si_4O_{10}(OH)_{7.294}$, a substantial departure from the ideal composition of chrysotile, $Mg_6Si_4O_{10}(OH)_8$. Greenalite and caryopilite (see Section 1.4.4) also have modulated structures and in general do not have regular periodicities and hence the concept of a structural formula, inferring as it does a regularly repeated structural unit, cannot apply. As the structures of clay minerals are studied in more detail further examples of such structural variants are likely to be discovered.

1.4 1:1 GROUP MINERALS

The layer silicate minerals with one tetrahedral sheet linked to one octahedral sheet in the layer structural unit are subdivided into two groups: the dioctahedral kaolin subgroup and the trioctahedral serpentine subgroup. The former are true clay minerals with particles predominantly $<2\,\mu$m in size, but the serpentines that are frequent constituents of ultrabasic rocks are sometimes found as crystals large enough for detailed structural study, although other minerals in this subgroup also occur in a fine-grained form.

1.4.1 The kaolin subgroup

Well-crystallized specimens of the three polytypes, kaolinite, dickite and nacrite, differ little in composition from the ideal formula $Al_4Si_4O_{10}(OH)_8$ (Table 1.7) and within the subgroup as a whole, the main chemical variations are relatively small deviations from this ideal. Halloysite (10 Å form), however, is notable in having water molecules intercalated between each 1:1 layer and a basal spacing of about 10 Å; the water is readily lost by air drying (see Chapter 5, Section 5.6.2), the basal spacing decreasing to about 7.2 Å. Following recent recommendations of the AIPEA Nomenclature Committee (Bailey 1980a) these two forms are called halloysite (10 Å) and halloysite (7 Å) respectively.

Most kaolin clays contain small amounts of accessory minerals like anatase and rutile, feldspar, iron oxides, mica, montmorillonite and quartz, which must be corrected for before it can be established that isomorphous substitution of an element into the kaolinite is occurring. Selective dissolution analysis and separation of kaolinite into narrow size fractions may help to decide how much of a constituent is impurity mineral and how much is substituted into the structure. These methods have certain limitations, however, and recent work has attempted to find confirmation by physical instrumental methods.

As shown in Table 1.7, Al:Si ratios in kaolinites differ from unity by only a few per cent. Jepson and Rowse (1975) used the electron microscope microprobe analyser (EMMA) to measure Al:Si ratios in many particles from two kaolinites from St Austell, Cornwall, and from Georgia, USA (analyses 3 and 4, Table 1.7). Ratios for individual particles varies somewhat, the variance being comprised of counting statistical and instrumental fluctuations and interparticle variations. Although the authors concluded that variation in particle composition could not be detected in their experiments, they showed that mean Al:Si ratios were significantly different from unity (0.962 ± 0.018) for the 0.9–1.0 μm fraction of the St Austell clay, but not significantly different from unity for the Georgia kaolinite (0.997 ± 0.019) for the 0.9–1.0 μm fraction. If Al were substituting for Si in the St Austell clay, the Al:Si ratio should have been greater than unity, so this hypothesis was rejected, and it was concluded that the kaolinite particles were covered with a surface coating of amorphous silica.

Jepson and Rowse also measured the ratios of minor components to Si on individual particles and confirmed the conclusion from Mossbauer and ESR measurements (see below) that Fe was present within kaolinite particles; Fe:Si ratios showed a wide and significant variation from particle to particle. About 12% of the total Ti in the Georgia kaolinite (i.e. $\sim 0.14\%$ TiO_2) was associated with the kaolinite particles as distinct from individual rutile and/or anatase particles, but the authors considered confirmatory evidence was needed to establish that this represented substitution of Ti into the structure, as was believed by Dolcater et al. (1970). Weaver (1976), however, presented much evidence that Ti in kaolinite was present as discrete surface-sorbed forms.

Statistically positive K:Si ratios were found for both the St Austell and Georgia kaolinite, sufficient to require about 1 mica layer per 250 kaolinite layers. Lee et al. (1975) have shown by HRTEM images that kaolinite can contain micaceous occlusions identified by a 10 Å lattice spacing defect in the predominantly 7 Å matrix. A mica concentration at this level is undetectable by X-ray diffraction but is sufficient to account for the K-levels in Jepson and Rowse's kaolinites.

The iron content of kaolinites has been the subject of some discussion (e.g. Angel and Hall 1973; Angel and Vincent 1978; Rengasamy 1976; Herbillon et al. 1976; Mestdagh et al. 1980; Cuttler 1981; Komusinski et al. 1981). Meads and Malden (1975) showed that several types of ESR signals were obtained from kaolinites, one attributed to Fe^{3+} substituting for octahedral Al, one attributed to trapped holes, and another due to Mn^{2+} and $(VO)^{2+}$. Herbillon et al. (1976) divided the forms of Fe in kaolinite into that in ancillary oxides, e.g. anatase and rutile, that in micas, and that in the kaolinite structure. Tropical soil kaolinite could contain up to 2% Fe^{3+} but there was increasing disorder as the iron content increased; the structural evidence for disorder has been studied in detail by Plançon and Tchoubar (1977). Mestdagh et al. (1980) further investigated this relationship and showed that a quantitative treatment of the ESR spectrum at $g_{eff} \simeq 4$ provided a method of estimating the structural Fe contents of kaolinites. Well-crystallized kaolinites have a narrow range of Fe^{3+} contents, but in poorly-crystallized kaolinites, the Fe^{3+} is partitioned equally into two sites characterized by their ESR spectrum. The difficulty of using selective dissolution analysis to define Fe content of kaolins is emphasized by the work of Komusinski et al. (1981) who showed that dithionite treatment did not totally remove all the "free" Fe oxides and that 9% HCl was sometimes needed; even this did not remove all the free oxides from some samples.

24 THE CHEMICAL CONSTITUTION OF CLAYS

TABLE 1.7. Analysis of kaolin subgroup minerals

	1	2	3	4	5	6
SiO_2	46.55	46.40	46.2	45.2	46.43	46.22
Al_2O_3	39.49	39.52	39.2	39.2	39.54	39.92
TiO_2	0	0	0.09	1.21	nil	
Fe_2O_3	0	0.09	0.23	0.17	0.15	
FeO	0	0				
Cr_2O_3	0					
MgO	0	0.15	0.07	0.08	0.17	
CaO	0	0.23	0.06	0.06	nil	
Na_2O	0	0.09	0.09	0.03	0.03	
K_2O	0	0.01	0.21	0.02	0.02	
H_2O^+	13.96	13.90	13.8	13.3	14.20	13.86
H_2O^-	0					
Total	100.00	100.39	99.95	99.27	100.54	100.00
Numbers of cations on basis of $O_{10}(OH)_8$						
Si	4	3.977	3.981	3.908	3.982	3.969
Al	0	0.023	0.019	0.092	0.018	0.031
Σ tet.	4	4.000	4.000	4.000	4.000	4.000
Al	4	3.970	3.962	3.902	3.979	4.010
Ti	0	0	0.006	0.079	0	
Fe^{3+}	0	0.006	0.015	0.011	0.010	
Fe^{2+}	0	0				
Cr^{3+}	0	0				
Mg	0	0.019	0.009	0.010	0.022	
Σ oct.	4	3.995	3.992	4.002	4.011	4.010
Ca	0	0.021	0.006	0.006	0	
Na	0	0.015	0.015	0.005	0.005	
K	0	0.001	0.023	0.002	0.002	
Al/Si	1	1.004	1.000	1.022	1.004	1.018
R^{3+}/Si	1	1.006	1.004	1.025	1.006	1.018

1. Ideal kaolinite.
2. Kaolinite, Keokuk, Iowa, USA (Keller et al. 1966).
3. Kaolinite, St Austell, Cornwall, UK. Contains 1–2% mica (Jepson and Rowse 1975).
4. Kaolinite, Washington County, Georgia, USA. Contains 1–2% anatase (Jepson and Rowse 1975).
5. Dickite, Barkly East, Cape Province, South Africa, 6–20 mμ (Schmidt and Heckroodt 1959).
6. Nacrite, Eureka Tunnel, St Peter's Dome, Colorado, USA (Blount et al. 1969). Recalculated after subtracting 0.68% CaF_2 from original analysis of Cross and Hillebrand (1885).
7. Kaolinite, b-axis disordered, Pugu, Tanganyika (Robertson et al. 1954). † Includes 0.28% dithionite extractable Fe_2O_3; ‡ Total includes 0.09% SO_3. Cation exchange capacity = 74 μeq g^{-1}.

The cation exchange properties of kaolinites have been investigated by many authors (Schofield and Samson 1954; Cashen 1959; Flegmann et al. 1969; Ferris and Jepson 1975; Bolland et al. 1976) and varying views have been expressed about the existence and origin of permanent negative surface charge, distinct from the pH-dependent positive and negative charges observed by most experimenters. Ferris and Jepson (1975) in a very detailed study concluded that cation retention decreased to zero as the electrolyte approached infinite dilution, this indicating no permanent negative charge. Bolland et al. (1976), however, believed that Ferris and Jepson reached this conclusion because they had not taken into account hydrolysis which led to hydroxy-Al species replacing the surface sorbed cation. Most authors seem to accept that kaolinites have a small pH-independent exchange capacity, approximately 10–80 μeq g^{-1},

7	8	9	10	11	12
44.59	48.98	40.09	46.20	44.7	40.55
38.12	36.83	35.38	39.84	28.1	24.56
1.38	0.33		0.02		
1.43†	0.09	trace	0.17	12.8	0.94
nil					
	0.69				11.72
0.06		trace	0.02	0.1	0.54
0.10		0.77	0.34	trace	0.46
0.12	0.03	0.10	0.01	1.7	0.20
0.08	0.02	trace	0.02	trace	0.10
13.91	13.06	15.00	14.00*	13.3	12.20
0.71	0.33	8.61			8.22
100.59‡	100.36	100.51	100.62	100.7	99.68
Numbers of cations on basis of $O_{10}(OH)_8$					
3.894	3.984§	3.907	3.957	4.029	4.025
0.106	0.016	0.093	0.043		
4.000	4.000	4.000	4.000	4.029	4.025
3.818	3.917	3.971	3.978	2.985	2.873
0.091	0.022		0.001		
0.076	0.006	0	0.011	0.868	0.070
0					
	0.049				0.920
0.008		0	0.003	0.013	0.080
3.993	3.994	3.971	3.993	3.866	3.943
0.009		0.080	0.031	0	0.049
0.020	0.005	0.019	0.002	0.297	0.038
0.009	0.002	0	0.002	0	0.013
1.008	0.987	1.040	1.016	0.741	0.714
1.029	1.001	1.040	1.019	0.956	0.960

8. Cr-dickite, sample 2378, Testic, Yugoslavia. Contains a small amount of kaolinite and about 5% quartz (Maksimovic et al. 1981).§ After subtraction of 5% free SiO_2.
9. Halloysite (10 Å), Wagon Wheel Gap, Colorado, USA (Larsen and Wherry 1917).
10. Halloysite (7 Å) (dehydrated), Djebel Debar, Morocco. (Garrett and Walker 1959). ‖ Water lost above 300°C.
11. Iron-rich halloysite (10 Å), weathered pumice bed, Uenae, Hokkaido, Japan, ¶ Analysis of sample pretreated with dithionite and 2% $NaCO_3$ and oven-dried (Wada and Mizota 1982).
12. Cr-halloysite, dark blue, Takova, Yugoslavia (Maksimovic and White 1973).

depending on particle size. The surface charge density is in the range 0.11 to 0.15 C m^{-2} (Bolland et al. 1976), which is about the same as that of smectites (see Chapter 5, Section 5.3.2). Thus it would be expected that the poorly crystalline kaolinites with a large specific surface would have a larger exchange capacity than the coarse grained samples; the cation exchange capacities as high as 250 μeq g^{-1} that have been reported by some authors seem too high, implying specific surfaces of the order of 250 m^2 g^{-1}, which are not observed for kaolinites, and Lim et al. (1980) have concluded that exchange capacities greater than 10 μeq g^{-1} were caused by impurity in the kaolins.

It is rare for kaolins to contain appreciable amounts of elements other than Si, Al and Fe, but Brookins (1973) described chromiferous kaolinite from The Geysers, California, and

Maksimovic et al. (1981) have reported Cr-kaolinite and Cr-dickite in the products of the hydrothermal alteration of ultramafic rocks in Yugoslavia. The kaolinite contains 0.7% Cr_2O_3 in the bulk analysis and electron microprobe study of the dickite grains shows a uniform distribution of 1.0% Cr throughout. The authors conclude that Cr is replacing Al in the structures and record changes in the infrared spectra that this causes.

1.4.1.1 Halloysites
The halloysite (10 Å) form differs from kaolinite and its polymorphs in having additional molecular water held between the aluminosilicate layers. Most of this water volatilizes at room temperature if the water vapour pressure is below saturation, but a little is retained to a higher temperature (250 °C: Brindley and Goodyear 1948).

Halloysite (10 Å) occurs in a variety of morphologies and these have been related to their chemical compositions (Tazaki 1982). The usual forms reported are short and long tubes that appear to form by scrolling of one or more 1:1 layers (Kirkman 1975, 1977); halloysites (10 Å) with these morphologies mostly contain rather little iron. Higher Fe contents are found in halloysites (10 Å) with spheroidal morphology, and halloysite in the form of a "crinkly film" (Tazaki 1982) or with "crumpled lamellar" morphology (Wada and Mizota 1982) contains from 4 to 15% Fe_2O_3. Halloysites appear to have a much wider range for the ratio (Al + Fe)/Si than kaolinites, some sources quoting ratios mostly greater than unity (Weaver and Pollard 1973), while others (Tazaki 1982) quote ranges with an excess of silica. These wide ranges probably reflect Al:Si ratios in the rocks from which the halloysites are formed, and the nature of the alteration processes.

The exchange capacity of halloysites has been the subject of contradictory observations. Quite large values, 500 μeq g^{-1}, have been reported by some authors, but smaller values of about 100 μeq g^{-1} are also obtained. The origin of these differences appears to be salt absorption into the interlayer where it may be trapped during the washing stage if a dehydrating solvent is used (Garrett and Walker 1959). Salts of the alkali metals, which have a low hydration energy, are readily sorbed into the interlamellar volume and are desorbed when the clay is exchanged with another electrolyte. Carr et al. (1978) analysed such salt-treated halloysite (10 Å) and showed that both cations and anions were sorbed; they also extended the list of salts known to be intercalated into halloysites (10 Å). Dehydrated halloysites (7 Å) are less well able to intercalate salts, but ammonium and potassium acetates are effective in concentrated solution. The proven true exchange capacity of halloysite appears to be about 90–95 μeq g^{-1} (Garrett and Walker 1959), though Wada and Mizota (1982) report a value of 580 μeq g^{-1} for iron-rich halloysite (10 Å), this being attributed to substitution of Fe^{2+} for Al (analysis 11, Table 1.7). Some confirmation of the exchange capacities of Fe-halloysites (10 Å) seems to be needed, because dehydrated halloysites (7 Å) have small exchange capacities (Garrett and Walker 1959).

As with kaolinite and dickite, halloysites (10 Å) with a high Cr content are formed in the alteration zone of ultramafic rocks in Yugoslavia (Maksimovic and White 1973). As much as 11.7% Cr_2O_3 is reported on one sample (analysis 12, Table 1.7), much larger than the substitution in kaolinite and dickite; this appears to parallel the iron contents in the two types of mineral, with only 2% Fe_2O_3 substituting in kaolinite but as much as 15% Fe_2O_3 in halloysites.

1.4.2 Minerals of the serpentine subgroup

The ideal formula of the trioctahedral serpentine minerals is $(R_6^{2+})(Si_4)O_{10}(OH)_8$ where R^{2+}

may be Mg, Fe^{2+}, Mn^{2+} or Ni^{2+}, but substitution by trivalent cations occurs in octahedral and tetrahedral sites and deficiencies of octahedral cations are common. These compositional differences are intimately related to differences in the structure and morphology of the layers.

The 1:1 layers in the serpentine subgroup may be of three kinds, planar layers, layers with continuous curvature producing rolled or tubular structures, and layers in which the direction of curvature changes about the mean plane of the layers forming overall planar structures. Which of these three types is formed depends on the degree of misfit, if any, between the lateral dimensions of the octahedral and tetrahedral sheets of the 1:1 layers and these in turn vary with their chemical constitution. The lateral dimensions of layers of octahedral and tetrahedral sheets must match at their shared anion plane to form continuous planar layers. When tetrahedral sheets are larger than octahedral sheets, alternate twisting of tetrahedra in their ideally hexagonal rings can readily decrease their dimensions. When tetrahedral sheets are smaller than octahedral sheets stretching the tetrahedral sheet and shrinking the octahedral sheet is much more difficult. This situation is found in those serpentine group minerals in which the trioctahedral sheet is entirely populated by divalent cations and the tetrahedral sheet contains Si only, i.e. compositions with no trivalent cations. Substitution of trivalent cations decreases the size of the octahedral sheet and enlarges the tetrahedral sheet leading to more similar lateral dimensions.

The overall picture is clear. Serpentine group minerals containing substantial amounts of Al^{3+} and/or Fe^{3+} form planar structures (lizardites, berthierines, amesite, cronstedtite). In the absence of trivalent cations, layers may either be planar or curved. Highly magnesian compositions in which misfit is relatively small occur as planar layers (Mg-lizardites), as layers curved to form rolls (chrysotiles) or as layers in which the direction of curvature changes periodically in the form of an alternating wave (antigorites); conditions of formation may account for these differences in form. When larger divalent cations (Fe^{2+}, Mn^{2+}) predominate (greenalites and caryopilites), the larger misfit is accommodated by the formation of small domed or saucer-like regions about 18–25 Å in diameter. The wavy antigorite type and the domed structures allow fewer octahedra to articulate with a tetrahedral sheet than the planar or rolled structures. Nickel-rich serpentine group minerals, on the basis of ionic radii, would be expected to behave like magnesian varieties and indeed a rolled fibrous mineral pecoraite (Faust et al. 1969), the Ni-analogue of chrysotile, has been described. Nickel-containing varieties also occur in what appears to be crumpled aggregates of indefinite morphology (nepouites, rich in Ni) which Brindley and Wan (1975) consider to be part of a continuous series, via Ni-lizardites, with the planar lizardites.

The main sources of analyses for this section were those of the Mg-serpentines that Whittaker and Wicks (1970) considered to be of high quality, for berthierines those from Brindley (1982) and the electron microprobe analyses of Guggenheim et al. (1982) for greenalites, caryopilites and tosalites and the analysis of tosalite (Yoshimura 1967); for Ni-serpentines the data of Brindley and Wan (1975) were the main source.

In all 83 modern analyses were considered. Structural formulae were calculated for all of these on the basis of the ideal $O_{10}(OH)_8$ anion framework of the serpentine structure. For many this results in formula with more than 4.00 Si (up to 4.4); for these, formulae were recalculated on the basis of 4Si per formula unit, which implies that the anionic charge is less than 28. Formulae which show 4 or less Si per $O_{10}(OH)_8$ anion can probably be taken to represent the distribution of cations in the different structural sites and can therefore be called structural formulae. Clearly formulae based on the ideal serpentine structure cannot apply to antigorites, greenalites and caryopilites; formulae that had to be recalculated on the 4.0 Si basis should therefore be

28 THE CHEMICAL CONSTITUTION OF CLAYS

TABLE 1.8. Average formulae of serpentine group minerals

Mineral	No. of analyses	Octahedral cations	Tetrahedral cations	Anion
Chrysotile	5	$Mg_{5.89}Fe^{2+}_{0.01}Al_{0.02}Fe^{3+}_{0.03}$	$Si_{3.99}Al_{0.02}$	$O_{10}(OH)_8$
Lizardite	9	$Mg_{5.73}Fe^{2+}_{0.02}Al_{0.02}Fe^{3+}_{0.19}$	$Si_{3.86}Al_{0.08}Fe^{3+}_{0.06}$	$O_{10}(OH)_8$
9-layer serpentine	1	$Mg_{4.17}Fe^{2+}_{0.15}Al_{1.46}$	$Si_{2.97}Al_{1.03}$	$O_{10}(OH)_8$
Amesite	1	$Mg_{3.30}Fe^{2+}_{0.67}Al_{2.00}$	$Si_{2.01}Al_{1.99}$	$O_{10}(OH)_8$
Tosalite†	5	$Fe^{2+}_{3.91}Mn^{2+}_{0.79}Mg_{0.54}Al_{0.55}$	$Si_{3.87}Al_{0.13}$	$O_{10}(OH)_8$
Berthierine	17	$Fe^{2+}_{3.29}Mg_{0.59}Mn_{0.01}Al_{1.54}Fe^{3+}_{0.30}$	$Si_{2.71}Al_{1.29}$	$O_{10}(OH)_8$
Cronstedtite	1	$Fe^{2+}_{4.76}Fe^{3+}_{1.42}$	$Si_{2.24}Fe^{3+}_{1.62}Al_{0.14}$	$O_{10}(OH)_8$
Antigorite	5	$Mg_{5.53}Fe^{2+}_{0.10}Al_{0.14}Fe^{3+}_{0.05}$	Si_4	$(O,OH)_{18}$
Greenalite†	7	$Fe^{2+}_{4.53}Mg_{0.60}Mn^{2+}_{0.17}Al_{0.03}$	Si_4	$(O,OH)_{18}$
Caryopilite†	4	$Mn^{2+}_{4.50}Mg_{0.27}Fe^{2+}_{0.07}Al_{0.17}$	Si_4	$(O,OH)_{18}$
Nepouite	4	$Ni_{4.03}Mg_{0.69}Fe^{3+}_{0.12}Al_{0.06}$	Si_4	$(O,OH)_{18}$

† All Fe treated as FeO.

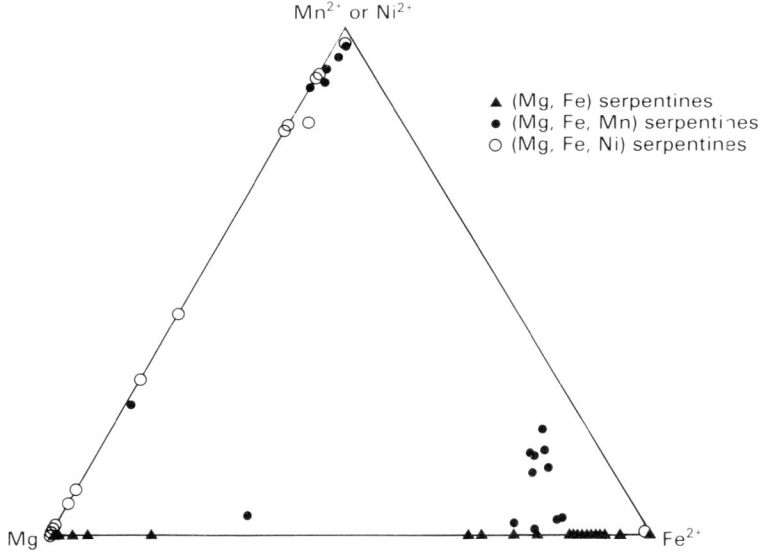

FIG. 1.1. Serpentine group minerals: relative proportions of Mg, Fe^{2+} and Mn^{2+} or Ni^{2+}.

regarded not as structural formulae but as a convenient way to express cationic ratios for comparison with the other serpentine minerals. Average formulae for the more important serpentine minerals, calculated on $O_{10}(OH)_8$ or $Si=4$, are given in Table 1.8.

The triangular diagram (Fig. 1.1) shows the relative abundance of the different species of divalent cations, Mg, Fe^{2+} and Mn^{2+} or Ni, in serpentine minerals. All four can be represented on a triangular diagram because Mn and Ni (and Fe and Ni) do not occur together in the same mineral. A notable feature is the rarity of compositions that are not on, or very close to, the edges; only six analyses were found in which three different cation species (Fe^{2+}, Mg and Mn) occur in appreciable proportions, and all of these have more than 70% of the dominant cation, Fe^{2+}

1.4.3 Serpentine subgroup minerals with planar layers

Minerals in this category, all of which give acceptable structural formulae when calculated on $O_{10}(OH)_8$ basis, are lizardites (including in this term all planar Mg–Al serpentines), berthierines and brindleyite, a Ni-rich berthierine (Maksimovic and Bish 1978), amesite and its Mn-analogue, kellyite (Peacor *et al.* 1974), and cronstedtite. Despite appearing to have excess Si the minerals named Ni-lizardites and nepouites are also discussed in this section because Brindley and Wan (1975) believe they form a continuous series with the planar lizardites.

Structural formulae of all of these calculated on the basis of a total anionic charge of 28 ($O_{10}(OH)_8$ formula unit) contain Si $\leqslant 4.0$ within analytical error. They can be represented (as can chrysotile) by the general formula

$$(R^{2+}_{6-x-y}R^{3+}_x \square_y)(Si_{4-x+2y}R^{3+}_{x-2y})O_{10}(OH)_8$$

where R^{2+} and R^{3+} represent the sums of the divalent and trivalent cations, respectively. Their $(R^{2+}+R^{3+})/Si$ ratios range upwards from 1.5 to about 1.7 for lizardites, through the range 2 to 3.5 for berthierines, brindleyite and the 9-layer aluminian serpentine (Jahanbagloo and Zoltai 1968), to about 4 for amesite and kellyite. The relation between the amounts of R^{3+} in octahedral and tetrahedral sites is shown in Fig. 1.2, in which the diagonal lines indicate the number of vacant octahedral sites. The minerals occur in two distinct groups, those with $R^{3+}_{tet}<0.5$ and $R^{3+}_{oct}<1.0$ and those with $R^{3+}_{tet}>0.75$ and $R^{3+}_{oct}>1.15$. In the former group are chrysotiles, lizardites (excepting 9-layer serpentine), tosalites, and some antigorites; the latter group are mainly berthierines, with the 9-layer serpentine, brindleyite, amesite, kellyite and cronstedtite.

1.4.3.1 Lizardites
This name is used here, as proposed by Wicks and Whittaker (1975), for all planar Mg-rich serpentines, but excluding the well-established mineral amesite (following Bailey (1980b). Lizardite analyses are given in Table 1.9. Excepting the unusual 9-layer aluminian serpentine (Jahanbagloo and Zoltai 1968), lizardites contain between 0.06 and 0.3 R^{3+} in tetrahedral sites, up to 0.42 R^{3+} in octahedral sites, and have few octahedral vacancies, usually less than 0.1 per formula unit. Their $R^{3+}_{oct}/R^{3+}_{tet}$ ratios lie between 0.9 and 2.55 (mean 1.81). One sample from Unst, Shetland Isles (analysis 6, Table 1.9) however contains very little R^{3+}_{oct} and has a ratio of 0.05.

The 9-layer serpentine (analysis 8, Table 1.9) of Jahanbagloo and Zoltai (1968) is intermediate in composition between amesite (analysis 9, Table 1.9) and the lizardites except that it has more octahedral vacancies.

1.4.3.2 Nepouites and Ni-lizardites
According to Brindley (1980) nepouite is a nickel-rich lizardite; when Ni constitutes less than half the octahedral cation total, the name Ni-lizardite is recommended. Fine-grained hydrous Ni–Mg silicates for which the name garnierite can be used (Faust 1966) are common in many lateritic ores and are almost invariably found to be intimate mixtures of serpentine-like minerals and trioctahedral 2:1 layer silicates. The nature of garnierites has been studied by chemical

30 THE CHEMICAL CONSTITUTION OF CLAYS

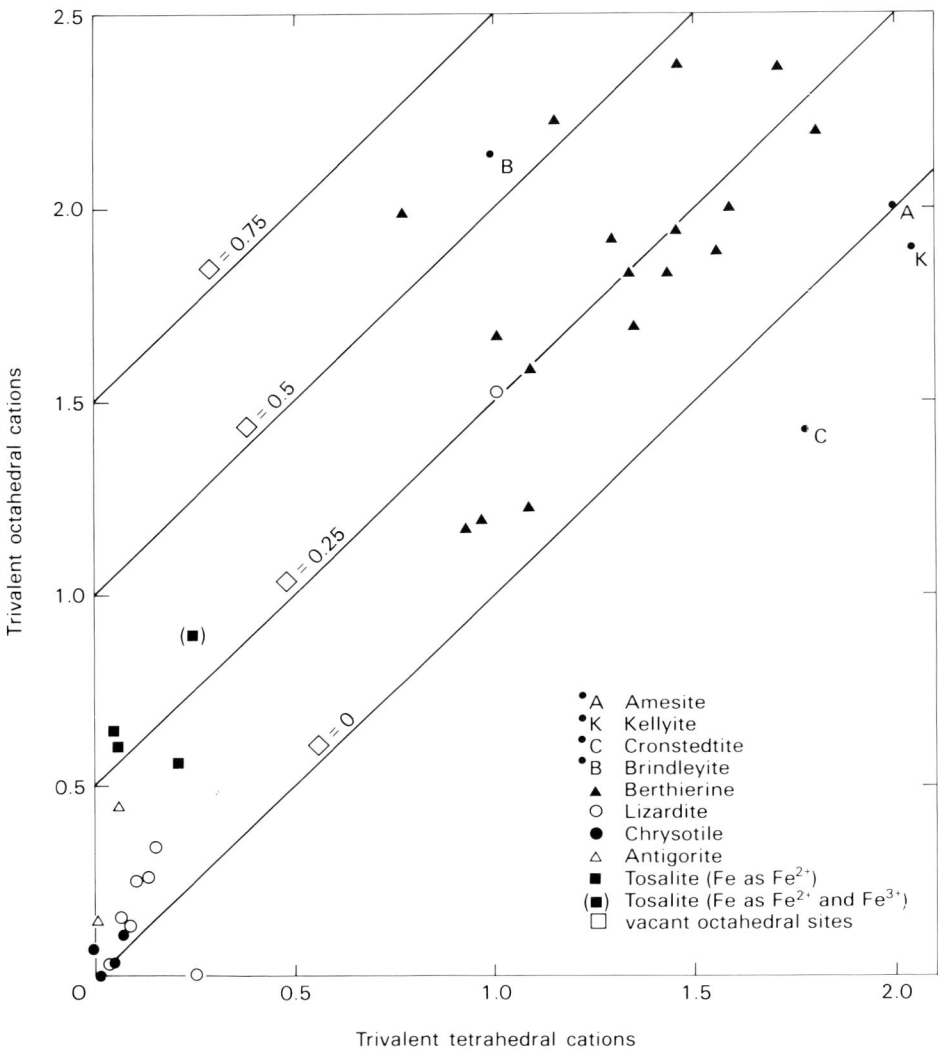

FIG. 1.2. Serpentine group minerals: R^{3+} (octahedral) v. R^{3+} (tetrahedral) for the $O_{10}(OH)_8$ formula unit.

analysis, X-ray diffraction and HRTEM by Brindley and collaborators (Brindley and Pham Thi Hang 1973; Uyeda *et al.* 1973), who showed that they were mixtures of varying amounts of serpentine-like (7 Å) and talc-like (10 Å) phases with additional adventitious quartz in some samples.

Brindley and Wan (1975) examined 14 dominantly 7 Å (serpentine-like) garnierites including four previously shown to be impure. Their R^{3+} content is always small, less than 0.5 atoms per 4Si and in this respect they resemble lizardites. All but one have $(R^{2+} + R^{3+})$/Si ratios between 1.13 and 1.33, much less than Mg-lizardites (1.53–1.58). The ratio for one sample from Valojoro, Madagascar, containing 1.28% NiO, is 1.44. Brindley and Wan (1975) attribute these small ratios to the presence of admixed colloidal silica even though Ni-bearing samples from the same localities, rich in 2:1 layer silicates, do not have unusually small $(R^{2+}+R^{3+})/Si$ ratios (Brindley et al. 1979) and that it has been shown previously that four of the samples contained quartz and/or 2:1 silicates. It seems likely that all the nepouites and Ni-lizardites examined by Brindley and Wan (1975) are impure and the silica in excess of that required for a serpentine structure is present as 2:1 layers, either in a discrete component or as layers interstratified or intergrown in a mainly 7 Å-layer matrix. These 2:1 layers are clearly shown as spacings $\geqslant 10$ Å in the HRTEM images of Uyeda et al. (1973). The existence of pure homogeneous nepouites and Ni-lizardites must therefore be considered still unproven. No analyses are quoted here because those in the literature almost certainly relate to samples with a large proportion of impurity. The sample from Valojoro, which incidentally gives the clearest X-ray powder pattern (Fig. 1(b), Brindley and Wan 1975) is probably the least contaminated. Its structural formula

$$(Mg_{5.646}Ni_{0.097}Fe^{3+}_{0.067}Al_{0.029}Ti_{0.001})(Si_{4.05})O_{10}(OH)_8$$

is closer to those of antigorites than lizardites.

1.4.3.3 Berthierine

Formerly called chamosite, a Fe^{2+}-rich serpentine group mineral is a common constituent of iron ores (see analyses 1–10, Table 1.10). It is distinguished from greenalite (Section 1.4.4.3) by its larger R^{3+} content (dominantly Al^{3+}) in octahedral (1.18 to 2.38 atoms) and tetrahedral (0.77 to 1.805 atoms) sites; octahedral vacancies range from 0.09 to 0.64 per $O_{10}(OH)_8$ formula unit. Brindley's (1982) contention that $R^{3+}_{oct} \simeq 1.3 R^{3+}_{tet}$ is not borne out by the formulae (Table 1.10 and Fig. 1.2); the ratio ranges from 1.2 to 2, though for most samples it is less than 1.6. Brindleyite (analysis 11, Table 1.10) was described by Maksimovic and Bish (1978) as a Ni-berthierine and its formula

$$(Ni_{2.662}Mg_{0.520}Fe^{2+}_{0.106}Al_{2.214}Cr^{3+}_{0.015})(Si_{3.01}Al_{0.99})O_{10}(OH)_8$$

clearly shows its resemblance to berthierines.

1.4.3.4 Amesite

Amesite with formula close to

$$(R^{2+}_4 Al_2)(Si_2Al_2)O_{10}(OH)_8$$

where R^{2+} is dominantly Mg, is notable for its large Al content and may be considered the Al-rich end-member of the lizardite series. An Mn-rich analogue, with formula

$$(Mn_{3.667}Mg_{0.494}Al_{1.715}Fe^{3+}_{0.183}Ti_{0.004})(Si_{1.96}Al_{2.04})O_{10}(OH)_8$$

has been described by Peacor et al. (1974) and given the name kellyite. Analyses of amesite and kellyite are in Table 1.9.

1.4.3.5 Cronstedtite

This serpentine group mineral is notable for its large content of Fe^{3+} in tetrahedral sites and its structural formula, derived from Hendricks (1939) (analysis 12, Table 1.10) is

$$(Fe^{2+}_{4.759}Fe^{3+}_{1.417})(Si_{2.232}Fe^{3+}_{1.624}Al_{0.144})O_{10}(OH)_8$$

32 THE CHEMICAL CONSTITUTION OF CLAYS

TABLE 1.9. Analyses of lizardites, amesite and kellyite

	1	2	3	4	5
SiO_2	41.64	40.97	41.25	39.92	41.65
Al_2O_3	0.72	none	0.54	0.44	0.10
TiO_2	none	none	0.02	none	none
Fe_2O_3	1.96	5.55	1.32	5.68	2.88
FeO	0.18	trace	0.09	trace	0.16
MnO	0.03		0.07		0.05
MgO	41.55	40.43	41.84	39.10	41.06
CaO	none	none	0.02	none	none
Na_2O	0.10				
K_2O	none				
H_2O^+	13.74	} 13.80	13.68	13.80	13.10
H_2O^-	0.36		0.97		
F	none				
Total	100.28	100.75	99.80	98.84	100.12

Numbers of cations on basis of $O_{10}(OH)_8$

	1	2	3	4	5
Si	3.91	3.86	3.92	3.85	3.94
Al	0.08		0.06	0.05	0.01
Fe^{3+}	0.01	0.14	0.02	0.10	0.05
Σ tet.	4.00	4.00	4.00	4.00	4.00
Al					
Ti					
Fe^{3+}	0.13	0.26	0.07	0.31	0.15
Fe^{2+}	0.01		0.01		0.01
Mn			0.01		
Mg	5.82	5.68	5.92	5.62	5.79
Σ oct.	5.96	5.94	6.01	5.93	5.95

1. Lizardite, Dognacska, Hungary (Faust and Fahey 1962, sample F-23); contains 0.018 Na per $O_{10}(OH)_8$.
2. Lizardite, Tyrol, Austria (Faust and Fahey 1962, sample F-47).
3. Lizardite, Transvaal, South Africa (Deer et al. 1962).
4. Lizardite, near Bellows Falls, Vermont, USA (Faust and Fahey 1962, sample F-46).
5. Lizardite (six-layer serpentine), Shetland Islands, UK (Brindley and von Knorring 1954).
6. Lizardite (six-layer serpentine), Shetland Islands, UK (Brindley and von Knorring 1954).

1.4.4 Serpentine subgroup minerals with non-planar layers

In this category are chrysotiles, antigorites, greenalites, and the Mn-rich minerals allocated to the serpentine group by Bailey (1980b), Bayliss (1981) and Guggenheim et al. (1982) under the name caryopilite. Tosalites are also included here because, despite their larger Al contents and larger $(R^{2+} + R^{3+})/Si$ ratios, they have affinities with greenalites and caryopilites.

1.4.4.1 Chrysotile

This has composition near the ideal serpentine formula

$$Mg_6Si_4O_{10}(OH)_8$$

(analyses 1–5, Table 1.11), and deviations from this are small. Although the average compositions of chrysotile and lizardite differ (Table 1.8), some chrysotiles and lizardites have

6	7	8	9	10
38.40	38.50	31.1	20.95	17.60
0.10	4.09	22.2	35.21	28.55
none	0.07			<0.05
3.42	3.52			2.18
none	1.29	1.9	8.28	
0.05	0.05			38.84
41.91	37.75	29.3	22.88	2.97
none	none	0.3	0.58	
	0.17	1.0		
15.03	13.79		13.02	
			0.23	
1.26	0.94			
100.17	100.22	85.8	100.15	90.14
	Numbers of cations on basis of $O_{10}(OH)_8$			
3.75	3.70	3.00	2.01	1.96
0.01	0.30	1.00	1.99	2.04
0.24				
4.00	4.00	4.00	4.00	4.00
	0.16	1.52	2.00	1.72
0.01	0.26			0.18
	0.10		0.67	
				3.67
6.10	5.41	4.21	3.28	0.49
6.11	5.93	5.73	5.95	6.06

7. Lizardite (six-layer serpentine), Thompson Lake, Quebec, Canada. Total includes 0.05% Cr_2O_3 (Olsen 1961).
8. Lizardite (nine-layer serpentine), North Shore Beaches, Lake Superior, Minnesota, USA. CaO and Na_2O omitted for structural formula calculation (Jahanbagloo and Zoltai 1968).
9. Amesite, Chester, Massachusetts, USA (Gruner 1944a); contains 0.08 Ca per $O_{10}(OH)_8$.
10. Kellyite, Bald Knob, North Carolina, USA. Analysis by electron microprobe; all iron as Fe_2O_3 (Peacor *et al.* 1974).

the same composition (see Section 1.4.6). Pecoraite (Faust *et al.* 1969) has been described as the Ni-analogue of chrysotile and its analysis is in Table 1.11.

1.4.4.2 Antigorite

This is a magnesium-rich serpentine (analyses 7–11, Table 1.11) characterized by a superlattice in the X-direction that arises from the alternating curvature of the 1:1 layers required to accommodate the misfit between its octahedral and tetrahedral sheets (see Bailey 1980b). At the inversion points of the waves octahedral cations and OH groups are omitted and a general formula

$$m[Mg_{3(1-1/m)}Si_2O_5(OH)_{1+3(1-2/m)}]$$

has been given by Bailey (1980b) where m is the number of tetrahedra in the X direction in the full-wave of the superstructure. Superlattice repeats ranging from about 20 to 110 Å ($m=8$ to 43)

34 THE CHEMICAL CONSTITUTION OF CLAYS

TABLE 1.10. Analyses of berthierine, brindleyite and cronstedtite

	1	2	3	4	5	6
SiO_2	19.08	22.86	21.40	23.72	26.24	22.47
Al_2O_3	26.88	26.66	25.40	24.33	22.17	21.82
TiO_2						
Fe_2O_3	4.29	3.85	0.25	3.75	6.62	0.22
Cr_2O_3						
FeO	34.52	32.30	37.40	29.64	32.54	37.24
MnO			0.05		0.04	
MgO	1.55	0.97	2.04	7.13	1.58	2.57
NiO						
CaO						
Na_2O						
K_2O						
H_2O^+	}11.18	}10.48	}12.02	}11.00	}10.81	}10.07
H_2O^-						
Total	99.25	97.02	98.56	99.57	100.00	94.39
	Numbers of cations on basis of $O_{10}(OH)_8$					
Si	2.20	2.55	2.45	2.55	2.85	2.65
Al	1.80	1.45	1.55	1.45	1.15	1.35
Fe^{3+}						
Σ tet.	4.00	4.00	4.00	4.00	4.00	4.00
Al	1.84	2.05	1.87	1.64	1.70	1.68
Ti						
Fe^{3+}	0.37	0.32	0.02	0.30	0.54	0.02
Cr^{3+}						
Fe^{2+}	3.32	3.00	3.58	2.67	2.96	3.67
Mn						
Mg	0.27	0.16	0.35	1.14	0.26	0.45
Ni						
Σ oct.	5.80	5.53	5.82	5.75	5.46	5.82

1 to 9 Berthierines, see Brindley (1982) for sources of analyses and localities.
10. Berthierine, Ayrshire, Scotland. $CO_2 = 0.40$, $SO_3 = 0.27$, $P_2O_5 = 0.18$, organic = 0.03% included in total (Brindley 1951). TiO_2 omitted in calculation of formula.

have been reported for different antigorites (Zussman *et al.* 1957); commonly the superlattice periodicity lies between 33 and 44 Å ($m = 13$ to 17) and different values are frequently found in different crystals of the same sample. Antigorites would therefore be expected to be deficient in R^{2+} (mainly Mg) and OH relative to the ideal formula and although the differences are small, the most careful studies indicate this trend (see Whittaker and Wicks 1970 and the analyses in Table 1.11). Unfortunately high quality analyses and superlattice periodicities determined on the same crystals that would provide a quantitative test of the differences predicted by Bailey's (1980b) formula are lacking.

A mineral from Orange County, New York, reported by Frondel (1962) to have a powder pattern identical to that of antigorite from Val Antigorio, contains unusually large amounts of ferrous iron for a dominantly magnesian serpentine (analysis 12, Table 1.11). It is included among the antigorites here because the author has named it ferroan antigorite, but no clear evidence was presented to confirm that it has a superlattice.

1.4.4.3 Greenalite

Greenalite is an iron-rich serpentine group mineral differing from berthierine and cronstedtite

7	8	9	10	11	12
23.92	26.9	25.5	22.03	27.45	16.42
20.55	18.4	14.9	22.91	24.09	0.90
			3.63	0.99	
5.50			0.46		29.72
			0.05	0.17	
35.00	36.8	35.6	36.68	1.15	41.86
		1.0	0.04		
2.54	8.15	5.6	1.91	3.18	
				30.18	
			0.07	0.07	1.32
			0.08		
			0.03		
}11.41	}9.58	}17.4	10.65		10.17
			0.63		
98.92	99.83	100.00	100.05	87.28	100.39
Numbers of cations on basis of $O_{10}(OH)_8$					
2.71	2.91	3.07	3.23	3.01	2.23
1.29	1.09	0.93	0.77	0.99	0.15
					1.62
4.00	4.00	4.00	4.00	4.00	4.00
1.46	1.26	1.18	1.95	2.12	
0.47			0.03		1.42
				0.01	
3.32	3.33	3.58	3.09	0.11	4.76
		0.10			
0.43	1.32	1.01	0.29	0.52	
				2.66	
5.68	5.91	5.87	5.36	5.42	6.18

11. Brindleyite, Marmara karsitic bauxite deposits, Greece. Analysis 3 (by electron microprobe) of Maksimovic and Bish (1978). Lanthanides = 0.35% omitted from total. TiO_2 and CaO omitted in calculation of formula.
12. Cronstedtite, Kisbanya, Hungary (Hendricks 1939). CaO omitted in calculation of structural formula.

by having a much smaller R^{3+} content. Iron is the dominant octahedral cation ($>70\%$) and Fe/Si ratios range from 0.92 to 1.35. Greenalites contain little Al^{3+} (<0.12 atoms per 4Si); the only other trivalent cation found is Fe^{3+}. Probably the most reliable analyses are those of Guggenheim et al. (1982) by electron microprobe (analyses 1–5, Table 1.12); this technique has the advantage of decreasing the impurity content of the material analysed but the disadvantage of failing to distinguish Fe^{2+} and Fe^{3+}. Analyses by Gruner (1936) and Guggenheim et al. (1982) by wet chemical and Mössbauer spectroscopy respectively yield Fe^{3+}/Fe^{2+} ratios of about 1:4 to 1:5 so all the iron determined by microprobe analyses has been treated as Fe^{2+}. Even if one-quarter of the total iron is present as Fe^{3+}, the R^{3+}/Si ratios are still small, about 0.2 to 0.35, compared with 0.7 to 1.8 for berthierines, and 1.4 for cronstedtite. All the microprobe analyses show a considerable excess of Si (4.11–4.26 atoms) when formulae are calculated on the 28 charge basis, and the ratio $(R^{2+}+R^{3+})/Si$ lies between 1.285 and 1.388 (mean = 1.33) compared to 1.5 for ideal serpentine structures. These ratios are consistent with the structural model proposed by Guggenheim et al. (1982) which can be regarded as a two-dimensional analogue of the one-dimensional wave modulations of the antigorite structure in response to the need to

TABLE 1.11. Analyses of chrysotile, pecoraite, antigorite and ferroan antigorite

	1	2	3	4	5	6
SiO_2	42.02	41.90	41.33	42.44	43.04	31.0
Al_2O_3	0.52	none	0.80	0.64	none	1.4
TiO_2	none		0.02	none	0.02	
Fe_2O_3	0.19	0.34	1.29	0.19	0.87	
Cr_2O_3						
FeO	0.11	none	0.08	0.03	0.16	0.7
MnO	0.03	0.04	0.04	0.03	0.15	
MgO	41.44	42.58	41.39	42.76	40.64	0.5
NiO						51.5
CaO	none	none	trace		0.85	0.4
Na_2O				0.06	0.07	
K_2O				0.08	0.01	
H_2O^+	14.04	13.92	13.66	13.58	13.76	9.7
H_2O^-	1.64	0.99	1.57	0.50	0.41	4.1
F		none				
Total	99.99	99.97	100.18	100.31	100.77	99.3

Numbers of cations on basis of $O_{10}(OH)_8$

	1	2	3	4	5	6
Si	4.00	3.97	3.93	3.96	4.07	4.03
Al			0.07	0.04		
Fe^{3+}		0.02				
Σ tet.	4.00	3.99	4.00	4.00		4.03
Al	0.06		0.02	0.03		0.21
Ti						
Fe^{3+}	0.01		0.09	0.01	0.06	
Cr^{3+}						
Fe^{2+}	0.01		0.01		0.01	0.08
Mn					0.01	
Mg	5.88	6.02	5.86	5.94	5.74	0.10
Ni						5.39
Σ oct.	5.96	6.02	5.98	5.98	5.82	5.78
Ca						0.06
Na				0.01	0.01	
K				0.01		

1. Chrysotile, Gila County, Arizona, USA (Faust and Fahey 1962, sample F-20).
2. Chrysotile, Montville, New Jersey, USA (Faust and Fahey 1962, sample F-24).
3. Chrysotile, Transvaal, South Africa. (Brindley and Zussman 1957). Note that what appears to be the same analysis in Deer et al. (1962) has $SiO_2 = 41.83\%$ and $Al_2O_3 = 0.30\%$.
4. Chrysotile, Montville, New Jersey, USA (Faust and Fahey 1962, sample F-25).
5. Chrysotile, Balmant Corners, New York, USA. Total includes $CO_2 = 0.74\%$ and $Cl = 0.05\%$ (Faust and Fahey 1962, sample F-56). CaO, CO_2 and Cl omitted from calculation of formula.
6. Pecoraite, formed in the Wolf Creek meteorite (Faust et al. 1969).

articulate the even larger Fe^{2+}-rich octahedral sheet with an entirely siliceous tetrahedral sheet. The small $(R^{2+} + R^{3+})/Si$ ratios of greenalites are in accord with a defective 1:1 layer structure in which some octahedra are omitted around the domed islands of six-membered rings of tetrahedra.

A mineral designated manganoan lizardite (analysis 10, Table 1.12) has affinities with greenalites and caryopilites in having a modulated layer structure and a deficiency of octahedral cations relative to Si. It has $(R^{2+} + R^{3+})/Si = 1.225$, much smaller than those of lizardites which have ratios between 1.53 and 1.70 but similar to those of greenalites, 1.285–1.388, and caryopilites, 1.16–1.33.

	7	8	9	10	11	12
	44.52	43.45	43.60	44.50	41.50	37.84
	none	0.81	1.03	1.41	3.18	0.48
		0.02	0.01	none		
	0.12	0.88	0.90	none	1.81	
	0.07	0.69	0.81	0.06	0.28	
	0.002	none	0.04	0.35	4.10	21.03
	41.88	41.90	41.00	none	0.09	
				41.56	35.44	25.00
				0.095	0.16	2.53
	0.64	0.04	0.05	0.02	0.10	
		0.05	0.01	none		
		0.02	0.03	none		
	12.64	12.29	12.18	12.36	12.60	13.29
	0.06	0.04	0.08	none	0.80	
	0.31					
	100.24	100.19	99.82	100.36	100.06	100.17
	Numbers of cations on basis of $O_{10}(OH)_8$					
	4.09	3.99	4.02	4.05	3.94	3.97
		0.01			0.06	0.03
	4.09	4.00	4.02	4.05	4.00	4.00
		0.08	0.11	0.15	0.29	0.03
	0.01	0.06	0.06		0.13	
					0.02	
	0.01	0.05	0.06	0.03	0.33	1.84
					0.01	0.23
	5.74	5.73	5.63	5.64	5.01	3.91
				0.01	0.01	
	5.76	5.92	5.86	5.83	5.80	6.01
	0.06				0.01	
		0.01				

7. Antigorite, Smithfield, Rhode Island, USA (Faust and Fahey 1962, sample F-8). Total corrected for $O \equiv F = 100.11$.
8. Antigorite, Mikonui, New Zealand (Zussman 1954).
9. Antigorite, Caracas, Venezuela. Total includes 0.02% Cr_2O_3 (Hess et al. 1952).
10. Antigorite, State Line Pits, Low's Mine, Pennsylvania, USA (Faust and Fahey 1962, sample F-1).
11. Antigorite, Val Antigorio, Piedmont, Italy (Faust and Fahey 1962, sample F-15).
12. Ferroan antigorite (jenkinsite), Monroe, Orange County, New York, USA (Frondel 1962).

1.4.4.4 Caryopilite

Bailey (1980b) and Bayliss (1981) include caryopilite, a Mn-rich hydrous layer silicate, in the serpentine group. The name, however, is also currently applied to another Mn-mineral structurally related to the friedelite family (Peacor and Essene 1980; Dunn et al. 1981). In this chapter we are concerned with the serpentine-like mineral only, for which we use the name caryopilite.

Guggenheim et al. (1982) have shown that caryopilite has a two-dimensionally modulated layer structure very similar to that of greenalite except that the diameter of the domed tetrahedral islands is smaller, about 17 Å in caryopilite compared with 23 Å in greenalite. Their

TABLE 1.12. Electron microprobe analyses of greenalite, caryopilite, manganoan lizardite and tosalite

	1	2	3	4	5	6	7
SiO_2	34.7	36.5	34.4	36.5	36.3	35.4	35.4
Al_2O_3	0.90	0.25	0.05	0.16	0.10	0.78	2.38
$FeO^\dagger$	47.3	40.2	55.5	46.7	50.4	0.72	0.95
MnO	0.15	8.71	0.18	1.50	0.06	52.1	49.2
MgO	4.98	3.75	0.26	4.53	3.80	0.58	1.23
Na_2O	0.00	n.d.	0.08	n.d.	0.17	0.15	0.05
K_2O	0.00	n.d.	0.00	n.d.	0.04	0.02	0.03
CaO	0.00	n.d.	0.00	0.02	0.00	0.00	0.16
Total	88.03	89.41	90.47	89.41	90.87	89.75	89.40

Numbers of cations on basis of Si = 4.00 for 1–10; on basis of $O_{10}(OH)_8$ for 11–14

	1	2	3	4	5	6	7
Si	4.00	4.00	4.00	4.00	4.00	4.00	4.00
Al							
Σ tet.	4.00	4.00	4.00	4.00	4.00	4.00	4.00
Al	0.12	0.03	0.01	0.02	0.01	0.10	0.32
Fe^{2+}	4.56	3.68	5.40	4.28	4.65	0.07	0.09
Mn	0.02	0.81	0.02	0.14	0.01	4.99	4.70
Mg	0.86	0.61	0.04	0.74	0.62	0.10	0.21
Σ oct.	5.56	5.13	5.47	5.18	5.29	5.26	5.32
Na			0.02		0.04	0.03	0.01
K					0.01		
Ca							0.02

† All iron reported as FeO.
1. Greenalite, Biwabik iron formation, Minnesota, USA.
2. Greenalite, Bluebell mine, Riondel, British Columbia, Canada.
3. Greenalite, near Glenluce, Wigtownshire, Scotland.
4. Greenalite, Sokoman iron formation, Knob Lake, Labrador, Canada.
5. Greenalite, Gunflint iron formation, Ontario, Canada.
6. Caryopilite, Hurricane Claim, Olympic Peninsula, Washington, USA.

electron microprobe analyses (analyses 6–9, Table 1.12) which treat all manganese as Mn^{2+}, have $(R^{2+} + R^{3+})/Si$ ratios between 1.16 and 1.33 (mean = 1.25), even smaller than those of greenalites. The smaller ratios are possibly in response to the need to adjust for the larger misfit between Mn^{2+}-rich octahedral sheet and the siliceous tetrahedral sheet.

1.4.4.5 Tosalite

The name tosalite was suggested (Yoshimura 1967; see Fleischer 1970) for a serpentine-like mineral

$$(Fe^{2+}_{3.403}Mn_{0.977}Mg_{0.356}Al_{0.531}Fe^{3+}_{0.364}Ti_{0.004})(Si_{3.760}Al_{0.240})O_{10}(OH)_8$$

from the Matsuo mine, Japan. Bayliss (1981) regards it as manganoan greenalite and tosalite an unnecessary name. Guggenheim et al. (1982) have analysed four tosalites by electron microprobe (analyses 11–14, Table 1.12). Unlike greenalites, satisfactory structural formulae can be calculated on the basis of an $O_{10}(OH)_8$ anion framework; these have $R^{3+}_{tet}(Al^{3+})$ between 0.05 and 0.21, octahedral Al between 0.30 and 0.64 and $(R^{2+} + R^{3+})_{oct}$ between 5.64 and 5.92. The average formula derived from the microprobe analyses, which treat all iron as Fe^{2+}, is

$$(Fe^{2+}_{3.918}Mn^{2+}_{0.741}Mg_{0.587}Al_{0.526})(Si_{3.888}Al_{0.112})O_{10}(OH)_8$$

	8	9	10	11	12	13	14
	37.9	38.9	45.1	32.8	32.9	34.7	34.6
	0.78	1.44	0.51	3.05	5.65	5.09	4.96
	0.94	0.48	0.65	43.6	40.6	38.5	39.7
	47.0	47.0	16.25	5.87	6.54	7.97	8.52
	2.78	2.15	27.0	4.16	4.14	3.37	2.82
	0.00	0.00	0.00	0.00	0.00	0.22	0.05
	0.00	0.12	0.00	0.10	0.00	0.08	0.02
	0.27	0.00	0.13	0.00	0.00	0.22	0.16
	89.67	90.09	89.64	89.49	89.83	90.15	90.83

Numbers of cations on basis of Si = 4.00 for 1–10; on basis of $O_{10}(OH)_8$ for 11–14

	8	9	10	11	12	13	14
	4.00	4.00	4.00	3.87	3.79	3.95	3.94
				0.13	0.21	0.05	0.06
	4.00	4.00	4.00	4.00	4.00	4.00	4.00
	0.10	0.17	0.05	0.30	0.56	0.64	0.61
	0.08	0.04	0.05	4.30	3.91	3.67	3.78
	4.20	4.09	1.22	0.73	0.64	0.77	0.82
	0.44	0.33	3.57	0.59	0.71	0.57	0.48
	4.82	4.63	4.89	5.92	5.82	5.65	5.69
						0.05	0.01
		0.02				0.01	
	0.03		0.01			0.03	0.02

7. Caryopilite, Olympic Mountains, Washington, USA.
8. Caryopilite, Fallota, Grisons, Switzerland.
9. Caryopilite, Ichinomata mine, Kunamoto Prefecture, Japan.
10. Manganoan lizardite (?), Harstigen mine, Persberg district, Värmland, Sweden.
11, 12, 13, 14. Tosalite granules from Matsuo mine, Kochi Prefecture, Japan.
All analyses from Guggenheim et al. 1982.

Tosalites appear to differ in two respects from the other serpentine minerals. Their $R^{3+}_{oct}/R^{3+}_{tet}$ ratios are larger than any of the planar serpentines. This is true even for the samples analysed by electron microprobe in which none of the iron is treated as Fe^{3+} (Fig. 1.2); secondly, of the six analyses which plot appreciably away from the edges of the triangle in Fig. 1.1, five are tosalites; the other is a greenalite (analysis 2, Table 1.11).

Even though tosalites have larger Al and smaller Si contents than greenalites, Guggenheim et al. (1982) regard them as manganoan greenalites. They comment, however, that "It is unlikely that a large amount of Al_2O_3 can be incorporated in a true greenalite structure" but do not suggest a limiting Al content.

1.4.5 Relations between serpentine-group and chlorite-group minerals

Hydrothermal syntheses by Yoder (1952), Nelson and Roy (1958) and Gillery (1959) have shown that for compositions at which both structural types are formed, the chlorite is the higher temperature form. In terms of the formula

$$(R^{2+}_{6-x-y}R^{3+}_x \square_y)(Si_{4-x+2y}Al_{x-2y})O_{10}(OH)_8$$

chlorites are formed above 500 °C for $0.5 < x - 2y < 2$ and serpentine minerals below 450 °C for $0 < x - 2y < 2$, although in natural materials there are few, if any, serpentines with $0.25 < x - 2y < 0.75$ (Fig. 1.2). Serpentines whose compositions overlap those of chlorites are berthierines, amesite, kellyite, cronstedtite and the nine-layer serpentine described by Jahanbagloo and Zoltai (1968).

1.4.6 Chemical differences among serpentine-group minerals

The serpentine group minerals can be broadly classified in terms of the predominant divalent cation into magnesian serpentines (chrysotiles, antigorites, lizardites, amesite), ferroan serpentines (berthierines, tosalites, cronstedtites), manganoan serpentines (kellyite, caryopilites), and nickel serpentines (pecoraite and the 7 Å-phases in garnierites which have been named nepouites). Figure 1.1 shows that in all but the Mg–Ni system (and there must be some doubt about the purity of these specimens) the dominant cation makes up more than 70% of the total divalent cation content.

There has been discussion of whether the magnesium-rich crystallographically distinct varieties chrysotile, antigorite, and lizardite are "polymorphs". Page (1968) concluded that they differed chemically. Whittaker and Wicks (1970) critically surveyed the data in the literature. They found few samples they believed had been correctly identified and accurately analysed. From a consideration of analyses of well-authenticated specimens they concluded that:

1. Antigorite has a larger SiO_2 and smaller MgO and H_2O weight per cent.
2. Chrysotile has a small Al_2O_3 content.
3. Lizardite has a large Fe_2O_3/FeO ratio and a small FeO content.
4. Chrysotile and lizardite contain H_2O^+ in excess of the ideal formula.
5. Antigorite has the largest $FeO/(FeO + Fe_2O_3 + Al_2O_3)$ ratio and lizardite the lowest.
6. The amount of substitution by Al and Fe tends to be in the order chrysotile < lizardite < six-layer serpentine though the ranges overlap and substitution in antigorite extends over the range of the other highly magnesian serpentines.

The number of samples is however small – there being 7 chrysotiles, 5 lizardites, 3 six-layer serpentines and 7 antigorites that Whittaker and Wicks consider to be sufficiently well characterized for a study of this kind and because the differences are small, conclusions about compositional differences between varieties are still tentative. It is clear however that chemical composition alone cannot be used to identify individual samples because, for example, chrysotile (analysis 3, Table 1.11) from a vein and adjacent lizardite from the matrix (analysis 3, Table 1.9) have almost identical compositions as have a similar pair of samples from Aboutville, New York, USA (Deer *et al.* 1962, p. 176, analyses 1 and 5, Table 29).

1.5 THE TALC–PYROPHYLLITE GROUP

Ideally talc and pyrophyllite consists of electrically neutral 2:1 layers; talc is trioctahedral, $Mg_6Si_8O_{20}(OH)_4$, and pyrophyllite is dioctahedral, $Al_4Si_8O_{20}(OH)_4$. Recently however, it has been shown that this mineral group is more complex than it was formerly thought to be.

A pyrophyllite rich in Fe^{3+} has been reported (Chukhrov *et al.* 1979), the possible coupled substitution $R^{3+} + H^+ \rightleftharpoons Si$ has been proposed in synthetic talc (Forbes 1969) and pyrophyllite (Rosenberg 1974), limited substitution of Al in natural (McKie 1959; Spear *et al.* 1981) and synthetic talcs (Fawcett and Yoder 1966) has been reported, and a material named sodian aluminian talc has been described (Schreyer et al. 1980). A nickel-rich talc has been identified (De Waal 1970a), and complete solid solution of Ni for Mg in fine-grained disordered talc-like minerals has been authenticated (Brindley et al. 1977, 1979). Minnesotaite, formerly believed to be the ferroan analogue of talc has been shown to have a structure more complex than that of talc (Guggenheim and Bailey 1982) and a high-pressure "talc hydrate" has been synthesized by Bauer and Sclar (1981).

1.5.1 Pyrophyllite

There are few analyses of well-characterized pyrophyllites but a selection representative of the range of natural compositions presented in Table 1.13 shows Si/R^{3+} ratios ranging from about 1.75 to 1.97 compared to the ideal ratio of 2.0. Structural formulae calculated on the basis of an $O_{20}(OH)_4$ anion framework show tetrahedral Al in the range 0.05 to about 0.5 atoms. Excepting ferripyrophyllite (vide infra), Al ions make up between 96.5% and 98.5% of the octahedral cations and octahedral cation totals lie between 3.9 and 4.1. Alkali and alkaline earth cations are recorded in almost all pyrophyllite analyses up to a maximum of about 0.2 cations and 0.25 equivalents per formula unit. Most pyrophyllites have fewer large cations, usually less than 0.05 per formula unit. Interlayer cations are not essential because pyrophyllites have been synthesized from starting materials containing Al_2O_3, SiO_2 and H_2O only (Rosenberg 1974). Rosenberg and Cliff (1980) have shown by EMMA analyses (Section 1.2.2.5) that pyrophyllite having $Si/Al = 1.722$ can be synthesized from gels and this seems to be the minimum ratio found in the pyrophyllite structure. To explain correlations between basal spacings, which range from near 9.19 Å in natural pyrophyllites to about 9.33 Å for some synthetic materials, and the intensity of OH-stretching bands in infrared absorption spectra, Rosenberg (1974) proposed that the coupled substitution $Al^{3+} + OH^- \rightleftharpoons Si^{4+} + O^{2-}$ occurs leading to the structural formula

$$(Al_4)(Si_{8-x}Al_x)O_{20-x}(OH)_{4+x}$$

and supported this hypothesis from the apparent inverse relationship between both $R_2O_3\%$ and $H_2O^+\%$, and $SiO_2\%$ in natural pyrophyllite analyses. If Rosenberg's suggested coupled substitution occurs in natural materials there should be a linear 1:1 correspondence between tetrahedral R^{3+} and "extra" OH (total OH $-$ 4.00) in structural formulae calculated on the basis of 44 anions, making use of determined H_2O^+ contents. The results in Table 1.13 and those for other pyrophyllites (Deer et al. 1962, p. 118; Iwao and Udagawa 1969, p. 82) do not show this relationship although there is a trend towards larger amounts of extra OH^- with larger amounts of tetrahedral Al. The failure to find an exact relationship may well arise because the allocation of weight lost or water lost between 105 °C and 1000 °C (H_2O^+ of conventional analyses) to structural hydroxyls is unreliable even for relatively coarse-grained materials (cf. Section 1.2.1.2).

Unlike synthetic pyrophyllites, natural pyrophyllites show little variation in basal spacing but a wide range of Si/R^{3+} ratios. The basal spacing of triclinic pyrophyllites from Brazil (Lee and Guggenheim 1981) and New Zealand (Swindale and Hughes 1968; Wardle and Brindley 1972) with $Si/R^{3+} = 1.962$ and about 1.76 respectively, are 9.1903 Å and 9.1970 Å; monoclinic pyrophyllites from USA (Rayner and Brown 1966) and Japan (Brindley and Wardle 1970) have $Si/R^{3+} = 1.951$ and 1.83–1.84 respectively, and basal spacings, 9.1938 Å and 9.1961 Å. Even if

42 THE CHEMICAL CONSTITUTION OF CLAYS

TABLE 1.13. Analyses and structural formulae of pyrophyllite

	1	2	3
SiO_2	64.88	66.04	65.96
Al_2O_3	28.64	28.25	28.25
TiO_2	0.02		trace
Fe_2O_3	0.48	0.64	0.18
FeO			
MnO	0.02	0.02	
MgO	0.08	} 0.12	
CaO	0.03		
Na_2O	0.03	} 0.06	
K_2O	0.04		
H_2O^+	5.47	5.02	5.27
H_2O^-	0.09		0.14
Total	99.78	100.13	99.80

Numbers of ions calculated on basis of (a) $O_{20}(OH)_4$, (b) $(O,OH)_{22}$

	a	b	a	b	a	b
Si	7.88	7.81	7.94	7.94	7.97	7.93
Al	0.12	0.19	0.06	0.06	0.03	0.07
Σ tet.	8.00	8.00	8.00	8.00	8.00	8.00
Al	3.98	3.87	3.94	3.94	3.99	3.93
Ti						
Fe^{3+}	0.04	0.04	0.06	0.06	0.02	0.02
Fe^{2+}						
Mg	0.01	0.01	0.02	0.02		
Mn						
Σ oct.	4.03	3.92	4.02	4.02	4.01	3.95
Ca						
Na	0.01	0.01				
K	0.01	0.01	0.01	0.01		
Interlayer charge	0.02	0.02	0.01	0.01		
OH^-	4	4.39	4	4.02	4	4.23
O^{2-}	20	19.61	20	19.98	20	19.77
Σ charges	44	43.61	44	43.98	44	43.77
$Si/(Al^{3+}+Fe^{3+})$	1.902		1.955		1.973	

1. Pyrophyllite, Nasum parish, Kristianstad County, Sweden (Deer *et al.* 1962, analysis 2, Table 22).
2. Pyrophyllite, Moore County, North Carolina, USA (Deer *et al.* 1962, analysis 4, Table 22).
3. Pyrophyllite, Tres Cerritos, Mariposa County, California, USA (Deer *et al.* 1962, analysis 6, Table 22).
4. Pyrophyllite, pale blue, Honami mine, Nagano Prefecture, Japan (Iwao and Udagawa 1969, analysis 1, Table 10).

Rosenberg's (1974) structural formula applies to synthetic pyrophyllites (and there is little doubt that there is a relationship between structural OH and basal spacing in these), on currently available evidence it does not apply to natural pyrophyllites because wide ranges of Si/R^{3+} ratios give almost identical basal spacings for both triclinic and monoclinic forms.

1.5.1.1 Ferripyrophyllite

This name has been approved by the IMA Commission on New Minerals and New Mineral Names for an inferred component of mixtures of non-expanding (pyrophyllite-like) and expanding (smectite-like) minerals (Chukhrov *et al.* 1979). The authors do not report chemical

	4		5		6		7
	63.57		64.02		62.7		66.04
	29.25		29.60		29.7		28.15
	0.04		0.09				
	0.10		0.16		1.0		0.64
	0.12		0.08				
	none		none				
	0.37		0.45				0.04
	0.38		0.12				0.01
	trace		trace		0.2		0.04
	0.02		0.01				
	5.66		5.45		5.8		5.27
	0.66		0.84		0.6		
	100.17		100.82		100.0		100.19

Numbers of ions calculated on basis of (a) $O_{20}(OH)_4$, (b) $(O,OH)_{24}$

a	b	a	b	a	b	a	b
7.77	7.67	7.76	7.69	7.70	7.61	7.95	7.92
0.23	0.33	0.24	0.31	0.30	0.39	0.05	0.08
8.00	8.00	8.00	8.00	8.00	8.00	8.00	8.00
3.98	3.83	3.98	3.88	3.99	3.84	3.95	3.89
		0.01	0.01				
0.01	0.01	0.01	0.01	0.10	0.10	0.06	0.06
0.01	0.01	0.01	0.01				
0.07	0.07	0.08	0.08			0.01	0.01
4.07	3.92	4.09	3.99	4.09	3.94	4.02	3.96
0.05	0.05	0.02	0.02				
				0.05	0.05	0.01	0.01
0.10	0.10	0.04	0.04	0.05	0.05	0.01	0.01
4	4.55	4	4.37	4	4.53	4	4.21
20	19.45	20	19.63	20	19.47	20	19.79
44	43.45	44	43.63	44	43.47	44	43.79
1.840		1.829		1.755		1.962	

5. Pyrophyllite, pale green, Honami mine, Nagano Prefecture, Japan (Iwao and Udagawa, 1969, analysis 2, Table 10).
6. Pyrophyllite, near Tairua, Coromandel Peninsula, New Zealand, <2 μm fraction (Swindale and Hughes 1968). Structural formulae calculated after correction for 3% kaolinite and 1% quartz.
7. Pyrophyllite, Ibitiara, Bahia, Brazil (Lee and Guggenheim 1981).

analyses but quote structural formulae,

$$Ca_{0.10}(Fe^{3+}_{3.92}Mg_{0.22})(Si_{7.60}Al_{0.26}Fe^{3+}_{0.14})O_{20}(OH)_4 \cdot 2H_2O$$

and

$$Ca_{0.36}(Fe^{3+}_{3.94}Mg_{0.04})(Si_{3.58}Al_{0.46})O_{20}(OH)_4 \cdot 3H_2O$$

for mixtures from Strassenschacht, German Democratic Republic, containing 15–20% expanding interlayers, and from Mount Tologay, central Kazakhstan, containing 40–50% expanding interlayers respectively. The small Al^{3+}/Fe^{3+} ratios and the virtual absence of other cations in these dioctahedral materials can probably be taken as evidence that a non-expanding Fe-rich pyrophyllite-like mineral exists as a component of these mixtures.

44 THE CHEMICAL CONSTITUTION OF CLAYS

TABLE 1.14. Analyses and structural formulae of talc, willemseite, minnesotaite, kerolite and pimelite

	1	2	3	4	5	6	7
SiO_2	63.22	63.90	51.83	55.95	52.89	51.26	50.22
Al_2O_3		0.03	0.38	0.06	1.45	0.97	1.49
TiO_2		0.10					
Fe_2O_3	0.33	0.21	1.77				
FeO	0.89		0.31	22.54†	26.48†	34.25†	38.75†
MgO	30.45	31.49	7.09	16.18	11.45	6.67	3.35
NiO			34.55				
CoO			0.46				
MnO				0.05	0.14	0.06	0.03
CaO		0.08	0.28	0.03	0.17	0.02	0.05
Na_2O	0.02	0.02					
K_2O	0.05	0.01		0.02	0.56	0.35	0.23
F	0.06						
H_2O^+	4.78	4.86	3.61				
H_2O^-	0.20		0.05				
Total	100.00	100.70	100.33	94.83	93.14	93.58	94.12

Numbers of cations calculated on the basis of $O_{20}(OH)_4$ formula unit

	1	2	3	4	5	6	7
Si	8.04	8.02	7.83	7.94	7.85	7.88	7.85
Al		0.07		0.01	0.15	0.12	0.15
Fe^{3+}			0.10				
Σ tet.	8.04	8.02	8.00	7.95	8.00	8.00	8.00
Al					0.10	0.06	0.13
Ti		0.01					
Fe^{3+}	0.03	0.02	0.10				
Fe^{2+}	0.09		0.04	2.67†	3.29†	4.40†	5.07†
Mg	5.77	5.89	1.60	3.42	2.53	1.53	0.78
Ni			4.19				
Co			0.06				
Mn				0.01	0.02	0.01	
Σ oct.	5.89	5.92	5.99	6.10	5.94	6.00	5.98
Ca		0.01	0.05		0.03		0.01
Na							
K	0.01				0.11	0.07	0.05
Interlayer charge	0.01	0.02	0.10		0.17	0.07	0.07

1. Talc, Harford County, Maryland, USA (Rayner and Brown, 1973); after correction for O≡F total=99.97.
2. Talc, fine-grained, from Manchuria (Brindley et al. 1977).
3. Willemseite, Scotia Talc Mine, Barberton, Transvaal, Union of South Africa (De Waal 1970a).
4. Talc, ferroan, Gunflint Iron Formation, Minnesota-Ontario. Electron microprobe analysis (Floran and Papike 1978, analysis 2).
5, 6, 7. Minnesotaite, Gunflint Iron Formation, Minnesota-Ontario. Electron microprobe analysis (Floran and Papike 1978, analyses 3, 4 and 9).
8. Talc, aluminian, Mautia Hill, Tanganyika (McKie 1959).

1.5.2 Talc subgroup

The trioctahedral subgroup of the talc–pyrophyllite group consists of the frequently macrocrystalline talc minerals, usually of hydrothermal or metamorphic origin and the fine-grained low-temperature minerals of the kerolite–pimelite series. For talc itself, almost all the data in mineralogical texts, with the exception of minnesotaite (*vide infra*), described by Gruner (1944b) as the ferroan analogue of talc, show only minor deviations from the ideal composition

	8	9	10	11	12	13	14
	63.49	60.76	55.3	58.63	61.55	53.3	50.6
	3.95	2.87	4.8	0.05	0.02	0.18	0.04
		0.06			0.10	0.01	0.08
				0.10	0.01	0.23	0.28
	$0.56^{\dagger}$	$4.25^{\dagger}$					
	26.66	27.34	29.8	30.84	29.44	18.9	6.58
				0.05	0.20	18.3	32.7
	0.06	0.03					
	0.64	0.01	0.06	0.33	0.32	0.08	0.24
		0.32	2.7		0.04	0.22	0.06
		0.17	0.09		0.10	0.16	0.04
		0.11					
	4.64			8.29	8.02	7.78	8.02
	0.12			3.87			
	100.12	95.92	92.75	102.16	99.80	99.16	98.64
			Numbers of cations calculated on the basis of $O_{20}(OH)_4$ formula unit				
	7.99	7.80	7.32	7.87	8.07	7.79	8.06
	0.01	0.20	0.68	0.01		0.03	
				0.01		0.03	
	8.00	8.00	8.00	7.89	8.07	7.85	8.06
	0.57	0.23	0.07				0.01
		0.01			0.01		0.01
							0.03
	$0.06^{\dagger}$	$0.46^{\dagger}$					
	5.00	5.23	5.88	6.17	5.75	4.12	1.56
				0.01	0.02	2.15	4.19
	0.01						
	5.64	5.93	5.95	6.18	5.78	6.27	5.80
	0.09		0.01	0.05	0.05	0.01	0.04
		0.08	0.69		0.01	0.06	0.02
		0.03	0.02		0.02	0.03	0.01
	0.18	0.11	0.73	0.10	0.13	0.11	0.11

9. Talc, sodian aluminian, Post Pond Volcanics, Vermont, USA. Electron microprobe analysis (Spear *et al.* 1981).
10. Talc, sodian aluminian, Derrag, Tell Atlas, Algeria. Electron microprobe analysis (Schreyer *et al.* 1980, analysis 11.2, Table 3).
11. Kerolite, Goles Mountain, Yugoslavia (Maksimovic, 1966; see also Brindley *et al.* 1977).
12. Kerolite, Kremze, Bohemia (Brindley *et al.* 1977).
13. Kerolite, nickeloan, Sua-Sua, Indonesia (Brindley *et al.* 1979, analysis 10).
14. Pimelite, Frankenstein, Lower Silesia (Brindley *et al.* 1979, analysis 18).
† All iron reported as FeO.

$Mg_6Si_8O_{20}(OH)_4$. Analyses 1 and 2 of Table 1.14 represent this type of material. In these classical talcs, there is little or no replacement of Si^{4+} in tetrahedral sites by Al^{3+} or Fe^{3+} ions, octahedral occupancy is close to six, Mg^{2+} is the overwhelmingly dominant octahedral cation (usually $>98\%$), alkali or alkaline earth cations (interlayer) are negligible and H_2O^+ values are close to the theoretical 4.75%.

In the last fifteen years or so it has become clear that the talc structure can accommodate considerable deviations from the ideal formula. De Waal (1970a) has described a nickel–talc (analysis 3, Table 1.14), which has been given the name willemseite. Forbes (1969, 1971) has shown by hydrothermal syntheses at temperatures between 375 and 675 °C, pressures of 1 and 2 kbar, and a range of oxygen fugacities, that iron can be taken into the talc structure, but under the conditions he used, the amount of substitution was limited. The largest amount, $Fe/Fe + Mg = 0.2$, occurred in products of synthesis at the lowest oxygen fugacity (magnetite–iron buffer) where most of the iron is presumably in the ferrous state. At the largest oxygen fugacity studied (hematite–magnetite buffer) in which much of the iron is likely to be Fe^{3+}, the maximum $Fe/Fe + Mg$ ratio was 0.062. Forbes did not determine Fe^{2+}/Fe^{3+} ratios in his products and he suggested that the following substitutions were possible:

$$Mg^{2+} \rightleftharpoons Fe^{2+}, \qquad Fe^{3+} + H^+ \rightleftharpoons Si^{4+}, \qquad Fe^{2+} + OH^- \rightleftharpoons Fe^{3+} + O^{2-}$$

Taking account of these Forbes (1969) proposed

$$(Mg_{6-z-y}Fe^{2+}Fe^{3+}_y)(Si_{8-x}Fe^{3+}_x)O_{20+y-x}(OH)_{4-y+x}$$

as a possible general structural formula for talcs synthesized in the $Mg_6Si_8O_{20}(OH)_4$–$Fe_6Si_8O_{20}(OH)_4$ system.

Discussion of Fe^{2+} for Mg substitution leads to consideration of minnesotaite, described by Gruner (1944b) as the ferroan analogue of talc. Floran and Papike (1978) have presented 10 microprobe analyses of ferroan talc and minnesotaites from the Gunflint Iron Formation which chemically appear to show a range of solid solution for $Fe/(Fe + Mg) = 0.43$ to $Fe/(Fe + Mg) = 0.87$ (analyses 4–7, Table 1.14). All the iron is treated as FeO in their analyses, which is probably a justified approximation because the analyses of minnesotaites by Gruner (1944b) and Blake (1965) show that more than 95% of the iron in their materials was in the ferrous state. The synthetic talcs of Forbes (1969) cover the $Fe/(Fe + Mg)$ range of 0 to 0.2 for synthesis temperatures greater than about 400 °C and extrapolation to lower temperatures (Forbes 1971) suggests that larger ratios up to at least 0.4 are possible. The results of Floran and Papike (1978) and Forbes (1969, 1971) appear to show that there is a complete range of solid solution between Mg and Fe in the talc–minnesotaite series. Recently however Guggenheim and Bailey (1982) have carefully examined a minnesotaite from the Cuyuna district, Minnesota, but give no analysis of the material they examined. Its composition is probably similar to that reported by Blake (1965) which yields the structural formula

$$Fe^{2+}_{4.80}Mg_{0.78}Al_{0.18}Fe^{3+}_{0.06})(Si_{7.91}Al_{0.09})O_{20}(OH)_4$$

Guggenheim and Bailey's studies showed that the structure of their sample differs from that of talc. The structural relation between talcs and minnesotaites is as yet unclear but chemically there appears to be a complete series of compositions from $Mg_6Si_8O_{20}(OH)_4$ to near $Fe^{2+}_6Si_8O_{20}(OH)_4$.

Trivalent cations are taken into the talc structure only to a limited extent. McKie (1959) described an aluminian talc from Tanganyika containing 3.95% Al_2O_3 (analysis 8, Table 1.14) in which almost all the Al is in octahedral sites, only 5.64 of which are occupied; this talc also contains 0.64% CaO, equivalent to an interlayer charge (if the Ca is in interlayer sites) of 0.18 equivalents per $O_{20}(OH)_4$ formula unit. The structural formula indicates that $2Al(oct) + \square(oct) \rightleftharpoons 3Mg$ is the dominant substitution in this specimen with a minor contribution only from the substitution $Ca(interlayer) + \square(oct) \rightleftharpoons Mg$. Fawcett and Yoder (1966) have synthesized talc in the system $MgO–Al_2O_3–SiO_2–H_2O$ assuming $Al(oct) + Al(tet) \rightleftharpoons Mg + Si$ substitution. They found that the maximum Al_2O_3 content of the starting material which produced talc was 4.0 wt% at water pressures up to 10 kbar. Spear *et al.* (1981) have made microprobe analyses of talc associated with wonesite (see Section 1.7.3.1) from Vermont. Their published analysis is given in Table 1.14 (analysis 9) but they added that the K_2O content of this specimen was unusually large. Other specimens from the same locality contained 2.5–3.7% Al_2O_3 and 0.3–0.6% Na_2O. In the analysed material there are about equal amounts of Al in octahedral and tetrahedral sites indicating that Al(oct) + Al(tet) ⇌ (Si + Mg) is dominant substitution in this specimen with small contributions from

$$2Al + \square \rightleftharpoons 3Mg \text{ and } 2Na \text{ (interlayer)} + \square\text{(oct)} \rightleftharpoons Mg$$

Schreyer *et al.* (1980) have published electron microprobe analyses of materials from Algeria, to which they gave the name sodian aluminian talc. Their analysis of an Na- and Al-rich specimen is given in Table 1.14 (analysis 10). Other analyses report 3.36–4.82% Al_2O_3 and 1.78–2.7% Na_2O. The average structural formula derived from 10 analyses is

$$(Na_{0.58}K_{0.01}Ca_{0.02})(Mg_{5.92}Al_{0.04})(Si_{7.41}Al_{0.59})O_{20}(OH)_4$$

The average composition of these 10 Na–Al talc analyses is compared with those of other trioctahedral layer silicates in Fig. 1.6 (Section 1.7.3.1). These Na–Al talc compositions lie in the field of the trioctahedral smectites with layer charges between about 0.5 and 0.75 per formula unit. Schreyer *et al.* are uncertain that there is genuine solid solution of Na and Al in these specimens and suggest that they "might represent disordered mixed layer phases of talc and Na-phlogopite". In view of the broad basal reflections reported by Schreyer *et al.*, it seems possible that these so-called sodian aluminian talcs may be complex intergrowths (Veblen 1983a, 1983b) of layer silicate units having layer charges between that of talc and that of mica. Determination of expandability in water and ethylene glycol, and exchangeability of the interlayer Na is required and, as suggested by Veblen (1983a), examination by HRTEM might clarify their status.

1.5.2.1 Kerolite–pimelite series

Most of the talcs discussed above were formed in hydrothermal or metamorphic environments. Talc-like minerals formed in low-temperature environments, usually by alteration of ultramafic rocks, have been known for many years. They frequently occur as intimate mixtures with fine-grained serpentine minerals in varying proportions and both components are poorly ordered; the names deweylite (Bish and Brindley 1978; Speakman and Majumbar 1971) and garnierite (Brindley and Pham Thi Hang 1973) can be used for Mg- and Ni-rich mixtures respectively. Only recently have relatively pure specimens of the components of these mixtures been studied.

Brindley *et al.* (1977, 1979) have shown that the structure and composition of the 2:1 layers in these materials, named kerolite and pimelite respectively, for Mg-rich and Ni-rich varieties, are similar to those in talc and willemseite. They are notably free of cations other than Mg and Ni, their layers are turbostratically stacked and they contain more water (H_2O^+) than talc and willemseite. Analyses and structural formulae of representative materials are given in Table 1.14 (analyses 11–14). The small excess of octahedral cations in some specimens can be accounted for by 5–10% serpentine impurity. Solid solution appears to be continuous between Mg and Ni end-members. Dehydration of kerolite occurs in two stages (Brindley *et al.* 1977); weight loss equivalent to the theoretical hydroxyl content, $2H_2O$ per $O_{20}(OH)_4$ formula unit, is lost between 700 and 850 °C, but additional weight loss equivalent to about 1.64–2.4 H_2O is reversibly lost between 105 and 700 °C. Pimelites lose weight similarly in two stages (Brindley *et al.* 1979). Most specimens of kerolite and pimelite contain a small proportion of interlayers that expand slowly in ethylene glycol but not in water. The kerolite from Goles Mountain, Yugoslavia, has been shown to contain 0.04 Ca and 0.06 Mg ions per $O_{20}(OH)_4$ formula unit (290 μeq g^{-1}) that are readily exchangeable (Maksimovic 1966). The water lost between 105 and 700 °C is believed to be located between the 2:1 layers and also on the large external surface, 196 m^2 g^{-1}, measured by N_2-adsorption for a kerolite from North Carolina (Brindley *et al.* 1977).

1.5.2.2 The 10 Å talc hydrate of Bauer and Sclar (1981)

A synthetic phase in the system MgO–SiO_2–H_2O, stable at pressures between 32 and 95 kbar and at temperatures between 375 and 535 °C, has been described by Bauer and Sclar (1981). This phase is a trioctahedral phyllosilicate that has unit cell dimensions similar to K-phlogopite. Chemically it resembles talc with Mg/Si = 0.75 but was shown to contain additional chemically bound water equivalent to about $2H_2O$ per $O_{20}(OH)_4$ formula unit. Weight loss on heating occurs in two distinct stages; approximately equal weights are lost (each $\equiv 2H_2O$) between 100 and 750 °C and between 750 and 900 °C and the first weight loss is accompanied by a decrease in basal spacing from 10 Å (mica-like) to 9.35 Å (talc-like). Bauer and Sclar believe that the water lost between 105 and 750 °C represents H_3O^+, formed by association of interlayer H_2O with structural hydroxyls and that the phase is best represented by the structural formula

$$(H_3O^+)_2(Mg_6Si_8)O_{22}(OH)_2$$

Chemically the composition of this phase is identical to ideal kerolite

$$Mg_6Si_8O_{20}(OH)_4 \cdot 2H_2O$$

despite their very different conditions of formation.

1.6 2:1 GROUP WITH INTERLAYER CATIONS, 0 < X < 1.9

Isomorphous substitution within the 2:1 layer structures of the smectite and vermiculite subgroup minerals is unbalanced so that the layers carry a net negative charge; this is balanced by cations interleaved between the 2:1 layers. The interlayer cations are often hydrated and readily exchangeable, the degree of hydration depending upon the humidity of the ambient atmosphere. Extreme desiccation will dehydrate the minerals almost completely but on exposure to atmospheric humidity the minerals rehydrate rapidly, the distance between the layers increasing to accommodate the sorbed water molecules.

The presence of interlayer cations presents a special problem in the analysis of these subgroup minerals. It is important to establish the negative charge on the 2:1 layers as this affects the swelling properties of the minerals. However, if the interlayer cations include those also present in the 2:1 layers (e.g. Mg, Al) it may not be possible to decide how they are distributed between the interlayer and the 2:1 layer. In many early studies of vermiculite, for instance, the natural mineral was analysed and as this usually contains both interlayer and structural Mg, a further assumption (e.g. that the octahedral sites were fully occupied: Foster 1963) was needed to calculate a structural formula. As the interlayer cations in vermiculite are exchangeable it is possible to displace them with another cation not present in the natural mineral, determining the displaced cations independently, and then analysing the mineral in its new cation form; the two estimates of interlayer charge should, of course, agree. In selecting analyses of smectites and vermiculites for inclusion in this section, preference has been given to those in which an independent estimate of interlayer cations has been made.

Isomorphous substitution is extensive, Si, Al and Fe^{3+} occupying tetrahedral sites and Al, Mg, Fe^{3+} and Fe^{2+} occupying octahedral sites. For smectites, the net negative charge on the layers arising from isomorphous substitutions is defined to lie between 0.50 and 1.20 but it is usually between 0.7 and 1.0 for an $O_{20}(OH)_4$ formula unit, and for vermiculites from 1.3 to 1.9 per formula unit (Bailey 1980a).

Both di- and trioctahedral smectites are common. In the vermiculite group only the trioctahedral variety is common; dioctahedral varieties probably exist but they have not been found free from other minerals so no analyses are available.

1.6.1 Dioctahedral smectites

Dioctahedral smectites can be subdivided chemically into the predominantly aluminian varieties, montmorillonite and beidellite, and the iron-rich variety, nontronite.

1.6.1.1 Aluminian dioctahedral smectites

Montmorillonites are distinguished from beidellites on the basis of the site of the negative charge on the layers. In montmorillonite, for which the ideal structural formula is

$$(Al_{3.15}Mg_{0.85})(Si_{8.00})O_{20}(OH)_4X_{0.85}nH_2O$$

the charge arises from divalent cations, usually Mg, in octahedral sites, whereas in beidellite

$$(Al_{4.00})(Si_{7.15}Al_{0.85})_{20}(OH)_4X_{0.85}nH_2O$$

it arises from Al^{3+} in tetrahedral sites; X is a monovalent interlayer cation. Minerals intermediate between these end-members are common and it is convenient to describe as montmorillonites those minerals in which more than half of the charge originates in octahedral sites, and as beidellites those in which tetrahedral charge predominates.

Many of the analyses in the literature are of impure materials; the analyses of montmorillonites and of beidellites given in the tables have been selected to represent carefully purified materials.

Grim and Kulbicki (1961) subdivided montmorillonites on the basis of their thermal properties into two types named Cheto and Wyoming. Schultz (1969) distinguished six different types of aluminous smectites and showed that differences in thermal properties can be correlated with chemical constitution. Disregarding possible differences in hydroxyl content, aluminous smectites can be divided chemically into four types. Using Schultz's names these are:

1. Wyoming type, characterized by a layer charge <0.85 per $O_{20}(OH)_4$ formula unit in which tetrahedral substitution causes 15 to 50% of the total layer charge (analyses 1–5, Table 1.15). In the sample from Aberdeen, Mississippi, analysis 5, 53% of the total layer charge is due to tetrahedral substitution, but if TiO_2 is treated as an impurity, the structural formula obtained

   $$Al_{2.88}Fe^{3+}_{0.68}Mg_{0.47})(Si_{7.71}Al_{0.29})O_{20}(OH)_4X_{0.67}$$

   has only 44% of the total charge due to tetrahedral substitution.
2. Otay type, with layer charge >0.85 and $<15\%$ of the charge attributable to tetrahedral substitution (analyses 6–8, Table 1.15).

TABLE 1.15. Analyses of sodium-saturated montmorillonites

	1	2	3	4	5	6
SiO_2	64.7	57.50	59.60	54.07	55.68	52.78
Al_2O_3	18.6	20.59	22.17	21.02	19.42	17.91
TiO_2		0.11	0.09	0.14	0.80	
Fe_2O_3	8.12	3.94†	4.32	3.72†	6.51†	0.06†
FeO	0.25					
MnO			trace		0.01	
MgO	4.32	2.45	2.73	2.09	2.30	3.56
CaO			0.14			
Na_2O	3.30	2.87	3.18	2.84	2.48	3.19
K_2O	0.04		0.03			
H_2O^-						
H_2O^+			6.02			
Total	99.33		100.17			

Numbers of cations on the basis of $O_{20}(OH)_4$

	1	2	3	4	5	6
Si	7.85	7.79	7.68	7.65	7.65	7.97
Al	0.15	0.21	0.32	0.35	0.35	0.03
Σ tet.	8.00	8.00	8.00	8.00	8.00	8.00
Al	2.51	3.07	3.05	3.15	2.80	3.16
Ti		0.01	0.01	0.01	0.08	
Fe^{3+}	0.74	0.40	0.42	0.40	0.67	0.01
Fe^{2+}	0.03					
Mg	0.78	0.49	0.52	0.44	0.47	0.80
Σ oct.	4.06	3.97	4.00	4.00	4.02	3.97
Ca			0.02			
Na	0.78	0.75	0.80	0.78	0.66	0.93
K	0.01					
Interlayer charge	0.79	0.75	0.84	0.78	0.66	0.93
% tet. charge	19	28	38	45	53	3

1. Montmorillonite, Redhill, Surrey, England, <0.2 μm fraction, ignited weight basis (Weir 1965).
2. Montmorillonite, Upton, Wyoming, USA, clay fraction, air dry sample (Foster 1953a).
3. Montmorillonite, Clay Spur, Wyoming, USA, 0.2–0.07 μm fraction, dried at 85 C for analysis; ‖ total includes 0.02% P_2O_5 (Earley et al. 1953).
4. Montmorillonite, Belle Fourche, South Dakota, USA, clay fraction, air dry sample (Foster 1953a).
5. Montmorillonite, Aberdeen, Mississippi, USA, clay fraction, air dry sample (Foster 1953a).
6. Montmorillonite, Tatatila, Mexico, clay fraction, air dry sample (Foster 1953a).
7. Montmorillonite, Otay, California, USA, 0.2–0.7 μm fraction, dried at 85 C for analysis; τ total includes 0.05% P_2O_5 (Earley et al. 1953).

3. Tatatila and Chambers type, with layer charge >0.85 and 15 to 50% of the charge caused by tetrahedral substitution (analyses 9–12, Table 1.15).
4. Beidellite type, with layer charge >0.85 and more than 50% of the layer charge arising from tetrahedral substitution (analyses 1–3, Table 1.16).

It would appear, therefore, that these types could be distinguished by their structural formulae. An example of the difficulties in doing so can be illustrated by considering the different structural formulae given in the literature for the Tatatila montmorillonite (analysis 6, Table 1.15).

Ross and Hendricks (1945) give the formula

$$(Al_{3.26}Mg_{0.84})(Si_{7.88}Al_{0.12})O_{20}(OH)_4X_{1.04}$$

2:1 GROUP WITH INTERLAYER CATIONS. 0<X<1.9

	7	8	9	10	11	12
	62.58‡	56.56	66.32	56.60	60.43‡	55.56
	18.44	17.18	22.42	21.24	20.38	22.48
	0.14	0.16		0.00	0.37	0.00
	1.20†	2.80†	3.30	2.0	3.66	0.20
	0.01			0.01	0.03	0.02
	7.30	4.85	4.07	3.46	4.52	3.30
	0.08		0.03		0.14	
	3.40	3.58	3.68	3.71	3.70	3.58
	0.02		0.06		0.04	
	6.47				6.40	
	100.15τ		99.88		100.43ϕ	

Numbers of cations on the basis of $O_{20}(OH)_4$

	7	8	9	10	11	12
	7.93	7.90	7.86	7.70	7.72	7.66
	0.07	0.10	0.14	0.30	0.28	0.34
	8.00	8.00	8.00	8.00	8.00	8.00
	2.68	2.72	2.99	3.10	2.80	3.32
	0.01	0.02			0.04	
	0.11	0.29	0.29	0.29	0.35	0.02
	1.38	1.01	0.72	0.70	0.86	0.68
	4.18	4.04	4.00	4.00	4.05	4.00
	0.01				0.02	
	0.84	0.97	0.85	0.98	0.92	0.96
			0.01		0.01	
	0.86	0.97	0.86	0.98	0.97	0.96
	8	10	16	31	29	35

8. Montmorillonite, Amargosa Valley, California, USA, clay fraction, air dry sample (Foster 1953a).
9. Montmorillonite, Camp Berteau, France, <2 μm fraction, ignited weight basis (Weir 1965).
10. Montmorillonite, Santa Rosa, Mexico, clay fraction, air dry sample (Foster 1953a).
11. Montmorillonite, Chambers, Arizona, USA, 0.2–0.07 μm fraction, dried at 85°C for analysis; ϕ total includes 0.03% P_2O_5 (Earley et al. 1953).
12. Montmorillonite, Greenwood, Maine, USA, clay fraction, air dry sample (Foster 1953a).
† Total Fe as Fe_2O_3.
‡ Value obtained after correction for free silica using Earley et al. (1953) values and Foster's (1953b) method of correction.

where X is a monovalent exchangeable cation. Foster (1951) showed that, unless precautions are taken, some of the magnesium found by chemical analysis may be exchangeable. Using the same analysis as Ross and Hendricks but supplemented by determining exchangeable cations and cation exchange capacity, and allocating only non-exchangeable Mg to the silicate layer, she calculated

$$(Al_{3.22}Mg_{0.80})(Si_{7.84}Al_{0.16})O_{20}(OH)_4 X_{0.90}$$

as the structural formula. Later Foster (1953a) found that the Na-saturated clay fraction of the Tatatila sample gave the analysis reported in Table 1.15 (analysis 6), after correction for free Al_2O_3. The structural formula she calculated from this analysis is

$$(Al_{3.16}Mg_{0.80})(Si_{7.98}Al_{0.02})O_{20}(OH)_4 X_{0.90}$$

52 THE CHEMICAL CONSTITUTION OF CLAYS

TABLE 1.16. Analyses of beidellites and nontronites

	1	2	3	4	5	6$^\tau$	7	8
SiO_2	59.30	55.80	64.00	53.12	42.40	45.8	52.30	49.29
Al_2O_3	36.11	28.60	29.00	0.36	5.60	0.7	4.43	13.47
TiO_2		0.26		<0.05			0.05	0.10
Fe_2O_3	0.50	0.41	0.21	29.69	32.53	45.3	22.73	13.45
FeO								
MnO				0.03			0.01	0.06
MgO	0.10	2.03	3.03	2.49	0.32	0.3	2.20	2.10
CaO	0.02	2.23		1.51			1.21	2.07
Na_2O	3.98	0.09	3.98	n.d.		3.55	1.26	0.13
K_2O	0.11	0.48	0.05	0.30	5.14†		0.06	0.67
H_2O^-								
H_2O^+		9.70		}12.5	}14.03	4.3	}15.75	}18.64
Total	100.12	99.60	100.27	100.05	100.02	99.95	100.00	99.98
Numbers of cations on the basis $O_{20}(OH)_4$								
Si	6.97	7.27	7.49	8.00	6.91	6.81	8.00	7.54
Al	1.03	0.73	0.51		0.77	0.13		0.46
Fe^{3+}					0.32‡	1.06		
Σ tet.	8.00	8.00	8.00	8.00	8.00	8.00	8.00	8.00
Al	3.98	3.66	3.49	0.05	0.29		0.80	1.98
Ti		0.03						
Fe^{3+}	0.04	0.04	0.02	3.40	3.67	4.01	2.62	1.56
Fe^{2+}								
Mg	0.02	0.39	0.53	0.55	0.04	0.07	0.50	0.48
Σ oct.	4.04	4.12	4.04	3.95	4.00	4.08	3.92	4.02
Ca		0.31		0.24			0.18	0.34
Na	0.91	0.02	0.90			1.02	0.38	0.04
K	0.02	0.08	0.01	0.06	1.13		0.02	0.14
Interlayer charge	0.93	0.72	0.91	0.54	1.13	1.02	0.76	0.86
% tet. charge	111	101	56	0	96	117	0	53

1. Beidellite, Black Jack Mine, Carson District, Owyhee County, Idaho, USA. Na-saturated sample (Weir 1965).
2. Beidellite, Castle Mountains, California, USA (Heystek 1963).
3. Beidellite, Unterrupsroth, Rhone, Germany. <2 μm fraction Na-saturated sample (Weir 1965).
4. Iron-rich montmorillonite, alteration product of hedenbergite, Silurian limestone, Giralang, Canberra, Australia (Eggleton 1977).
5. Nontronite, Garfield, Washington, USA. † Potassium saturated for analysis; ‡ tetrahedral composition based on structural and Mossbauer studies (Besson et al. 1983).
6. Nontronite, Clausthal, Zellerfeld, Germany. τ Analysis calculated from the structural formula and CEC given by Goodman et al. (1976); tetrahedral Fe^{3+} by Mössbauer spectroscopy.
7. Iron-rich montmorillonite, Passau, Germany. Analysis 2 given by Brigatti (1983).
8. Iron-rich smectite, Monte Brosimo, Vicenza, Italy. Analysis 5 given by Brigatti (1983).

Schultz (1969), using the same analysis as Ross and Hendricks, gave the formula

$$(Al_{3.15}Mg_{0.85}Fe^{3+}_{0.01})(Si_{7.80}Al_{0.20})O_{20}(OH)_4X_{1.04}$$

From Foster's (1953a) analysis we calculate the formula to be

$$(Al_{3.158}Mg_{0.801}Fe^{3+}_{0.01})(Si_{7.970}Al_{0.030})O_{20}(OH)_4X_{0.932}$$

and from Ross and Hendricks's analysis

$$(Al_{3.141}Mg_{0.848}Mn_{0.008}Fe^{3+}_{0.007})(Si_{7.794}Al_{0.206})O_{20}(OH)_4X_{1.052}$$

The proportion of tetrahedral charge ranges from 19% for Schultz's formula to 2% for Foster's (1953a) formula. The former places the sample in Schultz's Tatatila and Chambers group and the latter in the Otay group. The fact that these different structural formulae were derived from analyses of one material emphasizes the difficulty of differentiating clay types on the basis of relatively small differences in their structural formulae.

Recently Brigatti and Poppi (1981) re-examined by a type of cluster analysis the possibility of grouping dioctahedral smectites into a number of smaller groups. They concluded that the main compositional effects are: (1) substitution of tetrahedral Si by a trivalent cation; (2) substitution of divalent cations, particularly Mg, for octahedral trivalent cations; (3) montmorillonites and beidellites contained <0.6 Fe^{3+}; (4) non-ideal montmorillonites (Schultz 1969) and Fe-beidellites contained $0.6 < Fe^{3+} < 1.0$; (5) nontronites contained $Fe^{3+} > 3$. The divisions they distinguished broadly confirm those of Schultz (1969), which are now widely accepted.

1.6.1.2 Iron-rich smectites and nontronites

Brindley (1980) gives the "end-member" structural formula of nontronite as

$$(M_x^+ \cdot nH_2O)Fe_4^{3+}(Si_{4-x}Al_x)O_{20}(OH)_4$$

which corresponds to a beidellite with all octahedral Al replaced by Fe^{3+}. Until recently custom has been to use the term "nontronite" for dioctahedral smectites with octahedral $Fe^{3+} > 3$ per $O_{20}(OH)_4$. Montmorillonites and beidellites generally have octahedral $Fe^{3+} < 1$ (Brigatti and Poppi 1981), and there appeared therefore to be a composition gap separating them from nontronites. A survey by Brigatti (1983) has shown, however, that intermediate compositions do occur, so precise definitions may not be possible.

For practical descriptive purposes, it is arguable (Schultz 1969; Brigatti 1983) that montmorillonites can be distinguished from "non-ideal" montmorillonites on the basis of their dehydroxylation temperatures and Fe^{3+} content, montmorillonite *senso strictu* dehydroxylating in the region 600–700 C and having $Fe^{3+} < 0.5$ per $O_{20}(OH)_4$, whereas non-ideal montmorillonite has 0.5–1.2 Fe^{3+} per $O_{20}(OH)_4$ and dehydroxylates near 600 C. Montmorillonites with more Fe^{3+} than this are then described as ferrian (or Fe-rich) montmorillonites and the term nontronite is reserved for dioctahedral smectites with $Fe^{3+} > 3$ per $O_{20}(OH)_4$.

Five representative analyses of ferrian smectites and nontronites are given in Table 1.16. Some analyses when calculated as structural formulae give totals of octahedral cations significantly exceeding 4 per $O_{20}(OH)_4$ and it has been argued that this has a true structural foundation (Brigatti 1983). It has been shown earlier (Section 1.2.4), however, that the presence of non-siliceous hydrous oxide impurity or interlayering can yield such apparent structural formulae and it is best to apply additional tests, such as selective oxide dissolution or collapse of the interlayer spacing on heating, before reaching such conclusions. As Brigatti (1983) points out, the resolution of this problem is particularly difficult with Fe-rich smectites, which are very liable to be decomposed by pretreatments.

The compositions given in Table 1.16, which exclude those with total octahedral cations exceeding 4.1, show that Fe-rich analogues of both montmorillonite and beidellite are known, including two containing less than 1% Al_2O_3 (analyses 4 and 6). The nontronites in particular have a structural interest in that Fe^{3+} must be in tetrahedral sites in those smectites containing insufficient Si + Al to satisfy the common assumption that all tetrahedral sites are occupied. Further interest has centred on the question whether Fe^{3+} replaces Si even when there is sufficient Al present; this is relevant to the procedure used to allocate atoms to sites in writing a structural formula. Recently, a very detailed study (Besson *et al.* 1983) combined information from selected-area electron diffraction, oblique texture electron diffraction and X-ray diffraction with Mössbauer spectra to show that in the Garfield nontronite (analysis 5) the vacancies are in the *trans* octahedral positions and that Fe^{3+} is present in the tetrahedral sheet. The structural formula for this nontronite given in Table 1.16 gives the cation distribution determined by these authors and has Al and Fe in both octahedral and tetrahedral sheets.

1.6.2 Trioctahedral smectites

Of the trioctahedral minerals in the smectite subgroup, saponites have predominantly magnesium in the octahedral sheet and can be represented approximately by the formula

$$(Mg_{6-z}R^{3+}_z)(Si_{8-y}Al_y)O_{20}(OH)_4 \cdot X^+_x \cdot nH_2O$$

Net negative charge on the layers derives from Al-for-Si substitution in the tetrahedral sites but this is partially compensated by substitution of trivalent cations into the octahedral sites. However there is also a tendency for octahedral occupancy to be less than 6 per $O_{20}(OH)_4$ (see Table 1.17) making the substitution pattern somewhat similar to that in vermiculites (Section 1.6.4). In natural saponites the charge density is typically that for a smectite-like mineral, in the range 0.8 to 1.2 per $O_{20}(OH)_4$ but saponites with a higher charge density have been synthesized, though not analysed (Suquet *et al.* 1977) and this suggests that there is no chemical composition break between the two minerals. Vermiculites, however, occur as coarse-grained flaky crystals whereas saponites are fine-grained clays. The saponite from Kozakov, Czechoslovakia (analysis 5, Table 1.17) has been the subject of detailed X-ray and infrared investigations (Suquet *et al.* 1975, 1977). Granular and fibrous saponite clays from Orrock, Fife, Scotland, have also been investigated in detail by Cowking *et al.* (1983) but the interpretation of the chemical composition was hindered by the possible presence of hydroxidic interlayers.

Iron-rich saponites are known and described by Sudo (1978) who considers that they contain Fe^{2+} with a general formula of the type

$$(Fe^{2+},Mg)_6(Si_{8-y}Al_y)O_{20}(OH)_4 \cdot X^+_y \cdot nH_2O$$

However they readily oxidize on exposure to the atmosphere and the analyses usually contain a preponderance of Fe^{3+}. Calculation of a structural formula on the assumption that all the iron was originally in the Fe^{2+} form leads to a plausible result (analysis 8, Table 1.17).

Swinefordite is a lithium-containing smectite intermediate in composition between the dioctahedral and trioctahedral smectites and its compositional relation to them is thus similar to lepidolite in the mica group minerals. An average formula is given by Tien *et al.* (1975) as

$$(Al_{1.87}Fe^{3+}_{0.15}Fe^{2+}_{0.09}Mg^{2+}_{1.31}Li_{1.76})(Si_{7.66}Al_{0.34})O_{20}(F_{0.65}OH_{3.35})$$

The remaining varieties, hectorite and stevensite, contain only small amounts of trivalent cations. Hectorite (analyses 9 and 10) contains essential Li and F and, since Al^{3+} and Fe^{3+} are absent, the tetrahedral sites are almost all occupied by Si, and it is substitution of Li for divalent cations in octahedral sites that is responsible for the negative charge on the layers. Hectorite-like clays have been synthesized and produced on an industrial scale (analysis 11, Table 1.17). Hectorites appear to have a smaller layer charge density than most other smectites.

Stevensite is essentially a hydrous magnesium silicate. Although there is disagreement about its exact nature (Faust and Murata 1953; Brindley 1955; Faust et al. 1959; see Brindley 1980), it is closely related to the trioctahedral smectites; one analysis is given in Table 1.17. The negative charge on the layers of stevensite is less than half that of smectites and appears to be caused by a small deficiency of octahedral cations. It has been suggested that stevensite is an interstratification of talc and saponite (Shimoda 1971).

1.6.3 Smectites containing uncommon elements

The elements Cr, Ni, Cu and Zn occasionally substitute in major amounts into structures similar to smectites, but the mineralogical characteristics of these materials have not always been fully established.

1.6.3.1 Volkhonskoite
This is a chromian smectite and in a strict sense the name should be applied only to minerals in which Cr is the dominant octahedral cation, i.e. the mineral contains $>15\%$ Cr_2O_3; many chromian smectites contain less Cr than this. The chromian smectites are dioctahedral with net negative charge arising from either tetrahedral (beidellitic) or octahedral (montmorillonitic) substitution; a typical formula is given by Brindley (1980):

$$(Cr_{2.02}Fe^{3+}_{0.82}Al_{0.72}Mg_{0.14}Ni_{0.08})(Si_{6.76}Al_{1.24})O_{20}(OH)_4 \cdot Ca_{0.56}$$

An iron-free volkhonskoite from Jordan has recently been described by Khoury et al. (1984) that contains only Cr and Mg in the octahedral sites, with an octahedral occupancy of 4.7 per $O_{20}(OH)_4$. Although many chromian smectites may indeed have Cr substitution into the octahedral sheet, there is also the possibility that some may contain hydroxy-Cr interlayers like those prepared artificially by Brindley and Yamanaka (1979) (see Section 1.8.3). Mixed-layer chromian minerals are also known (Maksimovic and Brindley 1980).

1.6.3.2 Medmontite
This was at one time thought to be a copper smectite but is now considered to be a mixture of chrysocolla and mica (Chukhrov et al. 1969).

1.6.3.3 Sauconite
Sauconite is a trioctahedral smectite containing Zn in the octahedral sheet; the structural formula for a zinc-rich saponite from Coon Hollow Mine, Arkansas, USA was reported by Ross (1946) to be:

$$(Al_{0.09}Fe^{3+}_{0.04}Zn_{5.78}Mn_{0.02}Mg_{0.07})(Si_{6.68}Al_{1.32})O_{20}(OH)_4 \cdot X^+_{1.17}$$

1.6.3.4 Nickel-smectites
The existence of *Ni-rich smectites* does not seem to have been definitely established, as *pimelite*, formerly thought to be a Ni-smectite, is now believed to be a nickel analogue of kerolite (Section 1.5.2.1); however, some garnierite may contain a Ni-smectite-like mineral (Brindley and Maksimovic (1974).

1.6.3.4 Vanadium-smectite
A vanadium-rich smectite has been reported by Güven and Hower (1979) with the following

TABLE 1.17. Analyses of trioctahedral smectites

	1	2	3	4	5	6
SiO_2	53.88	43.62	50.01	54.74†	51.4	39.64
Al_2O_3	4.47	5.50	3.89	8.93†	9.0	9.05
TiO_2	0.25	0.00	<0.04			
Fe_2O_3	0.60	0.66	0.21	0.43†	5.4	7.32
FeO					4.8	7.83
MnO		0.06				
MgO	31.61	24.32	25.61	33.28†	26.1	15.80
CaO		2.85	1.31	2.65†	3.2	2.93
Na_2O	0.01	0.08			0.04	0.71
K_2O	0.05	0.04			0.12	none
Li_2O						
F						
H_2O^-		17.42‡	7.28	9.17		12.31
H_2O^+	9.28	5.48‡	12.02	9.93		4.90
Total	100.15	100.03	100.37		100.00	100.49
$O \equiv F$						
Total corrected						
Numbers of cations on basis of $O_{20}(OH)_4$						
Si	7.23	7.00	7.50	6.76	6.60	6.34
Al	0.71	1.00	0.50	1.24	1.36	1.66
Fe^{3+}	0.06				0.04	
Σ tet.	8.00	8.00	8.00		8.00	8.00
Al		0.04	0.19	0.06		0.05
Ti	0.03		0.00			
Fe^{3+}	0.00	0.08	0.02	0.04	0.48	0.88
Fe^{2+}					0.52	1.05
Mn		0.01				
Mg	5.97	5.81	5.72	5.90	5.00	3.77
Li						
Σ oct.	6.00	5.94	5.92	6.00	6.00	5.75
Ca		0.49	0.21	0.35	0.44	0.50
Mg	0.35			0.22	0.00	
Na	0.00	0.02			0.01	0.22
K	0.01	0.01			0.02	
Interlayer charge	0.71	1.01	0.42	1.14	0.91	1.22

1. Saponite, Krugersdorp, Transvaal, South Africa (Schmidt and Heystek 1953).
2. Saponite, Allt Ribbein, Skye, Scotland. Ca saturated; ‡ H_2O^+ = water lost above 300 C (Mackenzie 1957).
3. Saponite, Milford, Utah, USA (Cahoon 1954).
4. Saponite, Grosslattengrun, Fichtelgebirge, Germany. † Values on basis of ignited weight (Weiss et al. 1955).
5. Saponite, Kozakov, Czechoslovakia. Analysis calculated from the structural formula on an anhydrous basis (Suquet et al. 1975).
6. Saponite, iron-rich (griffithite), Griffith Pass, California, USA (Ross 1960).

percentage composition: SiO_2 40.18; V_2O_3 30.75; Al_2O_3 4.76; MgO 1.54; FeO 1.18; CaO 0.87; Na_2O 4.26; LOI 14.92; the oxidation states of Fe and V were not determined.

1.6.4 Vermiculites

Vermiculites are trioctahedral minerals that frequently occur as large crystals with a platy morphology like that of micas but are much softer and contain interlayer water. They are formed

7	8	9	10	11	12
48.96	39.68	55.86	55.02	66.03	57.30
7.30	3.93	0.13	1.12	0.30	none
0.20	0.37	none		0.02	
11.93	19.82	0.03		0.06	0.32
1.24	1.12		0.70		none
	0.19	none			0.21
23.39	11.21	25.03	24.89	29.03	27.47
2.42	2.37	trace	0.54	0.34	0.97
0.04		2.68	0.94	3.19	0.03
0.06		0.10	0.43	0.04	0.03
		1.05	0.36	0.98	
		5.96	3.22		none
	15.11	9.90	7.66		6.69
4.45	6.16	7.24	6.42		7.17
99.99	99.96	102.98	101.60$^\phi$	100.00	100.19
		2.51	1.36		
		100.47	100.24		

Numbers of cations on basis of $O_{20}(OH)_4$					
6.60	7.15	7.98	7.97	8.01	8.02
1.16	0.83	0.02	0.03		
0.24					
8.00	7.98	8.00	8.00	8.01	8.02
			0.16	0.04	
0.02	0.05				
0.97	0.00$^\tau$				0.03
0.14	2.86		0.08		
	0.03				0.02
4.70	3.01	5.33	5.38	5.25	5.73
		0.60	0.21	0.48	
5.83	5.95	5.93	5.79	5.78	5.78
0.35	0.46		0.03	0.04	0.15
0.01		0.74	0.26	0.75	0.01
0.01		0.02	0.08	0.01	0.01
0.72	0.92	0.76	0.40	0.84	0.32

7. Saponite, Winnweiler, Pfalz, Germany. Analysis calculated from structural formula (Quakernaat 1970).
8. Saponite, Green Tuff, Tertiary iron sand beds, Irisugawa, Gunma Prefecture, Japan (Sudo 1978). τ Structural formula calculated assuming all Fe_2O_3 originally present as FeO.
9. Hectorite, Hector, California, USA (Ross 1960).
10. Hectorite, Morocco. ϕ Total includes 0.30% CO_2 (Faust et al. 1959).
11. "Laponite", a synthetic hectorite; mean composition calculated from eight analyses (van Olphen and Fripiat 1979). The original analysis gave 8.3% F (Neumann 1965).
12. Stevensite, Springfield, NJ, USA (Faust et al. 1959).

mainly by the alteration of micas, particularly phlogopites and biotites, by a process in which interlayer potassium is replaced by a hydrated cation, usually magnesium (but see also Section 1.9.4). As this involves comparatively little disruption of the mica structure, the chemistry of vermiculites is quite closely related to that of the parent phlogopites or biotites. The two main distinguishing features are that the net negative layer charge is lower, in the range 1.2 to 1.9 per $O_{20}(OH)_4$, and that the iron is oxidized relative to the parent mica. Norrish (1973a) showed that these two differences are related, the lower net charge correlating closely with the oxidation of

TABLE 1.18. Analyses of calcium-saturated vermiculites

	1	2	3	4	5	6
SiO_2	45.56	41.87	42.27	43.35	44.00	44.80
Al_2O_3	15.82	19.20	18.30	16.48	5.96	14.03
TiO_2	0.38	0.36	0.34	2.18	1.36	0.05
Fe_2O_3	1.40	4.54	5.94	9.48	16.33	10.32
FeO		0.35		0.25	0.75	0.80
MnO	0.13	0.20	0.13	0.08	0.07	0.17
MgO	29.66	27.03	26.15	22.73	27.15	23.72
CaO	6.86	6.73	6.14	5.67	5.52	5.32
Na_2O						
K_2O	0.17	0.01	0.00	0.16	0.10	0.03
Total	99.98	100.29	99.27	100.38	101.24	99.24
Numbers of cations on the basis of $O_{20}(OH)_4$						
Si	5.79	5.38	5.48	5.61	5.83	5.87
Al	2.21	2.62	2.52	2.39	0.93	2.13
Fe^{3+}					1.24	
Σ tet.	8.00	8.00	8.00	8.00	8.00	8.00
Al	0.16	0.28	0.27	0.12		0.03
Ti	0.04	0.03	0.03	0.21	0.14	
Fe^{3+}	0.13	0.44	0.58	0.92	0.38	1.02
Fe^{2+}		0.04		0.03	0.08	0.09
Mn	0.01	0.02	0.01	0.01	0.01	0.02
Mg	5.62	5.17	5.05	4.38	5.36	4.63
Σ oct.	5.96	5.98	5.94	5.67	5.97	5.79
Ca	0.93	0.93	0.85	0.79	0.78	0.75
Na						
K	0.03			0.03	0.02	
Interlayer charge	1.89	1.86	1.70	1.61	1.58	1.50

1. Vermiculite, Carl Moss Ranch, Llano County, Texas, USA, Stop 12†.
2. Vermiculite, Santa Ollala, Spain, 7-1.
3. Vermiculite, Santa Ollala, Spain, 11-1.
4. Vermiculite, Busumbu, Uganda.
5. Vermiculite, Palabora, South Africa.
6. Vermiculite, Carl Moss Ranch, Llano County, Texas, USA, Stop 10†.
7. Vermiculite, Young River, Western Australia.

Fe^{2+} to Fe^{3+}. However, as pointed out by Foster (1963), there is not a 1:1 relation between the charge reduction below 2 per $O_{20}(OH)_4$ and the content of Fe^{3+} in the vermiculite, so that additional changes in composition must also occur. These include the coupled oxidation–reduction reaction:

$$Fe^{2+} + \text{structural } OH^- \rightarrow Fe^{3+} + \text{structural } O^{2-}$$

which causes no overall change in net charge balance, and ejection of Fe^{3+} from the structure when the octahedral sites become overpopulated with trivalent cations (Farmer et al. 1971).

The chemical compositions of vermiculites generally exemplify these relationships. However, because the interlayer cation is usually Mg, which also occurs in the octahedral sheet, the charge balance in vermiculites cannot be established unless the interlayer Mg is replaced by an index cation that is not included in the layer structure. The data given by Norrish (1973a) (Table 1.18) are for Ca-saturated vermiculites analysed by XRF and are therefore to be preferred to some earlier analyses in which the interlayer cations could not be so distinguished.

7	8	9	10	11	12
43.21	41.18	41.56	44.09	43.07	43.57
17.21	20.49	16.77	16.98	16.92	13.50
0.06	1.12	0.47	2.18	0.68	1.14
8.03	9.19	10.48	8.30	11.75	10.86
0.60	1.50	0.90		0.95	
0.21	0.08	0.35	0.03	0.36	0.09
25.71	20.92	23.21	23.64	22.63	26.07
5.10	5.02	4.63	4.76	4.66	4.39
0.05		0.06		0.06	
0.32	0.01	0.19	0.02	0.03	0.07
100.50	99.81	98.62	100.00	101.11	99.69
Numbers of cations on the basis of $O_{20}(OH)_4$					
5.57	5.40	5.51	5.67	5.57	5.68
2.43	2.60	2.49	2.33	2.43	2.08
					0.24
8.00	8.00	8.00	8.00	8.00	8.00
0.18	0.54	0.13	0.24	0.14	
0.01	0.11	0.04	0.21	0.07	0.12
0.78	0.90	1.04	0.80	1.14	0.82
0.06	0.16	0.10		0.10	
0.02	0.01	0.04		0.04	0.01
4.94	4.06	4.58	4.53	4.36	5.07
5.99	5.78	5.93	5.78	5.85	6.02
0.70	0.70	0.66	0.66	0.65	0.61
0.01		0.02			
0.05		0.03		0.02	0.01
1.46	1.40	1.37	1.32	1.32	1.23

8. Vermiculite, West Chester, Pennsylvania, USA.
9. Vermiculite, Young River, Western Australia.
10. Vermiculite, Beni Buxera, Morocco.
11. Vermiculite, Young River, Western Australia.
12. Vermiculite, Nyasaland.
   All analyses on ignited basis (from Norrish 1973a).
† Stop numbers refer to *10th National Clay Conference Guidebook of Excursions* (Folk et al. 1961).

Substitution of Al-for-Si in the tetrahedral sites generally exceeds 2 per $O_{20}(OH)_4$, so that an overall charge balance of less than 2 is achieved by replacement of some divalent octahedral cations by trivalent. In consequence the nominal charge on the octahedral sheet is positive, ranging from +0.29 to +1.22. This positive octahedral charge and the large negative charge (>2) on the tetrahedral sheet together appear to distinguish vermiculites from high-charge saponites.

1.6.5 Charge heterogeneity in swelling 2:1 layer silicates

Once methods for recognizing and determining interstratifications in layer silicates became available, it was observed that the ethylene glycol and glycerol complexes of some smectites did not represent regular sequences of layer spacings (Byrne 1954; McAtee 1958a, 1958b). These irregularities were attributed to inhomogeneity in the distribution of charge (Tettenhorst and

Johns 1965) among the different 2:1 layers. Vibration spectra likewise indicate heterogeneity of charge in smectites (Chourabi and Fripiat 1981).

Weiss *et al.* (1970) proposed determining the charge density sequence from the complexes with n-alkyl ammonium compounds of increasing chain length, a method that has since been applied to a large number of layer silicates (Lagaly 1981). Short-chain alkylammonium ions form complexes with smectites in which the chains are arranged in monolayers; long chain alkylammonium ions however form bilayers, the transition coming at the point where the chain length can no longer be accommodated as a monolayer at the given interlayer cation density. If the layer charge were homogeneous, the interlayer cation density would be expected to be the same in each interlayer and the transition from the monolayer with a 13.6 Å interlayer spacing to a bilayer with a spacing of 17.7 Å would be expected to occur at a critical chain length, n_c. Most smectites, however, do not show a sharp transition and this was interpreted to mean that the interlayer cation density was not uniform. The transition from the 13.6 Å monolayer to the 17.7 Å bilayer may, for instance, span a chain length from $n_c = 10$ to $n_c = 15$, corresponding to an interlayer cation density of from 0.66 to 0.48. By detailed interpretation of the transition region, frequency distributions for the charge density have been drawn up for 200 smectites (Lagaly and Weiss 1976) and for 25 vermiculites (Lagaly 1982); almost all are shown to have heterogeneous charge distribution.

This heterogeneity can be readily modelled in terms of possible structural formulae. Suppose a smectite intermediate in composition between beidellite and montmorillonite with an average structural formula

$$(Al_{4-p}Mg_p)(Si_{8-q}Al_q)O_{20}(OH)_4 \cdot X^+_{p+q}$$

Heterogeneity could be due to the sample containing m layers of montmorillonite type

$$(Al_{4-x}Mg_x)(Si_8)O_{20}(OH)_4 \cdot X^+_x$$

and n layers of beidellite type

$$(Al_4)(Si_{8-y}Al_y)O_{20}(OH)_4 \cdot X^+_y$$

the proportions being such that

$$p = mx/(m+n); \qquad q = ny/(m+n)$$

If $x \neq y$, there are three possible interlayer cation densities (x, $(x+y)/2$, and y) depending on whether two montmorillonite, one montmorillonite and one beidellite or two beidellite layers are adjacent to each other in the layer stack.

A further possibility is that each layer is itself heterogeneous, perhaps with a distribution of the type

$$(Si_4)(Al_{4-p}Mg_p)(Si_{4-q})O_{20}(OH)_4$$

and the net negative charge could then be $p/2$ on the "montmorillonitic" side and $(p/2 + q)$ on the "beidellitic" side, the interlayer cation densities being p, $p+q$, and $p+2q$ for the three combinations. If this type of heterogeneity is combined with different types of layers as in the first example, then more combinations are possible.

These studies serve to emphasize that structural formulae are a representation of the *average* composition of the layers, but not a formula for any *one* chosen layer unless the mineral is homogeneous.

Of the 200 smectites examined by Lagaly and Weiss (1976), only one was homogeneous, and 70% of the interlayer spaces had cation densities between 0.56 and 0.84 eq per $O_{20}(OH)_4$ and 40% between 0.62 and 0.78; the limits of the distribution were between 0.4 and 1.0 eq per $O_{20}(OH)_4$.

Vermiculites do not necessarily have the same broad range of charge heterogeneity as smectites. High-charge vermiculites (e.g. Llano County, analysis 1, Table 1.18) are characterized by a linear increase in spacing with increasing chain length (Lagaly 1982); this is because the chains are tilted by about 50° to the basal plane. Such "paraffin-type" structures are formed if the interlayer cation density is greater than 1.5 eq per $O_{20}(OH)_4$. However, with low-charge vermiculites transition zones can be distinguished and it is possible to estimate charge heterogeneity, e.g. for Beni-Buxero vermiculite (analysis 10, Table 1.18) the range is 1.14 to 1.4, mean 1.26 and for Young River vermiculite (analysis 9, Table 1.18) the range is 1.04 to 1.40, mean 1.22 (Lagaly 1982). It is likely therefore that charge heterogeneity in expanding 2:1 layer silicates is the rule rather than the exception.

1.6.6 Chemical composition and interlamellar swelling

Some 2:1 layer silicates swell in water, ethylene glycol and a wide range of similar compounds (MacEwan and Wilson 1980) by intercalation of molecules between 2:1 layers and this property is used as a diagnostic criterion in clay mineral identification by X-ray diffraction. There is no comprehensive account of the factors involved in swelling. It is generally agreed, however, that swelling, or lack of it, is controlled by the balance of attractive and repulsive forces between adjacent 2:1 layers.

The main attractive force is the electrostatic interaction of the negatively charged layers and positively charged interlayer material, frequently an alkali or alkaline earth cation. The size of this force depends on surface charge density and also on the separation of positive and negative charges, principally in the direction normal to the plane of the layers (dependent on layer separation) but also in the plane of the layers (dependent on distribution of substitutions in the layers).

When layers are in contact, or nearly so, there are van der Waals' attractive forces between atoms in the surfaces of adjacent layers and also residual electrostatic forces between charged or partially charged atoms in adjacent layers. The latter, like the overall electrostatic attraction between layers and interlayer material, are dependent on layer separation and structure.

The main repulsive force arises from the solvation of interlayer cations (in natural systems, hydration) supplemented by that from interaction of the solvate (intercalate) with the surface oxygens of the 2:1 layers; the latter seems to be less than the sum of van der Waals' and electrostatic attraction for layers in contact, because minerals of the talc–pyrophyllite group with near neutral layers and few, if any, interlayer cations, do not swell. In addition, there are repulsive forces between the hydrogens of hydroxyl groups linked to the octahedrally coordinated cations in the 2:1 layer, and the interlayer cations. The size of this repulsion depends on the proximity of the hydroxyl hydrogen to the seat of charge in the interlayer and is greatest when the O—H dipole is normal or nearly so, to the plane of the 2:1 layer. Replacement of OH^- by F^-, common in trioctahedral micas, decreases proportionately this repulsion.

Giese (1971, 1975a, 1975b, 1977, 1979) and Jenkins and Hartman (1979, 1980, 1982) have quantitatively evaluated electrostatic interactions for specific structures and compositions and these methods, when applied more widely, promise to give a better understanding of the crystal chemical factors that control intercalation and swelling. At present, however, Eberl's (1980) approach appears to be more generally useful as an indication of expandability of 2:1 layer structures. Interlayer cations are unavailable for exchange when they occur in irreversibly collapsed interlayers so the critical values for fixation that Eberl derives can be applied to judge relative expandability of layer silicates of different composition. Eberl considers only the forces due to attraction of interlayer cations to the silicate layer and repulsion due to cation hydration; these are probably the most important attractive and repulsive forces involved in expansion. Eberl uses the concept of an "equivalent anionic radius", of the 2:1 layer, r_a, given by

$$r_a = \frac{A}{4\pi C}$$

where C is the charge per formula unit, $O_{20}(OH)_4$ here and $A = (a \times b)$ Å^2. Minerals of different composition have different values of A because the cell dimensions a and b change with changes in composition. Using Eberl's critical values of r_a for fixation (i.e. failure to swell), the upper charge limits for swelling can be obtained. For K- and Na-clays in aqueous media, these are $C_K = 0.0329 A$ and $C_{Na} = 0.0366 A$ for structures with interlayer K and Na respectively, and hence critical upper charge limits for swelling in water can be calculated if the cell dimensions are known. Table 1.19 shows values of C_K and C_{Na} calculated for a range of compositions of 2:1 layers.

In general terms Table 1.19 suggests that a 2:1 layer silicate with interlayer Na would continue to expand at a layer charge at which that with interlayer K would not; trioctahedral micas should expand at larger layer charge than dioctahedral micas and less aluminous trioctahedral micas (phlogopites and biotites) should remain expandable at charge values that would prevent expansion of the more aluminous varieties (prieswerkite).

These charge limits for expansion in water take account of electrostatic attraction between the 2:1 layers and the interlayer cations and repulsion due to cation hydration. They will be modified by the effects of the abundance of structural F$^-$ for OH$^-$ replacement, the position of the hydrogen of these OH groups relative to the interlayer cation and any oxidation of OH$^-$ to O^{2-}.

These effects have been proposed to explain the observed differences in K-release characteristics of di- and trioctahedral micas in which the OH dipoles are oriented differently

TABLE 1.19. Upper limits of C_K and C_{Na}, the charges per $O_{20}(OH)_4$ unit, for expansion in water of K- and Na-micas of different chemical compositions

Micas		Reference[†]	A Å^2	C_K	C_{Na}
Dioctahedral,	K-aluminous (muscovite, illite)	1	46.8	1.54	(1.72)
	K-iron rich (glauconite, celadonite)	1	47.4	1.56	(1.74)
Trioctahedral,	K-magnesian (phlogopite)	2	48.9	1.61	(1.80)
	K-ferroan (annite)	2	50.5	1.66	(1.85)
	K-aluminous (aluminous biotite)	2	48.5	1.59	(1.78)
Dioctahedral,	Na-aluminous (paragonite, brammallite)	1	45.7	(1.50)	1.68
Trioctahedral,	Na-magnesian (Na-phlogopite)	3	48.4	(1.59)	1.78
	Na-aluminous (preiswerkite)	2, 4	47.2	(1.55)	1.73

[†] References for a and b dimensions used to calculate A: 1 – Bailey (1980b); 2 – Hewitt and Wones (1975); 3 – Carman (1974); 4 – Keusen and Peters (1980).

(Bassett 1960; Norrish 1973a) and the readier K-release from OH-phlogopites than from F-phlogopites (Newman 1969).

A comprehensive theory of expandability would require quantitative evaluation of the relative importance of various factors outlined above.

1.7 2:1 GROUP, X > 1.6: THE MICA MINERALS

The chemistry of the mica group minerals is complex and many types of isomorphous substitution are possible in their structures; these have been reviewed in detail by Foster (1956, 1960a, 1964, 1969) and Deer *et al.* (1962); the structures of specifically named micas are described by Bailey (1980b).

Representing the micas by the general formula $X_2Y_{4-6}Z_8O_{20}(OH,F)_4$, the group can be divided into dioctahedral and trioctahedral subgroups depending on whether the number of octahedral Y cations is close to 4 or to 6 respectively. Usually the interlayer cations (X) in the true micas are predominantly K or Na, but in brittle micas, Ca is the main interlayer cation. Lithium substitutes in the octahedral sheet but Li-micas can be intermediate between di- and trioctahedral and it is more convenient to consider them as a separate class.

The main substituents in the mica minerals are shown as ideal formulae in Table 1.20; actual composition may deviate considerably from the ideal. Besides the main constituents Si, Al, Fe, Al, Mg, K, Ca and Li, small amounts of Ti and Mn may be found in micas and are assigned to the octahedral sites, and Sr, Ba, Rb and Cs are sometimes present in the interlayer sites. Micas

TABLE 1.20. Simplified ideal cation compositions for some micas and brittle micas of general formula $[X_2][Y_{4-6}][Z_8]O_{20}(OH,F)_4$

	Interlayer cations, X	Octahedral cations, Y	Tetrahedral cations, Z
Dioctahedral micas			
Muscovite	K_2	Al_4	Si_6Al_2
Phengite	K_2	$(Al,Fe^{3+})_3(Mg,Fe^{2+})$	Si_7Al
Celadonite	K_2	$(Fe^{3+})_2(Mg)_2$	Si_8
Paragonite	Na_2	Al_4	Si_6Al_2
Dioctahedral brittle mica			
Margarite	Ca_2	Al_4	Si_4Al_4
Trioctahedral micas			
Phlogopite	K_2	Mg_6	Si_6Al_2
Biotite	K_2	$(Al,Fe^{3+})_y(Mg,Fe^{2+})_{6+(x/2)-(3y/2)}$	$Si_{6-x}Al_{2+x}$ $(x<1, y<2)$
Annite	K_2	Fe^{2+}_6	Si_6Al_2
Lepidomelane	K_2	$Fe^{3+}_2Fe^{2+}_3$	Si_6Al_2
Siderophyllite	K_2	$Al^{3+}_2Fe^{2+}_3$	Si_6Al_2
Trioctahedral brittle mica			
Clintonite	Ca_2	$Al_x(Mg,Fe^{2+})_{6-x}$	$Si_{4-x}Al_{4+x}$
Lithium micas			
Lithian muscovite	K_2	$Li_{0-1.5}Al_{4-3.5}$	Si_6Al_2
Trilithionite	K_2	Li_3Al_3	Si_6Al_2
Polylithionite	K_2	Li_4Al_2	Si_8
Zinnwaldite	K_2	$LiAl.Fe^{2+}_{2.5}$	Si_6Al_2
Taeniolite	K_2	Li_2Mg_4	Si_8

containing unusual amounts of Cr, V, Mn, Ni, Co, Cu, Ba, Be and B are mentioned in Section 1.7.5.

The structural chemistry of the micas has been derived mainly from studies on crystals large enough for the determination of optical properties and structure by single crystal X-ray diffraction. Many of the principles so derived have been used by analogy in the study of the fine-grained mica-like minerals commonly found in argillaceous sediments and soils, and in hydrothermal alteration products. It is therefore necessary to describe the chemistry of the coarse-grained micas of igneous and metamorphic rocks before going on to clay micas and mica-like sedimentary clays. The latter are usually dioctahedral and have their own distinct structural and chemical features so they are considered after the sections on dioctahedral micas. Interstratified minerals with micaceous components are included in Section 1.9.

1.7.1 Dioctahedral micas $X_2Y_4Z_8O_{20}(OH)_4$: General considerations

The chemistry of dioctahedral micas with predominantly monovalent interlayer cations (Na and K) can be described in terms of the ideal formulae:

$$K_2(R^{3+})_4(Si_6Al_2)O_{20}(OH)_4 \text{ (``trisilicic mica'')}$$

and

$$K_2(R^{3+})_2(R^{2+})_2(Si_8)O_{20}(OH)_4 \text{ (``tetrasilicic mica'')}$$

where R^{3+} is Al or Fe^{3+} and R^{2+} is Mg or Fe^{2+} (Schaller 1950).

In this trisilicic–tetrasilicic series, so-called from the number of Si atoms in the half-formula unit, tetrahedral aluminium in trisilicic muscovite is progressively replaced by Si, and a corresponding replacement of octahedral Al by Mg maintains the net layer charge at 2 per formula unit. These coupled replacements are represented by the equation

$$Al(\text{IV}) + Al(\text{VI}) \rightleftharpoons Si(\text{IV}) + Mg(\text{VI})$$

where the Roman numerals indicate the coordination of the cation. In proportion to the extent of this replacement, the wholly tetrahedral charge of muscovite decreases while the octahedral charge correspondingly increases until the net layer charge is wholly octahedral in origin in the tetrasilicic mica.

Superimposed upon this, substitution of Al and Mg by Fe^{3+} and Fe^{2+} introduces further variables in the substitutional pattern, but with the exception of the lithium micas, there is little tendency for the number of octahedral cations to exceed 4 per 6 sites. Consequently, tetrahedral Al in the common dioctahedral micas does not exceed 2 per eight sites, a substitutional range distinctly different from that in the trioctahedral micas (Section 1.7.3).

A further substitution series, in which Si replaces Al(IV) without a corresponding change in the octahedral composition decreases the net layer charge towards montmorillonite and pyrophyllite. This substitution seems in part responsible for the smaller layer charge of hydrous micas and may be represented by the equation

$$Si(\text{IV}) \rightleftharpoons Al(\text{IV}) + K(\text{XII})$$

The reverse of this reaction, with Al + K replacing Si in smectites is considered to occur in illite diagenesis (Eberl 1978).

In general, both of these substitutional patterns occur together in 2:1 layer silicates and can be represented reasonably accurately on a ternary diagram similar to that used by Yoder and

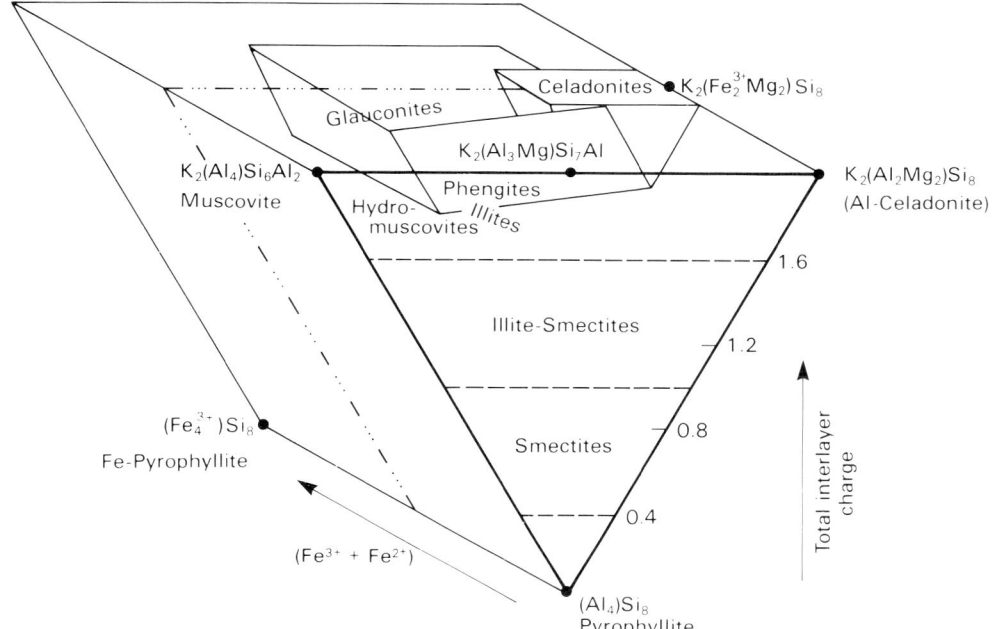

FIG. 1.3. Dioctahedral 2:1 layer silicates: some ideal end-member compositions for the $O_{20}(OH)_4$ formula unit, and the compositional ranges for some micas, illites, illite–smectites and smectites.

Eugster (1955). If net negative charge in the tetrahedral sheet arising from trivalent cations replacing Si is plotted in one direction, and net negative charge in the octahedral sheet from replacement of trivalent cations by divalent is plotted in a second direction, 120° to the first, total layer charge is represented in the third direction bisecting the first two vectors. This is shown in Fig. 1.3.

The trisilicic end-member mica, muscovite, $K_2Al_4(Si_6Al_2)O_{20}(OH)_4$, is represented in one corner; the tetrasilicic end-member, Al-celadonite, $K_2Al_2Mg_2(Si_8)O_{20}(OH)_4$, is represented in the second corner; and pyrophyllite, $Al_4Si_8O_{20}(OH)_4$, in the third. A further axis for total Fe converts the planar figure into a prism within which the compositions of most 2:1 dioctahedral layer silicates can be represented as a point. Minerals with otherwise similar compositions but different Fe^{2+}/Fe^{3+} ratios are not separated but on the whole little information is lost. However, important differences could be masked in the celadonite and glauconite groups and this should be remembered when using this representation.

A few mineral analyses generate data that fall outside the prism. This can arise when the net charge in the octahedral or tetrahedral sheets apppears to be positive, for example, if the Si occupancy of the tetrahedral sheet is apparently greater than 8 per eight sites or the octahedral sheet in a dioctahedral mineral apparently contains more than 4 trivalent cations per six sites. Both instances suggest impurity in the sample, irregularities in the structure or error in the analysis, but positive octahedral charge can also arise if the clay contains hydroxide-like interlayers, as for example, in glauconites (Thompson and Hower 1975) although in this example a positive octahedral charge did not result.

A relationship between Fe content and the distribution of charge between tetrahedral and

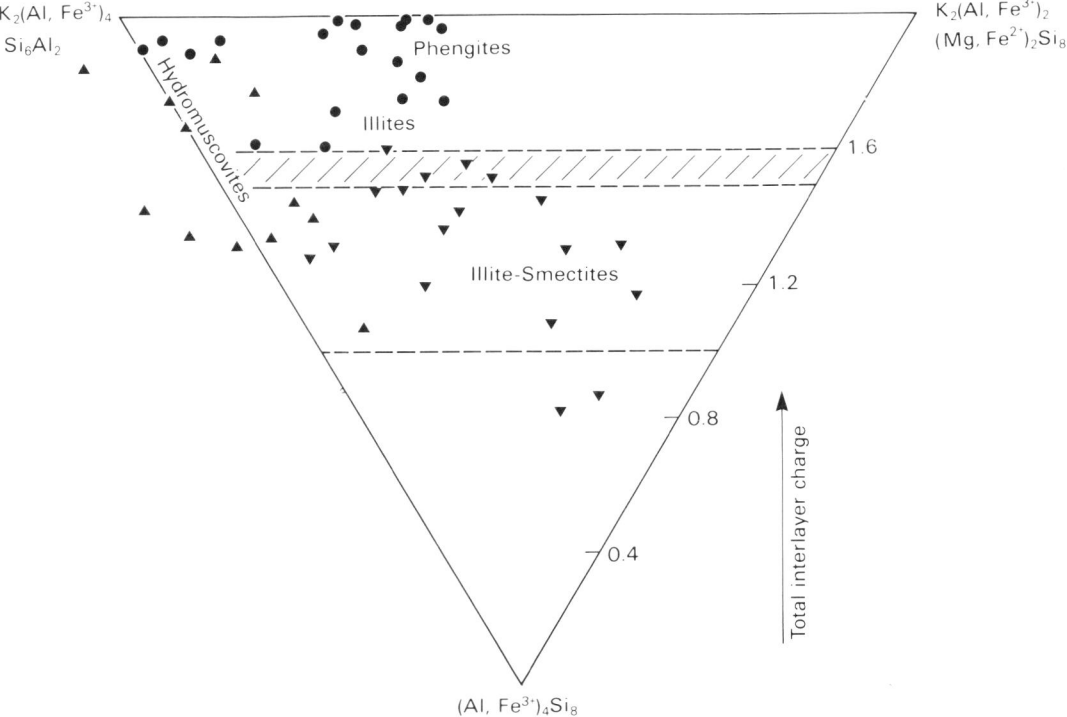

FIG. 1.4. Dioctahedral 2:1 layer silicates: analyses with Fe < 1 per $O_{20}(OH)_4$ projected on a prism face of Fig. 1.3. ● muscovites and phengites, ▲ hydromuscovites; ▼ illites and illite–smectites. Shading indicates a range of charge between illite *sensu stricto* (<5% expanding layers) and illite–smectites.

octahedral sites can be demonstrated for analyses of natural 2:1 dioctahedral minerals by grouping into those for which Fe < 1 per four octahedral sites and those for which Fe > 1 per four sites. Over 150 analyses taken from the references cited in this section were so grouped and their projections on a prism face (Figs. 1.4 and 1.5) show two distinct fields of composition, the low iron samples occurring towards the trisilicic, muscovite corner, and the high iron samples towards the tetrasilicic, celadonite corner. Despite this grouping, it seems that in the solid diagram there is a continuum of compositions from celadonite–glauconite to phengite–muscovite and illites (Velde 1977). Nevertheless, it is convenient to divide dioctahedral micas into two groups, the muscovite–phengite group, for which Fe < 1/4, and the celadonite–glauconite group with Fe > 1/4. The latter group also differs from the former in that they are essentially fine-grained and therefore come under the heading clay micas, Section 1.7.2.

The other general point illustrated in Figs. 1.4 and 1.5 is that within the groups of minerals generally described under the names hydromuscovite, illite and glauconite, only occasional examples have total net negative charge as high as 1.8 per $O_{20}(OH)_4$, and that in general net charge grades downwards towards that of smectites, 0.4 to 0.8 per $O_{20}(OH)_4$. The X-ray diffraction patterns of such clays often show that they are structurally inhomogeneous with interstratification by smectite increasing as the smectite composition is approached and these minerals are therefore discussed in Section 1.9 on mixed-layer minerals. The work of Hower and Mowatt (1966) on illite–smectites and Thompson and Hower (1975) on glauconites shows,

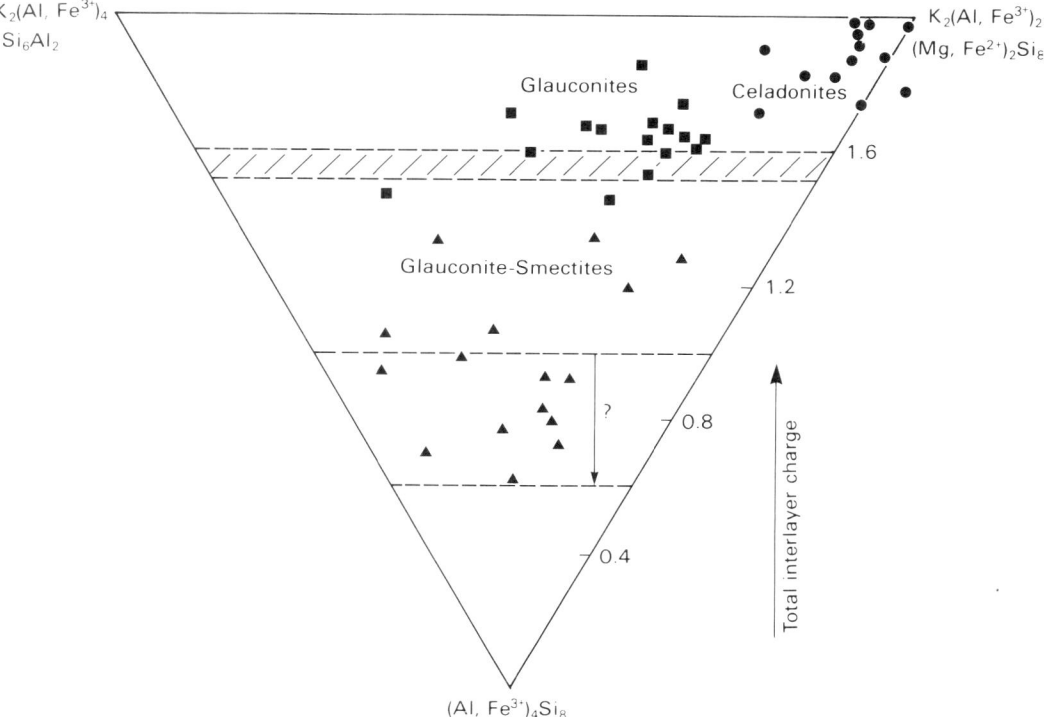

FIG. 1.5. Dioctahedral 2:1 layer silicates: analyses with Fe > 1 per $O_{20}(OH)_4$ projected on a prism face of Fig. 1.3. ● celadonites; ■ glauconites; ▲ glauconite–smectites. The arrow indicates the possible interlayer charge range for Fe-rich smectites.

however, that both groups tend towards an occupancy of 1.6K per two interlayer sites when no smectite layers can be detected by X-ray diffraction. Therefore analyses that give (K + Na) > 1.5 per two interlayer sites are included in the section on clay micas, Section 1.7.2, as such minerals are likely to be identified as "illites" by X-ray diffraction for Gaudette et al. (1966) defined illites as giving a 10 Å basal spacing.

1.7.1.1 Muscovite–phengite micas

As shown in Fig. 1.4 there is a continuous range of composition between

muscovite: $K_2(Al,Fe^{3+})_4(Si_6Al_2)O_{20}(OH)_4$

and

phengite: $K_2[(Al,Fe^{3+})_3(Mg,Fe^{2+})](Si_7Al)O_{20}(OH)_4$

where total Fe < 1 per four octahedral sites. Of the analyses examined, only that of alurgite quoted by Schaller (1950) as

$$K_{1.93}Na_{0.09}(Al_{2.58}Fe^{3+}_{0.12}Mn_{0.11}Mg_{1.21})(Si_{7.20}Al_{0.80})O_{20}(OH)_4$$

falls outside this field. In muscovite, tetrahedral composition is close to Si_6Al_2 and the octahedral sheet contains more than 3Al per four sites and very little Mg (e.g. analysis 1, Table 1.21).

68   THE CHEMICAL CONSTITUTION OF CLAYS

TABLE 1.21.   Analyses of dioctahedral micas

	1	2	3	4	5	6	7
SiO_2	45.55	47.08	49.92	50.45	44.41	47.0	29.84
Al_2O_3	36.89	24.46	24.38	25.77	40.09	39.1	50.18
TiO_2	0.26		1.70	0.12	0.22	0.02	0.00
Fe_2O_3	0.39	8.02	3.30	0.99	1.72	0.78	1.35
FeO	0.86	2.52	2.01	1.60	0.28		0.66
MnO	0.02	0.08	0.05	0.05	0.02	0.02	0.00
MgO	0.58	1.86	3.64	5.02	0.16	0.10	0.05
Li_2O	0.01			0.01			0.21
CaO	0.04		0.10	0.29	0.67	0.24	10.69
Na_2O	0.80	0.11	0.59	0.56	5.80	7.5	2.15
K_2O	10.17	10.6	9.87	10.11	2.22	0.81	0.48
H_2O^+	4.59	4.00	4.13	4.29	4.45	4.3	4.78
H_2O^-	0.03		0.29	0.09			0.03
F	0.04	0.14		0.00	0.08		
Total	100.27	98.87	99.98	99.70	100.12	99.87	100.51

Numbers of cations on the basis $O_{20}(OH)_4$

	1	2	3	4	5	6	7
Si	6.03	6.57	6.72	6.76	5.74	6.00	3.98
Al	1.97	1.43	1.28	1.24	2.26	2.00	4.02
Σ tet.	8.00	8.00	8.00	8.00	8.00	8.00	8.00
Al	3.79	2.58	2.59	2.84	3.85	3.89	3.88
Ti	0.03	0.00	0.17	0.01	0.02	0.00	0.00
Fe^{3+}	0.04	0.84	0.33	0.10	0.17	0.08	0.14
Fe^{2+}	0.10	0.29	0.23	0.18	0.03		0.07
Mn	0.00	0.01	0.01	0.006	0.00	0.00	0.00
Mg	0.11	0.38	0.73	1.00	0.03	0.02	0.01
Li	0.00			0.005			0.11
Σ oct.	4.07	4.10	4.06	4.14	4.10	3.99	4.21
Ca	0.01		0.01	0.06	0.09	0.03	1.53
Na	0.20	0.03	0.15	0.14	1.45	1.86	0.56
K	1.72	1.88	1.70	1.73	0.37	0.13	0.08
Σ interl.	1.93	1.91	1.86	1.93	1.91	2.02	2.17
Si/Al	1.05	1.64	1.74	1.66	0.94	1.02	0.50
Tetrahedral charge	−1.97	−1.43	−1.28	−1.24	−2.26	−2.00	−4.02
Octahedral charge	+0.03	−0.48	−0.59	−0.75	+0.26	−0.06	+0.32
Interlayer charge	−1.94	−1.91	−1.87	−1.99	−2.00	−2.06	−3.70

1. Muscovite, from pegmatite, Georgia (Ernst 1963). Total includes 0.04% BaO.
2. Iron-rich muscovite, Grandfather Mountain, North Carolina (Foster et al. 1960). $a_0 = 5.249$, $b_0 = 9.064$, $c_0 = 19.99$, $\beta = 95°\ 45'$, SG = 2.879. Corrected for 0.20% SiO_2, 0.30% TiO_2 and 0.14% CaO.
3. Phengitic mica, blueschist, Eastern Shikoku, Japan (Ernst 1964).
4. Phengitic mica, glaucophane schist, Tiburon Peninsula, California (Ernst 1963; Guven 1971). Total includes 0.32% Ba and 0.03% minor constituents.
5. Paragonite, chlorite–sericite schist, Glebe Mountain, Vermont (Rosenfeld 1956; E-an Zen and A. L. Albee 1964).
6. Paragonite, quartz–kyanite vein, Tasch Valley, Zermatt, Switzerland (E-an Zen et al. 1964). Total Fe as Fe_2O_3.
7. Margarite, Chester, Massachusetts (Velde 1971). Total includes 0.09% P_2O_5.

In a survey of 168 pegmatite muscovites, Heinrich et al. (1953) found that the total iron content varied between 0.63 and 5.1% Fe_2O_3 with a mean of 2.9%, and the MgO content between 0.01 and 1.9%, mean 0.93%. Most published analyses of muscovites fall within these limits; the muscovite described by Foster et al. (1960; analysis 2, Table 1.21) is exceptional for its large iron content ($Fe_{1.1}$) relative to the tetrahedral composition ($Si_{6.5}Al_{1.5}$).

Phengitic micas from metamorphic rocks contain more Si, Fe and Mg; the tetrahedral compositions lie between $(Si_{6.3}Al_{1.7})$ and (Si_7Al), with total Fe between 2.5 and 6% Fe_2O_3 and MgO between 2.5 and 4% (analyses 3 and 4, Table 1.21).

1.7.1.2 Paragonite

Muscovites commonly contain 0.5 to 1.5% Na_2O, with a maximum of about 2%; the sodium analogue of muscovite, paragonite (analysis 6, Table 1.21), may likewise contain up to 2% K_2O (analysis 5, Table 1.21), but these are the limits of the solid solutions. Although the structures of paragonite and muscovite are similar, local strains around "impurity" alkali ions in either structure are too severe to allow more than limited solid solution (Radoslovich 1963). Published analyses reviewed by Henley (1970) show that nearly 20 atoms per cent Ca can substitute into paragonite. The Na–K solid solution limits are temperature dependent and the composition of coexisting paragonite and muscovite may be used as a geothermometer (Henley 1970).

Paragonite is a relatively scarce mineral found in alumina-rich metamorphic rocks (Guidotti 1968) and most analyses are slightly silica deficient relative to muscovite (Kienast and Velde 1968) with a tetrahedral composition between $(Si_{5.75}Al_{2.23})$ and (Si_6Al_2).

Sodium-rich clay micas found in some Carboniferous sediments are known as brammallite (Bannister 1943) and appear to have the same relation to paragonite as illite does to muscovite–phengite; the layer charge is smaller and H_2O^+ is larger than in paragonite and it is therefore included in Section 1.7.2.

The mineral rectorite (Brown and Weir 1963) has a regular mixed-layer structure, non-expanding mica-like interlayers rich in sodium alternating with expanding smectite-like interlayer; potassium and calcium analogues are also known and both types are included in the section on mixed-layer clays, Section 1.9.

1.7.1.3 Margarite

The rare brittle mica margarite occurs mainly in highly aluminous metamorphic rocks and has the ideal formula $Ca_2(Al_4)(Al_4Si_4)O_{20}(OH)_4$, the layers having twice the net charge of the common micas. Many analyses are close to the ideal composition (analysis 7, Table 1.21), but Na substitutes for Ca (Afanasev and Aidinyan 1952) and this may be accompanied by substitution of Si for tetrahedral Al according to the equation

$$Si(IV) + Na(or\ K)(XII) \rightleftharpoons Ca(XII) + Al(IV)$$

this decreases the layer charge and alters the composition towards that of paragonite.

1.7.2 Clay micas

The AIPEA Nomenclature Committee has written that "the status of *illite* (or hydromica), *sericite*, etc., must be left open at present because it is not clear whether or at what level they would enter the Table [Table 1 in Bailey, 1980a]: many materials so designated may be interstratified".

At present, therefore, there are no definitions that enable the compositions of this very important group of clay minerals to be discussed in terms of end-member compositions, and furthermore there remain uncertainties about their structural composition.

For convenience only in the context of the present chapter, we propose using terms with the following meaning:

1. *Sericite:* a field term for a fine-grained white material of platy morphology, often a hydrothermal alteration product and usually but not always containing potassium.
2. *Hydrous mica:* a non-specific name for a mineral that contains mica-like layers but with analyses that show more water and less potassium than a true mica; sometimes hydrous mica may denote minerals with a mixed-layer structure.
3. *Hydromica:* synonymous with hydrous mica.
4. *Illite:* Gaudette et al. (1966) used the term illite for clay minerals that have a 10 Å basal repeat unit, show no evidence of interstratification but nevertheless contain less K and more H_2O^+ than true mica. To a first approximation an illite layer is structurally similar to a mica layer and it is therefore legitimate to describe a mineral as an "interstratified illite–smectite".

Thompson and Hower (1975) showed that glauconite is structurally similar to illite and likewise contains less K and more H_2O^+ than mica but differs from illite in containing much more Fe. By analogy with illite, they described structurally inhomogeneous minerals as, for example, "interstratified glauconite–smectite".

On the basis of the grouping demonstrated in Figs. 1.4 and 1.5 we shall describe as *illites* all clay micas having a 10 Å repeat unit with no evidence of interstratification, and with total Fe < 1 per four octahedral sites, and a *glauconites* and *celadonites* all clay micas with no evidence of interstratification but with total Fe > 1 per four octahedral sites.

1.7.2.1 Hydromuscovites

We include in this group those hydrous micas with Si/Al ratios between unity (the ratio for the ideal muscovite composition) and 1.5, and with less than 10% interstratification with smectite layers (analyses 1 and 2, Table 1.22). These fine-grained micas are reported mainly from Japan where they are found in regions of hydrothermal alteration and frequently contain rather little Mg and Fe (<2% MgO and <3% Fe_2O_3). The octahedral sheet has only a small net negative charge or even a slight positive charge from octahedral cations in excess of 4 per six sites. In clays where the octahedral net charge is positive and there is also an octahedral occupancy significantly greater than 4, it is possible that the clays contain aluminium hydroxy compounds, either sorbed on the surface or as interlayers, as found by Thompson and Hower (1975) for glauconites.

The hydrous micas contain H_2O^+ (i.e. water evolved on heating from 110 to 1000 °C) in excess of that for ideal muscovite 4.6%; the structural significance of this additional water is not yet certain. Some minerals in the group contain expanding layers and hydrated interlayer cations, so that some excess water is present in these expanding layers. Others are apparently structurally homogeneous yet contain 1% or more excess water. Kodama and Brydon (1968) found that the size fraction 0.5–2 μm of a microcrystalline muscovite contained 6.4% H_2O^+ whereas the fraction 2–10 μm contained 4.7% H_2O^+, close to the theoretical amount. They showed however that the additional water in the fine fraction was released between 100 and 300 °C in a temperature range below the main dehydroxylation range and that H_2O^+, 400 °C, was similar in all size fractions.

It is possible therefore that excess H_2O^+ in hydrous micas is sorbed water retained to temperatures higher than 105 °C conventionally used to distinguish between sorbed and structural water, but this has yet to be confirmed by systematic thermogravimetric study.

Regular and random interstratifications are common in this group of minerals and such minerals containing more than 10% expansible layers are discussed in Section 1.9.

1.7.2.2 Illites

Hower and Mowatt (1966) made a detailed study of the structure and composition of illites and showed that many so-called illites contained interstratified smectite interlayers. Nevertheless, when a trend diagram was constructed of interlayer $(K + Na)$ against the proportion of expandable layers, this extrapolated to 0% expandable layers at $1.5(K + Na)$ per O_{22}, not $2(K + Na)$ as might have been expected. Coupled with this deficit of $(K + Na)$ is an excess of water retained above 100 °C, H_2O^+, and these two observations had led Brown and Norrish (1952) to suggest that H_3O^+ was substituted for K^+ in the illite structure. Hower and Mowatt (1966) argue against this explanation on the grounds that the structural formulae so calculated are not consistent and that the trend line implies a constant H_3O^+/K^+ ratio, a diagenetically unlikely situation. They suggest alternatively that the illite layer charge is lower than that of muscovite or phengite and that the excess water is held at vacant interlayer sites as H_2O.

Recently, Norrish and Pickering (1983) reopened the discussion about the inverse relation between the K_2O and H_2O contents of illites that apparently contain no smectite interlayers. They argued that expansion of the illite by exchange of K^+ would also exchange H_3O^+ if present in the interlayers and so as exchange proceeds, the sum of exchangeable cations and K^+ should increase. They quote previously unpublished analyses of eight illites from which K^+ was extracted with sodium tetraphenylboron and show that the sum of interlayer cations and K^+ does not increase as would be expected if H_3O^+ was being displaced. They point out that as there is a continuous loss of water on heating, dividing the water evolved into hygroscopic and structural water is impossible. They also considered the possibility that the extra water might be present as randomly interstratified kaolinite layers in too small a concentration to detect by X-ray diffraction, but no conclusions are given.

Norrish and Pickering also note that there is an inverse relation between K_2O content and the cation exchange properties of illites but that this cannot be explained by the proportionately increased external surface of very small illite particles. They suggest that the excess exchange capacity might originate from replacement of K^+ in an outer annulus of disc-like illite particles.

However, whichever method is used to calculate the structural formulae of illites, the main bulk of analyses indicate a smaller substitution of Al in the tetrahedral sites than in muscovite, compositions tending towards that of phengite, Si_7Al, but with less substitution of Mg for Al in the octahedral layer, so that the total layer charge is significantly less than 2 per O_{22} (analyses 3–6, Table 1.22). Total iron contents are nearly always less than 1 per four octahedral sites, and this distinguishes illites from most glauconites, though the aluminian glauconites described by Berg-Madsen (1983) do overlap into the illite composition range (analysis 8, Table 1.23) and three very fine-grained illites from Australia, two thought to be of pedogenic origin (analyses 7 and 8, Table 1.22) and one of lacustrine origin (analysis 9), have much larger Fe contents than illites in hard sedimentary rocks.

There appears to be a sodium analogue of illite that occurs in some Carboniferous sediments and it was termed "brammallite" by Bannister (1943), though this name has not been officially recognized. A similar clay for which a more complete analysis has been given (analysis 10, Table 1.22) occurs in the Carboniferous deposits of the Dniepr–Donetz depression (Karpova 1966). The proportion of Na to K in this analysis lies outside the usual solid solution limits, suggesting that this sample may be a physical or structural mixture, like that reported by Frey (1969).

Natural ammonium illites recently described by Sterne et al. (1982) have NH_4^+ substituting significantly ($>50\%$) for K^+ in the interlayer.

THE CHEMICAL CONSTITUTION OF CLAYS

TABLE 1.22. Analyses of hydromuscovites, illites and paragonitic hydromica

	1	2	3	4
SiO_2	47.65	48.44	51.5	51.0
Al_2O_3	37.03	33.84	28.5	28.7
TiO_2	0.10	trace	0.77	0.27
Fe_2O_3	0.01	0.49	0.61	0.59
FeO	trace	0	0.91	0.39
MnO	trace	trace		
MgO	0.04	0.95	3.0	2.31
CaO	trace	0.11	0.05	0.30
Na_2O	0.76	0.50	0.10	0.15
K_2O	9.02	9.40	9.07	7.78
H_2O^+	4.97	5.42	5.5	5.96
H_2O^-	0.73	0.60	0.7	1.99
Total	100.31	99.75	100.71	99.44
Number of cations on the basis $O_{20}(OH)_4$				
Si	6.26	6.45	6.81	6.88
Al	1.74	1.55	1.19	1.12
Σ tet.	8.00	8.00	8.00	8.00
Al	3.99	3.76	3.25	3.45
Ti	0.01	0	0.08	0.03
Fe^{3+}	0	0.05	0.06	0.06
Fe^{2+}	0	0	0.10	0.04
Mn	0	0		
Mg	0.01	0.19	0.59	0.46
Σ oct.	4.01	4.00	4.08	4.04
Ca	0	0.01	0.01	0.04
Na	0.19	0.13	0.02	0.04
K	1.51	1.60	1.53	1.34
Σ interl.	1.70	1.74	1.56	1.42
Si/Al	1.09	1.21	1.53	1.51
Tetrahedral charge	−1.74	−1.55	−1.19	−1.12
Octahedral charge	+0.04	−0.20	−0.38	−0.35
Interlayer charge	−1.70	−1.75	−1.57	−1.47

1. Sericite, $2M_1$ polymorph, Yoji Pass, Gunma Prefecture, Japan; analysis YO(-01) in Table 1.10 of Sudo (1978).
2. Sericite, Seshido Mine, Fukushima Prefecture, Japan; analysis SS in Table 1.10 of Sudo (1978).
3. Illite, Interlake, well core, Williston Basin, Montana, USA. CEC = 120 μeq g^{-1}; < 10% expandable layers (Hower and Mowatt 1966).
4. Steamboat Springs illite; 10% expandable layers (Hower and Mowatt 1966).
5. Illite, Burnt Bluff Group, Fond du Lac Co., Wisc., USA (Gaudette 1965). Usually called "Marblehead Illite", but also known as Burnt Bluff Illite (Hower and Mowatt 1966).
6. Illite, Silver Hill, Jefferson Canyon, Montana, USA. CEC = 150 μeq g^{-1}; < 10% expandable layers (Hower and Mowatt 1966).

1.7.2.3 Celadonite–glauconite micas

For many years there has been considerable controversy over the structure, composition and terminology associated with this group of fine-grained mica minerals. Although their modes of occurrence were quite distinct, glauconites being found in recent and fossil sediments whereas celadonites were formed during the alteration of volcanic, particularly basaltic, rocks, they appeared to form a continuous isomorphous replacement series (Foster 1969). Recent work by Buckley et al. (1978) has however suggested that they are mineralogically distinct and can be

	5	6	7	8	9	10
	52.87	55.1	54.8	60.4	55.2	47.46
	24.91	22.0	20.1	16.4	17.9	27.65
	1.02	0.63	1.1	1.44	0.32	
	0.78	5.28	13.5	11.8	11.9	3.77
	1.19	1.34			0.85	1.48
	3.60	2.8	3.1	3.67	4.33	1.52
	0.69	0.02	1.86	2.25	0.84	0.42
	0.22	0.08			0.0	4.34
	7.98	8.04	5.4	3.98	7.88	4.40
	6.73	6.40				6.35
	2.56	1.03				2.60
	99.61	102.7	99.86	99.94	99.22	99.99

Number of cations on the basis $O_{20}(OH)_4$

	5	6	7	8	9	10
	7.07	7.28	7.01	7.56	7.18	6.57
	0.93	0.72	0.99	0.44	0.82	1.43
	8.00	8.00	8.00	8.00	8.00	8.00
	2.99	2.71	2.03	1.98	1.93	3.07
	0.10	0.06	0.11	0.14	0.03	
	0.08	0.52	1.30	1.11	1.17	0.39
	0.13	0.15			0.09	0.17
	0.72	0.55	0.54	0.68	0.84	0.31
	4.02	3.99	4.03	3.91	4.06	3.94
	0.10	0.0	0.25	0.30	0.12	0.06
	0.06	0.02				1.16
	1.36	1.36	0.88	0.64	1.31	0.78
	1.52	1.38	1.13	0.94	1.43	2.00
	1.80	2.12	2.31	3.12	2.62	1.46
	−0.93	−0.72	−0.99	−0.44	−0.82	−1.43
	−0.68	−0.66	−0.40	−0.80	−0.73	−0.63
	−1.61	−1.38	−1.39	−1.24	−1.55	−2.06

7. Illite, <2 μm fraction of Willalooka sand, a *solodized–solonetz* soil. Hundred of Laffer, South Australia (Norrish and Pickering 1983). Corrected for 20% kaolinite.
8. Illite, <2 μm fraction from *red-brown earth*, Tharbogang, New South Wales (Norrish and Pickering 1983). Corrected for 33% kaolin and 2.1% dithionite-soluble Fe_2O_3.
9. Illite, <2 μm fraction from lacustrine green clay, Muloorina Station, South Australia (Norrish and Pickering 1983).
10. Paragonite hydromica, tonstein, Carboniferous, Dniepr–Donetz depression, USSR (Karpova 1966).

differentiated by a combination of chemical and structural criteria. These criteria have been adopted by the AIPEA Nomenclature Committee (Bailey 1980a).

According to these definitions, *celadonite* is a dioctahedral mineral of ideal composition $K_2(Mg_2Fe_2^{3+})(Si_8)O_{20}(OH)_4$ but allowing up to 0.4Al substitution in the tetrahedral sheet, having $d(060) < 1.510$ Å and sharp OH stretching bands in the infrared region 3610–3530 cm^{-1}. *Glauconite* is defined as an Fe-rich dioctahedral mica with tetrahedral substitution greater than 0.4Al per 8Si and octahedral R^{3+} greater than 2.4 atoms per four sites. Glauconite has $d(060) > 1.510$ Å and usually broader absorption in the 3610–3530 cm^{-1} region. Microprobe analysis has established glauconite as an essentially single-phase mica, and the name glauconite

TABLE 1.23. Analyses of celadonites and glauconites

	1	2	3	4	5	6	7	8	9
SiO_2	55.61	53.40	52.58	50.12	49.11	51.19	56.98	53.8	49.03
Al_2O_3	0.79	2.51	5.28	2.41	9.41	9.33	13.33	18.5	17.93
TiO_2		0.04	0.20	0.02	0.04	0.13	0.08	0	1.06
Fe_2O_3	17.19	15.88	14.61	20.33	21.00	18.15	8.78	8.0	13.11
FeO	4.02	3.53	3.37	2.07	2.71	1.78	7.05		1.31
MnO	0.09					0.01		0.0	
MgO	7.26	6.45	6.35	6.35	3.06	3.34	5.35	4.7	2.79
CaO	0.21	0.07	0.68	0.16	0.35	0.58	0.05	0.0	0.39
Na_2O	0.19	0.05	0.31	0.02	0.03	0.02	0.23	0.3	0.10
K_2O	10.03	10.28	7.69	8.19	8.68	7.98	8.21	9.0	7.84
H_2O^+	4.88					5.21	†		6.00
H_2O^-						1.95	†		
Total	100.27	92.21	91.07	89.67	94.39	99.67	100.06	94.3	99.56

Number of cations on basis $O_{20}(OH)_4$

	1	2	3	4	5	6	7	8	9
Si	7.99	7.91	7.74	7.65	7.14	7.42	7.51	7.29	6.87
Al	0.01	0.09	0.26	0.35	0.86	0.58	0.49	0.71	1.13
Σ tet.	8.00	8.00	8.00	8.00	8.00	8.00	8.00	8.00	8.00
Al	0.13	0.35	0.65	0.09	0.75	1.01	1.58	2.24	1.84
Ti	0	0	0.02	0	0	0.01	0.01	0.00	0.11
Fe^{3+}	1.86	1.77	1.62	2.34	2.30	1.98	0.87	0.82	1.38
Fe^{2+}	0.48	0.44	0.42	0.26	0.33	0.21	0.78		0.15
Mn	0.01					0.00		0.00	
Mg	1.55	1.43	1.39	1.45	0.66	0.72	1.05	0.95	0.58
Σ oct.	4.03	3.99	4.10	4.14	4.04	3.93	4.29	4.01	4.06
Ca	0.03	0.01	0.11	0.03	0.05	0.09	0.01	0.00	0.06
Na	0.05	0.01	0.09	0.01	0.01	0.01	0.06	0.08	0.03
K	1.84	1.94	1.44	1.59	1.61	1.47	1.38	1.55	1.40
Σ interl.	1.93	1.96	1.64	1.63	1.67	1.57	1.45	1.63	1.49
Si/Al	57.1	18.0	8.5	17.4	4.43	4.67	3.66	2.47	2.31
Tetrahedral charge	−0.01	−0.09	−0.26	−0.35	−0.86	−0.58	−0.49	−0.71	−1.13
Octahedral charge	−1.95	−1.90	−1.49	−1.29	−0.87	−1.08	−0.96	−0.92	−0.42
Interlayer charge	−1.96	−1.99	−1.75	−1.64	−1.73	−1.66	−1.45	−1.63	−1.55

1. Celadonite, vesicular basalt, Reno, Nevada (Hendricks and Ross 1941).
2. Celadonite, massive, locality unknown (sample D of Buckley et al. 1978).
3. Celadonite, vein in crack, ocean-floor basalt, 15° 01′ N. 73° 24′ W., 2029 m (sample C of Buckley et al. 1978).
4. Glauconite, dark green botryoidal pellets, 31° 21.5′ N. 10° 19.5′ W., 500 m (sample 29D of Buckley et al. 1978).
5. Glauconite, olive green pellets, Upper Bracklesham Beds, Shepherds Gutter, Hampshire (sample 21 of Buckley et al.
6. Glauconite, Bohemian Upper Cretaceous, Praha-Prosek (sample 49 of Cimbalnikova 1971).
7. Glauconite, Ordovician, Basal Arenigian, Stora Brottet, Latorp, Sweden (sample G294 of Thompson and Hower 1975: 5% smectite, ISII ordered; † analysis of ignited sample).
8. Glauconite, Kalby Member, *Exsulans* Limestone Formation, Bornholm, Denmark (sample C4 of Berg-Madsen 1983).
9. Glauconite, Morrison Formation, Colorado, USA (Keller 1958).

does not have any genetic connotation: a green fecal pellet from a marine environment that meets the mineralogical definition for celadonite should be described as celadonite, not glauconite. Specimens that are structural mixtures with smectite can be described as interstratified glauconite–smectite.

Glauconite aggregates and pellets are often inhomogeneous (Bentor and Kastner 1965; Buckley et al. 1978) and inclusions of calcite, apatite and free iron oxides have been reported.

Bulk analyses may not therefore represent the chemical composition of the layer silicate particles, and distortion of the apparent isomorphous substitution pattern may occur. Free iron oxides for instance increase the apparent occupancy of the octahedral sheet (Kelley 1945).

Celadonites mostly have a net negative charge close to 2 per O_{22} but glauconites have a net charge significantly less than this, and are often interstratified with smectite. Thompson and Hower (1975) showed that there is a curvilinear inverse relationship between per cent expanding layers and interlayer (Na + K) and that this extrapolated to about 1.6(Na + K) at 0% expanding layers, a similar extrapolation point to that for illites. There seems therefore to be a close analogy between glauconite *senso strictu* and illite in respect to low K and high H_2O^+ contents.

Although glauconites are undoubtedly Fe-rich, the criterion $Fe^{3+} \gg Al$ suggested by AIPEA (Bailey 1980a) does not always hold and several glauconite analyses have $Fe^{3+} < Al$ (analyses 7–8, Table 1.23); as already noted, the Al-glauconite of analysis 8 more properly falls within the illite composition field. It seems therefore that a wide range of octahedral compositions is possible in minerals that have been called glauconite.

A substantial number of glauconite analyses suggest that some "trioctahedral" character is present, the octahedral contents exceeding 4.2 per O_{22} (analysis 7, Table 1.23). Foster (1969) suggested this was due to interlayer Mg and distributed cations between the sites accordingly. Thompson and Hower (1975) tested the hypothesis by acid extraction of two glauconites showing this feature and measuring the rate of release of Mg, Fe and Al. Their results showed that a proportion of Mg, Fe and Al was extracted at a rate that suggested it was not coordinated within the octahedral sheet. They considered that some Mg, Fe and Al was present as interlayer hydroxy complexes, which would increase the apparent occupancy of the octahedral sheet.

1.7.3 Trioctahedral micas $X_2Y_{5-6}Z_8O_{20}(OH,F)_4$

The trioctahedral micas are common constituents of igneous and metamorphic rocks and many hundreds of analyses of them have been published. Reviewing over 200 analyses in the literature, Foster (1960a) was able to interpret the wide range of composition in terms of a few simple types of substitution, and her interpretation and terminology is used here.

Compared with dioctahedral micas, the trioctahedral micas not only have a much larger divalent cation content but also contain less SiO_2 and the tetrahedral sites contain more Al; occasionally it is also necessary to assign Ti (analysis 3, Table 1.24) or Fe^{3+} (for example, the ferriphlogopite investigated by Steinfink, 1962) to the tetrahedral sheet. The principal cations in the octahedral sheet are Mg, Fe^{2+}, Al and Fe^{3+} with smaller proportions of Mn, Ti and Li; micas containing larger proportions of Li are described separately in the next section. Potassium is usually the dominant interlayer cation in the common micas; an exception is the brittle mica clintonite which has interlayer Ca (Forman *et al.* 1967; analysis and structural formulae in Tables 1.2, 1.3 and 1.5). Also, sodium is occasionally a dominant interlayer cation; these structures are discussed in Section 1.7.3.1.

A substantial amount of fluorine is present in many trioctahedral micas and conveys important properties on the structure, such as resistance to weathering (Rausell Colom *et al.* 1965; Newman 1969) and hardness (Bloss *et al.* 1959); the basal spacing decreases with increasing fluorine content (Rousseaux *et al.* 1973). In the absence of lithium, natural micas may contain 6% F (analysis 1, Table 1.24); synthetic fluorophlogopite contains over 9% F.

The end-member of the trioctahedral series is phlogopite, $K_2Mg_6(Si_6Al_2)O_{20}(OH,F)_4$; natural phlogopites with total Fe less than 1% FeO are rare. Simple replacement of Mg by Fe^{2+}

TABLE 1.24. Analyses of trioctahedral micas

	1	2	3	4	5
SiO_2	42.98	40.95	40.78	40.65	37.05
Al_2O_3	12.90	17.28	10.95	19.88	15.04
TiO_2	0.33	0.82	8.97	0.33	2.81
Fe_2O_3	0.91	0.43	2.18	0.00	4.12
FeO	2.70	2.38	3.73	7.89	5.10
MnO	0.08	trace	trace	0.12	0.28
MgO	25.93	22.95	19.66	15.66	19.74
Li_2O				0.28	
CaO	0.06	0.00	0.11	0.14	0.33
BaO		0.03	0.35		1.26
Na_2O	0.25	0.16	0.11	0.83	0.23
K_2O	10.63	9.80	10.59	9.46	9.55
H_2O^+	1.70	4.23	1.87	} 2.90	4.15
H_2O^-	0.70	0.48	0.19		0.30
F	6.04	0.62	0.66	2.57	0.40
Total	105.21	100.13	100.15	101.08	100.36
$O \equiv F$	−2.54	−0.26	−0.28	−1.08	−0.16
Total corrected	102.67	99.87	99.87	100.00	100.20

Number of cations and F atoms on the basis $O_{20}(OH,F)_4$

	1	2	3	4	5
Si	5.97	5.75	5.74	5.78	5.41
Al	2.03	2.25	1.81	2.22	2.59
Ti			0.45		
Σ tet.	8.00	8.00	8.00	8.00	8.00
Al	0.08	0.60	0.00	1.11	0.00
Ti	0.03	0.09	0.50	0.04	0.31
Fe^{3+}	0.10	0.05	0.23	0.00	0.45
Fe^{2+}	0.31	0.28	0.44	0.94	0.62
Mn	0.01	0.00	0.00	0.01	0.03
Mg	5.37	4.80	4.12	3.32	4.30
Li				0.16	
Σ oct.	5.90	5.82	5.29	5.58	5.71
Ca	0.01	0.00	0.02	0.02	0.05
Ba		0.002	0.02		0.07
Na	0.07	0.04	0.03	0.23	0.07
K	1.88	1.75	1.90	1.72	1.78
Σ interl.	1.96	1.79	1.97	1.97	1.97
F	2.65	0.27	0.31	1.16	0.18

1. Phlogopite, Ontario. New analysis by XRF (G. Brown); FeO and Na_2O from Newman (1967) and F from Newman (1969).
2. Phlogopite, from marble. Anxiety Point, Nancy Sound, New Zealand (Hutton 1947).
3. Titaniferous phlogopite, phlogopite–leucite lamproite, Howes Hill, Western Australia (Prider 1939).
4. Magnesian biotite, Monte Summa, Tuscany (Foster 1960a; analysis 37). Total includes 0.37% Rb_2O (0.03 atoms).
5. Magnesian biotite, biotite–augite–peridotite, Upper Kabirenge lava, Uganda (Coombe and Holmes 1945).

is never found and natural micas always contain octahedral Al^{3+}, Fe^{3+} and Ti (grouped together as R^{3+}); the additional positive charge from these polyvalent cations in the octahedral sheet is balanced in two ways: (1) Al in excess of that required to balance interlayer charge substitutes for Si in the tetrahedral sheet up to about $Si_{5.3}Al_{2.7}$ and balances the equivalent excess charge in the octahedral sheet (analysis 6, Table 1.24); (2) R^{3+} replaces R^{2+}, not atom for

6	7	8	9	10
38.04	39.17	35.55	31.64	33.60
15.77	11.24	16.16	15.34	22.36
0.39	2.33	2.63	0.33	0.06
2.62	1.86	4.34	8.38	1.44
11.13	16.58	17.19	31.78	28.54
0.10	0.89	0.61	0.28	0.33
18.35	13.51	8.95	0.90	0.10
0.00	0.18	0.32		0.32
0.91	0.20	0.46		0.14
0.15	0.05			
0.27	0.62	0.90	0.67	0.46
9.23	9.29	8.77	8.70	8.90
2.24	1.64	3.58	} 1.93	1.66
0.20	0.09	0.08		0.08
0.26	3.46	0.41	0.13	0.89
100.15	101.28	100.19	100.08	98.88
−0.11	−1.48	−0.17	−0.05	−0.37
100.04	99.80	100.02	100.03	98.51
Number of cations and F atoms on the basis $O_{20}(OH)_4$				
5.53	5.91	5.41	5.17	5.29
2.47	2.00	2.59	2.83	2.71
	0.09			
8.00	8.00	8.00	8.00	8.00
0.23	0.00	0.31	0.12	1.44
0.04	0.18	0.30	0.04	0.01
0.29	0.21	0.50	1.03	0.17
1.35	2.09	2.19	4.34	3.76
0.01	0.11	0.08	0.04	0.05
3.97	3.04	2.03	0.22	0.02
	0.11	0.20		0.20
5.89	5.74	5.61	5.79	5.65
0.14	0.03	0.07		0.02
0.01	0.003			
0.08	0.18	0.27	0.21	0.14
1.71	1.80	1.73	1.81	1.79
1.94	2.01	2.07	2.02	1.95
0.12	1.70	0.20	0.07	0.44

6. Magnesian biotite, granite pegmatite, Tollgate quarry, Malvern (Tha Hla 1945). Total includes 0.49% minor constituents.
7. Biotite, North Burgers, Ontario (Rausell Colom et al. 1965). Total includes 0.12% Cl and 0.05% Rb_2O.
8. Ferroan biotite, G1 granite, Keadew Strand, Donegal (Hall 1969). Total includes 0.24% Rb_2O (0.02 atoms).
9. Lepidomelane, French River, Sudbury district, Ontario (Foster 1960a; analysis 124).
10. Siderophyllite, Yagenyama, Japan (Foster 1960a; analysis 135).

atom, but in the proportion $2R^{3+}:3R^{2+}$ so that some octahedral sites remain unfilled (analysis 7, Table 1.24). In many formulae, both types of charge compensation occur.

These two modes of substitution imply a much wider composition range than is actually found in natural micas: Hazen and Wones (1972) point out that the octahedral Al content increases as Fe^{2+} replaces Mg and summarize the structural limitations on the substitutional

TABLE 1.25. Possible end-member compositions of siderophyllite and lepidomelane

Tetrahedral composition	Ratio of Fe^{2+} to R^{3+}	
	4:1	3:2
Si_6Al_2	$Fe^{2+}_{4.36}R^{3+}_{1.09}$	$Fe^{2+}_3R^{3+}_2$
$Si_{5.3}Al_{2.7}$	$Fe^{2+}_{4.62}R^{3+}_{1.15}$	$Fe^{2+}_{3.17}R^{3+}_{2.12}$

pattern. A plot of natural composition on the triangular diagram bounded by the proportions of Mg, ($Fe^{2+} + Mn^{2+}$) and R^{3+} (representing the sum of trivalent cations) in the octahedral sheet (Foster 1960a) shows that the proportion of Fe^{2+} to R^{3+} is mostly between 4:1 and 3:2 with a central value of 7:3. When combined with Al-for-Si substitution in the tetrahedral sheet of between Si_6Al_2 and $Si_{5.3}Al_{2.7}$, the end-members with Mg entirely absent (siderophyllites and lepidomelanes) should have formulae within the limits shown in Table 1.25.

Formulae in which the number of octahedral R^{3+} exceeds 1.8 per six sites are rare, and the number of unfilled octahedral sites is seldom greater than 0.8, which is evidence against the existence of a complete series between the dioctahedral and trioctahedral micas except when lithium is also present.

Foster (1960a) found that most of the analyses of the trioctahedral micas formed three natural groups and suggested the following terminology. Micas in which more than 70% of the occupied octahedral sites contained Mg^{2+} she termed phlogopites (analyses 1, 2, 3 and 5, Table 1.24), whereas in biotites 20–60% of the occupied sites contained Mg; the biotites were further subdivided into *magnesian biotites* (analyses 4, 6 and 7, Table 1.24) with between 40 and 60% Mg and *ferroan biotites* (analysis 8, Table 1.24) with 20–40% Mg. Finally, the micas with Mg in less than 10% of the occupied sites were divided into *siderophyllites* in which the dominant trivalent cation is Al (analysis 10, Table 1.24), and *lepidomelanes* in which Fe^{3+} is the main trivalent cation (analysis 9, Table 1.24).

The determined hydroxyl and fluorine contents of trioctahedral micas are frequently equivalent to less than the theoretical $(OH,F)_4$ (Foster 1964) and sometimes this is because the determination of structural hydrogen is inaccurate (see Section 1.2.1.2). If analyses that are deficient in H_2O^+ are calculated to structural formulae based on 24(O,F), the number of tetrahedral and octahedral cations frequently exceeds 14 and the number of interlayer cations exceeds 2. Such formulae are impossible structurally and should be regarded as unsatisfactory; should the same analyses calculated to $O_{20}(OH,F)_4$ give normal site occupancy, the water and/or the fluorine determinations are likely to be inaccurate. Eugster and Wones (1962) considered that some biotites are hydrogen deficient from internal oxidation of Fe^{2+} to Fe^{3+} expressed by the equation

$$K_2(Fe^{2+}_4 Mg_2)(Al_2Si_6)O_{20}(OH)_4 \rightleftharpoons K_2(Fe^{3+}_4 Mg_2)(Al_2Si_6)O_{24} + 2H_2$$

Foster (1964) found that internal oxidation explained hydroxyl deficiency in some but not all Fe^{2+}-dominant micas, and that other instances must represent real deficiency not coupled with oxidation or analytical error.

Trioctahedral micas are readily weathered so that trioctahedral analogues of the hydrous dioctahedral micas and illites are rarely found; the only reported example of a trioctahedral illite (Walker 1950) was too impure for its chemical composition to be established. Vermiculite deposits often contain hydrobiotite which has an interstratified structure composed of expanding vermiculite interlayers and potassium-rich mica-like interlayers; these are discussed in the section on interstratified minerals, Section 1.9.

1.7.3.1 Trioctahedral sodium micas

There have been three recent reports of naturally-occurring sodium-rich trioctahedral micas, preiswerkite (Keusen and Peters 1980), sodium phlogopite (Schreyer *et al.* 1980) and material named wonesite (Spear *et al.* 1981). They show marked differences in amounts of interlayer charge, and of substitution of Al^{3+} for Si^{4+} and of R^{3+} for R^{2+} in octahedral sites. The structural formulae of these minerals recalculated from published analyses are

$$(Na_{1.956}K_{0.033}Ca_{0.013})(Mg_{3.826}Fe^{2+}_{0.189}Fe^{3+}_{0.119}Al^{3+}_{1.831})(Si_{4.104}Al_{3.896})O_{20}(OH)_4$$

for preiswerkite,

$$(Na_{1.558}K_{0.086}Ca_{0.030})(Mg_{5.875}Al_{0.116}Ti_{0.050})(Si_{5.999}Al_{2.001})O_{20}(OH)_4$$

for sodium phlogopite, and

$$(Na_{0.794}K_{0.145}Ca_{0.006})(Mg_{4.394}Fe^{2+}_{0.783}Mn_{0.005}Al_{0.628}Cr_{0.008}Ti_{0.077})(Si_{6.470}Al_{1.530})O_{20}(OH)_4$$

for the material named wonesite.

Veblen (1983a, 1983b) has examined wonesite by high resolution transmission electron microscopy and shown that it is not homogeneous but is an intimate intergrowth of a talc-like material and what is referred to as "sodium biotite". The 2:1 layers are continuous across the two components which exist as fine lamellae inclined at an angle of 37° to the plane of the layers. From Veblen's (1983a) analytical electron microscopy examination of the two components, and assuming full occupancy of octahedral sites, the structural formula of the "Na-biotite" is calculated to be about

$$(Na_{1.20})(R^{2+}_{5.28}R^{3+}_{0.72})(R^{4+}_{6.08}R^{3+}_{1.92})O_{20}(OH)_4$$

Veblen (1983a) has discussed the nomenclature of Na-rich trioctahedral micas and presented their compositions on a triangular diagram, the vertices of which represent ideal talc $(\square_2)(Mg_6)(Si_8)O_{20}(OH)_4$, ideal phlogopite, $(K_2)(Mg_6)(Si_6Al_2)O_{20}(OH)_4$, and ideal preiswerkite, $(Na_2)(Mg_4Al_2)(Si_4Al_4)O_{20}(OH)_4$; the vertices indicate the distribution of the total cationic charge (+44) among interlayer, octahedral and tetrahedral sites. Using these vertices Fig. 1.6 compares the charge distribution in natural preiswerkite, natural Na-phlogopite, bulk wonesite and the associated "Na-biotite" with those of vermiculites (analyses in Table 1.18), trioctahedral smectites (analyses in Table 1.17), trioctahedral K-micas (analyses in Table 1.24), talcs, willemseite and minnesotaites (analyses in Table 1.14) and Na–Al talc (average of 10 microprobe analyses by Schreyer *et al.* 1980). Some analyses plot outside the triangle because the structural formulae lie outside the limits set by the vertices, e.g. interlayer charge >2.00, tetrahedral charge >32.00, octahedral charge <12.00. It is clear that bulk wonesite and the "Na-biotite" lie in the field of trioctahedral smectites and vermiculites and, following the definitions of layer silicate mineral groups (Bailey 1980a), should not be classified as micas. Preiswerkite has a full complement of interlayer cations and the layer charge of the Na-phlogopite, 1.704, places it near the upper limit (1.8) defined for vermiculite (Bailey 1980a). Compositionally bulk wonesite corresponds to a high-charge trioctahedral smectite and the associated "Na-biotite" to a low-charge vermiculite.

80 THE CHEMICAL CONSTITUTION OF CLAYS

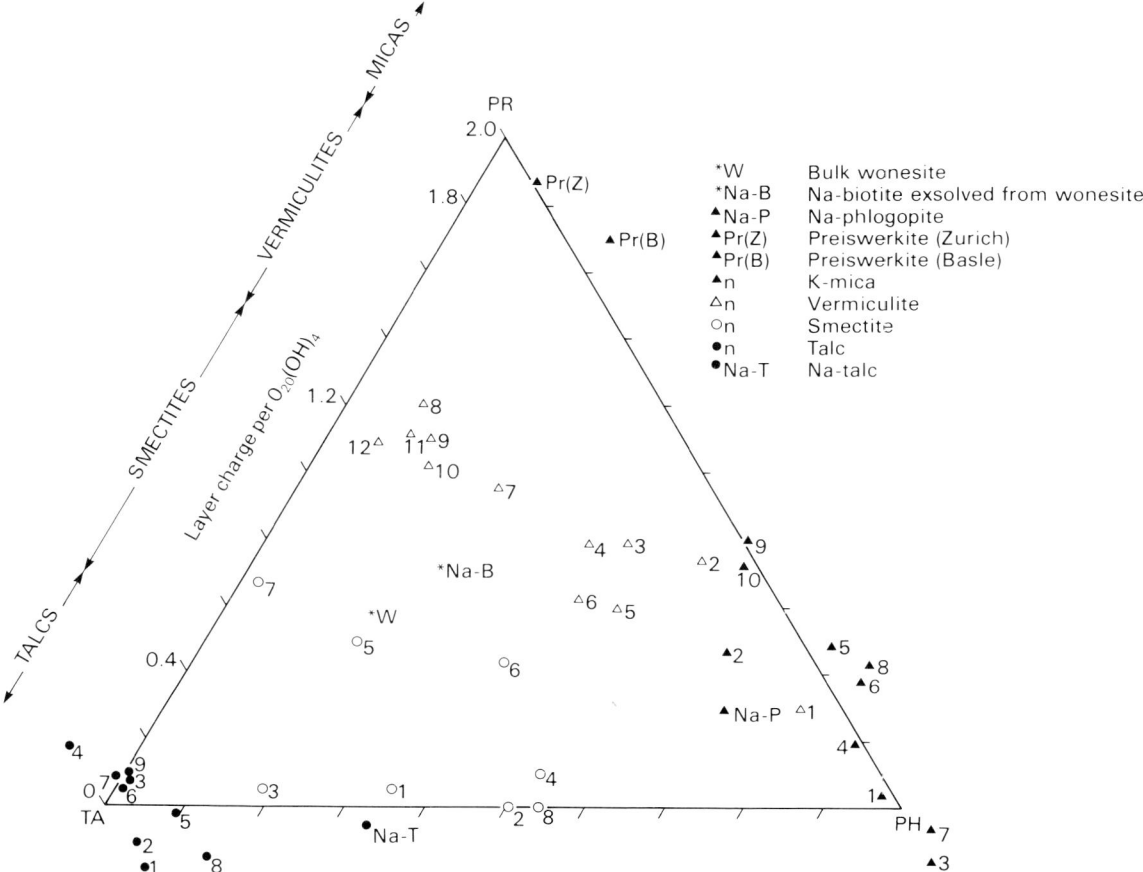

FIG. 1.6. Trioctahedral 2:1 layer silicates: compositions expressed in terms of distribution of positive charge among interlayer, octahedral and tetrahedral sites. Vertices present the distributions in ideal preiswerkite, PR, $(Na_2)(Mg_4Al_2)(Al_4Si_4)O_{20}(OH)_4$, phlogopite, PH, $(K_2)(Mg_6)(Al_2Si_6)O_{20}(OH)_4$, and ideal talc, TA, $(\square_2)(Mg_6)(Si_8)O_{20}(OH)_4$. Subscripts ($n$) to symbols indicate analyses numbers in Tables 1.17, 1.18 and 1.24.

The hydration properties of sodium micas have also been discussed by Veblen (1983a) but in terms of the factors influencing expandability in water discussed in Section 1.6.6, the only unexpected results are the reported expansion of two synthesized materials, the preiswerkite of Hewitt and Wones (1975) and the Na-phlogopite of Carman (1974). It must be remembered however that the compositions of synthetic products are usually inferred from the composition of the starting materials. If conversion is incomplete, the crystalline products may not have the inferred compositions. This may explain why Hewitt and Wones's (1975) synthetic preiswerkite and Carman's (1974) synthetic Na-phlogopite expand. From its overall interlayer charge, about 1.20 per $O_{20}(OH)_4$ formula unit, the "Na-biotite" in wonesite would be expected to expand. Its failure to do so can probably be attributed to the continuity of its 2:1 layers with those of the non-expanding, talc-like phase.

1.7.4 Lithium micas

Two series are known, the lithium aluminium micas or lepidolites, and the lithium iron micas which include the zinnwaldites. Starting from muscovite on one hand and siderophyllite on the other, both series converge with increasing Li content towards the composition of polylithionite, $K_2(Li_4Al_2)(Si_6)O_{20}(OH,F)_4$. Fluorine is a usual constituent of both series and in the lepidolites particularly, the lithium and fluorine contents are strongly correlated (Foster 1960b). Lithium micas generally contain appreciable amounts of Rb_2O and Cs_2O.

Lepidolites are essentially Si–Al–Li–K micas and contain only small amounts of Fe, Mg and Mn. If the compositions of the natural micas are plotted on a triangular diagram bounded by muscovite, polylithionite and trilithionite, $K_2(Al_3Li_3)(Si_6Al_2)O_{20}(OH,F)_4$, none is found close to trilithionite, and only a few approach the composition of polylithionite (analysis 1, Table 1.26). In most analyses, Li replaces octahedral Al in a ratio intermediate between 2Li:Al for trilithionite and 3Li:Al for polylithionite (Munoz 1968), the charge balance being maintained by substitution of Si for Al in the tetrahedral sheet; the end-member for this natural series would have the approximate composition $K_2(Al_{2.5}Li_{3.5})(Si_7Al)O_{20}(OH,F)_4$. Although there is a continuous series of chemical compositions between muscovite and this end-member composition, the series is structurally discontinuous. The muscovite structure persists up to about 1.5Li per six octahedral sites (analysis 2, Table 1.26) and such micas are usually called lithian muscovites, whereas when Li occupies more than 2.5 sites, the lepidolite (1-layer, 3-layer or 6-layer) structures are formed (analyses 4 and 5, Table 1.26). The intermediate compositions from $Li_{1.5}$ to $Li_{2.5}$ are a mixture of both structural types (analysis 2, Table 1.26).

Zinnwaldites contain ferrous iron, sometimes in large amounts, in addition to Si, Al, Li, K and F (analyses 6 to 9, Table 1.26) but only minor amounts of Mg and Mn. Foster (1960b) and Rieder (1970) showed that the compositions of zinnwaldites plot on or adjacent to the siderophyllite–polylithionite join, but the substitution mode in the series is complex. Starting from a mean ideal formula $K_2(Al_{1.5}^{3+}Fe_4^{2+})(Si_{5.5}Al_{2.5})O_{20}(OH,F)_4$ for siderophyllite (cf Table 1.25), Li substitutes for Fe^{2+} approximately atom for atom, octahedral Al increases and tetrahedral Al decreases until a composition approaching $K_2(Al_{2.67}Li_3)(Si_7Al)O_{20}(OH,F)_4$ is reached, similar to the intermediate composition in the muscovite–polylithionite series. The compositions of zinnwaldites seem to be quite well expressed by the general formula

$$K_2(Al_{(1.5+7x/18)}Fe^{2+}_{(4-4x/3)}Li_x)(Si_{(5.5+x/2)}Al_{(2.5-x/2)})O_{20}(OH,F)_4$$

where $0 \leqslant x \leqslant 3$. The name protolithionite is sometimes used for micas with $x < 1$, and cryophyllite for micas with $x > 2$, reserving zinnwaldite for micas with x between 1 and 2.

Taeniolite is a mica with the ideal formula $K_2(Mg_4Li_2)(Si_8)O_{20}(OH,F)_4$; only one analysis of a mica with this composition seems to have been published (analysis 10, Table 1.26).

1.7.5 Micas with unusual compositions

In laboratory syntheses micas have been prepared with Co, Ni, Cu and Fe^{2+} replacing Mg in octahedral sites and with B, Be, Ge and Ga replacing Si and Al in tetrahedral sites (Roy and Roy 1954; Klingsberg and Roy 1957; Stubican and Roy 1962; Hazen and Wones 1972). In natural occurrences, it is rare for micas to contain appreciable amounts of elements other than the main constituents listed earlier, but micas have been described that contain major or significant amounts of Cr, V, Mn, Ba, Be and Zn.

In the dioctahedral subgroup, the chromian micas fuchsite and mariposite have Cr^{3+}

TABLE 1.26. Analysis of the lithium micas, lepidolite, zinnwaldite and taeniolite

	1	2	3	4	5
SiO_2	60.40	45.18	50.20	50.92	52.80
Al_2O_3	12.00	35.76	28.18	26.92	19.94
TiO_2	0.37	0.15	trace	0.02	
Fe_2O_5	0.20	0.00		0.02	0.38
FeO	0.72	1.52	0.04	0.01	0.06
MnO		0.11	0.28	0.04	0.88
MgO	0.18	0.07	0.00	0.01	0.63
Li_2O	7.45	0.73	3.81	71	5.91
CaO	0.18	0.00	trace	')3	0.11
Na_2O	0.10	0.88	0.64	¦3	0.26
K_2O	11.80	9.95	9.91	10.04	9.00
Rb_2O	1.30	0.57	1.55	1.53	1.36
Cs_2O		trace	0.11		0.75
H_2O^+	0.16	4.48	2.18	1.59	2.85
H_2O^-	0.08	0.38	0.18	0.31	
F	8.10	0.88	4.97	6.71	7.07
Total	103.04	100.66	102.05	103.16	102.00
$O \equiv F$	−3.41	−0.37	−2.09	−2.82	−2.98
Total corrected	99.63	100.29	99.96	100.34	99.02
Numbers of cations and F atoms per $O_{20}(OH,F)_4$					
Si	7.96	6.06	6.67	6.74	7.19
Al	0.04	1.94	1.33	1.26	0.81
Σ tet.	8.00	8.00	8.00	8.00	8.00
Al	1.83	3.71	3.08	2.94	2.39
Ti	0.04	0.02	0.00	0.002	
Fe^{3+}	0.02	0.00		0.002	0.04
Fe^{2+}	0.08	0.17	0.004	0.001	0.01
Mn		0.01	0.03	0.005	0.10
Mg	0.04	0.01	0.00	0.002	0.13
Li	3.95	0.39	2.04	2.51	3.23
Σ oct.	5.92	4.31	5.15	5.46	5.90
Ca	0.025	0.00	0.00	0.00	0.02
Na	0.025	0.23	0.16	0.08	0.07
K	1.98	1.70	1.68	1.70	1.56
Rb	0.11	0.05	0.13	0.13	0.12
Cs		0.00	0.006		0.04
Σ interl.	2.14	1.98	1.98	1.91	1.81
F	3.38	0.37	2.09	2.81	3.04

1. Polylithionite, albitized granite, Ulkan pluton, South East Siberia (Gamaleya 1968).
2. Lithian muscovite, Varutrask, Sweden (Foster 1960b; analysis 12, Table 6) 2M structures.
3. Lepidolite, Chihuahua Valley, San Diego County, California (Foster 1960b; analysis 24, Table 6). Transition structure.
4. Lepidolite, Brown Derby pegmatite, Gunnison County, Colorado (Heinrich 1967; analysis 511) 6M structure.

replacing Al^{3+}; fuchsite is the chromian analogue of muscovite with tetrahedral Al:Si = 2:6, whereas mariposite has Al:Si = 1:7, analogous with phengite (Heinrich 1965). The chromian mica described by Martin-Ramos and Rodriguez-Gallego (1982), for instance, is a mariposite with the analysis: SiO_2 44.51; Al_2O_3 27.60; TiO_2 0.17; Cr_2O_3 3.35; Fe_2O_3 1.76; FeO 0.07; Mg 8.59; Li_2O 0.03; Rb_2O 0.01; Ba 0.01; CaO 1.71; Na_2O 0.40; K_2O 7.86; H_2O 4.21 = 100.28.

2:1 GROUP, X > 1.6: THE MICA MINERALS 83

6	7	8	9	10
45.51	40.08	35.38	33.60	58.82
20.59	22.34	20.00	22.36	1.29
0.24	0.06	0.10	0.06	0.11
1.51	2.53	3.90	1.44	0.40
9.20	17.14	22.50	28.54	0.24
0.53	0.57	0.53	0.33	
0.30	0.14	1.02	0.10	19.18
3.66	1.66	0.66	0.32	3.10
0.27	0.28	0.84	0.14	
0.60	0.24	0.16	0.46	0.64
9.68	8.36	8.00	8.90	10.44
	0.98	0.70		
	0.025	0.029		
1.90	3.99	4.23	1.66	0.59
1.21	0.08	0.11	0.08	0.09
7.45	3.85	4.04	0.89	8.56
102.65	102.33	102.20	98.88	103.46
−3.14	−1.62	−1.70	−0.37	−3.60
99.51	100.71	100.50	98.51	99.86
Numbers of cations and F atoms per $O_{20}(OH,F)_4$				
6.57	5.99	5.59	5.29	7.95
1.43	2.01	2.41	2.71	0.05
8.00	8.00	8.00	8.00	8.00
2.07	1.93	1.32	1.45	0.15
0.03	0.01	0.01	0.01	0.01
0.16	0.28	0.46	0.17	0.04
1.11	2.14	2.98	3.76	0.03
0.07	0.07	0.07	0.04	
0.06	0.03	0.24	0.02	3.86
2.12	1.00	0.42	0.20	1.69
5.62	5.46	5.50	5.65	5.78
0.04	0.04	0.14	0.02	
0.17	0.07	0.05	0.14	0.17
1.78	1.60	1.62	1.79	1.80
	0.09	0.07		
	0.002	0.002		
1.99	1.80	1.88	1.95	1.97
3.40	1.82	2.02	0.46	3.64

5. Lepidolite, Lithium pegmatite, Biskupice, nr Hrotovice, Czechoslovakia (Rieder 1970; analysis 41).
6. Zinnwaldite, albitized granite, Cinovec, Czechoslovakia (Rieder 1970; analysis 22).
7. Zinnwaldite, topaz–quartz–fluorite vein, Krupka, Czechoslovakia (Rieder 1970; analysis 4).
8. Zinnwaldite, Krupka, Czechoslovakia (Rieder 1970; analysis 7).
9. Zinnwaldite, topaz vein in granite, Yagenyama, Japan (Foster 1960b; analysis 2, Table 7).
10. Taeniolite, Magnet Cove, Arkansas (Foster 1960b; analysis 1, Table 5).

Alurgite is a muscovite containing 1–2% MnO (manganoan muscovite: Heinrich and Levinson 1955a). Roscoelite is a vanadiferous dioctahedral mica with V replacing octahedral Al; small amounts of V are accommodated in the muscovite structure, but micas with up to 17% V_2O_3 adopt the $1M$ structure; some vanadiferous micas also contain much interlayer Ba (9.9% BaO: Heinrich and Levinson, 1955b), and are known as chernykhite (Rozhdestvenskaya and Frank-Kamenetskii 1974; Bailey 1980b).

84 THE CHEMICAL CONSTITUTION OF CLAYS

In the trioctahedral subgroup, Zn and Mn can replace Mg and Fe in phlogopite and such micas are known as hendricksite if Zn > Mn; some samples contain as much as 21.6% ZnO and 15.3% MnO (Frondel and Ito, 1966; Frondel and Einaudi 1968). The analogue of phlogopite with Mn > Mg does not seem to have been described, but a number of manganoan phlogopites have been analysed (Frondel and Ito 1966).

Some brittle micas contain small amounts of Be (0.2–0.3% BeO: Schaller et al. 1967) and bityite is a rare brittle mica with Be substituting for tetrahedral Al (Bailey 1980b). Anandite is a trioctahedral mica with Fe in both tetrahedral and octahedral sites, and Ba as the predominant interlayer cation (Pattiaratchi et al. 1967). Although ephesite (ideal formula $Na_2(Li_2Al_4)(Al_4Si_4)O_{20}(OH)_4$) was originally considered to be a member of the margarite brittle mica group (Schaller et al. 1967) it is now considered to be more analogous to a true mica with charge compensation within each 2:1 layer (Bailey 1980b).

1.8 CHLORITES

The chlorite structure unit comprises a 2:1 layer whose negative charge is balanced by an interlayer consisting of a positively charged octahedrally coordinated hydroxide sheet. There are therefore two octahedral sheets in the structure unit, one in the 2:1 layers and another between them. These octahedral sheets may either be trioctahedral (brucite-like) or dioctahedral (gibbsite-like). In trioctahedral chlorites both octahedral sheets are trioctahedral, in dioctahedral chlorites both sheets are dioctahedral, and di,trioctahedral chlorites are dioctahedral in their 2:1 layers and trioctahedral in the interlayer sheets (Bailey 1980a). No well-characterized natural chlorites are known that are trioctahedral in the 2:1 layer and dioctahedral in the interlayer sheet.

1.8.1 Trioctahedral chlorites

The majority of minerals with the chlorite structure belong to this subgroup (Bailey 1975). Their compositions are ideally represented by the general structural formula

$$[(R^{2+},R^{3+})_6(Si,Al)_8O_{20}(OH)_4] \quad [(R^{2+}R^{3+})_6(OH)_{12}]$$

2:1 layer   interlayer hydroxide sheet

where R^{2+} represents the sum of divalent cations and R^{3+} the sum of trivalent cations; the two sets of octahedral cations, those in the 2:1 layer and those in the interlayer hydroxide sheet, cannot be distinguished by chemical analysis. The tetrahedral cations are almost invariably Si^{4+} and Al^{3+}. In the octahedral sheets Mg and Fe^{2+} are the most common divalent cations and substantial amounts of Mn^{2+} and Ni^{2+} occur in some chlorites. Trivalent cations may also be present in the octahedral sheets, commonly Al^{3+} and Fe^{3+}; Cr^{3+} may also occur.

In this chapter the recommendations of the AIPEA Nomenclature Subcommittee on chlorite (Bailey 1980a) are followed. Trioctahedral chlorites are named according to the dominant divalent cation. Clinochlore, chamosite, nimite and pennantite are the recommended names for Mg-, Fe^{2+}-, Ni- and Mn^{2+}-dominant chlorites with ideal end-member formulae

$$(Mg_{10}Al_2)(Si_6Al_2)O_{20}(OH)_{16}, \quad (Fe^{2+}_{10}Al_2)(Si_6Al_2)O_{20}(OH)_{16},$$
$$(Ni_{10}Al_2)(Si_6Al_2)O_{20}(OH)_{16}, \quad \text{and} \quad (Mn^{2+}_{10}Al_2)(Si_6Al_2)O_{20}(OH)_{16}$$

respectively. All other varietal and species names are discarded because subdivisions according to octahedral or tetrahedral compositions appear to have little or no structural significance (Bayliss 1975). Adjectival modifiers are used to indicate either important octahedral cations or unusual tetrahedral compositions. In this chapter the adjectival modifiers magnesian, ferroan, nickeloan and manganoan are used when there are 1.5 or more of the corresponding divalent cations per 12 available octahedral sites; the prefix aluminian is used when tetrahedral Al^{3+} is three or more per eight sites, i.e. one in excess of the ideal value. Trioctahedral chlorites containing more than 3% Cr_2O_3 are called chromian chlorites.

The main source of analytical data for trioctahedral chlorites is Foster's (1962) compilation of 154 analyses. This showed that tetrahedral cation composition almost invariably lay between (Si_7Al_1) and $(Si_{4.6}Al_{3.4})$. In some chlorites the negative charge on the tetrahedral sheets, arising from substitution of Al^{3+} for Si^{4+}, is balanced simply by the inclusion of an equal number of trivalent cations, Al^{3+}, Fe^{3+}, Cr^{3+}, in the octahedral sheets, the remaining sites being occupied by divalent cations. When this occurs the total number of octahedral cations is close to 12 for an $O_{20}(OH)_{16}$ formula unit. Often, however, trioctahedral chlorites have an incomplete complement of octahedral cations and in these there are more trivalent octahedral cations than are needed to balance the charge arising from trivalent cations in tetrahedral sites. Foster (1962) showed that the number of vacant octahedral sites, found to range up to about 1.1 per 12 sites, was closely equal to one-half the excess of trivalent cations in octahedral sites over the number of trivalent cations in tetrahedral sites.

The other important substitutions in trioctahedral chlorites involve mutual replacement of different species of divalent and trivalent cations. For divalent cations the commonest are $Mg^{2+} \rightleftharpoons Fe^{2+}$ which appear to replace each other in any proportion and are found in most trioctahedral chlorites (Table 1.27). Trioctahedral chlorites also occur in which Mn^{2+} (pennantite, analysis 11, Table 1.27) and Ni^{2+} (nimite, analysis 12, Table 1.27) are the dominant divalent cations and smaller amounts of Mn^{2+} are frequently present (Table 1.27). In addition, Al^{3+}, Fe^{3+} (analyses 2, 4, 6, 8, 9, 10, Table 1.27) and Cr^{3+} (analyses 13, 14, Table 1.27) are also found in octahedral sites up to a total of about 1.5 to 2 atoms per 12 sites. Lapham (1958) suggested that in some chromium-bearing chlorites, all or most of the Cr was in tetrahedral sites. However Bish (1977) and Phillips *et al.* (1980) have shown conclusively that Cr^{3+} occurs in octahedral sites only, and in addition in the specimens examined by the latter authors, all the heavy atoms (Cr^{3+}, Fe^{3+}) appear to be interlayer octahedral sites, the octahedral sites in the 2:1 layer being occupied by Mg only.

The structural formulae in this section have all been calculated on the basis of the ideal $O_{20}(OH)_{16}$ anion content, because in general little reliance can be placed on recorded H_2O^+ contents as a measure of structural hydroxyl, and calculations based on the ideal contents are sufficient for a general comparison of chlorites. When the H_2O^+ content, or some other measurement, provides a reliable value of structural hydroxyl, structural formulae can be calculated taking this into account.

1.8.2 Chlorites containing dioctahedral sheets

Chlorites also occur in which there are dioctahedral sheets. Many of these are fine-grained clay-sized materials and some of the analyses in the literature are recognized to be of impure material. Although Belov (1950) and Cerny (1970) have suggested possible structural sites for the small amounts of Ca, Na and K often reported in analyses of chlorites, their greater frequency and

86   THE CHEMICAL CONSTITUTION OF CLAYS

TABLE 1.27. Analyses of trioctahedral chlorites

	1	2	3	4	5	6	7
SiO_2	36.10	24.92	29.73	24.00	27.55	30.76	27.88
Al_2O_3	10.62	25.91	17.95	23.24	19.74	12.12	15.81
TiO_2		0.13		0.12			
Fe_2O_3	0.95	3.42	0.68	3.14	0.66	9.12	1.77
Cr_2O_3							
FeO	1.77	7.27	10.05	15.80	20.68	22.76	31.92
MnO	0.03	0.12	8.24	0.94	0.23	1.24	0.51
MgO	38.61	26.22	21.84	18.92	18.42	12.36	9.52
NiO							
CoO							
CaO				0.05		trace	0.20
Na_2O							
K_2O							
H_2O^+	11.92	10.43	11.90	}12.80	11.39	9.76	11.97
H_2O^-	0.38	1.65	0.42		0.31	1.80	0.15
Total	100.38	100.07	100.81	99.01	98.98	99.82	99.73
Numbers of cations on the basis $O_{20}(OH)_{16}$							
Si	6.69	4.81	5.95	4.98	5.69	6.54	6.14
Al	1.31	3.19	2.05	3.02	2.31	1.46	1.86
Σ tet.	8.00	8.00	8.00	8.00	8.00	8.00	8.00
Al	1.01	2.71	2.18	2.65	2.49	1.57	2.24
Ti		0.02		0.02			
Fe^{3+}	0.13	0.50	0.10	0.49	0.10	1.46	0.29
Cr^{3+}							
Fe^{2+}	0.27	1.17	1.68	2.74	3.57	4.05	5.87
Mn^{2+}		0.02	1.40	0.17	0.04	0.22	0.10
Mg	10.66	7.55	6.51	5.85	5.67	3.92	3.12
Ni							
Co							
Σ oct.	12.07	11.97	11.88	11.92	11.87	11.22	11.62
Ca				0.01			0.05
Na							
K							

1. Clinochlore (penninite†), Appennine Region, Italy (Foster 1962, analysis 77).
2. Aluminian clinochlore (sheridanite), Chester, Massachusetts, USA (Foster 1962, analysis 1).
3. Ferroan clinochlore, Kumanohata mine, Shiga Prefecture, Japan (Shirozu 1958).
4. Ferroan aluminian clinochlore (ripidolite), Chester, Massachusetts, USA (Foster 1962, analysis 82).
5. Ferroan clinochlore (brunsvigite), Old West Mine, Penhalonga, Southern Rhodesia (Foster 1962, analysis 124).
6. Magnesian chamosite (diabantite), Challis, Custer County, Idaho, USA (Foster 1962, analysis 134).
7. Magnesian chamosite (brunsvigite) Radauthal, Harz, Germany (Foster 1962, analysis 132).
8. Ferrian chamosite (ripidolite), Tateyama Hot Springs, Jodojama, Toyama Prefecture, Japan (Foster 1962, analysis 88).

abundance in analyses of di- and di,trioctahedral chlorites probably reflects the presence of impurities.

Ideal dioctahedral, di,trioctahedral and trioctahedral chlorites would have 8, 10 and 12 octahedral cations per formula unit. In fact trioctahedral chlorites have octahedral cation totals between 10.9 and 12.1 (Foster 1962) and dioctahedral minerals almost always have between 3.9 and 4.3 cations per octahedral sheet. On the basis of these values, dioctahedral chlorites would contain 7.8–8.6 octahedral cations, and di,trioctahedral chlorites would have 9.4–10.3. The structural formula of chlorites containing at least one dioctahedral sheet (Tables 1.28 and 1.29)

8	9	10	11	12	13	14
22.24	20.82	21.90	22.64	27.27	32.22	32.73
17.05	17.64	15.68	18.60	15.21	8.66	7.96
	trace					0.00
13.38	8.70	9.00	4.43	4.35	0.46	
					6.47	8.31
26.25	37.96	41.72		2.78	1.38	2.29
5.42			38.93	0.06		0.02
4.10	4.15	trace	1.48	10.13	35.06	34.88
				29.49		0.00
				0.38		
trace				0.38	1.04	
10.05	10.31	}11.13	9.40	10.48	13.82	
0.98	0.07			0.27	0.58	
99.48	99.65	99.43	96.81	100.80	99.69	86.19
Numbers of cations on the basis $O_{20}(OH)_{16}$							
5.09	4.83	5.26	5.36	5.95	6.33	6.39
2.91	3.17	2.74	2.64	2.05	1.67	1.61
8.00	8.00	8.00	8.00	8.00	8.00	8.00
1.69	1.66	1.70	2.54	1.87	0.34	0.23
2.30	1.52	1.63	0.79	0.71	0.07	
					1.01	1.28
5.02	7.37	8.38		0.51		0.37
1.05			7.80	0.01		
1.40	1.44		0.52	3.30	10.27	10.16
				5.18		
				0.07		
11.46	11.99	11.71	11.66	11.64	11.91	12.04
				0.09	0.22	

9. Aluminian ferrian chamosite (thuringite), Schmiedefeld, Thuringia, Germany (Foster 1962, analysis 136).
10. Ferrian chamosite (thuringite), Schmiedefeld, Eastern Thuringia, Germany (Foster 1962, analysis 146).
11. Pennantite, Benallt mine, Rhiw, Caernarvonshire, Wales (Smith *et al.* 1946, total includes BaO = 1.33%).
12. Magnesian nimite, Scotia Talc Mine, Barberton, South Africa (De Waal 1970b).
13. Chromian clinochlore, Erzincan province, Turkey (Lapham 1958).
14. Chromian clinochlore, Gumushane, Turkey. Electron microprobe analysis (Phillips *et al.* 1980). All iron treated as FeO.
† Names in parentheses are those used by Foster (1962).

have octahedral cation totals ranging from 8.29 to 10.33. Some in Table 1.29 (analyses 7, 8, 9, 10) have totals between 8.6 and 9.4, and therefore there must be some doubt as to the nature of these samples. One can only suggest that the analyses represent impure materials or that the analyses are in error, or that the assumed range of cation contents of di- and trioctahedral sheets are incorrect. It may not be without significance that modern sudoite analyses by electron microprobe, which are less likely to include impurities (see analysis 8, Table 1.28), and others by Kramm (1980), and those in which samples were likely to be pure (analyses 5 and 6, Table 1.28) have octahedral cation totals close to 10.

TABLE 1.28. Analyses of di, trioctahedral chlorites – sudoite and cookeite

	1	2	3	4	5	6	7
SiO_2	34.36	33.71	33.10	36.07	32.81	33.00	33.83
Al_2O_3	38.84	37.95	37.43	32.42	35.71	35.69	31.86
TiO_2			0.30		0.18	n.d.	0.34
Fe_2O_3	0.49	2.74	0.87	1.99	2.66	2.74	0.27
FeO		2.40			1.07	0.24	0.37
MnO					0.29	0.28	
MgO	10.21	12.08	13.12	15.94	13.54	14.07	18.86
CaO	0.07	1.49			0.03	n.d.	1.80
Li_2O							
Na_2O	0.06		0.15	} 0.44			0.58
K_2O	0.12		0.29		0.03	0.00	0.04
H_2O^+	13.98	13.07	13.32	13.05	12.95	13.83	12.06
H_2O^-	0.84		2.11				
Total	99.57	100.80	100.69	99.91	99.27	100.03	99.95
Numbers of cations on the basis $O_{20}(OH)_{16}$							
Si	6.25	5.93	6.03	6.48	5.99	6.02	6.08
Al	1.75	2.07	1.97	1.52	2.01	1.98	1.92
Σ tet.	8.00	8.00	8.00	8.00	8.00	8.00	8.00
Al	6.58	5.79	6.07	5.34	5.67	5.68	4.83
Ti			0.04		0.02		0.05
Fe^{3+}	0.07	0.36	0.12	0.27	0.37	0.38	0.04
Fe^{2+}		0.35			0.16	0.04	0.06
Mn					0.04	0.04	
Mg	2.77	3.17	3.56	4.27	3.68	3.82	5.05
Li							
Σ oct.	9.42	9.67	9.79	9.88	9.94	9.96	10.03
Ca	0.01	0.28			0.01		0.35
Na	0.23		0.05	0.15			0.20
K	0.03		0.07		0.01		0.01

1. Di, trioctahedral chlorite, sudoite, Kamikita mine, Akita Prefecture, Japan (Shirozu 1978).
2. Di, trioctahedral chlorite, sudoite, specimen from Kazanskii Mineral Museum, USSR (Alysheva *et al.* 1977).
3. Di, trioctahedral chlorite, sudoite, Niida, Odate, Akita Prefecture, Japan (Shirozu, 1978).
4. Di, trioctahedral chlorite, sudoite, Dnieper–Donetz depression, USSR (Alysheva *et al.* 1977).
5. Di, trioctahedral chlorite, sudoite, Ottre, Ardennes, Belgium (Kramm 1980).
6. Di, trioctahedral chlorite, sudoite, Venn-Stavelot Massif, Ardennes, Belgium (Fransolet and Bourguignon 1978).
7. Di, trioctahedral chlorite, sudoite, Novo-zolotushinski formation, USSR (Aleysheva *et al.* 1977).

The analytical data on chlorites containing dioctahedral sheets have been subdivided at a total octahedral cation content of 9.11, those with less are in Table 1.29, those with more in Table 1.28. Some in Table 1.29 (analyses 1–6), are dioctahedral chlorites for which the name donbassite has been recommended (Bailey 1980a). Others (analyses 7 to 10) as discussed may be di- or di,trioctahedral chlorites. The structural formulae of the minerals in Table 1.29 have tetrahedral cation contents ranging from $(Si_{5.38}Al_{2.62})$ to $(Si_{7.52}Al_{0.48})$. Two specimens (analyses 6 and 9) show appreciable Li_2O (1.00 and 0.53%) but in others Li may not have been sought. Small percentages of Li_2O make a considerable contribution in terms of atoms of Li, 1% Li_2O being equivalent to about 0.75 Li atoms per $O_{20}(OH)_{16}$ formula unit.

All the analyses in Table 1.28, with the possible exception of analysis 1, represent di,trioctahedral chlorites and, as recommended by Bailey (1980a), low-lithian minerals are named sudoite and high-lithian minerals, cookeite. The sudoites (analyses 1–9, Table 1.28) have

8	9	10	11	12	13	14
32.45	31.17	34.70	34.92	33.40	33.31	36.1
34.88	37.49	48.40	46.29	47.47	44.16	44.9
					trace	0.13
	1.04	0.10	0.07	0.00	1.72	
3.48†	1.95			0.71	1.69	
0.41			0.13	trace	0.18	
13.56	14.25	0.05	0.24	0.20	0.37	0.00
	0.26	0.12	0.30	0.45	0.00	0.40
	n.d.	2.45	2.67	3.12	3.40	3.90
	n.d.	0.10	0.14	} 0.09	0.70	
	n.d.				0.67	
	14.10	14.10	14.12	14.98	13.19	13.90
		0.28	0.23	0.46		
84.78	100.26	99.93	99.16	100.65	99.85	99.33
Numbers of cations on the basis $O_{20}(OH)_{16}$						
6.06	5.71	6.06	6.19	5.90	5.95	6.32
1.94	2.29	1.94	1.81	2.10	2.05	1.68
8.00	8.00	8.00	8.00	8.00	8.00	8.00
5.73	5.80	8.03	7.85	7.79	7.25	7.57
						0.02
	0.14	0.01	0.01		0.23	
0.54†	0.30			0.10	0.25	
0.06			0.02		0.03	
3.77	3.89	0.01	0.06	0.05	0.10	
	1.72	1.90	2.22	2.44	2.74	
10.10	10.13	9.77	9.84	10.16	10.30	10.33
	0.05	0.02	0.06	0.09		0.08
			0.05	0.03	0.24	
					0.15	

8. Di, trioctahedral chlorite, sudoite, Harz Mountains, German Democratic Republic. Electron microprobe analysis; † all Fe as FeO (Kramm 1980).
9. Di, trioctahedral chlorite, sudoite, Berezovska, Urals, USSR (Drits and Lazarenko 1967).
10. Di, trioctahedral chlorite, cookeite, North Little Rock, Arkansas, USA (Cerny 1970, analysis 10).
11. Di, trioctahedral chlorite, cookeite, Manono, Katanga. (Cerny 1970, analysis 9).
12. Di, trioctahedral chlorite, cookeite, Kalvinski Range, USSR (Cerny 1970, analysis 7).
13. Di, trioctahedral chlorite, cookeite, North-western USSR (Cerny 1970, analysis 4).
14. Di, trioctahedral chlorite, cookeite, Auburn, Maine, USA (Norrish, 1952).

octahedral cation totals between 9.42 and 10.13, and tetrahedral cation contents ($Si_{5.71}Al_{2.29}$) and ($Si_{6.48}Al_{1.52}$). The cookeites (analyses 10–14, Table 1.28) have 9.77 to 10.33 octahedral cations and tetrahedral Si ranges from 5.90 to 6.32. As pointed out by Cerny (1970), octahedral cation totals and Li contents in cookeite tend to be linearly related, larger totals being found with larger Li contents.

1.8.3 Chlorites with imperfect or incomplete interlayer hydroxide sheets

No account of the chemistry of clays would be complete without some discussion of minerals with imperfect hydroxidic interlayers. Their natural occurrence, identification, properties, the nature of the interlayer material and the consequences of their formation were comprehensively

TABLE 1.29. Analyses of dioctahedral chlorites – donbassite

	1	2	3	4	5
SiO_2	35.12	34.65	35.36	38.0	36.7
Al_2O_3	48.16	46.03	47.35	47.44	49.5
TiO_2				trace	trace
Fe_2O_3	n.d.	0.55	0.52		
FeO	n.d.	n.d.	0.07		
MnO					
MgO	n.d.	1.58	0.67	0.8	0.0
CaO	0.61	1.82	0.45		
Li_2O		n.d.	n.d.		
Na_2O	} 1.98	} 1.08	} 0.40		
K_2O					
H_2O^+	14.01	13.96	14.28	13.8	13.8
H_2O^-					
Total	99.88	99.67	99.10	100.0	100.0
Numbers of cations on the basis $O_{20}(OH)_{16}$					
Si	6.23	6.19	6.31	6.61	6.39
Al	1.77	1.81	1.69	1.39	1.61
Σ tet.	8.00	8.00	8.00	8.00	8.00
Al	8.29	7.89	8.26	8.33	8.54
Ti					
Fe^{3+}		0.07	0.07		
Fe^{2+}			0.10		
Mn					
Mg		0.42	0.18	0.21	
Li					
Σ oct.	8.29	8.38	8.52	8.54	8.54
Ca	0.12	0.35	0.09		
Na	0.68	0.37	0.14		
K					

1. Dioctahedral chlorite, donbassite (locality not given); Na_2O+K_2O treated as Na in structural formula (Drits and Lazarenko 1967).
2. Dioctahedral chlorite, donbassite, Nagolnaya Tarasovska, Donbass, USSR; Na_2O+K_2O treated as Na in structural formula (Drits and Lazarenko 1967).
3. Dioctahedral chlorite, donbassite, Novaya Zemlya, USSR (Drits and Lazarenko 1967).
4, 5. Dioctahedral chlorite, donbassite, Kesselberges bei Triberg, Schwartzwald, Germany (Muller 1963).
6. Dioctahedral chlorite, donbassite, from a pyrophyllite deposit, Itaya, Okayama Prefecture, Japan. Contains small amounts of kaolin and pyrophyllite (Shirozu 1978).
7. Dioctahedral chlorite(?), donbassite, Saint Paul-de-Fenouillet, Pyrenees-Orientales, France.

reviewed by Rich (1968). Since 1968 there has been continued interest in these materials and recently structurally similar materials have been produced synthetically by precipitation of incomplete hydroxidic interlayers in a smectite substrate with a view to their use as catalysts and molecular sieves (see Chapter 6, Section 6.7).

Minerals in this subgroup may be considered as intermediates between the expandable smectites and vermiculites, in which the interlayer material is readily exchangeable cations or hydrated cations, and the non-expanding chlorites in which the interlayers are continuous octahedrally coordinated gibbsite- or brucite-like sheets. They are of widespread occurrence in soils and sediments and have been variously referred to as dioctahedral vermiculites, dioctahedral chlorites, swelling chlorites, chlorite-like minerals and intergradient chlorite–

6	7	8	9	10
43.43	29.5	39.01	35.97	35.63
39.21	52.0	32.15	48.22	34.87
	1.2	0.47		none
0.17	0.5	0.90	0.53	5.01
	0.3	0.10	0.09	0.43
				0.05
0.13	trace	10.14	0.69	8.63
0.47	1.6	0.54		1.13
1.43			0.76	
1.41		0.10		0.24
0.49		1.52		0.46
11.59	13.9	14.15	13.74	12.24
1.47				1.91
99.91	100.00	99.50	100.00	100.60
Numbers of cations on the basis $O_{20}(OH)_{16}$				
7.52	5.38	7.12 (6.48)	6.28	6.50
0.48	2.62	0.88 (1.52)	1.72	1.50
8.00	8.00	8.00	8.00	8.00
7.53	8.56	6.03 (6.08)	8.20	5.99
	0.16	0.06		
0.02	0.07	0.12 (0.07)	0.07	0.69
	0.05	0.02	0.10	0.07
				0.01
0.03		2.76 (3.19)	0.18	2.35
1.00			0.53	
8.58	8.84	8.99 (9.34)	8.99	9.11
0.09		0.11 (0.12)		0.22
0.47		0.04 (0.03)		0.08
0.11		0.35 (0.32)		0.11

Total includes CaO = 1.6% and P_2O_5 = 1.0%, believed to be present as a calcium phosphate impurity. (Caillere et al 1962).

8. Dioctahedral chlorite(?), Furutobe Mine, Akita Prefecture, Japan. Total includes S = 0.42%, disregarded in calculation of structural formula. Contains quartz = 6.3%, illite 4.3% and pyrite. Numbers in brackets are structural formula obtained by Sudo and Sato (1966) after correction for impurities.
9. Dioctahedral chlorite(?), donbassite, Novaya Zemlya, USSR. Analysis recalculated from the structural formula of Drits and Alexandrova (1968).
10. Dioctahedral chlorite(?), Kamikita Mine, Aomori Prefecture, Japan. Contains a small amount of an interstratified mineral (Sudo and Sato 1966).

vermiculites. When they occur naturally they are almost invariably so intimately mixed with other clay-sized components that meaningful chemical analysis is not possible.

Many natural occurrences suggest that the interlayers were formed by precipitation and growth of Al-hydroxy (or Mg-hydroxy) polymers in the space between the layers of smectites vermiculites. They also appear to be formed by the replacement of interlayer potassium of micas and of the brucite-like interlayer in trioctahedral chlorites by Al-hydroxy species in the interlayer space. Almost invariably the resultant interlayer is incomplete, occurring as islands of gibbsite-like Al-hydroxy polymers. This results in interlayers that collapse less readily than those of smectites and vermiculites on heating but more readily than the complete interlayer sheets in chlorites. Some have interlayers that expand when treated with ethylene glycol and glycerol and these have been called swelling chlorites.

In acid soils, pH 4 to 6, hydroxy-Al is the principal interlayer material whereas hydroxy-Mg is probably the dominant cation in alkaline and most marine environments. Other cationic species, Fe^{3+} and Fe^{2+}, may also be present in situations where these ions are available (Rich 1968; Carstea et al. 1970; Herrera and Peech 1970). Some glauconites contain readily acid-extractable Al, Fe and Mg that may be present as interlayer hydroxides (Thompson and Hower 1975; see Section 1.9.3.3). Bassett (1958) has described a copper-bearing vermiculite-like mineral from Northern Rhodesia that contained up to 7% copper. The evidence strongly suggests that this mineral contains Cu-hydroxy interlayers.

Materials with incomplete Al-hydroxy interlayers, similar in most respects to the natural minerals, have been produced artificially (see Rich 1968; Sawhney 1968; Brown and Newman 1973; Nagasawa et al. 1974; Brindley and Sempels 1977; Lahav et al. 1978; Shabtai et al. 1984). The composition of the interlayer varies with the degree of polymerization of the products of hydrolysis, ranging from hydrated Al^{3+} species via material with a range of OH/Al ratios commonly up to a maximum of OH/Al of about 2.5 (Brown and Newman 1973).

Chlorite-like materials have also been synthesized with cations other than Al in the interlayer. Gupta and Malik (1969) and Yamanaka and Brindley (1978) have described nickel-hydroxy interlayered materials. Bassett (1958) produced material containing up to 10% copper, similar to his natural copper vermiculite from vermiculite and biotite simply by ion exchange with molar cupric chloride solution at 100°C for 144 hours.

Structurally related synthetic products that differ from those described above in that their basal spacing is greater than the 14 Å characteristic of chlorites, have been prepared with Cr-hydroxy (Brindley and Yamanaka 1979), Bi-hydroxy (Yamanaka et al. 1980) and Zr-hydroxy interlayers (Yamanaka and Brindley 1979). Brown and Newman (1973), Lahav et al. (1978) and Shabtai et al. (1984) have shown that materials with interlayer separations larger than the 5 Å of chlorite-like materials can also be produced with Al when appropriate conditions are used to develop incomplete Al-hydroxy interlayers.

1.9 INTERSTRATIFIED CLAY MINERALS

The 2:1 and 1:1 layers of clay minerals are strongly bonded internally but relatively weakly bonded to each other, and because the surface planes of the different kinds of layers and the hydroxide interlayer sheets of chlorites are geometrically similar, layers with different internal arrangements articulate well at their interfaces and can stack together in many different ways. An example already considered is that of the chlorite minerals in which hydroxide interlayers are regularly interposed between 2:1 aluminosilicate layers, but other multiple and irregular sequences are now widely recognized to be common, particularly in minerals formed at lower temperatures. Such minerals are structurally and chemically more complex than those that appear to have only one kind of structure unit, although even these, as noted earlier (Section 1.6.5), may show some heterogeneity when examined by the alkylammonium treatment technique.

The number of sequences that can be generated by stacking layers is very large; the structure units known in mixed sequences include 1:1 dioctahedral structure units, talc units, smectite units both di- and trioctahedral, mica and chlorite units. In addition it has become evident that in 2:1 layers Al-for-Si substitution in the tetrahedral sites may not be symmetrical on opposite sides of the octahedral sheet and that as a consequence the 2:1 layers may be asymmetrical. This would have the effect that the properties of the interlayers might vary, depending on how the 2:1 layers with different substitutions are superimposed.

Layer sequences can be fully ordered, e.g. ... ABABAB ..., partly ordered with a probability of B following A being less than 1 but greater than 0.5, or random. Minerals that are fully ordered give an X-ray diffraction pattern in which the basal reflections occur at d-spacings that are exact submultiples of the first order, that is, they comprise a rational series of basal reflections. The commonest interstratified minerals depart in varying degrees from this ideal rationality, ranging from minerals that give a first-order reflection that is close to the sum of the component layer spacings but with somewhat irrational higher orders, to minerals in which the irrational basal reflections are intermediate between the reflections for each component when it is not interstratified. The work of Reynolds and Hower (1970) and of Reynolds (1980) should be consulted for detailed interpretation of the X-ray diffraction patterns of interstratified minerals.

Minerals with regular sequences of layers (regular interstratifications) are given names by international agreement (see Bailey 1982); a statistical test based on the variation of the $l \times d(00l)$ is used to define whether the mineral can be described as a "regular interstratification". Partly ordered or random sequences are described in in terms of the component layers, their relative proportion and the type and degree of ordering. Thus a mineral might be described as an interstratified illite–smectite, 90% illite:10% smectite with maximum ISII ordering, indicating that all the smectite units (S) in a stack are preceded and followed by an illite unit (I). For convenience, this long description can be shortened to "illite–smectite (90:10) with maximum ISII ordering", and this convention will be used in the remaining parts of this section. A full account of the statistical parameters used to describe interstratified minerals is given by Reynolds (1980).

Calculation of a possible structural formula for interstratified minerals requires more structural information and often involves additional assumptions or inferences, compared with the calculations for minerals with one kind of structure unit. Besides the need to identify the kinds of structure units present, their relative proportions must also be estimated before a model can be proposed for the composition of the component structure units and the atoms allocated to the appropriate sites. If the layers are of different types, e.g. an interstratification of 2:1 layers with 1:1 layers or chlorite units, the anion content cannot be assumed; in the absence of the necessary information, all that can be calculated is the atomic ratios. When the layers are all of the 2:1 type, a mean structural unit composition can be calculated and the atoms allocated to "average" tetrahedral and octahedral sites. Such an allocation does not, however, explain differences between interlayers.

The difficulties of devising a convincing structural formula for interstratified clay minerals can be well illustrated with the very detailed data available for Na-rectorite. As shown by Brown and Weir (1963), Na-rectorite is a regularly interstratified mineral with equal proportions of mica-like and smectite-like interlayers. Sodium, Ca, K and Mg can be present both in the mica-like interlayers (non-exchangeable or fixed) and the smectite-like interlayers (exchangeable). Therefore in addition to the usual chemical analysis, the composition of the two kinds of interlayers must be determined by partial analysis of two different cation-exchanged forms, e.g. Sr- and Ca-saturated samples (Kodama 1966). This calculation makes the highly probable assumption that all the exchangeable cations come from one kind of interlayer, presumably swelling, and that all the fixed cations are situated in the non-swelling interlayer.

Using Kodama's analyses as an example, the numbers of cations in the structural unit are calculated for a unit containing two 2:1 layers giving a total anion content of $O_{40}(OH)_8$ (analysis 3, Table 1.30). The exchangeable interlayer composition is $(Ca+Sr)_{0.10}Mg_{0.19}K_{0.02}Na_{0.13}$ with

94 THE CHEMICAL CONSTITUTION OF CLAYS

TABLE 1.30. Analyses of regularly interstratified minerals

	1	2	3	4	5	6
SiO_2	46.21	53.15	54.11	51.98	52.62	47.58
Al_2O_3	14.44	3.48	40.38	39.35	29.08	36.36
TiO_2			0.01		0.12	
Fe_2O_3		3.48	0.15	1.44	0.19	0.47
FeO					0.52	0.28
MnO		0.03	0.00		0.06	
MgO	37.85	27.40	0.78	0.25	0.77	0.30
CaO	0.07	1.10	0.52	3.46	0.51	0.59
Na_2O	1.38	1.18	3.87	2.43	0.10	2.41
K_2O	0.06		0.29	1.08	4.44	3.52
Li_2O				0.03		
H_2O^+		} 10.18			5.52	5.79
H_2O^-					6.25	3.25
Total	100.10	100.00	100.24†	100.02	100.18	100.55

Numbers of cations

	1	2	3	4	5	6
Anion	O_{40}	O_{40}	O_{40}	O_{40}	O_{40}	O_{40}
Basis	$(OH)_{20}$	$(OH)_8$	$(OH)_8$	$(OH)_8$	$(OH)_8$	$(OH)_8$
Si	13.14	14.66	12.84	12.54	14.28	12.62
Al	2.86	1.13	3.16	3.46	1.72	3.38
Fe^{3+}		0.21				
Σ tet.	16.00	16.00	16.00	16.00	16.00	16.00
Al	1.98		8.12	7.73	7.58	7.99
Ti			0.00		0.02	
Fe^{3+}		0.51	0.03	0.26	0.04	0.09
Fe^{2+}					0.12	0.06
Mn		0.01	0.00		0.01	
Mg	16.05	11.26	0.09	0.09	0.31	0.12
Li						
Σ oct.	18.03	11.78	8.24	8.08	8.08	8.26
Total Ca	0.02	0.33	0.14	0.89	0.15	0.17
Total Na	0.76	0.63	1.78	1.14	0.05	1.24
Total K	0.02		0.09	0.33	1.54	1.19
Total	0.80	0.96	2.01	2.36	1.74	2.60
Exch. Ca			0.29‡	0.21		0.18
Exch. Na			0.13			0.24
Exch. K			0.01	0.01		
Charge			0.72	0.43		0.60

1. Kulkeite, low-grade dolomitic rock, Derrag, Tell Atlas, Algeria (Schreyer *et al.* 1982).
2. Alliettite, serpentine, Nure Valley, Piacenza Province, Italy (Allietti and Mejsner 1980).
3. Na-rectorite, Fort Sandeman, Baluchistan, Pakistan (Kodama 1966). † Includes exchangeable Ca, Mg and Sr; total Mg = 0.28 atoms per O_{88}. ‡ Total includes 0.13% SrO.
4. Ca-rectorite, Tooho mine, Aichi Prefecture, Japan (Nishiyama and Shimoda 1981).
5. K-rectorite, sample OP-79-2, Tulameen coalfield, British Colombia (Pevear *et al.* 1980).
6. "Tarasovite", selvages of quartz veins, Nagolnyi Tarasov, Donbass, USSR (Lazarenko and Korolev 1970).
7. Tosudite, Takatama mine, Fukushima Prefecture, Japan (Shimoda 1969).

a total charge of 0.73 per $O_{40}(OH)_8$ and the fixed interlayer contains $(Ca+Sr)_{0.04}K_{0.07}Na_{1.65}$ with a total charge of 1.80, or if fixed Mg (0.09 atoms) is also included in the interlayer, with a total charge of 1.98. Each of these interlayer compositions is consistent with a smectite-like (swelling) character and a mica-like (non-swelling) character, respectively.

7	8	9	10	11	12
42.14	36.96	41.60	39.74	35.95	33.95
37.38	32.09	36.40	35.87	15.13	19.20
	0.34		0.01	0.30	trace
0.30	1.57	1.82	0.98	6.65	0.71
	trace		2.77	9.15	0.69
				0.38	
0.08	8.20	0.29	3.08	14.76	26.31
1.65	2.21	0.38	0.06	1.86	0.70
0.15	0.16	0.14	0.12	1.13	0.74
1.40	0.23	0.38	0.62	1.13	0.05
		1.04	0.51		
11.22	12.71	11.12	} 15.70	7.30	11.26
6.16	6.12	6.87		6.62	6.55
100.48	100.59	100.04	99.46	100.36	100.16
		Numbers of cations			
O_{40}	O_{40}	O_{40}	O_{40}	O_{40}	O_{40}
$(OH)_{20}$	$(OH)_{20}$	$(OH)_{20}$	$(OH)_{20}$	$(OH)_{20}$	$(OH)_{20}$
13.72	12.50	13.60	13.03	12.80	11.79
2.28	3.50	2.40	2.97	3.20	4.21
16.00	16.00	16.00	16.00	16.00	16.00
12.05	9.29	11.63	10.89	3.15	3.65
	0.09		0.00	0.08	
0.07	0.40	0.45	0.24	1.78	0.19
			0.76	2.72	0.20
				0.12	
0.04	4.13	0.14	1.51	7.84	13.62
		1.37	0.67		
12.16	13.91	13.59	14.07	15.69	17.66
0.58	0.80	0.13	0.02	0.71	0.26
0.09	0.11	0.09	0.08	0.78	0.50
0.58	0.10	0.16	0.26	0.51	0.02
1.25	1.01	0.38	0.36	2.00	0.78
			0.49^τ		
			0.03		
			1.01		

8. Tosudite, Kuroko region, Niida, S. Odate, Akita Prefecture, Japan (analysis 4, Table 8.6: Sudo and Shimoda 1978).
9. Li-tosudite, Roseki deposit, Tooho mine, Aichi Prefecture, Japan (Nishiyama *et al.* 1975).
10. Li-tosudite, hydrothermal alteration product of Llanvirnian slates, Huy, Belgium (Brown *et al.* 1974). τ Denotes exchangeable Mg, not Ca.
11. Trioctahedral chlorite/saponite, almost regularly interstratified (see text), Green Tuff, Yamakata, Ibaragi Prefecture, Japan (analysis 3, Table 8.7: Sudo and Shimoda 1978).
12. Trioctahedral chlorite/saponite, almost regularly interstratified, Kuroko-type deposit, Noto mine, Ishikawa Prefecture, Japan (analysis 2, Table 8.7: Sudo and Shimoda 1978).

The structural unit of two 2:1 layers contains four tetrahedral sheets and two octahedral sheets. If the Si atoms are distributed equally among the four tetrahedral sheets, their mean composition would be $12.84/4 = 3.21$Si and hence $(4 - 3.21) = 0.79$Al per sheet; the remaining non-interlayer cations may then be allocated equally between the two octahedral sheets. The average structure derived in this way is the following:

			Net charge
Layer 1	T(1)	$Si_{3.21}Al_{0.79}$	-0.79
	O(1)	$Al_{4.06}Fe^{3+}_{0.014}$	$+0.22$
	T(2)	$Si_{3.21}Al_{0.79}$	-0.79
Interlayer 1 Non-swelling (mica-like)		$M^+_{1.98}$	$+1.98$
Layer 2	T(3)	$Si_{3.21}Al_{0.79}$	-0.79
	O(2)	$Al_{4.06}Fe^{3+}_{0.014}$	$+0.22$
	T(4)	$Si_{3.21}Al_{0.79}$	-0.79
Interlayer 2 Swelling (smectite-like)		$M^+_{0.73}$	$+0.73$
Layer 1'	T(1)	$Si_{3.21}Al_{0.79}$	-0.79
	O(1) etc. ...		

In this representation, T(1), T(2) and O(1) are the cations in the tetrahedral and octahedral sheets of layer 1 and T(3), T(4) and O(2) are those for layer 2; layer 1' is necessarily the same as layer 1 because the mineral is a 1:1 regular interstratification.

This structure can be seen to be unsatisfactory in two ways. First, the charge balance is very unequal locally, the total net negative layer charge being the same around each interlayer despite the different interlayer cation populations, and secondly, it does not explain why interlayer 2 swells and interlayer 1 does not.

Another hypothesis is that alternate 2:1 layers have a mica-like and a smectite-like negative charge. If again it is assumed that the cations in sheets O(1) and O(2) are identical, layers 1 and 2 should have the cation compositions and distributions $(Al_{4.06}Fe^{3+}_{0.014})(Si_{2.90}Al_{1.10})_2O_{20}(OH)_4$ and $(Al_{4.06}Fe^{3+}_{0.014})(Si_{3.52}Al_{0.48})_2O_{20}(OH)_4$ and charges of -1.98 and -0.74 respectively. However, this distribution of cations seems equally improbable because the charge distribution about interlayer 1 is the same as that about interlayer 2 and so the alternation of swelling and non-swelling interlayers remains unexplained.

A third possibility is that the Si occupancy of the two tetrahedral sheets within each 2:1 layer is asymmetric, that is, $Si_{T(1)} \neq Si_{T(2)}$ and $Si_{T(3)} \neq Si_{T(4)}$. This type of allocation enables the alternation of interlayer charge to be accounted for, but makes an actual structural formula indeterminate with the present data, even with the assumption, as before, that the occupancy of O(1) = O(2). There are many possible ways of allocating the Si between the four tetrahedral sheets while maintaining approximate local charge balance and alternation of high- and low-charge interlayers. Incidentally, this difficulty is not confined to interstratified minerals: if equal substitution in the two tetrahedral sheets in 2:1 layers is not a set condition, all structural formulae for composition with $Si < 8$ per $O_{20}(OH)_4$ are potentially indeterminate without additional evidence.

For Na-rectorite, one possible hypothesis is that, although $Si_{T(1)} \neq Si_{T(2)}$, $Si_{T(2)} = Si_{(3)}$ and $Si_{T(4)} = Si_{T(1)}$, that is, the substitution on opposite sides of each interlayer is equal and the composition and structure of all the 2:1 layers are identical except that alternate layers are inverted. An argument in support of this distribution is that it favours local symmetry of charge balance. The structure that this distribution generates is the following:

			Net charge	Kodama's formula[†]
Layer 1	T(1)	$Si_{3.52}Al_{0.48}$	-0.48	Si_3Al_1
	O(1)	$Al_{4.06}Fe_{0.014}$	$+0.22$	$Al_4Fe^{3+}_{0.02}Mg_{0.07}$[†]
	T(2)	$Si_{2.90}Al_{1.10}$	-1.10	Si_3Al_1
Interlayer 1 Non-swelling (mica-like)		$M^+_{1.98}$	$+1.98$	$M^+_{1.80}$[†]
Layer 2	T(3)	$Si_{2.90}Al_{1.10}$	-1.10	$Si_{3.42}Al_{0.58}$
	O(2)	$Al_{4.06}Fe_{0.014}$	$+0.22$	$Al_{4.12}Fe^{3+}_{0.01}Mg_{0.02}$[†]
	T(4)	$Si_{3.52}Al_{0.48}$	-0.48	$Si_{3.42}Al_{0.58}$
Interlayer 2 Swelling (smectite-like)		$M^+_{0.74}$	$+0.74$	$M^+_{0.74}$
	T(1)	$Si_{3.52}Al_{0.48}$	-0.48	Si_3Al_1
	O(1) etc. ...	$Al_{4.06}Fe_{0.014}$	$+0.22$	$Al_4Fe^{3+}_{0.02}Mg_{0.07}$[†]

[†] The formula devised by Kodama is given in the right-hand column; note that he included Mg in the octahedral sheets, whereas the formula on the left includes Mg in the interlayer, although either allocation seems allowable.

Although this structural formula is reasonable in that it reduces net charge separation and accounts satisfactorily for the alternation of high-charge and low-charge interlayers, other equally plausible formulae with $Si_{T(1)} = Si_{T(4)}$ and $Si_{T(2)} = Si_{T(3)}$ could be written, and it must be emphasized that the cation distribution given in the table is not unique.

In this particular example, $(Si + Al) = 24.12$ per $O_{40}(OH)_8$ thus establishing the essential dioctahedral nature of rectorite with close to four cations in both octahedral sheets. In other minerals containing more Mg substituting for Al in the octahedral sheets, or in minerals like tosudite with three or more non-equivalent octahedral sheets, allocation of cations between these sheets may be indeterminate with the information currently available.

A further problem is that the smectite interlayers themselves may have a range of net charge, or of Al-for-Si substitution in the 2:1 layers. Kodama recognized this in Na-rectorite and calculated that 75% of the layers were beidellitic and 25% montmorillonitic in nature, although his structural formulae do not account for the alternation of swelling and non-swelling interlayers. There is other evidence, however, that interlayer charge in the swelling minerals is not uniform and Lagaly (1979) has demonstrated heterogeneity of interlayer charge in three regularly interstratified mica–smectites by application of the alkylammonium technique.

This extended discussion of possible structural formulae for Na-rectorite has been given to bring out two important points:

1. Within one 2:1 layer, Si occupancy in the two tetrahedral sheets need not be equal, and in at least some interstratified minerals, is unlikely to be so.
2. If asymmetric occupancy of the tetrahedral sites in 2:1 aluminosilicate layers is possible, then only the average structural formula for the 2:1 layer can be obtained unless there is evidence additional to that normally available.

In the following sections, interstratified minerals with non-expanding layers are described first, and then the various combinations of expanding layers with other types; in each section named regularly interstratified minerals are considered before the less well-ordered or randomly interstratified minerals. As it is not possible in the space available to discuss each formula in the detail given above, only average structural formulae are given.

98   THE CHEMICAL CONSTITUTION OF CLAYS

1.9.1 Minerals with non-expanding interlayers

Several types are known or suspected. The mineral described in most detail is kulkeite (see below) which is a regular 1:1 talc–chlorite interstratification. Brindley and Gillery (1954) reported a "daphnite" as a "mixed-layer kaolin–chlorite" (i.e. interstratified serpentine–chlorite) but the model they proposed is somewhat ambiguous and the mineral could perhaps be an intimate intergrowth, rather in the manner of wonesite (Section 1.7.3.1). Tamura (1956) described a soil weathering sequence in which a mica–chlorite was formed by the weathering of mixed-layer vermiculite–illite, aluminium entering the expanding interlayer and becoming "chloritized" by hydrolysis. The formation of hydroxidic (chlorite-like) interlayers in laboratory alteration of mica with aluminium solutions has been described by Nagasawa *et al.* (1974) and this "chloritization" reaction seems to be well established. The hydroxide interlayers produced under low temperature/low pressure conditions are however less ordered and less stable than the interlayers in true chlorites (see Section 1.8.3).

1.9.1.1 Kulkeite
This is described by Schreyer *et al.* (1982) as a regular 1:1 chlorite–talc interstratification, formed in a low-grade metamorphic dolomitic rock (analysis 1, Table 1.30). The empirical formula given in the table is calculated to an anion content of $O_{40}(OH)_{20}$, which is the aggregate anion composition for one $O_{20}(OH)_4$ unit (talc) and one $O_{20}(OH)_{16}$ unit (chlorite). The mineral contains sufficient Al that about 1Al has to be allocated to one or more of the octahedral sheets. If it is assumed that the talc unit has a full complement of Mg_6 in its octahedral sheet and Si_8 in its tetrahedral sheets, the 2:1 layer in the chlorite component must have a tetrahedral composition of $Si_{5.14}Al_{2.86}$ and an octahedral composition of $Mg_{10.02}Al_{1.98}$ which is possible for a trioctahedral chlorite. However, because the mineral is regularly interstratified, this would place a highly charged tetrahedral chlorite sheet next to an uncharged talc tetrahedral sheet, leading to a locally highly unbalanced charge. As this seems unlikely, it is perhaps more reasonable to expect a more even distribution of Al-for-Si and Al-for-Mg substitutions throughout the structure. As the mineral contains Na additional to that normal for talc and chlorite, it is probable that substitution of the type $Na + Al \rightleftharpoons Si$ is also occurring, as suggested for sodian aluminian talc (Section 1.5.2).

1.9.2 Interstratification of smectite with talc-type units

This seems to be comparatively rare and the best characterized mineral is aliettite, a regular 1:1 talc–saponite.

1.9.2.1 Aliettite
This was first described by Alietti (1956), and subsequently by Veniale and van der Marel (1969) in serpentinized rocks of the Nure Valley, Piacenza Province, Italy. A detailed study by Alietti and Mejsner (1980) showed that it could be described as a regular 1:1 talc–saponite interstratification and established a chemical analysis for the mineral from the Taro Valley, Parma Province, Italy (analysis 2, Table 1.30). The structural formula given by Mejsner and Alietti apparently includes interlayer OH additional to the normal anion content $O_{20}(OH)_4$, but as no explanation of this was given, the formula in Table 1.30 has been calculated for $O_{40}(OH)_8$ the anionic content of a structural unit containing two 2:1 layers. The structure is a

regular alternation of swelling smectite-like interlayers presumably containing Ca and Na found by analysis, and non-swelling talc-like interlayers with few or no interlayer cations but the formula is an average one. Any further assignment of cations to the two octahedral and four tetrahedral cation sheets in the structural units would require reasoning of the kind outlined previously for rectorite.

1.9.3 Interstratification of smectite and mica units

This is the commonest interstratification of the clay minerals, and because the reaction

$$\text{smectite} + Al^{3+} + K^+ \rightarrow \text{illite} + Si^{4+}$$

occurs during burial metamorphism of argillaceous sediments (Hower *et al.* 1976) and its reverse during weathering this may indeed be one of the most abundant clay mineral types in the lithosphere. All portions of illite to smectite are known, and the cation in the non-expanding interlayers can be K, Na or Ca, or a mixture of all three.

Regular 1:1 mica–dioctahedral smectite interstratification is called rectorite, which has priority over the name allevardite (Bailey 1982). The latter name however is still used in the description "allevardite-like ordering" (Reynolds and Hower 1970; Hower *et al.* 1976; see also Chapter 9) for minerals with less than 30–40% smectite layers in which there is the maximum possible probability that an illite unit precedes and follows a smectite unit (IS ordering). If the proportion of expanding interlayers exceed about 40%, the sequence is usually random, and ordering of the ISII type is confined to samples with 10–15% expanding interlayers (Reynolds and Hower 1970). Thus in this subgroup of minerals, structure and composition are strongly related.

A series of interstratifications of glauconite and smectite units also occurs, analogous to that of the illite–smectite series but with an octahedral Fe content greater than one per $O_{20}(OH)_4$ (Thompson and Hower 1975).

1.9.3.1 Rectorite
The original comparison between rectorite and allevardite (Brown and Weir 1963) was done on samples in which Na was the dominant cation in the non-expanding layers, but other studies have characterized minerals that have K (Nemecz *et al.* 1963; Pevear *et al.* 1980) and Ca (Nishiyama and Shimoda 1981) as the dominant cation. These are described as Na-, K- or Ca-rectorites respectively; the expanding interlayers in rectorites may be either montmorillonitic or beidellitic in character. Na- and Ca-rectorites are found in Kuroko, Roseki and Toseki deposits in Japan (Sudo and Shimoda 1978), whereas K-rectorite has been described in an Eocene coalfield metamorphosed at temperatures > 160 °C (Pevear *et al.* 1980).

The allocation of cations to the various sites in Na-rectorite was discussed earlier. In general, unless the analysis includes estimates of exchangeable and fixed interlayer cations (Na, K, Ca and Mg), assumed to be located in the smectite-like and mica-like interlayers, respectively, only mean structural formulae for the rectorites can be calculated. As the rectorite structure unit consists of two 2:1 layers the cation contents have been calculated for an anion content of $O_{40}(OH)_8$, equivalent to 88 negative charges (analyses 3–5, Table 1.30).

The more frequently reported rectorites have Na > (Ca + K) in the micaceous, non-swelling interlayer (Brown and Weir 1963; Kodama 1966) with net layer charge derived from Al-for-Si substitution. In the example of Na-rectorite from Baluchistan (analysis 3, Table 1.30), possible substitutions are $2 \times (Al_{1.10}Si_{2.90})$ about the non-swelling interlayers and $2 \times (Al_{0.48}Si_{3.52})$ about the smectitic interlayers. There is a small net positive charge from the octahedral cations.

Nishiyama and Shimoda (1981) have reported a Ca-bearing rectorite formed by hydrothermal alteration of andesitic rock (analysis 4, Table 1.30) in which the non-swelling interlayer has a composition $Ca_{0.68}Na_{1.14}K_{0.33}$, which comprises a charge of $+2.83$, intermediate between paragonite-like and margarite-like interlayers. The implied Al-for-Si substitution in the adjacent tetrahedral sheets is $2 \times (Al_{1.45}Si_{2.55})$. Octahedral R^{3+} is 7.99 per $O_{40}(OH)_8$ and the mean charge $+0.08$ per sheet. The swelling interlayers have a mean charge of 0.43, which is small for a smectite.

Pevear et al. (1980) have reported a K-rectorite from the bentonites associated with the Tulameen coalfield (analysis 5, Table 1.30). No estimates were made of fixed interlayer cations so little can be deduced about the distribution of cations. However, overall Al-for-Si substitution was smaller than for Na- and Ca-rectorites but as the Mg content was not significantly higher, the fixed interlayer charge may be lower than for the other rectories.

1.9.3.2 "Tarasovite"
This mineral was studied and named by Lazarenko and Korolev (1970) who reported it as a "new dioctahedral ordered interlayer mineral". From their studies and its recent re-examination by Brindley and Suzuki (1983), it appears to be an interstratified illite–smectite consisting mainly of 3:1 illite–smectite units (ISII) with randomly added illite units making the overall illite to smectite ratio close to four. The sequence of basal reflections is not sufficiently regular to justify a specific mineral name, and it has been recommended that the name tarasovite be preserved pending discovery of a more regular interstratification of the same type (Bailey 1982). The ISII sequence also occurs in the "Kalkberg ordered" interstratifications but in these there are more additional interstratified illite units (Reynolds and Hower 1970; Reynolds 1980).

Because the mineral described above appears to be the most nearly 3:1 regularly interstratified illite–smectite discovered so far, its chemical composition is given in Table 1.30 (analysis 6); its small Mg and Fe contents and its interlayer cation population show it to be chemically related to the rectories.

1.9.3.3 Partially-ordered and randomly-interstratified mica–smectites
Many dioctahedral mica–smectites have a less-ordered structure than that of the rectories and they appear to fall into two broad groups: the magnesium- and iron-poor minerals, like those originating as hydrothermal alteration products of rocks in Japan (Shimoda 1978), contrast with the mica–smectites that result from burial metamorphism of argillaceous sediments (Hower et al. 1976). Chemically the former are closely related to the rectorites but the latter have a wide range of compositions, from a low-charge muscovite–phengite–beidellite type of substitution to a low-charge celadonite–montmorillonite composition. It seems probable that there is a continuum of compositions throughout the fields of substitutions shown in Fig. 1.3, Section 1.7.1.

Unique structural formulae cannot normally be given for these minerals and so the chemical analyses given in Table 1.31 (analyses 1 to 6) are calculated for a mean formula based on $O_{20}(OH)_4$ anions. However, if so wished, some tentative inferences can be made about the nature of the interlayers and their surrounding charge distributions if the assumptions discussed in the introduction (Section 1.9) to this section are used. These are:

1. That a mean octahedral composition is the same in both types of 2:1 layers.

2. That the substitutions in the tetrahedral sheets on either side of each type of interlayer minimize the local charge balance.

Non-exchangeable Mg is assumed to substitute in the octahedral sheet (but see the section on glauconite, Section 1.7.2.3, for discussion that some Mg may be present in partially chloritized interlayers), so the magnesium content is generally an indication of the place of the mineral on the beidellite–montmorillonite axis.

Iron is predominantly present as Fe^{3+} substituting for Al^{3+}; illite–smectites have total $Fe < 1$ per $O_{20}(OH)_4$ but glauconites–smectites (analyses 7 to 10, Table 1.31) have total $Fe > 1$, compositions with $Fe > 2$ per O_{20} being quite common. Many glauconite–smectites contain significant Fe^{2+} but Fe^{3+} remains in excess.

As pointed out by Reynolds and Hower (1970), the type of disorder found depends on the proportion of illite to smectite interlayers, which in turn is related to the cation exchange capacity and the potassium content of the illite–smectite (Hower and Mowatt 1966). Most illite–smectites with less than about 35% smectite interlayers are either IS- or ISII-ordered, the latter being restricted to minerals with less than about 15% smectite interlayers. As noted in the section on clay micas (Section 1.7.2.2), the potassium content of illite interlayers is less than that of the true micas, often averaging 1.4 to 1.6 K per $O_{20}(OH)_4$ layer. Thompson and Hower (1975) found with glauconite–smectites that a similar relation between ordering and proportion of expanding interlayers existed but that the potassium-containing interlayers had a lower charge density than in the illite–smectites.

Very many analyses of interstratified mica–smectites could be quoted; those given in Table 1.31 were chosen to illustrate the general points discussed above, using analyses of minerals for which an estimate of exchangeable and non-exchangeable cations has been made and for which the proportion of expanding to non-expanding interlayers has been estimated from the X-ray diffraction data.

1.9.4 Interstratification of smectite or vermiculite with chlorite

Minerals of this type are common weathering or hydrothermal alteration products and present considerable difficulties in characterization. Their presence is usually recognized by X-ray diffraction, using the changes in basal spacings that occur on treatment with ethylene glycol or glycerol and on heating in stages to 550 °C. With relatively few exceptions, these changes are less clear-cut than might be expected in ideal behaviour, apparently indicating not only interstratification but also imperfection in the interlayer hydroxide sheet of the chlorite-like units.

In interstratified chlorite–smectites (or vermiculites), the total charge on the expanding interlayers can be found by determining the exchangeable cations, but because the number of expanding interlayers is not known, this will not show whether the interlayer is smectite-like or vermiculite-like. Often little more than the cation ratios in the remainder of the mineral can be established. The structural components are 2:1 layers, expanding interlayers and interlayer hydroxidic sheets and in the general case, knowledge of the proportion of hydroxidic to expanding interlayers is needed to establish the anion content. In practice this means that regularly interstratified examples are the most favourable for further chemical characterization for in these the proportion of the two kinds of interlayer must be a simple rational fraction; this ratio is 1:1 in all known regularly interstratified chlorite–smectites and chlorite–vermiculites.

A 1:1 regularly interstratified chlorite–smectite (CS) structure unit contains four tetrahedral

TABLE 1.31. Analyses of interstratified illite–smectites and glauconite–smectites

	1	2	3	4	5
SiO_2	54.54	55.08	58.58	53.7	55.7
Al_2O_3	39.45	35.67	24.98	19.0	21.8
TiO_2	0.01	1.04		0.07	0.22
Fe_2O_3	0.07	0.28	7.24	3.04	2.19
FeO					
MgO	0.44	0.41	2.90	3.90	4.1
CaO	2.51	1.35	0.01	0.78	0.52
Na_2O	1.96	0.35	2.77	1.89	1.30
K_2O	1.99	5.92	3.63	2.30	4.64
H_2O^+				14.7	7.6
H_2O^-				3.3	4.3
Total	100.97	100.01	100.11	102.68	102.37
Numbers of cations on the basis $O_{20}(OH)_4$					
Si	6.47	6.68	7.21	7.69	7.51
Al	1.53	1.32	0.79	0.31	0.49
Total tet.	8.00	8.00	8.00	8.00	8.00
Al	3.99	3.79	2.83	2.90	2.98
Ti	0.001	0.09		0.01	0.02
Fe^{3+}	0.006	0.03	0.67	0.33	0.22
Fe^{2+}					
Mg	0.08	0.07	0.53	0.83	0.82
Total oct.	4.07	3.98	4.03	4.07	4.04
Ca	0.32	0.18	0.00	†	†
Na	0.45	0.08	0.66	0.52	0.34
K	0.30	0.92	0.57	0.42	0.80
Total interlayer	1.07	1.18	1.23	0.94	1.14
Exch. X^+	0.64	0.34	0.66	0.52	0.34

1. Mica–smectite, 50% mica with ordering, $P_{AB}=0.80$; Goto mine; Nagasaki Prefecture, Japan (Kodama et al. 1969). Contained 5.99% H_2O^- and 5.36% H_2O^+.
2. Mica–smectite, 55% mica with some ordering, $P_{AB}=0.25$; Yonago mine, Nagano Prefecture, Japan (Kodama et al. 1969). Contained 4.55% H_2O^- and 5.90% H_2O^+.
3. Illite–smectite, 55% illite, randomly interstratified; Denchworth series soil, Oxford Clay, England (Weir and Rayner 1974).
4. Illite–smectite, 60% smectite, randomly interstratified; Kinnekulle (KB-B), Ordovician K-bentonite, Sweden, Chasmops Fm (Hower and Mowatt 1966). † Indicates structural formula calculated by assuming CaO was impurity and 0.31% Na_2O for each 100 μeq g^{-1} exchange capacity.
5. Illite–smectite, 68% illite, maximum IS ordering; Kinnekulle (KB-A-2), Ordovician bentonite, Sweden Chasmops Fm. (Hower and Mowatt 1966). † See analysis 4.
6. Illite–smectite, 90% illite, ISII ordering; L. Devonian bentonite, Kalkberg, Cherry Valley, New York (Hower and Mowatt 1966). † See analysis 4.

sheets, two in each of the 2:1 layers, three octahedral sheets, one in each of the 2:1 layers and the other in the hydroxidic interlayer and one expanding interlayer containing the exchangeable cations. The implied anion content is $O_{40}(OH)_{20}$. As mentioned above the exchangeable cations may reasonably be allocated to the expanding interlayer. If the octahedral sheets all have their full complement of cations, these may be $12 (= 3 \times 4)$, $14 (= 2 \times 4 + 1 \times 6)$, $16 (= 1 \times 4 + 2 \times 6)$, or $18 (= 3 \times 6)$ octahedral cations depending on the number of di- or trioctahedral sheets in the CS unit and the most one can hope to deduce is whether the total octahedral cation content (Σ) suggests a predominantly dioctahedral ($\Sigma = 12$–14), trioctahedral ($\Sigma = 16$–18), or a mixed

	6	7	7A	8	9	10
	56.0	54.89		50.80	55.16	59.31
	24.4	18.62	16.63	6.26	16.71	14.40
	0.28	0.74		0.18	0.66	0.36
	1.04	14.92	7.42	17.93	17.24	12.05
		0.14		1.62		1.38
	4.1	5.39	3.92	4.45	4.07	6.93
	0.16	0.97		1.30	0.05	0.19
	0.96	1.61		0.09	0.56	0.37
	6.47	1.81		5.96	4.75	6.68
	7.0			6.84		
	1.7			3.94		
	102.11	99.09		99.66	99.20	101.67

Numbers of cations on the basis $O_{20}(OH)_4$

	6	7	7A	8	9	10
	7.34	7.04	7.67	7.66	7.14	7.48
	0.66	0.96	0.33	0.34	0.86	0.52
	8.00	8.00	8.00	8.00	8.00	8.00
	3.10	1.85	2.42	0.78	1.69	1.62
	0.03	0.07	0.08	0.02	0.06	0.03
	0.10	1.44	0.78	2.04	1.68	1.14
		0.02	0.02	0.20		0.15
	0.80	1.03	0.82	1.00	0.78	1.30
	4.03	4.41	4.12	4.04	4.21	4.24
	‡	‡	‡	0.12	‡	‡
	0.24	0.44	0.44	0.03	0.14	0.09
	1.08	0.30	0.32	1.15	0.78	1.07
	1.32	0.70	0.76	1.30	0.92	1.16
	0.24	0.40	0.44		0.14	0.09

7. Glauconite–smectite, 65% smectite, randomly interstratified; G68A, Eocene, Claiborne stage: Hurricane lentil in Landrum fm., Leon County, Texas, USA (Thompson and Hower 1975). ‡ Indicates structural formula was calculated by the authors by disregarding Na and Ca, and assuming that the cation exchange capacity is 80 μeq g^{-1} for each 10% smectite layers.

7A. Oxide determinations of analysis 7 corrected for Al, Fe and Mg dissolved in progressive acid treatment (see section 1.7.23).

8. Glauconite–smectite, ~25% smectite layers (ordering not recorded); sample 62, Bohemian Upper Cretaceous, Jikev, near Nymburk, Czechoslovakia (Cimbalnikova 1971). Total includes MnO = 0.02% and $P_2O_5 = 0.27\%$; structural formula calculated after subtraction of Ca and P_2O_5 equivalents for apatite.

9. Glauconite–smectite, 75–80% glauconite, IS ordered; Jurassic: Sundance fm., Thermopolis, Wyoming, USA (Thompson and Hower 1975). ‡ See analysis 7.

10. Glauconite–smectite, 85% glauconite, ISII ordered; Mississippian: Basal Barnett fm., Marble Falls, Texas, USA (Thompson and Hower 1975). ‡ See analysis 7.

dioctahedral–trioctahedral structure ($\Sigma = 14$–16). Any additional characterization involving allocation of cations to particular tetrahedral or octahedral sheets is impossible without additional structural information.

Minerals that are regularly interstratified and dioctahedral "on average", i.e. contain between 12 and 14 octahedral cations per $O_{40}(OH)_{20}$, are called tosudite; it has recently been suggested the name corrensite should only be used for the regular interstratification of trioctahedral chlorite with either trioctahedral smectite or trioctahedral vermiculite (Bailey 1982), and would be expected to have 16–18 octahedral cations per formula unit.

104   THE CHEMICAL CONSTITUTION OF CLAYS

1.9.4.1 Tosudite
Two samples (analyses 9 and 10, Table 1.30) meet the strict definition of tosudite for structural regularity (Bailey 1982), but both contain Li. Analysis 7 is for the originally described mineral; although it does not quite meet the criteria for regularity it is included as an example of a dioctahedral smectite–chlorite. Analysis 8 contains more MgO and, in common with analyses 9 and 10 with nearly 14 octahedral cations per $O_{40}(OH)_{20}$, probably has two dioctahedral and one trioctahedral sheet. Only a few analyses of tosudites are available and their composition range is not established. In their study of hydrothermally altered serpentine in Yugoslavia, Maksimovic and Brindley (1980) found a tosudite containing about 2% Cr_2O_3 but as it occurred in intimate mixture with Cr-kaolinite, its structural composition is not definitely known.

1.9.4.2 Corrensite
Originally corrensite was described as a regular 1:1 interstratification of chlorite and swelling chlorite, the latter being the imperfect type of chlorite structure that does not collapse to 10 Å on heating to 500 °C but expands with glycerol solvation to 18 Å. It is considered not appropriate, however, to give a species name to structures of this kind, and current usage of the name corrensite is for a regular 1:1 interstratification of trioctahedral chlorite with trioctahedral smectite or vermiculite. The Middle Keuper marls and shales are commonly a source of minerals of this type (Bailey 1982). Most occurrences are, however, not fully characterized mineralogically, and chemical analyses are frequently of impure materials. Analyses 11 and 12 in Table 1.30 are of minerals which do not strictly meet the requirement of regular interstratification, but otherwise could be described as corrensites; one is iron-rich and the other contains rather little iron. It should be noted however that analysis 11 in particular contains significant amounts of both Na and K, neither of which should be present in chlorite, so that the clay may contain mica impurity. Recently, Brigatti and Poppi (1984) described a number of "corrensite-like minerals" from Italy; although full structural data are not given, two of these seem to show a tendency towards regular interstratification. Assuming that the anion content is therefore $O_{40}(OH)_{20}$, the structural formula for sample 40 from Gotra, Taro Valley, Italy is

$$Na_{0.18}K_{0.02}Ca_{0.52}(Ti_{0.05}Al_{0.63}Fe^{3+}_{2.12}Mn_{0.04}Mg_{15.18})$$
$$(Si_{11.88}Al_{4.12})O_{40}(OH)_{20}$$

showing that the mineral is trioctahedral in all three octahedral sheets.

1.9.4.3 Less-regular interstratifications of chlorite with smectite or vermiculite
At least three types of structure can be recognized: in the least disordered, the first diffraction maximum occurs at a *d*-spacing that is the approximate sum of the component layers, but higher orders are irrational. Less ordered are minerals that give a first maximum at a spacing that is an approximate mean of the two components but gives a number of higher orders; least ordered of all are those that give only one or two orders of basal reflection.

An example of the latter is the mineral called swelling chlorite by Bain and Russell (1981), though swelling chlorite in the original description (Stephen and MacEwan 1950) gave rational higher orders of 14 Å when air dry and 17.8 Å when glycerol treated. Bain and Russell's mineral gave an unsatisfactory structural formula when calculated on a chlorite anion content, $O_{20}(OH)_{16}$, and they quoted a formula based on $O_{20}(OH)_{10}$, i.e. the same anion content as a 1:1

chlorite–smectite. The distinction between this type of mineral and a swelling chlorite, that is, a non-interstratified mineral with imperfect hydroxide sheets, is not at present well defined. Bayliss and James (1981) have described a similar type of mineral, but again there was no structural justification for calculating a formula on an $O_{20}(OH)_{10}$ basis.

A number of authors have described interstratified chlorite–smectites or vermiculites that give superlattice reflections based on 28 Å repeat distances when air dry or 31 Å when glycol treated. Assuming that calculations based on $O_{40}(OH)_{20}$ are justified, the mineral described by Bradley and Weaver (1956) was a trioctahedral chlorite–trioctahedral smectite interstratification, but the material investigated by Earley *et al.* (1956) appeared to have one dioctahedral and two trioctahedral sheets. April (1980) published an analysis of an iron-rich chlorite–vermiculite (octahedral Fe exceeding the other octahedral cations) with two trioctahedral and one dioctahedral sheet and of a chlorite–smectite with Mg the dominant octahedral cation in which all three sheets are trioctahedral (April 1981). None of these minerals was pure and considerable correction was needed for illite or quartz present.

These and other studies have shown that non-regularly interstratified chlorite–smectites or chlorite–vermiculites are not uncommon and probably have a wide composition range. Practical difficulties have often, however, prevented their characterization in the detail possible with other interstratified minerals.

1.9.5 Interstratification of smectite with kaolinite

Japanese bentonites and "acid clays" usually contain montmorillonite but Sudo and Hayashi (1956) showed that the structure of some acid clays is more complex and suggested that they might consist of interstratified kaolinite–smectite. Later, such interstratified kaolinite–smectites were found in Quaternary volcanic ash beds (Sudo *et al.* 1964) and in hydrothermal deposits in Florida where montmorillonite has altered to kaolinite (Altschuler *et al.* 1963); in the Yucatan peninsula, Mexico, volcanic ash has altered to montmorillonite and then to kaolinite–montmorillonite (Schultz *et al.* 1971). Srodon (1980) showed that interstratified kaolinite–smectite is formed when montmorillonite reacts with $AlCl_3$ under hydrothermal conditions. Most kaolinite–smectites contain a predominance of kaolinite layers, with irregular or partially segregated interstratification.

Although a number of analyses of kaolinite–smectites have been reported (Table 1.32), their expression in a rational form presents some difficulty. Calculation of a formula for the average structural unit requires a knowledge of the proportion of 1:1 and 2:1 layers and there may be some uncertainty about this (Schultz *et al.* 1971; Sakharov and Drits 1973). Here we have preferred to present the chemical compositions in the form of the ratios of Si, R^{3+} ($=Al+Fe^{3+}+Ti+\cdots$) and R^{2+} ($=Mg+Fe^{2+}+Mn+\cdots$) to their total number, S ($=Si+R^{3+}+R^{2+}$), assuming that the atoms Ca, Na and K are in the interlayer sites and not within the aluminosilicate layers. It is then possible to express the composition of kaolinite–smectites on a ternary diagram with Si, R^{3+} and R^{2+} in each of the corners (see for instance, Fig. 9.3 in Chapter 9). An enlarged part of this diagram is shown in Fig. 1.7 where the compositions of ideal kaolinite, $(Si_4)(Al_4)O_{10}(OH)_8$, beidellite, $(M^+)(Si_7Al)(Al_4)O_{20}(OH)_4$ and montmorillonite, $(M^+)(Si_8)(Al_3Mg)O_{20}(OH)_4$, are shown; it would be expected that the compositions of actual kaolinite–smectites should be intermediate between these three ideal compositions. In practice, the compositions of most of the analyses in Table 1.32 are outside this triangle, tending to lie on the Mg-rich side of the kaolinite–montmorillonite join. The significance of Mg is further

106 THE CHEMICAL CONSTITUTION OF CLAYS

TABLE 1.32. Analyses of interstratified kaolinite–smectites

	1	2	3	4	5	6	7
SiO_2	41.94	47.02	44.5	43.3	45.0	46.5	48.73
Al_2O_3	30.12	35.28	27.3	28.0	26.3	29.9	19.54
TiO_2	0.40	0.02	0.43	0.43	0.34	0.39	0.28
Fe_2O_3	2.42	0.58	4.3	4.2	4.3	0.94	2.07
FeO	0.21	0.03	0.04	0.08	0.04	0.27	0.13
MgO	1.52	0.96	1.6	1.2	1.8	1.03	2.96
CaO	0.32	0.41	0.38	0.92	0.59	0.76	0.72
Na_2O		0.08	0.47	0.15	0.50	0.43	0.32
K_2O		0.05	0.96	0.63	0.45	0.64	0.28
H_2O^+	11.10	13.04	}20.0	}21.0	}20.5	12.79	8.68
H_2O^-	12.88	2.53				6.87	15.59
Total	100.91	100.00	99.98	99.91	99.82	100.52	99.30
Si/S	0.513	0.520	0.538	0.530	0.547	0.550	0.624
R^{3+}/S	0.457	0.464	0.432	0.447	0.420	0.429	0.318
R^{2+}/S	0.030	0.016	0.029	0.023	0.033	0.021	0.058
CEC		270	528	495	628	374	1049
% smectite layers		14	35[†]	30[†]	40[†]	~25	100

1. Awazu acid clay, Ishikawa Prefecture, Japan (Sudo and Hayashi 1956).
2. Clay from veins in quartz-schists, Jeglowa, Poland (Wiewiora 1973).
3. Clay beds, weathered volcanic ash, Tepakan, Yucatan Peninsula, Mexico (Schultz et al. 1971). † Per cent smectite layers calculated by Sakharov and Drits (1973).
4. Clay beds, weathered volcanic ash, Ticul, Yucatan Peninsula, Mexico (Schultz et al. 1971). † See note to analysis 3.
5. Clay beds, weathered volcanic ash, Becal, Yucatan Peninsula, Mexico (Schultz et al. 1971). † See note to analysis 3.
6. Acid clay in Miocene sandstone, Nakamaruke, Niigata, Japan (Shimoyama et al. 1969); inferred to have formed by alteration of montmorillonite, analysis 7.
7. Montmorillonite, Miocene sandstone, Nakamaruke, Niigata, Japan (Shimoyama et al. 1969).

S represents the sum $(Si + R^{2+} + R^{3+})$;
CEC expressed as $\mu eq\, g^{-1}$.

emphasized in Fig. 1.8 (see p. 108), which relates the proportion of Mg to the cation exchange capacity; the linear relation suggests that net negative charge results primarily from replacement of Al by Mg. Whether such replacement occurs solely within the 2:1 layers, as is assumed by most authors, or may also occur in the 1:1 layers it not known.

1.9.6 Interstratification of mica with vermiculite

The name "hydrobiotite" has long been associated with interstratified biotite–vermiculite minerals, which are often found in commercial vermiculite deposits, but the name has not yet been internationally recognized. Recently, Brindley et al. (1983) showed the existence of well-ordered varieties that meet the specification for a special name but gave no chemical analyses and it may be that the name will be adopted for certain examples. The origin of these minerals, regular and non-regular, is still debated, there being some doubt as to whether they are formed by removal of K from biotite during hydrothermal alteration (Boettcher 1966; Brindley et al. 1983) or are formed when K is sorbed by vermiculite (Rhoades and Coleman 1967); both mechanisms may operate. Boettcher showed that the ordered varieties could be formed under hydrothermal conditions at temperatures in the region of 450 °C, whereas the products formed by sorption of K from solution are poorly or randomly ordered. Norrish (1973a) discussed the alteration of biotites to vermiculite and concluded that oxidation of Fe^{2+} to Fe^{3+} was a

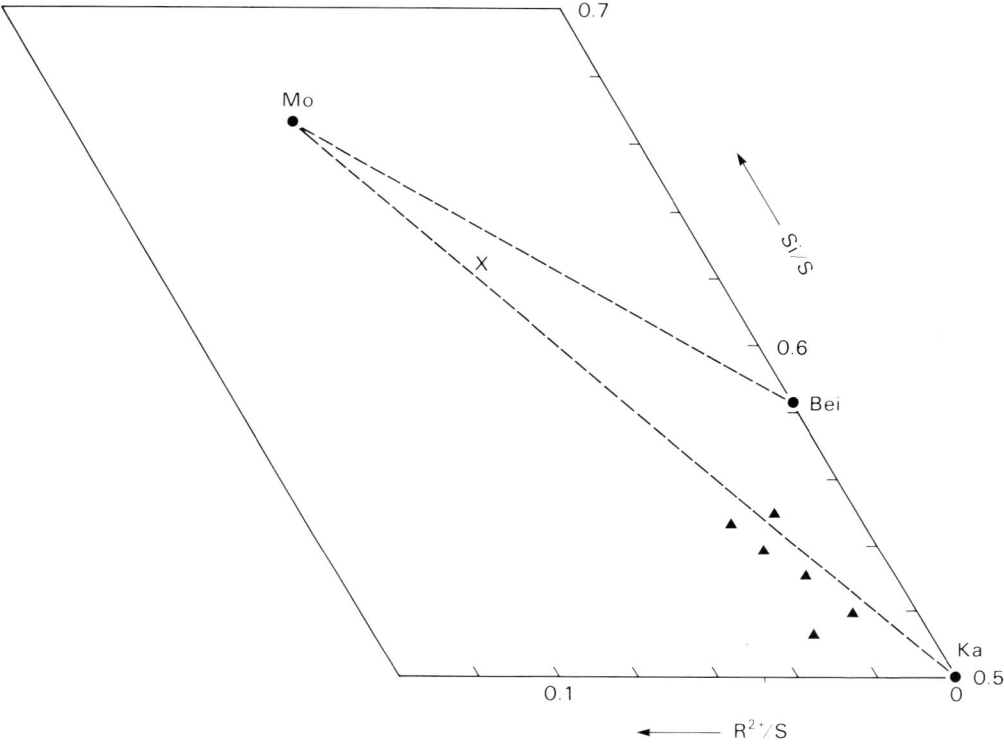

FIG. 1.7. Interstratified kaolinite–smectites: their relationship to ideal montmorillonite (Mo), ideal beidellite (Bei) and ideal kaolinite (Ka) on the Si–R^{3+}–R^{2+} composition diagram (S = Si + R^{3+} + R^{2+}). X represents the smectite (analysis 7, Table 1.32) from which a kaolinite–smectite was inferred to have formed (Shimoyama *et al.* 1969).

concomitant reaction leading to the decrease of net negative charge. Boettcher (1966) analysed biotite, vermiculite and hydrobiotite from adjacent sites and the evidence of his analyses is that the proportion of Fe^{2+}/Fe^{3+} in hydrobiotite is much closer to that in vermiculite than that in biotite (Table 1.33). It could therefore be argued that this hydrobiotite is not likely to be an intermediate stage in the alteration of biotite to vermiculite.

1.10 PALYGORSKITE AND SEPIOLITE

The structure of the fibrous minerals palygorskite (formerly also known as attapulgite) and sepiolite differs from that of other layer silicates in lacking continuous octahedral sheets. Both structures can be regarded as containing ribbons of 2:1 phyllosilicate structure, one ribbon being linked to the next by inversion of SiO_4 tetrahedra along a set of Si—O—Si bonds. This "corner" linking of the ribbons, shown diagrammatically in Chapter 7 (Fig. 7.5; see also Bailey 1980b), generates a relatively open framework structure containing channels that run parallel with the edges of the ribbons and the length of the fibre axes. As the octahedral sheets are discontinuous at each inversion of the tetrahedra, oxygen atoms of the octahedra at the edge of the ribbons are coordinated to cations on the ribbon side only; along the channel side,

108 THE CHEMICAL CONSTITUTION OF CLAYS

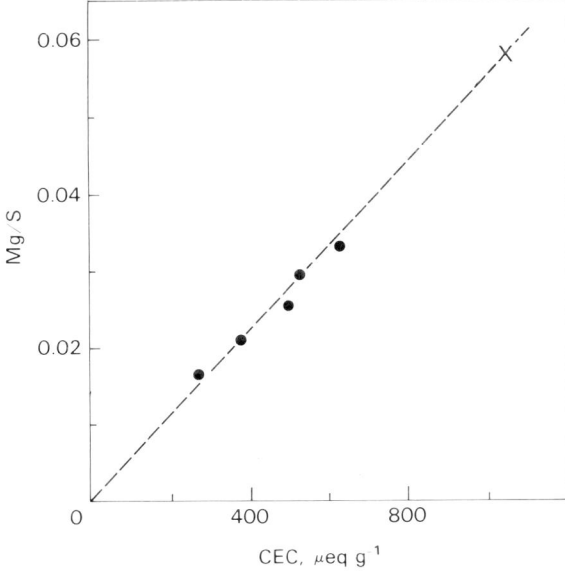

FIG. 1.8. Interstratified kaolinite–smectites: the relationship between cation exchange capacity (CEC) and the proportion of Mg present ($Mg/(Si+R^{2+}+R^{3+})$). X is the smectite of Shimoyama et al. (1969).

TABLE 1.33. Analyses of biotite, hydrobiotite and vermiculite

	Analyses				Structural formulae per $O_{20}(OH)_4$		
	1	2	3		1	2	3
SiO_2	39.10	36.77	35.57	Si	5.68	5.77	5.72
Al_2O_3	13.30	11.60	11.47	Al	2.27	2.14	2.17
TiO_2	1.21	1.02	1.06	Fe^{3+}	0.05	0.09	0.11
Fe_2O_3	2.56	8.19	7.49	Σ tet.	8.00	8.00	8.00
FeO	7.23	0.98	0.34	Ti	0.13	0.12	0.13
MnO	0.10	0.08	0.06	Fe^{3+}	0.23	0.88	0.80
MgO	21.55	20.04	22.57	Fe^{2+}	0.88	0.13	0.05
CaO	0.12	1.94	0.73	Mn	0.01	0.01	0.01
Na_2O	0.23	0.12	0.00	Mg	4.66	4.68	5.41
K_2O	10.05	3.84	0.96	Σ oct.	5.91	5.82	6.40
H_2O^+	3.74	6.69	9.01	Ca + Ba	0.04	0.34	0.13
H_2O^-	0.06	7.80	10.11	Na	0.06	0.04	
F	0.35	0.30		K	1.86	0.77	0.20
Total	100.26†	99.93‡	99.74τ	Σ interlayer cations	1.97	1.15	0.33

1. Biotite, Rainy Creek, nr Libby, Montana, USA (Boettcher 1966). † Total includes 0.25% Cr_2O_3, 0.02% NiO, 0.005% SrO, 0.35% BaO and 0.03% Rb_2O.
2. Hydrobiotite, Rainy Creek, nr Libby, Montana, USA (Boettcher 1966). ‡ Total includes 0.27% Cr_2O_3, 0.01% NiO, 0.01% SrO, 0.19% BaO, 0.0158% Rb_2O, 0.0005% Cs_2O and 0.06% P_2O_5.
3. Vermiculite, Rainy Creek, nr Libby, Montana, USA (Boettcher 1966). τ Total includes 0.18% Cr_2O_3, 0.02% NiO, 0.01% SrO, 0.10% BaO and 0.06% P_2O_5.

coordination and charge balance are completed by additional H, H_2O and a small number of loosely bound or exchangeable cations. The structure and composition of this ribbon-edge region, which has been difficult to describe precisely, is a much larger proportion of the total structure than in the other phyllosilicates and consequently contributes importantly to their chemistry and properties.

In palygorskite the ribbons are the width of two bridged siloxane chains, whereas in sepiolite the ribbon contains three bridged chains; in the former there are five octahedral sites per structural unit whereas sepiolite contains eight. Sepiolites contain mainly Mg in the octahedral sites but in palygorskites the ratio of Mg to trivalent cations can vary from 3:1 to 1:3, and there is a significant proportion of vacancies; sepiolites are therefore predominantly trioctahedral whereas palygorskites tend towards a dioctahedral composition.

Besides surface-sorbed water, palygorskites and sepiolites contain molecular or zeolitic water within the channels, water coordinated to the edge octahedral cations and the normal hydroxyl groups of 2:1 layer silicates at the centre of the ribbons. On heating in vacuum, zeolite water is lost between 100 and 300 °C together with about half of the coordinated water, and the loss of the water causes a part collapse or folding of the structure (Nagata *et al.* 1974; Van Scoyoc *et al.* 1979; see Fig. 7.5); at this stage the structures are readily rehydrated if water vapour is readmitted. Further heating to 500 °C drives off the remainder of the coordinated water and in the case of palygorskite also initiates dehydroxylation. Rehydration after this stage can only be achieved by heating with water under pressure. The interaction of water with palygorskite and sepiolite is discussed in more detail in Chapter 5.

From the failure to synthesize palygorskite and sepiolite under hydrothermal conditions it has been concluded that they are formed at low temperatures and pressures (Henin and Caillere 1975); Wollast *et al.* (1968) showed that a sepiolite-like phase was formed in the reaction between silica and sea water at earth surface conditions.

Sepiolites occur in veins or crusts as alteration products of serpentines, and also as products from other basic rocks; the deposit at Ampandrandava is associated with the alteration of phlogopite. Palygorskites are found in association with less basic rocks, e.g. in a syenite (Stephen 1954) and a weathered granite. It has been suggested that the presence of pyroxene or amphibole minerals provides a template for the formation of palygorskite and sepiolite and that calcite is also necessary (Hénin and Caillère 1975).

Both also occur in sedimentary rocks; sepiolite in particular is associated with limestones at Sommiere, Gard (France), Vallecas (Spain) and Eski Chehir, Asia Minor, and is also found in the marine Keuper formations. Palygorskite appears to be more widespread in its distribution, particularly in lacustrine deposits. The famous deposit near Attapulgus, Georgia, is also associated with limestone but the actual conditions of formation remain controversial (van Olphen and Fripiat 1979). Elsewhere, palygorskite is also formed in arid soils (Norrish and Pickering 1983).

1.10.1 Palygorskite

Based on the work of Drits and Aleksandrova (1966), who suggested a modification of the structure proposed by Bradley (1940), the approximate structural formula for palygorskite is

$$(Mg_{5-y-z}R^{3+}_y\square_z)(Si_{8-x}R^{3+}_x)O_{20}(OH)_2(OH_2)_4 \cdot R^{2+}_{(x-y+2x)/2}(H_2O)_4$$

Gard and Follett (1968) proposed an alternative structure for palygorskite in which the ribbon widths are alternately one and three (Si—O—Si chains and this leads to a different structural

110 THE CHEMICAL CONSTITUTION OF CLAYS

TABLE 1.34. Analyses of palygorskite

	1	2	3	4	5
SiO_2	52.0	55.12	52.35	51.00	52.6
Al_2O_3	17.5	15.7	15.44	12.89	12.6
TiO_2					
Fe_2O_3	1.6	1.6	2.12†	0.09	3.8
FeO				2.68	0.8
MnO				0.08	
MgO	7.0	6.14	6.60	7.54	8.4
CaO	1.4	0.41	0.14		2.2
Na_2O				0.27	
K_2O				0.44	
H_2O^+	14.5	13.52	12.00	} 24.74	9.0
H_2O^-	6.0	7.08	10.32		10.6
Total	100.0	99.57	98.97	99.73	100.0
Number of cations on the basis of $O_{20}(OH)_4(OH_2)_2$					
Si	7.34	7.75	7.61	7.71	7.50
Al	0.66	0.25	0.39	0.29	0.50
Σ tet.	8.00	8.00	8.00	8.00	8.00
Al	2.25	2.35	2.26	2.00	1.62
Ti					
Fe^{3+}	0.17	0.17	0.23	0.01	0.41
Fe^{2+}				0.34	0.10
Mn				0.01	
Mg	1.47	1.29	1.43	1.70	1.78
Σ oct.	3.59	3.81	3.92	4.06	3.90
Ca	0.21	0.06	0.02		0.34
Na				0.08	
K				0.08	
Σ charge	0.42	0.12	0.04	0.16	0.68

1. Palygorskite, Taguenout-Hagueret, Algeria (Caillère and Rouaix 1958).
2. Palygorskite, Klesovo, Volynia, USSR (Dromashko 1953).
3. Palygorskite, Bakkasetter, Shetland Isles (Stephen 1954).
4. Palygorskite, Cabrach, Aberdeenshire, Scotland (Stephen 1954).
5. Palygorskite, Tafraout, Morocco (Caillère and Hénin 1957).
6. Palygorskite, Jezda, Kazahk Republic, USSR (Drits and Aleksandrova 1966). H_2O^+ is above 200 °C; H_2O^- is below 200 °C.
7. Palygorskite, Attapulgus, Georgia, USA (Bradley 1940).

formula from that given above. The Drits and Aleksandrova structure is however more widely accepted.

The general structural formula shows that $4H_2O$ molecules are present in the channels (zeolitic water) and $4H_2O$ are bound to octahedral cations. On heating these latter water moecules are evolved in two stages: during the first stage, $2H_2O$ are lost and the structure collapses by alternate rotation of the ribbons ("folding"), but in the second stage, when the remaining coordinated water is lost, some dehydroxylation also occurs (Van Scoyoc et al. 1979).

Structural formulae for palygorskites have been calculated for the Bradley–Drits–Aleksandrova structure which gives 21 oxygens in the dehydrated and dehydroxylated half-unit cell (Table 1.34). The formulae for analyses 6 and 9 contain more than 8 silicon atoms, too many for the eight available sites; the occupancies of 8.06 and 8.09 are however probably within normal limits of accuracy. Apart from these two, tetrahedral occupancy ranges from

6	7	8	9	10	11
53.47	55.03	53.75	55.59	59.09	53.57
11.27	10.24	10.23	9.17	6.84	2.18
				0.99	0.78
0.42	3.53	1.83		4.21†	0.96
0.02		0.26			3.85
0.65					
7.23	10.49	9.39	10.32	14.12	17.81
0.52		2.29	0.79	<0.01	1.12
0.14		trace	0.25		0.75
0.04	0.47	0.02	0.77		0.21
10.51	10.13	12.04	}23.72	14.18	10.30
11.5	9.73	10.16			8.08
99.10‡	99.62	99.97	100.71	99.43	99.61
Number of cations on the basis of $O_{20}(OH)_4(OH_2)_2$					
8.06	7.80	7.82	8.09	7.88	7.75
	0.20	0.18		0.12	0.25
8.06	8.00	8.00	8.09	8.00	8.00
2.00	1.51	1.57	1.57	0.95	0.12
				0.10	0.08
0.05	0.38	0.20		0.42	0.10
		0.03			0.47
0.08					
1.62	2.22	2.04	2.24	2.81	3.84
3.75	4.11	3.84	3.81	4.28	4.61
0.08		0.36	0.12		0.17
0.04			0.07		0.21
0.01	0.09		0.14		0.04
0.21	0.09	0.72	0.45		0.59

8. Palygorskite, Kuzun District, Tochigi Prefecture, Japan (Imai et al. 1969).
9. Palygorskite, Donetz Basin, USSR (in Drits and Aleksandrova 1966). H_2O^+ includes bound and hydroxyl water.
10. Palygorskite, Mt Flinders, Queensland, Australia (Rogers et al. 1954).
11. Palygorskite, Volhynia basalts, USSR (in Drits and Aleksandrova 1966). H_2O^+ indicates bound plus hydroxyl water.

† Total iron as Fe_2O_3.
‡ Total includes CO_2 0.60%, C 0.84%, SO_3 1.90%.

($Si_{7.88}Al_{0.12}$) to ($Si_{7.34}Al_{0.66}$). The sum of octahedral cations lies between 3.76 and 4.64 with a mean value of 4.00, indicating that the minerals should be classed as dioctahedral. Al, Fe^{3+} and Fe^{2+} are the main elements substituting for Mg; a manganoan variety has been called yofortierite (Perrault et al. 1975).

Based on a study of the pleochromism of the OH spectra in the infrared region, Serna et al. (1977) concluded that the octahedral cations are ordered with the vacant sites at the centre of the ribbons. Octahedral cations (Table 1.34) range from about $2\frac{1}{2}$ trivalent plus $1\frac{1}{2}$ divalent for samples with much Al to $1\frac{1}{2}$ trivalent with $2\frac{1}{2}$ divalent for the palygorskite from Queensland and even to $\frac{1}{2}$ trivalent with $4\frac{1}{4}$ divalent for the highly magnesian sample 11.

Exchange capacities of palygorskites appear to be small, though few are reported. The sedimentary palygorskite from Attapulgus, Georgia, and minerals from Shetland and Aberdeenshire have exchange capacities in the region of 200–300 μeq g^{-1} (Stephen 1954),

TABLE 1.35. Analyses of sepiolite

	1	2	3	4	5
SiO_2	52.50	53.97	54.83	52.36	54.56
Al_2O_3	0.60	0.86	0.28	0.23	0.99
TiO_2					
Fe_2O_3	2.99	0.70	0.45	0.40	1.56
FeO	0.70				0.88
Mn_2O_3		3.14			
MnO				0.01	3.02
MgO	21.31	22.50	24.51	22.60	21.72
CuO		0.87			
CaO	0.47		0.55	0.83	
Na_2O			0.35		0.01
K_2O			0.03		0.02
H_2O^+	9.21	9.90	10.74	9.34	9.23
H_2O^-	12.06	8.80	8.18	13.85	7.92
Total	99.84	99.74	99.92	99.62	99.91
Number of cations on the basis of $O_{30}(OH)_4(OH_2)_4$					
Si	11.81	11.61	11.84	11.95	11.77
Al	0.16	0.22	0.07	0.05	0.23
Fe^{3+}	0.03	0.12	0.07		
Σ tet.	12.00	11.95	11.98	12.00	12.00
Al				0.01	0.02
Fe^{3+}	0.47			0.07	0.25
Fe^{2+}	0.13				0.16
Mn^{2+}		0.52			0.55
Mg	7.14	7.35	7.89	7.69	6.98
Cu		0.14			
Σ oct.	7.74	8.01	7.89	7.77	7.96
Ca	0.11		0.13	0.20	
Na			0.15		
K			0.01		0.01
Σ charge	0.22		0.42	0.40	0.01

1. Sepiolite, Ampandrandava, Madagascar (Caillère and Hénin 1961).
2. Sepiolite, Little Cottonwood, Utah, USA (Nagy and Bradley 1955).
3. Sepiolite, Yavapai County, Arizona, USA (Kauffman 1943).
4. Sepiolite, Karasarva Mine, Tochigi Prefecture, Japan (Takahashi 1966).
5. Sepiolite, Akatani Mine, Niigata Prefecture, Japan (Otsuka et al. 1966).
6. Sepiolite, Kuzu District, Tochigi Prefecture, Japan (Imai et al. 1966).
7. Bergholz, Sterzing, Tyrol, Austria (Brauner and Preisinger 1956).

whereas the Queensland sample (Rogers et al. 1954) has a value of 57 μeq g^{-1}. It is possible that part of the Ca, K and Na reported in some analyses may be enclosed within the channel frameworks and be not readily exchangeable.

1.10.2 Sepiolite

Originally there were alternative suggestions about whether the tetrahedral inversion at the edge of the ribbons in sepiolite occurred along the middle of the zigzag Si—O—Si chains (Nagy and Bradley 1955) or along their edges (Brauner and Preisinger 1956). The latter structure is now widely accepted and gives the general structural formula

6	7	8	9	10	11
52.85	49.58	50.50	52.43	50.80	51.52
1.03	0.88	0.36	7.05	0.66	0.75
trace	trace			0.02	0.02
0.04	17.20	16.46	2.24	1.85	2.64
0.01	1.43	0.63	2.4	1.51	1.35
<0.01	0.30	0.17			
23.74	12.67	13.90	15.08	16.01	19.60
0.51	0.72	0.42			0.02
				8.16	0.06
				0.00	0.00
9.04	10.59	9.83	9.45	6.82	10.42
12.67	7.12	8.38	10.48	13.68	13.38
99.89	100.49	100.65	100.01†	100.10τ	99.76
Number of cations on the basis of $O_{30}(OH)_4(OH_2)_4$					
11.78	11.23	11.38	11.54		11.96
0.22	0.24	0.10	0.46		0.04
	0.53	0.52			
12.00	12.00	12.00	12.00		12.00
0.06			1.37		0.17
0.01	2.40	2.27	0.37		0.46
	0.27	0.12	0.44		0.26
	0.06	0.03			
7.89	4.28	4.67	4.95		6.78
7.96	7.01	7.09	7.13		7.67
0.12	0.17	0.10			
					0.03
					0.00
0.24	0.34	0.20	0.45‡		0.03

8. Xylotile, Sterzing, Tyrol, Austria (Brauner and Preisinger 1956).
9. Aluminium sepiolite, Tintinara, South Australia (Rogers *et al.* 1956). † Included in total; 0.88% $(NH_4)_2O \equiv 0.58\%$ NH_3. ‡ Total monovalent cations = 0.45, present as NH_4^+.
10. Loughlinite, Sweetwater County, Wyoming, USA (Fahey *et al.* 1960). τ Included in total: 0.38% dolomite, 0.21% magnesite.
11. Loughlinite (leached 58 days). (Fahey *et al.* 1960).

$$(Mg_{8-y-z}R_y^{3+}\square_z)(Si_{12-x}R_x^{3+})O_{30}(OH)_4(OH_2)_4 \cdot R_{(x-y+2z)/2}^{2+} \cdot (H_2O)_8$$

This formula shows that sepiolites may contain more zeolitic water than palygorskites but less, in proportion, water coordinated to the octahedral cations at the edges of the ribbons. As with palygorskites, coordinated water is evolved in two stages on heating, folding of the structure occurring when half of the coordinated water has been lost (Nagata *et al.* 1974; Rautureau and Tchoubar 1976).

The structural formulae for sepiolites have been calculated on the basis of the general formula given above, which provides 32 oxygens in the dehydrated and dehydroxylated structure. The formulae given (Table 1.35) show that tetrahedral occupancy ranges from $(Si_{11.96}Al_{0.05})$ to $(Si_{11.23}Fe_{0.53}^{3+}Al_{0.24})$ and the total number of octahedral cations ranges from 7.01 to 8.01; the smaller totals for Si and octahedral cations are found for materials with larger trivalent cation contents. Octahedral cations are predominantly Mg with some Mn, Fe^{2+}, Fe^{3+} and Al^{3+}; a nickel analogue with Ni > Mg is known and has been called falcondoite (Springer 1976) and iron-rich varieties are called xylotile (analysis 8, Table 1.35). An Na-rich variety, named loughlinite, has been shown to have 2Na substituting for 2Mg at the edges of the ribbons (Preisinger 1963), with 2Na in the channels.

1.11 ALLOPHANE AND IMOGOLITE

Allophane is a colloidal material lacking crystalline structure in the normally accepted sense, often gel- or glass-like in the natural moist state but drying to an earthy appearance. It is mainly composed of silica, alumina and water and is then white or colourless, though other ions or hydrous oxides present in small amounts may colour the samples (Wada 1978). Iron and Mg usually comprise less than 1% of the total but several per cent of CaO may be present; CuO, ZnO, PbO and other uncommon oxides have been reported in special circumstances (Weaver and Pollard 1973).

Allophanes are found in diverse environments, being formed in the tropical weathering of basic igneous rocks, in soils on more acid rocks during the development of podzols, and particularly in the weathering of volcanic ash. Allophanic soils often also contain much organic matter and have very low bulk densities and other unusual physical properties (Maeda *et al.* 1977). The chemical composition of natural allophanes can be difficult to deduce as allophanes are often relatively minor components in soils. They react with many of the solutions normally used as pretreatments to mineralogical study, so that the production of artefact during the preparation is always a possibility. In consequence most detailed studies have been made on allophanes formed in pumice, volcanic ash and other pyroclastic rocks, from which they may be dispersed relatively uncontaminated.

Because of their non-crystalline nature, their chemical constitution cannot be described as precisely as for the other clay minerals. However, one type known as imogolite occurs in a thread-like form that can be recognized in the electron microscope and has a more regular structure than other allophanes. This has enabled its constitution to be deduced in some detail, and it will be considered separately.

1.11.1 Disordered allophanes

The ratio SiO_2/Al_2O_3 is an important characteristic of disordered allophanes (i.e. lacking the characteristics of imogolite) varying from about 0.84 (or even as little as 0.2 for very silica-poor materials in stream deposits: Henmi 1979) up to nearly 2, the characteristic ratio for kaolinites. As shown by infrared study, the oxides are not just physically mixed but are chemically combined. Early theories suggested a chain-like sequence of alumina and silica (Wada 1967), but later studies (Brindley and Fancher 1969; Okada *et al.* 1975) have shown that a primitive kaolinite-like structure with aluminium in four- and six-fold coordination is more likely. Henmi and Wada (1976) made a comprehensive study of 16 well-known allophane samples by electron microscopy, chemical analysis, infrared spectroscopy, and X-ray fluorescence spectrometry, using the position of the Al K_α line to study the coordination number of the aluminium. They found that most of their samples were mixtures of hollow spherule-like particles of allophanes and thread-like particles of imogolite. The ratio SiO_2/Al_2O_3 was close to 1 for samples containing most imogolite, but with samples containing least, the ratio varied between 1 and 2.

This showed that imogolite has a nearly constant SiO_2/Al_2O_3 ratio of 1 but that in allophanes it can vary between 1 and 2. In imogolite aluminium is only in six-fold coordination but in allophanes it is both four- and six-fold coordinated, the proportion in four-fold coordination increasing as the ratio SiO_2/Al_2O_3 increases. However a recent NMR study of imogolite (see below) has suggested that these conclusions based on Al K_α fluorescence data may require modification. Aluminium in four-fold coordination chains might be expected to impart permanent net negative charge to the structure but Henmi and Wada (1976) did not find that CEC and Al(IV) were related. They suggested that positively charged hydroxy-aluminium species were strongly sorbed on the surfaces balancing these net negative charges.

It has been suggested (Henmi 1983) that allophanes are of two basic kinds, the first consisting of units with $SiO_2/Al_2O_3 = 1$, like the unit (gibbsite + monomeric SiO_2) of imogolite; the second kind, having a SiO_2/Al_2O_3 ratio = 2, contains polymerized SiO_2 in its structural unit. So far, however, evidence about the degree of polymerization of Si in allophanes of different kinds has been sparse.

Allophanes contain both structural and sorbed water, the structural water giving a ratio H_2O/Al_2O_3 of 2 or more; adsorbed water amounts to 10–25%. Structural water is present as OH bound to both Al and Si; as allophanes are alkaline dispersive, this presumably indicates that neutral $\equiv$SiOH groups become ionized at high pH. Hydroxyl ions in allophanes readily exchange with D_2O indicating that the OH groups are accessible to ambient solution. Thermal analysis of allophanes shows an endotherm at low temperatures (below 150 °C) and an exotherm near 900 °C, the endotherm at 150 °C being related by infrared study to the loss of structural OH.

The surface acidities of H(Al)-allophanes is weaker at high moisture contents than those of H(Al)-kaolinite or H(Al)-montmorillonite, but increases as the concentration of water decreases when the samples are dried (Henmi and Wada 1974). This is a consequence of the pH-dependent nature of the surface charge, which can change from about $+500\,\mu\text{eq g}^{-1}$ at pH 4 to $-700\,\mu\text{eq g}^{-1}$ at pH 8 (Wada 1977).

1.11.2 Imogolite

By contrast with other allophanes imogolite is acid dispersive and was first described by Yoshinaga and Aomine (1962), who extracted it from the Imogo pumice layer widespread in Kyushu Island, Japan; it was subsequently shown to occur in quite high concentrations in other pumice deposits (Wada 1977). According to Cradwick *et al.* (1972), its structure consists of a gibbsite sheet ($Al_4(OH)_{12}$) with isolated Si atoms bridging 3 oxygen atoms over the vacant octahedral site; coordination of Si is completed with OH ions directed away from the octahedral sheet. The short Si—O bonds impart curvature to the octahedral sheet, which forms a tubular structure with OH on the outside and Si—OH directed towards the centre of the tube. The ideal composition for this structure is

$$(OH)_6Al_4O_6Si_2(OH)_2$$

or

$$SiO_2 \cdot Al_2O_3 \cdot 2H_2O$$

Wada (1977) gives the range of composition of imogolites as $(SiO_2)_{1.0-1.2} \cdot Al_2O_3(H_2O)_{2.3-3.0}$ showing that additional water and silica may be present. It is not clear at present whether this range of composition is due to structural variation or to allophanic impurity.

The presence of isolated SiO_4 groups in imogolite was demonstrated by applying the trimethylsilylation technique of Götz and Masson (1971) to imogolite: 95% of the silica present was considered to be unpolymerized (Cradwick *et al.* 1972).

Recently, Wilson *et al.* (1984) have used ^{27}Al magic angle spinning NMR to study the coordination of Al in imogolite, allophane and non-clay samples extracted from a Vitric Andosol developed on rhyolitic pumice. They showed that whereas Al in the unweathered pumice is tetrahedrally coordinated, the fine clay consisting of imogolite contains octahedrally coordinated Al only; the coarse clay has Al in both four- and six-fold coordination. They note that there is some discrepancy between the average coordination numbers obtained by NMR and by X-ray fluorescence and suggest that the measurements of the position of the Al K_β are related to Al—O bond lengths rather than to average coordination number.

Imogolite has little or no permanent exchange capacity but absorbs salts. Anions like oxalate and phosphate are less strongly sorbed on imogolite than on gibbsite (Wada 1978). It is very susceptible to alteration by dry grinding (Henmi and Yoshinaga 1981).

Several methods are available for the synthesis of imogolite (Farmer *et al.* 1983), although at present the best yields are obtained in dilute solutions. Freshly formed hydroxy-aluminium species react with orthosilicate, forming a reactive gel that transforms to imogolite on heating to near 100 °C. Careful adjustment of pH during the preparation is necessary to keep the hydrolysed Al in the most reactive form and to prevent polymerization of the orthosilicate before it has reacted.

REFERENCES

AFANASEV G.D. & AIDINYAN N.KH. (1952) On natron margarite from northern Caucasus. *Bull. Acad. Sci. USSR Geol. Ser.* **No. 2**, 138, (MA, **12**, 140).

ALIETTI A. (1956) Il minerale a strati misti saponite-talco di Monte Chiara (Val di Taro, Appennino Emiliano). *Rend. Accad. Naz. Lincei* **VIII-21**, 201–207.

ALIETTI A. & MEJSNER J. (1980) Structure of a talc/saponite mixed-layer mineral. *Clays Clay Miner.* **28**, 388–390.

ALTSCHULER Z.S., DWORNIK E.J. & KRAMER H. (1963) Transformation of montmorillonite to kaolinite during weathering. *Science, N.Y.* **141**, 148–152.

ALYSHEVA E.I., RUSINOVA O.V. & CHEKVAIDZE V.B. (1977) On sudoites from the polymetal deposits of Rydnyy Altai. *Dokl. Acad. Nauk. SSSR* **236**, 722–724.

ANGEL B.R. & HALL P.L. (1973) Electron spin resonance studies of kaolins. *Proc. Int. Clay Conf. 1972, Madrid*, pp. 47–60. Division de Ciencias CSIC, Madrid.

ANGEL B.R. & VINCENT W.E.J. (1978) Electron spin resonance studies of iron oxides associated with the surface of kaolins. *Clays Clay Miner.* **26**, 263–272.

APRIL R.H. (1980) Regularly interstratified chlorite/vermiculite in contact metamorphosed Red Beds, Newark Group, Connecticut Valley. *Clays Clay Miner.* **28**, 1–11.

APRIL R.H. (1981) Trioctahedral smectite and interstratified chlorite/smectite in Jurassic strata of the Connecticut Valley. *Clays Clay Miner.* **29**, 31–39.

BAILEY S.W. (1975) Chlorites, Ch. 7 in *Soil Components.* Vol. 2. *Inorganic Components* (J.E. Gieseking, ed.). Springer-Verlag: Berlin, Heidelberg, New York.

BAILEY S.W. (1980a) Summary of recommendations of AIPEA Nomenclature Committee. *Clay Minerals* **15**, 85–93.

BAILEY S.W. (1980b) Structures of layer silicates. Ch. 1 in *Crystal Structures of Clay Minerals and their X-ray Identification* (G.W. Brindley and G. Brown, eds). Mineralogical Society, London.

BAILEY S.W. (1982) Nomenclature for regular interstratifications. *Clay Minerals* **17**, 243–248.

BAIN D.C. & RUSSELL J.D. (1981) Swelling minerals in a basalt and its weathering products from Morvern, Scotland: II Swelling chlorite. *Clay Minerals* **16**, 203–212.

BANNISTER F.A. (1943) Brammallite (sodium illite), a new mineral from Llandebie, South Wales. *Mineralog. Mag.* **26**, 304–307.

BASSETT W.A. (1958) Copper vermiculites from Northern Rhodesia. *Am. Miner.* **43**, 1112–1133.

BASSETT W.A. (1960) Role of hydroxyl orientation in mica alteration. *Bull. Geol. Soc. Am.* **71**, 449–456.

BAUER J.F. & SCLAR C.B. (1981) The "10 Å phase" in the system $MgO–SiO_2–H_2O$. *Am. Miner.* **66**, 576–585.

BAYLISS P. (1975) Nomenclature of the trioctahedral chlorites. *The Can. Mineralogist* **13**, 178–180.
BAYLISS P. (1981) Unit cell data of serpentine group minerals. *Mineralog. Mag.* **44**, 153–156.
BAYLISS P. & JAMES D.P. (1981) Di/dioctahedral chlorite–vermiculite–montmorillonite irregular mixed-layer mineral. *Clay Minerals* **16**, 213–215.
BELOV N. (1950) Research in the field of structural mineralogy. *Mineral. Sbornik L.G.O. Lvov* **4**, 21–34 (in Russian).
BENTOR Y.K. & KASTNER M. (1965) Notes on the mineralogy and origin of glauconite. *J. sedim. Petrol.* **35**, 155–166.
BERG-MADSEN V. (1983) High-alumina glaucony from the Middle Cambrian of Olana and Bornholm, Southern Baltoscandia. *J. sedim. Petrol.* **53**, 875–893.
BERNAS B. (1968) A new method for decomposition and comprehensive analysis of silicates by atomic absorption spectrophotometry. *Analyt. Chem.* **40**, 1682–1686.
BERRY, R. & JORGENSEN P. (1969) Separation of illite and chlorite by electromagnetic techniques. *Clay Minerals* **8**, 201–212.
BESSON G., BOOKIN A.S., DAINYAK L.G., RAUTUREAU M., TSIPURSKY S.I., TCHOUBAR C. & DRITS V.A. (1983) Use of diffraction and Mössbauer methods for the structural and crystallochemical characterization of nontronites. *J. appl. Crystallogr.* **16**, 374–383.
BISH D.L. (1977) A spectroscopic and X-ray study of the coordination of Cr^{3+} in chlorites. *Am. Miner.* **62**, 385–389.
BISH D.L. & BRINDLEY G.W. (1978) Deweylite, mixtures of poorly crystalline hydrous serpentine and talc-like minerals. *Mineralog. Mag.* **42**, 75–79.
BLAKE R.L. (1965) Iron phyllosilicates of the Cuyuna district in Minnesota. *Am. Miner.* **50**, 148–169.
BLOSS F.D., SHEKARCHI E. & SHELL H.R. (1959) Hardness of synthetic and natural micas. *Am. Miner.* **44**, 33–48.
BLOUNT A.M., THREADGOLD I.M. & BAILEY S.W. (1969) Refinement of the crystal structure of nacrite. *Clays Clay Min.* **17**, 185–194.
BOAR P.L. & INGRAM L.K. (1970) The comprehensive analysis of coal ash and silicate rocks by atomic-absorption spectrophotometry by a fusion technique. *Analyst, Lond.*, **95**, 124–130.
BOETTCHER A.L. (1966) Vermiculite, hydrobiotite and biotite in the Rainy Creek igneous complex near Libby, Montana. *Clay Minerals* **6**, 283–296.
BOLLAND M.D.A., POSNER A.M. & QUIRK J.P. (1976) Surface charge on kaolinites in aqueous suspension. *Aust. J. Soil Res.* **14**, 197–216.
BRADLEY W.F. (1940) The structural scheme of attapulgite. *Am. Miner.* **25**, 405–410.
BRADLEY W.F. & WEAVER C.E. (1956) A regularly interstratified chlorite–vermiculite clay mineral. *Am. Miner.* **41**, 497–504.
BRAUNER K. & PREISINGER A. (1956) Struktur und Enstehung des Sepioliths. *Tschermaks Miner. Petrogr. Mitt.* **6**, 120–140.
BRIGATTI M.F. (1983) Relationships between composition and structure in Fe-rich smectites. *Clay Minerals* **18**, 177–186.
BRIGATTI M.F. & POPPI L. (1981) A mathematical model to distinguish the members of the dioctahedral smectite series. *Clay Minerals* **16**, 81–89.
BRIGATTI M.F. & POPPI L. (1984) "Corrensite-like minerals" in the Taro and Ceno valleys, Italy. *Clay Minerals* **19**, 59–66.
BRINDLEY G.W. (1951) The crystal structure of some chamosite minerals. *Mineralog. Mag.* **29**, 502–525.
BRINDLEY G.W. (1955) Stevensite a montmorillonite-type mineral showing mixed layer characteristics. *Am. Miner.* **40**, 239–247.
BRINDLEY G.W. (1980) Order-disorder in clay mineral structures. Ch. 2 in *Crystal Structures of Clay Minerals and their X-ray Identification* (G.W. Brindley and G. Brown, eds). Mineralogical Society, London.
BRINDLEY G.W. (1982) Chemical compositions of berthierines – a review. *Clays Clay Miner.* **30**, 153–155.
BRINDLEY G.W., BISH, D.L. & WAN H-M. (1977) The nature of kerolite, its relation to talc and stevensite. *Mineralog. Mag.* **41**, 443–452.
BRINDLEY G.W., BISH D.L. & WAN H-M. (1979) Compositions, structures and properties of nickel-containing minerals in the kerolite–pimelite series. *Am. Miner.* **64**, 615–625.
BRINDLEY G.W. & FANCHER D. (1969) Kaolinite defect structures; possible relation to allophanes. *Proc. Int. Clay Conf. 1969, Tokyo*, Vol. 2, pp. 29–34. Israel Program for Scientific Translations, Jerusalem.
BRINDLEY G.W. & GILLERY F.H. (1954) A mixed-layer kaolin–chlorite structure. *Clays Clay Miner.* **2**, 349–353.
BRINDLEY G.W. & GOODYEAR J. (1948) X-ray studies of halloysite and metahalloysite. II The transition of halloysite to metahalloysite in relation to relative humidity. *Mineralog. Mag.* **28**, 407–422.
BRINDLEY G.W. & MAKSIMOVIC Z. (1974) The nature and nomenclature of hydrous nickel-containing silicates. *Clay Minerals* **10**, 271–278.
BRINDLEY G.W. & PHAM THI HANG (1973) The nature of garnierites. I Structures, chemical compositions and color characteristics. *Clays Clay Miner.* **21**, 27–40.
BRINDLEY G.W. & SEMPELS R.E. (1977) Preparation and properties of some hydroxy-aluminium beidellites. *Clay Minerals* **12**, 229–237.
BRINDLEY G.W. & SUZUKI T. (1983) Tarasovite, a mixed-layer illite–smectite which approaches an ordered 3:1 ratio. *Clay Minerals* **18**, 89–94.
BRINDLEY G.W. & VON KNORRING O. (1954) A new variety of antigorite (ortho-antigorite) from Unst, Shetland Islands. *Am. Miner.* **39**, 794–804.

BRINDLEY G.W. & WAN H-M. (1975) Compositions, structures, and thermal behavior of nickel-containing minerals in the lizardite–nepouite series. *Am. Miner.* **60**, 863–871.
BRINDLEY G.W. & WARDLE R. (1970) Monoclinic and triclinic forms of pyrophyllite and pyrophyllite anhydride. *Am. Miner.* **55**, 1259–1272.
BRINDLEY G.W. & YAMANAKA S. (1979) A study of hydroxy-chromium montmorillonites and the formation of hydroxy-chromium polymers. *Am. Miner.* **64**, 830–835.
BRINDLEY G.W. & YOUELL R.F. (1953) Ferrous chamosite and ferric chamosite. *Mineralog. Mag.* **30**, 57–70.
BRINDLEY G.W., ZALBA P.E. & BETHKE C.M. (1983) Hydrobiotite, a regular 1:1 interstratification of biotite and vermiculite layers. *Am. Miner.* **68**, 420–425.
BRINDLEY G.W. & ZUSSMAN J. (1957) A structural study of the thermal transformation of serpentine minerals to forsterite. *Am. Miner.* **42**, 461–474.
BROOKINS D.G. (1973) Chemical and X-ray investigation of chromiferous kaolinite ("miloschite") from The Geysers, Sonoma County, California. *Clay Clay Miner.* **21**, 421–422.
BROWN, G., BOURGUIGNON P. & THOREZ J. (1974) A lithium-bearing aluminian regular mixed layer montmorillonite–chlorite from Huy, Belgium. *Clay Minerals* **10**, 135–144.
BROWN G. & NEWMAN A.C.D. (1973) The reactions of soluble aluminium with montmorillonite. *J. Soil Sci.* **24**, 339–354.
BROWN G. & NORRISH K. (1952) Hydrous micas. *Mineralog. Mag.* **29**, 929–932.
BROWN G. & WEIR A.H. (1963) The identity of rectorite and allevardite. *Proc. Int. Clay Conf. Stockholm*, Vol. 1, pp. 27–35; Vol. 2, pp. 87–90.
BUCKLEY H.A., BEVAN J.C., BROWN K.M., JOHNSON L.R. & FARMER V.C. (1978) Glauconite and celadonite: two separate mineral species. *Mineralog. Mag.* **42**, 373–382.
BYRNE P.J.S. (1954) Some observations on montmorillonite–organic complexes. *Clays Clay Miner.* **2**, 241–253.
CAHOON H.P. (1954) Saponite from Milford, Utah. *Am. Miner.* **39**, 222–230.
CAILLERE S. & HENIN S. (1957) The sepiolite and palygorskite minerals. Ch. IX in *The Differential Thermal Investigation of Clays* (R.C. Mackenzie, ed.). Mineralogical Society, London.
CAILLERE S. & HENIN S. (1961) Sepiolite. Ch. VIII in *The X-ray Identification and Crystal Structures of Clay Minerals* (G. Brown, ed.). Mineralogical Society, London.
CAILLERE S., HENIN S. & POBEGUIN T. (1962) Presence d'un nouveau type de chlorite dans les bauxites de Saint Paul-de-Fenouillet (Pyrenees-Orientales), *C.r. hebd. Séanc. Acad. Sci., Paris* **254**, 1657–1658.
CAILLERE S. & ROUAIX S. (1958) Sur la presence de palygorskite dans le region de Taguenout-Hagueret (A.O.F.), *C.r. hebd. Séanc. Acad. Sci., Paris* **246**, 1442–1444.
CARMAN J.H. (1974) Synthetic sodium phlogopite and its two hydrates: stabilities, properties and mineralogic implications. *Am. Miner.* **59**, 261–273.
CARR R.M., CHAIKUM N. & PATTERSON N. (1978) Intercalation of salts in halloysite. *Clays Clay Miner.* **26**, 144–152.
CARSTEA D.D., HARWARD M.E. & KNOX E.G. (1970) Comparison of iron and aluminium hydroxy interlayers in montmorillonite and vermiculite. I Formation. *Soil Sci. Soc. Am. Proc.* **34**, 517–521.
CASHEN G. (1959) Electric charges on kaolin. *Trans. Faraday Soc.* **55**, 477–486.
CERNY P. (1970) Compositional variations in cookeite. *Can. Mineralogist* **10**, 636–647.
CHOURABI B. & FRIPIAT J.J. (1981) Determination of tetrahedral substitutions and interlayer surface heterogeneity from vibration spectra of ammonium in smectites. *Clays Clay Miner.* **29**, 260–268.
CHUKHROV F.V., ZVYAGIN B.B., DRITS V.A., GORSHKOV A.J., ERMILOVA L.P., GOILO E.A. & RUDNITSKAYA E.S. (1979) The ferric analogue of pyrophyllite and related phases. *Proc. Int. Clay Conf. 1978, Oxford*, pp. 55–64. Elsevier: Amsterdam, Oxford, New York.
CHUKHROV F.V., ZVYAGIN B.B., ERMILOVA L.P., GORSHKOV A.I. & RUDNITSKAYA E.S. (1969) The relation between chrysocolla, medmontite and copper-halloysite. *Proc. Int. Clay Conf. 1969, Tokyo*, Vol. 1, pp. 141–150. Israel Program for Scientific Translations, Jerusalem.
CIMBALNIKOVA A. (1971) Chemical variability and structural heterogeneity of glauconites. *Am. Miner.* **56**, 1385–1392.
CLEMENTS R.L., SERGEANT G.A. & WEBB P.J. (1971) The determination of fluorine in rocks and minerals by a pyrohydrolytic method. *Analyst, Lond.* **96**, 51–54.
CLIFF G. & LORIMER G.W. (1975) The quantitative analysis of thin specimens. *J. Microscopy* **103**, 203–207.
COOKE C.J. & DUNCOMB P. (1968) Performance analysis of a combined electron microscope and electron probe microanalyser "EMMA". *Fifth Congress on X-ray Optics and Microanalysis*, Tubingen, pp. 245–247. Springer, Berlin.
COOMBE A.D. & HOLMES A. (1945) The kalsilite-bearing lavas of Kabirenge and Lyakauli, South-West Uganda. *Trans. Roy. Soc. Edin.* **61**, 359–379.
COWKING A., WILSON M.J., TAIT J.M. & ROBERTSON R.H.S. (1983) Structure and swelling of fibrous and granular saponitic clay from Orrock Quarry, Fife, Scotland. *Clay Minerals* **18**, 49–64.
CRADWICK P.D.G., FARMER V.C., RUSSELL J.D., MASSON C.R., WADA K. & YOSHINAGA N. (1972) Imogolite, a hydrated aluminium silicate of tubular structure. *Nature Phys. Sci., Lond.* **240**, 187–189.
CREER M.H., HARDY J.B.C., ROOKSBY H.P. & STILL J.E. (1971) Some applications of thermohygrometric analysis to the study of clay and associated minerals. *Clay Minerals* **9**, 19–34.
CROSS W. & HILLEBRAND W.F. (1885) Contributions to the mineralogy of the Rocky Mountains. *Bull. U.S. Geol. Surv.* **20**, 113 pp.

CUTTLER A.H. (1981) Further studies of ferrous iron doped synthetic kaolinite dosimetry of X-ray induced effects. *Clay Minerals* **16**, 69–80.
DEAN J.A. (1960) *Flame Photometry*. McGraw-Hill, London.
DEAN J.A. & RAINS T.C. (eds) (1969) *Flame Emission and Atomic Absorption Spectrometry*. Vol. 1, *Theory*. Marcel Dekker, London.
DEER W.A., HOWIE R.A. & ZUSSMAN J. (1962) *Rock Forming Minerals*. Vol. 3. *Sheet Silicates*. Longman, London.
DE WAAL S.A. (1970a) Nickel minerals from Barberton, South Africa: III Willemseite, a nickel-rich talc. *Am. Miner.* **55**, 31–42.
DE WAAL S.A. (1970b) Nickel minerals from Barberton, South Africa. II Nimite, a nickel-rich chlorite. *Am. Miner.* **55**, 18–30.
DOLCATER D.L., SYERS J.K. & JACKSON M.L. (1970) Titanium as free oxide and substituted forms in kaolinite and other soil minerals. *Clays Clay Miner.* **18**, 71–79.
DRITS V.A. & ALEKSANDROVA V.A. (1966) On the crystallographic nature of palygorskite. *Zap. vses. Miner. Obshch.* **95**, 551–560.
DRITS V.A. & ALEKSANDROVA V.A. (1968) The structure of a mineral of the donbassite group – dioctahedral chlorite from Novaya Zemlya. *Mineral Sbornik, Lvov* **22**, 162–167.
DRITS V.A. & LAZARENKO E.K. (1967) Structural–mineralogical characteristics of donbassites. *Mineral Sbornik, Lvov* **21**, 40–48.
DROMASHKO S.G. (1953) Comparative characteristics of palygorskite, talc, and pyrophyllite. *Mineral. Sbornik, Lvov* **7**, 191–212; *Mineral. Absts*, 1956, **13**, 60.
DUNCOMB P. (1962) An electron-optical bench for microscopy, diffraction and X-ray microanalysis, *Proc. 5th Int. Conf. for Electron Microscopy*, paper KK4. Academic Press, New York.
DUNN P.J., PEACOR D.R., NELEN J.A. & NORBER J.A. (1981) Crystal-chemical data for schallerite, caryopilite and friedelite from Franklin and Sterling Hill, New Jersey. *Am. Miner.* **66**, 1054–1062.
EARLEY J.W., BRINDLEY G.W., MCVEAGH W.J. & VAN DEN HEUVEL R.C. (1956) A regularly interstratified montmorillonite–chlorite. *Am. Miner.* **41**, 258–267.
EARLEY J.W., OSTHAUS B.B. & MILNE I.H. (1953) Purification and properties of montmorillonite. *Am. Miner.* **38**, 707–724.
EBERL D. (1978) The relation of montmorillonite to mixed-layer clay: the effect of interlayer alkali and alkaline earth cations. *Geochim. et Cosmochim. Acta* **42**, 1–7.
EBERL D.D. (1980) Alkali cation selectivity and fixation by clay minerals. *Clays Clay Miner.* **28**, 161–172.
EGGLETON R.A. (1977) Nontronite: Chemistry and X-ray diffraction. *Clay Minerals* **12**, 181–194.
EL-ATTAR H.A., JACKSON M.L. & VOLK V.V. (1972) Fluoride loss from silicates on ignition. *Am. Miner.* **57**, 246–252.
ERNST W.G. (1963) The significance of phengitic micas from low-grade schists. *Am. Miner.* **48**, 1357–1373.
ERNST W.G. (1964) Petrochemical study of coexisting minerals from low-grade schists, Eastern Shikoku, Japan. *Geochim. Cosmochim. Acta* **28**, 1631–1668.
EUGSTER H.P. & WONES D.R. (1962) Stability relations of the ferruginous biotite, annite. *J. Petrology* **3**, 82–125.
EVANS W.H. & SERGEANT G.A. (1967) The determination of small amounts of fluorine in rocks and minerals *Analyst, Lond.* **92**, 690–694.
FAHEY J.J., ROSS M. & AXELROD J.M. (1960) Loughlinite, a new hydrous magnesium silicate. *Am. Miner.* **45**, 270–281.
FARMER V.C., ADAMS M.J., FRASER A.R. & PALMIERI F. (1983) Synthetic imogolite: properties, synthesis and possible applications. *Clay Minerals* **18**, 459–472.
FARMER V.C., RUSSELL J.D., MCHARDY W.J., NEWMAN A.C.D., AHLRICHS J.L. & RIMSAITE J.Y.H. (1971) Evidence for loss of protons and octahedral iron from oxidized biotites and vermiculites. *Mineralog. Mag.* **38**, 121–137.
FAUST G.T. (1966) The hydrous nickel–magnesian silicates – The garnierite group. *Am. Miner.* **51**, 279–298.
FAUST G.T. & FAHEY J.J. (1962) The serpentine group minerals, *Prof. Pap. U.S. Geol. Surv.* **384-A**, 92 pp.
FAUST G.T., FAHEY J.J., MASON B. & DWORNIK E.J. (1969) Pecoraite, $Ni_6Si_4O_{10}(OH)_8$, nickel analog of clinochrysotile, formed in the Wolf Creek meteorite. *Science, N.Y.* **165**, 59–60.
FAUST G.T., HATHAWAY J.C. & MILLOT G. (1959) A restudy of stevensite and allied minerals. *Am. Miner.* **44**, 342–370.
FAUST G.T. & MURATA K.J. (1953) Stevensite redefined as a member of the montmorillonite group. *Am. Miner.* **38**, 973–987.
FAWCETT J.J. & YODERS H.S. (1966) Phase relationships in the system $MgO–Al_2O_3–SiO_2–H_2O$. *Am. Miner.* **51**, 353–380.
FERRIS A.P. & JEPSON W.B. (1975) The exchange capacities of kaolinite and the preparation of homoionic clays. *J. Colloid Interface Sci.* **51**, 245–259.
FLEGMANN A.W., GOODWIN J.W. & OTTEWILL R.H. (1969) Rheological studies on kaolinite suspensions. *Proc. Brit. Ceram. Soc.* **13**, 31–45.
FLEISCHER M. (1970) New mineral names. *Am. Miner.* **55**, 1070–1071.
FLORAN R.J. & PAPIKE J.J. (1978) Mineralogy and petrology of the Gunflint Iron Formation, Minnesota-Ontario: Correlation of compositional and assemblage variations at low to moderate grade. *J. Petrology* **19**, 215–288.
FOLK R.L., HAYES M.O., BROWN T.E., EARGLE D.H., WEEKS A.D., BARNES V.E. & CLABAUGH S.E. (1961) Field Excursion Central Texas; *10th National Clay Conference*. University of Texas Guidebook No. 3.
FORBES W.C. (1969) Unit-cell parameters and optical properties of talc on the join $Mg_3Si_4O_{10}(OH)_2–Fe_3Si_4O_{10}(OH)_2$. *Am. Miner.* **54**, 1399–1408.

FORBES W.C. (1971) Iron content of talc in the system $Mg_3Si_4O_{10}(OH)_2$–$Fe_3Si_4O_{10}(OH)_2$. *J. Geology* **79**, 63–74.
FORMAN S.A., KODAMA H. & MAXWELL J.A. (1967) The trioctahedral brittle micas. *Am. Miner.* **52**, 1122–1128.
FOSTER M.D. (1951) The importance of exchangeable magnesium and cation exchange capacity in the study of montmorillonitic clays. *Am. Miner.* **36**, 717–730.
FOSTER M.D. (1953a) Geochemical studies of clay minerals. II Relation between ionic substitution and swelling in montmorillonite. *Am. Miner.* **38**, 994–1006.
FOSTER M.D. (1953b) Geochemical studies of clay minerals. III The determination of free silica and alumina in montmorillonite. *Geochim. Cosmochim. Acta* **3**, 143–154.
FOSTER M.D. (1956) Correlation of dioctahedral potassium micas on the basis of their charge relations. *Bull. U.S. Geol. Surv.* **1036-D**, 57–67.
FOSTER M.D. (1960a) Interpretation of the composition of trioctahedral micas. *Prof. Pap. U.S. Geol. Survey* **354-B**, 11–49.
FOSTER M.D. (1960b) Interpretation of the composition of lithium micas. *Prof. Pap. U.S. Geol. Survey* **354-E**, 115–147.
FOSTER M.D. (1962) Interpretation of the composition and classification of the chlorites. *Prof. Pap. U.S. Geol. Survey* **414-A**, 33 pp.
FOSTER M.D. (1963) Interpretation of the composition of vermiculites and hydrobiotites. *Clays Clay Miner.* **10**, 70–89.
FOSTER M.D. (1964) Water content of micas and chlorites. *Prof. Pap. U.S. Geol. Survey* **474-F**.
FOSTER M.D. (1969) Studies of celadonite and glauconite. *Prof. Pap. U.S. Geol. Survey* **614-F**.
FOSTER M.D., BRYANT B. & HATHAWAY J. (1960) Iron-rich muscovitic mica from the Grandfather Mountain area, North Carolina. *Am. Miner.* **45**, 839–851.
FRANCIS C.W. (1973) Adsorption of polyvinylpyrrolidone on reference clay minerals. *Soil Sci.* **115**, 40–53.
FRANCIS C.W., BONNER W.P. & TAMURA T. (1972) An evaluation of zonal centrifugation as a research tool in soil science: I Methodology. *Soil Sci. Soc. Am. Proc.* **36**, 366–372.
FRANCIS C.W. & TAMURA T. (1972) An evaluation of zonal centrifugation as a research tool in soil science: II Characterization of soil clays. *Soil Sci. Soc. Am. Proc.* **36**, 372–376.
FRANSOLET A-M & BOURGUIGNON P. (1978) Di/trioctahedral chlorite in quartz veins from the Ardennes, Belgium. *Can. Mineralogist* **16**, 365–373.
FREY A. (1969) A mixed-layer paragonite–phengite of low-grade metamorphic origin. *Contr. Mineral. Petrol.* **24**, 63–65.
FRIPIAT J.J., CHAUSSIDON J. and TOUILLAUX R. (1960) Study of dehydration of montmorillonite and vermiculite by infrared spectroscopy. *J. phys. Chem. Ithaca* **64**, 1234–1241.
FRONDEL C. (1962) Ferroan antigorite (Jenkinsite). *Am. Miner.* **47**, 783–785.
FRONDEL C. & EINAUDI M. (1968) Zinc-rich micas from Sterling Hill, New Jersey. *Am. Miner.* **53**, 1752–1754.
FRONDEL C. & ITO J. (1966) Hendricksite, a new species of mica. *Am. Miner.* **51**, 1107–1113.
GAMALEYA, YU.N. (1968) Polylithionite from granites of the Ulkan pluton and the conditions of its formation. *Dokl. Akad. Nauk. SSSR* **182**, 1186–1188.
GARD J.A. & FOLLETT E.A.C. (1968) A structural scheme for palygorskite. *Clay Minerals* **7**, 367–369.
GARRETT W.G. & WALKER G.F. (1959) The cation exchange capacity of hydrated halloysite and the formation of halloysite–salt complexes. *Clay Miner. Bull.* **4**, 75–80.
GAUDETTE H.E., EADES J.L. & GRIM R.E. (1966) The nature of illite. *Clays Clay Miner.* **13**, 33–48.
GAUDETTE H.E., GRIM R.E. & METZGER C.F. (1966) Illite: a model based on the sorption behaviour of cesium. *Am. Miner.* **51**, 1649–1656.
GIESE R.F. (1971) Hydroxyl orientation in muscovite as indicated by electrostatic energy calculations. *Science, N.Y.* **172**, 263–264.
GIESE R.F. (1975a) Interlayer bonding in talc and pyrophyllite. *Clays Clay Miner.* **23**, 165–166.
GIESE R.F. (1975b) The effects of F/OH substitution on some layer silicate minerals. *Z. Kristallogr.* **141**, 138–144.
GIESE R.F. (1977) The influence of hydroxyl orientation, stacking sequence, and ionic substitution on the interlayer bonding in micas. *Clays Clay Miner.* **25**, 102–104.
GIESE R.F. (1979) Hydroxyl orientations in 2:1 phyllosilicates. *Clays Clay Miner.* **27**, 213–233.
GILLERY F.H. (with HILL V.G.) (1959) The X-ray study of synthetic Mg–Al serpentines and chlorites. *Am. Miner.* **44**, 143–152.
GOODMAN B.A. (1981) Mössbauer spectroscopy. Ch. 5 in *Advanced Techniques for Clay Mineral Analysis* (J.J. Fripiat, ed.), pp. 113–137. Elsevier, Amsterdam.
GOODMAN B.A., RUSSELL J.D., FRASER A.R. & WOODHAMS F.W.D. (1976) A Mössbauer and infrared spectroscopic study of the structure of nontronite. *Clays Clay Miner.* **24**, 53–59.
GOTZ J. & MASSON C.R. (1971) Trimethylsilyl derivatives for the study of silicate structures, Pt II Orthosilicate, pyrosilicate and ring structures. *J. chem. Soc. A* 686–688.
GREENFIELD S., JONES I.L. & BERRY C.T. (1964) High-pressure plasmas and spectroscopic emission sources. *Analyst, Lond.* **89**, 713–720.
GRIM R.E. & KULBICKI G. (1961) Montmorillonite: high temperature reactions and classification. *Am. Miner.* **46**, 1329–1369.
GROVES A.W. (1937) *Silicate Analysis*. T. Murby & Co., London.
GRUNER J.W. (1936) The structure and chemical composition of greenalite. *Am. Miner.* **21**, 449–455.
GRUNER J.W. (1944a) The kaolinite structure of amesite and additional data on chlorites. *Am. Miner.* **29**, 422–430.

GRUNER J.W. (1944b) The structure and composition of minnesotaite, a common iron silicate in iron formations. *Am. Miner.* **29**, 363–372.
GUGGENHEIM S. & BAILEY S.W. (1982) The superlattice of minnesotaite. *Can. Mineralogist* **20**, 579–584.
GUGGENHEIM S., BAILEY S.W., EGGLETON R.A. & WILKES P. (1982) Structural aspects of greenalite and related minerals. *Can. Mineralogist* **20**, 1–18.
GUIDOTTI C.V. (1968) On the relative scarcity of paragonite. *Am. Miner.* **53**, 963–974.
GUPTA G.C. & MALIK, W.U. (1969) Transformation of montmorillonite to nickel–chlorite. *Clays Clay Miner.* **17**, 233–239.
GUVEN N. (1971) The crystal structures of $2M_1$ phengite and $2M_1$ muscovite. *Z. Kristallogr.* **134**, 196–212.
GUVEN N. & HOWER W.F. (1979) A vanadium smectite. *Clay Minerals* **14**, 241–245.
HALL A. (1969) The micas of the Rosses granite complex, Donegal. *Sci. Proc. Roy. Dublin Soc.*, series A **3**, 209–217.
HALMA G. (1969) The separation of clay mineral fractions with linear heavy liquid density gradient columns. *Clay Minerals* **8**, 59–69.
HANNAKER P., HAUKKA M. & SEN S.K. (1984) Comparative study of ICP-AES and XRF analysis of major and minor constituents on geological materials. *Chem. Geology* **42**, 319–324.
HAYASHI H., AITA S. & SUZUKI M. (1978) Semi-quantitative chemical analysis of asbestos fibres and clay minerals with an analytical electron microscope. *Clays Clay Miner.* **26**, 181–188.
HAZEN R.M. & WONES D.R. (1972) The effect of cation substitutions on the physical properties of trioctahedral micas. *Am. Miner.* **57**, 103–129.
HEINRICH E.W. (1965) Further information on the geology of chromian muscovites. *Am. Miner.* **50**, 758–762.
HEINRICH E.W. (1967) Micas of the Brown Derby pegmatites, Gunnison County, Colorado. *Am. Miner.* **52**, 1110–1121.
HEINRICH E.W. & LEVINSON A.A. (1955a) Studies in the mica group: mangan muscovite from Mattkarr, Finland. *Am. Miner.* **40**, 1132–1135.
HEINRICH E.W. & LEVINSON A.A. (1955b) Studies in the mica group: X-ray data on roscoelite and barium muscovite. *Am. J. Sci.* **253**, 39–43.
HEINRICH E.W., LEVINSON A.A., LEVANDOWSKI D.W. & HEWITT C.H. (1953) Studies in the natural history of micas. University of Michigan Engineering Research Institute, Project M.978; final report.
HENDRICKS S.B. (1939) Random structures of layer minerals as illustrated by cronstedtite. Possible iron content of kaolin. *Am. Miner.* **24**, 529–539.
HENDRICKS S.B. & ROSS C.S. (1941) The chemical composition and genesis of glauconite and celadonite. *Am. Miner.* **26**, 683–708.
HENIN S. & CAILLERE S. (1975) Fibrous minerals. Ch. 9 in *Soil Components*. Vol. 2. *Inorganic Components* (J.E. Gieseking, ed.), pp. 335–349, Springer-Verlag, Berlin.
HENLEY K.J. (1970) Application of the muscovite–paragonite geothermometer to a staurolite-grade schist from Sulitjelma, north Norway. *Mineralog. Mag.* **37**, 693–704.
HENMI T. (1979) The occurrence of allophane in a stream-deposit from Ehime-Prefecture, Japan. *Clay Minerals* **14**, 333–338.
HENMI T. (1983) Structural changes of allophanes during dry grinding: dependence on SiO_2/Al_2O_3 ratio. *Clay Minerals* **18**, 101–107.
HENMI T. & WADA K. (1974) Surface acidity of imogolite and allophane. *Clay Minerals* **10**, 231–246.
HENMI T. & WADA K. (1976) Morphology and composition of allophane. *Am. Miner.* **61**, 379–390.
HENMI T. & YOSHINAGA N. (1981) Alteration of imogolite by dry grinding. *Clay Minerals* **16**, 139–149.
HERBILLON A.J., MESTDAGH M.M., VIELVOYE L. & DEROUANE E.G. (1976) Iron in kaolinite with special reference to kaolinite from tropical soils. *Clay Minerals* **11**, 201–220.
HERRERA R. & PEECH M. (1970) Reaction of montmorillonite with iron(III). *Soil Sci. Soc. Am. Proc.* **34**, 740–742.
HESS H.H., SMITH R.J. & DENGO G. (1952) Antigorite from the vicinity of Caracas, Venezuela. *Am. Miner.* **37**, 68–75.
HEWITT D.A. & WONES D.R. (1975) Physical properties of some synthetic Fe–Mg–Al trioctahedral biotites. *Am. Miner.* **60**, 854–862.
HEY M.H. (1982) The determination of ferrous and ferric iron in rocks and minerals; and a note on sulphosalicylic acid as a reagent for Fe and Ti. *Mineralog. Mag.* **46**, 111–118; 512–513.
HEYSTEK H. (1963) Hydrothermal rhyolitic alteration in the Castle Mountains, California. *Clays Clay Min.* **11**, 158–168.
HILLEBRAND W.F. (1919) The analysis of silicate and carbonate rocks. *Bull. U.S. Geol. Surv.* **700**.
HOWER J., ESLINGER E.V., HOWER M.E. & PERRY E.A. (1976) Mechanism of burial metamorphism of argillaceous sediment: I Mineralogical and chemical evidence. *Bull. Geol. Soc. Amer.* **87**, 725–737.
HOWER J. & MOWATT T.C. (1966) The mineralogy of illites and mixed-layer illite/montmorillonites. *Am. Miner.* **51**, 825–854.
HUANG W.G. & JOHNS W.D. (1967) Simultaneous determination of fluorine and chlorine in silicate rocks by a rapid spectrophotometric method. *Analyt. Chim. Acta* **37**, 508–515.
HUGHES J.C. (1982) High gradient magnetic separation of some soil clays from Nigeria, Brazil and Colombia. I The interrelationships of iron and aluminium extracted by acid ammonium oxalate and carbon. *J. Soil Sci.* **33**, 509–519.
HUTTON C.O. (1947) Contributions to the Mineralogy of New Zealand. Part 3. *Trans. Roy. Soc. New Zealand* **76**, 481–491.
IANICELLI J. (1976) High extraction magnetic filtration of kaolin clay. *Clays Clay Miner.* **24**, 64–68.

IMAI N., OTSUKA R., KASHIDE H. & HAYASHI H. (1969) Dehydration of palygorskite and sepiolite from the Kuzuu district, Tochigi Prefecture, Central Japan. *Proc. Int. Clay Conf.* 1969, Tokyo, Vol. 1, pp. 99–108. Israel Universities Press, Jerusalem.

IMAI N., OTSUKA R., NAKAMURA T. & INOUE, H. (1966) A new occurrence of well-crystallized sepiolite from the Kuzu district, Tochigi Prefecture, Central Japan, Nendo Kagaku. *J. Clay Sci. Soc. Japan* **6**, 30–40; *Min. Absts* 1968, **19**, 308.

INGAMELLS C.O. (1962) The application of an improved steam distillation apparatus to the determination of fluoride in rocks and minerals. *Talanta* **9**, 507–516.

INGAMELLS C.O. (1966) Absorptiometric methods in rapid silicate analysis. *Analyt. Chem.* **38**, 1228–1234.

INGAMELLS C.O. (1970) Lithium metaborate flux in silicate analysis. *Analyt. Chim. Acta* **52**, 323–334.

INGRAM B.L. (1970) Determination of fluorine in silicate rocks without separation of aluminium using a specific ion electrode. *Analyt. Chem.* **42**, 1825–1827.

IUPAC (International Union of Pure and Applied Chemistry) (1979) *Manual of Symbols and Terminology for Physicochemical Quantities and Units* (prepared by D.H. Whiffen) 41 pp. Pergamon Press, Oxford.

IWAO S. & UDAGAWA S. (1969) Pyrophyllite and "Roseki" clays. Ch. II in Part II *The Clays of Japan* (S. Iwao, ed. in chief). Geological Survey of Japan.

JACKSON M.L. (1956) *Soil Chemical Analysis – Advanced Course.* Published by the author. Dept. of Soils, University of Wisconsin, Madison, USA.

JAHANBAGLOO I.C. & ZOLTAI T. (1968) The crystal structure of a hexagonal Al-serpentine. *Am. Miner.* **53**, 14–24.

JEFFERY P.G. (1970) *Chemical Methods of Rock Analysis.* Pergamon Press, Oxford.

JEFFERY P.G. & WILSON A.D. (1960) Closed-circulation systems for determining water, carbon dioxide and total carbon in silicate rocks and minerals. *Analyst, Lond.* **85**, 749–755.

JENKINS H.D.B. & HARTMAN P. (1979) A new approach to the calculation of electrostatic energy relations in minerals: The dioctahedral and trioctahedral phyllosilicates. *Phil. Trans. Roy. Soc. Lond.* **A293**, 169–208.

JENKINS H.D.B. & HARTMAN P. (1980) Application of a new approach to the calculation of electrostatic energies of expanded di- and trioctahedral micas. *Phys. Chem. Minerals*, **6**, 313–325.

JENKINS H.D.B. & HARTMAN P. (1982) Calculations on a model intercalate containing a single layer of water molecules: A study of potassium vermiculite $K_{2x}Mg_6(Si_{4-x}Al_x)O_{20}(OH)_4 \cdot (H_2O)_4$ for $1 \leqslant x \leqslant 0$. *Phil. Trans. Roy. Soc. Lond.* **A304**, 397–446.

JEPSON W.B. & ROWSE J.B. (1975) The composition of kaolinite – an electron microscope microprobe study. *Clays Clay Miner.* **23**, 310–317.

KARPOVA G.V. (1966) Paragonite hydromicas in terrestrial rocks from the Great Donbasin. *Dokl. Akad. Nauk SSSR* **171**, 443–445.

KAUFFMAN A.J. JR. (1943) Fibrous sepiolite from Yavapai County, Arizona. *Am. Miner.* **28**, 512–520.

KELLER W.D. (1958) Glauconitic mica in the Morrison Formation in Colorado. *Clays Clay Miner.* **5**, 120–128.

KELLER W.D., PICKETT E.E. & REESMAN A.L. (1966) Elevated dehydroxylation temperature of the Keokuk geode kaolinite – a possible reference mineral. *Proc. Int. Clay Conf.* Jerusalem, Vol. 1, pp. 75–85. Israel Program for Scientific Translations, Jerusalem.

KELLEY W.P. (1945) Calculating formulas for fine grained minerals on the basis of chemical analysis. *Am. Miner.* **30**, 1–26.

KEUSEN H.R. & PETERS TJ. (1980) Preiswerkite, an Al-rich trioctahedral sodium mica from the Geisspfad ultramafic complex (Penninic Alps). *Am. Miner.* **65**, 1134–1137.

KHOURY H.N., MACKENZIE R.C., RUSSELL J.D. & TAIT J.M. (1984) An iron-free volkhonskoite. *Clay Minerals* **19**, 45–57.

KIENAST J.R. & VELDE B. (1968) Sur l'existence probable d'un nouveau type de substitution ionique affectant les paragonites. *C.r. hebd. Séanc. Acad. Sci., Paris* **267D**, 1909–1912.

KIRKMAN J.H. (1975) Clay mineralogy of some tephra beds of Rotorua Area, North Island, New Zealand. *Clay Minerals* **10**, 437–449.

KIRKMAN J.H. (1977) Possible structure of halloysite disks and cylinders observed in some New Zealand rhyolitic tephras. *Clay Minerals* **12**, 199–216.

KLINGSBERG C. & ROY R. (1957) Synthesis, stability and polytypes of nickel and gallium phlogopites. *Am. Miner.* **42**, 629–634.

KODAMA H. (1966) The nature of the component layers of rectorite. *Am. Miner.* **51**, 1035–1055.

KODAMA H. & BRYDON J.E. (1968) Dehydration of microcrystalline muscovite. *Trans. Faraday Soc.* **64**, 3112–3119.

KODAMA H., SHIMODA S. & SUDO T. (1969) Hydrous mica complexes: their structure and chemical composition, *Proc. Int. Clay Conf.* 1969, Tokyo, Vol. 1, pp. 185–196.

KOMUSINSKI J., STOCH L. & DUBIEL S.M. (1981) Application of electron paramagnetic resonance and Mössbauer spectroscopy in the investigation of kaolinite-group minerals. *Clays Clay Miner.* **29**, 23–30.

KRAMM U. (1980) Sudoite in low-grade manganese-rich assemblages. *Neues Jb. Miner. Abh.* **138**, 1–13.

LAGALY G. (1979) The "layer charge" of regular interstratified 2:1 clay minerals. *Clays Clay Miner.* **27**, 1–10.

LAGALY G. (1981) Characterization of clays by organic compounds. *Clay Minerals* **16**, 1–21.

LAGALY G. (1982) Layer charge heterogeneity in vermiculites. *Clays Clay Miner.* **30**, 215–222.

LAGALY G., FERNANDEZ GONZALEZ M. & WEISS A. (1976) Problems in layer-charge determination of montmorillonites. *Clay Minerals* **11**, 173–187.

LAGALY G. & WEISS A. (1976) The layer charge of smectitic layer silicates. *Proc. Int. Clay Conf. 1975, Mexico*, pp. 157–172. Applied Publishing Ltd, Wilmette, Ill., USA.
LAHAV N., SHANI U. & SHABTAI J. (1978) Cross-linked smectites. I Synthesis and properties of hydroxy-aluminium montmorillonite. *Clays Clay Miner.* **26**, 107–115.
LANGMYHR F.J. & PAUS P.E. (1968) The analysis of inorganic siliceous materials by atomic absorption spectrophotometry and the hydrofluoric acid decomposition technique. *Analyt. Chim. Acta* **43**, 397–408.
LANGMYHR F. J. & PAUS P.E. (1969) Hydrofluoric acid decomposition-atomic absorption analysis of nine silicate universal and rock reference samples. *Analyt. Chim. Acta* **47**, 371–373.
LANGMYHR F.J. & PAUS P.E. (1970) A bomb for the hydrofluoric decomposition of inorganic materials. *Analyt. Chim. Acta* **49**, 358–359.
LANGMYHR F.J. & SVEEN S. (1965) Decomposability in hydrofluoric acid of the main and some minor trace minerals of silicate rocks. *Analyt. Chim. Acta*, **32**, 1–7.
LANGSTON R.B. & PASK J.A. (1968) The nature of anauxite. *Clays Clay Miner.* **16**, 425–436.
LAPHAM D.M. (1958) Structural and chemical variation in chromium chlorite. *Am. Miner.* **43**, 921–956.
LARSEN E.S. & WHERRY E.T. (1917) Halloysite from Colorado. *J. Washington Acad. Sci.* **7**, 178–180.
LAZARENKO E.K. & KOROLEV YA.M. (1970) Tarasovite, a new dioctahedral ordered interlayered mineral. *Zapiski Vses. Obschch* **99**, 214–224.
LEE J.H. & GUGGENHEIM S. (1981) Single crystal X-ray refinement of pyrophyllite – 1Tc. *Am. Miner.* **66**, 350–357.
LEE S.Y., JACKSON M.L. & BROWN J.L. (1975) Micaceous occlusions in kaolinite observed by ultramicrotomy and high-resolution electron microscopy. *Clays Clay Miner.* **23**, 125–129.
LIM C.H., JACKSON M.L., KOONS R.D. & HELMKE P.A. (1980) Kaolins: sources of differences in cation exchange capacities and cesium retention. *Clays Clay Miner.* **28**, 223–229.
LINGANE J.J. (1967). A study of the lanthanum fluoride membrane electrode for end-point detection in titrations of fluoride with Th, La and Ca. *Analyt. Chem.* **39**, 881–887.
LONG J.V.P. (1977) Electron probe microanalysis. Ch. 6 in *Physical Methods in Determinative Mineralogy* (J. Zussman, ed.). Academic Press: London, New York, San Francisco.
LORIMER G.W. & CLIFF G. (1976) Analytical electron microscopy of minerals. Ch. 7 in *Electron Microscopy in Mineralogy* (H.-R. Wenk, coord. ed.). Springer-Verlag: Berlin, Heidelberg, New York.
LORIMER G.W., RAZIK N.A. & CLIFF G. (1973) The use of the analytical electron microscope EMMA-4 to study solute distribution in thin foils: some applications to metals and minerals. *J. Microscopy* **99**, 153–164.
MACEWAN D.M.C. & WILSON M.J. (1980) Interlayer and intercalation complexes of clay minerals. Ch. 3 in *Crystal Structures of Clay Minerals and their X-ray Identification* (G.W. Brindley and G. Brown, eds). Mineralogical Society, London.
MACKENZIE R.C. (1957) Saponite from Allt Ribhein, Fiskavaig Bay, Skye. *Mineralog. Mag.* **31**, 672–680.
MAEDA T., TAKENAKA H. & WARKENTIN B.P. (1977) Physical properties of allophanic soils. In *Advances in Agronomy* (N.C. Brady, ed.), Vol. 29, pp. 229–264. Academic Press, New York.
MAKSIMOVIC Z. (1966) The kerolite–pimelite series from Goles Mountain, Yugoslavia. *Proc. Int. Clay Conf. 1966*, Jerusalem, pp. 97–105, Vol. 1. Israel Program for Scientific Translations, Jerusalem.
MAKSIMOVIC Z. & BISH D.L. (1978) Brindleyite, a nickel-rich aluminous serpentine mineral analogous to berthierine. *Am. Miner.* **63**, 484–489.
MAKSIMOVIC Z. & BRINDLEY G.W. (1980) Hydrothermal alteration of a serpentine near Takova, Yugoslavia, to chromium-bearing illite/smectite, kaolinite, tosudite and halloysite. *Clays Clay Miner.* **28**, 295–302.
MAKSIMOVIC Z. & WHITE J.L. (1973) Infrared study of chromium-bearing halloysites. *Proc. Int. Conf. 1972*, Madrid, pp. 61–73. Division de Ciencias. CSIC, Madrid.
MAKSIMOVIC Z., WHITE J.L. & LOGAR M. (1981) Chromium-bearing dickite and chromium-bearing kaolinite from Teslic, Yugoslavia. *Clays Clay Miner.* **29**, 213–218.
MARTIN-RAMOS J.D. & RODRIGUEZ-GALLEGO M. (1982) Chromian mica from Sierra Nevada, Spain. *Mineralog. Mag.* **46**, 269–272.
MAVRODINEANU, R. (ed.) (1970) *Analytical Flame Spectroscopy: Selected Topics*. Macmillan, London.
MAXWELL J.A. (1968) *Rock and Mineral Analysis*. Interscience, London.
MCATEE J.L. (1958a) Heterogeneity in montmorillonite. *Clays Clay Miner.* **5**, 279–288.
MCATEE J.L. (1958b) Random interstratification in organophilic bentonites. *Clays Clay Miner.* **5**, 308–317.
MCKIE D. (1959) Yoderite, a new hydrous magnesium iron aluminosilicate from Mautia Hill, Tanganyika. *Mineralog. Mag.* **32**, 282–307.
MEADS R.E. & MALDEN P.J. (1975) Electron spin resonance in natural kaolinites containing Fe^{3+} and other transition metal ions. *Clay Minerals* **10**, 313–345.
MEDLIN J.H., SUHR N.H. & BODKIN J.B. (1969) Atomic absorption analysis of silicates using $LiBO_2$ fusion. *Atomic Abs Newsl.* **8**, 25–29.
MESTDAGH M.M., VIELVOYE L. & HERBILLON A.J. (1980) Iron in kaolinite. II. The relationship between kaolinite crystallinity and iron content. *Clay Minerals* **15**, 1–13.
MULLER G. (1963) Zur Kenntnis di-octaedrischer Vierschicht-Phyllosilikate (Sudoit-Reihe der Sudoit-Chlorit-Gruppe), *Proc. Int. Clay Conf.* Stockholm, Vol. 1, pp. 121–130. Pergamon Press, Oxford.
MUNOZ J.L. (1968) The physical properties of synthetic lepidolites. *Am. Miner.* **53**, 1490–1512.

NAGASAWA K., BROWN G. & NEWMAN A.C.D. (1974) Artificial alteration of biotite into a 14 Å layer silicate with hydroxy-aluminium interlayers. *Clays Clay Miner.* **22**, 241–252.

NAGATA H., SHIMODA S. & SUDO T. (1974) On dehydration of bound water of sepiolite. *Clays Clay Miner.* **22**, 285–293.

NAGY B. & BRADLEY W.F. (1955) The structural scheme of sepiolite. *Am. Miner.* **40**, 885–892

NARDOZZI M.J. & LEWIS L.L. (1961) Pyrolytic separation and determination of fluoride in raw materials. *Analyt. Chem.* **33**, 1261–1264.

NELSON B.W. & ROY R. (1958) Synthesis of the chlorites and their structural constitution. *Am. Miner.* **43**, 707–725.

NEMECZ E., VARJU G. & BARNA J. (1965) Allevardite from Kiralhegy, Tokaj Mountains, Hungary, *Proc. Int. Clay Conf.* 1963, Stockholm, Vol. 2, pp. 51–67. Pergamon Press, Oxford.

NEUMANN B.S. (1965) Behaviour of a synthetic clay in pigment dispersions. *Rheologica Acta* **4**, 250–255.

NEWMAN A.C.D. (1967) Changes in phlogopites during their artificial alteration. *Clay Minerals* **7**, 215–227.

NEWMAN A.C.D. (1968) A simple apparatus for separating fluorine from aluminosilicates by pyrohydrolysis. *Analyst, Lond.* **93**, 827–831.

NEWMAN A.C.D. (1969) Cation exchange properties of micas. I The relation between mica composition and potassium exchange in solutions of different pH. *J. Soil Sci.* **20**, 357–373.

NEWMAN A.C.D. & BROWN G. (1966) Chemical changes during the alteration of micas. *Clay Minerals* **6**, 297–310.

NISHIYAMA T. & SHIMODA S. (1981) Ca-bearing rectorite from Tooho mine, Japan. *Clays Clay Miner.* **29**, 236–240.

NISHIYAMA T., SHIMODA S., SHIMOSAKA K. & KANAOKA S. (1975) Lithium bearing tosudite. *Clays Clay Miner.* **23**, 337–342.

NORRISH K. (1952) A determination of the crystal structure of cookeite, Part 2. Ph.D. Thesis. University of London.

NORRISH K. (1973a) Factors in the weathering of mica to vermiculite. *Proc. Int. Clay Conf.* 1972, Madrid, pp. 417–432. Division de Ciencias, Madrid.

NORRISH K. (1973b) Forces between clay particles. *Proc. Int. Clay Conf.* 1972, Madrid, pp. 375–383. Division de Ciencias, Madrid.

NORRISH K. & CHAPPELL B.W. (1977) X-ray fluorescence spectrometry. Ch. 5 in *Physical Methods in Determinative Mineralogy* (J. Zussman, ed.). Academic Press: London, New York, San Francisco.

NORRISH K. & PICKERING J.G. (1983) Clay minerals. Ch. 22 in *Soils: an Australian Viewpoint*, pp. 281–308. Division of Soils, CSIRO. CSIRO: Melbourne/Academic Press: London.

OKADA K., MORIKAWA S., IWAI S., OHIRA Y. & OSSAKA J. (1975) A structure model of allophane. *Clay Sci.* **4**, 291–303.

OLSEN E.J. (1961) Six-layer ortho-hexagonal serpentine from the Labrador Trough. *Am. Miner.* **46**, 434–438.

OTSUKA R., IMAI N. & NISHIKAWA M. (1966) On the dehydration of sepiolite from the Akatani Mine, Niigata Prefecture, Japan. *J. Chem. Soc. Japan* (Industr. Chem. Soc.) **66**, 1677–1680; *Miner. Absts* 1967, **18**, 155.

PAGE N.J. (1968) Chemical differences among the serpentine "polymorphs". *Am. Miner.* **53**, 201–215.

PATTIARATCHI D.B., SAARI E. & SAHAMA T.G. (1967) Anandite, a new barium iron silicate from Wilagedera, North Western Province, Ceylon. *Mineralog. Mag.* **36**, 1–4.

PAUS P.E. (1971) New equipment for decomposition of inorganic materials. *Atomic Abs Newsl.* **10**, 44.

PEACOR D.R. & ESSENE E.J. (1980) Caryopilite – a member of the friedelite rather than the serpentine group. *Am. Miner.* **65**, 335–339.

PEACOR D.R., ESSENE E.J., SIMMONS W.B. JR. & BIGELOW W.C. (1974) Kellyite, a new Mn–Al member of the serpentine group from Bald Knob, North Carolina, and new data on grovesite. *Am. Miner.* **59**, 1153–1156.

PECK L.C. & SMITH V.C. (1964) Spectrophotometric determination of fluorine in silicate rocks. *Talanta*, **11**, 1343–1347.

PERRAULT G., HARVEY Y. & PERBSWSKY R. (1975) La yofortierite, un nouveau silicate hydrate de manganese de St.-Hilaire, P.Q. *Can. Mineralogist* **13**, 68–74.

PEVEAR D.R., WILLIAMS V.E. & MUSTOE G.E. (1980) Kaolinite, smectite and K-rectorite in bentonites: relation to coal rank at Tulameen, British Columbia. *Clays Clay Miner.* **28**, 241–254.

PHILLIPS T.L., LOVELESS J.K. & BAILEY S.W. (1980) Cr^{3+} coordination in chlorites: a structural study of 10 chromian chlorites. *Am. Miner.* **65**, 112–122.

PLANÇON A. & TCHOUBAR C. (1977) Determination of structural defects in phyllosilicates by X-ray powder diffraction. II Nature and proportion of defects in natural kaolinites. *Clays Clay Miner.* **25**, 436–450.

PREISINGER A. (1963) Sepiolite and related compounds: its stability and application. *Clays Clay Miner.* **10**, 365–371.

PRIDER R.T. (1939) Some minerals from the leucite-rich rocks of the West Kimberley area, Western Australia. *Mineralog. Mag.* **25**, 373–387.

QUAKERNAAT J. (1970) A new occurrence of a macrocrystalline form of saponite. *Clay Minerals* **8**, 491–493.

RADOSLOVICH E.W. (1963) The cell dimensions and symmetry of layer-lattice silicates. V. Composition limits. *Am. Miner.* **48**, 348–366.

RAMIREZ-MUNOZ J. (1968) *Atomic Absorption Spectroscopy and Analysis by Atomic Absorption Flame Photometry*. Elsevier, London.

RAUSELL COLOM J.A., SWEATMAN T.R., WELLS C.B. & NORRISH K. (1965) Studies in the artificial weathering of mica. In *Experimental Pedology* (E.G. Hallsworth and D.V. Crawford, eds.). *Proc. 11th School Agric. Sci., Nottingham.* Butterworth, London.

RAUTUREAU M. & TCHOUBAR C. (1976) Structural analysis of sepiolite by selected area diffraction – relations with physico-chemical properties. *Clays Clay Miner.* **24**, 43–49.

RAYNER J.H. & BROWN G. (1966) Structure of pyrophyllite. *Clays Clay Miner.* **13**, 73–84.

RAYNER J.H. & BROWN G. (1973) The crystal structure of talc. *Clays Clay Miner.* **21**, 103–114.
RENGASAMY P. (1976) Substitution of iron and titanium in kaolinites. *Clays Clay Miner.* **24**, 265–266.
REYNOLDS R.C. (1980) Interstratified clay minerals. Ch. 4 in *Crystal Structures of Clay Minerals and their X-ray Identification* (G.W. Brindley and G. Brown, eds), pp. 249–303. Mineralogical Society, London.
REYNOLDS R.C. & HOWER J. (1970) The nature of interlayering in mixed-layer illite–montmorillonites. *Clays Clay Min.* **18**, 25–36.
RHOADES J.D. & COLEMAN N.T. (1967) Interstratification in vermiculite and biotite produced by potassium sorption: I Evaluation by simple X-ray diffraction pattern inspection. *Soil Sci. Soc. Am. Proc.* **31**, 366–372.
RICH C.I. (1968) Hydroxy interlayers in expansible layer silicates. *Clays Clay Miner.* **16**, 15–30.
RIEDER R. (1970) with chemical analyses by M. Huka, D. Kucerova, L. Minarik, J. Obermajer and P. Povondra). Chemical composition and physical properties of lithium–iron micas from the Krusne hory Mts (Erzgebirge). *Contr. Miner. Petrol.* **27**, 131–158.
ROBERTSON R.H.S., BRINDLEY G.W. & MACKENZIE R.C. (1954) Mineralogy of kaolin clays from Pugu, Tanganyika. *Am. Miner.* **39**, 118–138.
ROGERS L.E.R., MARTIN A.E. & NORRISH K. (1954). The occurrence of palygorskite, near Ipswich, Queensland. *Mineralog. Mag.* **30**, 534–540.
ROGERS L.E.R., QUIRK J.P. & NORRISH K. (1956) Occurrence of an aluminium–sepiolite in a soil having unusual water relationships. *J. Soil Sci.* **7**, 177–184.
ROSENBERG P.E. (1974) Pyrophyllite solid solutions in the system Al_2O_3–SiO_2–H_2O. *Am. Miner.* **59**, 254–260.
ROSENBERG P.E. & CLIFF G. (1980) The formation of pyrophyllite solid solutions. *Am. Miner.* **65**, 1217–1219.
ROSENFELD J.L. (1956) Paragonite in the schist of Glebe Mountain, Southern Vermont. *Am. Miner.* **41**, 144–147.
ROSS C.S. (1946) Sauconite – a clay mineral of the montmorillonite group. *Am. Miner.* **31**, 411–424.
ROSS C.S. (1960) Review of the relationships in the montmorillonite group of clay minerals. *Clays Clay Miner.* **7**, 225–229.
ROSS C.S. & HENDRICKS S.B. (1945) Minerals of the montmorillonite group; their origin and relation to soils and clays. *Prof. Pap. U.S. Geol. Surv.* **205-B**, 23–79.
ROUSSEAUX J.M., NATHAN Y., VIELVOYE L.A. & HERBILLON A. (1973) The vermiculization of trioctahedral micas. II Correlations between the K level and crystallographic parameters. *Proc. Int. Clay Conf.* 1972, Madrid, pp. 449–456. Division de Ciencias, Madrid.
ROY D.M. & ROY R. (1954) An experimental study of the formation and properties of synthetic serpentines and related layer silicate minerals. *Am. Miner.* **39**, 957–975.
ROZHDESTVENSKAYA I.V. & FRANK-KAMENETSKII V.A. (1974) The structure of the dioctahedral mica chernykhite. In *Crystal Chemistry and Structure of Minerals*, pp. 18–23. Izdat. Nauka, Leningrad.
RUBESKA I. & MOLDAN B. (1969) *Atomic Absorption Spectrophotometry* (translated by P.T.Woods). Iliffe, London.
SAKHAROV B.A. & DRITS V.A. (1973) Mixed-layer kaolinite–montmorillonite: a comparison of observed and calculated diffraction patterns. *Clays Clay Miner.* **21**, 15–17.
SAWHNEY B.L. (1968) Aluminium interlayers in layer silicates. Effect of OH/Al ratio of Al solution, time of reactions, and type of structure. *Clays Clay Miner.* **16**, 157–163.
SCHAFER H.N.S. (1966) The determination of iron(II) oxide in silicate and refractory materials. *Analyst, Lond.* **91**, 755–762.
SCHALLER W.T. (1950) An interpretation of the composition of high-silica sericites. *Mineralog. Mag.* **29**, 406–415.
SCHALLER W.T., CARRON M.K. & FLEISCHER M. (1967) Ephesite, $Na(LiAl_2)Al_2Si_2O_{10}(OH)_2$, a trioctahedral member of the margarite group, and related brittle micas. *Amer. Min.* **52**, 1689–1696.
SCHMIDT E.R. & HECKROODT R.O. (1959) A dickite with elongated habit and its dehydroxylation. *Mineralog. Mag.* **32**, 314–323.
SCHMIDT E.R. & HEYSTEK H. (1953) A saponite from Krugersdorp district, Transvaal. *Mineralog. Mag.* **30**, 201–210.
SCHOFIELD R.K. & SAMSON H.R. (1954) Flocculation of kaolinite due to the attraction of oppositely-charged crystal faces. *Disc. Faraday Soc.* **18**, 138–145.
SCHREYER W., ABRAHAM K. & KULKE H. (1980) Natural sodium phlogopite coexisting with potassic phlogopite and sodian aluminian talc in a metamorphic evaporite sequence from Derrag, Tell Atlas, Algeria. *Contr. Miner. Petrol.* **74**, 223–233.
SCHREYER W., MEDENBACH O., ABRAHAM K., GEBERT W. & MULLER W.F. (1982) Kulkeite, a new metamorphic phyllosilicate mineral: ordered 1:1 chlorite/talc mixed-layer. *Contr. Miner. Petrol.* **80**, 103–109.
SCHULTZ L.G. (1969) Lithium and potassium absorption, dehydroxylation temperature and structural water content of aluminous smectites. *Clays Clay Miner.* **17**, 115–149.
SCHULTZ L.G., SHEPARD A.O., BLACKMON P.D. & STARKEY H.C. (1971) Mixed-layer kaolinite–montmorillonite from the Yucatan Peninsula, Mexico. *Clays Clay Miner.* **19**, 137–150.
SCHULZE D.G. & DIXON J.B. (1979) High gradient magnetic separation of iron oxides and other magnetic minerals from soil clays. *Soil Sci. Soc. Am. J.* **43**, 793–799.
SCHWERTMANN U. (1979) Dissolution Methods. In *Data Handbook for Clay Materials and Other Non-metallic Minerals* (H. van Olphen and J.J. Fripiat, eds), pp. 163–176. Pergamon Press, Oxford.
SERNA C., VAN SCOYOC G.E. & AHLRICHS J.L. (1977) Hydroxyl groups and water in palygorskite. *Am. Miner.* **62**, 784–792.

SHABTAI J., ROSELL M. & TOKARZ M. (1984) Cross-linked smectites. III Synthesis and properties of hydroxy-aluminium hectories and fluorhectorites. *Clays Clay Miner.* **32**, 99–107.
SHAPIRO L. & BRANNOCK W.W. (1955) Rapid determination of water in silicate rocks. *Analyt. Chem.* **27**, 560–562.
SHIMODA S. (1969) New data for tosudite. *Clays Clay Miner.* **17**, 179–184.
SHIMODA S. (1971) Mineralogical studies of a species of stevensite from the Obori mine, Yamagata Prefecture, Japan. *Clay Minerals* **9**, 185–192.
SHIMODA S. (1978) Interstratified minerals, Ch. 8 In *Clays and Clay Minerals of Japan* (T. Sudo and S. Shimoda, ed), pp. 265–322. Elsevier, Amsterdam.
SHIMOYAMA A., JOHNS W.D. & SUDO T. (1969) Montmorillonite–kaolin clay in acid clay deposits from Japan. *Proc. Int. Clay Conf. 1969, Japan*, Vol. 1, pp. 225–231. Israel Program for Scientific Translations, Jerusalem.
SHIROZU H. (1958) X-ray powder patterns and cell dimensions of some chlorites in Japan, with a note on their interference colours. *Miner. J. Japan* **2**, 209.
SHIROZU H. (1978) Chlorite minerals. Ch. 7 in *Clays and Clay Minerals of Japan* (T. Sudo and S. Shimoda, eds), pp. 243–262. Elsevier, Amsterdam.
SLAVIN W. (1968) *Atomic Absorption Spectroscopy*. Interscience, London.
SMITH W.C., BANNISTER F.A. & HEY M.H. (1946) Pennantite, a new manganese-rich chlorite from Benallt mine, Rhiw, Caernarvonshire. *Mineralog. Mag.* **27**, 217–220.
SPEAKMAN K. & MAJUMBAR A.J. (1971) Synthetic "deweylite". *Mineralog. Mag.* **38**, 225–234.
SPEAR F.S., HAZEN R.M. & RUMBLE D. III (1981) Wonesite: a new rock-forming silicate from the Post Pond volcanics, Vermont. *Am. Miner.* **66**, 100–105.
SPRINGER G. (1976) Falcondoite, nickel analogue of sepiolite. *Can. Mineralogist* **14**, 407–409.
SRODON J. (1980) Synthesis of mixed-layer kaolinite/smectite. *Clays Clay Miner.* **28**, 419–424.
STEINFINK H. (1962) Crystal structure of a trioctahedral mica, phlogopite. *Am. Miner.* **47**, 886–896.
STEPHEN I. (1954) An occurrence of palygorskite in the Shetland Isles. *Mineralog. Mag.* **30**, 471–480.
STEPHEN I. & MACEWAN D.M.C. (1950) "Swelling chlorite". *Geotechnique* **2**, 82–83.
STERNE E.J., REYNOLDS R.C. & ZANTOP H. (1982) Natural ammonium illites from black shales hosting a stratiform base metal deposit, DeLong Mountains, Northern Alaska. *Clays Clay Miner.* **30**, 161–166.
STUBICAN V. & ROY R. (1962) Boron substitution in synthetic micas and clays. *Am. Miner.* **47**, 1166–1173.
SUDO T. (1978) An outline of clays and clay minerals in Japan. Ch. 1 in *Clays and Clay Minerals of Japan* (T. Sudo and S. Shimoda, eds), pp. 1–103. Elsevier, Amsterdam.
SUDO T. & HAYASHI H. (1956) A randomly interstratified kaolin–montmorillonite in acid clay deposits in Japan. *Nature, Lond.* **178**, 1115–1116.
SUDO T., KURABAYASHI S., TSUCHIYA T. & KANNEKO S. (1964) Mineralogy and geology of Japanese volcanic ash soils. *Trans. 8th Int. Congr. Soil Sci.* Bucarest, Romania, Vol. 3, pp. 1095–1104.
SUDO T. & SATO T. (1966) Dioctahedral chlorite, *Proc. Int. Clay Conf. 1966*, Jerusalem, Vol. 1, pp. 33–39. Israel Program for Scientific Translation, Jerusalem.
SUDO T. & SHIMODA S. (eds) (1978) *Clays and Clay Minerals of Japan*. Kodansha/Elsevier, Amsterdam.
SUHR N.H. & INGAMELLS C.O. (1966) Solution technique for analysis of silicates. *Analyt. Chem.* **38**, 730–734.
SUQUET H., DE LA CALLE C. & PEZERAT H. (1975) Swelling and structural organization of saponite. *Clays Clay Miner.* **23**, 1–9.
SUQUET H., IIYAMA J.T., KODAMA H. & PEZERAT H. (1977) Synthesis and swelling properties of saponites with increasing layer charge. *Clays Clay Miner.* **25**, 231–242.
SWEATMAN T.R. & LONG J.V.P. (1969) Quantitative electron-probe microanalysis of rock-forming minerals. *J. Petrology* **10**, 332–379.
SWINDALE L.D. & HUGHES I.R. (1968) Hydrothermal association of pyrophyllite, kaolinite, diaspore, dickite, and quartz in the Coromandel area, New Zealand. *J. Geology and Geophysics, New Zealand* **11**, 1163–1183.
TAKAHASHI H. (1956) Occurrence of sepiolite from the Karasawa Mine, Tochigi Prefecture. *J. Jap. Ass. Min. Petr. and Econ. Geol.* **56**, 187–190; *Miner. Absts.* 1967, **18**, 242.
TAMURA T. (1956) Weathering of mixed-layer clays in soils. *Clays Clay Miner.* **4**, 413–422.
TAZAKI K. (1982) Analytical electron microscopic studies of halloysite formation processes – morphology and composition of halloysite. *Proc. Int. Clay Conf. 1981*, Italy, pp. 573–584. Elsevier, Amsterdam.
TETTENHORST R. & JOHNS W.D. (1965) Interstratification in montmorillonite. *Clays Clay Miner.* **13**, 85–93.
THA HLA (1945) Electrodialysis of mineral silicates: an experimental study of rock-weathering. *Mineralog. Mag.* **27**, 137–145.
THOMPSON G.R. & HOWER J. (1975) The mineralogy of glauconite. *Clays Clay Miner.* **23**, 289–300.
THOMPSON M. & WALSH J.N. (1983) *A Handbook of I.C.P. Spectrometry*. Black and Son Ltd.
TIEN P-L., LEAVENS P.B. & BELEN J.A. (1975) Swinefordite, a dioctahedral–trioctahedral Li-rich member of the smectite group from Kings Mountain, North Carolina. *Am. Miner.* **60**, 540–547.
UYEDA N., PHAM THI HANG & BRINDLEY G.W. (1973) The nature of garnierites. II Electron-optical study. *Clays Clay Miner.* **21**, 41–50.
VAN LOON J.C. & PARISSIS C.M. (1969) Scheme of silicate analysis based on the lithium metaborate fusion followed by atomic-absorption spectrophotometry. *Analyst, Lond.* **94**, 1057–1062.

VAN OLPHEN H. & FRIPIAT J.J. (eds) (1979) *Data Handbook for Clay Materials and Other Non-Metallic Minerals.* Pergamon Press, Oxford.

VAN SCOYOC G.E., SERNA C.J. & AHLRICHS J.L. (1979) Structural changes in palygorskite during dehydration and dehydroxylation. *Am. Miner.* **64**, 216–223.

VEBLEN D.R. (1983) Exsolution and crystal chemistry of the sodium mica, wonesite. *Am. Miner.* **68**, 554–565.

VEBLEN D.R. (1983b) Microstructures and mixed layering in intergrown wonesite, chlorite, talc, biotite, and kaolinite. *Am. Miner.* **68**, 556–580.

VELDE B. (1971) The stability and natural occurrence of margarite. *Mineralog. Mag.* **38**, 317–323.

VELDE B. (1977) *Clay and clay minerals in natural and synthetic systems.* Elsevier, Amsterdam.

VENIALE F. & VAN DER MAREL H.W. (1969) Identification of some 1:1 regular interstratified trioctahedral clay minerals. *Proc. Int. Clay Conf. 1969, Tokyo,* Vol. 1, pp. 233–244. Israel Program for Translations, Jerusalem.

VOLBORTH A. (1969) *Elemental Analysis in Geochemistry. Part A, Major elements.* Elsevier Publ. Co., London.

WADA K. (1967) A structural scheme for soil allophane. *Am. Miner.* **52**, 690–708.

WADA, K. (1977) Allophane and imogolite. Ch. 16 in *Minerals in Soil Environments* (J.B. Dixon, S.B. Weed, J.A. Kittrick, M.H. Milford and J.L. White, eds). Soil Science Society of America, Madison, Wisconsin, USA.

WADA K. (1978) Allophane and imogolite. Ch. 4 in *Clays and Clay Minerals of Japan* (T. Sudo and S. Shimoda, eds), pp. 147–187. Elsevier, Amsterdam.

WADA S-I. & MIZOTA C. (1982) Iron-rich halloysite (10 Å) with crumpled lamellar morphology from Hokkaido, Japan. *Clays Clay Miner.* **30**, 315–317.

WALKER G.F. (1950) Trioctahedral minerals in the soil clays of north-east Scotland. *Mineralog. Mag.* **29**, 72–84.

WALSH J.N. (1980) A simultaneous determination of the major, minor and trace constituents of silicate rocks using inductively coupled plasma spectrometry. *Spectrochim. Acta* **35B**, 107–111.

WALSH J.N. & HOWIE R.A. (1980) An evaluation of the performance of an inductively coupled plasma source spectrometer for the determination of major and trace constituents of silicate rocks and minerals. *Mineralog. Mag.* **43**, 967–974.

WARDLE R. & BRINDLEY G.W. (1972) The crystal structure of pyrophyllite, 1Tc and its dehydroxylate. *Am. Miner.* **57**, 732–750.

WASHINGTON H.S. (1918) *Manual of the Chemical Analysis of Rocks* (3rd ed.). Wiley, New York.

WEAVER C.E. (1968) Electron microprobe study of kaolin. *Clays Clay Miner.* **16**, 187–189.

WEAVER C.E. (1976) The nature of TiO_2 in kaolinite. *Clays Clay Miner.* **24**, 215–218.

WEAVER C.E. & POLLARD L.D. (1973) *The Chemistry of Clay Minerals.* Elsevier, Amsterdam.

WEIR A.H. (1965) Potassium retention in montmorillonite. *Clay Minerals* **6**, 17–22.

WEIR A.H. & RAYNER J.H. (1974) An interstratified illite–smectite from Denchworth series soil in weathered Oxford clay. *Clay Minerals* **10**, 173–187.

WEISS A., BECKER H.O. & LAGALY G. (1970) Determination of charge density sequence in regular interstratified mica-type silicates by means of their n-alkylammonium derivatives: I Layer charge sequence in allevardite from Hungary. *Proc. Int. Clay Conf. 1969, Tokyo,* Vol. II, pp. 67–73. Israel Program for Scientific Translations, Jerusalem.

WEISS A., KOCH G. & HOFFMANN U. (1955) Zur Kenntnis von Saponit. *Berichte der deutschen keramischen Gesellschaft* **32**, 12–17.

WHITTAKER E.J.W. & WICKS F.J. (1970) Chemical differences among the serpentine "polymorphs": a discussion. *Am. Miner.* **55**, 1025–1047.

WICKS F.J. & WHITTAKER E.J.W. (1975) A reappraisal of the structures of the serpentine minerals. *Can. Mineralogist* **13**, 227–243.

WIEWIORA A. (1973) Mixed-layer kaolinite–smectite from Lower Silesia, Poland: final report, *Proc. Int. Clay Conf. 1972, Madrid,* pp. 75–88. Division de Ciencias, Madrid.

WILLARD H.H. & WINTER O.B. (1933) Volumetric method for determination of fluorine. *Ind. Engng Chem. Analyt. Edn.* **5**, 7–10.

WILSON A.D. (1962) The determination of total water in rocks by a simple diffusion method. *Analyst, Lond.* **87**, 598–600.

WILSON M.A., BARRON P.F. & CAMPBELL A.S. (1984) Detection of aluminium coordination in soils and clay fractions using ^{27}Al magic-angle spinning n.m.r. *J. Soil Sci.* **35**, 210–217.

WINEFORDNER J.D., SVOBODA V. & CLINE L.J. (1970) A critical comparison of atomic emission, atomic absorption and atomic fluorescence flame spectrometry. *Critical Rev. Analyt. Chem.* **1**, 223–274.

WOLLAST R., MACKENZIE F.T. & BRICKER O.P. (1968) Experimental precipitation and genesis of sepiolite at earth-surface conditions. *Am. Miner.* **53**, 1645–1662.

YAMANAKA S. & BRINDLEY G.W. (1978) Hydroxy-nickel interlayering in montmorillonites by titration method. *Clays Clay Miner.* **26**, 21–24.

YAMANAKA S. & BRINDLEY G.W. (1979) High surface area solids obtained by reaction of montmorillonite with zirconyl chloride. *Clays Clay Miner.* **27**, 119–124.

YAMANAKA S., YAMASHITA G. & HATTORI M. (1980) Reaction of hydroxybismuth polycations with montmorillonite. *Clays Clay Miner.* **28**, 281–284.

YODER H.S. (1952) The MgO–Al$_2$O$_3$–SiO$_2$–H$_2$O system and the related metamorphic facies. *Am. J. Sci.* Bowen volume, 559–627.

YODER H.S. & EUGSTER H.P. (1955) Synthetic and natural muscovites. *Geochim. cosmochim. Acta* **6**, 157–185.

YOSHIMURA T. (1967) *Part 1, Manganese mineralization, minerals and ores* (Supplement to Manganese Ore Deposits of Japan). *Scientific Reports of the Faculty of Science, Kyushu University*, **9**, Series D, 485 pp.

YOSHINAGA N. & AOMINE S. (1962) Imogolite in some Ando soils. *Soil Sci. Plant Nutr.* **8**, 6–13.

YULE J.W. & SWANSON G.A. (1969) A rapid method for decomposition and the analysis of silicates and carbonates by atomic absorption spectrophotometer. *Atomic Abs Newsl.* **8**, 30–33.

ZEN E-AN & ALBEE A.L. (1964) Coexistent muscovite and paragonite in pelitic schists. *Am. Miner.* **49**, 904–925.

ZEN E-AN, ROSS M. & BEARTH P. (1964) Paragonite from Tasch valley near Zermatt, Switzerland. *Am. Miner.* **49**, 183–190.

ZUSSMAN J. (1954) Investigation of the crystal structure of antigorite. *Mineralog. Mag.* **30**, 498–512.

ZUSSMAN J., BRINDLEY G.W. & COMER J.J. (1957) Electron diffraction studies of serpentine minerals. *Am. Miner.* **42**, 133–153.

Chapter 2

Non-Silicate Oxides and Hydroxides

R. M. TAYLOR

	Page		Page
2.1 INTRODUCTION	129	2.4.4 Association of aluminium oxides with other soil components	172
2.2 GENERAL PROPERTIES	130	2.4.5 The dissolution of aluminium oxides	173
2.2.1 Structure	130	2.5 THE MANGANESE OXIDE GROUP OF MINERALS	173
2.2.2 Surface charge	133	2.5.1 Formation of manganese oxides	175
2.2.3 Ion adsorption	135	2.5.2 The manganese oxides present in soils	176
2.2.4 Isoelectric point	140	2.5.2.1 Birnessite	177
2.2.5 General discussion	140	2.5.2.2 Vernadite	178
2.3 THE IRON OXIDE GROUP OF MINERALS	141	2.5.2.3 Lithiophorite	179
2.3.1 Individual oxides	142	2.5.2.4 Todorokite	179
2.3.1.1 Goethite	142	2.5.2.5 Hollandite, cryptomelane and coronadite	181
2.3.1.2 Ferrihydrite	147	2.5.2.6 Pyrolusite	181
2.3.1.3 Feroxyhite	148	2.5.2.7 Amorphous manganese oxides	182
2.3.1.4 Hematite	148	2.5.3 Cation and anion adsorption on manganese oxide surfaces	182
2.3.1.5 Lepidocrocite	149	2.5.4 Dissolution of manganese oxides	185
2.3.1.6 Green rusts	151	2.6 THE TITANIUM GROUP OF MINERALS	186
2.3.1.7 Magnetite and maghemite	152	2.6.1 Titanium minerals in soils	186
2.3.1.8 Akaganeite	155	2.6.1.1 Rutile	187
2.3.2 Adsorption by soil iron oxides	155	2.6.1.2 Anatase	187
2.3.3 Dissolution treatments for iron oxides	159	2.6.1.3 Brookite	187
2.3.4 Association of iron oxides with other soil components	160	2.6.1.4 Perovskite	188
2.4 THE ALUMINIUM OXIDE GROUP OF MINERALS	161	2.6.1.5 Ilmenite	188
2.4.1 Forms of aluminium oxide in soils	163	2.6.2 Properties of soil titanium oxide minerals	188
2.4.1.1 Gibbsite (hydrargillite)	163	NOTES	189
2.4.1.2 Bayerite	165	REFERENCES	190
2.4.1.3 Nordstrandite	166		
2.4.1.4 Boehmite	167		
2.4.1.5 Diaspore	168		
2.4.1.6 Corundum	169		
2.4.2 Compositional variations within the soil aluminium oxides	169		
2.4.3 Adsorption by soil aluminium oxides	170		

2.1 INTRODUCTION

In recent years the group of minerals referred to as "soil oxides" has received increasing attention because their presence in a profile often gives some insight into the pedogenic conditions prevailing during their formation, and, more importantly, because of the realization of their contribution to the retention and release of necessary nutrient elements. The term "oxide" is taken here, and in most other studies, to include metal hydroxides, oxy-hydroxides and hydrous oxides (where non-stoichiometric water is included in the structure). They are not always simple compounds like rutile (TiO_2), or goethite (α-FeOOH), but sometimes include an essential second cation of the same element in a different valence state, as in magnetite ($Fe^{2+}Fe_2^{3+}O_4$), or birnessite in which Mn may exercise three valencies. The second metal cation may also be of a different element, e.g. ilmenite, $Fe^{2+}Ti^{4+}O_3$. In other minerals "foreign" metal cations in minor amounts are often necessary for the formation and stabilization of a particular mineral, an example being Al and Li in lithiophorite, $(Al,Li)MnO_2(OH)_2$. These foreign elements may not even appear in the mineral formula, and in other instances, for example lithiophorite, there is no assignment of the relative amounts present.

Because of the variability of the chemical environment during the formation of these soil oxides, and their frequent ability to incorporate foreign ions into their structure, these soil minerals are generally impure and differ from pure specimen minerals or synthesized samples not only in composition but also frequently in morphology. Although these minerals may have a low overall concentration in the soil, they generally exhibit a relatively high surface area, which enhances their adsorption capacities and their susceptibility to dissolution when changes in environmental Eh and pH occur. In many cases this high surface area may arise from the restriction of crystal growth caused by the incorporation or adsorption of minor amounts of foreign ions.

In this chapter the oxides of Fe, Mn, Al and Ti will be discussed. Generally, many polymorphic forms of these oxides have been recognized, some being high temperature phases, while others have only been produced synthetically under conditions far removed from those encountered during weathering and pedogenesis. From the point of relevance to this chapter, only mineral forms that are common in soils will be discussed in detail, and because of the variation in their relative abundancies and the ease with which the different oxides can be isolated or synthesized, more attention will naturally be given to the most common mineral forms and those which have been studied more extensively. Certain general properties like structure and surface adsorption apply to this group of minerals as a whole and are outlined in the following section.

2.2. GENERAL PROPERTIES

2.2.1 Structure

If we consider a two-dimensional plane[1] of oxygen and/or hydroxyls in their closest packed arrangement (e.g. layer A in Fig. 2.1(a)), and overlay it with a similar layer, octahedrally coordinated interstices are created when the two-plane structure assumes its minimum volume. This double layer, with a proportion of the octahedral vacancies occupied by a smaller metallic cation, is the basic structural unit of the *simple* oxides discussed. The stacking arrangement of subsequent oxygen/hydroxyl planes, the continuity of the structure in the direction of stacking and the extent and manner in which the vacancies are occupied give rise to the different minerals.

The second oxygen/hydroxyl plane (B), see Fig. 2.1, in coming to the position of minimum separation from layer A will have the centres of its constituent atoms above either type 1 or type 2 triangular interstices in the top surface of layer A, without any real difference in the resultant structure. Suppose we chose type 1 interstice. The third plane (C) can again be positioned with the centres of its atoms sitting above either type of interstice in the top surface of layer B. If the atoms occupy type 2 interstice, Fig. 2.1(a), they are then directly above the oxygens in the A plane, and the sequence of the A, B and C planes in fact becomes ABAB.... This is seen in Fig. 2.1(a) by the dashed circle shown in plane A above which would lie atom Z from plane C. This stacking sequence gives hexagonal close packing (hcp). If, however, the atoms in plane C again occupied type 1 interstices in the top surface of B we would have an ABCABC... sequence, constituting a cubic close packing (ccp). This is seen in Fig. 2.1(b) where atom Z in layer C now

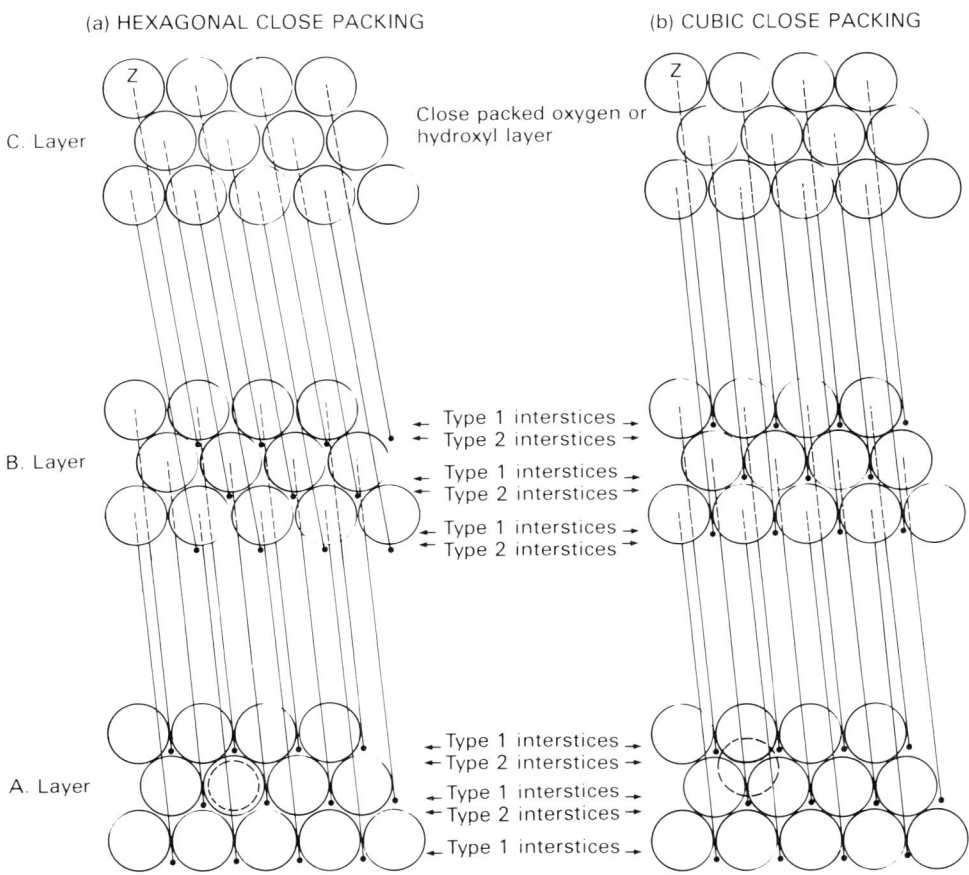

FIG. 2.1. Translocation of atom centres in bringing oxygen layers together to form (a) hexagonal close packing, and (b) cubic close packing.

occupies a different site (dashed circle in layer A) to that for the hcp. The two types of packing differ little in lattice energies and represent an equally economical filling of space (Evans 1964).

From Table 2.1 it can be seen that most of the minerals to be discussed in this chapter have a close packed oxygen/hydroxyl octahedral sheet as their basic unit, but the degree and ordering of sites occupied by the metal cations creates differences between the structures. This occupancy is determined by the valence exercised by the cation and whether hydroxyl substitute in part or totally for oxygen atoms. This can be demonstrated by reference to hematite, α-Fe_2O_3, and goethite, αFeOOH, both of which have trivalent Fe octahedrally coordinated in hcp oxygen layers. However, in goethite Fe is surrounded by 3 (OH) and 3 O rather than 6 O as in hematite (Busing and Levy 1958), resulting in 12 and 16 Fe atoms per 24 oxygen atoms for the respective two minerals. In hematite each oxygen is coordinated to 4 Fe, whereas in goethite, replacement of an O by OH reduces its ability to be coordinated to a fourth neighbouring Fe. This causes a reduction and resultant re-ordering of the octahedral occupancy, but bonding through the structure is continued by hydrogen bonding to a non-hydroxyl oxygen neighbour. These effects alter bond lengths and cause slight distortion of the octahedral layers.

TABLE 2.1. Formulae, packing sequences and occurrences of common soil oxides, hydroxides and oxy-hydroxides of Fe, Al, Mn and Ti

Oxide group	Name	Formula	Packing	Isostructural minerals and remarks
Fe	Goethite	α-FeOOH	hcp	Diaspore, ramsdellite, groutite (α-MnOOH)
	Lepidocrocite	γ-FeOOH	ccp	Boehmite
	Akaganeite	β-FeOOH		Hollandite, cryptomelane
	Ferrihydrite	$H_1Fe_5O_8 \cdot 4H_2O$	hcp	
	Feroxyhite	δ-FeOOH	hcp	
	Hematite	α-Fe$_2$O$_3$	hcp	Corundum, ilmenite
	Maghemite	γ-Fe$_2$O$_3$	ccp	Magnetite
	Magnetite	Fe$_3$O$_4$	ccp	Maghemite
Al	Diaspore	α-OOH	hcp	Goethite, ramsdellite and groutite (α-MnOOH)
	Boehmite	γ-AlOOH	ccp	Lepidocrocite
	Bayerite	α-Al(OH)$_3$	hcp	Not normally found or formed in soils
	Gibbsite	γ-Al(OH)$_3$		Alternate double layers of hcp and open packed oxygens
	Nordstrandite	Al(OH)$_3$		Mixed layers of gibbsite and bayerite layers
	Corundum	α-Al$_2$O$_3$	hcp	Hematite. Not regarded as being formed in soils
Mn	Ramsdellite	MnO$_2$	hcp	Diaspore, goethite. Not found in soils
	Pyrolusite	β-MnO$_2$	hcp	Rutile. Not considered to form in soils
	Hollandite	Ba$_2$Mn$_8$O$_{16}$		Akaganeite, cryptomelane (tunnel structure)
	Cryptomelane	K$_2$Mn$_8$O$_{16}$		Hollandite, akaganeite (tunnel structure)
	Lithiophorite	(Al,Li)MnO$_2$(OH)$_2$		Layered structure
	Birnessite	variable		Layered structure
	Vernadite	δ-MnO$_2$		Layered structure, often classed as poorly crystalline birnessite
	Todorokite	variable		Mixed layers, structure unknown
Ti	Rutile	TiO$_2$	hcp	Pyrolusite
	Anatase	TiO$_2$	ccp	
	Brookite	TiO$_2$	ccp-hcp	
	Ilmenite	FeTiO$_3$	hcp	Hematite. Generally a residual mineral
	Perovskite	CaTiO$_3$		

The superimposition of close packed oxygen planes also creates tetrahedrally coordinated interstices in both ccp and hcp structures, e.g. the space beneath atom Z when it is positioned on the top surface of layer B, Fig. 2.1. These tetrahedral sites can also accommodate small metal cations giving the possibility of further variation in mineral form. These sites may also be possible locations for the inclusion of foreign atoms, for example, the Si or P often present as impurities in many of the oxide minerals.

Magnetite has an inverse spinel structure and is an example of tetrahedral site occupancy. It has 8 Fe(III) atoms in tetrahedral positions in one sheet and 8 Fe(III) and 8 Fe(II) in octahedral coordination in the adjacent sheet, giving a total of 24 Fe per 32 O in the unit cell. On oxidation to its isostructural analogue, maghemite (γ-Fe$_2$O$_3$) further octahedral vacancies are created to satisfy the oxidation of Fe(II) → Fe(III) giving a resultant $21\frac{2}{3}$ Fe sites for the same 32 O locations (in normal spinels, the divalent cation usually occupies the tetrahedral sites, for example, in hausmannite where Mn^{2+} is in tetrahedral and Mn^{3+} in octahedral coordination).

In some minerals the lower octahedral occupancy caused by the oxidation of constituent cations to a higher valence can so disrupt the structure as to cause its breakdown and reversion to a more stable mineral form. This is particularly so in the case of the Mn oxides. The ability for topotactic transformations, especially when heat processes are involved, relies on the ease of cation diffusion to achieve the necessary rearrangement of occupied sites.

Structural rearrangements also occur when dehydroxylation or dehydration occurs. Referring again to goethite and hematite as typical examples, the difference between the hcp structures of these Fe(III) compounds arises when 3 OH in the coordination sphere of the Fe atoms in goethite are eliminated according to the reaction:

$$2\alpha\text{-FeOOH} \xrightarrow[c.\ 300\ °C]{\text{heat}} \alpha\text{-Fe}_2\text{O}_3 + \text{H}_2\text{O}$$

resulting in the rearrangement of occupied octahedral sites to accommodate the change from OH$^-$ to O^{--}.

Greek notation is used in the British system for different polymorphs of Fe and Al compounds, e.g. diaspore, α-AlOOH, and boehmite, γ-AlOOH, the α-form is usually hcp whereas the γ form is ccp, see Table 2.1. (Gibbsite, γ-Al(OH)$_3$ is an exception and its structure will be discussed separately.) According to Evans (1964) the difference between hcp and ccp may also be responsible for some variation in mechanical properties. For the Fe and Al minerals, heating a ccp γ-form yields a dehydrated α-structure, sometimes by way of an intermediate dehydrated γ-form.

There also exists a degree of isotropy between many of the γ- and α-forms of the oxides and oxy-hydroxides of the different elements (Table 2.1) although some of these forms are not naturally occurring or do not form under pedogenic conditions. It is an interesting observation that, whereas one mineral may be very common in soils of most climatic regions, its isostructural analogue from another element may not readily form under soil conditions. Goethite and its Al analogue, diaspore, demonstrate this; diaspore is not common in soils even though it can exist in partial solid solution with goethite as is seen in the frequent occurrence of Al substituted goethites in soils (Norrish and Taylor 1961).

Structurally, the Mn oxides are generally not as simple as the oxides of the other elements discussed and the Greek prefixes used do not conform with the above connotation, the γ-phase minerals generally being structural intergrowths (Giovanoli 1969).

2.2.2 Surface charge

The basal surfaces of the layer lattice clay minerals are unique in that their surface O atoms in the tetrahedral (smectites) and tetrahedral + octahedral (kandites) sheets are fully coordinated. Broken bonds occur only at edge surfaces, which become hydroxylated by adsorption of H$^+$ or OH$^-$ to achieve full coordination of the surface atoms. These hydroxylated surfaces can develop an electrical charge (whose sign is dependent on environmental conditions) by donation or acceptance of protons. However, the area of these edges is generally small in relation to that of the basal surfaces, in the case of the smectites about 1% (Dyal and Hendricks 1950), so that the net charge on the structure is largely determined by the extent of isomorphous substitution that has occurred in the tetrahedral and octahedral sheets. Generally metal cations of lower valence substitute for the Si and Al in these sheets (Pauling 1930), so that a permanent net negative charge is generally imparted to the structure.

In marked contrast to these clay minerals is the development of charge on the surface of the oxide minerals, where it arises from unshared O and OH resulting from broken surface bonds. Although isomorphous substitutions of both higher and lower valent cations do occur in these oxides, their contribution to the net surface charge is generally negligible in relation to the pH-dependent surface charge due to the generally high specific surface of these minerals.

The origin and nature of the variable or pH-dependent (amphoteric) charge on the surface of oxide minerals have been discussed by many workers, and readers are referred to the reports of Parks and de Bruyn (1962), Atkinson et al. (1967), Parks (1967) and Bowden et al. (1973, 1977). A recent review by Gast (1977) explains the surface charge characteristics of these minerals in a general way and gives examples of typical calculations.

Generally no O atoms will persist at the surface of these oxides under normal environmental conditions if exposure to gaseous or liquid water has occurred. Whereas O atoms in the bulk of the structure are shared by more than one metal cation, those in the surface are not fully coordinated, and hold a residual negative charge. A reaction with the H^+ and OH^- from dissociated water gives rise to a fully hydroxylated surface, even for those minerals without hydroxyl groups in their structure. Breeuwsma (1973) represents the hydroxylation at a dehydrated hematite surface by step 1 in the following reaction:

Breeuwsma states that this hydroxylation occurs readily and its presence has been confirmed using analyses of water isotherms (Jurinak 1966), infrared spectroscopy (Blyholder and Richardson 1962), heats of immersion (Zettlemoyer and McCafferty 1969) and dielectric relaxation (McCafferty et al. 1970) techniques. This OH is said to be chemisorbed in contrast to the physical sorption of subsequent water layers through hydrogen bonding to these structural hydroxyl groups, depicted by Breeuwsma in step 2 of the above reaction. The first layers of physically adsorbed water are strongly bound to the hydroxylated surface and are more dissociated than subsequent layers further from the surface (Parks 1967).

Charges develop on these hydroxylated surfaces through a process which may be interpreted as either an amphoteric dissociation of the surface hydroxyls, or as adsorption of either H^+ or OH^- ions. For this reason the hydroxylated surfaces are said to function as Brønsted[2] or protonic acids. Parks and de Bruyn (1962) represent these reactions for an octahedrally coordinated trivalent metal cation in the following way:

$$3H_2O + \begin{array}{c}\text{Fe}\\ \text{O}\\ \text{Fe}\end{array}\begin{array}{c}OH_2\\ OH_2\\ OH_2\end{array}\bigg|^{+3} \xleftarrow{3H_3O^+} \begin{array}{c}\text{Fe}\\ \text{O}\\ \text{Fe}\end{array}\begin{array}{c}OH\\ OH\\ OH\end{array}\bigg|^{0} \xrightarrow{3OH^-} \begin{array}{c}\text{Fe}\\ \text{O}\\ \text{Fe}\end{array}\begin{array}{c}O\\ O\\ O\end{array}\bigg|^{-3} + 3H_2O$$

From either viewpoint, adsorption or desorption of H^+ or OH^- ions is mainly responsible for the establishment of a surface charge, σ_0, and these ions are referred to as potential determining ions (PDI),[3] since an electrical surface potential, ψ_0, arises from σ_0 because of the finite distance between the surface and the counter-ions attracted from the surrounding electrolyte to balance this charge. Gast (1977) in his review gives explicit examples of how to calculate the surface potential on an oxide surface when the net surface charge is altered by changing the pH of the surrounding solution. The pH at which, in the absence of specific adsorption, the surface has no net charge is referred to as the isoelectric point (IEPS).

The pH value at which the net surface charge is zero is referred to as the zero point of charge (ZPC). At this value coagulation and consequently sedimentation rates are greatest, and the anion and cation exchange capcities (adsorption of counter-ions from solution to balance the charge) are equal and at their minimal value (Parks 1967). Possibly of more importance than the effect of surface charge on the physical behaviour of these colloids is its effect on the chemical environment. Adsorption of counter-ions from the enveloping soil solution to balance charge can modify the composition and pH of the solution (Parks 1967).

2.2.3 Ion adsorption

The retention and release of nutrient trace elements by soil oxides are often controlled by adsorption and desorption of anions and cations on these charged surfaces. Readers are referred to the following recent reviews on the general facets of adsorption at oxide surfaces and the adsorption characteristics of specific minerals and adsorbates: Reisenauer et al. (1962), Grimme (1968), Hingston et al. (1968, 1972, 1974), Bowden et al. (1973), Breeuwsma and Lyklema (1973), Loganathan and Burau (1973), Murray (1975), Gast (1977), Hsu (1977), McKenzie (1977), and Barrow et al. (1980a,b). Sufficient references are given in this group to earlier fundamental research in these fields.

When conditions are more acid than the ZPC, the oxide surface is positively charged, and anions or the negative regions of a polarizable molecule are attracted to, and electrostatically held adjacent to, the surface. Under conditions more alkaline than the ZPC the same attraction applies to cations or their equivalents. However, this pure electrostatic bonding is not the only factor responsible for ion adsorption. Bowden *et al.*, in presenting a general model for ion adsorption at oxide surfaces, suggesting that the free energy of adsorption, ΔG_{ads}, can be arbitrarily considered to be composed of three components, namely ΔG_{coul}, ΔG_{int} and ΔG_{chem}.

ΔG_{coul} is the pure electrostatic (coulombic) bonding described above, representing the interaction between a theoretical point charge and an electrical field, and depends only on the ionic charge. If this were the only component of the adsorption, ions of similar charge would be adsorbed strictly in proportion to their activity in solution.

ΔG_{int} arises from specificity in bonding of different ions due to their charge, size and polarizability, so that adsorption may no longer be in proportion to the activity of an ion in solution. However, it is dependent on the attraction of oppositely charged counter ions and hence will be zero for an uncharged surface. If the ZPC is known the surface potential developed by these minerals in an indifferent electrolyte at any other pH can be approximated by the formula: $\psi_0 = (RT/F)\ln[(a_H^+)/(a_H^+)_{ZPC}]$ giving a value round 59 mV per pH unit deviation from the ZPC (for small deviations). Here ψ_0 is the electrical surface potential, RT has its usual connotation and a_H^+ is the hydrogen ion activity at a particular pH. It should be noted that the Nernst equation given above relates to situations not far removed from the ZPC, i.e. for low values of ψ_0, and is in fact a limiting form of equation (14) in Bowden *et al.* (1977). The surface potential is not influenced by the concentration of the electrolyte (Fig. 2.2), although at distances removed from the surface, the potential decreases more rapidly the more concentrated is the solution.

The surface charge which gives rise to this potential can be calculated from theoretical

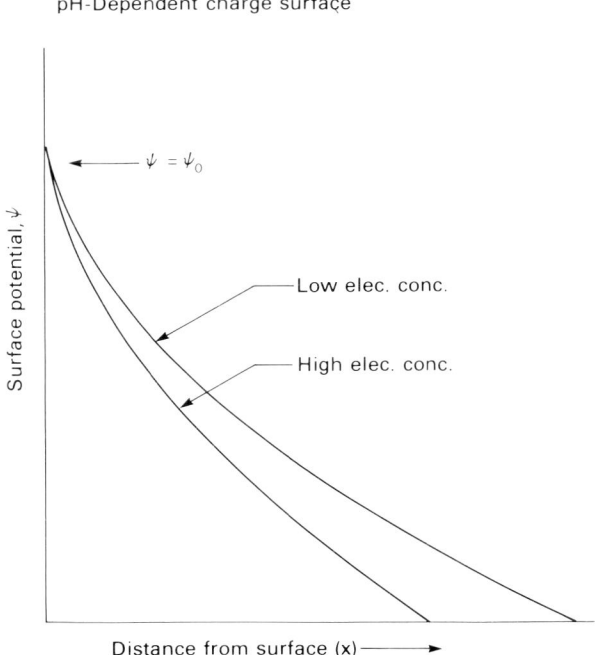

FIG. 2.2. Variation in potential at surfaces with a pH-dependent charge as a function of distance from surface. (Reproduced from *Minerals in Soil Environments*, 1977, p. 33, by permission of the Soil Science Society of America.)

GENERAL PROPERTIES   137

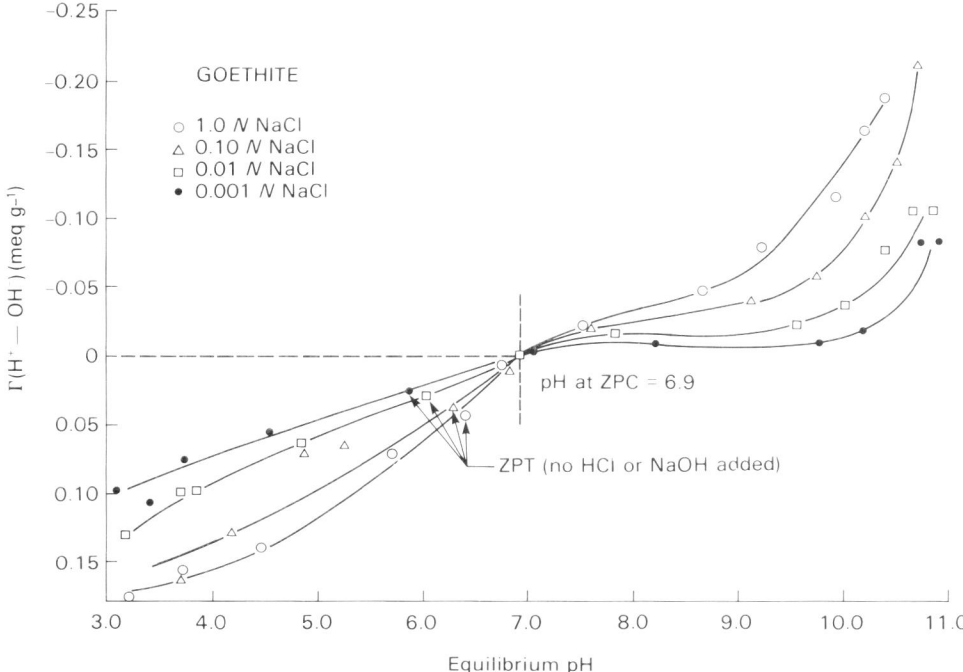

FIG. 2.3. pH titration curve for synthetic goethite showing variations in surface charge with pH for various electrolyte concentrations. (Reproduced from *Minerals in Soil Environments*, 1977, p. 38, by permission of the Soil Science Society of America.)

relationships between surface potential and electrolyte concentration, or can be measured directly as a function of pH and electrolyte concentration from pH titration curves or from the measurement of anion and cation retention values (Gast 1977). A typical surface charge v. pH relationship for varying concentrations of an indifferent electrolyte is shown in Fig. 2.3, and it is seen that both increases in pH and electrolyte concentration cause an increase in surface charge.

ΔG_{chem} is due to a specific interaction between an ion and the surface, and can be +ve or −ve or zero depending on whether adsorption, repulsion or no interaction occurs. It arises from the electronic nature of the ions of the surface and the electrolyte, and is composed of coordination, van der Waals' and polarization forces, so that an ion may be adsorbed by an uncharged or even a similarly charged surface. If ΔG_{chem} is small, so that ions are adsorbed predominantly on oppositely charged surfaces, the ions in solution are termed "indifferent" and their adsorption is referred to as non-specific (Hingston et al. 1967). In contrast, some anions may be specifically adsorbed ($\Delta G_{chem} > 0$) regardless of the surface charge or their activity in solution (Hingston et al. 1972) and this phenomenon has been termed "specific adsorption" or "ligand exchange" (Hingston et al. 1967). Such specific adsorption of anions affects the surface potential and lowers the ZPC to more acid values (Hingston et al. 1970). Conversely, specific adsorption of cations raises the ZPC to more alkaline values.

This specific adsorption is usually associated with the anions of weakly dissociated acids, although some fully dissociated acids also exhibit the phenomenon on certain oxide surfaces (Hingston et al. 1972). The mechanism is seen as a direct coordination of the surface metal cation with the oxy-anions, or replacement of the surface hydroxyl group by an anion directly

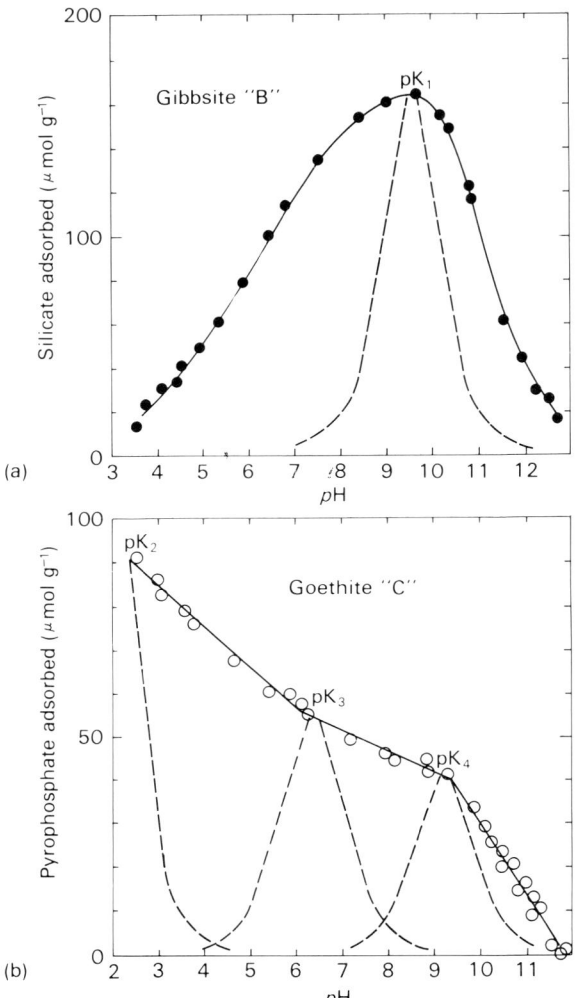

FIG. 2.4. (a) Adsorption of silicate on gibbsite. Initial silicate concentration in solution 33 p.p.m. Si; 0.149 g gibbsite in 25 ml 0.1 HCl. (b) Adsorption of pyrophosphate on goethite. (Data from Hingston *et al.* 1970.) Broken lines indicate changing nature and proportions of anionic species in solution.

coordinated to the surface metal cation. The specific adsorption of an anion from a monobasic acid shows a maximum near the pH equal to the pK_a of the acid. With the polyvalent anions there is a change in the slope in the pH–adsorption relationship, rather than a maximum, at pH values corresponding to those at which the anionic species in solution changes. These two types of adsorption envelope (the variation in adsorption with pH) are shown in Fig. 2.4 with those predicted by the early model proposed by Hingston *et al.* (1970).

Predicted adsorption of anions and cations at different pH and electrolyte concentrations often does not fit observed data over a full range of conditions, mainly due to the limitations in the earlier models used for the derivation of these relationships. Bowden *et al.* (1973, 1977) proposed models to explain most of the principal features of anion and cation adsorption, both

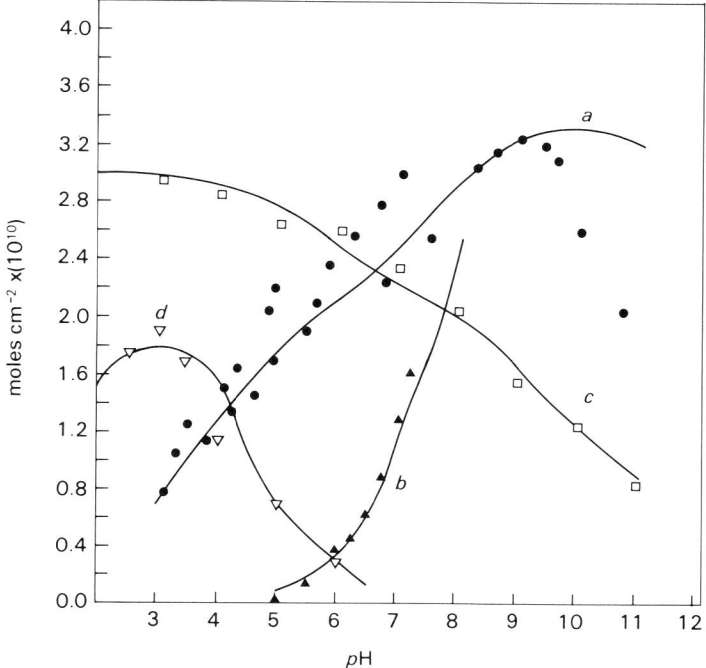

FIG. 2.5. Adsorption on goethite as a function of pH. a Silicate, 8×10^{-4} M (in 0.1M NaCl). b Zinc, 1×10^{-4} M (in 0.1M NaCl). c Phosphate, 3.2×10^{-4} M (in 0.1M NaCl). d 2,4-D, 3×10^{-4} M (in 0.01M NaCl). Continuous lines give adsorptions predicted by model of Bowden *et al.* (1973).

specific and non-specific, on these variable charged oxide surfaces. Figure 2.5 compared predicted curves from their modified theory with actual adsorption data.

Their model allows for a change in binding energy as the surface charge itself varies with pH and ion adsorption. For example, an increase in charge at constant pH makes subsequent specific adsorption more difficult, even before all available adsorption sites are occupied. This is due to repulsion of like-charge and is considered responsible for the plateau values attained for the specific adsorption at different pH values. The model also provides for the changing nature of the surface and the adsorbing ionic species changing with pH and therefore gives a modified explanation for, and a better prediction of, the specific adsorption maxima or discontinuities around the pK values of the relevant ionic species. From the earlier work of Bowden *et al.* (1973, 1977) Barrow *et al.* (1980a,b) presented an objective method for fitting models of ion adsorption on variable charge surfaces. Their techniques enabled them to obtain predictions on the charging behaviour of a goethite surface that closely resembled the observed values. Equation parameters are found from titration and surface area measurements for a particular sample, and from these parameters the surface charge can be calculated for any other pH or ionic strength. Subsequently, McKenzie (1981) applied this model to predict the surface charge on manganese oxides at different pH values and electrolyte concentrations. For cryptomelane there was good agreement between observed and predicted values, but the model could not be applied to birnessites. Cabrera *et al.* (1977) produced a model to predict phosphate adsorption by various oxides based on the chemical potential of the species in solution. Davis *et al.* (1978) and Davis and Leckie (1978, 1980) presented further models for the adsorption of ions at the oxide/water interface and Balistrieri and Murray (1979) using this model found that in sea water Na, K and Cl appeared to be associated with the goethite surface mainly by electrostatic attraction, whereas the association with Mg, Ca and SO_4 suggested stronger or more specific bonds.

A practical application of specific anion adsorption is its effect of increasing the CEC in soils containing colloids with pH-dependent charges. Wann and Uehara (1978) suggest that in the more weathered soils of the warm humid regions, added phosphate not only serves as a nutrient but acts as an amendment to increase the CEC due to a combination of three possible factors: (1) a shift in the ZPC to lower pH values; (2) neutralization of positive charges; and (3) electrolyte inhibition (Thomas 1960).

2.2.4 Isoelectric point

In addition to the ZPC, there exists a similar value, the isoelectric point (IEPS) at which the net surface charge is also zero. This value is defined as those solution conditions resulting in zero net charge in the absence of specific adsorption effects, i.e. the surface charge is established only by the PDI H_3O^+ and OH^- or any metal–hydroxo complexes existing in solution from the dissolution of the lattice cation (Bowden *et al.* 1977). Under these conditions the ZPC and the IEPS are equivalent.

Since the IEPS of an oxide will determine whether non-specific anion or cation adsorption occurs in a particular soil environment, any factor influencing this value must in turn influence the adsorption characteristics conferred on a soil by these components. Van Schuylenborgh (1950) notes that the isoelectric point of synthetic iron oxides is lower when:

(a) the concentration of the initial system is lower;
(b) the formation of the oxides is slower;
(c) the samples are dried to remove non-stoichiometric water; and
(d) the final products are more crystalline.

Parks (1967) confirmed some of these effects on the IEPS, and also that the inclusion of ionic impurities had an effect, anions lowered the IEPS while cations often raised the value. Because of the difficulties in ascertaining the trace impurities in either a natural or synthetic sample it would appear to be unwise to compare IEPS values for the same mineral oxide derived from different sources. However, it would appear from the above effects that natural samples, which are generally formed by combinations of slow oxidation, hydrolysis and dehydration (not necessarily in this order) from dilute soil solutions, could be expected to exhibit lower IEPS than materials formed by rapid precipitation in the laboratory. This could result in a lower IEPS for natural samples and allow their surfaces to be susceptible to both anion and cation adsorption following small fluctuations in pH about values that could be expected in many soils.

2.2.5 General discussion

Impurities commonly encountered in pedogenic oxides can alter, both directly and indirectly, the properties that these oxides exercise in soils. For example, the incorporation of a foreign ion into the metal oxide structure can directly influence (a) lattice distortion, (b) crystal growth or development in a certain direction, (c) the IEPS and consequently the charge developed under a certain set of conditions, and (d) the degree and components of specific adsorption due to variation in the nature of the surface metal cations. The charge developed per unit area can obviously be further modified by the indirect influence of the structural variations (a) and (b) above. For these reasons the isomorphous substitution of Al in soil goethites as a possible influence on the phosphate adsorption was suggested by Norrish and Taylor (1961). The incorporation of foreign elements can also affect physical properties of oxides. For example, Al incorporation within maghemite, γ-Fe_2O_3, increases the temperature at which the cubic structure is transformed into the rhombohedral α-Fe_2O_3 (Michel and Pouillard 1949), and certain magnetic properties of hematite can be affected or inhibited by the inclusion of small amounts of the trivalent elements, Al, Cr, Ga and In (Svab and Kren 1979).

Therefore it is important to regard each pedogenic oxide as a product of a particular environment, and to expect variations within the same mineral species formed under different pedogenic conditions. Many adsorption and surface charge characteristics of soil oxides are inferred from studies carried out on synthetic samples, but before this inference is accepted, it should be established that synthetic samples and their natural pedogenic analogues, with possible variations in morphology and composition, exhibit similar charge characteristics.

2.3 THE IRON OXIDE GROUP OF MINERALS

All the Fe oxides listed in Table 2.1 can form in soils. They form under humid conditions in all climatic zones (Segalen 1971a; Schwertmann and Taylor 1977) in which wide variations in parent materials and other soil-forming factors are encountered. Each mineral form, however, is favoured by particular environmental factors, so that its presence in a soil may suggest a defined or restricted climatic or chemical environment during its formation, for example, hematite is not considered to be forming at present in humid temperate climates (Schwertmann 1971). To understand better the variations in composition and properties between these different oxides a brief review of possible methods of their pedogenic formation will be given; an earlier review was given by Oades (1963).

Originally the soil Fe existed mainly as divalent Fe in primary minerals from which it was released during weathering most probably as a soluble Fe(II) species. Fe oxides in sediments and rocks formed during an earlier weathering cycle may also give rise to soluble Fe(II) species under suitable reducing conditions, or to soluble Fe(III) species through complex formation with organic compounds (Bloomfield 1958; McHardy et al. 1974; Chukhrov et al. 1976) or dissolution under acidic soil conditions. Although much work has been done on the influence in Fe(III) solutions of concentration, ageing times, rate and degree of hydrolysis and ageing conditions (pH and chemical environment) on the final product (Schellman 1959; Schwertmann 1959a, 1970; Feitknecht et al. 1973, 1975; Knight and Sylva 1974; Murphy et al. 1975, 1976), soil iron oxides should be considered to have formed initially and predominantly via soluble Fe(II) species, through hydrolysis–precipitation–oxidation or oxidation–hydrolysis–precipitation reactions, which can be influenced by the chemical micro-environment. Under pH conditions normally encountered in soils, greater mobility and solubility exist for the Fe(II) than for the trivalent species.

However the microbial decomposition of mobilized Fe(III) complexes and their subsequent excretion as ferrihydrite, a precursor of goethite or hematite (Fischer and Ottow 1972; Chukhrov et al. 1976), is an important process in many soils. Lepidocrocite can form from the hydrolysis of particular Fe(III) salts under conditions not expected in soils (Fordham 1970; Murphy et al. 1976), but its formation and that of the other ccp γ-soil oxide, maghemite, appear to require a precursor "green rust" stage which needs both Fe(II) and Fe(III) (Feitknecht and Keller 1950;

Bernal *et al.* 1959). The hcp α-oxides are generally derived from the hydrolysis of Fe(III) species in solution although they can also form from the oxidation transformations of "green rust" phases under particular environments (Taylor 1980; Taylor and McKenzie 1980).

Syntheses from an Fe(II) system have shown that each of the common Fe oxides can be formed at pH 7 and ambient temperatures by slight variations in the rate of aeration, the Fe(II) concentration and the chemical environment. Thus small variations in the micro-environment can be seen to be responsible for intimate associations of the various oxide phases considered to have formed contemporaneously in soils. Taylor and Schwertmann (1974a) demonstrated this in a chloride system in which slow oxidation rates and higher initial Fe(II) concentrations favoured maghemite formation, whereas the converse conditions favoured lepidocrocite. Where some oxidation was allowed to precede hydrolysis (i.e. higher initial Fe(III)/Fe(III) + Fe(II) ratios) goethite or ferrihydrite was favoured, and ferrihydrite was further enhanced by higher rates of aeration. In more recent experiments in the Fe(II) system Taylor and Schwertmann (1978), working with a bicarbonate system, which is more realistic in terms of most soils, found that goethite is favoured by low rates of aeration and/or higher Fe(II) concentrations, whereas lepidocrocite is favoured by opposite conditions. At much higher aeration rates, ferrihydrite is formed at pH 7. In all systems the direction of mineralization may be further controlled by the influence of foreign ions which may override the effects of concentration and aeration rates by enhancing, retarding or even inhibiting necessary precursors for the formation of a particular mineral phase (Schellmann 1959; Schwertmann 1959a; Krause and Borkowska 1963; Gastuche *et al.* 1964; Inouye *et al.* 1972; Nalovic 1974; Detournay *et al.* 1975; Taylor and Schwertmann 1978).

The impurities in soil iron oxides can be present either as adsorbed species on crystal surfaces or within the crystal structure. Nalovic (1974) stated that trivalent cationic impurities would tend to occupy lattice positions, whereas divalent cations can be associated in either manner. Koons *et al.* (1980) also found evidence for the effect of the environment on the composition of the iron oxides that developed during soil formation. They observed that Co, Cr, Mn, Sc, Th, U and Zn (and the heavy rare earth elements) were associated with iron oxides formed during the weathering of the basic diabase, and, with the exception of Mn and the inclusion of As, the same elements were associated with Fe oxides formed on granites. Anionic impurities may also be present in either manner and this is discussed later. Commonly several mineral forms are associated together, each exhibiting a similar hydroxylated surface. As it is impossible differentially to dissolve *only* one phase, the preference of a particular foreign ion for, or even its association with, a particular mineral variety is difficult to determine. Even though an ion which can favour the formation of a particular phase is present, it may not be incorporated within that phase, as its influence may be on the kinetics of the reaction rather than a direct inhibition or enhancement of the necessary intermediate phase formation. Since many of the influences that control the direction of mineralization are to some extent interrelated (Stumm and Lee 1961) it is again emphasized that the formation, and consequently the properties of a soil oxide reflect its particular formation environment.

2.3.1 Individual oxides

2.3.1.1 Goethite
This α-form (hcp structure) is one of the most widely occurring soil Fe oxides and is stable under normal soil conditions. There is no satisfactory evidence that it will dehydrate to form the red

hematite under soil conditions. Transformations however do occur but these are reconstructive transformations, occurring by dissolution and re-precipitation under changed environmental conditions (Schwertmann and Taylor 1977). The markedly acicular morphology exhibited by most synthetic samples is not common in soil goethites, possibly due to isomorphous substitutions and the interference to crystal development by foreign ions in the soil environment. Nakai and Yoshinaga (1980) identified fibrous goethite as a common constituent in some Japanese and Scottish soils and attributed this morphology to structural incorporation of Al and possibly Si and other ions. Al is the most common impurity in soil goethites and can be present in higher concentrations than other foreign ions (up to approximately 33 mol %), and reduces the unit cell, particle size and acicularity (Fig. 2.6). Analysis of the data of these and other workers has shown that incorporation of the smaller Al^{3+} cation does not produce uniform or consistent contractions of the unit cell in all axial directions; the b and c axes contracted linearly with the degree of substitution but the a axis behaved irregularly, exhibiting both contraction and expansion (Schulze 1980). The $d(110)$ X-ray diffraction line of goethite is the most dependent on the a axis and its non-systematic variation with Al substitution is in marked contrast to the linear displacement of the $d(130)$ and $d(111)$ (Taylor and Schwertmann 1978), suggesting that the synthesis conditions can have a large influence on the a-axis dimension. These ideas have been confirmed by the studies of Schulze (1982) on the behaviour of Al-substituted goethites.

The marked association of Al with goethite and its common occurrence may be related because the presence of Al in an initial Fe(II) system causes the formation of α-oxides (goethite or hematite via ferrihydrite, depending on the concentrations and rates of oxidation) under conditions which in the absence of Al lead to the γ-oxides, lepidocrocite and maghemite (Taylor and Schwertmann 1978; see Fig. 2.6). This influence of Al is seen also in soils; Taylor and Schwertmann (1974b) found that in morphologically similar magnetic and non-magnetic iron oxide soil concretions, the magnetic samples contained hematite and maghemite and very little layer silicate clay (a source of Al) whereas the non-magnetic samples contained goethite and hematite and much more clay.

Thiel (1963), Jonas and Solymar (1970), Lewis and Schwertmann (1979a,b) and others have produced Al-goethites from Fe(III) systems but under conditions far removed from those encountered in soils. Goethite is believed to be formed following its nucleation from the species $[Fe(OH)_2)]^+$ or $[Fe_2(OH)_4]^{2+}$ (Knight and Sylva 1974) the formation of which would be favoured by the slow oxidation and hydrolysis of dilute Fe(II) solutions. A recent study by Taylor and McKenzie (1980) on the formation of Fe(II)Al(III) hydroxy-sulphates, -chlorides and -carbonates gives a further path by which Al-substituted goethites may form under conditions that could be expected in the soil environment, and which offers an explanation for the upper limit of replacement that is found in nature.

Goethite also forms from ferrihydrite via solution (Chukhrov et al. 1973) probably through nucleation from $Fe(OH)_2^+$ species in equilibrium with the precipitate. Cr^{3+} and Ti^{4+} behave like Al in directing mineralization of an Fe(II) system towards goethite during oxidation–hydrolysis.[4] Ni^{2+} is also found in association with goethite,[5] occurring within the lattice and also, because it can be partially leached without dissolution of the goethite, on crystal surfaces (Nalovic 1974). Norrish (1975) and Schellmann (1976) have analysed a series of goethites from different sources and their compositions are listed in Tables 2.2(a) and 2.2(b). Goethite pseudomorphs after pyrites have also been reported to contain higher concentrations of trace elements such as Co, Cu, Ni, Pb and Au (Bell and Hornig 1972), again reflecting the influence of the environment on the composition of the oxides. A similar effect was noted in a soil goethite

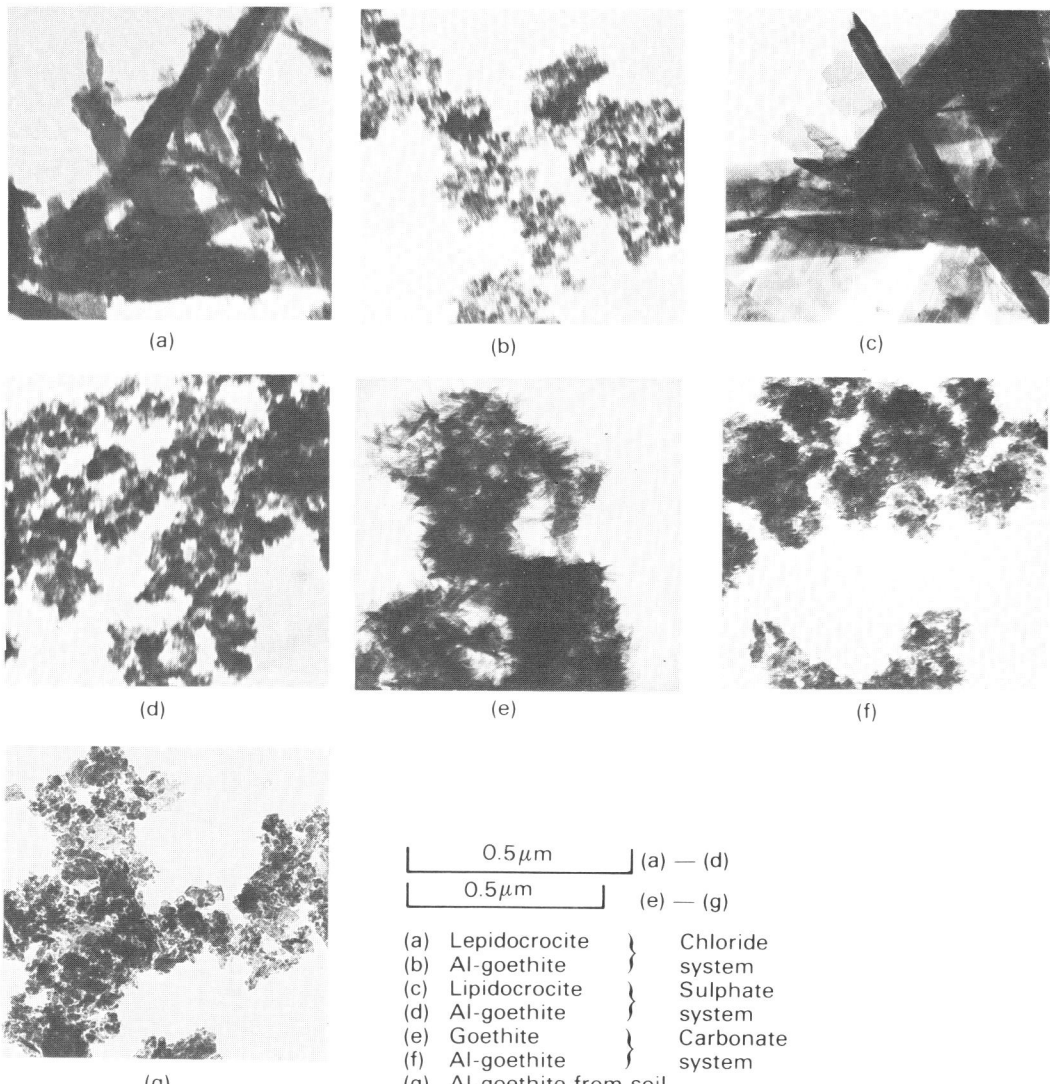

FIG. 2.6. Electron micrographs showing influence of Al during oxidation of Fe(II) chloride, carbonate and sulphate systems and the EM of a natural Al-goethite extracted from a soil. (a) From oxidation of 150 ml 0.032M Fe(II) chloride at pH 5.5 with 1.5 ml air/min. (b) As in (a) but Al added to give molar ratio Al/Al + Fe = 0.29. (c) As in (a) but 0.032M Fe(II) sulphate used. (d) As in (c) but Al added to give molar ratio Al/Al + Fe = 0.29. (e) From oxidation of 100 ml 0.02M Fe(II) carbonate with 5 ml air/min. pH ⩾ 6. (f) As in (e) but 200 ml 0.02M Fe(II) carbonate and Al added to give Al/Al + Fe = 0.13. (g) Al goethite extracted from soil H 165/10 < 2 μm (Norrish and Taylor 1961).

which had developed above an old lead mine: the goethite was found to contain up to 0.95% Pb and 1.5% Zn.[6]

Impurities such as SiO_2 were previously considered to be admixed separate phrases, but the inability to remove freshly precipitated silica by NaOH without dissolution of the co-formed goethite (Schwertmann and Taylor 1972a) suggests that it could be present in the structure,

TABLE 2.2(a). Electron microprobe analyses of goethites

	Concentration, %			
	1	2‡	3	4
Fe_2O_3	73.9	88.5	81.3	73.7
Al_2O_3	8.68		0.30	2.96
TiO_2	1.53	0.7	0.02	0.01
V_2O_5		3.0		
MnO			0.34	0.65
ZnO		2.15		2.93
CuO		0.10		
MoO_3		1.42		
P_2O_5	2.16	2.66	4.43	0.03
SO_3	0.7		0.08	
SiO_2	1.2		0.30	5.02
Crystal size† Å	300		1000	500

Sample 1 is a collection of single crystals from Queensland. 2 Goethite-rich V shale, composition calculated by difference between analysis before and after goethite removal. 3 Goethite from gossan in Cobar district, New South Wales. 4 A goethite from near Broken Hill, New South Wales (after Norrish 1975).
† Crystal size estimated from X-R-D line broadening.
‡ Ignited basis; Fe_2O_3 in pure FeOOH = 89.87%.

TABLE 2.2(b). Chemical analyses of lateritic goethites

	Concentration, %				
	1	2	3	4	5
Fe_2O_3	72.7	74.8	69.9	71.7	59.2
Al_2O_3	4.17	2.60	3.71	3.61	11.0
SiO_2	3.27	2.40	1.99	2.69	4.88
MgO	0.63	0.40	0.81	0.43	1.42
Cr_2O_3	2.93	2.92	3.51	3.60	3.31
MnO	0.65	0.33	0.85	1.53	1.15
NiO	1.27	1.01	1.92	1.49	1.86
CoO	0.11	N.D.	N.D.	0.03	0.11
Loss on ignition	13.3	13.3	17.0	14.5	16.0
Surface area $m^2 g^{-1}$		64.2			101.7

Samples 1–3 from New Caledonia, samples 4 and 5 from Philippines. (Quoted by permission of W. Schellmann, Institut für Geowissenschaften und Rohstoffe, Hannover.)

possibly in the tetrahedral interstices formed from the stacking of close packed oxygen layers. Phosphorus and similar sized amphoteric elements could also occupy such interstitial positions. Whether Si occupies these sites or whether it is strongly bound to surfaces and occluded by further crystal development is not clear. Norrish (1975 and priv. comm.), in his electron microprobe analyses of goethites, found that the silica was homogeneously distributed in apparently well-crystallized goethites which would support either hypothesis. However, in one goethite that appeared to be a single crystal, the microstructure was revealed as a mosaic of rhombic-shaped aggregates, the boundaries of which are clearly defined and at which occlusion of silica could occur (see Fig. 2.7; Smith and Eggleton, 1983).

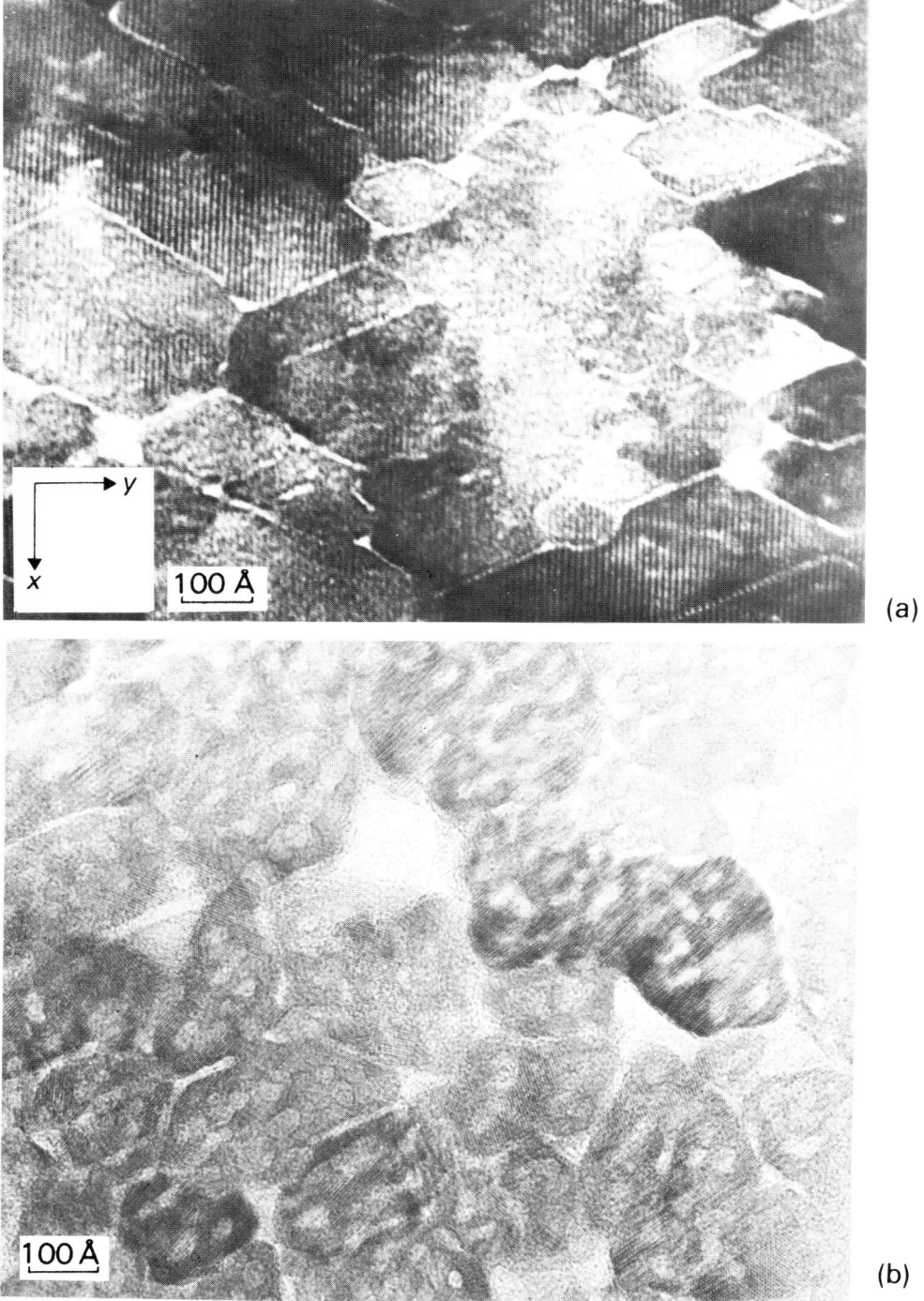

FIG. 2.7. Electron micrographs of goethites showing boundaries between crystallites: (a) on apparent single crystal; (b) fibrous massive goethite looking along fibre axis direction. Note apparent holes in structure (pers. comm. Dr A. Eggleton, Australian National University, Canberra).

2.3.1.2 Ferrihydrite

Colloidal ferric hydroxide has in the past been termed amorphous, but even in freshly precipitated "ferric hydroxide" there is a small degree of short range order that results in at least two defined bands in X-ray diffraction. The degree of ordering increases with ageing under appropriate conditions, the material developing into what is now called ferrihydrite (Chukhrov *et al.* 1973). It was previously given the formula $5Fe_2O_3 \cdot 9H_2O$, but Russell (1979) has shown that structural OH groups are present and has suggested that the formula be amended to $Fe_2O_3 \cdot 2FeOOH \cdot 2.6H_2O$. Its colour can vary from yellow–brown to brown–red depending on its method of formation. It is a precursor of hematite and a possible source of goethite. It is formed from the fast hydrolysis of Fe(III) salts if its solubility product (10^{-37}–10^{-39}) is exceeded, or from a sufficiently high rate of oxidation of a soluble Fe(II) species at a pH where rapid hydrolysis occurs. Many naturally occurring ferrihydrites are rich in organic carbon compounds, which from their adsorption on the surface can often inhibit further crystallization by preventing the aggregation required (Schwertmann and Fischer 1966) for hematite formation or the dissolution required for transformation to goethite. These carbon-rich samples may arise from the excretion of micro-organisms following the ingestion of Fe(III)–organic complexes (Chukhrov *et al.* 1976; Fischer and Ottow 1972).

Rapid oxidation of Fe(II) in bicarbonate solutions also causes ferrihydrite formation (Taylor and Schwertmann 1978). The presence of inorganic and organic ions not only stabilizes the ferrihydrite by inhibition of further crystallization (Scheffer *et al.* 1957; Schellmann 1959; Schwertmann 1966), but their presence may be the cause of the ferrihydrite formation. For example, 0.150 dm^3 of a 0.016 mol dm^{-3} FeSO$_4$ solution on oxidation with 5 ml air/min at pH 7 yields lepidocrocite with a trace of goethite, but the addition[4] of silicate to the system (5.6×10^{-4} mol dm^{-3}) causes only ferrihydrite to form. Excess Al also present during the oxidation–hydrolysis of Fe(II) chloride or bicarbonate can cause ferrihydrite to be the preferred end product (Taylor and Schwertmann 1978). Krause *et al.* (1970) demonstrated that foreign divalent cations Ca, Mg, Co and Pb also caused ferrihydrite formation during the oxidation of an Fe(II) carbonate solution. This material possesses a very high specific area and it is impossible to say whether in natural samples high surface adsorption after its formation stabilized the mineral, or whether high concentration of foreign ions in the environment caused its formation and resultant high specific surface.

Ferrihydrites formed from Fe(II) systems in the presence of high concentrations of foreign cations or from microbial decomposition of Fe(III) complexes often display a yellow–brown colour which is common in natural samples, as opposed to the dark red–brown colour displayed on rapid hydrolysis of an Fe(III) solution. With such a poorly crystalline and, as yet, little studied mineral, very little has been published about the incorporation of foreign ions into the structure. However, Lubner and Malone (1974) noted that the precipitates formed from bacterial oxidation often contained trace elements originally associated with the Fe-oxidizing bacteria.

In the transformation of ferrihydrite to goethite by dissolution at high pH, goethite quickly attains its maximum development in the crystallographic *a*-axis direction and subsequent growth is in the *b* and *c* axial directions (Lewis and Schwertmann 1980). Ferrihydrite co-precipitated with Al in M KOH crystallizes more readily to goethite than pure ferrihydrite allowed to age for a period before Al addition (Lewis and Schwertmann 1979a). Rapidly precipitated ferrihydrite transforms more effectively to hematite at higher temperatures and higher concentrations of added Al, whereas increased pH favours the transformation to goethite by dissolution (Lewis and Schwertmann 1979b).

The action of organic matter on ferrihydrite appears to depend on the relative amounts, the types of organic matter involved and the pH. For example Kodama and Schnitzer (1977) observed that, for ageing of ferrihydrite in fulvic acid in the pH range 4–10, a decrease in the Fe(III)/fulvic acid ratio from ∞ to 6 favoured the formation of hematite over goethite. At higher fulvic acid concentrations the formation of these more crystalline species was inhibited. These results agree with the observations of Fischer and Schwertmann (1975) on the behaviour of oxalic acid in favouring hematite formation. The presence of fulvic acid or other organic compounds added *before* the precipitation of ferrihydrite may possibly produce a different end product by controlling the rate of hydrolysis so that the solubility product of ferrihydrite is not exceeded. It is an interesting observation of Kodama and Schnitzer (1977) that the hydroxy–carboxylic (polybasic) compounds had a greater inhibiting influence on crystallization. This may possibly be due to the more specific adsorption of these hydroxy compounds by the highly hydroxylated ferrihydrite surface.

2.3.1.3 Feroxyhite

Feroxyhite, designated δ-FeOOH, is a poorly crystalline material resembling ferrihydrite in being composed of poorly ordered 4Fe(O,OH) octahedral units. However, feroxyhite does not display the characteristic 1.97 Å X-ray diffraction line of ferrihydrite, and the (100) diffraction peak of feroxyhite at around 2.54 Å is generally sharper than the ferrihydrate (110) line at the same spacing. In both minerals this line is asymmetrical and tails off towards lower 2θ values (Carlson and Schwertmann 1980, 1981; Chukhrov *et al.* 1973). A similarity may be drawn between these two minerals and the birnessite and vernadite forms of δ-MnO$_2$ in the degree of order and missing diffraction lines. Carlson and Schwertmann (1980) identified feroxyhite, in association with goethite and lepidocrocite, as the dominant Fe species in some rusty precipitates in recently opened gravel pits in Finland. They synthesized similar samples by rapid oxidation of FeCl$_2$ solutions between pH 5 and 8 with 30% H$_2$O$_2$ and suggested that feroxyhite forms in nature from the rapid oxidation of Fe(II) in solution. Chukhrov *et al.* (1977) suggest that silica can favour the formation of feroxyhite when Fe(II) solutions are rapidly oxidized around pH 6–6.5. Taylor (1980) proposed that Fe(II)Fe(III) hydroxy-carbonate (isostructural with hydrotalcite) could be an intermediate compound in the formation of many common soil iron oxides. The washed and dried green–blue compound turned a yellow–brown after a few minutes' exposure to air and, after 31 days, gave an X-ray diffraction pattern with a sharp asymmetrical peak around 2.54 Å, suggesting feroxyhite synthesized under conditions that could be expected in nature.

Carlson and Schwertmann (1980) also established that feroxyhite is less soluble in acid ammonium oxalate at pH 3 (in the dark) than is ferrihydrite, suggesting that, although it lacks some of the spacings of the latter mineral, it is possibly more crystalline, as would be suggested by the sharper diffraction peaks.

2.3.1.4 Hematite

This hcp structure, α-Fe$_2$O$_3$, forms at low temperature through the aggregation and subsequent dehydration of its poorly crystalline ferrihydrite precursor. Acceleration of aggregation by higher temperatures or denser concentrations of ferrihydrite (Schwertmann and Fischer 1966), or the influence of foreign ions, e.g. oxalate (Fischer and Schwertmann 1975), Ca and Mg (Schellmann 1959) favour its formation. Collepardi *et al.* (1973) aged ferrihydrites at 100 °C and pH values between 5.3 and 12, and it appears that at pH 5.3 salts which formed stable complexes with Fe(III) generally accelerated the formation of hematite rather than goethite. Although goethite will dehydroxylate to hematite, soil occurrences are not considered to form in this way.

Thrierr-Sorel et al. (1978) presented a kinetic study on the thermal transformation between these two phases. The presence of even small amounts of Al (5%) in a rapidly precipitated Fe(III) hydroxide inhibited goethite formation (Gastuche et al. 1964) and instead favoured hematite. Schwertmann et al. (1979) confirmed this observation and further suggested that the resultant Al-hematites form via ferrihydrite in which Al is already substituting for Fe. Al-hematites possess a disordered structure (Perinet and Lafont 1972) and Schwertmann et al. (1977) have demonstrated that differential disorder, as seen from line broadening and peak shifts to higher 2θ values in X-ray diffractograms, is similar for soil occurrences and synthetic Al hematites.

This observation in the inhibition of goethite formation by Al may however give further support to alternative formation paths for hematite. The frequent intimate association of goethite and hematite in soil profiles, coupled with the existence in soils of Al-substituted hematites (Beneslavsky 1957; Janot and Gilbert 1970; Janot et al. 1971; Perinet and Lafont 1972, and Schwertmann et al. 1977), seem inconsistent with the more common occurrence of Al goethites. This, however, may be a further argument for the importance of Fe(II) in the formation of soil iron oxides, since an increase in the proportion of Al present in an Fe(II)–Al system causes Al-ferrihydrite (which transforms to hematite rather than goethite) to form during oxidation–hydrolysis (Taylor and Schwertmann 1978). Moreover, Chukhrov (1973) states that hematite in red beds appears to derive from ferrihydrite formed from bacterial oxidation of soluble Fe(II) species but not from organically complexed Fe(II) compounds.

Little is known about the compositional variations in soil hematites, possibly due to its frequent association with other Fe oxides. TiO_2 is often present in small amounts, but it is difficult to determine whether intergrowths of ilmenite have occurred. In a pure sample this could be determined by the composition of successive dissolution extracts. D'yachkova (1971) reports that hematite may exhibit isomorphous substitutions like magnetite, where Fe^{3+} is placed by Cr^{3+}, Ti^{3+} and other elements, but in samples developed from metasomatic rocks, hematite contains only a fraction of the V^{3+} (approx. 1%) found in magnetites. Small amounts of divalent Fe (1–2%) have also been found in mineral hematites.

Adsorbed silica inhibits hematite crystallization from ferrihydrite (Chukhrov et al. 1973) and certainly in the analyses of mineral hematites, the SiO_2 contents (0.16% average of 4) are much lower than in mineral goethites (1.1% average of 5) (Deer et al. 1962) where the SiO_2 is considered an admixture. The average SiO_2 contents of soil goethites are even higher, 1.6% (average of 4: Table 2.2(a)) and 3.0% (average of 5: Table 2.2(b)). Although these figures are not conclusive it is suggested that hematites may not contain as much SiO_2 impurity as goethite.

From the work of Nalovic et al. (1975) it would appear that Mössbauer spectroscopy can indicate the purity of fine-grained soil hematites and possibly whether foreign transition element ions are incorporated in the structure. These techniques could probably be applied to other soil Fe oxides.

2.3.1.5 Lepidocrocite
This is the ccp γ-form of FeOOH. It is generally less common than either goethite or hematite. A recent investigation by Christensen and Christensen (1978) produced a useful projection of the lepidocrocite structure, showing the location of H bonds and the coordinations of the several types of surface oxygens. The occurrences of lepidocrocite are almost exclusively limited to hydromorphic soils where microbial reduction under anaerobic conditions leads to the presence of soluble Fe(II) species. As soon as it has been partly oxidized, the resultant Fe(III) hydroxy species reacts with Fe(II) hydroxy species in solution to form a highly insoluble Fe(II)–Fe(III)

hydroxy compound, a "green rust" phase (Feitknecht and Keller 1950; Bernal *et al.* 1959) which oxidizes under a suitable environment to lepidocrocite (Taylor and Schwertmann 1974a).

Oxidation of an Fe(II) system is considered the most likely pathway for lepidocrocite formation and is supported by the morphological similarity of natural soil samples and synthetic samples produced by these techniques (Schwertmann 1973). McHardy *et al.* (1974), however, have shown that the hydrolysis of unaged Fe(III) chelates of compounds that could be expected in soils can also lead to lepidocrocite formation. The conditions required for lepidocrocite formation from Fe(III) solutions and sometimes the nature of these solutions preclude them as being likely paths for pedogenic formation.

From soluble $Fe(HCO_3)_2$, which is probably a species of mobile Fe in soils, lepidocrocite formation is favoured over goethite by lower Fe(II) concentrations and higher rates of subsequent aeration (Taylor and Schwertmann 1978). Lepidocrocite formation can however be inhibited by the presence in the soil solution of small amounts of Al and Si, which can be expected in most soils. These effects may be to some extent causal factors in the infrequent occurrences of lepidocrocite, especially as the only oxide form present. The influence of Al has been discussed in Section 2.3.1.1, Goethite (see Fig. 2.6). Schwertmann and Thalmann (1976) found that silica in solution caused ferrihydrite formation until all the silica had been removed from solution, after which lepidocrocite again formed. The competition between the soluble silica and the Fe(II) hydroxy species present in solution for reaction or adsorption with the oxidized Fe(III) species formed may inhibit the formation of the green rust precursor. Although the influence of Al suggests that Al-lepidocrocite should not form under normal soil conditions, its existence was proposed by de Villiers and Van Rooyen (1967). However, the inhibition of lepidocrocite (and also of maghemite) by Al only applies under oxidizing conditions. If allowed to react under strictly anaerobic conditions, Al-hydroxy species can induce the hydrolysis of Fe(II) at pH 7 with the subsequent formation of green rust complexes (Taylor and McKenzie 1980), to be discussed later. Oxidation in water of the Fe(II)Al(III) hydroxy-chloride gives lepidocrocite with an absent or markedly reduced (020) basal reflection (Taylor and Schwertmann 1980), whereas oxidation in air of the dried green complex gives akaganeite, β-FeOOH. The presence of this disordered lepidocrocite could explain why it was not detected by de Villiers and Van Rooyen in their work mentioned above.

Thermodynamically lepidocrocite is metastable under soil conditions. Van Oosterhout (1967) showed that it could transform via solution to goethite. Schwertmann and Taylor (1972a) showed that Si in the soil solution retarded goethite nucleation and so retarded this transformation. However its persistence in soils was shown to be due to its low rates of dissolution under normal conditions (Schwertmann and Taylor 1972b). Detournay *et al.* (1975) found that in the oxidation of an Fe(II) solution, under conditions which normally lead to goethite formation, the presence of large amounts of divalent Co or Mn $(M/(M+Fe) > 0.05)$ caused lepidocrocite also to form. They suggested this was due to the ease with which the octahedra of lepidocrocite are deformed compared with those in the more compact goethite. Small amounts of Cu(II)(Cu/Fe = 0.1) in an Fe(II) system normally forming lepidocrocite causes a rapid formation of maghemite, due presumably to catalytic oxidation of the Fe(II) at pH 7.[4]

Lepidocrocite transforms on heating initially to γ-Fe_2O_3 and at higher temperatures in an oxidizing medium this converts to α-Fe_2O_3. Giovanoli and Brütsch (1974) have suggested that, in order to form γ-Fe_2O_3, the corrugated sheets of edge- and corner-shared FeO_6 octahedra in γ-FeOOH, and linked by H bonding to similar sheets, must collapse and fit together. The front of such a coalesced region would be under strain, and therefore be a suitable nucleus for further transformation. When the particle size of the γ-FeOOH is small, the ordering of the resultant γ-Fe_2O_3 is suppressed.

Because of the difficulty of obtaining pure samples, there is no available literature over compositional variations expected in lepidocrocites.

2.3.1.6 Green rust phases

Feitknecht and Keller (1950) found that the dark green precipitate formed during the oxidation of Fe(II) chloride solutions had a well-defined X-ray diffraction pattern. This compound is very susceptible to oxidation, especially if moist, and becomes yellow–brown after a few minutes' exposure to air. Bernal *et al.* (1959) called this and the similar dark green–blue complex formed in the presence of sulphate "green rust" and proposed that they transformed topotactically on oxidation to lepidocrocite or magnetite. These compounds are now recognized to have a double octahedral layer structure with essential anions and water occupying the interlayer region, a structure similar to the pyroaurite group of minerals (Brindley and Bish 1976). According to Taylor (1973) these types of compound form readily at neutral or slightly acid pH by the interaction of di- and trivalent metal cations in the presence of suitable anions.

Taylor (1980) formed a dark green Fe(II)–Fe(III) hydroxy-carbonate at pH 7 that was isostructural with hydrotalcite, the $Mg_6Al_2(OH)_{16}(CO_3) \cdot 4H_2O$ member of the pyroaurite group. Under different environmental conditions this Fe(II)–Fe(III) basic carbonate transformed on oxidation to goethite, lepidocrocite, maghemite or ferrihydrite or mixtures of these phases. The CO_3^{2+} anion is common in most soils, especially under anaerobic conditions where Fe is also often present in the mobile divalent state and a similar formation of this form of green rust could be expected. The blue–green colour of the mottled material in gleyed horizons of some soils and the rapidity with which this colour becomes yellow–brown suggest that these natural formations may contain green rust compounds which transform to more common and stable oxide phases on exposure to air.

The presence of Al-hydroxy species in a medium containing Fe^{2+} cations and carbonate anions around pH 7 led to the formation of an Fe(II)–Al(III) analogue of the Fe(II)–Fe(III) hydroxy-carbonate when oxidation was inhibited (Taylor and McKenzie 1980). This compound was also a blue–grey to dark green and rapidly changed colour to the yellow–brown on exposure to air. Continued exposure of the freeze-dried compound to air gave ferrihydrite with possible Al substitution, whereas allowing the hydroxy-carbonate to oxidize in water caused Al-substituted goethite to form. In these double hydroxide phases the divalent/trivalent cation ratio can vary between 2 and 3. At the lower value it can be seen that conversion to aluminous goethite would give a 33 mol% substitution, around the maximum degree of replacement suggested to occur in nature. The oxidative transformation of the Fe(II)–Fe(III) hydroxy-carbonate to various α- or γ-phase Fe oxides, depending on the conditions, as opposed to transformation of the Fe(II)–Al(III) analogue most commonly to goethite or ferrihydrite may explain the earlier observation of Taylor and Schwertmann (1978) that the presence of Al-hydroxy species inhibits γ-phase oxide formation in favour of the α-phase during the oxidation of Fe(II) systems at neutral pH and room temperatures.

The rapidity with which these compounds alter structurally on oxidation creates a difficulty in their identification by X-ray diffraction or other technique. This behaviour could also have contributed to the previous inability to identify the mineral composition of the green–blue mottles in gleyed horizons. Taylor (1982) described a method for stabilizing these synthetic hydroxy-carbonates against oxidation and proposed that the technique should be suitable for

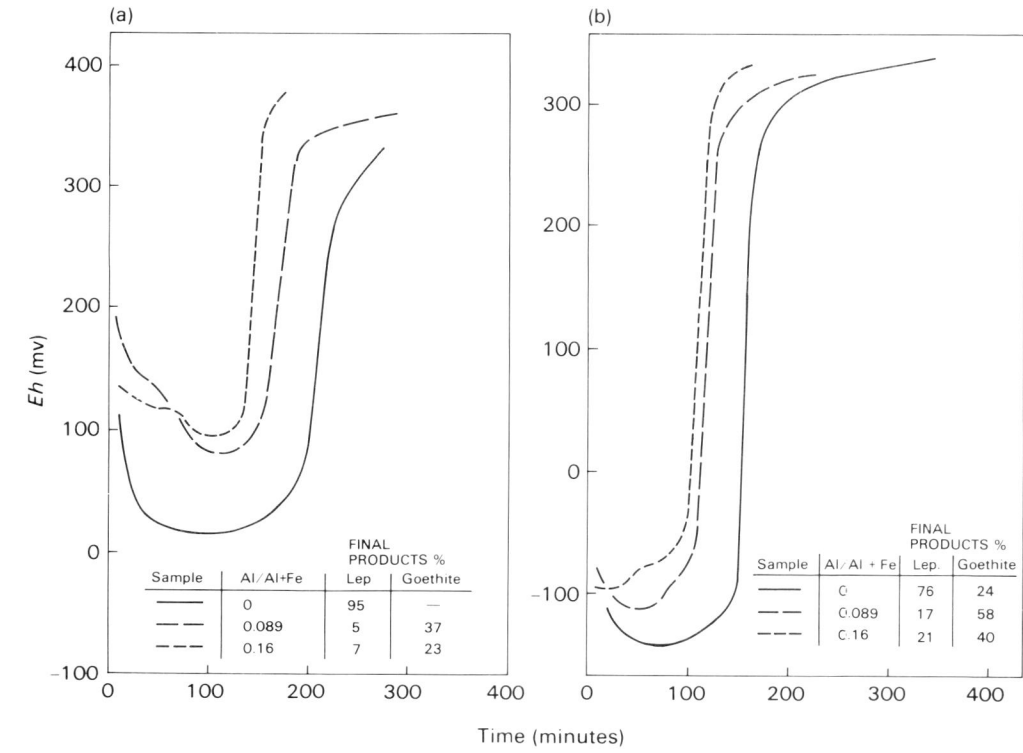

FIG. 2.8. The influence of Al-hydroxy species on Eh during the oxidation of 150 ml 0.016M $FeCl_2$ with 1 ml air/min: (a) at pH 6; (b) at pH 7.

similar naturally occurring material. The technique has been used by Murad and Taylor (1984) to follow the oxidation of Fe(II)–Fe(III) and Fe(II)–Fe,Al(III) hydroxy-carbonates by Mössbauer spectroscopy.

During the oxidation of Fe^{2+} in the presence of Al-hydroxy species in the formation of a green rust or its oxidation product, the actual presence of the Al causes a marked increase in the Eh of the system[4]. This is demonstrated in Fig. 2.8, and is considered to be due to the higher concentrations of Fe(III) in solution, possibly arising from the formation of more soluble mixed Fe(III)–Al(III) hydroxy species.

2.3.1.7 Magnetite and Maghemite

Before discussing magnetite and maghemite occurrences in soils, some arbitrary definition is required to differentiate between these two isostructural minerals. In soil maghemites the Fe is not all trivalent, Taylor and Schwertmann (1974b) found an average of about 10% Fe(II) in maghemites examined from Australian soil concretions.

For this work the definition of Fasiska (1967) will be used to differentiate between these two compounds, although it would appear that the products of pedogenesis should always fall into the maghemite classification. According to Fasiska, the Fe/O ratio in magnetites extends only over a short range, 0.744–0.75 and the Fe(III)/Fe(II) is close to 2.0. At lower Fe(II) contents as the Fe/O ratio decreases from 0.72 to 0.67, the oxide is designated as maghemite, even though ideally γ-Fe_2O_3 contains no Fe(II). Because of the ease of oxidation of moist fine-grained magnetite (Feitknecht 1959) and the tendency for an intermediate product to oxidize towards γ-Fe_2O_3 (Annersten and Hafner 1973), most pedogenic magnetitic oxides formed from solution would fall into the maghemite range.

Because soil maghemites contain Fe(II) and are associated with other Fe oxides in soils, it is likely that they are generally formed from solution like the associated oxides. This expresses the views of Oades and Townsend (1963) who suggested that neither the firing of soils (thermal dehydration and transformations of goethite and lepidocrocite to maghemite) nor the oxidation of primary magnetite could explain the common occurrence of maghemite in many soils. However, maghemite formation from burning of ferruginous soils is not uncommon. For example, the surface soil surrounding a burnt *Xanthorrhoea Australis*, a member of the lily family rich in combustible resin, contained maghemite and hematite. Non-magnetic goethite ferricrete taken from 1 m beneath the surface was ground with powdered resin from the plant and ignited and maghemite and hematite were formed in approximately the same relative amounts as in the topsoil.[4] Formation from primary magnetite also occurs and Fitzpatrick and Leroux (1976) report occurrences of titano-magnetites in South African soils that derived from the weathering of titano-maghemites. Mullins (1977) states that titano-maghemites formed in this way are more common than previously thought. In other formations he agrees that maghemite appears to be a product of repeated oxidation–reduction cycles in a soil and is more common in the upper horizons.

During oxidation of some green rust phases under particular conditions there is a topotactic transformation to a magnetic phase somewhere between the composition of magnetite and maghemite. Bernal *et al.* (1959) and Misawa *et al.* (1973) state that low rates of aeration of these compounds favoured formation of maghemite over lepidocrocite. Taylor and Schwertmann (1974a) confirmed this effect and showed that from an Fe(II) chloride system all common oxides could form around pH 7 by slight modification of other parameters (see earlier section in this chapter), and maghemite was favoured by higher total Fe concentrations, lower rates of oxidation, higher temperatures (30 °C rather than 20 °C) and small initial amounts of Fe(III) present before the commencement of hydrolysis. Taylor (1984) further extended the earlier work of Taylor and Schwertmann (1974a) to form magnetite particles of about 0.1 μm diameter and these transformed after 1 hour at 105 °C to maghemites which displayed superstructure X-ray diffraction peaks much broader than the associated normal diffraction lines. Under other environments reactions between stoichiometric amounts of ferrihydrite and unoxidized $Fe(OH)_2$ can lead to a rapid precipitation of magnetic material without a green phase being observed.

Moreover, as has been mentioned in the section on goethite, maghemite formation, under conditions that normally lead to its synthesis, is inhibited by small amounts of Al (Taylor and Schwertmann 1978) or Cr^{3+},[5] and natural occurrences of maghemite supported these observations. However, Al-maghemites can be made by adding NH_4OH to a boiling solution of ferrous sulphate and aluminium sulphate, oxidizing the precipitate, sintering at 1000 °C and then reducing at 350 °C (De Boer and Selwood 1954), and Al is said to stabilize the structure. The effect of foreign ions is demonstrated in Fig. 2.9 where the linear relationship between the a-axis dimension and the Fe(II) content depended on whether NaOH or NH_4OH was used to maintain pH during synthesis. In both cases the a-axis dimension was larger than would be predicted from the Fe(II) content asssuming linear changes of the axial dimension between Fe_3O_4 and γ-Fe_2O_3. The influence of the Na^+ and NH_4^+ ions is not clear although Michel (1949) reported an increase in maghemite cell size with Na^+ content.

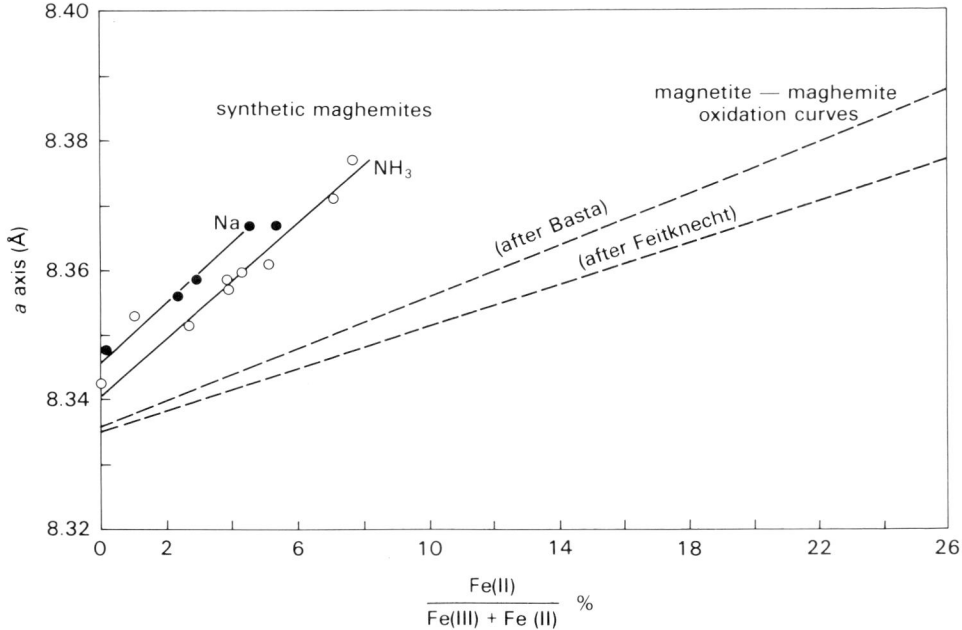

FIG. 2.9. Variation in an axis dimension with Fe(II) content in maghemites produced by thermal oxidation of magnetites and by syntheses using NaOH or NH$_4$OH to control pH at 7 during aerial oxidation of 10% Fe(III)–90% Fe(II) chloride solutions with 5 ml air/min.

Low temperature synthesis could be expected to yield a disordered form of maghemite[7] and Annersten and Hafner (1973) have stated that the structure of maghemite is dependent on its mode of preparation. Creation of octahedral vacancies on oxidation of a magnetite phase to maghemite produces X-ray diffraction superstructure reflections. Synthetic magnetites formed at pH 7 by the method of Taylor and Schwertmann (1974a) on oxidation gave superstructure lines much broader than the main reflections, whereas in some natural maghemites the reflections appear equally sharp. These variations may be the result of compositional differences, the temperature of formation, or pedogenic precipitation, all processes which may influence the distribution of Fe(II) between octahedral and tetrahedral sites. With new techniques a relationship may be found between the initial Fe(II) coordination in the magnetite phase and the final degree of vacancy ordering as measured by the sharpness and intensity of superstructure reflections.

Although inclusions of foreign ions in spinels is common (Lewis 1970), the compositional changes in pedogenic maghemites must be controlled by the environment. During synthesis divalent Mn, Co and Ni did not suppress maghemite formation and it is supposed therefore that they could occur in soil maghemites. Silica (Shcheka et al. 1977) and calcium (De Sitter et al. 1977) have been found in magnetites and could therefore be expected to be incorporated into the similar maghemite lattice or persist in the structure if the maghemite forms from oxidation of magnetite. Random distribution of vacancies in both the tetrahedral and octahedral sites

(Annersten and Hafner 1973) should allow flexibility of incorporation of foreign ions during formation. The behaviour of the trace elements Co, Ni, Zn, Cu, Mn and Cr in synthetic magnetites during their thermal alteration to maghemite (220 °C) and to hematite (650 °C) was studied by Sidhu et al. (1980). In the original magnetites Co, Ni and Zn were uniformly distributed, whereas Cr was more concentrated near the centre of grains and Cu and Mn were more concentrated near the surface. This distribution was retained during the transformation to maghemite, although there was a dilution in the surface regions due to outward migration of Fe. On transformation to hematite, Mn and Cr were retained but much of the Co, Ni, Zn and Cu migrated to the surface regions.

2.3.1.8 Akaganeite

Akaganeite, β-FeOOH, is not formed under normal soil conditions but does occur in nature (Mackay 1962; Chukhrov et al. 1976). Chloride or fluoride appears necessary for its formation and initially occupies tunnel positions in a structure similar to hollandite. The halide can be removed by subsequent leaching (Bernal et al. 1959), and in some natural samples no Cl is detected (Mackay 1962). However, this is doubted by Ellis et al. (1976) who state that chloride ion is specifically adsorbed and stabilizes the structure. Chukhrov et al. (1976) however have found Cl in some samples formed from waters rich in Cl and F. Synthetically it can form from $FeCl_2.4H_2O$; the brown insoluble material often noted on the surface of AR grade $FeCl_2.4H_2O$ generally gives a pure β-FeOOH X-ray diffraction pattern and a sample of this β-FeOOH has not altered after 15 years in water and shows streaming in suspension. It is also formed hydrolytically by boiling ferric solutions containing Cl^- ions. The Fe(II)–Al(III)–Cl compound of the pyroaurite group prepared by Taylor and McKenzie (1980) (isostructural with the Fe(II)–Fe(III) "green rust 1" of Bernal et al. 1959) alters to akaganeite if oxidized in the dry state in air.

2.3.2 Adsorption by soil iron oxides

For non-specific adsorption all the oxides should behave similarly, i.e. adsorbing sufficient anions or cations to balance the respective pH-dependent surface charge. Variations in exchange capacity of different samples of the same mineral reflect variations in IEPS, which in turn is controlled by composition and mode of formation.

Only certain OH groups on the substrate surface may be involved in ligand exchange. By investigating isotopic exchange processes Atkinson et al. (1972) concluded that the inert character of Fe(III)–phosphate complexes is due to the formation of a bridging ligand structure of binuclear complexes on the predominantly exposed (100) goethite crystal faces (see also Atkinson et al. 1971). They further indicate that there is only one site per two unit cell areas on the (100) face where steric limitations allow these binuclear complexes to form. Binuclear sites on the (100) and (010) faces were due to unshared apices of adjacent Fe octahedra (Atkinson et al. 1974), and the processes involved in adsorption were further investigated by Parfitt et al. (1975, 1976). In the case of phosphate adsorption by goethite, the phosphate ion replaces two singly coordinated surface OH groups. Two of the O atoms in the phosphate ion became coordinated with Fe^{3+} resulting in a binuclear surface complex as represented below.

$$\text{H–O–Fe} \quad \text{H–O–Fe} \diagdown_O \diagup + HPO_4^{2-} \longrightarrow \begin{array}{c} O \diagdown \diagup OH \\ P \\ O \diagup \diagdown O \\ | \quad\quad | \\ Fe \quad Fe \\ \diagdown O \diagup \end{array} + 2OH^-$$

These reactive singly coordinated surface OH groups also occur in pairs on selected faces of hematite, lepidocrocite and akaganeite (Parfitt *et al.* 1975) allowing binuclear bridging complexes to form with such anions as phosphate and changing the surface charge of these oxides to more negative values. Parfitt and Russell (1977) and Parfitt and Smart (1977) showed that sulphate is similarly adsorbed on goethite surfaces, though less strongly than is phosphate. Parfitt and Smart (1978) presented evidence that this complex is also formed at surfaces of lepidocrocite, akaganeite, ferrihydrite and hematite.

The OH groups in the goethite surface may be coordinated to either 1, 2 or 3 Fe atoms, and the energy of adsorption may be insufficient to exchange OH groups with higher coordinations. Ionic size or steric interaction between adsorbing species may limit adsorption. In some cases monodentate bonding occurs, and transitions between bidentate and monodentate can occur with increasing concentration of the adsorbing species (Parfitt *et al.* 1977a), allowing further adsorption.

The observation that different iron oxides exhibit some degree of specific adsorption towards the same ionic species suggests that the results obtained for synthetic goethites, on which most studies have been made, may be applicable to the other minerals in this group. Moreover, Schwertmann and Taylor (1977) showed that the adsorption of phosphate per unit surface by synthetic goethites (from the results of Atkinson *et al.* 1972) was similar to that displayed by natural goethites (Taylor and Schwertmann 1974c), about 0.08 mg P m^{-2} of goethite surface. Ryden *et al.* (1977a) also demonstrated that the phosphate sorption exhibited by many soils is similar to that shown by the Fe oxides. However the same workers (Ryden *et al.* 1977b) found that adsorption of P from solution approached equilibrium more slowly in a soil or in natural goethite and ferrihydrite systems than on synthetic goethite surfaces. They suggest that a transition from physi- to chemi-sorption occurs and that the time-dependent sorption may involve the diffusion of P into structurally porous short range order material present in the natural samples.

There has been much controversy as to whether Fe or Al oxides fix phosphate more efficiently in soils, the arguments being based on the uptake by synthetic samples of each oxide group. In an autoradiographic and electron microprobe analysis of many Australian soils, the Fe oxides were found to be the most reactive, even in the presence of large amounts of gibbsite (see Norrish and Rosser 1983). Contributing to this effect would be the relatively larger particle size of the Al oxides compared to the Fe oxide crystallites.

Schwertmann and Taylor (1977) also tabulated the values obtained by various workers for the specific adsorption of anions and metal cations by different goethite, hematite and ferrihydrite samples. Unfortunately, comparisons cannot be made between the adsorption values obtained per unit surface area of the different minerals because of the wide variations in

TABLE 2.3. References to adsorption of anions by iron oxide minerals

Adsorbate	Fe oxide		
	Goethite	Hematite	Ferrihydrite
Phosphate	Hingston et al. (1968)	Breeuwsma (1973)	
Silicate	McKeague and Cline† (1963)	McKeague and Cline (1963)	
	Hingston et al. (1974)	Beckwith and Reeve (1963)	
Molybdate	Hingston et al. (1974)	Reyes and Jurinak (1967)	Reisenauer et al. (1962)
Sulphate	Hingston et al. (1974)	Aylmore et al. (1967)	Parfitt and Smart (1978)
	Parfitt and Smart (1977)	Breeuwsma and Lyklema (1973)	
		Parfitt and Smart (1978)	
Sulphate (in sea water)	Balistrieri and Murray (1979)		
Chloride (in sea water)	Balistrieri and Murray (1979)		
Borate			Sims and Bingham (1968a)
Fulvic and Humic Acids	Parfitt et al. (1977b)		
Oxalate	Parfitt et al. (1977a)		
Benzoate	Parfitt et al. (1977a)		
Selenite	Parfitt and Russell (1977)	Hamdy and Gissel-Nielsen (1977)	

† McKeague and Cline (1963) also found lepidocrocite could remove monosilicic acid from solution.

experimental conditions used (pH, equilibrium concentrations of adsorbing species and the presence, nature and concentration of the indifferent electrolyte). A list of some inorganic and organic anions specifically adsorbed on iron oxide surfaces and pertinent references are given in Table 2.3.

Although some anions are more strongly adsorbed than others, competition for available sites exists between the various organic and inorganic species capable of being specifically adsorbed. This competition can increase phosphate availability by reducing its adsorption on the oxide surfaces (Hingston et al. 1968). Nagarajah et al. (1970) showed that organic substances expected to be exuded by plant roots can similarly reduce phosphate adsorption on goethite and thereby increase phosphate availability. Large organic molecules possess various functional groups through which adsorption at the oxide surface can occur, and Greenland (1971) has observed additional hydrogen bonding and various types of electrostatic interactions. Such multiple contacts result in stronger associations shown for example in the adsorption of fulvic acid by goethite (Parfitt et al. 1977b).

Krishna Murti et al. (1976) proposed that negatively charged mixed Si–Al polymers can absorb on Fe oxide surfaces, probably behaving similarly to large organic anions. Baird et al. (1977) refluxed synthetic Fe oxy hydroxides with silica at pH 10 for up to 7 days but detected no penetration of Si into the lattice by infrared or electron microscopy although it has been shown for example that lepidocrocite can remove monosilicic acid from solution (McKeague and Cline 1963). Baird et al. attributed the lack of penetration to the H bonding in these compounds or the formation of adsorbed surface phases. The earlier comments on the presence of Si within the oxide structures may reflect its presence in low concentrations in the environment during the formation of particular oxide phases.

The specific adsorption of divalent metal cations on Fe oxide surfaces has also been widely investigated because of its importance in pollution control as well as trace element fixation and availability. Adsorption of the metal as a hydroxy species is generally considered to occur and can be represented diagrammatically by the following bidentate formation as suggested by Forbes et al. (1974):

$$\begin{array}{c} \text{H} \\ | \\ \text{O} \\ | \\ \text{Fe} \end{array} \quad \begin{array}{c} \text{H}^+ \\ | \\ \text{O} \\ | \\ \text{Fe} \end{array} \diagdown_{\text{O}} \diagup + \text{MOH}^+ \longrightarrow \begin{array}{c} \text{OH} \\ | \\ \text{M} \\ \diagup \diagdown \\ \text{O} \quad \text{O} \\ | \quad | \\ \text{Fe} \quad \text{Fe} \end{array} \diagdown_{\text{O}} \diagup + 2\text{H}^+$$

Grimme (1968) found that on goethite specific adsorption followed the sequence $Cu > Zn > Pb > Mn$ whereas Forbes et al. (1976) found the order to be $Cu > Pb > Zn > Co > Cd$. The adsorption of these transition elements under conditions expected in a marine environment was examined by Takematsu (1979) who found the order of uptake to be $Cu > Zn > Ni > Co > Mn$, a similar order to that found above for goethite. Hildebrand and Blum (1974) showed Pb was specifically adsorbed on hematite and ferrihydrite, as well as on the goethite surface, i.e. cations or their positively charged hydroxy species are adsorbed by a similarly charged surface, as discussed earlier. Zn (Kinniburgh and Jackson 1974), Ca (Kinniburgh et al. 1975) and Cu McLaren and Crawford 1974) were found to be specifically adsorbed by ferrihydrite, and Breeuwsma (1973) reported specific adsorption of Li and Mg by hematite. Tewari et al. (1972) investigated the adsorption of Co on magnetite surfaces, and Music et al. (1979) examined its adsorption by hematite in the presence and absence of certain amino acids. The adsorption of cations can be influenced by the presence of inactive salts in the environment, for example, Idzikowski (1971) reported that in the presence of Na, K or Mg nitrates the adsorption of Ag^+, Cu^{2+} and Al^{3+} on hematite was increased. It was subsequently shown that adsorption of these cations from sulphate solutions was three times greater than from the nitrate or chloride systems (Idzikowski 1973).

The interaction of ethylene, propylene and 1-butylene with hematite surfaces leads to the oxidation of the olefins by surface-sorbed oxygen and the formation of bound surface complexes with carbonate or carboxylate structures (Kuznetsov et al. 1977). Irreversible chemisorption of ethanol and butanol also leads to the formation of surface compounds with hematite (Yankovskaya et al. 1979). IR spectra after sorption and desorption of NH_3 on hematite surfaces showed that there was an interaction at adsorption sites to give surface amines or oxides or nitrogen (Belokopytov et al. 1979). IR spectroscopy was also used by Rochester and Topham (1979a) to investigate the adsorption of NH_3, pyridine, NO, CO_2 and methanol on hematite surfaces. The adsorption of NH_3 and pyridine depended on the prior heat treatment which could remove the Lewis acid sites at which adsorption occurred. There was no evidence for adsorption of NO and only a single weak absorption peak at 1320 cm^{-1} was found for CO_2. Rochester and Topham (1979b) used similar techniques to investigate the adsorption of methanol, pyridine, CO_2 and NO on goethite surfaces. Pyridine formed H bonds with the goethite hydroxyl groups and was also adsorbed at Lewis acid sites, while CO_2 and NO were chemisorbed to give HCO_3^-, CO_3^{2-}, NO_2 and NO_3^- surface groups respectively.

2.3.3 Dissolution treatments for iron oxides

As has been stated earlier, crystalline Fe oxides in soils cannot be separated from one another by differential dissolution techniques, as nearly all deferration treatments are based on the reduction of the Fe^{3+} in the oxide to the divalent state. Moreover, the relative rate of dissolution of two different mineral forms in one soil may differ from that of the same two oxides present in the same proportions in another soil. This is because each mineral in a soil has properties affecting its rate of dissolution imparted to it from the environment in which it was formed, e.g. crystal size, deviations from ideal stoichiometry, surface blocking of active sites, different fractions of different crystallographic faces exposed, porosity and the density of surface imperfections.

The sodium dithionate reduction technique of Deb (1950) has been modified in various ways to achieve more satisfactory extractions in various soils. One of the most commonly used modifications is that developed by Mehra and Jackson (1960) which contains bicarbonate buffer and citrate complexing additives. Whereas the unmodified treatment only slightly attacks gibbsite, kaolinite and illite (Segalen 1971b), the citrate addition increases dissolution of clay surfaces and Al oxides. Segalen (1971b) reviewed the various deferration treatments used in soils and investigated the efficiency of the unmodified Deb treatment. He found that for a given number of extractions it depended on the total free iron content, but was not influenced by the pedological character of the soil, except where this affected the total Fe content. He stated that the treatment could also remove Fe from octahedral positions in montmorillonite. Maghemite is less sensitive to dithionite dissolution than hematite with which it is commonly associated but dissolves more readily in hot 0.2M oxalic acid (Taylor and Schwertmann 1974b).

Another successful deferration treatment is the use of ultraviolet light to increase photo-reduction in an acid ammonium oxalate solution (De Endredy 1963). Segalen (1970) showed that for all but the Fe-rich red ferrallitic soils 1-2 hour extraction under these conditions gave results comparable with the multiple treatments of the unmodified dithionite extraction. Furthermore, only trace amounts of Si and Al were extracted. The variations in efficiencies of the various treatments suggest that prior identification of the Fe and other minerals present may help in choosing a treatment most suited to that particular soil fraction, and investigations along these lines could be rewarding.

Although the crystalline forms cannot be effectively separated by dissolution techniques, a partition of different forms of Fe in the soil can be made by a series of extractions. Treatment with 0.1M sodium pyrophosphate will remove the Fe–organic complexes (Aleksandrova 1960) and a subsequent extraction with 0.3M ammonium oxalate at pH 3 in the dark, which prevents photochemical reduction by ultraviolet light and dissolution of the better crystallized oxides also, was designed by Schwertmann (1959a, 1964) to remove the ferrihydrite. A final treatment with dithionite then removes the crystalline forms and analysis gives the type and distribution of Fe in the sample. These sequential treatments have been used successfully by McKeague *et al.*

(1971) and Weber *et al.* (1974). There are however limitations to these techniques. The pyrophosphate also removes Al–organic complexes, and the acid ammonium oxalate can also remove crystalline Fe oxides if a source of Fe^{2+} is present, e.g. siderite (Schwertmann 1959b) or maghemite, which has been shown to be susceptible to this treatment (McKeague *et al.* 1971; Pawluk 1972). A solution of 2M NaOH and ethylene glycol in the ratio of 10:1 has been found to bring ferrihydrite into solution without any obvious dissolution of the more crystalline iron oxides.[4] Dithionite will also extract Mn oxides which can contain Fe and so creates further uncertainties.

Karapetyan and Petrova (1972), during acid dissolution experiments on hematite and magnetite ores, observed that isomorphous impurities as well as crystal morphology and particle size can affect the rate of dissolution. Norrish and Taylor (1961) reported that the goethite fractions with the highest degree of Al substitution were the most difficult to solubilize by dithionite. Cornell *et al.* (1974) have studied the acid dissolution of synthetic goethites with various morphologies and their results may give some insight to the stability and solubility of soil Fe oxides. Anisotropic dissolution occurs with greater attack on those crystal faces where there are greater proportions of singly or doubly coordinated OH groups. For example there are only singly coordinated OH on the (001) face, the (010) face having singly and doubly coordinated groups. The (100) surface, having 1, 2 and 3 coordinated groups, shows lowest susceptibility to dissolution. If during pedogenesis foreign ions, which can restrict crystal development in certain directions, are present, the stability of the mineral could be altered due to a change in the proportion of the most vulnerable sites.

2.3.4 Association of iron oxides with other soil components

Reactions between the Fe oxides and other soil components can modify both the physical and chemical properties of the soil. Organic materials other than those discussed under adsorption may have an affinity for the normally positively charged oxide surfaces, and by blocking adsorption sites may reduce the phosphate or similar nutrient anion retention (Greenland 1971). The association may retard decomposition of the organic matter and be responsible for the lack of positive charge on the free oxides in surface soils (Deshpande *et al.* 1964a). Interaction between Fe oxides and organic matter may also be responsible for aggregation under certain conditions. Greenland (1971) states that ferrihydrite might be expected in very young soils, especially if actively weathering Fe-rich minerals are present (Oades 1963), or under conditions of periodic solution and reprecipitation of Fe. Coulombic attachment of the ferrihydrite to negatively charged clay surfaces may allow organic polymers with their multiple points of contact to effect a bridging between neighbouring "coated" clay particles. Since the organic matter can stabilize the ferrihydrite against crystallization (Schwertmann 1966), this mechanism may be an important factor in aggregation in some soils or horizons.

There are however opposing views on the ability of iron oxides to promote aggregation purely by interaction with inorganic soil constituents. Deshpande *et al.* (1964b), working with highly weathered soils with total Fe_2O_3 contents from 2 to 15%, found Fe was generally present as discrete mineral particles and did not influence the stability of aggregates by cementation. This discrete nature of the Fe minerals, even when they were present on the surfaces of clay minerals has also been noted by Fripiat and Gastuche (1952) and Greenland *et al.* (1968). On the other hand Kemper (1966) obtained a correlation between Fe content and the water stability of aggregates. Blackmore (1973) and Schahabi and Schwertmann (1970) found that freshly precipitated ferrihydrite promoted aggregate stability more than the more crystalline oxides, and Greenland and Oades (1968) found that hydrolysis of a $FeCl_3$ solution at pH 3 bonded kaolin particles into larger masses.

Such effects could occur on a smaller scale by hydrolysis following concentration of soluble Fe during a drying out period. The more notable and obvious effects of cementation involved in the formation of pisolites, lateritic crusts and bog-iron ores are due to crystal growth. Schnitzer and Kodama (1977) suggest that interaction between organic material might be involved to some extent in the hardening of laterites. This is presumably due to their ability to dissolve and mobilize Fe(III) from less crystalline oxides. Subsequent destruction of the organic matter in the presence of the more crystalline oxides will accelerate further crystalline precipitation and so increase induration.

2.4. THE ALUMINIUM OXIDE GROUP OF MINERALS

As opposed to the Fe minerals listed in Table 2.1, only two of the Al oxides listed commonly form under soil conditions. These are gibbsite (hydrargillite), γ-$Al(OH)_3$, correctly called trihydroxide but often referred to as Al hydroxide, and boehmite, γ-AlOOH, the oxy-hydroxide. The trihydroxides bayerite and nordstrandite, further polymorphic forms of $Al(OH)_3$, have also occasionally been identified in soils. Diaspore, α-AlOOH, and corundum, α-Al_2O_3, are occasionally found in bauxite deposits, but are rare in normal soil profiles. At times this may be due to insufficient concentrations in the soil to allow positive identification by the normal X-ray diffraction techniques. This is in marked contrast to the ubiquitous occurrences of their isostructural Fe analogues, goethite and hematite.

As with the Fe oxide minerals, each Al oxide is favoured by particular environmental conditions and in some cases the conditions of syntheses can be seen to have some relevance to natural occurrences. According to Segalen (1971a) the occurrence of free Al oxides in soils is mainly limited to the warm and humid climates, particularly in wetter areas with good drainage. Some metastable varieties have been found in arid conditions which probably inhibited transformation to the more stable minerals, e.g. the occurrence of bayerite in Hatrurim, Israel (Bentor et al. 1963). Hsu (1977) has published a detailed review which covers the crystallography, IR, X-ray diffraction and DTA data, and possible formation paths of the different Al oxides. Readers are also referred to the work of Schoen and Roberson (1970) dealing with the variation in structures between the mineral varieties and its geochemical implications. The natural occurrences of the Al oxide minerals and possible formation paths are also discussed. However, the detailed chemistry of Al oxide formation in soils remains obscure, mainly due to the uncertainties in the formation and type of hydroxy polymers involved.

In weatherable primary minerals Al exists in the trivalent state, generally in complex aluminosilicates, and its release into the soil system stems from the breakdown of these silicate structures. According to Segalen (1971a) the alkali and alkali earth metals are removed by solution and replaced by the smaller H^+ ion. This weakens the structure and affects the Si—O—Al bonds and subsequent breakdown causes the separation of Si and Al. This reaction proceeds from a continuous supply of H^+ arising from the conversion of dissolved CO_2 into HCO_3^-.

Depending on the environmental conditions (e.g. pH, drainage, the presence or absence of complexing organic or inorganic anions) the breakdown products can be (i) simultaneously removed through leaching, (ii) react together and form pedogenic clay minerals, or (iii) undergo differential removal, e.g. the precipitation of Al while Si is removed as a soluble species. This differential removal is possibly due to the large differences in the respective solubilities of their breakdown products ($Si(OH)_4$ and some hydrolysed Al species), and the rates at which these solubilities decrease[8] as the pH increases from say 4 to 6. At pH 4 pSi is approximately 2.57 (75 mg Si litre^{-1}) whereas pAl is around 3.7 (6 mg Al litre^{-1}). On increasing the pH to 6, pSi drops to 2.7 (50 mg Si litre^{-1}) whereas pAl is near 6 (0.03 mg Al litre^{-1}).

With a change in environment, the clay minerals previously formed by condition (ii) above may be subject to weathering conditions favouring the accumulation of Al oxides. The conditions for this accumulation are largely determined by [Si] in the soil solution. Gardner (1970) states that for the formation of gibbsite from the dissolution of kaolin, the infiltrating water must have a pH $\geqslant 4.2$ and at this pH must contain $< 10^{-4.6}$ mol litre^{-1} of dissolved silica. Kittrick (1969) states that this concentration is controlled by (i) the rates of dissolution of unstable silicates, (ii) the dilution and removal of silica charged soil solution, (iii) the rate of precipitation of neoform stable silicates, and (iv) the uptake of SiO_2 by plants. Moreover this Si concentration in the soil solution cannot arise from the solubility of quartz.

Other environmental factors can influence the relative Si and Al concentrations at a particular pH level by complex formation or a form of selective dissolution. For example, Huang and Keller (1970) studied the solubilities of Si, Al, Fe, Ca and Mg derived from various primary silicates in pure H_2O, CO_2 charged water and in the presence of citric, salicylic and tartaric acids, and noted that the [Al] increased markedly due to complex formation.

Given conditions that permit free Al oxide accumulation in soils, the actual processes leading to its precipitation and crystallization are not fully understood and many different hydrolysed species have been suggested to be involved. Below pH 4 Al is present as the hydrated $\{Al(OH_2)_6\}^{3+}$ but at slightly higher pH the soluble species $Al(OH)^{2+}$ or its dimer $\{Al_2(OH)_2\}^{4+}$ is expected. Initial hydrolysis of the hydrated Al^{3+} cation in acid solution is classically represented by

$$\{Al(OH_2)_6\}^{3+} \rightleftharpoons \{Al(OH_2)_5OH\}^{2+} + H^+$$

Various polymers resulting from subsequent hydrolysis have been suggested by different workers. Matijevic et al. (1961) suggested that octamer $\{Al_8(OH)_{20}\}^{4+}$ as the principal species in aqueous solution whereas Brosset et al. (1954) proposed a hexamer $\{Al_6(OH)_{15}\}^{3+}$ because it possessed a structure similar to gibbsite, an ultimate product of polymerization. Hayden and Ruben (1974) suggested also $\{Al_7(OH)_{17}\}^{4+}$ as a possible species but concluded that the principal complexes in acid solution were the monohydroxy $Al(OH)^{2+}$ ion and the octamer which became dominant just prior to precipitation. Akitt and Farthing (1981) suggested that the various species postulated may result from different conditions and times of hydrolysis. They stated that these various species may exist but not in freshly prepared solutions that are rapidly hydrolysed. Using nuclear magnetic resonance techniques Akitt and Farthing found that the original monomeric aquo complex $\{Al(OH_2)_6\}^{3+}$ decreased continuously to zero as the ratio OH/Al increased to 2.5, and that the ions $\{Al_2(OH)_2\}^{4+}$ and $\{AlO_4Al_{12}(OH)_{24}(OH_2)_{12}\}^{7+}$ formed sequentially as hydrolysis proceeded. This latter species can be rewritten $\{Al_{13}(OH)_{32}(OH_2)_8\}^{7+}$ which does not appear to be very different from the species proposed by other workers. The above formula can be reduced to $\{Al(OH)_{2.46}^{0.54+}\}_{13}$ which is very similar to the $\{Al(OH)_{2.56}^{0.44}\}_p$ species (p is an integer) proposed by Brown and Newman (1973) from studies on the adsorption of partially hydrolysed Al species on Wyoming bentonite. Akitt and Farthing's large ion can be thought of as an $Al(OH)_4^-$ nucleus surrounded by apically bound $\{Al_2(OH)_2\}^{4+}$ ions.

Environmental anions, although not affecting the initial hydrolysis (Akitt and Farthing 1981), influence polymerization because their incorporation during ageing may require a higher OH subsequently to replace the anion and induce precipitation. Ross and Turner (1971) found that the inhibition of crystallization by foreign anions was related to their ability to be incorporated in the hydroxy species, the more readily incorporated having the greater influence.

Besides the isomorphous replacement of Fe by Al in the Fe oxides already discussed, the complementary effect has also been observed for the Al oxides (Beneslavsky 1957). Also by analogy with their effects on the formation on Fe oxides, foreign ions could be expected to exert an influence on the formation conditions, and the direction of mineralization of the Al oxides. For example, Gastuche and Herbillon (1962) noted that inorganic salts present during the ageing of $Al(OH)_3$ precipitates slowed down crystallization and Hsu (1977) gives references to the effects of other foreign ions. However, Hsu states that the influence of organic acids capable of complexing Al may be more important than inorganic soil components, and quotes the inhibitory effects of citric acid on the crystallization as reported by Kwong and Huang (1975) (cf. the influence of citrate on the crystallization of Fe oxides, Schwertmann *et al.* 1968). Such effects may be important in the formation of boehmite as will be discussed later. However, in summarizing their results and those of other workers Schoen and Roberson (1970) emphasize that most *natural* foreign ions only slow down but do not change the direction of mineralization on which pH exerts the controlling influence. This was also demonstrated by Violante and Violante (1980) who found that the effectiveness of inhibition of crystallization increases with the concentration of the anion and lower pH values. The formation of moderate or strong chelating bonds with some organic anions gives stable pseudoboehmite or amorphous material as final products.

The subsequent evolution of a solid phase precipitate evolves from the condensation of the soluble polymerized species, leading to a reduction in surface charge thus reducing repulsion of further polymeric species (Hayden and Rubin 1974) and allowing the development of sheets of close packed hydroxyls with Al in octahedral coordination. Using different initial polymeric species Hsu (1977, p. 120) graphically depicts the condensation of these polymers to give a sheet structure of composition approaching $Al(OH)_3$ with further hydrolysis. The stacking sequence of these basic sheets then gives rise to the various crystalline polymorphs of $Al(OH)_3$, and presumably to the replacement of some OH by O leading to the formation of boehmite.

Above the pH range of minimum solubility the rapidly formed $Al(OH)_3$ flocculant precipitate reacts with the $Al(OH)_4^{-1}$ ion which is considered to be the dominant soluble species in the alkaline range. According to the equation

$$Al(OH)_3 + Al(OH)_4^{-1} \rightarrow 2Al(OH)_3 + OH^{-1}$$

the precipitate is subject to continual growth at the expense of the aluminate ions during ageing and crystallization, with an accompanying increase in pH.

2.4.1 Forms of aluminium oxide in soils

2.4.1.1 Gibbsite (hydrargillite)
This is the most common pedogenic form of $Al(OH)_3$ and occurs in many different soil types. It is often the principal mineral in bauxite deposits and commonly occurs in laterites. Bauxite can form from a great variety of parent materials and requires a warm humid climate with a high rainfall and a suitable topography for its accumulation (Schoen and Roberson 1970).

Segalen (1971a) stated that, in addition to warm conditions and high rainfall, good drainage was also required for gibbsite formation. Often its occurrence in soils of cooler climates may be attributed to an earlier formation under a warmer environment (e.g. Wilke and Schwertmann 1977). Kittrick (1969), from thermodynamic studies, concluded that gibbsite should not occur with montmorillonite, but should be found in association with kaolinite. An idea of the distribution of gibbsite in various soil types and of the minerals with which it is commonly associated is given in the bibliography published by the Commonwealth Bureau of Soils (1957–1967). Generally these occurrences support the requirements for the climatic conditions mentioned, and the association with kaolin rather than montmorillonite.

Gibbsite consists of sheets of hcp layers with open packing between successive sheets (Fig. 2.10). In the lateral extension of the hcp sheets each Al is octahedrally coordinated to six OH and each OH is coordinated to two Al cations, leaving one octahedral interstice vacant. According to Schoen and Roberson (1970), this distribution causes a certain degree of polarization which influences the direction of the surrounding O—H bonds, displacing the protons towards the vacant sites. Additionally, two positively charged Al cations can further polarize their shared OH, inducing areas of positive and negative charge on the outer surfaces of the hcp sheet. Bonding with oppositely charged areas of the next hcp sheet is such that the OH of neighbouring sheets sit directly above one another and not in the position of closest packing.

Although gibbsite can be produced in the laboratory under highly alkaline conditions, Hsu (1977) emphasizes the anomaly that in nature it is confined to highly weathered and acidic soils.

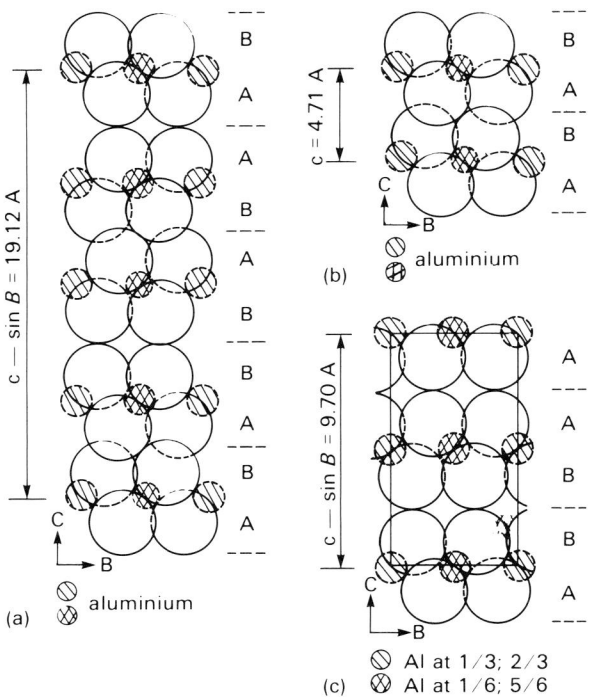

FIG. 2.10. Stacking arrangement of octahedral layers in (a) nordstrandite, (b) bayerite and (c) gibbsite. (After Schoen and Roberson 1970.)

However, other laboratory syntheses (Barnhisel and Rich 1965); Hem and Roberson 1967) confirm its formation at acid pH values with the appearance of its polymorph bayerite at higher pH. Hsu (1966) obtained similar results and believed the source of OH ions was the determining factor. Whereas at higher pH there exists a ready source of OH resulting in an immediate precipitation of the amorphous Al(OH)$_3$ which would age to bayerite, the hydroxyls at the lower pH were derived from the dissociation of water and favoured the much slower hydrolysis and subsequent formation of gibbsite. Hsu (1979) also showed that foreign anions inhibited the crystallization of precipitated Al hydroxide to gibbsite, and noted that the influence decreased in the order phosphate > silicate > sulphate > chloride > nitrate > perchlorate. This is similar to the effect of some of these anions in inhibiting the crystallization of ferrihydrite to hematite or goethite. Bardossy and White (1979) state that the carbonate anion also has an important influence in inhibiting crystallization as is evidenced by the smaller grain size of gibbsite in karstic bauxite deposits as compared to lateritic bauxites. According to Violante and Violante (1978) montmorillonite could be expected to favour the formation of stable pseudoboehmite, which is easily formed in clay-free systems in the presence of certain organic anions capable of complexing the Al.

The results of Gastuche et al. (1964) suggest that the removal of foreign ions through dialysis accelerates the formation of gibbsite and this agrees with its formation under natural conditions of high rainfall and good drainage (Segalen 1971a). In addition the results of Gastuche et al. showed an interesting feature in that during the ageing of mixed Al–Fe(III) gels at pH 4.6, water was absorbed by the dialysis bag in the highly aluminous samples (Al/Al + Fe ⩾ 0.9), whereas in the more ferruginous samples water was expelled. The affinity of the aluminous samples for water is interesting when considered in conjunction with the observations that under pedogenic conditions the anhydrous α-Al$_2$O$_3$ is rare relative to its Fe analogue, and the trihydroxide species of Al are common whereas in the Fe oxides only the oxy-monohydroxide is found.

This retention of a higher degree of hydroxyl coordination in the Al minerals in relation to the Fe oxides may be due to many causes. A possible explanation may be the higher solubility of the Al-hydroxy species above a certain degree of hydrolysis and the greater rate of subsequent polymerization above pH 5 of these Al species.[9]

2.4.1.2 Bayerite
This polymorphic form of Al(OH)$_3$ is readily synthesized under alkaline conditions by the rapid hydrolysis of Al salts. It develops a pyramidal crystal form in contrast to the platy habit of gibbsite. It possesses an hcp structure, and the process of its crystal development is discussed by Schoen and Roberson (1970) who attribute the difference between its structure and that of gibbsite (Fig. 2.10) to the degree of polarization to which OH in the respective structures is subjected. At the lower pH where Al(OH)$^{2+}$ species give rise to polymers resulting in gibbsite, the full polarizing power of the Al^{3+} is directed towards a single OH, giving a greater degree of polarization and leading to the distortions discussed earlier. With bayerite, formed at higher pH, the presence of the Al(OH)$_4^-$ species results in a lower degree of polarization of the coordinated OH groups in the subsequent polymerized species. Intensive grinding of bayerite causes a structural breakdown and dehydration and the resultant amorphous material transforms directly to α-Al$_2$O$_3$ (Kacsalova 1979).

As mentioned earlier there have been reported occurrences of bayerite in nature, but not in soils. In Hatrurim, bayerite reported by Bentor et al. (1963) occurs in veins associated with calcite and gypsum, suggesting higher pH conditions which are known to favour its formation. Schoen and Roberson also refer to a recent report by Khorosheva (1968) who identified bayerite in association with gibbsite and diaspore in the weathering crust of amphibolites and serpentine. This and the Hatrurim occurrence are considered to represent meta-stable stages, its persistence in the Hatrurim deposit being attributed to the present arid conditions of the area. Under an aqueous alkaline environment it would transform to nordstrandite which is considered to be the more stable polymorph (Schoen and Roberson 1970). Violante and Violante (1980) found that with increasing stability of the complexes formed between Al and various organic acids, the formation of bayerite was inhibited in favour of gibbsite and/or nordstrandite, and, with stronger complexing, pseudoboehmite or amorphous phases.

2.4.1.3 Nordstrandite
The third polymorph of $Al(OH)_3$, is also rare in nature although readily formed during synthesis, generally in association with gibbsite and/or bayerite under more alkaline conditions than favour gibbsite alone. Structurally (Fig. 2.10) it consists of a stacking arrangement of the basic octahedral sheets that can be interpreted as an alternation of gibbsite and bayerite layers. Schoen and Roberson (1970) attribute its formation to the influence of pH, which, by altering the hydroxy species present, affects the polarizing effect of the Al^{3+} cation. Hsu and Bates (1964) found an influence of anion, obtaining gibbsite with or without nordstrandite from the hydrolysis of Al sulphate solutions, but only pure bayerite from chloride solutions (cf. the formation of TiO_2 polymorphs depending on the anion present). The preparation of nordstrandite from inorganic solutions often results in the formation together of bayerite and gibbsite, depending on the pH of the medium. However Violante and Jackson (1979) found that it can be crystallized pure in the pH range 9–11 from citrate systems with an Al/citrate ratio greater than 10. They also state that montmorillonite favours the formation of nordstrandite under these conditions by inhibiting the formation of bayerite or pseudoboehmite. Ross and Turner (1971), using 80% hydrolysed systems with different anions, found that nordstrandite formed as an early crystalline precipitate but transformed in time to gibbsite. From the arguments of Schoen and Roberson and from many results of syntheses, its formation is seen to be favoured under intermediate pH conditions.

It has been found in nature as crystals radiating into solution cavities in limestone in Guam by Hathaway and Schlanger (1965), and also in Borneo as small pellets in a terra rossa soil (Wall *et al.* 1962), again associated with limestone. It is seen that the pH expected in these natural occurrences is supported by the conditions suggested by laboratory synthesis

The differences between the results of syntheses regarding reaction paths, stabilities, transformations and influences of foreign ions for these three forms of $Al(OH)_3$ may be in part explained by the differences in experimental conditions, e.g. concentrations, rates and degree of hydrolysis, ageing environments, etc. By contrast gibbsite is relatively common in nature, whereas bayerite and nordstrandite are in fact rarities. Furthermore, gibbsite forms by slow hydrolysis in slightly acid conditions where the solubility of amorphous silica is near its minimum. At the higher pH values which favour bayerite and nordstrandite, the solubility of silica is much higher and may possibly be a factor in preventing their formation.

However, bayerite and nordstrandite may be more prevalent in nature than would be indicated by their very few reported occurrences. Where they have been identified, they have generally occurred in pure or concentrated deposits of Al oxides. Their apparent rarity in nature could possibly reflect difficulties in identification, particularly at low concentrations, arising from the similarity of their X-ray diffraction patterns to that of the more common gibbsite, especially for the two diagnostic highest spacings.

2.4.1.4 Boehmite

This mineral, γ-AlOOH, is isostructural with lepidocrocite (γ-FeOOH) and is more common in nature than bayerite or nordstrandite, but much less common than gibbsite, especially in soils. It often occurs in bauxite deposits in association with gibbsite from which it is considered to form through partial dehydration. Rooksby (1961) showed that both gibbsite and bayerite will dehydrate to boehmite if their crystallite sizes are large enough, the finer particles ($<0.2\ \mu$m) dehydrating completely to an Al_2O_3 phase. Harder (1952) considered heat from the sun may be sufficient to transform gibbsite to boehmite in tropical bauxites. Schoen and Roberson (1970) suggests that although temperatures above 130 °C are required for this transformation in the laboratory, the presence of salts in the soil solution may reduce the activity of the water sufficiently to facilitate partial dehydration at lower temperatures. Its occurrence therefore in bauxites that have undergone diagenesis or hydrothermal alteration is expected.

The mechanism of formation by a slow thermal dehydration is supported by the observation of Bridge (1952) who noted that Cenozoic bauxites consisted predominantly of gibbsite with small amounts of boehmite, whereas mesozoic deposits are mainly boehmite; diaspore, α-AlOOH, is generally confined to the older Palaeozoic. This agrees with the thermodynamic studies of Kittrick (1969) which give the stability of gibbsite > boehmite > diaspore in contrast with the order gibbsite < boehmite = diaspore previously accepted.

Occurrences of boehmite in normal soils are rare. Segalen (1971a) states that it has been found in soils derived from limestone, and in Australia it has been identified in earthy and siliceous sands, and in lateritic and gleyed podzolic soils (Stace *et al.* 1968). However it is difficult to identify positively boehmite in a soil, especially if it is poorly crystalline or present in small amounts.

Lippens and Steggerda (1970) believe that many identified boehmite occurrences in European bauxites may actually be pseudoboehmite. This material possesses a short-range order which gives rise to broad diffraction reflections which roughly correspond to those of the more crystalline boehmite, hence the name. However, the diagnostic (020) reflection of boehmite at 6.11 Å is considerably displaced in pseudoboehmite to 6.6–6.7 Å (Rooksby 1961). Tettenhorst and Hofmann (1980) synthesized a series of boehmites which ranged in crystallite size from a few unit cells to about 65 in the b-axis direction (perpendicular to the octahedral sheets), and showed a relationship between the shifts to higher d values of the (020) reflection with increased line broadening. They suggest that boehmite and pseudoboehmite are continuous in their structure, but that the latter is restricted in its crystalline development in the b-axis direction and contains higher amounts of this intercalated water. This poorly crystalline material forms in less well drained areas than does gibbsite, and Segalen (1971a) attributes its formation to the presence of foreign anions which inhibit gibbsite crystallization. Hsu (1967) also observed this effect of salts during synthesis. Since pseudoboehmite alters to crystalline $Al(OH)_3$ rather than to boehmite on removal of the foreign ions from the environment, the occurrence of boehmite in soils other than those derived from bauxites may be less common than believed.

The basic structural unit in boehmite is a double sheet of oxygen octahedra whose centres are partially occupied by Al cations. The stacking arrangement of the three oxygen layers comprising the double octahedral layer is ccp. There are two types of oxygen coordination in this structure. Each O in the middle layer is shared by four octahedra, while those in the two outer layers are shared by only two octahedra. These outer O are additionally H bonded to two similarly coordinated O in the adjacent basic structural units above and below. In a recent refined crystal structure of synthetic boehmite Christoph *et al.* (1979) found that the edge lengths of shared and unshared Al octahedra agree with those of layered Al silicates. On heating, boehmite dehydrates at about 300 °C to γ-Al$_2$O$_3$ which has a spinel structure (cf. lepidocrocite alteration to maghemite). In view of this thermal dehydration the natural conditions required for the transformation of boehmite to the more stable diaspore, suggested by the composition of bauxites of different ages, remain obscure. However, in the next section dealing with the formation of diaspore, further observations suggest that the transformation may be via solution, and this may be facilitated when the boehmite is more poorly crystallized, i.e. when the amount of water intercalated between the octahedral sheets is higher.

2.4.1.5 Diaspore (α-AlOOH)
Isostructural with goethite, diaspore consists of hcp O sheets with Al filling two-thirds of the octahedral sites. The occupied octahedra are linked together by edge sharing to form double chains in the *c*-axis direction. These units are in turn joined to similar units above and below by sharing apical oxygen atoms. Diaspore differs from its ccp polymorph, boehmite, in the coordination of its O, each of which is H bonded to another O. The protons are however more closely associated with one of these oxygen pairs (Hill 1979). It has generally been considered a product of diagenetic rather than pedogenic weathering. However, unless very high pressures are involved, its formation from boehmite without dehydration would be difficult (Ervin and Osborn 1951). Kennedy (1959) discussed diaspore occurrences where high temperatures and pressures were unlikely to have been involved, and concluded from laboratory and field studies that high pressures and temperatures may not always be required for its formation. He stated that topotactic alteration of boehmite to diaspore would not occur, and that the transformation to the more stable polymorph takes place via solution. This occurs in the case of the Fe analogues of these minerals, lepidocrocite transforms to goethite via solution where the reaction is facilitated by higher temperatures but can be retarded or completely inhibited by the ionic environment (Schwertmann and Taylor 1972b).

Further observations supporting the low temperature–pressure formation of diaspore are given by the work of Khorosheva (1968) who observed diaspore, gibbsite and nordstrandite in the weathering crusts on amphibolites and serpentinites, suggesting environmental factors were involved. Gout (1974) investigated bauxite diaspores with 3–4% Fe(III) substitution for Al and which had corresponding increases in unit cell parameters. On heating to 300–500 °C the diaspore transformed to corundum and the lattice Fe was liberated as free hematite. From syntheses Wefers (1967) obtained diaspore under normal pressures at temperatures around 100 °C from an Al$_2$O$_3$–Fe$_2$O$_3$–H$_2$O system, and suggests diaspore could form initially as microscopic outgrowths from its isostructural analogue, goethite. Keller (1978) identified diaspore as a surface weathering product formed by the desilicification of a parent kaolinitic flint clay. He also notes that minor amounts of goethite were present which might have catalysed diaspore formation. Biais *et al.* (1972) synthesized Fe substituted diaspore in an Fe–Al system and also concluded that high temperatures and pressures are not necessary for diaspore formation in bauxites. Rengasamy[9] has also recently synthesized diaspore under ambient conditions in the following way. Nitrate solutions of different Fe/(Fe+Al) ratios were hydrolysed by the addition of NaOH up to a mole ratio of OH/M = 2.5. The precipitates were then dialysed for 6 hours against distilled water and then allowed to age at room temperature at pH 7 for 30 days. For Fe/(Fe+Al) ratios in the range 0.29–0.37 diaspore was obtained in association with bayerite and gibbsite (at 0.29) and bayerite and goethite (at 0.37).

In view of the above evidence, it is suggested that diaspore formation may not require high temperatures and pressures, and could be present in soils derived from parent materials that have not been subjected to diagenesis. Given that gibbsite with a distinctive diagnostic X-ray diffraction pattern cannot generally be detected in soils at concentrations lower than 5% (Jackson 1969), the identification of small amounts of diaspore, with its major diffraction lines among those of common soil components, would be even more difficult.

Since free Fe minerals (presumably oxides) were associated with the diaspores investigated by Gout (1974), and Fe would most certainly be present in the weathering of the amphibolites and serpentines investigated by Khorosheva (1968), this element could have some influence on the formation of diaspore at lower temperatures and normal pressures. This possible influence of iron in diaspore formation is supported by its observed association with other minerals. Kiskyras et al. (1978) found it associated with chamosite and magnetite in bauxite deposits, and suggest that reducing conditions were possibly involved in its formation. They also note that kaolin, which is associated with boehmite, prevents the formation of diaspore. Sijaric (1978) also found diaspore in bauxites associated with hematite and goethite, as well as boehmite. However, other transition metals besides Fe in the environment could have an influence on diaspore formation. Caillere et al. (1978) reported that the mineral was found in bauxites on or within nodules of takovite (a Ni–Al hydroxy-carbonate of the pyroaurite group), erythrite (an hydrated Co arsenate) and sphalerite (zinc sulphide), although these minerals could all possibly contain small amounts of Fe substitution.

2.4.1.6 Corundum
This is an anhydrous hcp oxide and is usually associated with high grade aluminous metamorphic rocks or certain igneous rocks under-saturated with respect to silica (Deer et al. 1962) but is very rarely present in clays. Being a highly resistant mineral, it can persist as a clay-sized particle during subsequent weathering (Jackson et al. 1948). However it has also been found in pedogenic materials, for example, lateritic bauxites of Western Australia.[10] It is suggested by Jackson et al. (1948) that such occurrences are associated with hydrothermal activity, but there may be some particular, and as yet undefined, conditions under which corundum will form directly during pedogenesis, and by its nature, persist in the clay fraction.

2.4.2 Compositional variations within the soil aluminium oxides

Because of the rarity of many of the minerals in soils, and the difficulty in isolating them for analysis, very few data are available as to their typical composition or the trace elements they commonly incorporate. In contrast to the Fe oxides, where slight variations in environmental conditions can possibly account for the occurrence of all common species (see earlier sections of this chapter), the Al oxides appear to need distinct and different conditions for their respective formation. Therefore, a foreign cation, whose ionic radius would permit its incorporation, may be included in a particular structure, but be excluded from a polymorph requiring a higher pH for its formation. This could be due to a change in the surface charges reducing the tendency for adsorption which might be a necessary step in the incorporation, or, at the higher pH, the cation may have been effectively removed from the environment through hydrolysis and precipitation.

Fe(III) is commonly incorporated in the lattice of some of the Al oxides (Beneslavsky 1957; Biais et al. 1972), but the degree of isomorphous substitution is not as high as in the opposite case where the smaller and similarly coordinated Al substitutes in Fe oxides. For example goethite can contain up to about 30 mol% Al substitution for Fe (Norrish and Taylor 1961) whereas diaspore can contain up to only 5% Fe (Deer et al. 1962). A manganoan variety of diaspore with 4.3% Mn_2O_3 has also been analysed and given in this last reference. Deer et al. also give analyses for two boehmites with trace amounts of Fe, Ti, Ca, Mg and Ga (all less than 1%) and 2–3 SiO_2. For gibbsite these authors quote SiO_2, Ti, Fe, Mg, Mn and Ca as having been found but note that some may be present as admixed oxide impurities.

A comprehensive trace element association for the gibbsite and boehmite phases in the Weipa (Australia) bauxite deposit has been given by Jepson and Schellmann (1974). In the sediments on which the bauxite developed, TiO_2 and Ga are very highly correlated ($r=0.97$) with Al_2O_3 which occurs almost exclusively as kaolinite in these horizons. In the weathering bauxite horizon these correlations are less, that for Ga more so than Ti. Moreover, in this zone Cr and Al are more positively correlated ($r=0.86$) than in the underlying sediments. Their examination of the actual mineral phases in which these elements were located showed that boehmite was highly correlated with Ti ($r=0.81$), Cr ($r=0.72$), Ni ($r=0.73$), Y ($r=0.73$) and Zr ($r=0.66$). The gibbsite with which the boehmite was associated in the same horizons was moderately to highly correlated with Fe ($r=0.52$), Cr ($r=0.66$), Ga ($r=0.68$), Sn ($r=0.50$) and Th ($r=0.72$). Hematite in the same horizons was also correlated with Al, V, Cr, Ga, Sn and Th, and was more correlated with gibbsite ($r=0.75$) than it was with boehmite ($r=0.52$). It may be justified to extend their results and state that if these elements were present during the formation of the same two minerals in other situations, they could also be incorporated. Cr has also been found in boehmite (Abdulvaliev et al. 1978), and Be in diaspore–boehmite of Russian bauxites (Lavrenchuk and Eremeev 1979).

Interpretation of these statistical associations lends further confusion to the possible formation paths of the Al oxides in nature. If boehmite forms from the partial dehydration with time from gibbsite, we should not expect to find such differences in the trace element associations of the two minerals from the one profile. It is suggested by Schellmann[5] that boehmite may form directly in zones of higher concentrations of soluble salts. This is in accord with the experiments of Pedro (1964) who weathered basalt for two years in a Soxhlet extractor and obtained boehmite in the zone above the water level, and gibbsite in the zone below. The same effect in nature could be achieved by seasonal variations in the water content in a particular weathering horizon which could give rise to the two mineral phases in intimate association.

2.4.3. Adsorption by soil aluminium oxides

In comparison with the Fe oxides, adsorption of relatively few ionic species and molecules by Al oxide surfaces has been investigated. Generally these studies have been limited to synthetic preparations of gibbsite, bayerite and amorphous precipitated $Al(OH)_3$. For a few ionic species, adsorption by a soil known to contain free Al oxides has been studied before and after removal of these oxides. Here, however, quantitative assessments of adsorption and retention are not always possible because of the possible changed characteristics of the remaining soil minerals after the oxide removal.

Cation adsorption by Al oxides has been little studied, but because of the similarity of their hydroxylated surfaces to those of the iron oxides, any cationic species that is specifically adsorbed by the iron minerals should display a similar affinity towards the Al oxide surfaces, although possibly not to the same degree. For example, Kalbasi et al. (1978) found that α-Fe_2O_3 specifically adsorbed greater amounts of Zn^{2+} from solution than α-Al_2O_3 at the same pH. Voznesenskii et al. (1958) found that Ce and Ru were adsorbed by positively charged amorphous $Al(OH)_3$ surfaces, and Kinniburgh et al. (1975) reported a similar behaviour for Ca and Sr. Subsequently Kinniburgh et al. (1976) showed that there was selective adsorption of alkaline earth elements with $Mg > Ca > Sr > Ba$. The heavy metals were even more strongly adsorbed. At pH values well below the measured isoelectric point of the precipitated $Al(OH)_3$, Pb, Cu, Zn, Ni, Co and Cd were adsorbed selectively in the order given. Tewari et al. (1972) examined the adsorption of Co(II) from solution by Al_2O_3 and suggested that hydrolysis of the Co(II) may be responsible for the pH and temperature dependence of the reaction.

The specific adsorption of anions and molecular species has received much more attention. Phosphate adsorption in particular has been extensively investigated because of its agricultural implications, and the relative importance of the Al as opposed to the Fe oxides in soils in the fixation of phosphate has been widely discussed. Bromfield (1965), from a study of the correlations between P and both Fe and Al in soils, suggests that the Al oxides were more effective. In contrast, Fordham and Norrish,[11] using electron microprobe techniques, found that soil Fe oxides were more reactive even in the presence of a large amount of gibbsite. One reason advanced by these authors is the relatively small particle sizes of the Fe oxides compared to those of soil gibbsites. This difference in particle size agrees with the influences of foreign ions on the formation of the respective oxides discussed earlier; the iron oxides more readily incorporate foreign ions which inhibit crystal growth whereas gibbsite only forms under conditions free from the influence of dissolved impurities which can completely inhibit its crystallization. Hsu (1977) states that controversy also exists as to whether the reaction between Al oxides and phosphate is one of adsorption or precipitation, as phosphate in high concentrations can cause decomposition of the $Al(OH)_3$. The evidence of Parfitt et al. (1977c) confirms by IR analysis that, at low concentrations, singly coordinated OH and H_2O groups on the edge faces of gibbsite undergo ligand exchange with the phosphate similar to the reaction with goethite where singly coordinated (OH) groups are present on all faces. A similar effect for singly coordinated groups on the surfaces of the other Al oxides could be expected. This high affinity of edge sites for phosphate at low concentrations was first suggested by Muljadi et al. (1968a,b) in a study on its adsorption by gibbsite and pseudoboehmite. Readers are referred to the bibliography published by the Commonwealth Bureau of Soils (1950–1960) for earlier work on the fixation of phosphate by Al oxides.

Monosilicic acid is specifically adsorbed from solution by the Al oxides. Jones and Handreck (1963) investigated the adsorption on poorly crystalline boehmite (possibly pseudoboehmite), boehmite and amorphous $Al(OH)_3$. Increasing adsorption with decreasing crystallinity was observed, and for samples of comparable crystallinity the Al oxides were more effective than Fe oxides. McKeague and Cline (1963) showed also that $Si(OH)_4$ adsorption was related to the specific surface of the oxide and found strong adsorption for the anhydrous Al_2O_3. Hingston and Raupach (1967) found that the first layer of silicic acid was rapidly adsorbed by Al hydroxide mineral surfaces, and their results suggested that the slower adsorption of subsequent layers could have formed by polymerization of silicic acid. Hingston et al. (1972), in extending their work on the definition of specific adsorption, showed that the maximum adsorption near pH 9.5, as found by other workers, was at the pH of the dissociation constant (pK_a) of the acid as predicted by theory.

Specific adsorption of sulphate has been widely studied and increases with decreasing pH. Sulphate is less tightly bound than phosphate which can completely replace it on these surfaces (Harward and Reisenauer 1966). However, anions such as chloride which are not specifically adsorbed have little effect in displacing the sulphate.

Al and Fe oxides were considered by Jacobs et al. (1970) to be the soil components mainly responsible for arsenate adsorption. Sims and Bingham (1968b) suggest that these oxides are also the dominant factors in boron retention by soils. Hatcher et al. (1967) noted the decrease in available B after the liming of acid soils and attributed the effect to adsorption by Al(OH)$_3$ precipitated by the rise in pH. They found that B retention by precipitated amorphous Al(OH)$_3$ decreased with the time of equilibration in borate solution, suggesting a decrease in the specific area available for adsorption with time. Sims and Bingham (1968a) also noted that B retention by amorphous Al(OH)$_3$ depended on pH and the degree of ageing, and retention by Al oxides was found to be an order of magnitude higher than for the Fe oxides. The adsorption of molecular HBO$_3$, a possibility suggested by Hingston (1964) and Hatcher et al. (1967) was not indicated by their results, as the uptake decreases in the acid pH ranges. Moreover, the maximum uptake in the alkaline range did not occur at the same pH for both the Fe and Al amorphous precipitates, so that other factors besides ligand exchange, which would produce a maximum adsorption at the pK_a value, could be involved. These authors suggested from their results that precipitation of Al-hydroxy borates might occur.

Parfitt et al. (1977b) demonstrated that fulvic acid is adsorbed on gibbsite by a ligand exchange process involving the singly coordinated OH groups on the (100) edge face, water being formed from the protons of the carboxylic acid and the displaced hydroxyl groups. Sodium fulvate is also specifically adsorbed with the liberation of OH into the solution. Whereas all the singly coordinated OH can be replaced by phosphate, steric limitations in the case of fulvate reduce the adsorption. Multiple contacts with the surface due to the number of carboxylic groups result in a stronger adsorption at lower concentrations of fulvate, but some points of contact may be given up for further adsorption as the concentration increases. The doubly coordinated OH on the (001) face of gibbsite, although unreactive towards phosphate or fulvate in neutral solutions, becomes coated with presumably un-ionized fulvic acid at lower pH, probably due to H bonding. These authors also found that specific adsorption occurred when Na humate was added to gibbsite as indicated by the liberation of OH into the solution.

Oxalate and benzoate adsorption on gibbsite has also been investigated by Parfitt et al. (1977c). At low concentrations and near neutral pH, oxalate is more strongly adsorbed as a bidentate complex on the (100) edge of gibbsite than on the similarly coordinated OH groups of goethite. Benzoate is also adsorbed in a bidentate form at lower concentrations and at higher concentrations further physical sorption occurs possibly through an interaction between the already adsorbed benzoate and the aromatic rings of the acid molecule. Kavanagh et al. (1976) found that the higher molecular weight polyvinyl alcohols were adsorbed to a greater extent than the lower molecular weight polymers. They found no influence of pH or concentration on the adsorption which they attributed to H bonding or van der Waals' dispersion forces.

2.4.4 Association of aluminium oxides with other soil components

Hsu (1977) considers that Al–OH polymers act as small fragments of solid Al hydroxide when they occupy interlayer positions of expandable clay minerals. Here they can be more tightly bound than the cations they replace and their presence leads to a reduction in swelling characteristics as well as aggregation produced from the bonding of adjacent clay particles. Fe oxides generally do not enter these interlayer positions and Deshpande et al. (1964b) consider Al oxides to be more important in promoting aggregation in soils than the oxides of Fe.

The greater effect of Al oxides was also confirmed by Saini et al. (1966) in podzol B horizon material where a higher correlation was found between aggregation and Al oxide than Fe oxide content. In such soils interactions between Al oxides, present as surface deposits on other minerals, and organic polymers would cause bridging effects which would contribute to aggregate stability as in the case of the Fe oxides (see Section 2.3.4). In some cases the COOH groups on the organic matter can cause the dissolution and mobilization of Fe and Al oxides (Schnitzer and Skinner 1963; Schnitzer and Kodama 1977). El Swaify and Emerson (1975), in precipitating Fe and Al hydroxides on to illite and illite–kaolinite mixtures, suggested that the greater effect of Al may be associated with its ability to form films over larger surface areas than the Fe which appears to precipitate at specific hydroxyl sites on the clay surface (Rengasamy and Oades 1977).

As discussed earlier, Al hydroxy species can also inhibit to some extent the formation of γ-phase Fe oxides during the oxidation–hydrolysis of Fe^{2+} solutions around neutral pH values (Taylor and Schwertmann 1978). The ability of Al-hydroxy species to induce the hydrolysis of Fe^{2+} discussed earlier (Taylor and McKenzie 1980) may influence the pedogenesis of other soil minerals. The increase in Eh caused by Al in such systems, see Fig. 2.8, could possibly cause the oxidation and precipitation of mobile Mn^{2+} at lower pH values than would be required in a pure Mn(II) system. This may contribute to the common association of Mn and Fe oxides in soils and offer some explanation for the Al content of lithiophorites which are generally found in slightly acidic soils (Taylor et al. 1964).

2.4.5 The dissolution of aluminium oxides

There are no specific treatments designed to remove only Al oxides. Many Fe oxide removal treatments utilizing citrate, oxalate or other complexing agents may also dissolve some of the Al oxides present, and reagents such as NaOH that attack the Al oxides and have minimal effect on the Fe oxides, can cause partial dissolution of the associated aluminosilicates. Janekovic et al. (1979) reported that acetylacetone dissolved the Fe and Al oxides commonly encountered in soils at a rate depending on the M—O bond, so that some distinction can be made between the minerals of each group.

2.5. THE MANGANESE OXIDE GROUP OF MINERALS

Mn oxides generally constitute $<1\%$ of soil minerals and form from the oxidation or precipitation and oxidation of divalent Mn dissolved during the weathering of primary minerals mainly by the action of carbonic acid. During weathering both Fe(II) and Mn(II) are released together but due to the greater ease of oxidation of Fe(II) at lower pH and Eh values, a separation of these two cations often occurs during pedogenesis.

The Mn oxides are generally modifications of MnO_2 in which substitution of Mn(II) and Mn(III) for the Mn(IV) can produce over 150 variations with compositions ranging from $MnO_{1.2}$ to MnO_2 (Dubois 1963). The reversible substitutions of Mn(II) or Mn(III) for Mn(IV) may be achieved by a topochemical reduction or oxidation without a change in position of the Mn ions in a structure (Feitknecht et al. 1960). The substitutions cause a change in the Mn—O bond lengths and involve a conversion of some O^{2-} to OH^- to maintain electrical neutrality. The partial substitution of lower valence Mn also allows incorporation of foreign ions, e.g. Na^+, K^+, Ca^{2+}, Mg^{2+}, Ba^{2+}, etc., in vacant sites (Ross et al. 1976). After a certain degree of substitution or in situ reduction or oxidation, the structure can become unstable and reverts to a more stable species (Feitknecht et al. 1960). In nature Mn exists only in the $2+$, $3+$ and $4+$ states (Bricker

TABLE 2.4. Some manganese oxide minerals (after McKenzie 1972a)

Mineral name	Other names	Composition
Pyrolusite	βMnO_2, polianite	MnO_2
Ramsdellite		MnO_2
Nsutite	γ-MnO_2, (ρ-MnO_2)	$Mn^{4+}_{(1-x)}Mn^{2+}_{x}O_{(2-2x)}(OH)_{2x}$
Birnessite	δ-MnO_2, manganous manganite	$\{ Na_4Mn_{14}O_{27} \cdot 9H_2O$ $Mn_7O_{13} \cdot 5H_2O$
Todorokite		see text
Cryptomelane	α-MnO_2	$K_2Mn_8O_{16}$
Hollandite	α-MnO_2	$Ba_2Mn_8O_{16}$
Lithiophorite		$(Al,Li)MnO_2(OH)_2$
Groutite	α-MnOOH	MnOOH
Manganite	γ-MnOOH	MnOOH
Partridgeite	α-Mn_2O_3	Mn_2O_3
Hausmannite		Mn_3O_4
Vernadite	δ-MnO_2	$MnO_2 \cdot m(R_2O,RO,R_2O_3) \cdot nH_2O$
Rancieite		$(Ca,Mn^{2+})Mn^{4+}_4O_9 \cdot 3H_2O$

1965), and in soils Wadsley and Walkley (1951) consider that the mean valency is between 3 and 4 which allows quite varied compositional ranges in terms of O, and OH and included foreign cations, see Table 2.4.

Giovanoli (1969) discussed the polymorphism of the various MnO_2 modifications and stated that only two stoichiometric ideal forms exist, viz. pyrolusite and ramsdellite. The latter is rare in nature but pyrolusite is quite common and has been found in soils. Other Mn oxides found in soils are less crystalline than pyrolusite and generally contain relatively high concentrations of foreign cations, some of which are necessary for the particular structure.

The literature on the formation, stabilities, chemical compositions and interrelationships between the Mn oxides is voluminous, but in this present work references will be restricted to general studies or to particular minerals found in soils. Readers are particularly referred to recent reviews on Mn oxides in soils by McKenzie (1972a, 1977, 1978a), Burns and Burns (1975a), Burns et al. (1975) and Ross et al. (1976) and to reviews on marine Mn minerals, where the structure and mineral associations are applicable in many cases to the soil oxide phases (Burns and Burns 1979).

About 20 oxides have been identified as separate minerals (Burns et al. 1975), and reports by Hewett and Fleischer (1960) and Hariya (1961) deal with the parent material, composition and conditions of formation of Mn oxide mineral deposits. Of these only about six are considered to form during pedogenesis or to be commonly found in soils (Taylor et al. 1964; Taylor 1968; Chukhrov et al. 1980); these species have been identified in soils of many countries and climatic conditions and are also common in manganiferous marine nodules. The Mn oxides in soils are often reported to be amorphous, e.g. McKeague et al. (1968), though in many cases this could be due to difficulty in identification. Their generally poor crystallinity, small particle sizes and low concentrations, and the fact that they are often intimately associated with the more crystalline soil minerals, make identification by X-ray diffraction techniques almost impossible without concentration procedures (Taylor et al. 1964; Ross et al. 1976). Structural modification of some of these minerals is common (Turner and Buseck 1981) and is possibly responsible for the periodic description of a claimed new species formed under a particular environment. There is still much controversy over the discrete nature of some of these phases and IR spectroscopy is providing a useful technique in this respect (Potter and Rossman 1979).

2.5.1 Formation of manganese oxides

Due possibly to the mechanisms involved in their precipitation (to be discussed later) Mn oxides occur generally in discrete sites rather than in a more widely dispersed form. Accumulations are often associated with coatings on ped faces, root channel cementation, pore infills and very commonly occur as concretions (this term does not necessarily imply concentric layering, although the type of formation is quite common) of variable size and matrix composition. Taylor et al. (1964) did not observe any relationship between the mode of occurrence and mineralogy which is often governed by the composition of the environment.

Synthesis of the common soil oxides through oxidation of soluble Mn^{2+} with air is not readily achieved at neutral or slightly acid pH values of the environment in which these minerals are sometimes found. Although it is accepted that bacterial oxidation influences the precipitation of Mn(IV) (Krauskopf 1957; Hem 1964; Schweisfurth and Gattow 1966; Sokolova 1968; Mustoe 1981), there is no indication that biological processes are always involved. Ross and Bartlett (1981) state that many previous experiments demonstrating the microbial oxidation of Mn are not specific. Inhibition of microbial activity did not affect the oxidation of Mn in soils, which was faster than could be expected from the bacterial strains present. Thermodynamically, divalent Mn should not be stable in solutions in contact with oxygen in the pH ranges of most natural waters (Morgan and Stumm 1965) although Hem (1964) states that direct precipitation at pH 6 from a solution containing 0.1 p.p.m. Mn^{2+} would require a redox potential of 0.7 volt, which is higher than normally obtained from a solution of atmospheric oxygen. Hem showed that Mn^{2+} adsorbed from solution on other mineral surfaces is oxidized on subsequent drying, suggesting that oxidation is catalysed by such surfaces. On the introduction of further soluble Mn^{2+} the process is repeated. Acceleration of this process may be achieved through the concentration of the Mn^{2+} in solution during evaporation resulting in the soil solution occupying increasingly smaller voids (Drosdoff and Nickiforoff 1940). Oxidation–precipitation may also be induced by specific adsorption at the surface of a MnO_2 phase present, regardless of how this phase was originally formed. Above pH 5–6 specific adsorption could lead to a reversal of surface charge, and consequently an increased OH^- concentration near this surface. This would favour oxidation under the existing concentration of dissolved atmospheric O_2 (Ross and Bartlett 1981).

Similarly, the surface micro-environment of weathering potassium feldspar could provide a sufficiently high pH for oxidation and precipitation and so provide a nucleation site for further deposition. Taylor and Bond (1979) have shown that where dissimilar minerals are in contact and are bathed by an electrolyte solution, an electrolytic current flows between the non-contiguous surfaces, and the surface of one of the minerals acquires a potential that would enhance oxidation at lower pH values. Also the process of oxidation may be accelerated by "valence inductivity" by which Selwood (1948) found that the valence of the substrate metal can exert some influence on the oxidation state of the adsorbed Mn. This interaction was also suggested by Morgan and Stumm (1964) who stated that in order to interpret the kinetic data on the oxygenation of Mn(II) it was necessary to infer that it reacts with the solid oxidation products to form the non-stoichiometric higher valence Mn oxides with O/Mn ratios between 1.1 and 1.8. Another factor which may contribute to the precipitation of Mn(II) is its ability at low concentrations rapidly to flocculate suspensions of MnO_2 within pH ranges of normal stability. Morgan and Stumm (1964) found that 10^{-4}–10^{-5}M Mn^{2+} could cause this flocculation in a 10^{-3}M MnO_2 suspension.

Precipitated ferrihydrite may also assist the oxidation of Mn^{2+} (Hem 1964) and the

176 NON-SILICATE OXIDES AND HYDROXIDES

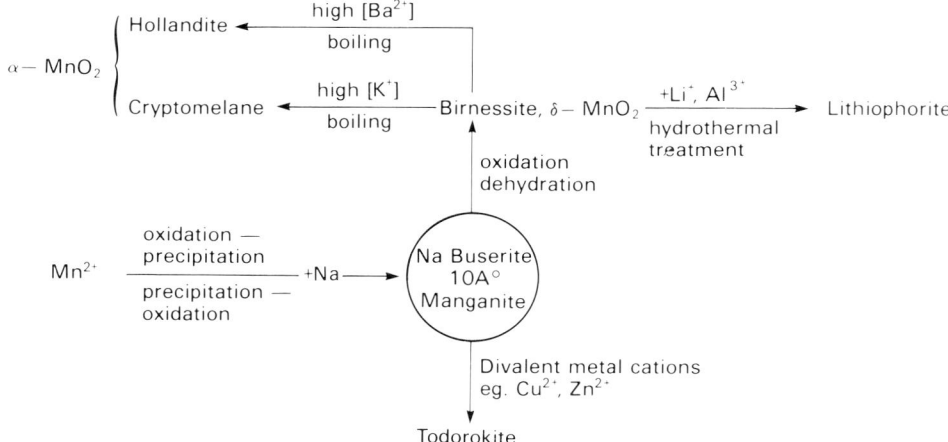

FIG. 2.11. Formation and transformation paths for synthetic Mn oxides. (After McKenzie 1971; Giovanoli et al. 1973; and Wadsley 1950a, b.)

frequent association of the oxides of Fe and Mn in soils and nodules supports this possibility. The reaction

$$2Fe(OH)_3 + Mn^{2+} + 2H^+ = MnO_2 + 4H_2O + 2Fe^{2+}$$

is normally displaced to the left but if the activity of Fe^{2+} is low enough and that of Mn^{2+} sufficiently high (as would occur during the easier oxidation and precipitation of Fe^{2+} at lower pH and Eh) the reaction could shift to the right. The greater adsorption by ferrihydrite of Fe^{2+} than Mn^{2+} would also assist in this catalytic oxidation since it would tend to lower the activity of Fe^{2+}. For this reaction $K = 10^{-5.67}$, $\Delta G° = 7.73$ and $\Delta G = 7.73 + 1.364 \log ([Fe^{2+}]^2/[Mn^{2+}]) + 2.728$ pH (Hem 1964; McKenzie 1978a). Burns and Burns (1975b) suggest that MnO_2 and ferrihydrite are structurally similar and that epitaxial growth of the two species inhibits recrystallization of the ferrihydrite and birnessite and initiates nodule growth. They suggest however that post-depositional crystallization may produce birnessite and todorokite.

Although the possibility of inorganic oxidation and precipitation under conditions that could be expected in soils is suggested from the above observations, the crystalline minerals found in soils have generally been synthesized under non-pedogenic environments and conditions for their natural formation are uncertain. The high concentrations of foreign ions in the soil environment, which preclude the formation of the more crystalline minerals such as pyrolusite, may play some part in the acceleration of oxidation and precipitation of the mineral species in which they are commonly incorporated, often as necessary constituents. Readers are referred to the work of McKenzie (1971) on the synthesis of some soil Mn oxides. Some information and transformation paths found by different workers during syntheses of Mn oxides are shown in Fig. 2.11.

2.5.2 The manganese oxides present in soils

Birnessite, lithiophorite, todorokite, the hollandite group, and, to a lesser extent, pyrolusite, psilomelane, vernadite and rancieite are the most common crystalline Mn oxides found in soils

(Taylor et al. 1964; Taylor 1968; Gallaher et al. 1973; Ross et al. 1976; Chukhrov et al. 1980). Apart from pyrolusite which is stoichiometric MnO_2, the compositions of these other minerals are quite variable.

2.5.2.1 Birnessite

This mineral is the naturally occurring form of the synthetic $\delta\text{-}MnO_2$ or manganous manganite. Other names which have been applied to members of this group include 7 Å manganite, manganese(III) manganate(IV) and manganous(II) manganate(IV). It was first reported in nature in a Scottish soil by Jones and Milne (1956) and analysis gave its formula as $(Na_{0.7}Ca_{0.3})Mn_7O_{14} \cdot 2.8H_2O$. Synthetic preparations by Giovanoli et al. (1970a,b) give typical formulae as $Na_4Mn_{14}O_{27} \cdot 9H_2O$; $Mn_7O_{13} \cdot 5H_2O$ and $Mn_7O_{12} \cdot 6H_2O$. Analyses of soil birnessites have been made on ignited hydrogen peroxide extracts (pH 3) and a comparison of the mean composition of samples from five Australian soils (Taylor et al. 1964) and six soils from Indiana (Ross et al. 1976) is shown in Table 2.5. Carlson et al. (1977) analysed a birnessite from Finland and detected, in addition to the common impurities listed in Table 2.5, trace amounts of Ti, Sn, Pb, Cr, Ni, Zn, Cu, Cd and Rb.

TABLE 2.5. Compositions of some soil and mineral formations of Mn oxides

	Birnessite		Lithiophorite			Hollandite		Vernadites$^\phi$	
	Australian† soils	Indiana‡ soils	Mineral sample$^\tau$ 2	3	Australian† soils	Australian† soils		Kurchatov	Lepkhe-Nelm.
Mn_3O_4	83.8	87.7	58.2	58.3	73.2	65.3	MnO_2	46.5	56.5
							MnO	1.1	4.8
SiO_2	1.38	0.99	1.12	0.79	3.8	6.7		0.8	1.3
Al_2O_3	1.0	2.21	26.72	26.32	11.5	5.7		1.0	1.0
Fe_2O_3	0.94	0.95	0.37	0.53	4.7	6.6		10.5	7.0
CaO	3.16	1.93		0.18	0.20	5.7		2.2	5.2
MgO	1.13	0.63		0.02	0.25	1.03		2.6	0.3
BaO	4.5	5.0	0.66	0.06	3.6	6.0			1.9
Li_2O	0.06	0.012	1.85	1.74	0.16	0.12		ND	ND
K_2O	0.22	0.19	0.29	0.24	0.32	0.37		0.6	0.2
Na_2O	0.09	0.26	0.12	0.14	0.20	0.14		2.3	0.1
CoO	0.57	ND	3.30	2.52	1.53	1.4		3.4	
NiO	0.43	ND	1.47	1.79	0.13	0.31		1.0	

$\dagger$ Taylor et al. (1964). Mean values of analyses on ignited sample extracts.
$\ddagger$ Ross et al. (1976). Mean value of analyses on ignited samples.
τ Fleischer and Faust (1963) samples 2 and 3, converted to H_2O free (ignited) basis.
ϕ Analyses not on ignited samples. Complete analyses given by Chukhrov et al. (1980).
ND = Not determined.

It possesses a double layer structure (Giovanoli et al. 1970a) with the main layer consisting of Mn(IV)–O octahedra. In this layer every sixth octahedral site is vacant, and Mn^{2+} and Mn^{3+} ions situated in the second layer, which is occupied by H_2O, OH and foreign cations, are considered to lie above or below these unoccupied sites. These ions are coordinated both to the O in the main octahedral layer and to H_2O and OH^- groups in the intermediate layer. The position of Na or other alkali and alkaline earth cations in this second layer is uncertain (Burns and Burns 1975a). These foreign cations are not essential to the structure but are the result of environmental impurities present during formation of natural phases.

The original differentiation between forms of birnessite and δ-MnO_2 was made on the basis of the degree of oxidation, the δ-MnO_2 was considered to have an O/Mn ratio greater than 1.9 (Feitknecht and Marti 1945; Buser et al. 1954). Originally the two varieties were distinguished by the absence of basal reflections around 7–7.2 Å and 3.5–3.6 Å in X-Ray diffraction pattern of the δ-MnO_2. Bricker (1965) considered the presence of only two diffraction lines from δ-MnO_2 rather than the four displayed by the more crystalline birnessite was due to the small particle sizes of the former, while other workers attribute the lack of basal reflections to disorder as well as particle size. Chukhrov et al. (1980), however, consider this poorly ordered structure to be a separate mineral, vernadite, which is discussed later. Another variation is the mineral rancieite, and Chukhrov et al. (1980) have confirmed that the difference between this variety and birnessite is due to the dissimilar interlayer cations. However, Perseil and Giovanoli (1979), in reinvestigating the original type rancieite mineral, suggest that it should no longer be regarded as a valid separate mineral species as they consider it to be a mixture of buserite at various stages of dehydration to birnessite. By contrast, Potter and Rossman (1979) compiled IR data on most of the MnO_2 phase minerals, and state that the spectra of well-ordered birnessite and rancieite suggest different structures, and the two should be regarded as distinct minerals. These workers found that there was no structural differences between the 10 Å buserite phases and their dehydration products and support Giovanoli et al. (1971) in the conclusion that the buserite forms should be regarded as hydrates of the 7 Å material.

Many reported soil occurrences of birnessite are associated with a sandy textured matrix. Ross et al. (1976) stated that horizons presenting most evidence for birnessite or a precursor were sandy in texture and had developed in zones of fluctuating water table where the oxide filled voids around sand grains. Carlson et al. (1977) described a deposit formed within the last twenty years which occurs with Fe oxide cementing glaciofluvial sands. Burns and Burns (1975b) suggested that Mn marine nodules nucleated similarly on various mineral surfaces or hard skeletal remains of organisms which had been previously coated with Fe oxide. In Australian soils birnessite was generally associated with near neutral to slightly alkaline horizons, and was differentiated from the other commonly occurring oxide, lithiophorite, by a significantly lower Li_2O/CaO ratio, viz. 0.022 v. 0.80 (mean values, Taylor et al. 1964), due mainly to the higher Ca contents of birnessites. However, Ross et al. (1976) did not find this association with alkaline horizons and found lower Ca and Mg content. Although the alkali and alkaline earth cations are commonly present in birnessites they are not considered essential (McKenzie 1971).

Birnessite may be synthesized by oxidation of $Mn(OH)_2$ under alkaline conditions. If the pH is controlled by NaOH, the oxidation leads initially to the formation of Na buserite (Giovanoli et al. 1970a), also referred to as 10 Å manganite, which dehydrates in air to birnessite without alteration of the Mn octahedral framework. KOH control of pH during oxidation leads instead directly to birnessite formation (McKenzie 1971).

Hariya and Kikuchi (1964) claimed bacteria were responsible for the oxidation and precipitation of Mn^{2+} from mineral spring waters with a pH of 6.8 and a Mn^{2+} concentration of 2.8–4.7 p.p.m. The freshly precipitated material was amorphous but older formations had recrystallized to birnessite and pyrolusite.

2.5.2.2 Vernadite

This mineral was described by Betekhtin (1940) before the more crystalline birnessite was reported by Jones and Milne (1956). It is reported in the ASTM X-ray diffraction files, but its composition and diffraction data vary, and many workers believe it to be a disordered birnessite lacking regular stacking in the c-axis direction. According to Chukhrov *et al.* (1978) the mineral resembles δ-MnO_2 and may be expressed by the formula $MnO_2 . m(R_2O, RO, R_2O_3) . nH_2O$, where R represents mono-, di- and tri-valent cations. Although vernadite and δ-MnO_2 have X-ray diffraction spacings at approximately 2.4 and 1.4 Å, Chukhrov *et al.* (1980) state that vernadite lacks the basal spacings around 7–7.2 and 3.5–3.6 Å.

Burns and Burns (1979) represent the vernadite structure as a two-layer hexagonal packing of oxygen and water with the octahedral vacancies partially filled by Mn^{4+}, the contents of water and other cations determining the extent of Mn^{4+} occupation. However, Potter and Rossman (1979) concluded that the IR spectra of vernadites are not sufficiently different from those of birnessites for it to be regarded as a different mineral species.

Chukhrov *et al.* (1980) consider that most natural vernadites are precipitated following microbial oxidation of Mn^{2+}, and they have synthesized the compound by similar techniques. They consider the very rapid oxidation of Mn^{2+}, precluding the formation of more crystalline and stable minerals, is the important factor in vernadite formation. Some chemical analyses of naturally occurring vernadite samples are given in Table 2.5.

2.5.2.3 *Lithiophorite*

This mineral, which contains Al and Li as essential constituents, occurred in a variety of Australian soils in nodules, crack infills and small cemented sandy aggregates (Taylor *et al.* 1964). It was subsequently identified associated with a dark clay in a solution pipe in a quarry at Lenham, England (Taylor 1968) and in nodules from Hawaiian soils.[12] In the Australian soils it was generally associated with slightly more acid soils than was birnessite. This is to be expected because of the increased levels of Al in the soil solution at the lower pH values. Ross *et al.* (1976) found no evidence for lithiophorite in the soils they examined, and attributed its absence to the low levels of Li. Fleischer and Faust (1963) analysed mineralized lithiophorite from two localities and found compositions approximating to the general formula proposed by Wadsley (1952), $(Al, Li)MnO_2(OH)_2$. The average of these mineral analyses is compared in Table 2.5 with the mean composition of soil lithiophorites. The marked differences from birnessite appear to be the higher levels of Al, Ni, Co and Li in lithiophorites.

Structurally, lithiophorite consists of alternating sheets of two different types of octahedral layers (Wadsley 1952). One layer is composed of continuous MnO_6 linked octahedra while the second is composed of $[Al, Li(OH)_6]$ layers without vacancies. This leads to a positive charge which is countered by the substitution of lower valent Mn in the first layer (Burns and Burns 1975a). In synthesis experiments Giovanoli *et al.* (1973) noted that Li^+ was essential and could not be completely replaced by Na^+. They also concluded that the composition of lithiophorite was not variable over a wide range. Their ideal formula, $[Mn_5^{4+}Mn^{2+}O_{12}]^{2-} . [Al_4Li_2(OH)_{12}]^{2+}$, gives the Al/Li ratio as 2, much lower than in the naturally occurring sample, Table 2.5, where it is assumed substitutions could occur in both octahedral layers. In their syntheses experiments Giovanoli *et al.* (1973) used hydrothermal conditions (300°C and 90–130 atm) but little is known about the processes involved in its pedogenic formation.

2.5.2.4 *Todorokite*

Todorokite has been found in soils and marine nodules in association with birnessite but some controversy exists over the distinction between this mineral and 10 Å manganite or Na buserite. According to Burns *et al.* (1975a) some mineralogists recognize no distinction, Giovanoli *et al.* (1971) concluding that todorokite was a product of partial decomposition of buserite. Support to the separate identity of these phases is given by the observations of McKenzie (1971) that contact with solutions of divalent heavy metals, e.g. Zn^{2+}, Cu^{2+}, stabilized the structure of 10 Å Na buserite and gave the X-ray diffraction pattern of todorokite for the product which does not dehydrate to birnessite. Todorokite formation however does not occur if the more crystalline birnessite is treated with heavy metal solutions. Burns *et al.* (1974) found that the more disordered and finer grained δ-MnO_2 transformed under pressure in water to give the 10 Å material whereas the more crystalline birnessite was unaltered by this treatment.

It is suggested therefore that this 10 Å material can be the precursor of both birnessite and todorokite, the final product being determined by the environment. Under marine conditions the high Na^+ concentrations and the continual presence of water would tend to stabilize the 10 Å structure as todorokite especially in the presence of heavy metals, whereas in a soil environment it may dehydrate more readily to birnessite.

Potter and Rossman (1979) interpreted IR spectra of todorokite as indicating either a layer structure, supported by the ease of transformations described above, or a tunnel structure of wide dimensions, favoured by Burns and Burns (1979) from morphological and chemical data. Chukhrov *et al.* (1979) found polymorphs with variable a axes in naturally occurring todorokites, and this variation in cell dimensions is interpreted as reflecting the number of edge-shared octahedral chains in the a-axis direction (Fig. 2.12). Turner and Buseck (1981) claim that many of the variations between the tunnel structured MnO_2 oxides are determined by the number of edge-shared octahedral chains in the directions perpendicular to the tunnel.

Analyses of mineral deposits (Straczek *et al.* 1960) led to a proposal for the formula $(Ca,Na,Mn^{2+},K)(Mn^{4+},Mn^{2+},Mg)_6O_{12}.3H_2O$. Larson (1962) described a todorokite with 5% Zn, and McKenzie (1972b) prepared samples containing respectively 15.4% Cu, 14.4% Co and 8.9% Ni, concentrations generally higher than those determined in naturally occurring samples (Frondel *et al.* 1960; Eckhardt and Schellmann 1962). Burns *et al.* (1975) give a general formula $(Na,Ca,Mn^{2+})_2Mn_5^{4+}O_{12}.3H_2O$, and suggest that only Na, Ca and Mn^{2+} are essential constituents. The occurrences described by Straczek *et al.*, Larson, and Eckhardt and Schellmann were each associated with limestone or carbonate deposits, and Eckhardt and Schellmann noted that todorokite appears to be found especially in calcareous paragenesis. The only sample found in the survey of Australian soils (Taylor *et al.* 1964) occurred as a black

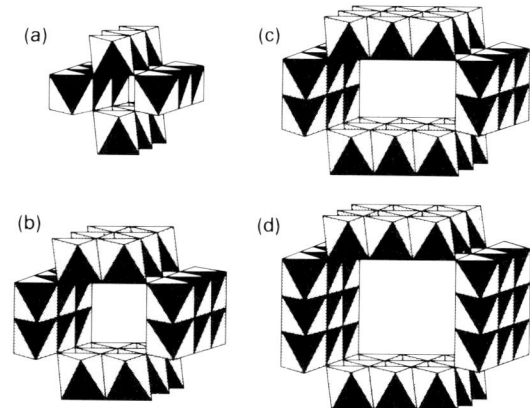

FIG. 2.12. Structural variations between some Mn oxides: (a) pyrolusite group; (b) hollandite group; (c) romancheite group; (d) todorokite group. (After Turner and Buseck 1981. Reproduced from *Science* (1981), **212**, p. 1024.)

coating on and in pockets in hardened aggregates of sepiolite and dolomite which agrees with the above observations.

Assuming that the tunnel structure of todorokite is correct, its ability to form in the presence of heavy metal cations from the layer-structured 10 Å phase needs further investigation.

2.5.2.5 Hollandite, cryptomelane and coronadite

These minerals are members of the α-MnO_2 group of oxides and differ from one another in the nature of the incorporated large cation that is essential for their particular structure. Ba, K and Pb are the respective cations that are necessary for the three minerals, although as with the other soil oxides, extensive further substitutions occur. Byström and Byström (1951) give a general formula for the group as $(Ba,K,Pb)_{2-y}(Mn,Fe)_{8-z}(O,OH)_{16} \cdot (H_2O)_u$. where y is close to unity but may be somewhat less for the Pb member coronadite, z is equal or close to zero (<0.5), and u is probably small and cannot exceed unity. According to Fleischer (1964) it is not certain whether a complete series of cation replacements exists.

MnO_6 octahedra share edges to form a double chain extending in the z-axis direction. Sharing of corners of octahedra with similar double chained octahedra gives rise to a three-dimensional structure with tunnels extending in the z direction (Fig. 2.12b). It is in these tunnels that the large cations are held, being coordinated to eight O atoms of the surrounding octahedra. The structure is discussed by Burns and Burns (1975a). McKenzie (1977) states that the tunnel cation sites are never more than half-filled, but a minimum occupation is necessary to prevent structural collapse.

Synthetic birnessite with low K contents was converted to cryptomelane by boiling (Buser *et al.* 1954), but McKenzie (1971) found that with higher concentrations of K, birnessites required heating to 400 °C before transformation occurred. He stated that there existed an upper limit (approx. 7%) for the K content of birnessite which still permitted transformation to cryptomelane on boiling, as well as the lower limit (0.25–2.5%) found by Buser and associates.

Again, as with todorokite, the tunnel structure is produced from a layer-structured phase reacting with large cations in solution. These syntheses, although far removed from natural conditions, suggest that the easily formed δ-MnO_2 or the hydrated 10 Å phase could be a possible precursor for the formation of cryptomelane.

However, only the Ba member of the group has to date been reported in soils (Taylor *et al.* 1964) and an average composition derived from three occurrences in Australian soils is given in Table 2.5. The identification of hollandite rather than cryptomelane in soils may however give some further insight to the chemical environment and formation paths responsible for its occurrence. Since there would be little difference in the relative ease of identification of these two minerals, and K is generally more plentiful than Ba in the soil environment, cryptomelane could be expected to be the more common oxide. Furthermore, K is completely soluble over the whole pH range in soils whereas Ba is relatively insoluble in the presence of carbonate. If adsorption of the relevant cation by an amorphous precipitated Mn oxide or precursor of birnessite was involved, cryptomelane should occur more frequently. The occurrence instead of hollandite would suggest that formation may occur by the association of Mn^{2+} with already precipitated Ba. This is consistent with the observations that (a) these cations are a necessary part of the structure, and (b) they are enclosed within the tunnel structure formed by the surrounding MnO_6 octahedra.

2.5.2.6 Pyrolusite

This oxide is common in Mn ores and is the most stable of the oxides in the supergene environment. Its formation can be inhibited by small amounts of foreign ions present during crystallization which would explain its uncommon formation in soils (McKenzie 1972a, 1977). Its more frequent occurrences in mineral deposits may be due to the removal of foreign ions during leaching, allowing a recrystallization of pyrolusite. Taylor et al. (1964) reported only one occurrence of pyrolusite in the suite of manganiferous soils they examined. This occurred at a depth of 3 m, and was an irregular-shaped aggregate free of contamination from other soil minerals. There is no evidence that this was formed in the particular soil which was derived from alluvium and its pedogenic origin is therefore questioned. Sherman et al. (1949) discussed the origin and composition of "pyrolusite" concretions in Hawaiian soils. Analyses suggest high Al and Fe contents but no evidence is given to support the identification as pyrolusite. Sung[12] identified lithiophorite in concretions from Hawaiian soils which would be in agreement with the high Al contents found by Sherman and colleagues.

It is structurally similar to rutile and consists of single chains of MnO_6 octahedra which are crosslinked to adjacent chains extending in the same direction by the corner sharing of oxygen atoms in adjacent octahedra (Fig. 2.12a).

2.5.2.7 Amorphous manganese oxides
Manganite, the trivalent oxy-hydroxide, γ-MnOOH, has been identified by Eswaran and Raghu Moran (1973) in voids in a very localized area of a lateritic concretion and cannot be regarded as a normal form of Mn oxides in soils. In this sample the majority of the Mn was present as unidentified or amorphous oxides. This amorphous form may be the most common phase in soils, and be a precursor for the formation of the more crystalline varieties with which it is associated. Where some crystalline phase has been found or concentrated from a soil, the chemical analyses, e.g. those given by Taylor et al. (1964), could represent the results of wide variations in the composition of the identified variety as well as an unidentified amorphous phase which would have been neglected. When it is realized that the crystalline oxides that were identified by Taylor and his associates had particle sizes of 0.1 μm for lithiophorites to 0.02 μm for the birnessites it is not surprising that "less crystalline" phases or fractions would make positive identification difficult.

Many early experiments on Mn availability utilized the more crystalline oxides from mineral deposits or synthetic products which may not indicate the higher reactivity exhibited by the few usually fine-grained oxides that actually occur in soils. As with the Fe oxides, solution, mobilization and reprecipitation of the Mn oxides are associated with their susceptibility to reducing and oxidizing environments. These conditions influencing the solubility of the oxides can also affect the availability of essential nutrient cations, some of which may be predominantly associated with these Mn minerals. Conversely, during dissolution of the oxides under anaerobic conditions toxic levels of both Mn and heavy metals associated with the minerals may result. Therefore it is seen that the Mn oxides, although present only in small total concentrations in a soil, may have important agronomic consequences. For example, the amounts of Pb, Co and Ni incorporated into subterranean clover grown on contaminated soils can be reduced by the application of birnessite or cryptomelane to the soils. There was, however, no effect for Cu or Zn (McKenzie 1978b).

2.5.3 Cation and anion adsorption on manganese oxide surfaces

Possessing hydroxylated surfaces these oxides exhibit pH-dependent surface charges. In studies on synthetic Mn oxides, the ZPC was found to increase with increasing crystallinity, ranging from 1.5 for birnessite to about 7.3 for pyrolusite (Healy *et al.* 1966; Morgan and Stumm 1964). These studies suggest that most pedogenic Mn oxides would have values in the range 1.5–4.6 (McKenzie 1972a), resulting in negatively charged surfaces in most soils and thus differing from the common soil oxides of Fe and Al. This is a further factor which would aid the oxidation–precipitation of Mn^{2+} at lower pH, as discussed earlier. McKenzie (1981) found that for cryptomelane the variations in charge density with pH and ionic concentration closely followed values predicted by the model proposed by Barrow *et al.* (1980a) for goethite. The charge variations were almost parallel to that obtained for a goethite, but the isoelectric points were displaced to much lower pH values. However, the model could not predict the much higher charge densities for birnessite samples. The low ZPC is possibly the reason why anion adsorption on Mn oxide surfaces has received so little attention, most studies being concerned with the high levels of cation adsorption, especially those of the heavy metals.

The hydroxylated surfaces could also be expected to give rise to specific adsorption of anions. Moreover, the generally small particle sizes and poorly expressed morphology in these soil oxides would minimize variations in the proportions of active sites for specific adsorption per unit surface area of these minerals (e.g. the singly coordinated hydroxyl groups involved in phosphate adsorption at Fe and Al oxide surfaces). In this regard, McKenzie[13] (priv. comm.) has found that the adsorption of molybdate per unit surface area fits the same linear relation as found for the Fe oxides. Platonov *et al.* (1982) found that in the adsorption of polyvinyl alcohol at a MnO_2 surface strong donor–acceptor bonds as well as H bonding was involved and adsorbed surface H_2O was displaced.

The adsorption of Co^{2+} by Mn oxides has received much more attention than other divalent metal cations, some of which may be initially taken up in higher amounts and/or at faster rates (McKenzie 1967, 1970, 1972b; Loganathan and Burau 1973; Murray 1975; Burns 1976). Although the relative enrichment of Co in Mn oxide phases has been long observed, Taylor and McKenzie (1966) and McKenzie (1975) showed that about 80% of the total soil Co was associated with the Mn oxides when they were also present, the mineralogy of the oxides having no significant influence. Subsequently, Co deficiency could be related to the Mn status of soils, and added Co fertilizer was observed to become rapidly associated with the Mn oxide phases present (Adams *et al.* 1969).

McKenzie (1970) found that after the initial fast adsorption of Co^{2+} there followed a slower adsorption during which Mn ions were released into the solution at the same rate as Co was taken up. This phase of adsorption was considered to be due to the replacement of Mn^{3+} which was substituting for Mn^{4+} in the oxide lattice. Electron transfer oxidized the Co^{2+} to Co^{3+} at the surface which, because of its higher crystal field stabilization energy (CFSE), replaced the Mn^{3+} and released it into the solution as Mn^{2+} on a mole to mole basis. This would explain why preferential replacement of lattice Mn by Co occurred in birnessite and in cryptomelane but not in pyrolusite where the Mn exists only in the Mn(IV) state.

Because of the variation between the ionic radii of octahedrally coordinated low spin Co^{3+} (0.53 Å) and high spin Mn^{3+} (0.65 Å), both being the forms found in Mn minerals, Burns (1976) considers the above explanation unlikely. He has suggested a modified mechanism for Co incorporation whereby hydrated Co^{2+} initially adsorbed at the surface is oxidized by Mn^{4+} in edge-shared MnO_6 octahedra at the surface and replaces the resultant Mn^{3+} which is released into solution. He considers that this replacement would be favoured by the similarity of the atomic radii of Co^{3+} (0.53 Å) and Mn^{4+} (0.54 Å). However, this modified theory does not explain the non-incorporation by pyrolusite. Since McKenzie (1970) suggested that the Co^{3+} replaced Mn^{3+} which was actually occupying a Mn^{4+} site, his theory does in principle agree with the modified version of Burns.

Loganathan and Burau (1973), in extending McKenzie's studies to the adsorption of Zn^{2+} on $\delta\text{-}MnO_2$, found that a small amount of Mn was released to the solution at pH 4 but this did not occur during the adsorption of Ca^{2+} or Na^+. They suggested that Mn^{2+} could be replaced by Zn^{2+} and that the greater adsorption and Mn release shown by Co^{3+} was due to its ability to replace both Mn^{2+} and Mn^{3+} in the lattice. Some of the Mn released into solution may arise from reduction of MnO_2 by water, which can occur below pH 3.5 (Murray 1974).

The order of ability to replace K from some synthetic samples of birnessite and cryptomelane was $Cu > Co > Ni$, but the order for replacement of lower valence Mn was $Co > Cu > Ni$ (McKenzie 1970). The initial fast adsorption of divalent metal cations was due to the balanced replacement of exchangeable Mn^{2+}, K^+ and H^+ and McKenzie (1967) established that in the samples used, the relative uptakes of Co^{2+}, Cu^{2+} and Ni^{2+} were in the ratios 2:5:1.

Because of the high degree of substitution of lower valence Mn^{2+} and Mn^{3+} for Mn^{4+} in soil Mn oxides there is a resultant high surface charge in addition to the pH-dependent charge common to all this group of oxide minerals. Adsorbed Co^{2+} diffuses from the ionic double layer to the solid surface where it replaced Mn^{2+} or Mn^{3+} after being oxidized by electron transfer. The released Mn ions then diffuse to the interface where they remain held until subsequent oxidation again incorporates them into the structure. McKenzie (1970) believes this process explains the observed scavenging action of these soil oxides for Co.

Murray (1975) showed that Co was adsorbed in proportion to the surface area, and that Ca^{2+} and Mg^{2+} in concentrations encountered in sea water could reduce uptake of Co^{2+} from dilute concentrations (5×10^{-8} M) over the whole pH range. He also observed that continued adsorption of Co produces reversal of the surface charge at about pH 6 which may reflect both the specific adsorption of Co and the precipitation at the surface of $Co(OH)_2$ at a pH below which this normally occurs. Murray's experiments suggested that adsorption at the ZPC of an oxide could not be explained by replacement of structural Mn by divalent cations. Here normal ligand exchange occurred with one proton released for each Co^{2+} ion adsorbed. Specific adsorption below the isoelectric point occurred with Co^{2+}, Cu^{2+} and Ni^{2+} ions but not with the monovalent Na^+ or K^+ cations (Murray et al. 1968).

The soil Mn oxides usually contain trace amounts of many other metals, besides those given in Table 2.5, which may have been incorporated through an initial adsorption process.

TABLE 2.6. Trace metal concentration ranges in peroxide extracts of Mn oxide accumulations in 19 Australian soils

	% range
Co	0.02–1.4
Ni	0.007–0.14
Mo	<0.0003–0.003
V	0.002–0.43
Ga	0.0002–0.001
Pb[†]	<0.01–0.25
Cr	<0.003–0.24
Zn	0.0019–0.12

† Norrish (1975) using the electron probe microanalyser has found Pb concentrations up to 2% in soil Mn oxides (normalized to 70% MnO_2).

Concentration ranges for some of these elements in the peroxide extracts of Mn oxide accumulations found in about 19 Australian soils are given in Table 2.6 from the results of Taylor and McKenzie (1966).

2.5.4 Dissolution of manganese oxides

Soil Mn oxides, because of their generally small particle sizes and ease of reduction, can be brought into solution by a variety of reducing agents. However, because of the normally high amounts of foreign cations usually incorporated, care must be taken in choosing a reagent that does not cause their precipitation. Furthermore, the reducing agents may also cause the dissolution of the more prevalent Fe oxides, or a complexing agent to retain the liberated foreign cations in solution may also partially attack other mineral phases.

Taylor et al. (1964) developed techniques both to concentrate the manganese phases, and also selectively to dissolve them with minimal attack on the associated non-manganiferous minerals. Mn oxides catalytically decompose H_2O_2 and the oxygen bubbles formed at the oxide surface can float these Mn-rich particles to the surface where they may be collected as the fluid surface overflows the vessel. Ross et al. (1976) slightly modified this technique with success. Better results are sometimes achieved when the Mn oxide is present as coatings and aggregates of silt size rather than associated with clay material. In acid solution the H_2O_2 causes a reduction and dissolution of the Mn oxides. Taylor et al. (1964) and Taylor and McKenzie (1966) used HNO_3 at pH 3 as most metals released as nitrates would be soluble under these conditions. Initial washes with the acid at this pH removed exchangeable cations and dissolved the carbonate minerals that may have been present, and the subsequent treatment in the presence of H_2O_2 gave solutions containing the dissolution products of the oxides. This dissolution treatment need not be carried out on the enriched Mn phase but is effective on the whole soil fractions (see also Ross et al. 1976).

Mn oxides adsorb Mn^{2+} more strongly than other cations commonly present in the soil solution and it is generally this adsorbed divalent Mn that provides readily available Mn necessary for plant nutrition. This supply is replenished under reducing conditions but its subsequent availability as measured by extractability with hydroquinone may be decreased in the presence of large amounts of organic matter which is also a site of strong retention (Heintze and Mann 1951).

Besides the microbiological reduction that can occur in the presence of organic matter, straight dissolution and complex formation by certain humic materials will affect the distribution of Mn oxides within a profile. Rosell and Babcock (1968) found that mixtures of fulvic and humic acids were effective in the dissolution of Mn^{3+} and Mn^{4+} oxides, and Baker (1973) found that at pH 3.4 humic acid alone could remove 70 times more Mn from pyrolusite than CO_2 charged water. Schnitzer and Kodama (1977) state that at the higher pH values humic complexes of Mn could be responsible for providing the necessary levels of soluble Mn for plant nutrition.

Because of the ease of dissolution and the fact that already precipitated Mn oxide surfaces strongly adsorb Mn ions from solution, the occurrence of these oxides as discrete aggregates, coatings and concretionary accumulations can be expected to be sinks for the mobile Mn. As the precipitates become more crystalline and grow in physical size, the level of available Mn, supplied either from exchange positions or from dissolution and complexing with humic materials will decrease.

2.6 THE TITANIUM OXIDE GROUP OF MINERALS

The three allotropic forms of TiO_2, rutile, anatase and brookite, and the double oxides ilmenite, $FeTiO_3$, and perovskite, $CaTiO_3$ (sometimes incorrectly referred to as titanates; Cotton and Wilkinson 1962), are the main Ti oxides found in soils. Leucoxene is not a separate phase but a generic term covering all fine-grained occurrences of these oxides. These minerals occur in soils either as relict grains of primary or metamorphic rocks, in which they commonly occur, or derive from the weathering of Ti-rich silicates such as biotites, augites, amphiboles and garnets. In these easily weatherable minerals Ti^{4+} often partially substitutes for octahedrally coordinated Al^{3+} and Fe^{3+} (Tillmans and Correns 1969). However, some of the Ti in these minerals may be present as Ti^{3+}, and there is difficulty in distinguishing between Ti(III) and Ti(IV) if Fe is also present in the di- and tri-valent states (Rankama and Sahama 1950). The formation of titania minerals during the weathering of these silicates suggests that Ti is released into solution, but the form of the soluble species is not definitely established.

Cotton and Wilkinson (1962) state that the high charge to radius ratio prevents hydration of the Ti^{4+} ion, and refute the existence of a definite $Ti(OH)_4$ hydroxide. They suggest that this commonly cited species is probably an hydrated form of TiO_2. However, charged hydrolysed species such as $Ti(OH)_3^+$ and $Ti(OH)_2^{2+}$ can exist under certain conditions and may be further hydrolysed and polymerized to give colloidal or precipitated hydrous TiO_2 by increases in pH (Cotton and Wilkinson 1962). These authors also state that there is no evidence for a titanyl (TiO^{2+}) ion so that the tri- or di-hydroxy species may be the main forms of mobile Ti during weathering and subsequent oxide mineral formation. Whether $(Ti(OH)_4$ is in fact an hydrated form of TiO_2 does not seriously influence the discussions on the formation of the oxides in soils. Loughnan (1969) states that hydrolysed Ti still persists in solution after Al has been precipitated and will not precipitate as $Ti(OH)_4$ until pH 5, and forms one of the titania polymorphs on subsequent dehydration. Sherman (1952), in a study of the Ti contents of Hawaiian soils, believed that this element was mobile as a colloidal hydrated oxide which dehydrated to TiO_2. McLaughlin (1954) considers that TiO_2 can be mobilized within a profile by rehydration facilitated by the action of humic acids. The transformation of amorphous hydrated Ti and Fe gels present is perennially wet forest soils into crystalline Ti and Fe–Ti oxide phases after deforestation supports this concept (Walker et al. 1969). Colloidal hydrated TiO_2 (100 Å particles) is extremely active and its presence can alter the direction of mineralization and be incorporated into the resultant Fe oxide phases during the oxidation of an Fe^{2+} solution.[4]

Rutile is the most common form of TiO_2 in nature (Cotton and Wilkinson 1962), and Jackson (1964) considers that anatase and rutile are the predominant Ti minerals in clay fractions of soils and found that they constituted 1–3% of the soil clays examined. Loughnan (1969) obtained similar values for some soils around Sydney (Australia). Being resistant minerals they can be concentrated during certain soil-forming conditions, e.g. tropical weathering, under which the concentrations can rise to 20–30% in certain parts of the profile (Sherman 1952; Loughnan 1969). Readers are referred to a review by Hutton (1977) on the Ti minerals in soil in which their origin, identification, quantitative estimation and distribution in soils, aspects not covered by this chapter, are discussed.

2.6.1 Titanium minerals in soils

The three polymorphic forms of TiO_2 are differentiated crystallographically by the manner in which the TiO_6 octahedra are linked. For example, in rutile octahedra share two of their edges with adjacent octahedra, while in brookite and anatase three and four edges respectively are shared in the development of the overall structure (Evans 1964). The manner of sharing edges can distort the octahedra and produce variations in the Ti—O bond lengths within a single octahedron. In describing the structures in this way it must be realized that the octahedra that share edges are the ones that are occupied by a Ti^{4+} cation. Since all three polymorphs consist of close packed oxygen layers, the variation in linking reflects only the distribution pattern of the Ti cations within the octahedral sites, since, to satisfy valencies, only one-half of the total possible octahedral sites are occupied.

2.6.1.1 Rutile

This hcp form of TiO_2 occurs in many igneous and metamorphic rocks and often persists as a residual mineral in the developed soils. It is also a common alteration product of weathering of other Ti-containing minerals, e.g. ilmenite, titaniferous magnetite, sphene and perovskite. Ziv (1956) reports conversion of ilmenite to rutile under supergene conditions. Although no data are available for the composition of pedogenically formed rutile, analyses of mineral deposits (Deer et al. 1962) show that it may contain divalent and trivalent Fe and some varieties contain Nb (26%) and Ta (36%) in major amounts. When pentavalent cations are incorporated in the structure further octahedral vacancies are created or divalent cations such as Fe(II) replace some Ti to maintain charge balance. Most reported syntheses have been carried out at high temperatures and cannot explain its formation in soils. However, there could be some environmental influence as Weiser and Milligan (1942) concluded from their synthesis experiments that, whereas hydrolysis of Ti sulphate leads to anatase, the chloride and nitrate salts under their experimental conditions favoured rutile.

2.6.1.2 Anatase

This is the ccp TiO_2 form and also occurs in igneous and metamorphic rocks and can persist as a residual mineral in soils. Loughnan (1969) states that anatase is generally formed from the titania released during the weathering of silicates containing Ti. The high amounts of Nb and Ta found in rutile are uncommon in anatase, although Fe is often a minor impurity (Deer et al. 1962). Anatase containing Fe(III) has been synthesized (Kiyama and Takada 1981) by the aerial oxidation at 95 °C of a strongly acidic solution of $TiCl_3$ and $FeSO_4$. Fitzpatrick et al. (1978) successfully synthesized anatase by ageing amorphous Ti hydroxide precipitated by addition of NH_3 to a chloride-free Ti nitrate solution. After 30 days at 25 °C at pH 5.5 the strongest X-ray diffraction lines of anatase were visible. Higher temperatures and longer periods of ageing increased the crystallinity which was reflected in a lower degree of solubility in ammonium oxalate in the absence of light. These authors showed that soils can contain variable amounts of this poorly crystalline, oxalate-soluble Ti, which may be related to the environmental soil-forming factors. Deelman (1979) synthesized anatase under conditions which he related to soil environments. Periodical immersion of a substrate in a hydrolysed TiO_2 colloidal suspension followed by air drying at 25 °C caused a transformation from amorphous to crystalline oxide. Sherman (1952) identified anatase as a secondary product of tropical weathering in Hawaiian soils and has related the TiO_2 contents to the horizon depth and climatic conditions of the soils.

2.6.1.3 Brookite

This is a combination of ccp and hcp layers and is less common than anatase or rutile. Hutton (1977) reports that this oxide is considered to be of secondary origin, occurring in metamorphic rocks. Its natural alteration product is rutile, and Fe(III) is regarded as an essential constituent

(Deer et al. 1962). This mineral also forms from the solution of Ti during the weathering of sphene and can also form in association with rutile and anatase during the weathering of ilmenite (Rankama and Sahama 1950). It has been synthesized by chemical transport reactions (Razumeenko et al. 1981) and its structure is stabilized by MgO.

If mobilized titania liberated during the weathering can form both anatase and rutile under pedogenic conditions, it is reasonable to expect that ionic environment can influence the direction of mineralization, as with the oxides of Fe, Al and Mn. The incorporation of foreign ions of lower or higher valence, and/or the level of this incorporation could affect the distribution of octahedral vacancies during the polymerization, dehydration and crystallization from colloidal or precipitated hydrous TiO_2.

2.6.1.4 Perovskite
Perovskite, $CaTiO_3$, often occurs as an accessory mineral in igneous rocks and in some thermally metamorphosed calcareous rocks in contact with basic igneous rocks. It often contains rare earth elements replacing Ca but is not regarded as forming under normal soil conditions. Any soil occurrences are probably relict. Structurally it consists of TiO_6 octahedra linked at their corners to form a cubic structure with the Ca ion occupying the central position between four TiO_6 octahedra and coordinated to twelve of the surrounding oxygen atoms.

2.6.1.5 Ilmenite
Ilmenite, $FeTiO_3$, is isostructural with hematite and occurs in soils as a residual resistate mineral. On weathering, TiO_2 minerals can form possibly via the intermediate pseudo-rutile, $FeTi_3O_9$ (Teufer and Temple 1966), which is sometimes detected in soils. White (1979) identified ilmenite as the cementing matrix in soil concretions, suggesting that ilmenite could form under pedogenic conditions. However, this phase might have been titaniferous hematite which resembles ilmenite in its X-ray diffraction pattern and can be readily synthesized from the ageing of mixed Ti–ferric hydroxide gels (Fitzpatrick and Le Roux 1976).

2.6.2 Properties of soil titanium oxide minerals

Hutton (1977) considers that the major importance of these oxides in soils is their value in soil development studies. The value of these studies would be enhanced if it could be established without doubt that the minerals were relict, or whether they had formed in soils. However, until synthesis experiments can elucidate the conditions which may be responsible for the formation of a particular oxide in preference to another, interpretation of the soil-forming processes leading to a particular mineral assemblage cannot be made. Because of their generally low concentrations in soils, retention and release of adsorbed nutrient anions and cations would not be expected to have much influence on the nutrient status of a soil. In tropical soils where their concentrations may be much higher, some effect may be demonstrated.

Many of the studies on the surface properties of TiO_2 have been initiated because of its importance in technology. These fine-grained TiO_2 materials are not mined as such, but are generally produced by the rutilization of ilmenite or other such processes, the enriched TiO_2 phase then being ground to the appropriate particle size range. In pigment technology, silica is often adsorbed on to the TiO_2 surface to alter its working properties, for example, its colour retention and gloss.

The broken edge bonds of the Ti oxides are hydroxylated and can be expected to exhibit pH-dependent charge characteristics. Parfitt (1976) found singly and doubly coordinated hydroxyl groups at the rutile surface (c.f. the surface hydrolysis of goethite and gibbsite discussed earlier). He found that the two sites differed in reactivity because while NH_3 is adsorbed at both sites, only one was sufficiently reactive to coordinate with weaker bases, e.g. pyridine. Graham et al. (1981) demonstrated that these surface OH groups were less perturbed by physical adsorption of n-heptane than by stronger H-bonding with some aromatic molecules. Day et al. (1979), working with rutile dehydroxylated at 673 °C, found that pretreatment with some alcohols reduced subsequent adsorption and sometimes caused a change in the orientation of adsorptives. Surface adsorption behaviour has been utilized to effect selective flocculation of rutile with tannic acid which is suggested to produce bridging between particles (Rinelli and Marabini 1981).

The surface activity of anatase also gives rise to some interesting reactions. Physisorption of aromatic hydrocarbons at its surface can produce red-shifts of about 3000 cm^{-1} in the fluorescence of the hydrocarbon, much greater than that resulting from sorption on γ-Al_2O_3 (Oelkrug and Radjaipour 1981). In primary aliphatic alcohols anatase is reduced under the action of UV to give Ti^{3+} ions and the corresponding aldehyde in solution (Malati and Seager 1981).

However, little work has been done on the adsorption of ions important in soil processes. Fuerstenau et al. (1981) investigated the adsorption of alkaline earth metal cations at the rutile–aqueous solution interface and found that although these species behave as indifferent cations at low concentrations (10^{-5} M), they exhibit strong specific adsorption at higher concentrations, and this is particularly significant at pH values greater than the ZPC. The affinity of these cations for the rutile surface decreases with decreasing ionic radii and consequently with the ease with which these ions dehydrate.

Ahmed and Maksimov (1969) found a shift in the ZPC of rutile towards more positive values with higher concentrations of KNO_3. This was attributed to specific adsorption of K^+ cations. Berube and DeBruyn (1968) found a preferential adsorption series among the monovalent cations with $Li > Na > Cs$. On the basis of their results they proposed a model for the rutile–solution interface to account for the adsorption of Li^+ as a hydrated cation, rather than as the dehydrated species as suggested by Breeuwsma (1973) for the preferential adsorption of Li^+ on hematite.

Cabrera et al. (1977) showed that anatase and rutile can absorb and retain phosphate and Fordham and Norrish (1979) showed that these minerals but not ilmenite could retain added arsenate. In microprobe studies an association between P and TiO_2 in some soils has been observed, however, only a small fraction of the total P was generally involved (Norrish and Rosser, 1983), possibly due to competition from other soil components. However, in a particular soil developed on the pallid zone of a laterite containing negligible free iron oxides, the association of P with TiO_2 accounted for approximately 40% of the total P.

NOTES

1. The terms "plane, sheet and layer" used here conform with the recommendations of the AIPEA, Nomenclature Committee adopted in the Madrid meeting, June 1972.
2. A Brønsted acid is a substance which can split off a proton, whereas Lewis acid sites are defined as those groups which combine with a base by sharing electrons with the formation of a coordination compound.

3. Metal–hydroxo complexes formed from the dissolution of the metal cations in the surface layers of the oxide can also be adsorbed by the surface and act as PDI, but since their concentrations are often quite small in relation to H^+ and OH^- they can be neglected (Atkinson *et al.* 1967; Parks and de Bruyn 1962).
4. Unpublished data R.M. Taylor.
5. W. Schellmann, Bundesanstalt fur Geowissenschaften und Rohstoffe, Hannover.
6. Prof. A. Herbillon, Université Catholique de Louvain, Louvain-la-Neuve, Belgium.
7. Dr R. Giovanoli, Inst. Inorganic Analytical and Physical Chem., Berne University.
8. Taken from the data of Raupach (1963) for Al oxides, and from Wilding *et al.* (1977) for Si. Similar values are obtained from studies on soils. Raupach (CSIRO Division of Soils, Adelaide) priv. comm.
9. Dr P. Rengasamy, Waite Agricultural Research Institute, University of Adelaide, S. Australia. Present address: Irrigation Res. Inst., Tatura, Victoria 3616.
10. K. Norrish, CSIRO Division of Soils, Adelaide, S. Australia. Priv. Comm.
11. A.W. Fordham and K. Norrish, CSIRO Division of Soils, Adelaide, S. Australia.
12. Priv. comm. from W. Sung, Mass. Inst. Technology, to R.M. McKenzie – see McKenzie (1977).
13. Unpublished data of R.M. McKenzie, CSIRO Division of Soils, Adelaide, S. Australia.

REFERENCES

ABDULVALIEV R.A., MEDVEDKOV B.E. & BOCHAROVA G.V. (1978) Chromium minerals in bauxites. *Kompleksn. Ispol's Miner. Syr'ya* **4**, 31–35.

ADAMS S.N., HONEYSETT J.L., TILLER K.G. & NORRISH K. (1969) Factors controlling the increase of cobalt in plants following the addition of a cobalt fertiliser. *Aust. J. Soil Res.* **7**, 29–42.

AHMED S.M. & MAKSIMOV D. (1969) Studies on the double layer on cassiterite and rutile. *J. Colloid Interface Sci.* **29**, 97–104.

AKITT J.W. & FARTHING A. (1981) Aluminium-27 nuclear magnetic resonance studies of the hydrolysis of aluminium(III). Part 4. Hydrolysis using sodium carbonate. *J. Chem. Dalton Proc.* 1617–1623.

ALEKSANDROVA L.N. (1960) The use of sodium pyrophosphate for isolating free humic substances and their organic mineral compounds from soil. *Soviet Soil Sci.* **2**, 190–197.

ANNERSTEN H. & HAFNER S.S. (1973) Vacancy distribution in synthetic spinels of the series Fe_3O_4–γFe_2O_3. *Z. Kristallog.* **137**, 321–340.

ATKINSON R.J., PARFITT R.L. & SMART R.ST.C. (1974) Infrared study of phosphate adsorption on goethite. *J. Chem. Soc. Faraday Trans.* 1, **70**, 1472–1479.

ATKINSON R.J., POSNER A.M. & QUIRK J.P. (1967) Adsorption of potential determining ions at the ferric oxide aqueous electrolyte interface. *J. Phys. Chem. Ithaca*, **71**, 550–558.

ATKINSON R.J., POSNER A.M. & QUIRK J.P. (1971) Kinetics of heterogeneous isotopic exchange reactions: derivation of an Elovich equation. *Proc. R. Soc. Lond.* **A324**, 247–256.

ATKINSON R.J., POSNER A.M. & QUIRK J.P. (1972) Kinetics of isotopic exchange of phosphate at the αFeOOH–aqueous solution interface. *J. Inorg. Nucl. Chem.* **34**, 2201–2211.

AYLMORE L.A.G., KARIM M. & QUIRK J.P. (1967) Adsorption and desorption of sulphate ions by soil constituents. *Soil Sci.* **103**, 10–15.

BAIRD T., FRYER J.R. & GALBRAITH S.T. (1977) Iron oxide and oxyhydroxide preparations and their reactions with silica. *Inst. Phys. Conf. Ser.* 36, Ch. 5, pp. 211–214.

BAKER W.E. (1973) Role of humic acids from Tasmanian podzolic soils in mineral degradation and metal mobilisation. *Geochim. Cosmochim. Acta* **37**, 269–281.

BALISTRIERI L. & MURRAY J.W. (1979) Surface of goethite (α-FeOOH) in seawater. *Amer. Chem. Soc. Sym. Ser. 93 Chem. Model Aqueous system*, 275–298.

BARDOSSY G. & WHITE J.L. (1979) Carbonate inhibits the crystallisation of aluminium hydroxide in bauxite. *Science* **203**, 355–356.

BARNHISEL R.I. & RICH C.I. (1965) Gibbsite bayerite and nordstrandite formation as affected by anions, pH and mineral surfaces. *Soil Sci. Soc. Am. Proc.* **29**, 531–534.

BARROW N.J., BOWDEN J.W., POSNER A.M. & QUIRK J.P. (1980a) An objective method for fitting models of ion adsorption on variable charge surfaces. *Aust. J. Soil Res.* **18**, 37–47.
BARROW N.J., BOWDEN J.W., POSNER A.M. & QUIRK J.P. (1980b). Describing the effects of electrolyte on adsorption of phosphate by a variable charge surface. *Aust. J. Soil Res.* **18**, 395–404.
BECKWITH R.S. & REEVE R. (1963) Soluble silica in soils. *Aust. J. Soil Res.* **1**, 157–168.
BELL H. & HORNIG C.A. (1972) Trace element content of iron oxide pseudomorphs of pyrites from Cherokee and York Counties, South Carolina. *S.C., Div. Geol. Notes* **16**, 49–58.
BELOKOPYTOV YU.V., KHOLYAVENKO K.M. & GEREI S.V. (1979) An infrared study on the surface properties of metal oxides. 2. The interaction of ammonia with the surface of iron(III) oxide, zinc oxide, mobybdenum(VI) oxide and vanadium (V) oxide. *J. Catal.* **60**, 1–7.
BENESLAVSKY S.I. (1957) New aluminium-bearing minerals in bauxites. *Dokl. Akad. Nauk. SSSR* **113**, 1130–1132.
BENTOR Y.K., GROSS S. & HELLER L. (1963) Some unusual minerals from the "mottled zone" complex, Israel. *Am. Miner.*, **48**, 924–930.
BERNAL J.D., DASGUPTA D.R. & MACKAY A.L. (1959) The oxides and hydroxides of iron and their structural interrelationships. *Clay Miner. Bull.* **4**, 15–30.
BERUBE Y.G. & DEBRUYN P.L. (1968) Adsorption on the rutile–solution interface. II. Model of the electrochemical double layer. *J. Colloid Interface Sci.* **28**, 92–101.
BETEKHTIN A.G. (1940) South Urals manganese deposits as a raw material base for the Magnitogorsk metallurgic plant. *Trudi Inst. Geol. Scien. Acad. Sci. USSR*, Ore deposit Ser. No. 4.
BIAIS R., BONNEMAYRE A., DE GRAMONT X., MICHEL M., GILBERT H. & JANOT C. (1972) Aluminium–iron substitutions in synthetic oxides and hydroxides. *Bull. Soc. Fr. Miner. Cristallogr.* **95**, 308–321.
BLACKMORE A.V. (1973) Aggregation of clays by the products of iron(III) hydrolysis. *Aust. J. Soil Res.* **11**, 75–82.
BLOOMFIELD C. (1958) Mobilisation of iron in podzol soils by aqueous leaf extracts. *Chem. Ind.* 259–260.
BLYHOLDER G. & RICHARDSON E.A. (1962) Infrared and volumetric data on the adsorption of ammonia, water and other gases on activated iron(III) oxide. *J. Phys. Chem. Ithaca* **66**, 2597–2602.
BOWDEN J.W., BOLLAND M.D.A., POSNER A.M. & QUIRK J.P. (1973) General model for anion and cation adsorption at oxide surfaces. *Nature, Lond., Phys. Sci.* **245**, 81–83.
BOWDEN J.W., POSNER A.M. & QUIRK J.P. (1977) Ionic adsorption on variable charge mineral surfaces. Theoretical charge development and titration curves. *Aust. J. Soil Res.* **15**, 121–136.
BREEUWSMA A. (1973) Adsorption of ions on hematite (α-Fe_2O_3). *Med. Landbouwhogeschool, Wageningen* 73-1.
BREEUWSMA A. & LYKLEMA J. (1973) Physical and chemical adsorption of ions in the electrical double layer on haematite (α-Fe_2O_3). *J. Colloid Interface Sci.* **43**, 437–448.
BRICKER O.P. (1965) Some stability relations in the system Mn–O_2–H_2O at 25 and one atmosphere total pressure. *Am. Miner.* **50**, 1296–1354.
BRIDGE J. (1952) Discussion in *Problems of Clay and Laterite Genesis. Am. Inst. Mining Metall. Engineers Symp.*, pp. 212–214.
BRINDLEY G.W. & BISH D.L. (1976) Green rust: a pyroaurite type structure. *Nature, Lond.* **263**, 353.
BROMFIELD S.M. (1965) Studies of the relative importance of iron and aluminium in the sorption of phosphate by some Australian soils. *Aust. J. Soil Res.* **3**, 31–44.
BROSSET C., BIEDERMANN G. & SILLEN L.G. (1954) Studies on the hydrolysis of metal ions. XI. The aluminium ion, Al^{3+}. *Acta. Chem. Scand.* **8**, 1917–1926.
BROWN G. & NEWMAN A.C.D. (1973) The reactions of soluble aluminium with montmorillonite. *J. Soil Sci.* **24**, 339–354.
BURNS R.G. (1976) The uptake of cobalt into ferromanganese nodules, soils, and synthetic manganese(IV) oxides. *Geochim. Cosmochim. Acta* **40**, 95–102.
BURNS R.G. & BURNS V.M. (1975a) Structural relationships between the Mn(IV) oxides. Ch. 16 in *Proc. Int. Symposium on Manganese Dioxide* (A. Kozawa and R.J. Brodd, eds). Electrochemical Society Publication, Cleveland.
BURNS R.G. & BURNS V.M. (1975b) Mechanism for nucleation and growth of manganese nodules. *Nature, Lond.* **255**, 130–131.
BURNS R.G. & BURNS V.M. (1979) "Manganese oxides" in Marine Minerals. Ch. 1. pp. 1–46 in Ribbe, P.D. (ed.), Mineralogical Society of America, Washington.
BURNS R.G., BURNS V.M., SUNG W. & BROWN B.A. (1974). Ferromanganese nodule mineralogy: suggested terminology of the principle manganese oxide phases. *Geol. Soc. Amer. Ann. Meeting, Abstr.* **6**, 1029–1031.
BURNS V.M., BURNS R.G. & ZWICKER W. (1975) Classification of natural manganese dioxide minerals. Ch. 15 in *Proc. Int. Symposium on Manganese Dioxide*. Electrochemical Society Publication, Cleveland.
BUSER W., GRAF P. & FEITKNECHT W. (1954) Beitrag zur Kenntnis der Mangan(II) – manganite und des δ-MnO_2. *Helv. Chim. Acta* **37**, 2322–2333.
BUSING W.R. & LEVY H.A. (1958) A single crystal neutron diffraction study of diaspore (AlO(OH). *Acta Crystallogr.* **11**, 798–803.
BYSTROM A. & BYSTROM A.M. (1951) The positions of the barium atoms in hollandite. *Acta Crystallogr.* **4**, 469.
CABRERA F., MADRID L. & DE ARAMBARRI P. (1977) Adsorption of phosphate by various oxides. Theoretical treatment of the adsorption envelope. *J. Soil Sci.* **28**, 306–313.
CAILLÈRE S., DIETRICH J.E., MAKSIMOVIC Z. & POBEGUIN T. (1978) Mineralogical and geochemical studies of the

complex nodule bearing bauxites of the Blanquette W deposit near Thoronet (Var). *4th Int. Cong. Study Bauxites, Alumina Alum.* **1**, 77–91.

CARLSON L., KOLJONEN T., LAHERMO P. & ROSENBERG R.J. (1977) Case study of a manganese and iron precipitate in a groundwater discharge in Somero, Southwestern Finland. *Bull. Geol. Soc. Finland* **49**, 159–173.

CARLSON L. & SCHWERTMANN U. (1981) Natural ferrihydrites of Finnish surface deposits and their association with silica. *Geochim. Cosmochim. Acta* (in press).

CARLSON L. & SCHWERTMANN U. (1980) Natural occurrence of feroxyhite (δ-FeOOH) Clays and Clay Miner. **28**, 272–280.

CHRISTENSEN H. & CHRISTENSEN A.N. (1978) Hydrogen bonds of γ-FeOOH. *Acta Chem. Scand.* **A32**, 87–88.

CHRISTOPH G.G., CARBATO C.E. & HOFMANN D.A. (1979) The crystal structure of boehmite. *Clays Clay Miner.* **27**, 81–86.

CHUKHROV F.V. (1973) Mineralogical and geochemical criteria in the genesis of red beds. *Chem. Geol.* **12**, 67–75.

CHUKHROV F.V., ERMILOVA L.P., ZVYAGIN B.B. & GORSHKOV A.I. (1976) Genetic system of hypergene iron oxides. *Proc. Int. Clay Conf. 1975, Mexico, pp. 275–286.* Applied Publishing Ltd., Wilmette, Illinois.

CHUKHROV F.V., GORSHKOV A.I., RUDNITSKAYA E.S., BEREZOVSKAYA V.V. & SIVTSOV A.V. (1978) On vernadite. *Izv. Akad. Nauk SSSR. Ser. Geol.* **1**, 86–94.

CHUKHROV F.V., GORSHKOV A.I., SIVTSOV A.V. & BEREZOVSKAYA V.V. (1979) New data on natural todorokites. *Nature, Lond.* **278**, 631–632.

CHUKHROV F.V., GORSHKOV A.I., RUDNITSKAYA E.S., BEREZOVSKAYA V.V. & SIVTSOV A.V. (1980) Manganese minerals in clays: a review. *Clays Clay Miner.* **28**, 346–354.

CHUKHROV F.V., ZVYAGIN B.B., ERMILOVA L.P. & GORSHKOV A.I. (1973) New data on iron oxides in the weathering zone. *Proc. Int. Clay Conf. 1972, Madrid,* Vol. 1, pp. 397–404. Division de Ciencias, CSIC, Madrid.

COLLEPARDI M., MASSIDDA L. & ROSSI G. (1973) Crystallisation of goethite and hematite by ageing iron hydroxide-gels. *Rend. Soc. Ital. Mineral. Petrologia* **29**, 251–270.

COMMONWEALTH BUREAU OF SOILS. Harpenden, England. Bibliography No. 1190, (1957–1967), *Gibbsite in Soils.*

COMMONWEALTH BUREAU OF SOILS. Harpenden, England. Bibliography No. 353, (1950–1960), *Phosphate Fixation by Aluminium, Iron and Clay.*

CORNELL R.M., POSNER A.M. & QUIRK J.P. (1974) Crystal morphology and the disolution of goethite. *J. Inorg. Nucl. Chem.* **36**, 1937–1946.

COTTON F.A. & WILKINSON G. (1962) *Advanced Inorganic Chemistry.* Interscience Publishers, London.

DAVIS J.A., JAMES R.O. & LECKIE J.O. (1978) Surface ionisation and complexion at the oxide/water interface. I. Computations of electrical double layer properties in simple electrolytes. *J. Colloid Interface Sci.* **63**, 480–499.

DAVIS J.A. & LECKIE J.O. (1978) Surface ionisation and complexion at the oxide/water interface. II. Surface properties of amorphous iron oxyhydroxide and adsorption of metal ions. *J. Colloid Interface Sci.*, **67**, 90–107.

DAVIS J.A. & LECKIE J.O. (1980) Surface ionisation and compexion at the oxide/water interface. III. Adsorption of anions. *J. Colloid Interface Sci.* **74**, 32–43.

DAY R.E., PARFITT G.D. & PEACOCK J. (1979) A comparison of the effect of pretreatment with hexanol, hexanol-1-ol and hexan-1:6-diol on the adsorption behaviour of rutile, alumina and silica. *J. Colloid Interface Sci.* **70**, 130–138.

DEB B.C. (1950) Estimation of free iron oxides in soils and clays and their removal. *J. Soil Sci.* **1**, 212–230.

DE BOER F.E. & SELWOOD P.W. (1954) The activation energy for the solid state reaction γ-Fe$_2$O$_3$ → αFe$_2$O$_3$. *J. Am. Chem. Soc.* 3365–3367.

DEELMAN J.C. (1979) Low-temperature synthesis of anatase (TiO$_2$). *Neues Jb. Miner. Monatsh* **6**, 253–261.

DE ENDREDY A.S. (1963) Estimation of free iron oxides in soils and clays by a photolytic method. *Clay Miner. Bull.* **29**, 209–217.

DEER W.A., HOWIE R.A. & ZUSSMAN J. (1962) *Rock Forming Minerals.* Vol. 5 – *Non Silicates* Longman, London.

DESHPANDE T.L., GREENLAND D.J. & QUIRK J.P. (1964a) Charges on iron and aluminium oxides in soils. *Trans. 8th Int. Congr. Soil Sci., Bucharest.* **3**, 1213–1225.

DESHPANDE T.L., GREENLAND D.J. & QUIRK J.P. (1964b) Role of iron oxides in the binding of soil particles. *Nature, Lond.* **201**, 107–108.

DE SITTER J., GOVAERT A., DE GRAVE E., CHAMBAERE D. & ROBBRECHT G. (1977) A Mössbauer study of calcium (2+)-containing magnetites. *Phys. Status Solidi A.* **43**, 619–624.

DETOURNAY J., GHODSI M. & DERIE R. (1975) Influence de la température et de la présence des ions étrangers sur la cenétique et la mécanisme de formation de la goethite en milieu aqueux. *Z. anorg. allg. Chem.* **412**, 184–192.

DE VILLIERS J.M. & VAN ROOYEN T.H. (1967) Solid solution formation of lepidocrocite–boehmite and its occurrence in soils. *Clay Minerals* **7**, 229–235.

DROSDOFF M. & NICKIFOROFF C.C. (1940) Iron–manganese concretions in Dayton soils. *Soil Sci.* **49**, 333–345.

DUBOIS P. (1963) Contribution a l'étude des oxydes du manganese. *Ann. Chim.* **5**, 411–482.

D'YACHKOVA I.B. (1971) Vanadium and iron isomorphism in iron oxides. *Geokhim. Gidroterm. Rudoobrazov. (V.L Barsukov, ed.), pp. 129–134.*

DYAL R.S. & HENDRICKS S.B. (1950) Total surface of clays in polar liquids as a characteristic index. *Soil Sci.* **69**, 421–432.

ECKHARDT F.J. & SCHELLMANN W. (1962) Eigenschaften und Beschreibung des Mangan-minerals Todorokit. *Geol. Jb.* **79**, 867–882.

ELLIS J., GIOVANOLI R. & STUMM W. (1976) Anion exchange properties of β-FeOOH. *Chimia* **30**, 141–144.
EL SWAIFY S.A. & EMERSON W.W. (1975) Changes in physical properties of soil clays due to precipitated aluminium and iron hydroxides: I. Swelling and aggregate stability after drying. *Soil Sci. Soc. Am. Proc.* **39**, 1056–1063.
ERVIN G. & OSBORN E.F. (1951) The system Al_2O_3–H_2O. *J. Geol.* **59**, 381–394.
ESWARAN H. & RAGHU MORAN N.G. (1973) The microfabric of petroplinthite. *Soil Sci. Soc. Am. Proc.* **37**, 79–82.
EVANS R.C. (1964) *Introduction to Crystal Chemistry* (2nd edn.). Cambridge University Press, Cambridge.
FASISKA E.J. (1967) Structural aspects of the oxides and oxyhydrates of iron. *Corrosion Sci.* **7**, 833–839.
FEITKNECHT W. (1959) Über die Oxydation von festen Hydroxyverbindungen des Eisens in wassrigen Lösungen. *Z. Elektrochem.* **63**, 34–43.
FEITKNECHT W., GIOVANOLI R., MICHAELIS W. & MÜLLER M. (1973) Über die Hydrolyse von Eisen(III) Salzlosungen. I. Die Hydrolyse der Lösungen von Eisen(III) chlorid. *Helv. Chim. Acta* **56**, 2847–2856.
FEITKNECHT W., GIOVANOLI R., MICHAELIS W. & MÜLLER M. (1975) ÜUber die Hydrolyse von Eisen(III) Salzlösungen. II. Die Alterung der Hydrolyseprodukte. *Z. anorg. allg. Chem.* **417**, 114–124.
FEITKNECHT W. & KELLER G. (1950) Über die dunkelgrünen Hydroxyverbindungen des Eisens. *Z. anorg. allg. Chem.* **262**, 61–68.
FEITKNECHT W. & MARTI W. (1945) Über die oxydation von Mangan(II)-hydroxid mit molekularem Sauerstoff. *Helv. Chim. Acta*, **28**, 129–148.
FEITKNECHT W., OSWALD H.R. & FEITKNECHT-STEIMANN V. (1960) Über die topochemische einphasige Reduktion von γ-MnO_2. *Helv. Chim. Acta* **48**, 1947–1950.
FISCHER W.R. & OTTOW, J.C.G. (1972) Abbau von Eisen (III)-citrat in durchlufteter wassriger Lösung durch Bodenbakterien. *Z. Pflanzenernähr. Bodenkd.* **131**, 243–253.
FISCHER W.R. & SCHWERTMANN U. (1975) The formation of hematite from amorphous iron(III)-hydroxide. *Clays Clay Miner.* **23**, 33–37.
FITZPATRICK R.W. & LE ROUX J. (1976) Pedogenic and solid solution studies on iron-titanium minerals. *Proc. Int. Clay Conf. 1975, Mexico City*, pp. 585–599. Applied Publishing Ltd, Wilmette, Illinois.
FITZPATRICK R.W., LE ROUX J. & SCHWERTMANN U. (1978) Amorphous and crystalline titanium and iron-titanium oxides in synthetic preparations, at near ambient conditions, and in soil clays. *Clays Clay Miner.* **26**, 189–201.
FLEISCHER M. (1964) Manganese oxide minerals. VIII. Hollandite. *Advancing Frontiers in Geol. and Geophysics*, pp. 221–232. Osmania University Press, Hyderabad.
FLEISCHER M. & FAUST G.T. (1963) Studies on manganese oxide minerals. VII. Lithiophorite. *Schweiz. Mineralogische u. Petrographische Mitteilungen.* **43**, 198–216.
FORBES E.A., POSNER A.M. & QUIRK J.P. (1974) The specific adsorption of inorganic Hg(II) species and Co(III) complex ions on goethite. *J. Colloid Interface Sci.* **49**, 403–409.
FORBES E.A., POSNER A.M. & QUIRK J.P. (1976) The specific adsorption of divalent Cd, Co, Pb and Zn on goethite, *J. Soil Sci.* **27**, 154–166.
FORDHAM A.W. (1970) Sorption and precipitation of iron on kaolinite. III *Aust. J. Soil Res.* **8**, 107–122.
FORDHAM A.W. & NORRISH K. (1979) Arsenate-73 uptake by components of several acidic soils and its implications for phosphate retention. *Aust. J. Soil Res.* **17**, 307–316.
FRIPIAT J.J. & GASTUCHE M.C. (1952) Physico-chemical study of clay surfaces. *Nat. Inst. Belg. Congo. Agromical Studies ci. Ser. No. 54.*
FRONDEL C., MARVIN U.B. & ITO J. (1960) New occurrences of todorokite. *Am. Miner.* **45**, 1167–1173.
FUERSTENAU D.W., MANMOHAN D. & RAGHAVAN S. (1981) The adsorption of alkaline-earth metal ions at the rutile aqueous solution interface. *Adsorpt. Aqueous Solutions* [*Proc. Symp.*] 1980 (P.H. Tewari, ed.), pp. 93–117. Plenum, New York.
GALLAHER R.N., PERKINS H.F., TAN K.H. & RADCLIFFE D. (1973) Soil concretions: II Mineralogical analysis. *Soil Sci. Soc. Am. Proc.* **37**, 469–472.
GARDNER L.R. (1970) A chemical model for the origin of gibbsite from kaolinite. *Am. Miner.* **55**, 1380–1389.
GAST R.G. (1977) Surface and Colloid Chemistry. Ch. 2 in *Minerals in Soil Environments* (J.B. Dixon and S.B. Weed, eds). Soil Science Society of America, Madison, Wisconsin.
GASTUCHE M.C., BRUGGENWERT T. & MORTLAND M.M. (1964) Crystallisation of mixed iron and aluminium gels. *Soil Sci.* **98**, 281–289.
GASTUCHE M.C. & HERBILLON A. (1962) Etude des gels alumine: Cristallisation en milieu desionisé. *Bull Soc. Chim. France* 1404–1412.
GIOVANOLI R. (1969) A simplified scheme for polymorphism in the manganese oxides. *Chimia* **23**, 470–472.
GIOVANOLI G. & BRÜTSCH R. (1974) Dehydration of γ-FeOOH: Direct mechanism. *Chimia* **28**, 181–184.
GIOVANOLI R., BÜHLER H. & SOKOLOWSKA K. (1973) Synthetic lithiophorite: electron microscopy and X-ray diffraction. *J. de Microscopie* **18**, 271–284.
GIOVANOLI R., FEITKNECHT W. & FISCHER F. (1971) Über Oxyhydroxide des vierwertigen Mangans mit Schichtengitter. 3. Reduktion von Mangan(III)–manganat(IV) mit Zimtalkohol. *Helv. Chim. Acta* **54**, 1112–1124.
GIOVANOLI R., STAEHLI E. & FEITKNECHT W. (1970a) Über Oxyhydroxide des vierwertigen Mangans mit Schichtengitter. 1. Natrium mangan (II, III) manganat (IV). *Helv. Chim. Acta* **53**, 209–220.
GIOVANOLI R., STAEHLI E. & FEITKNECHT W. (1970b) Über Oxyhydroxide des vierwertigen Mangans mit Schichtengitter. 2. Mangan(III)–manganat(IV). *Helv. Chim. Acta* **53**, 453–464.

GOUT R. (1974) Sur les ions ferriques présents dans le réseau des diaspores. 96ᵉ Congr. Natl. Soc. Savantes, Sect. Sci. 1971, **96**, 383–389.

GRAHAM J., ROCHESTER C.H. & RUDHAM R. (1981) Infrared study of the adsorption of hydrocarbons on rutile at the solid–vapour and solid–liquid interfaces. J. Chem. Soc., Faraday Trans. 1 **77**, 2735–2745.

GREENLAND D.J. (1971) Interactions between humic and fulvic acids and clays. Soil Sci. **111**, 34–40.

GREENLAND D.J. & OADES J.M. (1968) Iron hydroxides and clay surfaces. Trans. 9th Int. Cong. Soil Sci. Adelaide, **1**, 657–668.

GREENLAND D.J., OADES J.M. & SHERWIN J.W. (1968) Electron microscope observations of iron oxides in some red soils. J. Soil Sci., **19**, 116–122.

GRIMME H. (1968) Die Adsorption von Mn, Cu and Zn durch Goethit aus verdünnten Lösungen. Z. Pflanzenernähr, Bodenkd. **121**, 58–65.

HAMDY A.A. & GISSEL-NIELSEN G. (1977) Fixation of selenium by clay minerals and iron oxides. Z. Pflanzenernähr. Bodenkd. **140**, 63–70.

HARDER E.C. (1952) Examples of bauxite deposits illustrating variations in origin. In Problems of Clay and Laterite Genesis, pp. 35–64. Am. Inst. Mining Metall. Engineers, New York.

HARIYA Y. (1961) Mineralogical studies on manganese dioxide and hydroxide minerals in Hokkaido, Japan. J. Fac. Science. Hokkaido University. Ser. IV. Geol. & Mineral. **X**, 641–702.

HARIYA Y. & KIKUCHI T. (1964) Precipitation of manganese by bacteria in mineral springs. Nature, Lond. **202**, 416–417.

HARWARD M.E. & REISENAUER H.M. (1966) Reactions and movement of inorganic soil sulphur. Soil Sci. **101**, 326–335.

HATCHER J.T., BOWER C.A. & CLARK M. (1967) Adsorption of boron by soils as influenced by hydroxy aluminium and surface area. Soil Sci. **104**, 422–426.

HATHAWAY J.C. & SCHLANGER S.O. (1965) Nordstrandite ($Al_2O_3 \cdot 3H_2O$) from Guam. Am. Miner. **50**, 1029–1037.

HAYDEN P.L. & RUBEN A.J. (1974) Systematic investigation of the hydrolysis and precipitation of Al(III), Ch. 9 in Aqueous Environmental Chemistry of Metals, pp. 317–381. Ann. Arbor Science Publishers.

HEALY T.W., HERRING A.P. & FUERSTENAU D.W. (1966) The effect of crystal structure on the surface properties of a series of manganese dioxides. J. Colloid Interface Sci. **21**, 435–444.

HEINTZE S.G. & MANN P.J.G. (1951) A study of various fractions of the manganese of neutral and alkaline soils. J. Soil Sci. **2**, 234–242.

HEM J.D. (1964) Deposition and solution of manganese oxides. Chemistry of manganese in natural waters. Water-Supply Paper 1667-B, U.S. Geol. Surv., Washington.

HEM J.D. & ROBERSON C.E. (1967) Form and stability of aluminium hydroxide complexes in dilute solution. Water-Supply Paper 1827-A, U.S. Geol. Surv., Washington.

HEWETT D.F. & FLEISCHER M. (1960) Deposits of the manganese oxides. Econ. Geol. **55**, 1–55.

HILDEBRAND E.S. & BLUM W.E. (1974) Lead fixation by iron oxides. Naturwissenschaften **61**, 169–170.

HILL R.J. (1979) Crystal structure refinement and electron density distribution in diaspore. Phys. Chem. Minerals **5**, 179–200.

HINGSTON F.J. (1964) Reactions between boron and clays. Aust. J. Soil Res. **2**, 83–95.

HINGSTON F.J., ATKINSON R.J., POSNER A.M. & QUIRK J.P. (1968) Specific adsorption of anions on goethite. Trans. 9th Int. Cong. Soil Sci., Adelaide **1**, 669–678.

HINGSTON F.J., ATKINSON R.J. & QUIRK J.P. (1967) Specific adsorption of anions. Nature, Lond. **215**, 1459–1461.

HINGSTON F.J., POSNER A.M. & QUIRK J.P. (1970) Anion binding at oxide sufaces – the adsorption envelope. Search **1**, 324–327.

HINGSTON F.J., POSNER A.M. & QUIRK J.P. (1972) Anion adsorption by goethite and gibbsite. I. The role of the proton in determining adsorption envelopes. J. Soil Sci. **23**, 177–192.

HINGSTON F.J., POSNER A.M. & QUIRK J.P. (1974) Anion adsorption by goethite and gibbsite. II. Desorption of anions from hydrous oxide surfaces. J. Soil Sci. **25**, 16–26.

HINGSTON F.J. & RAUPACH M. (1967) The reaction between monosilicic acid and aluminium hydroxide. Aust. J. Soil Res. **5**, 295–309.

HSU PA HO (1966) Formation of gibbsite from aging hydroxy-aluminium solutions. Soil Sci. Soc. Am. Proc. **30**, 173–176.

HSU PA HO (1977) Aluminium hydroxides and oxyhydroxides. Ch. 4 in Minerals in Soil Environments, (J.B. Dixon and S.B. Weed, eds), pp. 99–143. Soil Science Society of America, Madison, Wisconsin.

HSU PA HO (1979) Effect of phosphate and silicate on the crystallisation of gibbsite from OH–Al solutions. Soil Sci. **127**, 219–226.

HSU PA HO & BATES T.F. (1964) Formation of X-ray amorphous and crystalline aluminium hydroxides. Mineralog. Mag. **33**, 749–768.

HUANG W.H. & KELLER W.D. (1970) Dissolution of rock forming silicate minerals in organic acids, simulated first stage weathering of fresh mineral surfaces. Am. Miner. **55**, 2076–2094.

HUTTON J.T. (1977) Titanium and zirconium minerals. Ch. 18 in Minerals in Soil Environments (J.B. Dixon and S.B. Weed, eds). Science Society of America, Madison, Wisconsin.

IDZIKOWSKI S. (1971) Adsorption of inorganic ions on α-iron(III) oxide from mixtures of strong electrolytes. I. Adsorption of silver, cupric and aluminium ion as a function of inactive salt concentration. Rocz. Chem. **45**, 1139–1149.

IDZIKOWSKI S. (1973) Adsorption of inorganic ions on α-iron(III) oxide from mixtures of strong electrolytes. IV. Adsorption of silver, cupric and aluminium sulphate in presence of indifferent electrolyte. *Rocz. Chem.* **47**, 231–238.

INOUYE K., ISHII S., KANEKO K. & ISHIKAWA T. (1972) The effect of copper(II) on the crystallisation of α-FeOOH. *Z. anorg. allg. Chem.* **391**, 86–96.

JACKSON M.L. (1964) Free oxides, hydroxides, and amorphous aluminosilicates. In *Methods of Soil Analysis*. Agronomy Monograph 9. American Society of Agronomy.

JACKSON M.L. (1969) *Soil Chemical Analysis – Advanced Course* (2nd edition). M.L. Jackson, Madison, Wisconsin, USA.

JACKSON M.L., TYLER S.A., WILLIS A.L., BOURBEAU G.A. & PENNINGTON R.P. (1948) Weathering sequence of clay-size minerals in soils and sediments. *J. Phys. Chem. Ithaca* **52**, 1237–1260.

JACOBS L.W., SYERS J.K. & KEENEY D.R. (1970) Arsenic sorption by soils. *Soil Sci. Soc. Am. Proc.* **44**, 750–754.

JANEKOVIC A., PRIBANIC M. & IVEKOVIC H. (1979) Dissolution of aluminium and iron minerals in acetylacetone. Correlation of digestion rate with mineral density. *Croat. Chem. Acta.* **52**, 25–28.

JANOT C. & GILBERT H. (1970) Les constituents du fer certains bauxites naturelles étudiées par effet Mössbauer. *Bull. Soc. Fr. Miner. Cristallogr.* **93**, 213–223.

JANOT C., GILBERT H., GRAMMONT X. & BIAS R. (1971) Étude des substitutions Al–Fe dans les roches latéritique. *Bull. Soc. Fr. Miner. Cristallogr.* **94**, 367–380.

JEPSON K. & SCHELLMANN W. (1974) Über den Stoffbestand und die Bildungsbedingungen der Bauxitlagerstatte Weipa, Australien. *Geol. Jahrb. Reihe D*, Mineralogie, Petrographie, Geochemie, Lagerstattenkunde, **7**, 19–106.

JONAS K. & SOLYMAR K. (1970) Preparation, X-ray, derivatographic and infrared study of aluminium-substituted goethites. *Acta Chim. Acad. Sci. Hung.* **66**, 383–394.

JONES L.H.P. & HANDRECK K.A. (1963) Effects of iron and aluminium oxides on silica in solution in soils. *Nature, Lond.* **198**, 852–853.

JONES L.H.P. & MILNE A. (1956) Birnessite, a new manganese oxide mineral from Aberdeenshire, Scotland. *Mineral. Mag.* **31**, 283–288.

JURINAK J.J. (1966) Surface chemistry of hematite: anion penetration effects on water adsorption. *Soil Sci. Soc. Am. Proc.* **30**, 559–562.

KACSALOVA L. (1979) Transformation of bayerite in α-aluminium oxide under mechanical effects. *Acta Chim. Acad. Sci. Hung.* **99**, 115–120.

KALBASI M., RACZ G.J. & LOEWEN-RUDGERS L.A. (1978) Mechanism of zinc adsorption by iron and aluminium oxides. *Soil Sci.* **125**, 146–150.

KARAPETYAN E.T. & PETROVA L.P. (1972) Effect of the composition and morphological features of magnetite and hematite on their solubility. *Obogasch. Rud.* **17**, 34–36.

KAVANAGH B.V., POSNER A.M. & QUIRK J.P. (1976) The adsorption of polyvinyl alcohol on gibbsite and goethite. *J. Soil Sci.* **27**, 467–477.

KELLER W.D. (1978) Diaspore recrystallised at low temperature. *Am. Miner.* **63**, 326–329.

KEMPER W.D. (1966) Aggregate stability of soils from Western United States and Canada. Tech. Bull. No. 1355, US Department of Agriculture.

KENNEDY G.C. (1959) Phase relations in the system $Al_2O_3–H_2O$ at high temperatures and pressures. *Am. J. Sci.* **257**, 563.

KHOROSHEVA D.P. (1968) O baierite boksitovogo gorizonta srednego pridneprov'ya. *Dokl. Akad. Nauk. SSSR* **182**, 434–436.

KINNIBURGH D.G. & JACKSON M.L. (1974) Zinc adsorption by iron hydrous oxide gels. *Agron. Abstracts* p. 122.

KINNIBURGH D.G., JACKSON M.L. & SYERS J.K. (1976) Adsorption of alkaline earth, transition and heavy metal cations by hydrous oxide gels of iron and aluminium. *Soil Sci. Soc. Am. J.* **40**, 796–799.

KINNIBURGH D.G., SYERS J.K. & JACKSON M.L. (1975) Specific adsorption of trace amounts of calcium and strontium by hydrous oxides of iron and aluminium. *Soil Sci. Soc. Am. Proc.* **39**, 464–470.

KISKYRAS D., CHORIANOPOULOU P. & PAPAZETI H. (1978) Some remarks about the mineralogical composition of Greek bauxites. *4th Int. Cong. Study Bauxites*, Alumina Alum. **1**, 409–433.

KITTRICK J.A. (1969) Soil minerals in the $Al_2O_3–SiO_2–H_2O$ system and a theory of their formation. *Clays Clay Miner.* **17**, 157–167.

KIYAMA M. & TAKADA T. (1981) Formation of anatase precipitates containing iron(III) by the air oxidation of strongly acid solutions. *Bull. Chem. Soc. Japan* **54**, 2960–2963.

KNIGHT R.J. & SYLVA R.N. (1974) Precipitation in hydrolysed iron(III) solutions. *J. Inorg. Nucl. Chem.* **36**, 591–597.

KODAMA H. & SCHNITZER M. (1977) Effect of fulvic acid on the crystallisation of Fe(III) oxides. *Geoderma* **19**, 279–291.

KOONS R.D., HELMKE P.A. & JACKSON M.L. (1980) Association of trace elements with iron oxides during rock weathering. *Soil Sci. Soc. Am. J.* **44**, 155–159.

KRAUSE A. & BORKOWSKA A. (1963) Der Einfluss von Fremdanionen auf die Luftoxydation von $Fe(OH)_2$ und die Struktur der Oxydations-produkte. *Z. anorg. allg. Chem.* **326**, 216–224.

KRAUSE A., IGNASIAK J. & KOSTRZEWA E. (1970) Über die Bildung von röntgenamorphem Eisen (III)-hydroxide durch Luftoxydation von Eisen (II)-hydroxide und Eisen (II)-carbonat. *Z. anorg. allg. Chem.* **378**, 210–212.

KRAUSKOPF K.R. (1957) Separation of manganese from iron in sedimentary processes. *Geochim. Cosmochim. Acta* **12**, 61–84.

KRISHNA MURTI G.S.R., SARMA V.A.K. & RENGASAMY P. (1976) Amorphous ferri-aluminosilicates in some tropical ferruginous soils. *Clay Minerals* **11**, 137–145.

KUZNETSOV V.A., GEREI S.V., GOROKHOVATSKII YA.B. & ROZHKOVA E.V. (1977) Study of the interaction of hydrocarbons with the surface of oxide catalysts. II. IR spectroscopic study of the interaction of olefins with nickel(II) oxide and iron(III) oxide. *Kinet. Katal.* **18**, 424–428.

KWONG NG KEE & HUANG P.M. (1975) Influence of citric acid on the crystallisation of Al hydroxide. *Clays Clay Miner.* **23**, 164–165.

LARSON L.T. (1962) Zinc-bearing todorokite from Phillipsburg, Montana. *Am. Miner.* **47**, 59–66.

LAVRENCHUK V.N. & EREMEEV A.F. (1979) Beryllium in diaspore–boehmite bauxites of the Ural region. *Geokhimiya* **2**, 247–253.

LEWIS D.G. & SCHWERTMANN U. (1979a) The influence of Al on iron oxides. III. Preparation of Al goethites in 1M KOH. *Clay Minerals* **14**, 115–126.

LEWIS D.G. & SCHWERTMANN U. (1979b) The influence of Al on the formation of iron oxide. IV. The influence of [Al], [OH], and temperature. *Clay Minerals* **27**, 195–200.

LEWIS D.G. & SCHWERTMANN U. (1980) The effect of [hydroxide] on the goethite produced from ferrihydrite under alkaline conditions. *J. Colloid Interface Sci.* **78**, 543–553.

LEWIS J.F. (1970) Chemical composition and physical properties of magnetite ejected plutonic blocks of the Soufriere volcano, St Vincent, West Indies. *Am. Miner.* **55**, 793–807.

LIPPENS B.C. & STEGGERDA J.J. (1970) Active alumina. In *Physical and Chemical Aspects of Adsorbents and Catalysts*, pp. 171–211. Academic Press, New York.

LOGANATHAN P. & BURAU R.G. (1973) Sorption of heavy metals by a hydrous manganese oxide. *Geochim. Cosmochim. Acta* **37**, 1277–1293.

LOUGHNAN F.C. (1969) *Chemical Weathering of the Silicate Minerals*. Elsevier, New York.

LUBNER K.E. & MALONE P.G. (1974) Trace element composition of bacterial iron oxide. *Geol. Soc. Am. Abstr. Programs* **6**, 1046.

MCCAFFERTY E., PRAVDIC V. & ZETTLEMOYER A.C. (1970) Dielectric behaviour of adsorbed water films on the α-iron(III) oxide surface. *Trans. Faraday Soc.* **66**, 1720–1731.

MCHARDY W.J., THOMPSON A.P. & GOODMAN B.A. (1974) Formation of iron oxides by decomposition of iron-phenolic chelates. *J. Soil Sci.* **25**, 471–482.

MACKAY A.L. (1962) β-Ferric oxyhydroxide-akageneite. *Mineral. Mag.* **33**, 270–280.

MCKEAGUE J.A., BRYDON J.E. & MILES N.M. (1971) Differentiation of forms of extractable iron and aluminium in soils. *Soil Sci. Soc. Am. Proc.* **35**, 33–38.

MCKEAGUE J.A. & CLINE M.G. (1963) The adsorption of monosilicic acid by soil and other substances. *Can. J. Soil Sci.* **43**, 83–96.

MCKEAGUE J.A., DAMMAN A.W.H. & HERINGA P.K. (1968) Iron–manganese and other pans on some soils of Newfoundland. *Can. J. Soil Sci.* **48**, 243–253.

MCKENZIE R.M. (1967) The sorption of cobalt by manganese minerals in soils. *Aust. J. Soil Res.* **5**, 235–246.

MCKENZIE R.M. (1970) The reaction of Co with manganese dioxide minerals. *Aust. J. Soil Res.* **8**, 97–106.

MCKENZIE R.M. (1971) The synthesis of birnessite, cryptomelane and some other oxides and hydroxides of manganese. *Mineral. Mag.* **38**, 493–502.

MCKENZIE R.M. (1972a) The manganese oxides in soils – A review. *Z. Pflanzenernahr. Bodenkd.* **133**, 221–242.

MCKENZIE R.M. (1972b) The sorption of some heavy metals by the lower oxides of manganese. *Geoderma* **8**, 29–35.

MCKENZIE R.M. (1975) An electron microprobe study of the relationships between heavy metals and manganese and iron in soils and ocean floor nodules. *Aust. J. Soil Res.* **13**, 177–188.

MCKENZIE R.M. (1977) Manganese oxides. Ch. 6 in *Minerals in Soil Environments* (J.B. Dixon and S.B. Weed, eds). Soil Science Society of America, Madison, Wisconsin.

MCKENZIE R.M. (1978a) The manganese oxides in soils. In *Geology and Geochemistry of Manganese* (I.M. Varentsov, ed.), Vol. 1, pp. 259–269. Hung. Acad. Science.

MCKENZIE R.M. (1978b) The effect of two manganese dioxides on the uptake of lead, cobalt, nickel, copper and zinc by subterranean clover. *Aust. J. Soil Res.* **16**, 209–214.

MCKENZIE R.M. (1981) The surface charge on manganese dioxides. *Aust. J. Soil Res.* **19**, 41–50.

MCLAREN R.G. & CRAWFORD D.V. (1974) Studies on soil copper. III. *J. Soil Sci.* **25**, 111–119.

MCLAUGHLIN (1954) Iron and titanium oxides in soil clays and silts. *Geochim. Cosmochim. Acta* **5**, 85–96.

MALATI M.A. & SEAGER N.J. (1981) Further investigation of the photo-induced oxidation of normal primary alcohols by anatase titanium dioxide. *J. Oil Colour Chem. Assoc.* **64**, 231–233.

MATIJEVIC E., MATHAI K.G., OTTEWILL R.G. & KERKER M. (1961) Detection of metal hydrolysis by coagulation. III. Aluminium. *J. Phys. Chem. Ithaca* **65**, 826.

MEHRA O.P. & JACKSON M.L. (1960) Iron oxide removal from soils and clays by dithionite–citrate system buffered with sodium bicarbonate. *Clays Clay Miner.* **7**, 317–327.

MICHEL A. (1949) Substitutions dans la magnetite et preparation de nouveaux sesquioxydes cubiques de fer. *Bull. Soc. Chem. France D* 128–131.

MICHEL A. & POUILLARD E. (1949) A new family of cubic iron sesquioxides. *C.r. hebd. Séanc. Acad. Sci., Paris*, **228**, 680–681.

MISAWA T., HASHIMOTO K. & SHIMODAIRA S. (1973) Formation of Fe(II), Fe(III) intermediate green complex on oxidation of ferrous iron in neutral and slightly alkaline sulphate solutions. *J. Inorg. Nucl. Chem.* **35**, 4167–4174.
MORGAN J.J. & STUMM W. (1964) Colloid-chemical properties of manganese dioxide. *J. Colloid Sci.* **19**, 347–359.
MORGAN J.J. & STUMM W. (1965) The role of multivalent metal oxides in limnological transformations as exemplified by iron and manganese. *Proc. Second Int. Water Pollution Research Conference, Tokyo 1964*, Pergamon Press, 103–123.
MULJADI D., POSNER A.M. & QUIRK J.P. (1966a) The mechanism of phosphate adsorption by kaolinite, gibbsite and pseudo-boehmite. Part I. The isotherms and the effect of pH on adsorption. *J. Soil Sci.* **17**, 212–228.
MULJADI D., POSNER A.M. & QUIRK J.P. (1966b) The mechanism of phosphate adsorption by kaolinite, gibbsite and pseudoboehmite. II. The location of the adsorption sites. *J. Soil Sci.* **17**, 230–237.
MULLINS C.E. (1977) Magnetic susceptibility of the soil and its significance in soil science – A Review. *J. Soil Sci.* **28**, 223–246.
MURAD ENVER & TAYLOR R.M. (1984) The Mössbauer spectra of hydroxycarbonate green rusts. *Clay Minerals* **19**, 77–83.
MURPHY P.J., POSNER A.M. & QUIRK J.P. (1975) Chemistry of iron in soils. Ferric hydrolysis products. *Aust. J. Soil Res.* **13**, 189–201.
MURPHY P.J., POSNER A.M. & QUIRK J.P. (1976) Characterization of hydrolysed ferric ion solutions. A comparison of the effects of various anions on the solutions. *J. Colloid Interface Sci.* **56**, 312–319.
MURRAY J.W. (1974) The surface chemistry of hydrous manganese dioxide. *J. Colloid Interface Sci.* **46**, 357–371.
MURRAY J.W. (1975) The interaction of cobalt with hydrous manganese dioxide. *Geochim. Cosmochim. Acta* **39**, 635–647.
MURRAY D.J., HEALY T.W. & FUERSTENAU D.W. (1968) The adsorption of aqueous metal on colloidal hydrous manganese oxide. In *Adsorption from Aqueous Solution. Adv. Chem. Series*, **79**, 74–81. American Chem. Soc., Washington.
MUSIC S., GESSNER M. & WOLF R.H.H. (1979) Sorption of small amounts of cobalt(II) and iron(III) oxide. *Mikrochim. Acta* **1**, 105–112.
MUSTOE G.E. (1981) Bacterial oxidation of manganese and iron in a modern cold spring. *Bull. Geol. Soc. Am.* **902**, 147–153.
NAGARAJAH S., POSNER A.M. & QUIRK J.P. (1970) Competitive adsorption of phosphate with polygalacturonate and other organic anions on kaolinite and oxide surfaces. *Nature, Lond.* **228**, 83–84.
NAKAI M. & YOSHINAGA N. (1980) Fibrous goethite in some soils from Japan and Scotland. *Geoderma* **24**, 143–158.
NALOVIC L. (1974) Geochemical research on transition elements in soils. An experimental study of the influence of trace elements on the behaviour of iron and the evolution of ferriferous compounds during pedogenesis. Summary of Thesis for "Doctorat d'Etat" Paris University VI.
NALOVIC L., PEDRO G. & JANOT C. (1975) Demonstration by Mössbauer spectroscopy of the role played by transitional trace elements in the crystallogenesis of iron hydroxides (III). *Proc. Int. Clay Conf. 1975, Mexico City*, pp. 601–610. Applied Publishing Ltd, Wilmette, Illinois.
NORRISH K. (1975) Geochemistry and mineralogy of trace elements. In *Trace Elements in Soil–Plant–Animal Systems* (D.J.D. Nicholas and A.R. Egan, eds), pp. 56–81. Academic Press Inc., New York.
NORRISH K. & ROSSER H. (1983) Mineral phosphate, Ch. 24 in *Soils: An Australian Viewpoint*, Division of Soils, CSIRO; CSIRO: Melbourne/Academic Press, London.
NORRISH K. & TAYLOR R.M. (1961) The isomorphous replacement of iron by aluminium in soil goethites. *J. Soil Sci.* **12**, 294–306.
OADES J.M. (1963) The nature and distribution of iron compounds in soils. *Soils and Fertilizers* **26**, 69–80.
OADES J.M. & TOWNSEND W.N. (1963) The detection of ferromagnetic minerals in soils and clays. *J. Soil Sci.* **14**, 179–187.
OELKRUG D. & RADJAIPOUR M. (1980) Fluorescence of adsorbed aromatic hydrocarbons. Molecular and donor–acceptor interactions with the surface of aluminium oxide, gallium(III) oxide and titanium dioxide. *Z. Phys. Chem.* (Wiesbaden) **123**, 163–172.
PARFITT G.D. (1976) Surface chemistry of oxides. *Pure and Appl. Chem.* **48**, 415–418.
PARFITT R.L., ATKINSON R.J. & SMART R.St.C. (1975) The mechanism of phosphate fixation by iron oxides. *Soil Sci. Soc. Am. Proc.* **39**, 837–841.
PARFITT R.L., FARMER V.C. & RUSSELL J.D. (1977a) Adsorption on hydrous oxides. I. Oxalate and benzoate on goethite. *J. Soil Sci.* **28**, 29–39.
PARFITT R.L., FRAZER A.R. & FARMER V.C. (1977b) Adsorption on hydrous oxides. III. Fulvic acid and humic acid on goethite, gibbsite and imogolite. *J. Soil Sci.* **28**, 289–296.
PARFITT R.L., FRAZER A.R., RUSSELL J.D. & FARMER V.C. (1977c) Adsorption on hydrous oxides. II. Oxalate, benzoate and phosphate on gibbsite. *J. Soil Sci.* **28**, 40–47.
PARFITT R.L. & RUSSELL J.D. (1977) Adsorption on hydrous oxides. IV. Mechanisms of adsorption of various ions on goethite. *J. Soil Sci.* **28**, 297–305.
PARFITT R.L., RUSSELL J.D. & FARMER V.C. (1976) Confirmation of the surface structures of goethite (α-FeOOH) and phosphated goethite by infrared spectroscopy. *J. Chem. Soc. Faraday Trans.* **72**, 1082–1087.
PARFITT R.L. & SMART R.St.C. (1977) Infrared spectra from binuclear bridging complexes of sulphate absorbed on goethite (α-FeOOH). *J. Chem. Soc. Faraday Trans.*, **73**, 796–802.

PARFITT R.L. & SMART R.ST.C. (1978) The mechanism of sulphate adsorption on iron oxides. *Soil Sci. Soc. Am. J.* **42**, 48–50.
PARKS G.A. (1967) Aqueous surface chemistry of oxides and complex oxide minerals. Isoelectric point and zero point of charge. In *Equilibrium Concepts in Natural Water Systems* (Symposium). Advances in Agronomy series 67, Am. Chem. Soc.
PARKS G.A. & DE BRUYN P.L. (1962) The zero point of charge of oxides. *J. Phys. Chem. Ithaca* **66**, 967–973.
PAULING L. (1930) The structure of micas and related minerals. *Proc. Natn. Acad. Sci. USA* **16**, 123–129.
PAWLUK S. (1972) Measurement of crystalline and amorphous iron oxides. *Can. J. Soil Sci.* **52**, 119–123.
PEDRO G. (1964) Contribution al'étude experimentale de l'alteration geochemique des roches cristallines. *Ann. agron.* **15**, 85–91.
PERINET G. & LAFONT R. (1972) Sur la présence d'hematite alumineuse désdordonée dans les bauxites du Var. *C.r. hebd. Séanc. Acad. Sci., Paris* **274**, 272–274.
PERSEIL E.A. & GIOVANOLI R. (1979) La genese des nodules de manganese. *Colloques Internationaux du Centre de la Recherche Scientifique*, 1978 No. 289, 369–377.
PLATONOV B.E., POLISHCHUK T.A. & TSENDROVSKII V.A. (1982) Study of poly(vinyl alcohol) adsorption on a manganese dioxide surface by multiple attenuated total internal reflectance. *Ukr. Khim. Zh.* **48**, 31–34.
POTTER R.M. & ROSSMAN G.R. (1979) The tetravalent manganese oxides; Identification, hydration, and structural relationships by infrared spectroscopy. *Am. Miner.* **64**, 1199–1218.
RANKAMA K. & SAHAMA TH.G. (1950) *Geochemistry*. Univ. Chicago Press, Chicago.
RAUPACH M. (1963) Solubility of simple aluminium compounds expected in soils. III. Aluminium ions in soil solutions and aluminium phosphates in soils. *Aust. J. Soil Res.* **1**, 46–54.
RAZUMEENKO M.V., GRUNIN V.S. & BOITSOV A.A. (1981) Growth of anatase and brookite single crystals by chemical transport reactions. *Kristallografiya* **26**, 650–652.
REISENAUER H.M., TABIKH A.A. & STOUT P.R. (1962) Molybdenum reaction with soils and the hydrous oxides of Fe, Al and Ti. *Soil Sci. Soc. Am. Proc.* **26**, 23–27.
RENGESAMY P. & OADES J.M. (1977) Interaction of monomeric and polymeric species of metal ions with clay surfaces. I. Adsorption of iron(III) species. *Aust. J. Soil Res.* **15**, 221–233.
REYES E.D. & JURINAK J.J. (1967) A mechanism of molybdate adsorption on α-Fe_2O_3. *Soil Sci. Soc. Am. Proc.* **31**, 637–641.
RINELLI G. & MARABINI A. (1981) A new reagent system for the selective flocculation of rutile *Dev. Miner. Process.* 1979, (Pub. 1981) 2. (Miner. Process., Part A), 316–345.
ROCHESTER C.H. & TOPHAM S.A. (1979a) Infrared studies of the adsorption of probe molecules onto the surface of hematite. *J. Chem. Soc., Faraday Trans.* **75**, 1259–1267.
ROCHESTER C.H. & TOPHAM S.A. (1979b) Infrared studies of the adsorption of probe molecules onto the surface of goethite. *J. Chem. Soc., Faraday Trans.* **75**, 872–882.
ROOKSBY H.P. (1961) Oxides and hydroxides of aluminium and iron. Ch. X in *The X-ray Identification and Crystal Structures of Clay Minerals* (G. Brown, ed.), 2nd edn. Mineralogical Society, London.
ROSELL R.A. & BABCOCK K.L. (1968) Precipitated manganese isotopically exchanged with [54]Mn and chelated by soil organic matter. In *Isotopes and Radiation in Soil Organic Matter Studies*, pp. 453–469. Int. Atomic Energy Agency, Vienna.
ROSS D.S. & BARTLETT R.J. (1981) Evidence for nonmicrobial oxidation of manganese in soil. *Soil Sci.* **132**, 153–160.
ROSS G.J. & TURNER R.C. (1971) Effect of different anions on the crystallisation of aluminium hydroxide in partially neutralised aqueous aluminium salt systems. *Soil Sci. Soc. Am. Proc.* **35**, 389–392.
ROSS S.J. (JR), FRANZMEIER D.P. & ROTH C.B. (1976) Mineralogy and chemistry of manganese oxides in some Indiana soils. *Soil Sci. Soc. Am. J.* **40**, 137–143.
RUSSELL J.D. (1979) Infrared spectroscopy of ferrihydrites: evidence for the presence of structural hydroxyl groups. *Clay Minerals* **14**, 109–114.
RYDEN J.C., MCLAUGHLIN J.R. & SYERS J.K. (1977a) Mechanisms of phosphate sorption by soils and hydrous ferric oxide gel. *J. Soil Sci.* **28**, 79–82.
RYDEN J.C., MCLAUGHLIN J.R. & SYERS J.K. (1977b) Time dependent sorption of phosphate by soils and hydrous ferric oxides. *J. Soil Sci.* **28**, 585–595.
SAINI G. R., MACLEAN A.A. & DOYLE J.J. (1966) The influence of some physical and chemical properties on soil aggregation and response to VAMA. *Can J. Soil Sci.* **46**, 155–160.
SCHAHABI S. & SCHWERTMANN U. (1970) Der Einfluss von synthetischen Eisenoxiden auf die Aggregation zweier Lossbodenhorizonte. *Z. Pflanzenernahr. Bodenkd.* **125**, 193–204.
SCHEFFER F., WELTE E. & LUDWIEG F. (1957) Zur Frage der Eisenoxihydrate in Boden. *Chem. Erde* **19**, 51–64.
SCHELLMANN W. (1959) Experimentelle Untersuchungen uber die sedimentare Bildung von Goethit und Hämatit. *Chem. Erde* **20**, 104–135.
SCHELLMANN W. (1976) Verteilung, Bindung und chemische Mobilisierung des Nickels in eisenreichen lateritischen Nickelerzen. *Abschlussbericht zum Forschungsvorhaben* Nr 2964.
SCHNITZER M. & KODAMA H. (1977) Reactions of minerals with soil humic substances. Ch. 12 in *Minerals in Soil Environments* (J.B. Dixon and S.B. Weed, eds). Soil Science Society of America, Madison, Wisconsin.

SCHNITZER M. & SKINNER S.I.M. (1963) Organo-metallic interactions in soils: 2. Reactions between different forms of iron and aluminium and the organic matter of a podzol B_H horizon. *Soil Sci.* **96**, 181–186.
SCHOEN R. & ROBERSON C.E. (1970) Structures of aluminium hydroxides and geochemical implications. *Am. Miner.* **55**, 43–77.
SCHULZE D.G. (1980) Unit cell dimensions of Al-substituted goethites. Fourth Meeting European Clay Groups, Freising, 1980. Abstracts 131–132.
SCHULZE D.G. (1982) The identification of iron oxides by differential X-ray diffraction and the influence of aluminium substitution on the structure of goethite. Ph.D. thesis, Lehrstuhl für Bodenkunde, Technische Universität München in Weihenstephan.
SCHWEISFURTH R. & GATTOW G. (1966) Untersuchungen über Röntgenstruktur und Zusammensetzung microbiell gebildeter Braunsteine. *Z. allg. Mikrobiol.* **6**, 303–308.
SCHWERTMANN U. (1959a) Über die Synthese definierter Eisenoxyde unter verschiedenen Bedingungen. *Z. anorg. allg. Chem.* **298**, 337–348.
Schwertmann U. (1959b) Die fraktionierte Extraction der freien Eisenoxyde in Böden, ihre mineralogischen Formen und ihre Enstehungsweisen. *Z. Pflanzenernähr. Dung. Bodenkd.* **84**, 194–204.
SCHWERTMANN U. (1964) Differenzierung der Eisenoxide des Bodens durch photochemische Extraction mit saurer Ammoniumoxalat-Losung. *Z. Pflanzenernähr. Bodenkd.* **105**, 194–202.
SCHWERTMANN U. (1966) Inhibitory effect of soil organic matter on the crystallization of amorphous ferric hydroxide. *Nature, Lond.* **212**, 645–646.
SCHWERTMANN U. (1970) Der Einfluss organischer Anionen auf die Bildung von Goethit und Hämatit aus amorphen Fe(III)-hydroxid. *Geoderma* **3**, 207–214.
SCHWERTMANN U. (1971) Transformation of hematite to goethite in soils. *Nature, Lond.* **232**, 624–625.
SCHWERTMANN U. (1973) Electron micrographs of soil lepidocrocites. *Clay Minerals* **10**, 59–60.
SCHWERTMANN U. & FISCHER W.R. (1966) Zur Bildung von α-Fe$_2$O$_3$ aus amorphen Eisen(III)-hydroxid. *Z. anorg. allg. Chem.* **346**, 137–142.
SCHWERTMANN U., FISCHER W.R. & PAPENDORF H. (1968) The influence of organic compounds in the formation of iron oxides. *Trans. 9th Int. Congr. Soil Sci., Adelaide* **1**, 645–655.
SCHWERTMANN U., FITZPATRICK R.W. & LE ROUX J. (1977) Al substitution and differential disorder in soil hematites. *Clays Clay Miner.* **25**, 373–374.
SCHWERTMANN U., FITZPATRICK R.W., TAYLOR R.M. & LEWIS D.G. (1979) The influence of aluminium on iron oxides. Part II. Preparation and properties of Al substituted hematites. *Clays Clay Miner.* **27**, 105–112.
SCHWERTMANN U. & TAYLOR R.M. (1972a) The influence of silicate on the transformations of lepidocrocite to goethite. *Clays Clay Miner.* **20**, 159–164.
SCHWERTMANN U. & TAYLOR R.M. (1972b) The *in vitro* transformation of soil lepidocrocite to goethite. In *Pseudogley and Gley Trans. Com. V and VI, Int. Soc. Soil Sci., Stuttgart*, pp. 45–54. Verlag Chemie, Weinheim.
SCHWERTMANN U. & TAYLOR R.M. (1977) Iron oxides. Ch. 5 in *Minerals in Soil Environments* (J.B. Dixon and S.B. Weed, eds). Soil Science Society of America, Madison, Wisconsin.
SCHWERTMANN U. & THALMANN H. (1976) The influence of [Fe(II)], [Si] and pH on the formation of lepidocrocite and ferrihydrite during oxidation of aqueous FeCl$_2$ solutions. *Clay Minerals* **11**, 189–200.
SEGALEN P. (1970) Extraction du fer libre des sols a sesquioxydes par la methode de De Endredy par irradiation a l'ultraviolet des solutions oxaliques. *Cah ORSTOM ser. Pedol.* **VIII**, 483–495.
SEGALEN P. (1971a) Metallic oxides and hydroxides in soils of the warm and humid areas of the world: formation, identification, evolution. *Natural Resources Research XI, Soils and Tropical Weathering Proc. Bandung Symposium* (UNESCO Paris), 15–24.
SEGALEN P. (1971b) La determination du fer libre dans les sols a sesquioxydes. *Cah. ORSTOM ser. Pedol.* **IX**, 3–27.
SELWOOD P.W. (1948) Valence inductivity. *J. Am. Chem. Soc.* **70**, 883.
SHCHEKA S.A., ROMANENKO I.M., CHUBAROV V.M. & KURENTSOVA N.A. (1977) Silica bearing magnetites. *Contr. Miner. Petrol.* **63**, 103–111.
SHERMAN G.D. (1952). The titanium content of Hawaiian soils and its significance. *Soil Sci. Soc. Am. Proc.* **16**, 15–18.
SHERMAN G.D., TOM, A.K.S. & FUJIMOTO C.K. (1949) The origin and composition of pyrolusite concretions in Hawaiian soils. *Pacific Science* **111**, 120–125.
SIDHU P.S., GILKES R.J. & POSNER A.M. (1980) The behaviour of cobalt, nickel, zinc, copper, manganese and chromium in magnetite during alteration to maghemite and hematite. *Soil Sci. Soc. Am. J.* **44**, 135–138.
SIJARIC G. (1978) Mineralogical investigations of the bauxites from Crvene Stijene (Bosnia). *4th Int. Cong. Study Bauxites, Alumina Alum.* **2**, 783–796.
SIMS J.R. & BINGHAM F.T. (1968a) Retention of boron by layer silicates, sesquioxides and soil materials. II. Sesquioxides. *Soil Sci. Soc. Am. Proc.* **32**, 364–369.
SIMS J.R. & BINGHAM F.T. (1968b) Retention of boron by layer silicates, sesquioxides and soil materials. III. Iron- and aluminium-coated layer silicates and soil materials. *Soil Sci. Soc. Am. Proc.* **32**, 369–373.
SMITH K. & EGGLETON R.A. (1983) Botryoidal goethite: a transmission electron microscope study. *Clays Clay Miner.* **31**, 392–396.
SOKOLOVA T.A. (1968) Iron manganese concretions from a strongly podzolic soil profile. *Trans. 9th Int. Cong. Soil Sci., Adelaide* **4**, 459–466.

STACE H.C.T., HUBBLE G.D., BREWER R., NORTHCOTE K.H., SLEEMAN J.R., MULCAHY M.J. & HALLSWORTH E.G. (1968) *A Handbook of Australian Soils*. Rellim Technical Publications, Adelaide, South Australia.

STEPHEN I. (1952) A study of rock weathering with reference to the soils of the Malvern Hills. *J. Soil Sci.* **3**, 20–33.

STRACZEK J.A., HOREN A., ROSS M. & WARSHAW C.M. (1960) Studies of the manganese oxides. IV. Todorokite. *Am. Miner.* **45**, 1174–1184.

STUMM W. & LEE G.F. (1961) Oxygenation of ferrous iron. *Ind. Engng. Chem.* **53**, 143–146.

SVAB E. & KREN E. (1979) Neutron diffraction study of substituted hematite. *J. Magn. Magn. Mater.* **14**, 184–186.

TAKEMATSU N. (1979) Sorption of transition metals on manganese and iron oxides and silicate minerals. *J. Oceanogr. Soc., Japan* **35**, 36–42.

TAYLOR H.F.W. (1973) Crystal structures of some double hydroxide minerals. *Mineralog. Mag.* **39**, 377–389.

TAYLOR R.M. (1968) The association of manganese and cobalt in soils – further observations. *J. Soil Sci.* **19**, 77–80.

TAYLOR R.M. (1980) Formation and properties of Fe(II)Fe(III) hydroxy-carbonate and its possible significance in soil formation. *Clay Minerals* **15**, 369–382.

TAYLOR R.M. (1982) Stabilisation of colour and structure in the pyroaurite-type compounds Fe(II)Fe(III)Al(III)-hydroxycarbonates. *Clay Minerals* **17**, 369–372.

TAYLOR R.M. (1984) The influence of chloride on the mineralisation of Fe from the Fe(II) system. I. Effects of Cl on the formation of magnetite. *Clays Clay Miner.* **32**, 167–174.

TAYLOR R.M. & BOND R.D. (1979) Electrolytic weathering effects in minerals. *Geoderma* **22**, 85–97.

TAYLOR R.M. & MCKENZIE R.M. (1966) The association of trace elements with Mn minerals in Australian soils. *Aust. J. Soil Res.* **4**, 29–39.

TAYLOR R.M. & MCKENZIE R.M. (1980) The influence of Al on iron oxides. Part VI. The formation of Fe(II)–Al(III) hydroxy-chlorides, -sulphates, and -carbonates as new members of the pyroaurite group and their possible significance in soils. *Clays Clay Miner.* **28**, 179–187.

TAYLOR R.M., MCKENZIE R.M. & NORRISH K. (1964) The mineralogy and chemistry of manganese in some Australian soils. *Aust. J. Soil Res.* **2**, 235–248.

TAYLOR R.M. & SCHWERTMANN U. (1974a) Maghemite in soils and its origin. II. Maghemite syntheses at ambient temperature and pH 7. *Clay Minerals* **10**, 299–310.

TAYLOR R.M. & SCHWERTMANN U. (1974b) Maghemite in soils and its origin. I. Properties and observations on soil maghemites. *Clay Minerals* **10**, 289–298.

TAYLOR R.M. & SCHWERTMANN U. (1974c) The association of phosphorus with iron in ferruginous soil concretions. *Aust. J. Soil Res.* **12**, 133–145.

TAYLOR R.M. & SCHWERTMANN U. (1978) The influence of Al on iron oxides. I. The influence of Al on Fe oxide formation from the Fe(II) system. *Clays Clay Miner.* **26**, 373–383.

TAYLOR R.M. & SCHWERTMANN U. (1980) The influence of aluminium on iron oxides. VII. Substitution of Al for Fe in synthetic lepidocrocite. *Clays Clay Miner.* **28**, 267–271.

TETTENHORST R. & HOFMANN D.A. (1980) Crystal chemistry of boehmite. *Clays Clay Miner.* **28**, 373–380.

TEUFER G. & TEMPLE A.K. (1966) Pseudo-rutile – a new mineral intermediate between ilmenite and rutile in the natural alteration of ilmenite. *Nature, Lond.* **211**, 179–181.

TEWARI P.H., CAMPBELL A.B. & LEE WOON (1972) Adsorption of Co^{2+} by oxides from aqueous solution. *Can. J. Chem.* **50**, 1642–1648.

THIEL R. (1963) Zum System α-FeOOH-α-AlOOH. *Z. anorg. allg. Chem.* **326**, 70–78.

THOMAS G.W. (1960) Effect of electrolyte inhibition upon cation exchange behaviour of soils. *Soil Sci. Soc. Am. Proc.* **24**, 329–332.

THRIERR-SOREL A., LARPIN J.P. & MOUGIN G. (1978) Etude cinetique de la transformation de la goethite α-FeOOH en hematite α-Fe_2O_3. *Annales de Chimie* **3**, 305–315.

TILLMANS E. & CORRENS C.W. (1969) Titanium. In *Handbook of Geochemistry* (K.H. Wederpohl, ed.), Vols. 11–12, Springer-Verlag, Berlin.

TURNER S. & BUSECK P.R. (1981) Todorokites: A new family of naturally occurring manganese oxides. *Science* **212**, 1024–1027.

VAN OOSTERHOUT G.W. (1967) The transformation of γ-FeO(OH) to α-FeO(OH). *J. Inorg. Nucl. Chem.* **29**, 1235–1238.

VAN SCHUYLENBORGH J. (1950) The electrokinetic behaviour of the sesquioxide hydrates and its bearing on the genesis of clay minerals. *Trans. 4th Int. Cong. Soil Sci., Amsterdam* **1**, 89–92.

VIOLANTE A. & JACKSON M.L. (1979) Crystallisation of nordstrandite in citrate systems and in the presence of montmorillonite. In *Proc. 6th Int. Clay Conf. 1978, Oxford* (M. Mortland and V.C. Farmer, eds). Elsevier Scientific Pub. Co., Amsterdam.

VIOLANTE A. & VIOLANTE P. (1978) Influence of carboxylic acids on the stability of chlorite-like complexes and on the crystallisation of $Al(OH)_3$ polymorphs. *Agrochimica* **XXII**, 335–343.

VIOLANTE A. & VIOLANTE P. (1980) Influence of pH, concentration, and chelating power of organic anions on the synthesis of aluminium hydroxides and oxyhydroxides. *Clays Clay Miner.* **28**.

VOZNESENSKII S.A., PUSHKAREV V.V. & BAGRETSOV V.F. (1958) Sorption of radio isotopes by aluminium hydroxides. *J. Inorg. Chem., USSR* **3**, 235–239.

WADSLEY A.D. (1950a) A hydrous manganese oxide with exchange properties. *J. Am. Chem. Soc.* **72**, 1881–1884.

WADSLEY A.D. (1950b) Synthesis of some hydrated manganese minerals. *Am. Miner.* **35**, 485–499.
WADSLEY A.D. (1952) The structure of lithiophorite (Al, Li) MnO$_2$(OH)$_2$. *Acta Crystallogr.* **5**, 676–680.
WADSLEY A.D. & WALKLEY A. (1951) The structure and reactivity of the oxides of manganese. *Rev. Pure Appl. Chem.* **1**, 203–213.
WALKER J.L., SHERMAN G.D. & KATSURA T. (1969) The iron and titanium minerals in the titaniferous ferruginous latosols of Hawaii. *Pacific Science* **23**, 291–304.
WALL J.R.D., WOLFENDEN E.B., BEARD E.H. & DEANS T. (1962) Nordstrandite in soil from West Sarawak, Borneo. *Nature, Lond.* **196**, 264–265.
WANN S.S. & UEHARA G. (1978) Surface charge manipulation of constant surface potential soil colloids: 1. Relation to sorbed phosphorus. *Soil Sci. Soc. Am. J.* **42**, 565–570.
WEBER, M.D., MCKEAGUE J.A., RAAD A.T., DE KIMPE C.R., CHANG WANG, HALUSCHAK P., STONEHOUSE H.B., PETTAPIECE W.W., OSBORNE V.E. & GREEN A.J. (1974) A comparison among nine Canadian Laboratories of dithionite-, oxalate-, and pyrophosphate-extractable Fe and Al in soils. *Can. J. Soil Sci.* **54**, 293–298.
WEFERS K. (1967) Das System Al$_2$O$_3$–Fe$_2$O$_3$–H$_2$O. *Z. Erzbergbau Metallhüttenw.* **20**, 13—19.
WHITE K.L. (1979) Chesterton soil concretions: Ilmenite and not iron-manganese cementing matrix, *Science* **204**, 1077–1078.
WEISER H.B. & MILLIGAN W.O. (1942) The constitution of inorganic sols. *Adv. Colloid Sci.* **1**, 227–246.
WILDING L.P., SMECK N.E. & DREES L.R. (1977) Silica in soils: Quartz cristobalite, tridymite and opal. Ch. 14 in *Minerals in Soil Environments* (J.B. Dixon and S.B. Weed, eds), pp. 471–552. Soil Science Society of America, Madison, Wisconsin, USA.
WILKE B.M. & SCHWERTMANN U. (1977) Gibbsite and halloysite decomposition in strongly acid podzolic soils developed from granite saprolite of the Bayerischer Wald. *Geoderma* **19**, 51–61.
YANKOVSKAYA A.K., PROPENKO V.A. & SIMUROV V.V. (1979) Adsorption of aliphatic alcohols and saturated acids on the surface of metal oxides. *Fiz.-Khim. Mekh. Promyuochnykh Tamponazhnykh Dispersii* (N.N. Kruglitskii, ed.) (Izd. Naukova Dumka: Kiev), pp. 59–65. 4th Mater. Resp. Konf. 1977.
ZETTLEMOYER A.C. & MCCAFFERTY E. (1969) Heat of immersion of α-ferric oxide in water. *Z. Phys. Chem.* **64**, 41–48.
ZIV E.F. (1956) Rutilisation of ilmenite under supergene conditions. *Izvest. Akad. Nauk. SSSR ser. geol.* **12**, 57.

Chapter 3

Dispersion and Flocculation

H. VAN OLPHEN

	Page		Page
3.1 GENERAL PRINCIPLES OF COLLOIDAL STABILITY	203	3.3.3 Born repulsion	213
3.1.1 Introduction	203	3.3.4 Protective action of hydrophilic colloids	213
3.1.2 Electrical double layer repulsion	204	3.3.5 Sensitizing effect of hydrophilic colloids	213
3.1.3 van der Waals' attraction between particles	205	3.4 QUALITATIVE TREATMENT OF COLLOIDAL STABILITY. ELECTRICAL DOUBLE LAYER THEORY	214
3.1.4 Net potential energy of particle interaction	205	3.4.1 Introduction	214
3.1.5 Preparation and physical properties of lyophobic colloidal systems	206	3.4.2 Potential and charge distribution in the single flat double layer	215
3.2 COLLOIDAL STABILITY OF CLAY SUSPENSIONS	207	3.4.3 Potential and charge distribution in interacting double layers	219
3.2.1 Effects of particle shape	207	3.4.4 Repulsive force and repulsive energy for interacting double layers	220
3.2.2 Constitution of the electrical double layers on clay particles	208	3.4.5 Specific effects in the electrical double layer	221
3.2.3 Flocculation of clay suspensions	209	3.4.6 Experimental analysis of potential and charge distribution in the double layer	222
3.2.4 Peptization of clay suspensions	210		
3.2.4.1 Edge charge reversal	210		
3.2.4.2 Reversal of face charge	211		
3.3 OTHER METHODS OF INFLUENCING COLLOIDAL STABILITY	212	3.5 VAN DER WAALS' ATTRACTION	223
3.3.1 Entropic repulsion	212	NOTE	223
3.3.2 Lyosphere repulsion	212	REFERENCES	223

3.1 GENERAL PRINCIPLES OF COLLOIDAL STABILITY

3.1.1 Introduction

Dispersions of clay particles in water are classified as "lyophobic colloidal systems". In contrast with "lyophilic colloidal systems", which are thermodynamically stable solutions of macromolecules or polyelectrolytes, the lyophobic systems are not in a thermodynamic equilibrium state: they tend to coarsen by recrystallization of the particles to minimize the Gibbs' interfacial energy. This process is called "ageing". Furthermore, at any given stage of ageing the interactions between particles and between particles and solvent are such that the Gibbs' energy of the systems is minimized by aggregation of the particles. This process is called "flocculation" or "coagulation". However, in these systems, generally an energy barrier for particle association exists, which retards the process of coagulation. Hence, the system may remain virtually unchanged for a considerable period of time, and thus appear to be stable, i.e. "colloidally stable" or "peptized".

In order to be reasonably stable, a colloidal system should be rather clean, i.e. contain only small amounts of electrolytes. In general, the addition of any "inert" electrolyte (an electrolyte which does not react chemically with the particle) promotes flocculation. In practice, the colloidal stability of a system is usually measured in terms of the "flocculation value" or the "critical coagulation concentration (c.c.c.)", which is the concentration of an electrolyte which increases the coagulation rate to an arbitrarily chosen value. The flocculation value can be determined simply by visual inspection of a series of suspensions containing different amounts of salt, after an arbitrarily chosen period of time ("flocculation series experiment"). The flocculating power of electrolytes varies according to the Schulze–Hardy rule: the higher the valance of the ion in the added electrolyte that has a sign opposite to the sign of the charge of the particle, the more powerful the electrolyte in its flocculating action. The valency of the ion of the same sign has has only a minor affect (Hardy 1900).

The most generalized theory of the stability of lyophobic colloids, which can explain the observed non-specific phenomena, considers two competing particle interaction forces: an electrical repulsion force and an attractive force due to the van der Waals'–London dispersion forces acting between all atoms of one particle and all atoms of another particle.

The synthesis of electrical double layer repulsion and van der Waals'–London attraction as a quantitative basis of the theory of the stability of lyophobic colloids was developed simultaneously in the early 1940s by Derjaguin and Landau (1945) and by Verwey and Overbeek (1948). Therefore, this theory is often referred to as the "DLVO Theory".

Some additional particle interaction forces will be discussed later.

3.1.2 Electrical double layer repulsion

A particle can obtain a surface charge in three different ways. (1) The adsorption of an ion for which the solid acts as a reversible electrode (e.g. I^- or Ag^+ for AgI particles, H^+ or OH^- for metal oxides). The potential of such a surface is determined by the activity of these ions – "potential determining ions" – in solution in accordance with the Nernst formula:

$$\psi_0 = \frac{RT}{nF} \ln \frac{a}{a_0}$$

in which a_0 is the activity of the potential determining ion at the point of zero charge (p.z.c.) of the surface. (2) The preferential adsorption (e.g. by chemisorption) of a cation or an anion from an electrolyte solution ("peptizing ions"). The charge of the surface varies with the electrolyte concentration as given by the adsorption isotherm for the peptizing ion and the solid. (3) An imbalance of charge within the solid; in this case the surface charge is a constant independent of the composition of the equilibrium liquid, and is determined by the average unit cell composition.

The surface charge forms one layer of the electrical double layer. The compensating charge of opposite sign is accumulated in the solution near the surface and is more or less diffuse as a result of competing electrostatic and diffusion forces. This layer of the electrical double layer contains an excess of ions of opposite charge ("counter-ions") and a deficit of ions ("co-ions") of the same charge as the surface charge. The sum of the excess counter-ion charge and the deficit of the co-ion charge is equivalent to the surface charge.

When two particles approach each other in solution due to Brownian motion, their diffuse double layers will overlap, which can be shown to result in an increase of the Gibbs' energy of the system, hence the particles will repel each other. This repulsion has a long-range character since a significant part of the diffuse double layer extends up to several hundred nanometres from the surface. The addition of electrolytes to the system will reduce the diffusion tendency of the ions; the diffusion double layer is therefore "compressed". The result is that the range of repulsion of the particles is reduced, and it is this effect of electrolytes on the range of repulsion which is responsible for the decreased colloidal stability of the system, as will be shown in the following analysis.

3.1.3 van der Waals' attraction between particles

The non-specific attraction force which is responsible for the flocculation phenomenon is the van der Waals'–London dispersion force between atoms of the two particles approaching each other by diffusion. Although the force between atoms decays rapidly with distance (the force is inversely proportional to the seventh power of the distance), the sum of all forces between all atom pairs of the particles has a much longer range, comparable with that of the double layer repulsion. The force between particles is inversely proportional to the third power of the particle distance up to distances of the order of 10 nm, and to the fourth power at higher distances due to "retardation" of the van der Waals' forces. Also the magnitude of the total attraction force is of the order of the double layer repulsion force. Direct measurements of the attractive forces between macroscopic objects have confirmed these facts, and the van der Waals' attraction is now firmly established as the cause of the flocculation phenomenon (Hamaker 1937; Vold 1954, 1961; van Silfhout 1966; Mahanty and Ninham 1977).

3.1.4 Net potential energy of particle interaction

The flocculation behaviour of colloidal systems and the effect of electrolytes can be qualitatively explained by combining the attractive and repulsive potential energy as a function of distance between the particles at different electrolyte concentrations. Figure 3.1 shows these potential curves for three electrolyte concentrations. The van der Waals' attraction is practically unaffected by electrolyte concentration, and is represented by the roughly hyperbolic curves below the abscissa. The electrical double layer repulsion is represented by the roughly exponential curves above the abscissa, and its range decreases with increasing electrolyte concentration. The net potential curves all show attraction at close approach, but at large particle distances repulsion may dominate when the electrolyte concentration is low. The resulting energy barrier for flocculation retards the flocculation which is maximal when an energy barrier is absent ("rapid coagulation"). The larger the barrier, the slower the coagulation process, and in very dilute electrolyte systems the rate becomes so small that the system remains practically unchanged for weeks or months.

A quantitative treatment of the repulsion and attraction forces allows the evaluation of the magnitude of the energy barrier at different electrolyte concentrations, which governs the colloidal stability of a colloidal system by retarding flocculation. The kinetics of flocculation has been treated as a diffusion problem in the absence of an energy barrier (rapid coagulation) by von Smoluchowski (1916, 1917), and in the presence of an energy barrier (slow coagulation) by Fuchs (1934). The factor W by which the coagulation rate of a given system is retarded by the energy barrier is called the "flocculation ratio". Colloidal stability is indeed not an absolute stability, but the difference between a stable and an unstable system is only gradual and is a matter of smaller or larger coagulation velocities. A "flocculation value" is the electrolyte concentration at which a system flocculates in an arbitrarily chosen time, hence, colloidal stability is an arbitrary parameter of the system.

The theory has been very successful in explaining the Schulze–Hardy rule governing the effect of ion valency. Also, the fact that W appears to depend strongly on electrolyte concentration explains why the simple flocculation test is so sensitive.

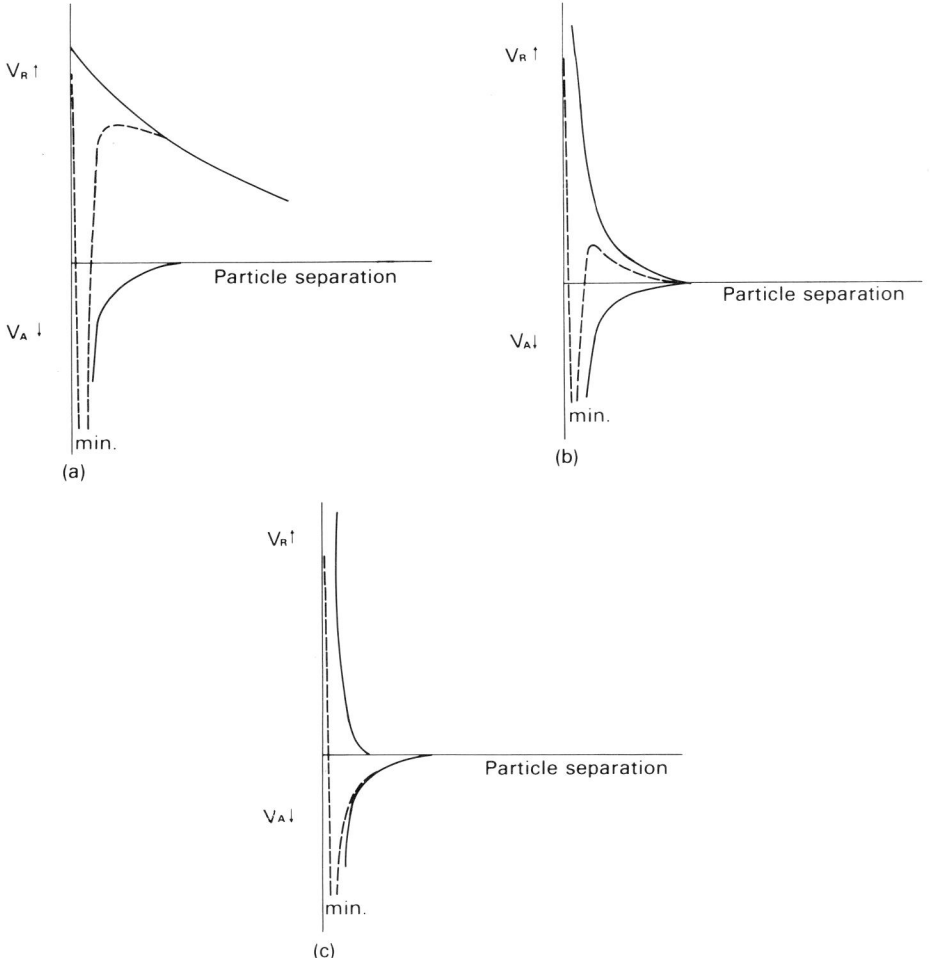

FIG. 3.1. Net interaction energy as a function of particle separation. (a) Low electrolyte concentration. (b) Intermediate electrolyte concentration. (c) High electrolyte concentration. (From H. van Olphen 1977.)

3.1.5 Preparation and physical properties of lyophobic colloidal systems

Lyophobic colloidal systems are not formed spontaneously. The proper size of the particles is obtained either by dispersion or by condensation methods. The latter involve precipitation from solution, and the ultimate particle size is determined by the course of the nucleation and growth processes, which can be influenced by variations in precipitation conditions. The dispersion methods comprise electric disintegration of metals in an arc (Bredig's method), ultrasonic dispersion, or mechanical disintegration. Clays usually can be dispersed directly in water since the particle size is already of colloidal dimensions as a result of natural processes.

In order to obtain stable suspensions, excess electrolyte must be removed by ultrafiltration, ultracentrifugation and washing, or by dialysis. The addition of an appropriate amount of peptizer is required in most cases.

The sign of the charge of the particles depends on the species of peptizing ion added. In those systems in which the particles act like reversible electrodes, the sign of the charge depends on the activity of the potential-determining ions in solution relative to their activity at the point of zero charge (p.z.c.). For example, an aluminium oxide sol is positively charged in acid solution, and negatively charged in alkaline solution. In the neighbourhood of pH 7 the system is uncharged and flocculates. The exact position of the p.z.c. depends on the crystal structure of the aluminium oxide. The sign of the charge can also be changed by specific chemisorption of anions or cations: for example, a positive alumina suspension becomes negative upon the addition of anions which react with Al to form complex anions, such as fluoride, metaphosphate and certain organic anions. A negative particle will also become positive when hydrolysable salts of three or four valent cations (e.g. Al or Th salts) are added. Hence, when a negative particle in suspension is treated with $AlCl_3$ solutions, the suspension first flocculates at a low concentration of $AlCl_3$ as is expected from the Schulze–Hardy rule, but with further addition of $AlCl_3$ the charge of the particle is reversed and a stable positive suspension is obtained. This positive suspension then flocculates with additional salt, primarily due to the flocculating action of the chloride ion. Such a sequence of events is called an "irregular series".

The bulk physical and mechanical properties of suspensions depend strongly on the state of flocculation or peptization of the suspended matter. Flocs are usually loose aggregates of particles and are heavier than the original particle and therefore they settle faster. The settling volume of the flocs is usually large because of the enclosed liquid. In contrast, the settling volume of the peptized system is small and the sediment is dense because the individual particles are able to attain the lowest possible position in the vessel by sliding along each other. The sediment of a peptized suspension is much more difficult to redisperse mechanically than that of a flocculated suspension. Analogous effects are observed when suspensions are filtered. The filter cake of a flocculated suspension is voluminous and permeable, but that of a peptized suspension is dense and rather impermeable. Such effects of the state of peptization and flocculation are important in many practical systems. Minor adjustments of the stability of the system by proper addition of electrolytes (flocculants or peptizers) allows the tailoring of the bulk properties of a suspension to suit a particular application. At sufficiently high concentrations flocculation may lead to a matrix of linked particles throughout the system, and a gel rather than a flocculated dispersion is obtained. Hence, gelation should be considered a special case of flocculation. The rheological properties of these gels depend largely on the attractive force between the particles and the number of particle links present.

3.2 COLLOIDAL STABILITY OF CLAY SUSPENSIONS

3.2.1 Effects of particle shape

The preceding general notes on the stability of lyophobic colloidal systems apply to clay suspensions as well. However, the colloidal phenomena in these suspensions are complicated by the fact that clay particles are usually plate like or lath shaped, and expose two crystallographically different surfaces to the solution, i.e. the face surfaces and the edge surfaces. As discussed below, the electrical double layers on these two crystal surfaces are different, hence the double layer interactions between faces, between edges, and between faces and edges are different. Also, the van der Waals' attraction will be different for particles associating face to face ("FF"), edge to face ("EF"), or edge to edge ("EE"), because of the different summation geometry of the forces between atoms of the two particles. Therefore, FF, EF, and EE flocculation is

208   DISPERSION AND FLOCCULATION

governed by different net potential curves of interaction, and depending on conditions, one or the other type of association may predominate. This is an important point to consider since the properties of the resulting flocculated systems will be quite different. EF and EE flocculation leads to loose card house structured flocs, or at a sufficiently high concentration to gels, whereas FF flocculation simply leads to the formation of thicker particles, although some chain formation due to partial overlap of associated faces may occur. In the clay literature it is customary to refer to EE and EF association as flocculation, whereas FF association is often called "aggregation", and the reverse processes are usually referred to as deflocculation or peptization, and "dispersion" respectively.

3.2.2 Constitution of the electrical double layers on clay particles

Clay minerals are characterized by the presence of crystal imperfections, and ion substitutions such as Al for Si or Mg for Al which result in an overall shortage of positive charge. This deficit of charge within the crystal is compensated by the adsorption of ions on the faces of each of the layers of smectite and vermiculite, or on the exterior crystal faces only in the case of illite and kaolinite particles. In suspension, the ions on the exterior surface assume a diffuse distribution; a diffuse electrical double layer is created with a negative surface charge, and a diffuse layer of compensating cations. This double layer has a constant charge independent of the composition of the equilibrium solution. As is the case with counter-ions in general, they can be exchanged for other ions. Since the number of exchangeable counter ions per unit weight of a clay is often relatively large, and the cation exchange capacity of a clay therefore a rather prominent property, the counter-ions are usually referred to as the exchangeable ions.

The situation at the edge surfaces is quite different from that on the faces. The edge surfaces are broken bond surfaces, and the double layers on these surfaces should be analogous to those on the surfaces of oxide particles, for which the double layer potential is determined by the activity of H^+ or OH^- ions in solution. By analogy with alumina, an edge surface at which aluminium–oxygen bonds are broken should carry a positive double layer in acid solution, and a negative double layer in alkaline solution, with a p.z.c. in the neighbourhood of pH 7 depending on the particular crystal structure. It appears that in neutral and acid solution the edges of clays do carry a positive double layer, which becomes negative at higher pH. This fact can be elegantly demonstrated by preparing electron micrographs from mixtures of kaolinite, and negative gold sols, which show exclusive deposition of the small gold particles on the edges, and not on the faces which repel the gold particles. In spite of the positive edge charge, the particle as a whole is electrophoretically negatively charged; apparently, the negative charge of the comparatively large crystal faces dominates. Nevertheless, the presence of a positive charge on the edge surfaces has important consequences for the stability behaviour of clay suspensions (van Olphen 1950, 1951a,b; Schofield and Samson 1954).

It was mentioned already that in kaolinite the face charge which compensates for shortage of positive charge within the crystal is located on the exterior basal surfaces of the rather thick layer stacks of which the particle consists. This charge will occur on both the siloxane face at one side of the particle and the hydroxyl face at the other side to achieve colloidal stability. Although the amount of exchangeable ions per unit weight is usually small, it is still large when based on unit face surface area, hence the surface charge density is comparable to that of a montmorillonite. At high pH the amount of exchange cations is enhanced by those located at the negative edges. A different picture of the surface condition of kaolinite was developed by

Ferris and Jepson (1975). Evidence was presented that for the particular kaolinite studied the particles are coated by polymeric Al—O—OH species so that cation and anion exchange as a function of pH is governed by the properties of these coatings.

3.2.3 Flocculation of clay suspensions

In the presence of a negative face double layer and a positive edge double layer, edge to face flocculation must occur in a clean suspension, since the van der Waals' attraction is not counteracted by electrical double layer repulsion, but enhanced by the electrostatic attraction between the oppositely charged surfaces. In principle, therefore, a clean neutral or acidic clay suspension cannot be stable. This phenomenon may be called "internal mutual coagulation" in analogy with coagulation taking place in mixtures of positive and negative colloids which is called mutual coagulation (van Olphen 1964). Although a clean dilute suspension of montmorillonite has the appearance of a homogeneous and colloidally stable suspension, at a higher concentration it becomes obvious that the system is indeed flocculated since a strong gel is obtained. The minimum concentration of particles required to create a gel can be calculated from the average size of the particles assuming that they associate in a cubic network. Such calculations give results in agreement with observations (van Olphen 1956, 1959, 1964).

The reason why in dilute montmorillonite suspensions the system has a stable appearance may be attributed to the formation of flocs consisting of only a few plates on the average, which are so light that they do not settle. When visualizing the association of three large plates edge to face, it is useful to think of a "double T" unit:

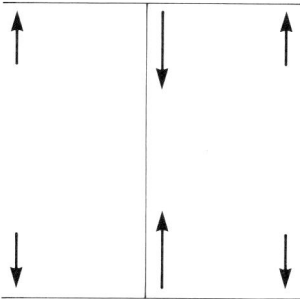

There would be a subtle balance in such flocs between edge to face attraction, and double layer repulsion between the faces, hence convection currents will easily disrupt such units so that the average floc size will remain small.

When salts are added to these apparently stable montmorillonite suspensions visual flocculation and sedimentation of flocs does occur, and flocculation values can be determined in the usual way. It then appears that the flocculation values follow the Schulze–Hardy rule for negative suspensions, correcting for ion exchange which takes place simultaneously. Apparently, salt flocculation is governed by the negative face double layers, and FF flocculation dominates, although EE and EF type flocculation will occur to some extent too. The flocculated appearance is now due to the creation of thicker FF associated units which make the flocs heavier, and cause them to settle. This is also apparent at higher suspension concentrations at which gels are created. The addition of electrolytes to the gels reduces their yield stress as a consequence of the reduction of the number of particles, and hence of the number of links in the matrix due to FF association.

3.2.4 Peptization of clay suspensions

The principle of peptizing clay suspensions by modification of the double layers is to eliminate internal mutual flocculation by reversing either the face charge or the edge charge, and to build up the reversed charge to a level sufficient to prevent flocculation by van der Waals' forces. The most common procedures are based on reversing the edge charge. Because of the relatively small area of the edge surfaces, edge charge reversal is usually economically more attractive than face charge reversal.

3.2.4.1 Edge charge reversal (van Olphen 1950)
There are two ways to reverse the edge charge. One is to increase the concentration of potential-determining OH ions by adding alkali, raising the pH, passing the p.z.c. into the negative region of the edge surface. The other is to add electrolytes containing anions which are chemisorbed to form complex anions with Al or other cations exposed at the broken bond edge surfaces. Examples of the latter are the fluoride ion, polymetaphosphate ions, and many other inorganic and organic anions.

The effectiveness of the peptizers can be conveniently measured in terms of the increase of the NaCl flocculation value of the system with the addition of a certain amount of peptizer. When salt flocculation value is plotted versus amount of peptizer added, "flocculation diagrams" of the kind shown in Fig. 3.2 are obtained.

The curves show a rapid increase of the salt flocculation value upon the addition of small amounts of peptizer until a maximum value is reached. Beyond this maximum, the effect of the peptizer diminishes since the cations of the peptizing salt exert a flocculating action on the flocculation clay particles, and finally the salt flocculation value is reduced to zero when the flocculation value for the peptizer by itself is reached.

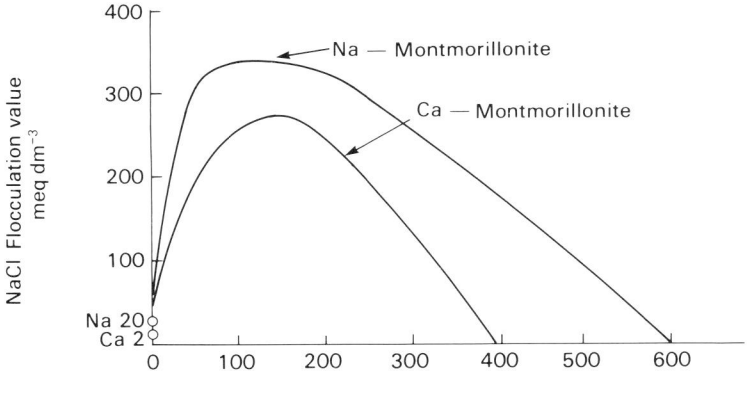

FIG. 3.2. Flocculation diagram of a sodium and a calcium montmorillonite sol with sodium polymetaphosphate. (From H. van Olphen 1977.)

Earlier it was thought that in general the peptizing effect of certain sodium salts would be due to the conversion of a calcium clay to the more stable sodium clay by ion exchange. However, from a comparison between the curves for sodium and calcium montmorillonite, it is obvious that ion exchange can only account for the small increase of stability from 2 mmol dm^{-3} for calcium montmorillonite to 20 mmol dm^{-3} for sodium montmorillonite. The major effect, which is due to edge charging by the metaphosphate anion, is of the same order of magnitude for both ion forms of the clay. Another contribution to this large effect is that in the presence of the polyanion the average activity, and therefore the flocculating power of cations in solution, is decreased. In the absence of this effect, a peptizer is less effective. For example, fluoride which is able to reverse the edge charge by the formation of AlF_6^{3-} ions increases the flocculation value of a sodium montmorillonite to a value of only about 30 mmol dm^{-3}. The same applies to the relatively small effect of peptization with NaOH, and also to the effect of a trivalent phenol, which is small compared to the large effect of polyvalent phenolates such as the natural tannates.

In summary, it appears that any salts containing anions which are able to become chemisorbed at the clay particle edges by complex anion formation with the exposed cations of the broken octahedral sheets will be potentially suitable as peptizers for clay suspensions. The term "complex anion formation" is used here in a broad sense, and includes the chemisorption of, for example, a dicarboxylic acid which for steric reasons becomes attached to the surface by means of one carboxylic group only, the other group providing the charge. Furthermore, these salts will be more effective if they significantly decrease the activity of the cations in the system.

The relative effectiveness of a group of peptizers depends on the type of clay treated; there is no unique scale of effectiveness. The reasons for differences in effectiveness are that in different clays different octahedral cations are exposed, and the complexing ability of certain anions varies with the type of cation. For example, the sequence of effectiveness of a number of anions is different for alumina and ferric oxide sols, and it has been shown (van Olphen 1951a) that the same differences are shown by clays with a small and with a large amount of Fe in octahedral position in addition to Al. Therefore, peptizers should be evaluated empirically for each individual system. Examples of commonly used peptizers are sodium salts of oxalic acid and other dicarboxylic acids, citrates, pyrophosphates and metaphosphates (not orthophosphates), natural tannates, and other polyphenolates such as alizarin dyes and lignosulphonates.

If the main purpose of the peptization is to reduce the yield stress of concentrated clay suspensions, the simultaneous promotion of FF association is very effective, as mentioned before. This can be achieved, for example, by adding calcium ions to the system by which the clay will be partially converted into a calcium clay which yields thicker particles. Only in those systems in which the sequestering ability of the anions is large enough to prevent EE and EF flocculation by the divalent cations can this method be applied. A well-known example is the treatment of drilling fluids by quebracho extract (a moderately high molecular weight phenolic polycondensate) and lime. The greater effect of Fe–Cr lignosulphonates compared with that of Na lignosulphonates may at least partially be explained in the same way.

Another example of the importance of the sequestering ability is the much greater effect of natural tannates compared with that of simple phenols. From a study of the effect of the latter it appeared that a phenolate is effective only if it contains at least three phenol groups, two of which must be adjacent, to achieve chemisorption.

3.2.4.2 Reversal of face charge
The face charge of a clay can be reversed by the adsorption of certain organic cations,

particularly the long chain quaternary ammonium salts. Upon the addition of these compounds, the organic cation is first adsorbed by ion exchange with a high preference of the surface for the organic cation. Since the cationic groups are attached to the surface, with the hydrocarbon chains pointed towards the solution, the double layer is completely condensed and the system flocculates. (The flocculation could also be regarded as the result of the reduction of solubility of the particle enveloped by the hydrocarbon chains.) With further addition of quaternary cations, these are now adsorbed through van der Waals' association of the hydrocarbon chains of the added quaternary compound with those on the surface, and the ionic groups of the second layer of cations point towards the solution. The positive particle thus created becomes colloidally stable with a reversed charge (van Olphen 1951c; Jordan 1963; Pham Ti Hang and Brindley 1970).

As mentioned before, this method of peptization is economically less attractive than edge charge reversal, particularly in the case of montmorillonites in which the quaternary ammonium ions are also adsorbed between the unit layers, requiring about equal weights of clay and organic peptizer for charge reversal.

3.3 OTHER METHODS OF INFLUENCING COLLOIDAL STABILITY

Although the electrical double layer repulsion and van der Waals' attraction are the most general forces operating in colloidal systems, there are some other special factors which influence stability, and which can be used to modify practical suspensions, both in water and in other media. Some such factors are given in the following paragraphs.

3.3.1 Entropic repulsion

Entropic repulsion may occur in systems in which long chain compounds are adsorbed on particle surfaces and are attached to the surface at one end of the compound, allowing the chain a certain degree of freedom of movement in the liquid medium. When two particles approach each other and the chains begin to interfere as the particle distance beomes smaller than twice the chain length, their motion is restricted and a decrease in entropy of the system results, which is manifested as a repulsion. Entropic repulsion has a comparatively short range and only creates stability where it can successfully compete with van der Waals' attraction, for instance, in systems containing small particles for which the van der Waals' attraction is comparatively small at distances less than twice the chain length. This mechanism of stabilization has been inferred to exist primarily in dispersions of particles in hydrocarbon liquids in which the formation of electrical double layers could not be assumed. However, in some dispersions in hydrocarbon liquids, ionized compounds do occur and in such systems, stability can be attributed to electrical double layers. When large particles are involved, stability in such organic systems can indeed only be achieved through electrical double layer repulsion (Mackor 1951; Koelmans and Overbeek 1954).

3.3.2 Lyosphere repulsion

When particles surrounded by an adsorbed layer of solvent molecules approach each other to distances smaller than twice the thickness of the adsorbed solvent layers, the required work of desorption is manifested as particle repulsion. Inasmuch as the thickness of these adsorbed layers may be expected to be limited to the thickness of one or two monomolecular layers only, this type of repulsion will be of a very short range. Its effect on the potential curves of particle interaction will be to limit the depth of the minimum in the net curve at close approach. The magnitude of this short-range repulsion can be estimated from the Gibbs' adsorption energy of the solvent which can be derived from vapour phase adsorption isotherms.

3.3.3 Born repulsion

At close approach of particles where the crystals come into contact, Born repulsion prevents further approach.

3.3.4 Protective action of hydrophilic colloids

"Hydrophilic colloids" are thermodynamically stable true solutions of macromolecules or polyelectrolytes. They do not precipitate at high electrolyte concentrations generally, although "salting out" sometimes occurs if the added salt reduces the solubility significantly. When such macromolecular solutions are added to hydrophobic colloidal systems, such as clay–water dispersions, these become protected against flocculation by salts to a much greater extent than can be achieved by peptizing chemicals of the types mentioned earlier (Fig. 3.3). The particles become enveloped by the macro-ions of the polyelectrolyte solutions. Several different mechanisms may be responsible for the increased stability, i.e. increase of surface charge, entropic repulsion, work of desorption, and possibly a reduction of the van der Waals' attraction due to the low density of the polymer hull attached to the particles (Healy and La Mer 1964; Hesselink *et al.* 1971).

3.3.5 Sensitizing effect of hydrophilic colloids

When polyelectrolyte solutions are added to hydrophobic colloids in relatively small amounts, they make the hydrophobic colloid more sensitive to flocculation by salt (Fig. 3.3). This effect is explained by the bridging of the particles by the long chain compounds, which at such relatively low concentrations may become adsorbed on two particles simultaneously. The presence of small amounts of salt in the system (less than required for flocculation) promotes the bridging effects since the presence of salt allows the particles to come closer together. The salt also facilitates the adsorption of the polymer on the particle surfaces. The effectiveness of this method depends on the relative rates of adsorption and of flocculation, and the sequence of addition of salt and polymer is also an important variable (Ruehrwein and Ward 1952).

In many applications involving clay and other suspensions, advantage is taken of either the protective or sensitizing action of polyelectrolytes. For example, polyelectrolytes such as modified starches are used in relatively high concentrations to stabilize drilling fluids under marine conditions. Used in relatively small amounts, polyelectrolytes aid flocculation and sedimentation of dispersed matter in water or waste treatment plants, and in "soil conditioning" they improve the permeability of the soil by promoting flocculation of the fine particle size fractions through the bridging mechanism.

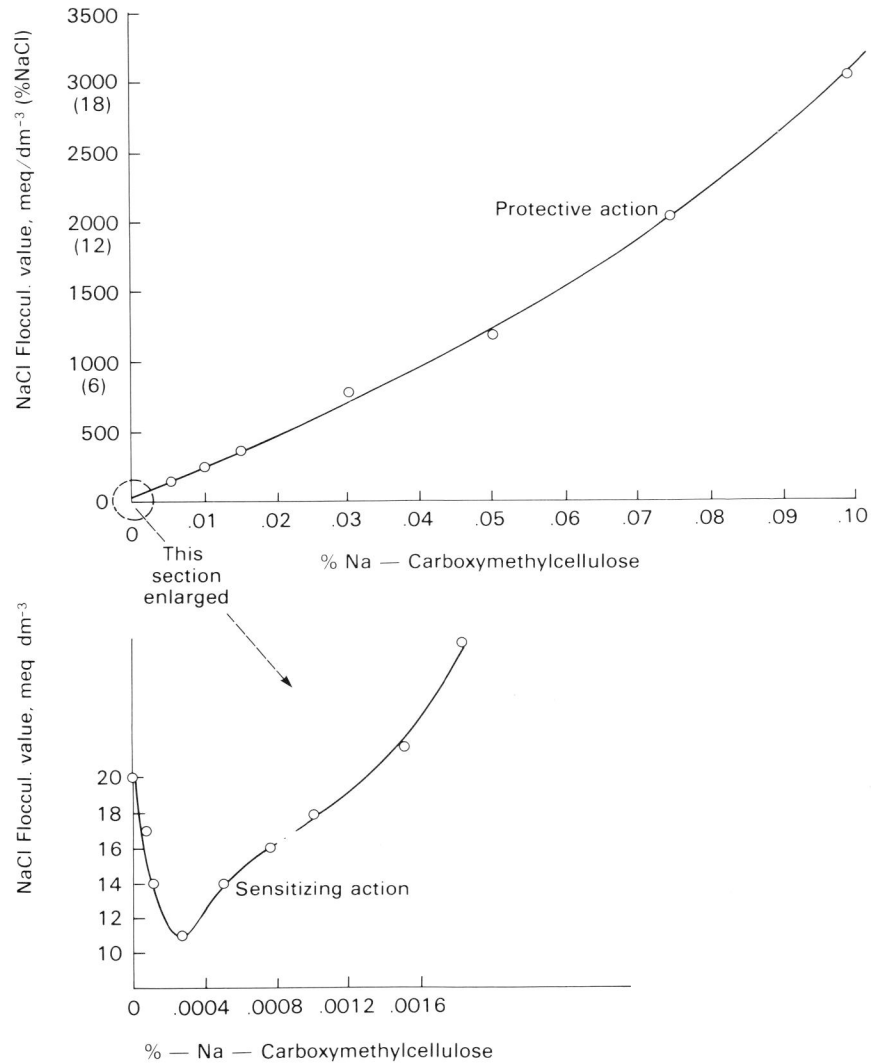

FIG. 3.3. Protective and sensitizing action of Na carboxymethylcellulose on a sodium montmorillonite sol. (From H. van Olphen 1977.)

3.4 QUANTITATIVE TREATMENT OF COLLOIDAL STABILITY. ELECTRICAL DOUBLE LAYER THEORY

3.4.1 Introduction

The colloidal stability of suspensions, osmotic swelling and ion exchange are governed by the structure of the diffuse electrical double layers on the particles in the suspension. These phenomena can be treated quantitatively on the basis of the diffuse electrical double layer theory. Gouy (1910, 1917) and Chapman (1913) were the first to develop a theory in which the ions were considered as point charges, hence, this theory is non-specific. Subsequently, many refinements have been introduced to incorporate effects of ion size, of specific interactions between the ions and the surface, and of mutual interactions between ions, solvent molecules and surfaces. The first two were first considered by Stern who introduced the concept of the distance of closest approach between counter-ion and surface which is equal to the radius of the counter-ion. In the resulting molecular condenser, the potential drops linearly and rather steeply ("Stern layer"). At the same time, Stern (1924) considered the effect of specific adsorption of the counter-ions as a possible contribution. Grahame (1947) refined Stern's model by distinguishing between the distance of closest approach of unhydrated counter-ions to the surface, and that of hydrated ions of the same sign as the surface charge (co-ions), which led to the concept of the inner and outer Helmholtz planes. In later years many further refinements have been introduced by various authors, but for practical purposes, the Gouy and the Stern models appear to be very useful as first approximations. Therefore, the following treatment deals with these approaches only.

In clay–water systems a quantitative treatment of the electrical double layer is most useful in dealing with those problems which involve the face double layers (e.g. osmotic swelling, counter-ion exchange, negative adsorption of co-ions). The following discussion will be limited to flat double layers (which simplifies the mathematic formulation), and the emphasis will be on the properties of double layers of constant charge, rather than of constant potential.

3.4.2 Potential and charge distribution in the single flat double layer

At a distance x from the surface, the local charge density ρ and the local electric potential ϕ are related by the *Poisson equation*:[1]

$$d^2\phi/dx^2 = -(4\pi/\varepsilon)\rho \tag{1}$$

in which ε is the dielectric constant of the medium, and ρ can be written

$$\rho = v_+ e n_+ - v_- e n_-$$

in which n_+ and n_- are the local concentrations of the cations and anions respectively, and v_+ and v_- the absolute values of their valencies, i.e. without regard to sign.

At equilibrium, the local concentrations of the ions are related to the local electric potential by the *Boltzmann equation*:

$$\begin{aligned} n_- &= n_-^0 \exp(v_- e\phi/kT) \\ n_+ &= n_+^0 \exp(-v_+ e\phi/kT) \end{aligned} \tag{2}$$

in which n_-^0 and n_+^0 are the concentrations of the anions and cations in the solution outside the range of the diffuse double layer. All concentrations are commonly expressed in numbers of ions per cubic centimetre, e is the electronic charge, k the Boltzmann constant and T the absolute temperature. For a negatively charged surface ϕ is negative, hence $n_+ > n_-$.

Combining (1) and (2) and considering symmetrical electrolytes so that $n_-^0 = n_+^0$ and $v_- = v_+ = v$ (since the ion of opposite sign is known to have a minor effect only, this simplification will not introduce a serious defect), the fundamental differential equation for the diffuse electrical double layer, the *Poisson–Boltzmann equation*, is obtained:

$$d^2\phi/dx^2 = (8\pi nve/\varepsilon)\sinh(ve\phi/kT) \tag{3}$$

It is convenient to introduce the following dimensionless quantities:

$$y = ve\phi/kT; \quad z = ve\phi_0/kT; \quad \xi = \kappa x \text{ in which } \kappa^2 = 8\pi n e^2 v^2/\varepsilon kT \text{ cm}^{-2}$$

which reduces equation (3) to

$$d^2y/d\xi^2 = \sinh y$$

Noting that the electrical potential is zero at infinity and has the value ϕ_0 at the particle surface, integrating twice with the boundary conditions that $dy/d\xi = 0$ and $y = 0$ for $\xi = \infty$, and that $\phi = \phi_0$ (or $y = z$) for $\xi = 0$ yields:

$$e^{y/2} = \frac{e^{z/2} + 1 + (e^{z/2} - 1)e^{-\xi}}{e^{z/2} + 1 - (e^{z/2} - 1)e^{-\xi}} \tag{4}$$

which represents an approximately exponential decay of the potential with the distance from the surface as sketched in Fig. 3.4.

For small surface potentials ($z \leqslant 1$ or $\phi_0 \leqslant 25$ mV) the Poisson–Boltzmann equation reduces

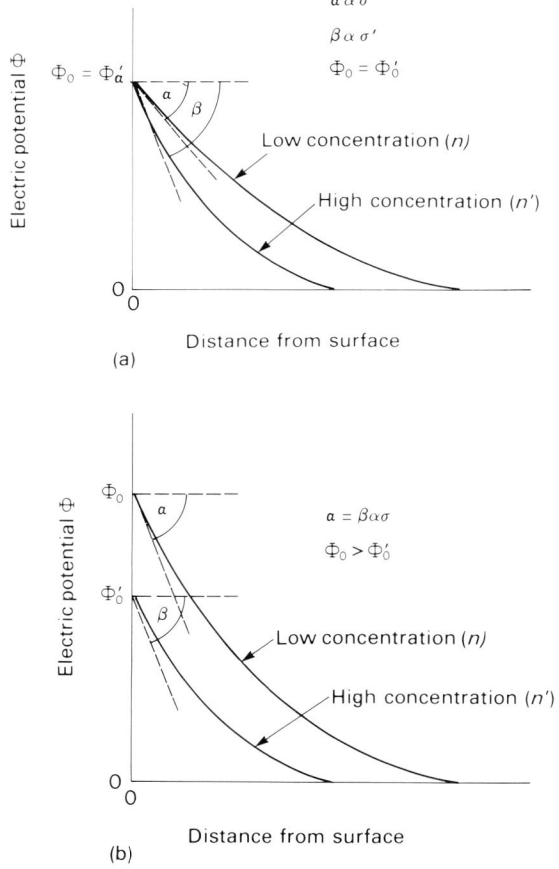

FIG. 3.4. Electric potential distribution in the diffuse layer at two electrolyte concentrations. (a) Constant surface potential. (b) Constant surface charge. (From H. van Olphen 1977.)

to $d^2\phi/dx^2 = \kappa^2\phi$ which yields upon integration:

$$\phi = \phi_0 \exp(-\kappa x) \qquad (4a)$$

representing a purely exponential decay of the potential with distance. The centre of gravity of the space charge is located at a distance $1/\kappa$ from the surface, hence $1/\kappa$ is a measure for the thickness of the diffuse double layer ($1/\kappa$ is the Debye characteristic length introduced in the Debye–Huckel theory for strong electrolytes).

The surface charge density can be calculated from the potential function as follows:

$$\sigma = -\int_\rho^\infty \rho\, dx = \frac{\varepsilon}{4\pi} \int_0^\infty \frac{d^2\phi}{dx^2}\, dx = -\frac{\varepsilon}{4\pi}\left[\frac{d\phi}{dx}\right]_{x=0}$$

The total surface charge is determined by the initial slope of the potential function, and equals:

$$\sigma = (n\varepsilon kT/2\pi)^{1/2}(2\cosh z - 2)^{1/2} = (2n\varepsilon kT/\pi)^{1/2}\sinh(z/2) \qquad (5)$$

which, for small surface potentials reduces to

$$\sigma = (n\varepsilon/2\pi kT)^{1/2}(v e \phi_0) = (\varepsilon\kappa/4\pi)\phi_0 = C\phi_0 \qquad (5a)$$

in which $C = \varepsilon\kappa/4\pi$ is the capacity of the double layer.

In Figs. 3.4 and 3.5 the general trends in potential and charge distribution at two electrolyte concentrations are sketched for surfaces of constant charge and of constant potential, in accordance with the above formulae. Note that in Fig. 3.4 the potential curves are displaced towards the surface when the electrolyte concentration is increased, which demonstrates the compression of the double layer. For constant potential surfaces the initial slope of the curves becomes steeper, as the double layer charge increases. For constant charge surfaces the surface potential decreases with increasing electrolyte concentration but the initial slope remains constant.

Figure 3.5 demonstrates the unequal distribution of cations and anions in the diffuse double layer. The large excess of counter-ions is proportional to the areas ABD, and the deficit of ions which are repelled by the surface is proportional to the areas CBD ("negative adsorption"). The sum of the excess and deficit charges is equal to the absolute value of the total surface charge. By integration the total deficit of charge (anion charge near a negative surface, for example) amounts to:

$$\sigma_- = (\varepsilon n kT/2\pi)^{1/2}[\exp(v e \phi_0/2kT) - 1]$$

and the ratio of the deficit of charge to the total surface charge is:

$$\frac{\sigma_-}{\sigma} = -\frac{[\exp(v e \phi_0/2kT) - 1]}{[\exp(-v e \phi_0/2kT) - \exp(v e \phi_0/2kT)]} \qquad (6)$$

Hence, for a surface of constant potential the ratio is constant, but for a surface of constant charge the ion deficit can be shown to increase with increasing electrolyte concentration, whereas the counter-ion excess decreases with increasing electrolyte concentration (Bolt and Warkentin 1958; Chaussidon 1958).

An interesting application is Schofield's method for the determination of the surface area per gram of colloidal particles in suspensions from the change of the amount of negative adsorption per gram as a function of the electrolyte concentration (Schofield and Talibudeen 1948).

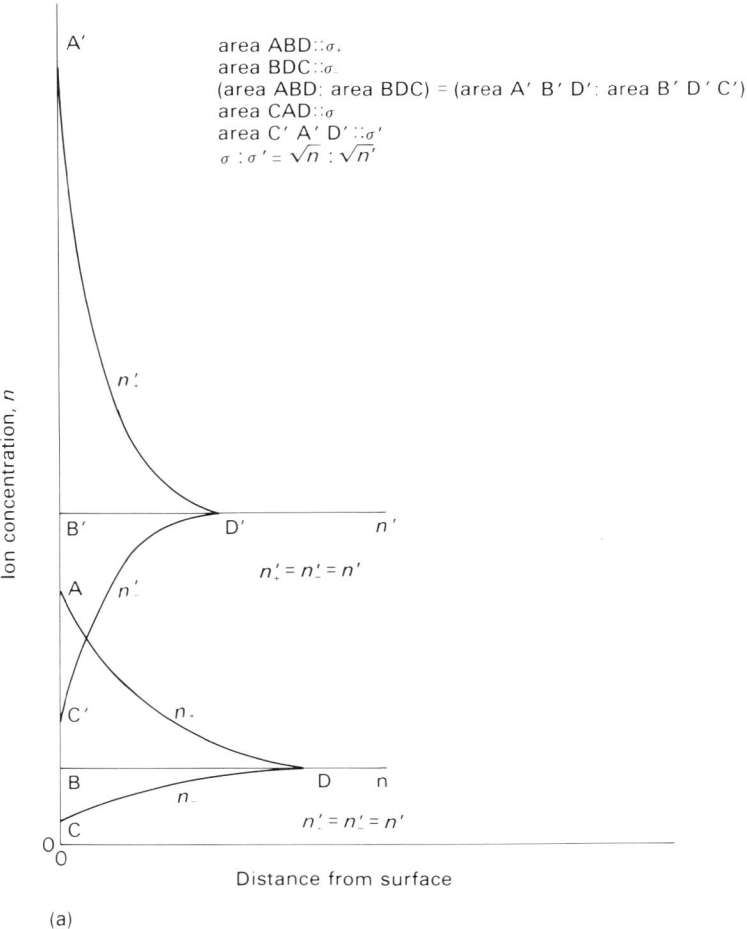

FIG. 3.5. Charge distribution in the diffuse layer of a negative particle at two electrolyte concentrations. (a) Constant surface potential.

The counter-ions can be exchanged by other ions of the same sign when available in solution. An *exchange equilibrium* is established whereby the ratio of ion species in solution and in the diffuse double layer is the same when the valencies of the two ions are the same. This follows directly from the non-specific Gouy theory. The theory also shows that ions of a higher valency are preferentially accumulated in the diffuse double layer.

The term *"counter-ion exchange capacity"* ("cation exchange capacity" or CEC for a negative surface) should be carefully defined. If it is determined by treating the suspension with an excess of a salt, for example an ammonium salt, and measuring the decrease of the ammonium ion concentration in the supernatant solution, one actually determines the counter-ion excess which differs from the total charge of the double layer by the amount of negative adsorption. If the CEC is determined by exhaustive treatment with an ammonium salt followed by separating the particles from the solution and measuring the total nitrogen content of the separated particles, the CEC would be equal to the total charge of the particle. Both definitions

QUANTITATIVE TREATMENT OF COLLOIDAL STABILITY. ELECTRICAL DOUBLE LAYER THEORY

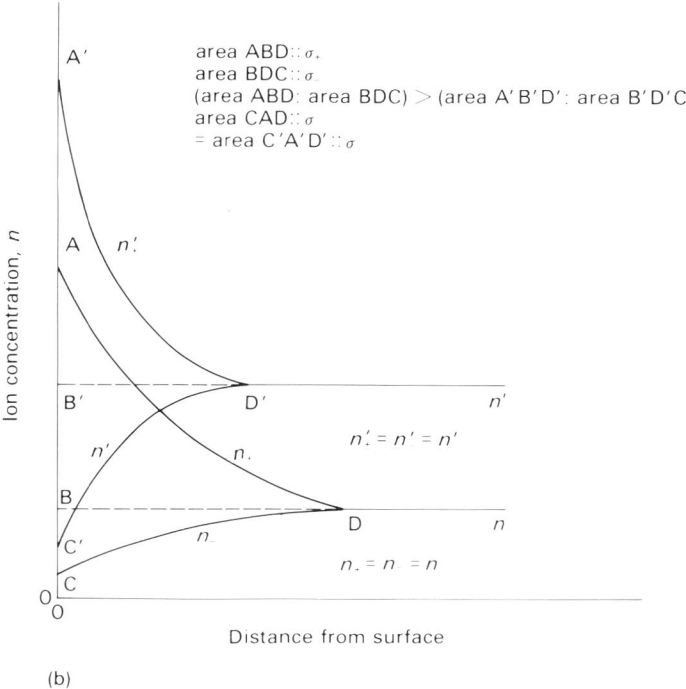

(b)

(b) Constant surface charge. (From H. van Olphen 1977.)

of CEC may be used, but for clays it is customary to quote a CEC value which is equivalent to the total surface charge. Therefore, when it is determined by the first method, negative adsorption should be measured at the same time and be added to the cation excess to obtain the surface charge. For clays which have a constant charge surface, the CEC is a material constant. For materials with constant potential surfaces, the CEC will vary with the composition of the medium, and is, therefore, not a material constant. In clays slight variations in CEC may occur, particularly for the thick kaolinite particle, due to the presence of the edge double layers which are of the constant potential type. Also, under conditions at which the edge charge is positive, the negative adsorption values should be corrected for anion adsorption at the edges. Bolt (1955) has introduced correction terms to the Boltzmann distribution function that makes the derived equations more realistic for equilibria between ions of the same valence but different size.

3.4.3 Potential and charge distribution in interacting double layers

The distribution of potential between the interacting surfaces is symmetrical with a minimum value midway between the surfaces; the potential in this plane at a distance d from either surface is called ϕ_d and $ve\phi_d/kT = u$.

Integration of the fundamental differential equation (3) with the boundary conditions that for $x=0$ and $x=2d$, $y=z$, and for $x=d$, $y=u$ and $dy/dx=0$, yields:

$$\int_z^u (2\cosh y - 2\cosh u)^{-1/2} dy = -\kappa d \tag{7}$$

The integral can be evaluated using appropriate tables. For practical purposes the evaluation of u is most important, and tables are available presenting sets of values for κd, z, and u. Tables which are convenient for constant potential double layers are given by Verwey and Overbeek (1948), and for constant charge systems by van Olphen (1977).

For small interaction, i.e. large values of κd, the midway potential will be approximately equal to the sum of the separate double layer potentials, i.e. $u = 2y_d$. Then, $y_d = 4\gamma \exp(-\kappa d)$ where $\gamma = (e^{z/2} - 1)/(e^{z/2} + 1)$ and one obtains:

$$u = 8\gamma \exp(-\kappa d) \tag{8}$$

In principle, the complete potential distribution between the surfaces, and therefore also the complete charge distribution, can be calculated. However, the ion concentration midway between the surface is again of primary importance. It can be derived from the midway potential by

$$n_- = n\, e^u \quad \text{and} \quad n_+ = n\, e^{-u} \tag{9}$$

When double layers of constant potential interact, the surface charge decreases, and when constant charge surfaces interact, the surface potential increases.

3.4.4 Repulsive force and repulsive energy for interacting double layers

The repulsive energy of interacting double layers can be evaluated directly from the change of the Gibbs' energy of the double layers with distance. Such calculations and results for systems of practical interest are presented by Verwey and Overbeek (1948) for interacting constant potential double layers.

Alternatively, the force between the double layers may be calculated and the interaction energy is then obtained by integration of the force–distance relation. As first proposed by Langmuir (1938), the force can be easily derived from the difference between the ion concentration midway between the double layers and that in the equilibrium solution, as an osmotic pressure:

$$p = 2nkT(\cosh u - 1) \tag{10}$$

By evaluating u, the repulsive force between the particles is easily calculated. This is the force dominating the swelling process of clays at interparticle distances where diffuse layers have developed.

Since u changes in a complicated way with the distance, the force equation is difficult to integrate. However, at small interaction, u changes purely exponentially with distance (equation (8)), and then the integration of equation (10) yields:

$$V_R = (64nkT/\kappa)\gamma^2 \exp(-2\kappa d) \quad \text{where } \gamma = (e^{z/2} - 1)/(e^{z/2} + 1)$$

This relation is a good approximation for κd values of less than 1.

3.4.5 Specific effects in the electrical double layer

On the basis of the Gouy–Chapman model, the quantitative interpretation of many colloid chemical phenomena has been quite successful, and the theory should be considered an excellent starting point for such interpretations. Then, when phenomena are encountered which cannot be explained satisfactorily with the Gouy model, specific interactions should be considered, and the Gouy model should be modified to match physical reality. One obvious defect of the Gouy model is that it leads to impossibly high local ion concentrations in the close proximity of a surface with a high potential, and this is inherent in considering the ions as point charges. Stern's treatment, referred to before, considers the actual size of the ions, and he developed the model of a molecular condenser in series with a diffuse double layer as sketched in Fig. 3.6. The potential governing the diffuse double layer properties is now greatly reduced due to the steep drop of the potential in the molecular condenser. If this effect is still not sufficient to explain the properties of certain suspensions, a still greater drop between surface potential and "Stern potential" (ϕ_δ) can be obtained by introducing a specific adsorption potential of the counter-ions at the surface. However, since other specific effects may be important (such as ion–solvent or ion–ion interactions) specific adsorption should not be resorted to indiscriminately, and should be physically justified (van Olphen 1954, 1962).

The qualitative treatment of the Stern double layer model is based on statistical considerations by which the distribution of the total charge in Stern and Gouy layer is derived from the number of available positions for ions in either layer and the local electric potentials, and, if applicable, the specific adsorption potential. Within the molecular condenser, the dielectric constant of water will be different from that of bulk water since the water is in the

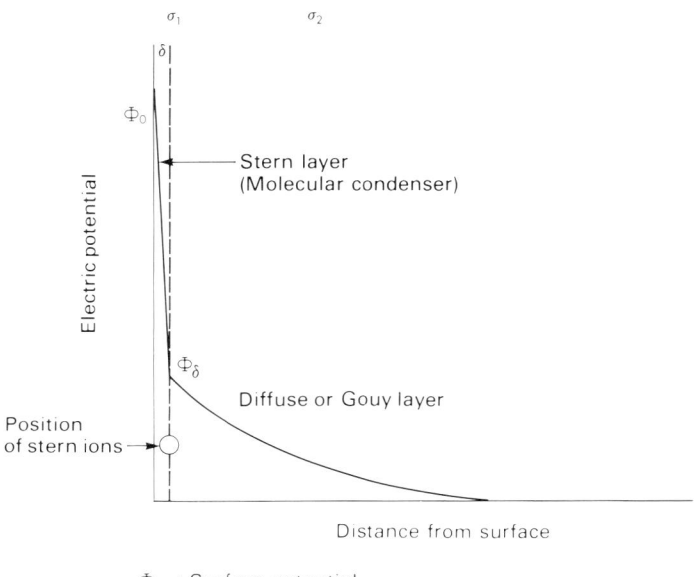

Φ_0 : Surface potential
Φ_δ : Stern potential
σ_1 : Net counter-ion charge of Stern layer
σ_2 : Net counter-ion charge of diffuse layer
Total charge = $\sigma_1 + \sigma_2$

Fig. 3.6. Stern's model of the potential distribution in the electrical double layer. (From H. van Olphen 1977.)

adsorbed state. A value of the order of $\varepsilon' = 3$ to 6 has been derived from double layer capacity data on compressed double layers on the mercury surface, obtained from electrocapillary curves.

The potential and charge distribution in a single flat Stern–Gouy negative double layer can be derived from the following equations (see Fig. 3.6):

$$-\sigma = \sigma_1 + \sigma_2 \tag{11}$$

$$\sigma = (\varepsilon'/4\pi\delta)(\phi_0 - \phi_\delta) \tag{12}$$

$$\sigma_2 = (2\varepsilon n k T/\pi)^{1/2} \sinh(v e \phi_\delta / 2kT) \tag{13}$$

$$\sigma_1 = \frac{N_1 v e M n}{N_A} \exp\left[-(v e \phi_\delta + \psi)/kT\right] \tag{14}$$

in which N_1 is the number of available positions on 1 cm^2 of surface, N_A is Avogadro's number, and M the molecular weight of the solvent. The specific adsorption potential is ψ. A reasonable figure for N_1 is about 10^{15} (10 Å^2 per position).

For interacting double layers of the Stern–Gouy type equation (13) should be substituted by

$$\sigma_2 = (nkT\varepsilon/2\pi)^{1/2}(2\cosh(v e \phi_\delta / kT) - 2\cosh(v e \phi_d / kT))^{1/2} \tag{13a}$$

and, since the ratio of available positions in Stern and Gouy layer may be taken as proportional to the respective volumes:

$$\sigma_1/\sigma_2 = [\delta/(d-\delta)] \exp\left[(v e \phi_\delta + \psi)/kT\right] \tag{14a}$$

although this treatment is by no means exact, it is sufficient to show trends.

3.4.6 Experimental analysis of potential and charge distribution in the double layer

The surface potential of suspended particles and the potential distribution in the double layer are, in principle, not accessible to experimental determination, because of the unknown response of the liquid junction between the suspension and the salt bridge connecting the suspension with a reference electrode. However, the total charge of the double layer can be determined from the adsorption isotherms for potential-determining or peptizing ions, or, for constant charge surfaces, from the ion exchange capacity as discussed before.

For constant potential surfaces, the magnitude of the surface potential in a given system can be estimated from the ratio of the concentration of potential-determining ions in the equilibrium liquid under the given conditions, to that at the point of zero charge, applying the Nernst equation, neglecting activity corrections. Hence, for constant potential surfaces, surface charge and approximate surface potential values can be obtained from which the capacity of the double layer can be calculated. From these data one can establish which model would best fit the data, either the Gouy model or the Stern model, with or without assuming specific counter-ion adsorption, or some other more refined model.

For other surfaces, the surface charge is the only known parameter, and surface potentials can only be derived by accepting a certain model. The possible reality of the selected model can be judged from its ability to explain phenomena which are dependent on the double layer structure. Of course, such applicability does not prove that the selected model is the correct one.

In clay–water systems, the determination of the swelling pressure allows evaluation of the potential midway between parallel clay plates. When the particle size is known and a parallel arrangement of the plates is achieved, the average particle distance can be calculated at various clay concentrations. Then, the midway potential will be experimentally known as a function of particle distance, allowing a check on the best fit of a double layer model.

The above considerations apply to equilibrium measurements. An alternative approach to probing the double layer is to use electrokinetic steady-state techniques. Electrokinetic phenomena are governed by the double layer potential at the hydrodynamic slipping plane which is located at some distance to the surface. Since the exact location of the slipping plane is unknown, and since the calculation of the electrokinetic potential (or "zeta-potential") contains a number of uncertainties, the zeta potential so calculated has only a qualitative significance. The value of the zeta potential as, for example, determined from the electrophoretic velocity of the particles is considerably lower than the surface potential and much closer to the Stern potential or the potential at the outer Helmholtz plane. Since the Stern model and the Grahame model have proved their usefulness in colloid chemical stability problems, it is not surprising that the zeta potential finds much practical use in spite of its somewhat ill-defined character (Hunter 1981).

3.5 VAN DER WAALS' ATTRACTION

The non-retarded van der Waals' attraction energy for two parallel plates of thickness at a distance $2d$ below about 10 nm is given by:

$$V_A = -\frac{A}{48\pi}\left[\frac{1}{d^2} + \frac{1}{(d+\Delta)^2} - \frac{1}{(d+\Delta/2)^2}\right] \quad (15)$$

in which the constant A (the "Hamaker constant") is of the order of a few times 10^{-5} J.

NOTE

1. In the classical literature on the electric double layer treatment, the electrostatic system was used. This practice is followed here. When the rationalized four quantity system of units is used the factor π only appears in equations for spherical particles.

REFERENCES

BOLT G.H. (1955) Analysis of the validity of the Gouy–Chapman theory of the electric double layer. *J. Colloid Sci.*, **10**, 206–218.
BOLT G.H. & WARKENTIN B.P. (1958) The negative adsorption of anions by clay suspensions. *Kolloid Z.* **156**, 41–46.
CHAPMAN D.L. (1913) A contribution to the theory of electrocapillarity. *Phil. Mag.* **25**, No. 6, 475–481.
CHAUSSIDON J. (1958) Adsorption négative des anions dans les suspensions d'argile. *Bull. Groupe Franc. Argiles* **10**, 31–35.
DERJAGUIN B. & LANDAU L.D. (1945). *J. Exp. Theor. Phys.*, USSR **15**, 662.
FERRIS A.P. & JEPSON W.B. (1975) The exchange capacities of kaolinite and the preparation of homoionic clays. *J. Colloid Interface Sci.* **51**, 245–259.
FUCHS M. (1934) Uber die Stabilität und Aufladung der Aerosole. *Z. Physik.* **89**, 736–743.
GOUY G. (1910) Sur la constitution de la charge électrique à la surface d'un electrolyte. *Ann. Phys. (Paris)*, Série 4, **9**, 457–468.
GOUY G. (1917) Sur la fonction électrocapillaire. *Ann. Phys. (Paris)* Série 9 **7**, 129–184.
GRAHAME D.C. (1947) The electrical double layer and the theory of electrocapillarity. *Chem. Rev.* **41**, 441–501.

HAMAKER H.C. (1937) The London–van der Waals attractions between spherical particles. *Physica* **4**, 1058–1072.
HARDY W.B. (1900) A preliminary investigation of the conditions which determine the stability of irreversible hydrosols. *Proc. Roy. Soc., Lond.* **66**, 110–125.
HEALY T.W. & LA MER V.K. (1964) The energetics of flocculation and redispersion by polymers. *J. Colloid Sci.* **19**, 323–332.
HESSELINK F.TH., VRIJ A. & OVERBECK J.TH.G. (1971) On the theory of the stabilization of dispersions by adsorbed macromolecules. *J. Phys. Chem., Ithaca* **75**, 2094–2103.
HUNTER R.J. (1981) *Zeta Potential in Colloid Science.* Academic Press, New York.
JORDAN J.W. (1963) Organophilic clay base thickeners. *Clays Clay Miner.* **10**, 299–308.
KOELMANS H. & OVERBEEK J.TH.G. (1954) Stability and electrophoretic mobility in non-aqueous media. *Disc. Faraday Soc.* **18** (Coagulation and Flocculation), 52–63.
LANGMUIR I. (1938) The role of attractive and repulsive forces in the formation of tactoids, thixotropic gels, protein crystals, and coacervates. *J. Chem. Phys.* **6**, 873–896.
MACKOR E.L. (1951) A theoretical approach of the colloid chemical stability of dispersions in hydrocarbons. *J. Colloid Sci.* **6**, 492–495.
MAHANTY J. & NINHAM B.W. (1977) *Dispersion Forces.* Academic Press.
PHAM TI HANG & BRINDLEY G.W. (1970) Methylene blue absorption by clay minerals. Determination of surface area and cation exchange capacities. *Clays Clay Miner.* **18**, 203–212.
RUEHRWEIN R.A. & WARD D.W. (1952) Mechanism of clay aggregation by polyelectrolytes. *Soil Sci.* **73**, 485–492.
SCHOFIELD R.K. & SAMSON H.R. (1954) Flocculation of kaolinite due to the attraction of oppositely charged crystal faces. *Disc. Faraday Soc.* **18** (Coagulation and Flocculation), 135–145; 220.
SCHOFIELD R.K. & TALIBUDEEN P. (1948) Measurement of internal surface by negative adsorption. *Disc. Faraday Soc.* **3**, 51–56.
STERN O. (1924) Zur Theorie der electrolytischen Doppelschicht. *Z. Elektrochem.* **30**, 508–516.
VAN OLPHEN H. (1950) Stabilization of montmorillonite sols by chemical treatment. *Rec. Trav. Chim.* **69**, 1308–1322.
VAN OLPHEN H. (1951a) Thesis, Delft.
VAN OLPHEN H. (1951b) Rheological phenomena of clay sols in connection with the charge distribution on the micelles. *Disc. Faraday Soc.* **11** (The Size and Shape Factor in Colloidal Systems), 82–84.
VAN OLPHEN H. (1951c) A tentative method for the determination of the base exchange capacity of small samples of clay minerals. *Clay Miner. Bull.* **1**, 169–170.
VAN OLPHEN H. (1954) Interlayer forces in bentonite. *Clays Clay Miner.* **2**, 418–438.
VAN OLPHEN H. (1956) Forces between suspended bentonite particles. *Clays Clay Miner.* **4**, 204–224.
VAN OLPHEN H. (1959) Forces between suspended bentonite particles. *Clays Clay Miner.* **6**, 196–206.
VAN OLPHEN H. (1962) Unit layer interaction in hydrous montmorillonite systems. *J. Colloid Sci.* **17**, 660–667.
VAN OLPHEN H. (1964) Internal mutual flocculation in clay suspensions. *J. Colloid Sci.* **19**, 313–322.
VAN OLPHEN H. (1977) *An Introduction to Clay Colloid Chemistry*, 2nd edn. Wiley – Interscience.
VAN SILFHOUT A. (1966) Dispersion forces between macroscopic objects. *Kon. Ned. Akad. Wetensch., Proc. Series B* **69**, 501–541 (46 references).
VERWEY E.J.W. & Overbeek J.TH.G. (1948) *Theory of the Stability of Lyophobic Colloids.* Elsevier.
VOLD M.J. (1954) van der Waals attraction between anisometric particles. *J. Colloid Sci.* **9**, 451–459.
VOLD M.J. (1961) The effect of adsorption on the van der Waals interaction of spherical colloidal particles. *J. Colloid Sci.* **16**, 1–12.
VON SMOLUCHOWSKI M. (1916) Drei Vorträge über Diffusion, Brownsche Molekularbewegung und Koagulation von Kolloidteilchen. *Physikalische Z.* **17**, 557–571, 585–599.
VON SMOLUCHOWSKI M. (1917) Versuch einer mathematischen Theorie der Koagulationskinetik kolloidaler Lösungen. *Z. Physik. Chem.* **92**, 129.

# Chapter 4

Cation Exchange Equilibria in Clays

H. LAUDELOUT

4.1 INTRODUCTION	225
4.2 DEFINITIONS	225
4.3 THERMODYNAMICS OF THE CATION EXCHANGE REACTION IN CLAYS	227
4.4 APPLICATIONS OF THE THERMODYNAMICS OF CATION EXCHANGE IN CLAYS	230
4.4.1 Statistics of cation exchange in clays	230
4.4.2 Simulation of cation exchange isotherms	231
4.4.3 Relation between cation exchange selectivity and properties of the clay	231
4.4.4 Cation properties and exchange selectivity on a given clay surface	233
4.4.5 Cation exchange equilibria with more than two counter-ions	233
REFERENCES	235

4.1 INTRODUCTION

The problems related to the exchange of ions in soils and clays have been the object of continued if not always intensive study for more than a century. The first studies were directed towards "the power of soil to retain manure" (Way 1852). Later, at the turn of the century, Dutch soil scientists investigated ion exchange processes fundamental to the reclamation of polder soils. More recent problems have been the disposal of radioactive waste and environmental contamination following nuclear accidents, and latterly there has been concern about eutrophication of river and lake waters which has again brought to the fore the study of ion exchange and movement in soils and clays.

Until quite recently, studies on ion exchange in clays and soils were directed towards obtaining empirical relationships between concentration and composition of the equilibrium solution and the cationic composition of the adsorbed phase on the clay surface. Most of these empirical relationships were later shown to be limiting cases of a more general treatment, thermodynamic in nature, and this demands that a precise definition of the process being studied be given first.

4.2 DEFINITIONS

If ion exchange is defined by the fact that a given cation has taken the place of a different one in the surface phase of the clay, this presupposes that an operational definition of the surface phase has been established. Furthermore, in stating that an adsorbed cation resides in the surface phase, the definition of the extent of the surface phase of a clay also defines what adsorption is. The logical problems involved in this were discussed many years ago by Gibbs and have been presented with great clarity by Prigogine and Defay (1951).

Without entering into details, we may state the problem of an operational definition of ion exchange as follows: "The surface charge of a clay entails a cation excess and an anion deficit at its surface, creating a zone of electrolyte exclusion in the surface phase which may be regarded as containing an excess of cations." Clearly any "surface excess" has to be referred to some agreed level, which could be the cations present in the bulk solution (e.g. supernatant or dialysate), but could also be the anion (i.e. electrolyte) concentration.

This convention is precisely the reverse of what is done usually in surface chemistry where the adsorption of the solute is referrred to that of the solvent which is consequently taken as non-adsorbed or absent from the surface phase.

There is a two-fold advantage of using a convention that defines cation adsorption in relation to solute in bulk solution. Firstly, the deficit or negative adsorption of anions at clay surfaces is "defined out" of the picture of cation exchange reactions, and secondly, an important parameter, the cation exchange capacity, is held constant irrespective of the ionic strength of the electrolyte solution.

Were it not so, the cation exchange capacity would vary from a figure proportional to half the value of the electrical surface charge at infinitely high salt concentration to the full value at infinitely low concentration.

The definition of a cation surface excess may be stated more precisely if we take the example of exchange between K^+ and Ca^{2+} carried out in the presence of a common anion, say Cl^-. We have

$$n_K^* = n_K - n_{Cl}(m_K/m_{Cl}) \quad (1)$$

$$n_{Ca}^* = n_{Ca} - n_{Cl}(m_{Ca}/m_{Cl}) \quad (2)$$

where n^* is the surface excess, m the molarity in the equilibrium solution and n is the number of cations or anions found in the clay–water–electrolyte system. In writing equations (1) and (2), the easily justifiable procedure has been used of apportioning the chloride to the two cations in direct relation to their concentrations in the dialysate, in which they are easily determined.

The cation exchange capacity is proportional to $n_K^* + 2n_{Ca}^*$ and from equations (1) and (2) we have

$$n_K^* + 2n_{Ca}^* = n_K + n_{Ca} - n_{Cl} \quad (3)$$

which shows that the total excess of cations in the surface phase is expressed with respect to the salt concentration in the clay–water–electrolyte system.

Before formulating an ion exchange reaction in a clay–water–salt system, it is necessary to clarify yet another arbitrary convention. Högfeldt (1953) pointed out many years ago, and it has been restated recently by Sposito and Mattigod (1977), that when formulating ion exchange reactions, it is immaterial whether cation concentrations are expressed in molar amounts or in equivalents. The exchange reaction between cations M and N with charges m^+ and n^+ with the clay surface, Z, may be represented in two ways for the molar convention,

$$nMZ_m + mN^{n+} \rightleftharpoons nM^{m+} + mNZ_n \quad (4)$$

and for the equivalent convention,

$$nmM_{1/m}Z + mN^+ \rightleftharpoons nM^{m+} + mnN_{1/n}Z \quad (5)$$

Although this distinction may seem trivial, its importance resides in the fact that the mass action relationships derived from one convention are not formally identical with those derived from the other even though the thermodynamic equilibrium constants derived from them are the same.

Summarizing, the definition of the ion exchange reaction involves a definition of the extent of the surface phase or, what amounts to the same, a definition of adsorption.

It also involves a definition of the meaning of the symbols used for representing the chemical species involved in the exchange such as clay-Ca or clay-K.

It is useful to emphasize that the method used for separating the clay suspension from its dialysate amounts to a definition of the extent of the surface phase. This implies that no experimental method can be the best one for that purpose unless its convenience in the laboratory is taken into account, but one method which has proven its value both in the field of protein physical chemistry and for adsorption studies on clays is the equilibrium dialysis method (Hughes and Klotz 1956).

Whatever the method used for establishing ion exchange equilibria in clays, there will always be a noticeable hysteresis, i.e. an enhancement of preference for the cation displaced from the clay, which diminishes as the ionic strength of the equilibrium solution is decreased. This supports the view that ion exchange hysteresis is due to restricted accessibility, the aggregation of the clay particles being less pronounced when electrostatic interactions between clay surfaces are the strongest, i.e. when the ionic strength is low.

Van Bladel and Laudelout (1967) have taken advantage of this observation for devising an extrapolation procedure which allows data to be obtained that are not only free from hysteresis effects but also correspond to the commonly chosen reference state of the homoionic clays.

4.3 THERMODYNAMICS OF THE CATION EXCHANGE REACTION IN CLAYS

Ion exchange measurements in clays are repetitious and consequently tedious and, furthermore, they can only be accurately carried out with fairly dilute clay suspensions in electrolyte solutions that are neither too dilute nor too concentrated. By contrast, practical applications often involve systems containing large amounts of clay with respect to the water they contain at salinities that are far above those experimentally practical in ion exchange studies. One aim of thermodynamic descriptions of ion exchange measurements is therefore to extend their applicability to actual practical conditions. Other possibilities of thermodynamics may also be utilized. For instance, the triangle rule may be invoked to calculate equilibria between two cations whose concentrations are difficult to measure experimentally. By involving each of them in turn in exchange reactions with a third, easily measured cation, the thermodynamic parameters for the exchange between the first two cations can be calculated.

The formulation of ion exchange reactions on clays requires some care, because they involve a two-phase system: the electrolyte solution, and the surface region of the clay, with equilibrium being established between the two.

The exchange reaction for the replacement of cation B by cation A on the clay surface may be expressed in words as follows. A number Z_A of cations B carrying Z_B positive charges which are adsorbed in the clay surface region where their hydration number is n_w^B are replaced by Z_B cations A of charge Z_A which have a surface hydration number n_w^A. Consequently an amount of $(n_w^B - n_w^A)$ moles of water will be released from the clay surface as a result of the difference of the two hydration numbers n_A and n_B.

The Gibbs–Duhem equation may then be written for the three constituents of the surface phase

$$n_A \, d\mu_A + n_B \, d\mu_B + n_w \, d\mu_w = 0 \qquad (6)$$

where the n's are numbers of moles in the surface region and the μ's are the chemical potentials in the surface region or, since equilibrium is established, the chemical potentials in the whole system: surface region + equilibrium solution.

Clearly, chemical potentials have thus been tacitly defined for ionic components A and B. Such a definition does not have operational significance since their concentrations cannot be varied independently. No logical inconsistencies will follow from this definition, however, since the condition of electroneutrality precludes arbitrary change in the amounts of cations present in the surface phase.

A further advantage of the constraint of electrical neutrality is that the composition of the surface region may be expressed in terms of equivalent fractions rather than as mole fractions.

The equivalent fraction will be defined as

$$N_A = \frac{Z_A n_A}{Z_A n_A + Z_B n_B} \tag{7}$$

$$N_B = \frac{Z_B n_B}{Z_A n_A + Z_B n_B} \tag{8}$$

The surface equivalent fraction for water or any neutral species will be zero.

The quantity $Z_A n_A + Z_B n_B$ will depend on the extent of the surface region which is being observed. There is no less in generality in taking it equal to unity, i.e. one equivalent of surface charge.

With the convention above, the Gibbs–Duhem equation takes the form

$$Z_B N_A \, d\mu_A + Z_A N_B \, d\mu_B + Z_A Z_B n'_w \, d\mu_w = 0 \tag{9}$$

where n'_w is the hydration number per equivalent of clay surface.

Before integrating this equation, it is necessary to fix the integration constants. In other words, since only *variations* in free energy or chemical potentials are of significance, it is necessary to choose reference levels from which the potentials will be measured.

As the choice of this level is purely a matter of convenience, it may seem that choosing identical reference levels in the surface region and in the equilibrium solution would be the easiest procedure. With this choice, the free energy change for the ion exchange reactions will always be zero. This is a logical consequence of the choice of reference levels which is nevertheless awkward. It must be remembered that this is precisely the choice which is the basis of the Donnan equilibrium for ion exchange reactions.

If another choice is made, we may then have

$$\mu_A = \mu_A^\circ + RT \ln a_A$$

$$\mu_B = \mu_B^\circ + RT \ln a_B$$

$$\mu_w = \mu_w^\circ + RT \ln a_w \tag{10}$$

a set of equations which define the activities of the three components in the surface phase. For the case of water, the same standard state may be taken in the surface phase and in the equilibrium solution.

We obtain for the free energy change

$$\Delta G = \Delta G_0 + RT \ln \frac{a_A^{Z_B} a_B^{Z_A}}{a_B^{Z_A} a_A^{Z_B}} \tag{11}$$

where ΔG_0 is given by

$$\Delta G_0 = Z_B \mu_A^\circ + Z_A \mu_B^\circ - Z_A \mu_B^\circ - Z_B \mu_A^\circ \tag{12}$$

It remains now to define the reference level from which chemical potentials will be measured in the surface region and in the electrolyte solution or in the usual phrase, the standard state. In the surface region, the one which has proved most convenient is the homoionic clay suspended in an infinitely dilute solution of the corresponding salt. Although this state is both inaccessible to experimental measurement and fictitious, this is not an obstacle to its use provided measurements of the chemical potential can be made "with respect to" that state, even though, for example, a Na-clay in an infinitely dilute solution will undergo rapid and irreversible hydrolysis leading to a Na-Al-clay. To show how this can be done, we shall describe the way an actual exchange process at a salt concentration suitable for making measurements can be referred to the standard state.

Starting from a homoionic A-clay in its standard state, we concentrate the solution to a finite molarity of the A-salt. Then A is exchanged for B at a finite molarity until homoionic B-clay is obtained and finally the solution of salt B is diluted to a vanishingly small concentration. Clearly the changes in the thermodynamic potentials along this path L are equivalent to the experimentally inaccessible transformations for the homoionic clays in the infinitely dilute solutions.

We then replace activities in equation (11) by the product of a concentration times an activity coefficient. Since in the surface region equivalent fractions are convenient and in the solution molarities are both usual and convenient, we have at equilibrium

$$\Delta G_0 = -RT \ln K = -RT \ln \frac{N_A^{Z_B} m_B^{Z_A} f_A^{Z_B} \gamma_B^{Z_A}}{N_B^{Z_A} m_A^{Z_B} f_B^{Z_A} \gamma_A^{Z_B}} \tag{13}$$

and

$$\Delta G_0 = -RT \ln K'_c \frac{f_A^{Z_B} \gamma_B^{Z_A}}{f_B^{Z_A} \gamma_A^{Z_B}} \tag{14}$$

In the above, K is the thermodynamic equilibrium constant, K'_c is a selectivity coefficient which can be experimentally determined, the γ are the mean activity coefficients for the salts and the f are the activity coefficients corresponding to the choice which has been made for the concentration units in the surface phase.

If we disregard the problem of finding the ratio of the activity coefficients γ in the solution, we have for the value of the thermodynamic equilibrium constant

$$\ln K = \ln K'_c + Z_B \ln f_A - Z_A \ln f_B \tag{15}$$

which after differentiation becomes

$$d \ln K'_c + Z_B d \ln f_A - Z_A d \ln f_B = 0 \tag{16}$$

When this equation is combined with the Gibbs–Duhem equation (6) in the form

$$Z_B d \ln f_A + Z_A d \ln f_B + Z_A Z_B n'_w d \ln a_w = 0 \tag{17}$$

it is possible to calculate the activity coefficients f, and after integration, the value of $\ln K$, i.e. the standard free energy change. The details of the calculation may be found in Gaines and Thomas (1953) and Laudelout and Thomas (1965).

The result of the integration along the path L described above is:

$$\ln K = (Z_B - Z_A) + \int_L (\ln K_c \, dN_B + n'_w \, d \ln a_w) \tag{18}$$

where the numerical term $Z_B - Z_A$ originates from the choice of the equivalent fraction as an expression of the concentration in the surface region; had one taken the mole fractions of the ions for expressing the concentrations, this would have been absent and of course the meaning of K_c would have been different. Needless to say the activity coefficients calculated by carrying the integration up to an incomplete exchange will be different according to the choice which has been made for expressing the concentrations in the surface region. The relationships between the various forms of the activity coefficients are easily calculated, see for instance Sposito (1981).

The importance of the choice made between various conventions or definitions which are equally logical and more or less convenient cannot be overemphasized. Many controversies which arise in this field could often be very easily resolved if the definitions of the quantities or properties such as ideality of the surface mixture of ions were carefully examined (Laudelout *et al.* 1971, 1972).

An important corollary of equation (18) may first be mentioned. Noticing that $\ln K$ is a point function and that the cross-differentiation rule may then be applied to the integrand, one obtains for a uni-univalent exchange:

$$\left(\frac{\partial \ln K_c}{\partial \ln a_w}\right)_{N_B} = \left(\frac{\partial n'_w}{\partial N_B}\right)_{a_w} \tag{19}$$

It is reasonable to suppose that the hydration of the clay is a linear function of the composition of the exchangeable cations, i.e.

$$n'_w = n_w^A N_A + n_w^B N_B \tag{20}$$

where n_w^A and n_w^B are the hydration numbers of ions A and B in the surface phase of the clay.
Since

$$\frac{dn'_w}{dN_B} = -n_w^A + n_w^B \tag{21}$$

we have

$$\left(\frac{\partial \ln K_c}{\partial \ln a_w}\right)_{B_B} = n_w^B - w_w^A \tag{22}$$

which means that the variation of the selectivity coefficient with water activity in the electrolyte solution can be calculated if the difference in the hydration number of the cations is known.

4.4 APPLICATIONS OF THE THERMODYNAMICS OF CATION EXCHANGE IN CLAYS

4.4.1 Statistics of cation exchange in clays

The thermodynamic formulation for various ion exchange reactions allows the values of the relative affinities of the clay surface for cations to be readily compared. A survey of the properties exhibited must then be followed by experimental and correlative work to relate the differences found to the surface or structural properties of the clay.

A recent survey by Bruggenwert and Kamphorst (1979) has filled a gap in the literature on this subject. The authors have tabulated "the data which characterize *briefly* the exchange properties of soil materials with respect to different cations". They have covered in depth 500 papers on the subject selected out of 2500 titles. Even though most of the references selected were in the field of soil science, many of the exchange studies were concerned with rather well-defined clay minerals, with the expected preponderance of montmorillonite type minerals.

The selectivity constant which has been used in these very useful tabulations is a fractional power of the one that has been defined by equation (13) namely

$$K'_N = \frac{N_A^{1/Z_A} m_B^{1/Z_B}}{N_B^{1/Z_B} m_A^{1/Z_A}} \tag{23}$$

Ranges of values for this selectivity constant have been tabulated and presumably averaged for calculating the equilibrium constant without weighting the experimental values according to the composition of the surface region.

It has been pointed out that due to the algebraic form of the selectivity constant, the error in its value becomes extremely large when the quantity absorbed is imprecisely known. The relative error on K_c may be calculated as (Laudelout et al. 1979)

$$\frac{\Delta K_c}{K_c} \simeq \frac{1}{\bar{N}_{Ca}} \frac{1 + \bar{N}_{Ca}}{1 - \bar{N}_{Ca}} \cdot \Delta \bar{N}_{Ca} \tag{24}$$

This error is minimal for $\bar{N}_{Ca} = 0.414$ and increases so rapidly in the ranges $0 < \bar{N}_{Ca} < 0.15$ and $0.75 < \bar{N}_{Ca} < 1.0$ that the numerical results become meaningless.

4.4.2 Simulation of cation exchange isotherms

Statistical studies like that quoted above summarize mean values of selectivity coefficients, and relations between the activity coefficients of adsorbed cations and the surface region composition. A primary practical use of such correlations is to facilitate the calculation, as accurately as necessary, of an exchange isotherm for any concentration and composition of the equilibrium solution and for any clay type, i.e. for any origin or density of the clay surface charge. Such a calculation is required whenever ion exchange chromatographic transport through soil or clay formation is being studied.

For this purpose, an algorithm for the calculation of ion exchange equilibria has to be incorporated in the numerical solution of the Fokker–Planck equation describing salt convection with hydrodynamic dispersion. A particularly simple model has been described by Laudelout et al. (1975) for that purpose. The algorithm is easily programmable and rests upon the relationship which is accepted between the surface region composition and the selectivity coefficient.

Quite often, instead of the third degree polynomial which, as shown by Cremers and Thomas (1968), quite accurately describes the relationship, a constant value of the selectivity coefficient may be taken without appreciably modifying the fit between the experimental data and the simulated isotherm.

This has been done by Robbins et al. (1980) for the four cations Ca^{2+}, Mg^{2+}, K^+ and Na^+.

4.4.3 Relation between cation exchange selectivity and properties of the clay

As stated by Eberl (1980) "one would like to predict the sorptive properties of a clay for all ions simply by characterizing the clay". There have been several approaches to this problem, of which we will mention only a few.

Clearly the subject of this section is outside the province of thermodynamics. It is, however, closely related to the use of thermodynamic methods since once the value of the free energy change has been obtained for a particular exchange reaction on a given clay mineral, knowledge of the influence of some of its properties on the exchange selectivity would avoid the repetition of particularly tedious experimental work.

One of the first attempts in that direction was the utilization of the surface charge density of the clay by way of the double layer theory. This description initiated by Eriksson (1952) was applied to complete thermodynamical and thermochemical data by Laudelout *et al.* (1968a) with the rather expected results that, since ions are considered as point charges in the double layer theory, the error increases with ion size. The problem has been reinvestigated by Maes and Cremers (1977, 1978) both on heterovalent and homovalent exchange equilibria.

Using different clays and reduced charged montmorillonites obtained by heating partially substituted Li-clay, these authors have observed that standard free energy changes for the Na^+–Ca^{2+} exchange were related to the surface charge density. A higher exchange capacity was accompanied by an increase in the calcium selectivity. Standard enthalpy and entropy changes also decreased with charge density.

In the case of heterovalent exchange, the double layer theory mentioned above predicts that

$$\ln K = 2\left(\ln \frac{\sqrt{\beta}}{2} - 1\right) + 2 \ln \Gamma \tag{25}$$

where Γ is the charge density and β a double layer constant.

Clearly the consequence of this is that the experimental values of $\ln K$ should increase linearly with the logarithm of charge density. Maes and Cremers (1977) found that this was indeed the case with a slope of 1.33. The experimentally found free energy loss is less dependent on charge density than predicted by the diffuse double layer theory.

Consequently, this theory, which has proven its value in other fields such as the calculation of electrolyte exclusion, can only be used for separating the coulombic effect from other mechanisms controlling ion exchange selectivity in heterovalent exchange. In fact, when such an analysis is done, it becomes obvious that non-coulombic forces play an important part in the overall selectivity pattern (Laudelout *et al.* 1968a; Maes and Cremers 1978).

In homovalent exchange, the surface charge density was shown to have a pronounced effect on the standard free energy of exchange as well as on the calorimetrically measured enthalpy of exchange. A linear relationship which was fairly accurate could be established without difficulty. In order to establish this relationship, it was necessary to use clays of varying charge density but identical charge localization, otherwise irregularities were observed.

There have been several other attempts to put ion exchange selectivity on a theoretical basis, for example, Shainberg and Kemper (1966, 1967), Ravina and Gur (1978), McBride (1979) and Eberl (1980).

The basis of the attempts was the calculation of the hydration energy of a cation on the one hand and, on the other hand, its attraction energy for the clay surface. The clay was considered as a spherical anion exerting the same force on the cation as the actual surface.

Regularities in behaviour seem fairly easy to observe and *ad hoc* theoretical interpretations not too difficult to find. At present, the predictive value of the latter seems limited.

4.4.4 Cation properties and exchange selectivity on a given clay surface

The concluding sentence of the last paragraph could very well serve as an opening sentence for this one, with some qualifications.

Martin and Laudelout (1963) have shown that linear relationships exist between the difference in polarizability of the two cations being exchanged and the free energy and calorimetrically determined enthalpy changes.

Maes and Cremers (1977) have pointed out that the compensatory effect between the enthalpy and entropy changes in heterovalent ion exchange reactions seems to be a common phenomenon.

A purely thermodynamic relationship has already been mentioned. If the difference in hydration between two cations is known, then the change of the selectivity coefficient with water activity in the equilibrium solution may be calculated. This relationship has been tested experimentally by Laudelout *et al.* (1972). The check consisted in determining the change of the selectivity coefficient and calculating the hydration differences of the two cations involved in the exchange.

The predictive value of the method can be tested in three ways. The determination of the change of ion exchange selectivity with solvent activity for three pairs of ions – A, B, and C – should yield a set of mutually consistent values for the changes in cation hydration, n_A-n_B, n_A-n_C and n_B-n_C. Furthermore, it is possible to convert observed basal spacings in clays to numbers of water molecules per exchange site. Finally, the determination of the electrolyte exclusion volume allowed another check on the amount of hydration water in the clay surface region when this contains alkali metal cations. Without entering into the details of the various consistency checks on the experimental results, it seems sufficient to quote the succession of values found for the exchanges $Li^+ \rightarrow Cs^+ \rightarrow Na^+ \rightarrow Rb^+ \rightarrow Li$, i.e. -37.2, 21.5, -21.6, 37.9 water molecules per equivalent for the hydration difference in the direction indicated by the arrows. The sum of these differences is zero as it should be, because the exchange cycle is closed.

Several other attempts have been made to explain ion exchange selectivity by incorporating properties of the cations such as their ion pair formation constants (Heald *et al.* 1964; Shainberg and Kemper 1967) into an amended diffuse double layer theory.

In this model, a fraction of the fixed charges of the clay surface is neutralized by the formation of ion pairs with Stern layer cations; the remainder of the surface charge is compensated by a diffuse double layer outside the Stern layer.

Each cation is characterized by the partition constant between the "ion-pair" state and the diffuse double layer.

4.4.5 Cation exchange equilibria with more than two counter-ions

The treatments described above were restricted to two counter-ions. Obviously, in studies on cation exchange in natural clay systems, more than two cations will be involved, namely Ca^{2+}, Mg^{2+}, K^+, Na^+, Al^{3+} and in a transient way H^+. A common situation is that of an irrigation water containing essentially calcium, magnesium and sodium percolating through an alluvial clay.

Comparatively few studies have been dedicated to the subject previous to the work of Elprince and Babcock (1975) and Wiedenfeld and Hossner (1978). The reason for this is easy to find: the most important practical case is of course the one mentioned above since the exchangeable sodium percentage will influence the physical properties of an irrigated soil on a clayey parent material.

In calculating the effect of exchange on the cation composition of a given irrigation water, long experience has shown that good results can be obtained by considering that the exchange behaviour of Ca^{2+} and Mg^{2+} was identical. The effect of the concentration and composition of irrigation water could then be expressed by the well-known sodium adsorption ratio (SAR) value. The predictive value of the latter can be noticeably improved (Sposito and Mattigod 1977) by using a comprehensive model of the equilibria involved in ion pair formation and solubilities to calculate the concentration of free ionic species.

In other situations, the short cut represented by the SAR parameter will no longer be valid and this expains the interest, albeit limited, in a general solution of the calculation of ternary of quaternary exchanges from data obtained on binary exchanges.

The approach has been to use the excess free energy of mixing

$$\Delta G^E = RT \Sigma B_i \ln f_i \qquad (26)$$

and relate it semi-empirically to a property of the ions mixing on the clay surface such as their volume fractions. Such expressions have the advantage that for n cations, calculations may be made from $n(n-1)$ parameters obtained from binary exchange equilibria (Wilson 1964).

The expression derived by Wilson (1964) and used by Elprince and Babcock (1975) and Weidenfeld and Hossmner (1978) is

$$\Delta G^E/RT = -\Sigma N_i \ln \left(1 - \sum_j N_j A_{ji}\right) \qquad (27)$$

or

$$\Delta G^E/RT = -\Sigma N_i \ln \lambda_{ij} N_j \qquad (28)$$

with $\lambda_{ij} = 1$ if $i = j$ and $\lambda_{ij} \neq \lambda_{ji}$

Since we have

$$\Delta G^E/RT = \Sigma N_i \ln f_i \qquad (29)$$

we equate the total differentials of equations (28) and (29) and add the constancy condition of the cation exchange capacity ($\Sigma dN_i = 0$) times a Lagrangian multiplier. Determining the latter gives:

$$\ln f_i = 1 - \ln\left(\sum_j N_j \lambda_{ij}\right) - \sum_k \frac{N^{k^{\lambda k_i}}}{\sum_j N_j \lambda k_j} \qquad (30)$$

If the λ_{ij} are known from binary exchange studies, the activity coefficients may then be calculated for a ternary exchange in which ion pair formation may also be taken into account. The computing involved which would have been prohibitive a few years ago is now well within the capabilities of present-day microcomputers.

Representing the excess free energy of mixing by an empirical or semi-empirical formulation may have its drawbacks, however. Laudelout et al. (1968b) have shown that, for several exchanges, Scatchard's formulation of the excess free energy of mixing for components having unequal volumes might give a very good fit at low temperatures (< 10 °C) but an extremely poor one at room temperature and above.

For this reason, it might be preferable to use the subregular model of mixed solid solutions as was done recently by Chu and Sposito (1981) and by Elprince *et al.* (1980).

REFERENCES

BRUGGENWERT M.G.M. & KAMPHORST A. (1979) Survey of experimental information on cation exchange in soil systems. In *Soil Chemistry: B. Physico chemical models* (G.H. Bolt, ed.). Elsevier Scientific, Amsterdam.

CHU S.Y. & SPOSITO G. (1981) The thermodynamics of ternary cation exchange systems and the subregular models. *Soil Sci. Soc. Am. J.* **45**, 1084–1089.

CREMERS A. & THOMAS H.C. (1968) Thermodynamics of sodium-cesium exchange on Camp Berteau montmorillonite: An almost ideal case. *Israël J. Chem.* **6**, 949–957.

EBERL D.D. (1980) Alkali cation selectivity and fixation by clay minerals. *Clays Clay Miner.* **28**, 161–172.

ELPRINCE A.M. & BABCOCK K.L. (1975) Prediction of ion exchange equilibria in aqueous systems with more than two counter ions. *Soil Sci.* **120**, 332–338.

ELPRINCE A.M., VANSELOW A.P. & SPOSITO G. (1980). Heterovalent ternary cation exchange equilibria: NH_4^+–Ba^{++}–La^{+++} exchange on montmorillonite. *Soil Sci. Soc. Am. J.* **44**, 964–968.

ERIKSSON E. (1952) Cation exchange equilibria on clay minerals. *Soil Sci.* **74**, 103–113.

GAINES G.L. & THOMAS H.C. (1953) Adsorption studies on clay minerals: II. *J. Chem. Phys.* **21**, 714–718.

HEALD W.R., FRERE M.H. & DE WIT C.T. (1964) Ion adsorption on charged surfaces, *Soil Sci. Soc. Am. Proc.* **28**, 622–627.

HÖGFELDT E. (1953) On ion exchange equilibria. II. Activities of the components in ion exchangers. *Arkiv for Kemi* **5**, 147–171.

HUGHES T.R. & KLOTZ, I.M. (1956) The equilibrium dialysis method. In *Methods of Biochemical Analysis* (D. Glick, ed.). Vol. III. Interscience Publishers, New York.

LAUDELOUT H., DUFEY J.E. & SHETA T.H. (1979) Ionic equilibria in semi-arid soils. In *Soils in Mediterranean Type Climates and their Yield Potential*, pp. 135–150. Publications I.P.I., Bern.

LAUDELOUT H., FRANKART R., LAMBERT R., MOUGENOT F. & PHAM MANH LE (1975) Modeling of solute interactions with soils. In *Modeling and Simulation of Water Resource Systems* (G.C. Vansteenkiste, ed.), pp. 361–366. North-Holland Publ. Co.

LAUDELOUT H. & THOMAS H.C. (1965) The effect of water activity on ion exchange selectivity. *J. Phys. Chem. Ithaca* **69**, 339–340.

LAUDELOUT H., VAN BLADEL R., BOLT G.H. & PAGE A.L. (1968a) Thermodynamics of heterovalent cation exchange reactions in a montmorillonite clay. *Trans. Faraday Soc.* **84**, 1477–1488.

LAUDELOUT H., VAN BLADEL R., GILBERT M. & CREMERS A. (1968b) Physical chemistry of cation exchange in clays. *Trans. 9th Int. Congr. Soil Sci., Adelaide, 1968* **1**, 565–575.

LAUDELOUT H., VAN BLADEL R. & ROBEYNS J. (1971) The effect of water activity on ion exchange selectivity. *Soil Sci.* **111**, 211–213.

LAUDELOUT H., VAN BLADEL R. & ROBEYNS J. (1972) Hydration of cations adsorbed on a clay surface from the effect of water activity on ion exchange selectivity. *Soil Sci. Soc. Am. Proc.* **36**, 30–34.

MAES A. & CREMERS A. (1977) Charges density effects in ion exchange. Part 1. Heterovalent exchange equilibria. *J. Chem. Soc., Faraday Trans.* **73**, 1807–1814.

MAES A. & CREMERS A. (1978) Charge density effects in ion exchange. Part 2. Homovalent exchange equilbria. *J. Chem. Soc., Faraday Trans.* **74**, 1234–1241.

MARTIN H. & LAUDELOUT H. (1963) Thermodynamique de l'échange des cations alcalins dans les argiles. *J. Chim. Phys.* **60**, 1086–1099.

MCBRIDE M.B. (1979) An interpretation of cation selectivity variations in M^+–M^{++} exchange on clays. *Clays Clay Miner.* **27**, 417–422.

PRIGOGINE I. & DEFAY R. (1951) *Tension superficielle et adsorption*. Liège, Desoer.

RAVINA I. & GUR Y. (1978) Application of the electrical double layer theory to predict ion adsorption in mixed ionic systems. *Soil Sci.* **125**, 204–209.

ROBBINS C.W., JURINAK J.J. & WAGENET R.J. (1980) Calculating cation exchange in salt transport model. *Soil Sci. Soc. Am. J.* **44**, 1195–1200.

SHAINBERG I. & KEMPER W.D. (1966) Hydration status of adsorbed cations. *Soil Sci. Soc. Am. Proc.*, **30**, 707–713.

SHAINBERG I. & KEMPER W.D. (1967) Ion exchange equilibria on montmorillonite. *Soil Sci.* **103**, 4–9.

SPOSITO G. (1977) The Gapon and the Vanselow selectivity coefficients. *Soil Sci. Soc. Am. J.* **41**, 1205–1206.

SPOSITO G. (1981) *The Thermodynamics of Soil Solutions*. Clarendon Press, Oxford; Oxford University Press, New York.

SPOSITO G. & MATTIGOD S.V. (1977) On the chemical foundation of the sodium adsorption ratio. *Soil Sci. Soc. Am. J.* **41**, 323–329.

VAN BLADEL R. & LAUDELOUT H. (1967) Apparent irreversibility of ion-exchange reactions in clay suspensions. *Soil Sci.* **104**, 134–137.

WAY J.T. (1852) On the power of soils to absorb manure. *J. Roy. Agric. Soc. Engl.* **13**, 123–143.

WIEDENFELD R.P. & HOSSNER L.R. (1978) Cation exchange equilibria in a mixed soil system containing three heterovalent cations. *Soil Sci. Soc. Am. J.* **42**, 709–713.

WILSON G.W. (1964) Vapor–liquid equilibrium XI. A new expression for the excess free energy of mixing. *J. Am. Chem. Soc.* **86**, 127–130.

Chapter 5

The Interaction of Water with Clay Mineral Surfaces

A.C.D. NEWMAN

		Page			Page
5.1	INTRODUCTION	237		5.7.2.1 Hydrogen ions	258
5.2	METHODS OF INVESTIGATION	238		5.7.2.2 Lithium and sodium ions	259
	5.2.1 The amount of water held by clay	238		5.7.2.3 Magnesium, calcium, strontium and barium ions	259
	5.2.2 Thermodynamics of water interaction	239		5.7.2.4 Potassium and ammonium ions	260
	5.2.3 Infrared spectroscopy	239		5.7.2.5 Rubidium and caesium ions	260
	5.2.4 Neutron scattering measurements	240		5.7.2.6 Aluminium	260
	5.2.5 Nuclear magnetic resonance	240	5.7.3	Vermiculites	260
	5.2.6 Electron spin resonance	241		5.7.3.1 Heats of immersion	262
	5.2.7 Other methods	241		5.7.3.2 Infrared studies	262
5.3	CLAY SURFACES	241		5.7.3.3 Nuclear magnetic resonance	262
	5.3.1 General considerations	241		5.7.3.4 Neutron scattering	263
	5.3.2 Surface charge density	242	5.7.4	Montmorillonites and beidellites	263
	5.3.3 Alteration of charge density	244		5.7.4.1 Sorption isotherms	263
	5.3.4 Interlamellar surface	244		5.7.4.2 Heats of immersion	265
	5.3.5 Special types of clay surface	245		5.7.4.3 Structure	266
5.4	SPECIFIC SURFACE OF CLAYS	245	5.7.5	Other smectites	267
5.5	THE REDUCED OR NORMALIZED ISOTHERM	247		5.7.5.1 Hectorites	267
5.6	1:1 GROUP MINERALS	249		5.7.5.2 Saponites	267
	5.6.1 Kaolinites	249		5.7.5.3 Mixed-layer minerals	268
	5.6.2 Halloysites	252	5.8	THE SEPIOLITE–PALYGORSKITE GROUP	268
5.7	THE 2:1 GROUP MINERALS	254		REFERENCES	270
	5.7.1 Micas	254			
	5.7.2 Swelling minerals: general considerations	255			

5.1 INTRODUCTION

In his discussion about the origin of life, Cairns-Smith (1971) describes the several states of clay under the title "The story of sloppy, sticky, lumpy and tough". These adjectives vividly convey the ability of clay to harden as it dries from a liquid slurry to a plastic material and eventually to a brittle solid. Reversing this process is more difficult and although on rewetting from the solid state a few types of clay will imbibe water spontaneously to form a slurry, generally mechanical work is needed. The origin of such differences between clays lies in the nature of the forces between the clay particles and how these are influenced by diffuse electrical double layers that form on their surfaces (Chapter 3). Further differences are attributable to the type of cation balancing negative charge on the clay surface, and also to specific effects such as the chemical composition of the mineral and the density of electric charge that forms as a result on the particle surfaces. The swelling of clays is a complex process that is still only partly understood (Norrish 1973a).

The development of films of water on clay surfaces is therefore an important fundamental process that has been studied since the early years of scientific investigation. In the late 1930s, definite structures were proposed for the arrangement of water on clay surfaces based on the pattern of hydrogen bonds that exist in ice (Hendricks and Jefferson 1938) but later studies disputed this. Graham (1964) has reviewed some of these controversies, and indeed certain aspects of these opposing views persist to the present day (Low 1982; Sposito and Prost 1982).

The application of modern physical techniques to the study of clay–water systems has greatly increased our knowledge about the state and properties of water on clay surfaces. Nuclear magnetic resonance (NMR), electron spin resonance (ESR) and infrared spectroscopy have all combined to define the diffusion, rotation and bonding of water near clay surfaces much more precisely. These studies suggest that the influence of the clay surface on the ordering of H_2O molecules does not extend much beyond the second or third layer of molecules, that is, it is restricted to a range of about 1 nm. Beyond this distance water appears to behave very like liquid water.

In this chapter an arbitrary choice has been made to restrict the review to the interaction of clays with small amounts of water, that is, to the water that is within the range of influence of the clay surface. Even with this restriction, the importance of the subject is reflected in the very large volume of literature on the subject. Fortunately this has been summarized from time to time in review articles; in addition to those mentioned above, there are articles by Martin (1962), Stocker (1969), van Olphen (1975), Farmer (1978) and MacEwan and Wilson (1980). The present chapter, in accordance with the aims of the monograph, presents this literature mineral by mineral, but begins with a brief review of the methods available for studying clay–water interaction, and then considers the different types of surfaces that clay minerals present. Because the literature on clay–water interaction is so large, the references quoted are illustrative rather than exhaustive, but where divergence between sources is evident, this is discussed.

5.2 METHODS OF INVESTIGATION

5.2.1 The amount of water held by clay

Knowing how much water is taken up by clay is clearly basic to many applications and may also help to characterize a clay, but it is important to make the measurements under well-defined conditions of ambient temperature and relative humidity. If in addition the equilibrium amount of water sorbed is measured as a function of water vapour pressure under isothermal conditions, the sorption isotherm so determined can be analysed further to make deductions about the extent of the clay surface and the free energy of its interaction with water.

Alternatively the amount of water retained by the clay as it is heated (thermogravimetric analysis, TGA) may also be used to characterize the clay. In this experiment, ideally the water vapour pressure should be specified and controlled but this is less often attempted as it is technically more difficult. A further variant measures the heat changes that occur as a clay is heated (differential thermal analysis, DTA), which at low temperatures generally reflect the loss of water from a hydrated clay. Both TGA and DTA, and their application to minerals are fully described elsewhere (Mackenzie 1957, 1970, 1981; Mackenzie and Caillère 1975) and are not further reviewed here.

The isothermal sorption and desorption of water by clays is usually measured either by the isopiestic (desiccator) method or with a sorption balance. In the former, the clay is placed in a weighing bottle in a vacuum desiccator over a source of known constant relative humidity, usually sulphuric acid solutions or saturated salt pastes, and weighed occasionally until it reaches constant weight. The isotherm is constructed by using a range of pastes or solutions in sequence. In the sorption balance method, the clay is weighed *in vacuo* and vapour pressure often measured directly with a manometer (Zettlemoyer *et al.* 1975). In general, the desiccator method is less accurate and the conditions less rigorously defined as the vacuum is broken at each weighing; it has the merit that it requires little special apparatus, and is reasonably accurate if 10 g samples can be used.

5.2.2 Thermodynamics of water interaction

Heats of adsorption are calculated from isotherms at two or more different temperatures, using the equation (analogous to the Clausius–Clapeyron equation)

$$\frac{\partial \ln p}{\partial (1/T)_n} = -\frac{q_i}{R} \tag{1}$$

where q_i is the isosteric heat of adsorption. The ratio $\Delta \ln p / \Delta(1/T_1 - 1/T_2)$ is evaluated from the isotherms for equal coverage n (or amount of water sorbed per unit mass of clay) and the differential heat of adsorption $q_d = q_i - RT$.

Heats of adsorption are also measured calorimetrically (Zettlemoyer *et al.* 1975), usually by immersion. The clay may be completely desiccated before immersion, or may contain presorbed water up to a chosen point on the sorption isotherm, so that the site–energy distribution can be mapped. Heat is evolved when clays are immersed in water and the enthalpy change is negative. Immersion calorimetry determines integral heats of adsorption, as all available sites are covered.

5.2.3 Infrared spectroscopy

By contrast with the previous measurements, which measure bulk properties on a long time scale (equilibrium properties), infrared and other spectral methods study short-term interactions and provide information on the electronic, vibrational, rotational and diffusional states of the sorbed water molecule. Sposito and Prost (1982) have classified techniques by their time scales: for absorption in the infrared region, molecular vibrations in the time scale 10^{-12}–10^{-15} s are sampled. The infrared spectra of minerals form the subject of a monograph containing chapters that describe techniques useful in studying clays (Russell 1974) and discuss the spectra of layer silicates (Farmer 1974). Only very brief description is included here and the reader is referred to the chapters mentioned above for more detail.

Water adsorbed on or in layer silicates absorbs radiation in the region 1630–1640 cm^{-1} due to the HOH bending vibration and this band is useful for studying dehydration qualitatively (Russell and Farmer 1964). The main OH vibrations absorb in the range 3300–3600 cm^{-1}, which overlaps with the vibration frequencies of the OH groups within the mineral structure. In order to avoid confusion with these structural vibrations, the hydration water is often replaced by D_2O, so that the water spectrum is displaced to frequencies 2400–2700 cm^{-1}, clear of the structural OH if not deuterated. The frequencies of the absorption maxima give information about the strength of hydrogen bond formation, and if the clay substrate is prepared as an oriented aggregate, the orientation of the water molecules can be deduced from the pleochroism of the absorption bonds (Farmer and Russell 1971).

5.2.4 Neutron scattering measurements

This is one of the most powerful techniques to have been developed in recent years and applied to the study of water in clays (Ross and Hall 1980; Hall, 1981). Four types of neutron scattering experiments can be distinguished:

1. *Neutron diffraction.* The elastic, coherent scattering of neutrons from regular structures is analogous to the diffraction of X-rays, the wavelength of thermal neutrons being in the same range as that of X-rays. The neutron scattering length for hydrogen is about the same magnitude as for other elements in clays, and in consequence the method is useful for locating structural hydrogen atoms (Rothbauer 1971; Rayner 1974). It has also been used to study the structure of interlayer water in montmorillonite (Hawkins and Egelstaff 1980).
2. *Small angle neutron scattering.* Measurements of the coherent diffraction intensities at small angles provide information about deviations from the periodic structure over distances in the range 5 to 500 nm. This is useful for studying voids and particle size in colloid materials, and has been used to investigate the effect of different cations on the association of montmorillonite particles (Cebula *et al.* 1980).
3. *Quasi-elastic neutron scattering.* As thermal neutrons have energies comparable with the diffusive motion of atoms, a small energy transfer occurs on interaction and the measurement of this energy transfer can be used to study translocation and orientation of molecules and groups. This type of experiment has particular relevance in the present chapter as it has been used to study the mobility of water associated with clays (Hall 1981).
4. *Inelastic neutron scattering.* This gives information about lattice vibrations similar to that found by infrared and Raman studies but is not limited by the selection rules governing the coupling of vibrations with electromagnetic radiation.

In this chapter, most attention will be given to quasi-elastic neutron scattering (QENS).

5.2.5 Nuclear magnetic resonance

Certain nuclei, including H, have a magnetic moment that will interact with an external field, and nuclear magnetic resonance (NMR) measures the frequency and energy of this interaction. As this interaction is influenced by local factors like chemical bonding, coupling with other magnetic moments, field distortions, and the population of spin states, it is able to sample the immediate environment of the nucleus in several different ways (Fripiat 1980). For the study of water in clays, the most frequently used techniques have been the analysis of the resonance line shape and the study of relaxation times (Stone 1981).

In liquid water, although a molecule experiences an instantaneous anisotropic interaction potential, this anisotropy is time averaged by molecular motion and a single proton resonance line results. If, however, the water is adjacent to a surface, the dipole–dipole coupling is no longer averaged to zero and the line splits into a doublet. In clay–water systems, the splitting is always less than in ice or in crystalline hydrates, indicating that some time averaging is occurring, and the preferential orientation of water molecules at the clay surface is described as "a dynamic ordered state". If a powdered sample is used, the resultant signal is also space averaged, but the use of oriented clay films has enabled the orientation of the interaction to the surface to be studied. The interpretation however is complicated by proton–proton exchange and interpretation is clarified by replacement of H_2O by D_2O.

Under equilibrium conditions, the population of the nuclear spin orientation states is that defined by the Boltzmann distribution for the appropriate temperature and field conditions. This equilibrium can be displaced by perturbation, and if the perturbation is subsequently relaxed, the decay of the displacement to the original equilibrium state can be measured, leading to rate constants for the process. This relaxation can be characterized by T_1 the spin–lattice relaxation time, and T_2 the spin–spin relaxation time. In clay–water systems, the usual technique has been to measure T_1 by the "spin echo" (pulsed radiofrequency field) technique; this gives information about the exchange of energy between the nuclear spin system and its surroundings (the "lattice"), a term which includes random motions within the sample. Various T_1-detected motions have been found in clays (Fripiat 1980) but their interpretation is not always easy.

A limitation in the application of NMR to clays is the complicating effect of paramagnetic centres, which at present limits its application to samples with low iron content. Most studies have therefore been done on synthetic clays and a very restricted range of natural samples.

5.2.6 Electron spin resonance

Like many nuclei, an electron has a spin of 1/2 and an associated magnetic moment. As electrons are regularly paired, however, transitions require the high energy excitation of visible or ultraviolet light. The electron resonance experiment of microwave energies is therefore confined to atoms and groups of atoms with an unpaired electron. This limits greatly its application in the study of water sorption by clays, and such studies have been confined to examination of the hydration states of Cu^{2+}, Mn^{2+} and VO_2^{2+} clays. The results of these investigations have agreed with those obtained by QENS and NMR (Hall 1981).

5.2.7 Other methods

The relaxation of water dipolar orientation can be studied by measuring the complex dielectric permittivity as a function of frequency. Although dielectric loss maxima are observed in clay–water systems, their interpretation is complicated in powder samples by dielectric loss due to surface conduction, which can totally obscure dipolar relaxation effects at high water contents (Newman 1955). Dielectric studies have been made on kaolinites (Muir 1954; Nelson *et al.* 1969) and homoionic montmorillonites (Mamy 1973; Calvet 1975).

Recently, differential heat flow scanning calorimetry has been used to study the freezing of water in clays (Homshaw and Chaussidon 1978), soils (Homshaw and Cambier 1980) and porous media (Homshaw 1981). This method measures pore size distribution, that is, it studies water remote from the particle surface; however, it has shown that water to a depth of two water-monolayer thicknesses on the surface of Ca illite is "unfreezable" (Homshaw and Quirk 1980), and therefore indirectly confirms results obtained by the other techniques mentioned earlier.

5.3 CLAY SURFACES

5.3.1 General considerations

The atomic structure at the external surfaces of layer silicates is of three basic types, all of which are demonstrated by a 1:1 type mineral like kaolinite. One basal surface consists of the close-packed hydroxyl ions on the exposed side of the octahedral sheet, and the opposite basal surface is composed of the oxygen atoms coordinated to silicon in six-membered siloxane rings of the tetrahedral sheet. The third type of surface is at the lateral boundary of the sheet structure where the chemical composition and bonding of the bulk structure cannot be maintained without additional ions or atoms to complete the coordination spheres of the cations in accordance with Pauling's rules. The "broken bonds" at these "edge" surfaces are usually completed by H^+ or OH^-, which are therefore the potential-determining ions, the net charge balance being dependent on the pH of the medium in which the clay is immersed. Changes in the net charge balance have a fundamental effect on dispersion and flocculation behaviour, as discussed in Chapter 3.

Kaolinite and other 1:1 minerals have both hydroxyl and siloxane basal surfaces but 2:1 minerals have only siloxane basal surfaces. Gibbsite and other hydroxide minerals have only hydroxyl basal surfaces; the surface properties of these minerals are discussed in Chapter 2. Chlorites have the possibility of either siloxane or hydroxyl basal surfaces, or both; a recent study of two chlorites (Jones 1981) seems to leave open the question of which type actually occurs.

5.3.2 Surface charge density

For many layer silicates in which isomorphous replacement causes a net deficit of positive charge within the sheet, the deficit is balanced by interlamellar cations. On external basal surfaces of such minerals, e.g. illites, the balancing cations are accessible to the surrounding medium, and if this is an aqueous salt solution, the cation can exchange for other cations. Under these conditions, double layers develop whose nature and thickness depend on the balancing cation, on the surface density of charge and on the concentration of the salt solution. These phenomena are discussed in Chapters 3 and 4.

Surface charge density, σ, can be estimated in several ways. If the cation exchange capacity (CEC) and the specific surface (S) of the clay are measured, the charge density is given by

$$\sigma = CEC/S \tag{2}$$

CEC is conventionally expressed in the units milliequivalents per gram (meq g^{-1}); it does, however, represent a charge per unit mass, and in SI units should therefore be expressed in the units "coulombs per unit mass", a CEC of 1 meq g^{-1} (i.e. that of a medium charge montmorillonite) is 96.5 C g^{-1} in SI units. The units of charge density are then C m^{-2}, and assuming that the montmorillonite has a total specific surface of 760 m^2 g^{-1}, its charge density is $96.5/760 = 0.127$ C m^{-2}.

The measurement of the specific surface of clays (see Section 5.3.3 below) presents particular difficulties and it is useful to be able to calculate surface charge densities independently. This can be done from the structural formula and unit cell parameters alone in some instances, or also with the additional help of exchange capacity measurements; both calculations make assumptions that may require careful validation for individual minerals.

The simple calculation from the structural formula is

$$\sigma = e \cdot \sum (\text{Interlayer cation charge})/2ab \tag{3}$$

when e is the elementary charge, 1.6022×10^{-19} C, and a and b are the unit cell parameters. This calculation assumes that the composition of clay surface is the same as that of the bulk material and that the interlayer cations can be unequivocally identified in the structural formula. For the example of clintonite given in Chapter 1 (Section 1.3.2), the interlayer cations are identified as $Ca_{1.906} Sr_{0.011}$; as $a = 5.14$ Å and $b = 9.01$ Å,

$$\sigma = 1.6022 \times 10^{-19} \times 2(1.906 + 0.011)/(2 \times 5.14 \times 9.01 \times 10^{-20})$$
$$= 0.663 \text{ C m}^{-2}.$$

Clintonite is a brittle mica with a very large charge density; the charge density of muscovite is about half this value, and clay minerals are smaller still.

When the CEC is known, the calculation for swelling minerals is

$$\sigma = e \cdot CEC \cdot FW/(2000ab) \tag{4}$$

If the CEC is expressed in meq per gram ignited weight of clay, then FW is the formula weight per O_{22}. Taking, for example, the data on beidellite (Weir and Greene-Kelly 1962), the CEC = 1.27 meq g^{-1} (ignited weight), and the formula weight is 708.6 per O_{22}, so that

$$\sigma = 1.6022 \times 10^{-19} \times 1.27 \times 708.6/(2000 \times 5.14 \times 8.93 \times 10^{-20})$$
$$= 0.157 \text{ C m}^{-2}$$

The calculation from the interlayer cation contents, $Na_{0.90} K_{0.02}$, gives $\sigma = 0.161$ C m^{-2}, in good agreement with the calculation from the cation exchange capacity.

The calculations for mixed-layer minerals, however, may require considerable care in interpretation, as in the following example of an interstratified smectite–illite (Weir and Rayner 1974). In the sodium-treated form, it contained $Na_{0.66} K_{0.59}$, these cations being assigned to the interlayer sites. The spacing of the 060 reflection is 1.501 Å (G. Brown, private communication) so that $ab = 46.83$ Å^2, and from equation (3)

$$\sigma = 0.214 \text{ C m}^{-2}$$

However, if charge density is calculated from equation (4) using the determined CEC of 0.67 meq g^{-1} 110 °C weight basis, and a formula weight (per $O_{20}(OH)_4$) of 757.8, the calculated charge density is 0.0869 C m^{-2}.

The discrepancy arises because calculation by equation (4) assumes that the whole layer structure is accessible to the cation exchange process, whereas the interlayers contining potassium do not in fact take part. Equation (4) is therefore only valid for layer silicates in which the whole interlayer surface is accessible to cation exchange.

On the other hand, calculation by equation (3) assumes that despite K being present and non-exchangeable in some interlayers, the chemical composition and the consequent charge density are uniform throughout the sample. The charge density so calculated, 0.21 C m^{-2}, is very large for a smectite, being in the range expected for a vermiculite (see Chapter 1, Section 1.6.4). As the clay has properties associated with smectite (e.g. the expanding layers have a spacing of 16.9 Å on treatment with ethylene glycol) it may be that the assumption of uniformity is unjustified. If, pursuing this point, it is assumed that the non-expanding layers contain 1.5 atoms of (K + Na) per O_{22} (Hower and Mowatt 1966), the charge density of the expanding layers can be calculated from the equation

$$\sigma_m = (1 - \alpha)\sigma_1 + \alpha\sigma_2 \tag{5}$$

where σ_1 and σ_2 are the charge densities of the illite and smectite layers respectively, σ_m is the mean (bulk) charge density, and α is the proportion of expanding layers in the clay. For the

Oxford clay being considered, Weir and Rayner (1974) calculated from the X-ray diffraction profiles of the basal reflections that the sample contained 45% expanding layers. For $(K + Na) = 1.5$ per O_{22}, $\sigma_1 = 0.258$ C m^{-2}, and using this value for the non-expanding layers in equation (5) gives $\sigma_2 = 0.160$ C m^{-2} for the expanding layers, which as shown above, is similar to the charge density for beidellite. The charge densities σ_1 and σ_2 so derived and the interlayer cation contents of $Na_{0.66} K_{0.59}$ can be used to calculate the proportions of expanding to non-expanding layers assuming that only the layers containing Na expand. The result of this calculation is 63% smectite to 37% illite, which is clearly at variance with the X-ray diffraction calculations. This discrepancy, noted by Weir and Rayner (1974) and subsequently found with very fine clay fractions from other soils, has not so far been explained.

5.3.3 Alteration of charge density

Surface charge from atom replacement within the structure is often termed "permanent" charge, but it can in fact be modified in various ways.

Charge arising from isomorphous substitution in the dioctahedral sheets of 2:1 minerals (e.g. montmorillonite) can be decreased by treatment with Li and heating to 220–240 °C. It is believed that at this temperature some Li ions migrate into vacant sites in the octahedral sheet, decreasing the net negative charge (Hofmann and Klemen 1950; Greene-Kelly 1952; Brindley and Ertem 1971; see Chapter 7, Section 7.2.1).

The net charge may also be modified by oxidizing octahedral Fe^{2+} to Fe^{3+}, and this reaction is believed to be one mechanism by which micas weathered to vermiculite decrease charge density (Norrish 1973b). However, in laboratory experiments, there is not a stoichiometric relationship between decrease in charge density and oxidation of Fe^{2+} (Newman and Brown 1966), so it is believed that a simultaneous loss or gain of protons can also occur (Newman 1967; Farmer et al. 1971) and decrease charge density.

The effective charge density (and the nature of the surface itself) is also changed by the highly energetic sorption of polymeric species, in particular, the polynuclear hydroxy cations formed by the many cations that hydrolyse readily in aqueous solution. Generally the interaction is sufficiently strong to resist leaching by salt solutions. On non-swelling clays, the polynuclear cations give rise to pH-dependent charge, and enhance the sorption of certain anions and organic molecules (Colombera et al. 1971; Huang et al. 1978; Perrott 1981). The interaction with swelling clays leads to the formation of hydroxy interlayers and partial "chloritization" (Rich 1968; Nagasawa et al. 1974). Cations capable of reacting in this way are formed by Al, Fe^{3+}, Cr^{3+}, Mg, zirconyl, Ni, uranyl and probably others, and the important catalytic properties of the products are discussed in Chapter 6 (Section 6.7). The effect of such hydroxycation surfaces on the interaction of water with clays is likely to be considerable. Kidder and Reed (1972) showed that both the macroscopic swelling and the specific surfaces measured by the EGME method (Heilman et al. 1965) of montmorillonite clays were decreased by interlayering with hydroxyl-aluminium; no other studies seem to have been undertaken with the specific aim of investigating the effect of Al on bulk swelling properties.

5.3.4 Interlamellar surface

The uptake of water between the aluminosilicate layers of expanding clay minerals differs in important respects from the sorption of water on free external surfaces. Interlamellar water used to be described as "zeolitic" but the analogy with water in zeolites is not exact. The three-dimensional framework structures of the latter are comparatively little influenced by cation exchange or dehydration at low temperature, whereas the basal spacing of layer silicates is strongly dependent on the interlayer cation and its hydration properties. In particular, for each mineral type and interlayer cation, a series of hydration complexes is formed with characteristic basal spacings (MacEwan and Wilson 1980). These correspond broadly to zero, one two and three sheets of water between the layers, the basal spacings being approximately 10, 12–13, 14–15 and 19 Å respectively. Hydration from the anhydrous condition by increasing the ambient vapour pressure apparently occurs in steps, each complex having a characteristic range of stability over a vapour pressure range. This discontinuous hydration contrasts strongly with sorption on external surfaces, where the thickness of the water film increases smoothly up to saturation. These contrasting hydration properties have provided a means of distinguishing between interlamellar and external surfaces (see Section 5.4).

Interlamellar water has a structure that can be determined by diffraction methods (i.e. on a long time scale) and change from one complex to another is analogous to a phase change. The boundary between the two phases can be observed optically in vermiculite and the kinetics of the transformation suggest that the rate is controlled by diffusion of water through the interlamellar space (Walker 1956). As confirmed by NMR spin–lattice relaxation studies, the effect of the aluminosilicate surfaces is to confine water diffusion between two planes parallel with the surfaces.

Cation–water dipole interaction within the interlamellar space has a strongly polarizing effect on OH vibration, and results in interlamellar water being a stronger acid than water in a salt solution of equivalent concentration. The protonation reactions of interlamellar water with bases have been studied in detail by infrared absorption (Chapter 8, Section 8.4.3) and the acidity function has profound importance in the catalytic activity of clays (see Chapter 6, Section 6.2).

5.3.5 Special types of clay surface

The interlamellar surfaces of expanding clay minerals contain cations, the hydration of which provides the driving force for expansion. Halloysite contains one sheet of water between the 1:1 aluminosilicate layers but no interlayer cations. In consequence, the dehydration of halloysite is irreversible.

Palygorskite and sepiolite have rectangular channels in their structure that contain small numbers of exchangeable Ca^{2+} or Mg^{2+} cations and water molecules, some of which are relatively tightly held because they complete the coordination of Al and Mg in the octahedral ribbons. The remaining water is more truly zeolitic in character, though dehydration of sepiolite and palygorskite does induce considerable change in their structures (see below, Section 5.8, and Chapter 7, Section 7.4.9).

5.4 SPECIFIC SURFACE OF CLAYS

Determining the surface area per unit mass or specific surface of clays is beset by unique problems (Haynes 1961; van Olphen 1969). One of the most universally applied procedures for determining specific surface derives from the multilayer theory of adsorption of Brunauer, Emmett and Teller (BET theory: see Gregg and Sing 1982), which is used to analyse the sorption of non-polar gases on non-porous solids. The method requires the prior removal of all sorbed gases, usually by evacuation and gentle heating before measuring the sorption isotherm for the non-polar gas, often nitrogen. Applied to hydrous clays, this prior evacuation thins out water films between particles, bringing the particles closer. As most clays have a platy morphology, quasi-contact between particles extends over a non-negligible proportion of the surface, and some regions become inaccessible to non-polar molecules like N_2. Further, if the clay contains expansible interlayers, these collapse on evacuation, and the interlayer surface also becomes inaccessible to N_2.

It is clear, therefore, that the surface of clay determined by N_2 sorption is not a unique property of the clay but depends on its history and preparation, and also on the interlayer cation (Greene-Kelly 1964; Knudson and McAtee 1974).

In consequence, the practice has developed of using polar molecules that can penetrate between particles and into interlayer regions to measure the total specific surface. This has dubious validity from a theoretical standpoint, because polar molecules tend to group around charge centres, particularly cations countering negative charge on the clay surface, so that a true "monolayer" is unlikely to develop. In addition, BET theory makes the assumption that the energy of interaction of a molecule with the surface is very much higher than the interaction with an adjacent molecule, and this is not always valid for molecules like water that associate by hydrogen bonding.

Nevertheless, the use of polar molecules for measuring clay surface has had considerable empirical success, especially when it is calibrated by concurrent measurements on clay minerals for which the total surface can be calculated, for example, montmorillonite. From measurements made on the known surface, and assuming that conditions for completion of a monolayer are established, the effective cross-section area of the molecule for the clay surface is calculated. The mass of substance sorbed on the unknown surface at monolayer capacity is measured, and using the cross-section calculated previously, the specific surface is determined.

Most methods use ethylene glycol (Dyal and Hendricks 1950), glycerol (Diamond and Kinter 1958) or ethylene glycol monoethyl ether, EGME (Heilman *et al.* 1965), the latter having received a considerable degree of acceptance. All appear to suffer from the disadvantage that expanding layer silicates having a high charge density (e.g. Llano vermiculite and altered micas) form a single sheet complex at saturation, whereas lower charge density clays (e.g. smectites) form a double sheet (Eltantawy and Arnold 1974). Consequently, the apparent surface coverage by the molecule varies from clay to clay and a general value cannot be assumed for a sample of unknown composition.

A further difficulty is so controlling the experimental conditions that neither more nor less than an effective monolayer is formed on external surfaces. In some procedures, a free liquid surface is used (e.g. Martin 1955), in others, a $CaCl_2$-solvate (Bower and Goertzen 1959) and yet others, a dynamic system (Rawson 1969). Rigorous predrying of the clay is essential; best results seem to be obtained by wetting the clay with liquid and evaporating the excess off over $CaCl_2$ under vacuum until the clay attains constant weight.

Many attempts have been made to use water as a polar molecule to measure the specific surface of clays. Because the vapour pressure can be readily controlled, earlier work measured sorption isotherms and used the BET equation

$$x/x_m = cR/[1-R)(1+(c-1)R)] \tag{6}$$

to calculate the monolayer coverage, x_m; in equation (6) R is the relative vapour pressure, p/p_0, x

is the mass of water sorbed per unit mass of clay and c is a parameter related to the energy of sorption.

However, Mooney *et al.* (1952) showed that the specific value of montmorillonite so calculated was approximately half that estimated from the unit cell dimensions. This is because at the relative pressure for which a monolayer is completed according to equation (6), the interlayer contains only a single sheet water complex. The use of the BET equation to calculate the specific surface of clays from water sorption isotherms was thus discouraged (Quirk 1955).

Recently it was shown that for clay soils that are predominantly Ca-saturated, water sorption and ethylene glycol sorption are highly correlated measurements and it has been suggested that the specific surface of clays can be measured from the central portions of water sorption isotherms ($0.35 < R < 0.75$), where little change in interlamellar volume is occurring (Branson and Newman 1983). Such a procedure also enables internal and external surface to be distinguished (Ormerod and Newman 1983).

Another method of determining specific surface (Haynes 1961) measures the heat of wetting, that is, the heat evolved when a clay pre-equilibrated at a high relative humidity is immersed in water. The method is technically difficult and has not in consequence been used very often, but as it depends on the destruction of a large free water surface by immersion in water it is theoretically an absolute method.

Adsorption of a solute from solution by clay has also been used to determine the clay surface. Cationic dyes, particularly methylene blue, suffer disadvantages but long chain quaternary amine salts, particularly cetyl pyridinium bromide (CPB) (Greenland and Quirk 1964) appear to pack quite reproducibly on clay surfaces with a limited charge density range. In general, however, the packing changes with the charge density (Lagaly 1981) and so the method requires careful calibration for the clay being studied.

5.5 THE REDUCED OR NORMALIZED ISOTHERM

The sorption isotherm consists of a set of measurements at constant temperature of mass sorbed versus relative pressure. Comparison of one set with another is greatly helped if both are expressed on a common basis, and the simplest way to do this is in terms of the thickness of the sorbed layer. The multilayer (BET) theory of vapour sorption supposes that water in the layer adjacent to a surface requires a greater activation energy of desorption, E_1, than the second and subsequent layers, E_1 being a characteristic of the surface–sorbate interaction. This theory leads to the BET equation, from which the monolayer coverage of a simple external surface is readily calculated for type II (S-shaped) isotherms commonly found in the sorption of water by clay. If the amount of water sorbed is expressed, not as g H_2O per g clay, but as g H_2O per g H_2O held at monolayer capacity, the shape of the isotherm depends only on the BET "c" parameter which is closely related to E_1, the activation energy for desorption. Figure 5.1 shows three reduced normalized isotherms based on published data for the sorption of H_2O on oxide surfaces that Hagymassy *et al.* (1969) consider represent isotherms for $c = 10$–14.5, $c = 23$, and $c = 50$–200. From Fig. 5.1 it is clear that monolayer coverage occurs at $p/p_0 = 0.097$ for $c = 50$–200, but for $c = 10$–14.5 the monolayer is completed at $p/p_0 = 0.24$.

Commonly monolayer coverage is estimated by fitting isotherm data to the BET equation, using data in the range $0.05 < p/p_0 < 0.35$. Equation (6) can be recast in the form

$$R/(x(1-R)) = 1/(c \cdot x_m) + R(c-1)/(c \cdot x_m) \tag{7}$$

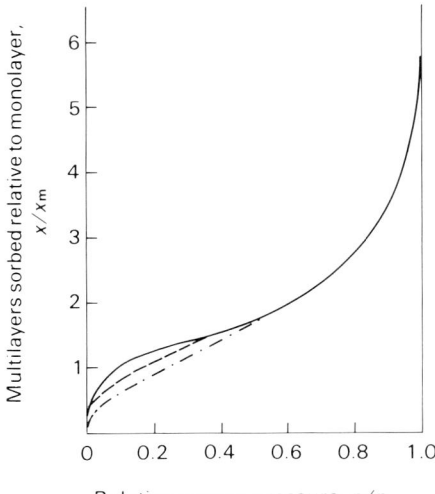

FIG. 5.1. Reference isotherms for water sorption on oxide surfaces (Hagymassy *et al.* 1969) for $c = 50$–200 (———), $c = 23$ (– – –) and $c = 10$–14.5 (– · – · –).

which is a linear equation of the form

$$z = ay + b$$

where $z = R/(x(1-R))$, $a = (c-1)/(c \cdot x_m)$, $y = R$, and $b = 1/(c \cdot x_m)$; a and b can readily be calculated by a least squares procedure, whence $c = a \cdot b + 1$ and $x_m = 1/b \cdot c$.

From the value of the monolayer coverage x_m so calculated, a table of x/x_m against p/p_0 is drawn up and can be plotted as a reduced isotherm for comparison with the reference isotherms shown in Fig. 5.1.

Equations other than the BET have been used to express multilayer sorption of water on clays. The best known of these is the Frenkel–Halsey–Hill (FHH) equation

$$\ln R = -k/((v/v_m)s) \qquad (8)$$

where v is the volume of sorbate at relative pressure R, v_m is the volume of sorbate at monolayer coverage, k is a parameter associated with the forces of adsorption, and s is related to the decay of the force field (Jurinak 1963). As this is a three-parameter equation, an independent estimate of v_m is needed to calculate k and s. This makes its application more difficult and often water sorption data for clay minerals fit the FHH equation less well than the BET equation.

For vermiculites and smectites, the BET equation is not directly applicable because the free energy conditions governing the development of interlamellar water complexes are different from those governing the development of multilayers on external surfaces. Consequently, a different method of analysis of the isotherm data is needed for all clays where the presence of interlamellar water sheets is suspected. Halloysites also present a special problem in the analysis of isotherm data. These examples are discussed in the appropriate mineral sections below.

Further information about the pore structure of a clay can be inferred if the reduced isotherm is plotted, not against relative pressure, but against the multilayer thickness (n) of water on the reference sorbent at each relative pressure. Such n-plots have several useful properties. If the observed isotherm corresponds *exactly* to the reference isotherm, a straight

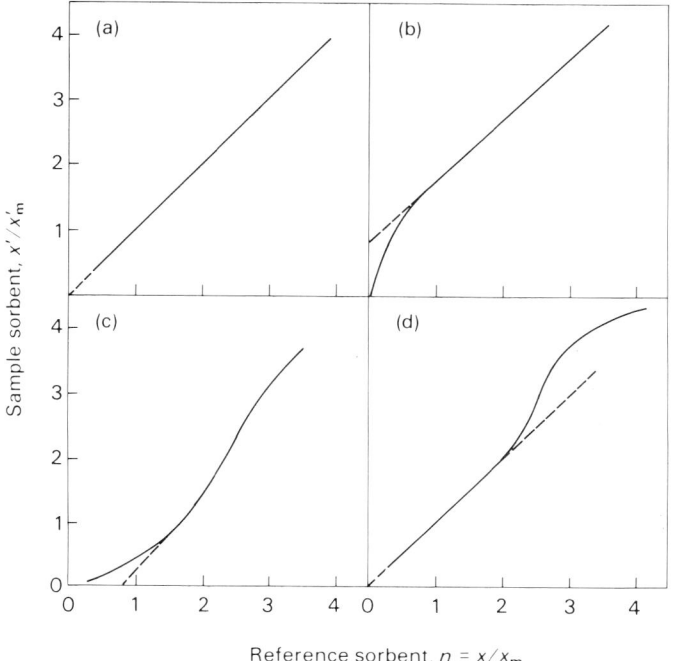

FIG. 5.2. Types of multilayer plots: (a) similar isotherms for sample and reference; (b) micropore filling up to $n=0.7$; (c) reference and sample have different "c" parameters; (d) capillary filling between $n=2$ and $n=3.5$.

line passing through the origin with unit gradient results (Fig. 5.2(a)). Usually, however, some deviation will be expected. If the clay contains micropores that fill at very low relative pressures (e.g. interlamellar pores) when the clay is treated with an appropriate cation, the graph is a straight line with a positive intercept b on the y-axis, the mass of water in the micropores being given by $b.x_m$ (Fig. 5.2(b)). A negative intercept (Fig. 5.2(c)) may indicate that an incorrect reference isotherm was chosen, or that the cation form of the clay was such that the filling of the interlamellar pores was held back relative to the formation of a monolayer on the external surface. If the early part of the plot is linear passing through the origin but the sections at higher relative pressure deviate upwards (Fig. 5.2(d)), this is usually taken to indicate condensation in capillaries only a little larger than the multilayer thickness at the corresponding relative pressure. In general, water sorption data for clays may combine any of the effects shown in Fig. 5.2.

5.6 1:1 GROUP MINERALS

5.6.1 Kaolinites

In principle, kaolinites should present a comparatively simple surface for water sorption. The particles are relatively large, often 50% or more greater than 0.5 μm, and easily measured in the electron microscope. The interlayer spacing is not affected by water sorption (unless the kaolinite has been intercalated, see Chapter 8) or by varying humidity. The sorption of water should therefore be analogous to that on other oxides and hydroxides.

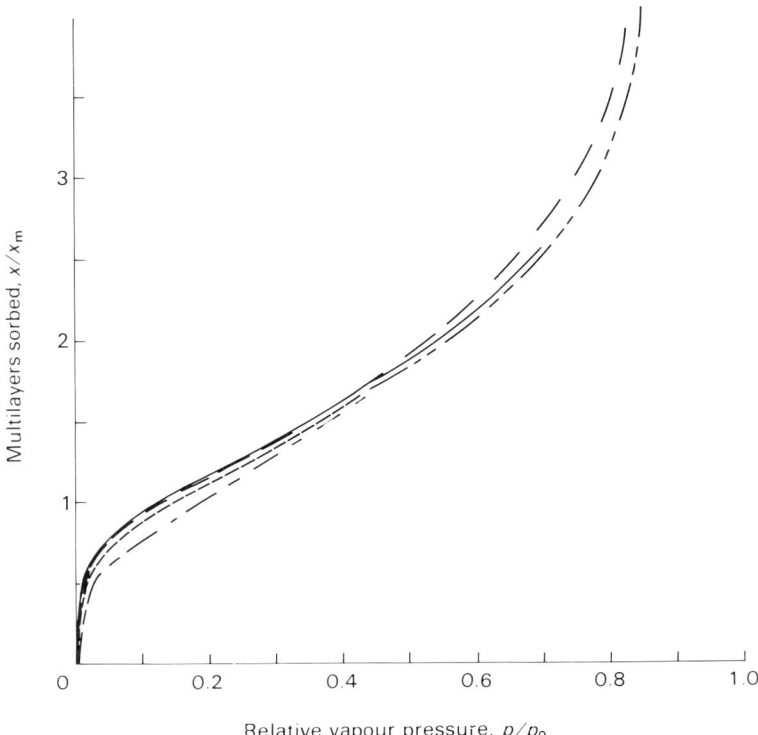

FIG. 5.3. Reduced isotherms for water sorption on several cation forms of kaolinite. ——— Mg (Martin 1959); ———
Li (Martin 1959); ----- Li (Keenan et al. 1951); —·—·— Ca (Keenan et al. 1951); —··—··— Ca (Jurinak 1963).

In detail, the sorption properties are not quite so simple. Although some observers (e.g. Martin 1959; Lloyd and Conley 1970) have calculated that the geometric surface is similar to the surface measured by N_2 sorption, others (Olivier and Sennet 1973) have found that the N_2 surface is considerably greater than that predicted from particle size measurements. Scanning electron microscope study showed that kaolinite particles consisted of booklets and that some of the additional porosity must be occurring between the leaves of the booklet.

Kaolinites have only a small permanent exchange capacity (permanent charge), but as the specific surfaces are also small, the charge density is in the same range as moderate to high charged smectites; most data calculate at 0.15 to 0.2 C m^{-2}.

Most kaolinites therefore have cations balancing the surface charge, and the effect of cation form on water sorption was investigated by Keenan et al. (1951), Martin (1959) and Jurinak (1963). The results of these investigators, calculated in accordance with the BET equation, and expressed as reduced isotherms, impressively demonstrate the concordance of the data (Fig. 5.3). The BET "c" parameter, which relates to the energy of interaction, is influenced by the surface cation, as is the monolayer coverage x_m (Table 5.1). Although the extent of variation of x_m with cation varies with investigator, all agree that Li-saturated kaolinites have the smallest monolayer coverage. This was originally explained (Keenan et al. 1951) by supposing that Li through its small size was able to fit into the structure so that it no longer influenced hydration of the surface, and subsequent work (Martin 1958; Jurinak 1961) tended to confirm this. Greene-

TABLE 5.1. Water sorption by kaolinites:BET parameters

	Peerless No. 2 <0.5 μm fract.[†]			Peerless ungraded[‡]		Georgia ungraded[§]		
	x_m mg/g	c	N_2 surface	x_m	c	x_m	c	N_2 surface
H	7.36	13.8	26.1					
Li	4.69	33.7	26.3	3.13	58.5	2.45	75.9	11.7
Na	6.03	18.8	27.8	3.50	41.3	2.66	75.8	11.7
K	6.03	18.8	28.9	3.22	36.2			
Rb	5.48	49.9	28.7					
Cs	5.65	19.5	30.2	3.40	30.6			
Mg				4.40	62.8			
Ca	7.88	20.6	27.6			3.28	50.6	12.0
Sr	7.16	18.8	28.5					
Ba	6.16	45.3	29.3			2.89	1.31	12.3
Al						3.84	57.2	12.6

[†] Keenan et al. (1951).
[‡] Martin (1959); $N_2 = 13.0 \, m^2 g^{-1}$.
[§] Jurinak (1963).

Kelly (1955) showed that Li introduced into the cation exchange positions was found to be non-exchangeable after drying by heating to 200 °C. It should be noted, however, that the effect of surface cation on x_m observed by the later workers was much less marked than in the original publication by Keenan et al.

Recent work has emphasized the arrangement and energy of interaction of water with the surface of kaolinite. Dielectric measurements on kaolinite-water systems (Muir 1954; Nelson et al. 1969) have shown that loss occurs in two frequency regions. The low frequency loss increasing with decreasing frequency below 1000 Hz is attributed to conduction effects, but the loss maxima in the region 10–100 kHz is believed to arise from relaxation of dipole rotation. The energy region indicates that the bonding of the water molecules is intermediate between that in free water and ice.

Heats of immersion and NMR measurements were made on Georgia kaolinite by Fripiat et al. (1982) and the NMR studies were extended to high pressure by Jonas et al. (1982). As the surface coverage before immersion is increased, the heat for immersion per unit surface declines steadily to a constant value of 119.5 mJ m^{-2} at average coverage of 2.6, indicating that the range of surface forces declines rapidly after the second multilayer is completed. In free water, the measurement by NMR of T_1 the spin-lattice relaxation time, has shown that increasing pressure distorts the random hydrogen bond pattern of the structure at atmospheric pressure whereas adjacent to the kaolinite surface this does not occur, the implication being the hydrogen bond pattern is already non-randomized by the adjacent surface. The activation energies for reorientation of the water molecules are lower than in free water and the influence of the surface extends for three to four layers of water (Jonas et al. 1982).

Studies of the freezing of water in kaolinite pastes (Anderson et al. 1973) show that a zone of water between the bulk ice and the clay surface remains unfrozen. In this zone, water molecules and solutes remain mobile at temperatures down to −30 °C.

All these studies on the structure and energy status of water adjacent to kaolinite surfaces seem to agree that the range of influence of the surface extends to between two and four molecular layers of water.

5.6.2 Halloysites

The halloysite structure consists of kaolinite-like aluminosilicate layers, interspersed in the fully hydrated mineral with one sheet of water molecules. Various morphologies have been described but the best known is tubular, one or more layers being rolled up spirally. The ideal formula unit is $Al_4Si_4O_{10}(OH)_8 \cdot 4H_2O$ and as the exchange capacity is small, about 0.10 meq g^{-1} (Garrett and Walker 1959), there are few cations to interact with the interlamellar water. In its fully hydrated state the interlayer spacing of halloysite is near 10.1 Å, but on dehydration it decreases to 7.2 Å; rewetting does not cause the structure to re-expand spontaneously to 10 Å. This combination of dehydration without rehydration has led to some confusion in nomenclature; in the present chapter, following the suggestion of Brindley (1980), these two forms of halloysite will be called "halloysite (10 Å)" and "halloysite (7 Å)".

The structural formula of halloysite (10 Å) given above corresponds to 0.1394 H_2O g^{-1} dehydrated clay; as with kaolinite, a further 14% water is lost by dehydroxylation at 550 °C. The area ab of one basal face of the unit cell is $5.14 \times 8.90 = 45.7$ Å^2 (Brindley 1980), which is covered by $4H_2O$, so that each H_2O covers 11.44 Å^2 on each aluminosilicate surface, but as two surfaces are adjacent to one H_2O, the aluminosilicate surface in contact with H_2O is 22.9 Å^2. The total surface of the basal planes is 1067 m^2 g^{-1}.

The dehydration–hydration isotherms of halloysite (Fig. 5.4) have the unusual characteristic of large irreversibility (Hughes 1966). During the primary dehydration, it appears

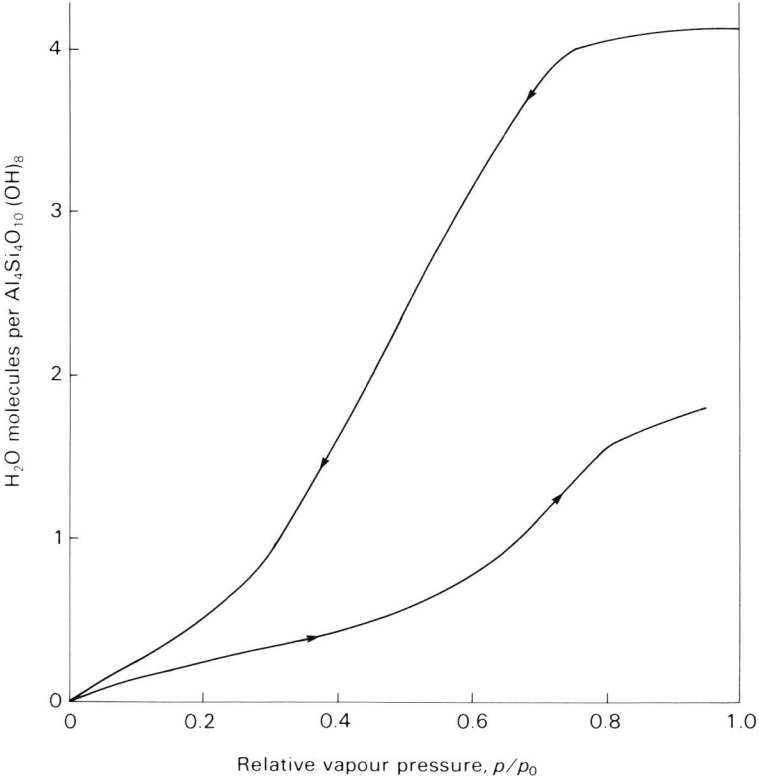

FIG. 5.4. Desorption of water from 10 Å-halloysite and its irreversibility on resorption.

that a multifaceted morphology develops with relatively large gaps between groups of layers in the spiral (Kirkman 1977). Such delaminations are observed in other instances where large changes in interlayer spacing have to be accommodated within an established morphology (e.g. the edge exchange of potassium from micas: Newman and Brown 1969) and is probably one source of irreversibility. However, the main reason why halloysites (7 Å) do not rehydrate seems to be that, in the absence of surface cations, the energy of interaction of water molecules with the surfaces alone is insufficient to overcome van der Waals' attraction between the layers.

Halloysites form interlayer complexes with a wide range of compounds (MacEwan and Wilson 1980), and this has suggested that halloysite might be rehydrated by using such a complex as an intermediate and washing out the interlayer compound with water. Churchman and Carr (1973) washed the hydrazine-, CsCl-, and potassium acetate (KOAc)-complexes of halloysite until no residue of the original complex was detected by X-ray diffraction. The products from the CsCl- and hydrazine-complexes appear to have been hydrated halloysite, but the authors state that "a fully hydrated halloysite which is identical to the original cannot be recovered by washing the KOAc complex of the dehydrated form of the mineral".

In view of the unusually irreversible nature of the halloysite–water interaction, a detailed analysis of the isotherms by the BET equation is probably inappropriate. However, it may be noted that the data for *sorption* in Fig. 5.4 conform very closely ($r=0.9997$) to the BET equation up to $p/p_0 = 0.7$, with $x_m = 12.7$ mg g^{-1} and $c = 4.26$, the small value of the latter parameter emphasizing qualitatively the weak interaction between water and the halloysite surface.

As shown in Fig. 5.4, the removal of water appears to be a smooth progression to a dehydrated state as the relative pressure is decreased. In an early study, Brindley and Goodyear (1948) suggested that dehydration did not begin unless $R < 30\%$ and that certain compositions (between 1.3 and 2.6 H$_2$O per Al$_4$Si$_4$O$_{10}$(OH)$_8$) were not found, but Churchman *et al.* (1972) confirmed the smooth dehydration observed by Hughes (1966). Churchman *et al.* (1972) found that the X-ray diffraction patterns of partially dried halloysites indicated interstratification of the 7 Å- and 10 Å-components, with a tendency to segregate into the two basic types. Nagasawa and Miyazi (1976), however, showed that halloysites differed in their readiness to dehydrate.

Even after prolonged desiccation, some residual molecular water remains in halloysite (Brindley and Goodyear 1948), and Cruz *et al.* (1978) found that, even after heating at 250 °C, a sample of halloysite (10 Å) from Indiana contained 11.7 mg H$_2$O/850 mg solid. Residual water is not allowed for in Fig. 5.4, which, like most isotherm data, conventionally assumes that after prolonged desiccation a clay is dehydrated, and this can sometimes cause significant discrepancies when comparing data from different sources (see Section 5.7.4).

Owing to the large interlamellar water content and lack of interlayer cations, halloysite (10 Å) is a useful material for studying H$_2$O–surface interaction. Cruz *et al.* (1978) used NMR together with infrared and specific heat measurements to study the structural arrangement of interlamellar water and also water in the pores enclosed by the morphological units. They confirmed previous observations by Yariv and Shoval (1975) that the vibrations of OH groups on the mineral surface are not appreciably perturbed by comparison with kaolinite, indicating that hydrogen bonding between water and surface OH is weak. Water molecules are hydrogen bonded to each other, though it was necessary to suppose that one OH is not used for H-bonding in order to account for a band at 3550 cm^{-1}. An analysis of the NMR relaxation time measurements showed that in halloysite (10 Å), the water molecules undergo anisotropic motion composed of reorientation about an axis along the bisector of H—O—H angle, and of reorientation of this axis around another perpendicular to the first, described as "tumbling". Reorientation around the bisector is faster than tumbling, and there is no evidence of preferential ordering of water relative to the halloysite surface.

5.7 THE 2:1 GROUP MINERALS

Differences between these minerals in respect to their interaction with water have their origin principally in the net negative charge on the aluminosilicate layers and in the presence of swelling interlayers, that is, interlayers that change in water content and interlamellar spacing as the partial pressure is varied. Pyrophyllite and talc have neither net negative charge nor swelling interlayer, so that the superficial surface is the bare basal oxygen surface of the siloxane network. Infrared study shows that water interacts weakly with this surface, and in the absence of charge-balancing cations the surface is, if anything, slightly hydrophobic; freshly cleaved talc is completely hydrophobic (Pashley 1981a).

So the main interest centres on the micas with a large negative charge but no swelling interlayers, and on vermiculites and smectites with a large interlayer surface accessible to water. Illites lie somewhere between these subgroups: they are often mixed-layer mica–smectites and their small particle size also complicates their behaviour. There has been little systematic study of the interaction of water with illites; by far the most attention has been given to the swelling minerals and to montmorillonite in particular.

5.7.1 Micas

Molecularly smooth surfaces can be prepared by cleaving sheets of micas, particularly muscovite mica, and these are ideal for studying surface properties of aluminosilicate layers. The presence of water vapour during cleavage decreases surface energy (Bailey and Kay 1967), and by studying the forces between mica surfaces as they are brought together in water and electrolyte solutions, Pashley (1981a,b) demonstrated that short-range repulsive forces attributed to hydration often prevented adhesive contact. However, a mica flake does not hydrate under unstressed conditions as the hydration energy of the interlayer K ions is insufficient to overcome the cooperative structural forces at the coherent edges of a cleavage surface (Newman and Brown 1969).

Therefore only the superficial surface of mica particles is accessible to water and micas should provide a useful substrate for measuring water sorption on an open charged surface. There have, however, been rather few measurements of sorption isotherms for micas. Orchiston (1955) used a fraction $<2\,\mu$m of a shale from Illinois, Mikhail et al. (1979) used Fithian illite also from Illinois, and Branson and Newman (1983) used a mica, two illites and an illitic soil clay, all Ca-saturated. Some of these data are given in reduced isotherms in Fig. 5.5. All are normal S-shaped (type II) isotherms that, with the exception of the mica, closely fit the BET equation up to $p/p_0 = 0.6$. This is demonstrated in Fig. 5.6 in which the multilayer thickness is plotted against the thickness on reference surfaces (n-plot: Section 5.5). The data points extrapolate through the origin except those for mica where the positive intercept indicates the presence of micropores. It is suggested that this arises from face-to-face contact of smooth mica particles. There may also be a small proportion of such micropores in the illites, but because their particle size is much smaller and the total specific surface much larger, their presence is masked by sorption on the free surfaces. Specific surfaces estimated by assuming that one water molecule covers 10.6 Å^2 are given in Table 5.2.

In general it can be said that water sorption by illites without mixed layering conforms quite closely with the BET equation even when the surface cation is not defined and that this sorption provides a reasonable basis for calculating specific surface.

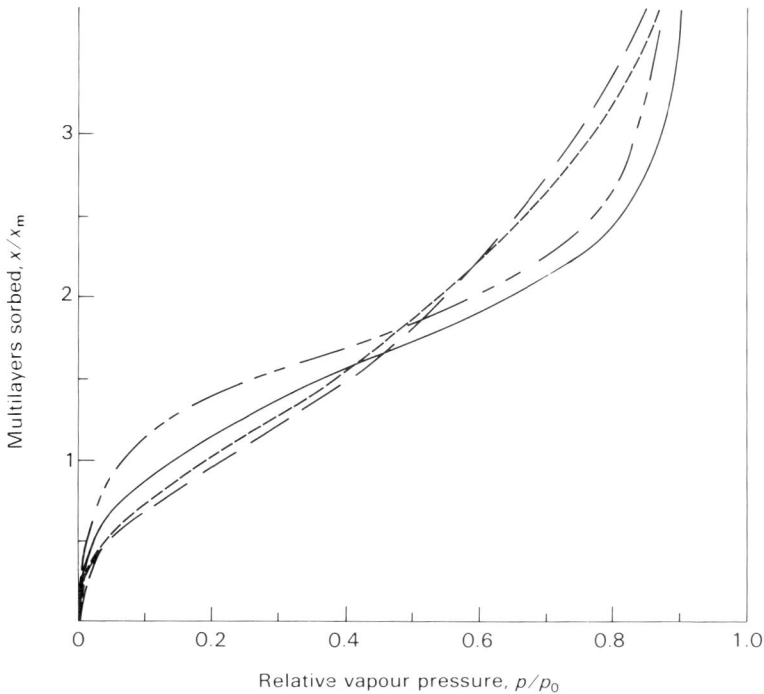

FIG. 5.5. Reduced isotherms for water sorption on illites and mica. ——— illite, Illinois (Orchiston 1955); ----- Fithian illite (Mikhail et al 1979); ——— Ca-Beavers Bend illite; —·—·— Ca-muscovite (Branson and Newman 1983).

5.7.2 Swelling minerals: general considerations

Although showing considerable chemical diversity, the minerals in this group, comprising vermiculites and smectites including montmorillonite, beidellite, hectorite, saponite and other varieties, have many features in common in their interaction with water. As they all contain a large proportion of swelling interlayers in which the water is arranged in a partly ordered structure around the interlayer cations, the cations themselves play a dominant role in their interaction with water. Not only does the interlamellar spacing depend on the cation but so also do the absolute amounts of water sorbed, the shape of the sorption isotherm and the acidity function (Chapter 6) of the water. All these properties are influenced by the cation size, its valency, its electronegativity and its hydration energy to a much greater extent than on free, non-interlamellar surfaces. Inter-mineral differences arise from the surface density of charge and its origin in the tetrahedral or octahedral sheets, and also on particle size, which determines the proportion of free surface to interlamellar surface.

The effects of cation type on the water sorption ability of montmorillonite were established by the now classical work of Hendricks et al. (1940) and Mooney et al. (1952). In both of these studies, the different cation forms were prepared by neutralizing the hydrogen-clay prepared by electrodialysis with an appropriate base. Although this procedure has been criticized (van Olphen 1970) on the grounds that H^+-clay is unstable and slowly becomes partly Al-saturated,

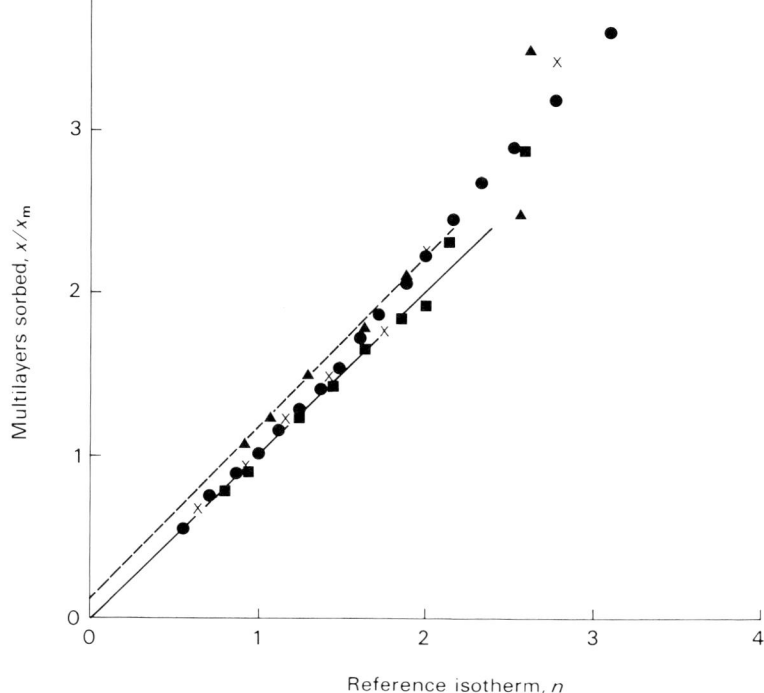

FIG. 5.6. Multilayer n–n' plots for the isotherms in Fig. 5.5. Note that the data for muscovite do not extrapolate through the origin. × = illite, Illinois; ● = Fithian illite; ■ = Beavers Bend illite; ▲ = muscovite.

TABLE 5.2. BET parameters for water sorption on mica and illites

	x_m (mg g^{-1})	S_{H_2O} (m^2 g^{-1})	c
Illite, shale, Illinois (Orchiston 1955)	23.7	84	13.7
Fithian illite (Mikhail et al. 1979)	15.9	53	18.6
Beavers Bend illite (Branson and Newman 1983)	9.2	31	38.6
Muscovite (1–4 μm fraction) (Branson and Newman 1983)	4.2	15	63.1

the results of Mooney et al. (1952) have since been verified, in part at least, and there seems little doubt that they are substantially correct.

An indication of how the isotherm shape varies with interlayer cation in Wyoming montmorillonite is given in Fig. 5.7, which also shows the interlamellar spacings at the various relative humidities. It is clear that the basal spacing changes in the regions where points of inflection occur on the isotherms. This is even more marked in the isotherm for Na-vermiculite which exhibits a steep rise as the spacing begins to change, and a plateau once a stable spacing

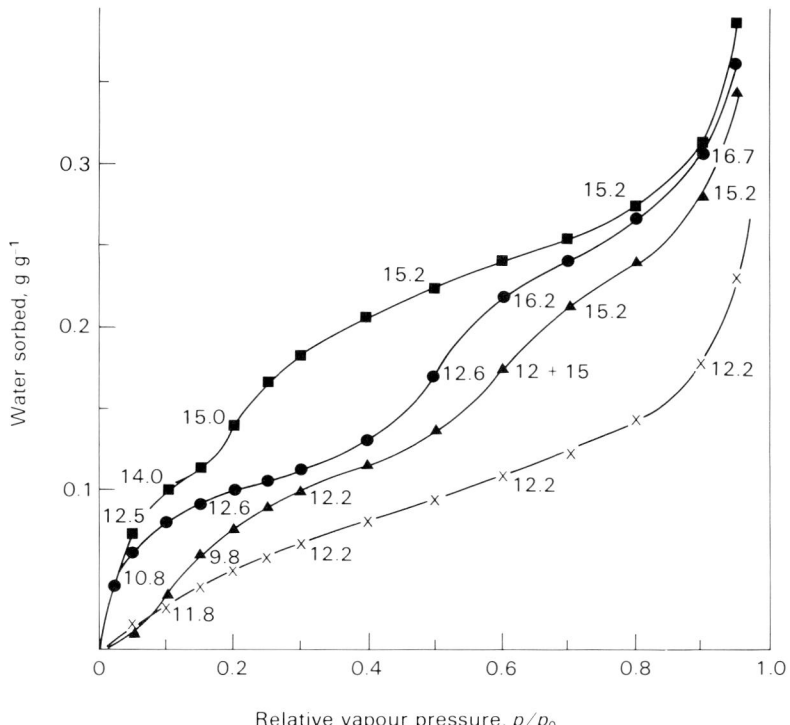

FIG. 5.7. Water sorption isotherms for several cation forms of Wyoming bentonite (Mooney *et al.* 1952): ■ Ca; ● Ba; ▲ Na; × Cs. The numbers indicate $d(001)$ spacings in Å at various points on the isotherms.

has been reached (Fig. 5.8). The sharpness of the step diminishes as the particle size becomes smaller, as shown by the isotherms for Na-Unterrupsroth beidellite, which consists of coarse clay, and Na-Camp Berteau montmorillonite, which is very fine grained (Glaeser and Mering 1968). Presumably as the aluminosilicate sheets become smaller they pack together in smaller stacks or less regularly so that the proportion of internal to external surface decreases and multilayer sorption obscures the interlayer hydration steps. The changes in interlayer hydration are still detectable in the isotherm for Na-montmorillonite, but are smoothed relative to Na-vermiculite.

Hydration of the interlamellar space proceeds therefore in steps, corresponding approximately to the intercalation of one, two and three sheets of water. The relative pressure range over which the interlamellar spacing changes is usually quite narrow and is followed by only a small change, or even none at all, as the relative pressure is increased up to the point where the next hydration complex forms. The relative pressures at which the spacing changes vary from cation to cation, and the ranges of humidity over which little change occurs were termed domains or zones of homogeneous hydration by Glaeser and Mering (1968).

It is possible to generalize to some extent about these zones of hydration and these are summarized in the following paragraphs; more complete data are given by MacEwan and Wilson (1980). It must be recognized, however, that exceptions are found, and in some instances, the data are scanty.

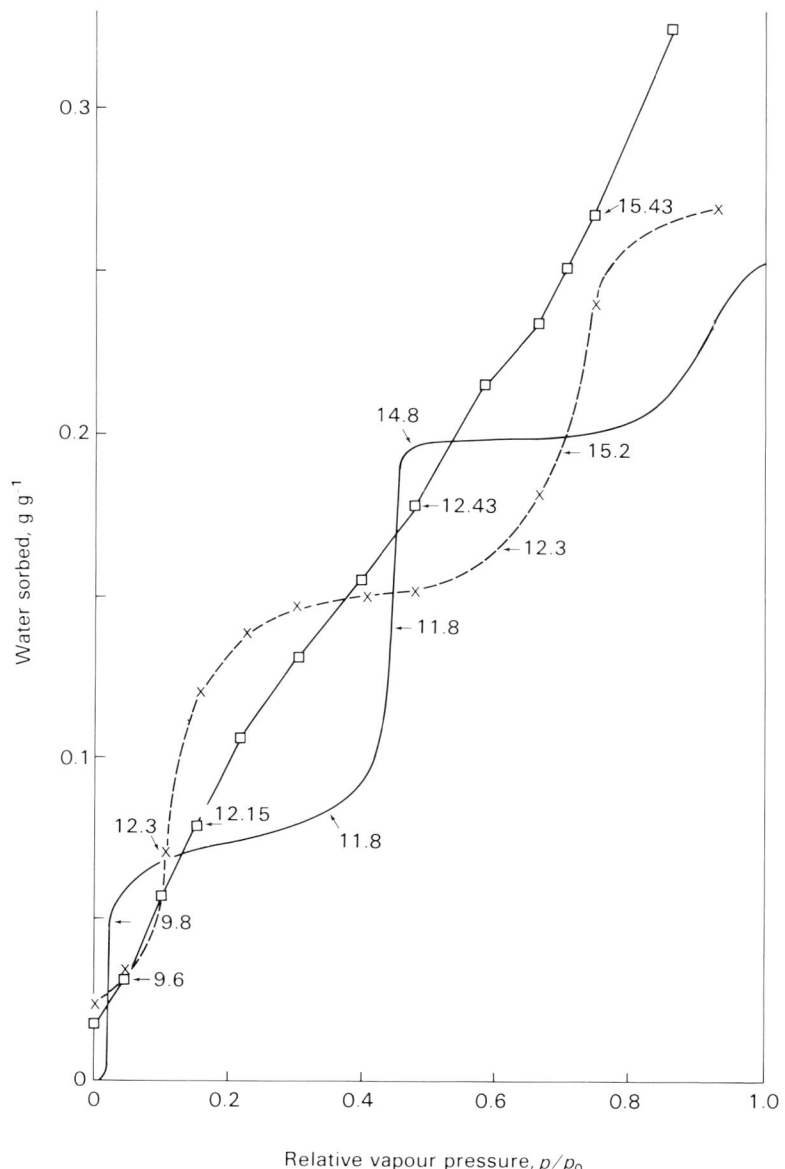

FIG. 5.8. Water sorption isotherms for Na-saturated clays: —— Llano vermiculite; □——□ Camp Berteau montmorillonite; ×----× Unterrupsroth beidellite. The numbers indicate $d(001)$ spacings in Å.

5.7.2.1 Hydrogen ions

H^+-saturated clays are unstable and tend to become Al-saturated (or in the case of trioctahedral vermiculites and smectites, Mg-saturated). In consequence, data on H^+-clays are subject to uncertainty. According to Norrish (1954) expansion up to a basal spacing of 22 Å is possible when immersed in water.

5.7.2.2 Lithium and sodium ions

Both cations form single sheet complexes ($d(001) = 12$–12.5 Å) at low humidities that expand to the two sheet complex ($d(001) = 15$ Å) at $p/p_0 = 0.5$–0.6. At very high humidities ($p/p_0 > 0.95$), most montmorillonites, but not vermiculites or some beidellites, form 19 Å (three sheet) complexes and when immersed in water, swell macroscopically. In this type of swelling, also known as osmotic or type 2 swelling, the mean interlamellar distance in aqueous solutions increases smoothly in proportion to $1/\sqrt{c}$ as the concentration, c, decreases (Norrish 1954). Li-vermiculite also shows osmotic swelling (Walker and Milne 1950; Norrish and Rausell-Colom 1963). Factors affecting whether osmotic swelling occurs are the total charge and particularly whether it is located in octahedral or tetrahedral sites. Table 5.3, adapted from Norrish (1973a), illustrates these factors.

5.7.2.3 Magnesium, calcium, strontium and barium ions

This sequence of cations illustrates the effect of decreasing hydration energy. Even under moderately dry conditions, Mg-clays hydrate to the double sheet complex and only rarely has evidence for the single sheet complex been found (van Olphen 1970). The single sheet becomes more stable through the sequence Ca → Sr → Ba, the transition to the double sheet complex of Ba occurring above $p/p_0 = 0.5$, as with Na. All give 19 Å when immersed in water but do not swell macroscopically.

TABLE 5.3. Swelling and dispersion of montmorillonites as affected by charge, its location and interlayer cation (after Norrish 1973a)

Sample	Cation[†] exchange capacity	Location[‡] of charge	Cation	Swelling in water[§]	
				$<2\,\mu m$	$>2\,\mu m$
1. Montmorillonite, Mt Binjour, Q	126	O	Li	D	D
			Na	D	D
2. Montmorillonite, Gummeracha, SA	126	O	Li	D	—
			Na	D	—
3. Montmorillonite, Chambers, Arizona	140	O	Li	D	—
			Na	19 + D	20 + D
4. Montmorillonite, Otay, California	156	O	Li	D	—
			Na	D	—
5. Beidellite, Unterrupsroth	140	$\frac{1}{2}$O + $\frac{1}{2}$T	Li	D	15.5
			Na	15.7 + D	15.5
6. Beidellite, Mt Binjour, Q	146	$\frac{1}{2}$O + $\frac{1}{2}$T	Li	D	18.5
			Na	D	18.5
7. Saponite, Mt Binjour, Q	125	T	Li	17 + D	15.5
			Na	15.9	15.2
8. Saponite, Groschlattengrun	130	T	Li	D	15.4
			Na	15.5	15.3
9. Beidellite, Drayton, Q	135	T	Li	D	—
			Na	15.6	16.0

† meq/100 g ignited weight basis.
‡ T = predominantly tetrahedral; O = predominantly octahedral.
§ XRD spacings observed from a wet paste; D = diffuse spacing greater than 40 Å.

TABLE 5.4. Free energies and enthalpies of hydration (kJ mol^{-1}) and Pauling atomic radii (Å) for monovalent cations (after Salomon 1970)

	$-\Delta G_{solv}$	$-\Delta H_{solv}$	r
Li	409	448	0.60
Na	303	336	0.95
K	230	252	1.33
Rb	212	222	1.48
Cs	196	193	1.69

5.7.2.4 Potassium and ammonium ions
The single ion free energy of hydration for K^+ is only a little over half that for Li (Table 5.4) and K^+ is easily dehydrated in the interlayer space. By preparing K-montmorillonite from the dispersed Na-clay, Norrish (1954) was able to obtain a spacing of 15 Å and some indication of swelling beyond this separation. However, most observers report a maximum spacing of 12–13 Å, except when immersed in water. Higher charge vermiculites have a spacing of 10–10.5 Å. In general, NH_4^+ behaves somewhat like K, with the additional complications that it may dissociate into $H^+ + NH_3$, especially on heating, and also can form H-bonds.

5.7.2.5 Rubidium and caesium ions
Both cations are large enough in the unhydrated state to prevent complete closure of the interlamellar space when swelling minerals are hydrated (Table 5.4), and the 12 Å interlamellar spacing of Cs-montmorillonite is almost independent of water content. The incomplete closure of the interlayer under vacuum allows some penetration of N_2 and nitrogen-specific surfaces of Rb- and Cs-montmorillonite are larger (64.2 and 108 m^2 g^{-1} respectively) than the surfaces of Li- and K-montmorillonite (41 and 48 m^2 g^{-1}) (Mooney *et al.* 1952).

5.7.2.6 Aluminium
The acidity function of water increases in the interlamellar space and interlayer Al^{3+} has a greater tendency to hydrolyse than Al in free solution. As a range of hydroxy-Al complexes can form, the basal spacing and hydration properties are strongly dependent on the degree of hydrolysis and also on the interlamellar charge density. At acid pH, Al^{3+}-montmorillonite gives a spacing of 15 Å when air-dry and 19 Å when wet. If the pH is raised and Al-OH polymers are formed in the interlayer, the interlamellar spacing increases to 19 Å air-dry and 22 Å when wet (Brown and Newman 1973). Vermiculite, however, forms only 14 Å complexes (Walker and Milne 1950; Brydon and Turner 1972). The water sorption properties of these various modifications have not been systematically studied, though Kidder and Reed (1972) showed that macroscopic swelling of two montmorillonites was greatly diminished following treatment with Al-OH polymers.

5.7.3 Vermiculites

Much information about the structures of vermiculites has been obtained by X-ray diffraction (Mathieson and Walker 1954; Mathieson 1958; Shirozu and Bailey 1966; Alcover *et al.* 1973) and quite detailed information about the interlayer regions has recently emerged for the two-sheet hydrates (Alcover and Gatineau 1980a; Fornes *et al.* 1980) and for the single- and zero-sheet hydrates (Alcover and Gatineau 1980b; Rausell-Colom *et al.* 1980).

The two-sheet hydrate of Mg-vermiculite has a well-ordered superstructure with a $3 \times a$ repeat unit, the water molecules coordinating the cation in inner and outer hydration shells. The Ba two-sheet hydrate also has a superstructure but with a $2 \times a$ repeat unit, but the two-sheet hydrates of Ca- and Sr-vermiculite are much less ordered and it was not possible to calculate H_2O–H_2O distances (Alcover and Gatineau 1980c).

In the one-sheet hydrates, Ba is displaced 1 Å from the central interlamellar plane, interacting with the structural OH groups and surrounded by five to six water molecules. Mg, Ca and Sr, however, lie on the central plane surrounded by three to four water molecules and do not perturb the structural OH vibrations (Rausell-Colom *et al.* 1980).

Although no relative pressure stability domains were established in these studies, the transformations from two-sheet to one-sheet hydrates were achieved by heating to 140 °C for Mg and Ni and to 90 °C for Ca and Sr; for Ba, the one-sheet hydrate is stable at ambient conditions (room temperature and 50% relative humidity). After intense desiccation, most cation forms of vermiculite are close to the dehydrated state, but Li-vermiculite retains much more water than the others. There is considerable evidence that because Li^+ has an unhydrated radius of only 0.6 Å, $Li + H_2O$ together occupy the cavities formed between layers by the coincidence of the quasi-hexagonal basal siloxane groups.

The phase changes between several Mg-vermiculite hydrates can be observed quite easily in macroscopic flakes, using a microscope. When a partly dried flake is allowed to rehydrate at atmospheric humidity, $\simeq 50\%$, a dark line is observed to enter at the edge, marking the boundary between the 11.59 Å and 13.82 Å phases, moving towards the centre of the flake as water diffuses through the interlayers (Walker 1956). This seems to indicate that the different hydrates of vermiculite can be regarded as separate phases in the thermodynamic sense and that at a given temperature, two hydrates can only coexist at one relative pressure. It should therefore be possible to map stability fields for the various hydrates in P/T space, and this is supported by detailed data for Na- and Mg-Llano vermiculite (van Olphen 1965, 1969), which exhibit transitions (steps) at definite partial pressures (Fig. 5.8).

How far particle size is itself a factor has yet to be established. The observation that water diffuses through macroscopic flakes at a relatively slow rate shows that particle size will affect the approach to equilibrium, but it is not known whether the position of equilibrium, that is, the partial pressure at which the transition occurs, is influenced by particle size. In two other related phenomena, the exchange of K from micas (Norrish 1973b), and the intercalation of kaolinites (Weiss *et al.* 1970), particle size has been shown to alter the ease with which the reaction occurs, and the hydration of vermiculite may be similarly affected.

Because, in vermiculites, the sorption of water and its interaction with the surface is dominated by interlayer cation hydration and expansion, analysis in terms of BET multilayer theory is inappropriate. Most of the surface is interlamellar, and the maximum sorption is one sheet on each interlamellar surface (two-sheet complex), and for this type of sorption the Langmuir equation is more suitable because it reaches a plateau maximum asymptotically. van Olphen (1965) found that the Langmuir equation adequately described each of the hydration steps in Na-vermiculite, where the stability fields are sufficiently separated for each to be considered independent.

The adsorption isotherm for Na-vermiculite does not follow the desorption isotherm but is displaced to higher relative pressures. It is believed that this is because the introduction of water into an established structure requires an activation energy to open the layer stacks. Once initial

TABLE 5.5. Heats of immersion, Q (J g^{-1}), and of adsorption per mole H$_2$O, H (J mol^{-1}), of Na- and Mg-Llano vermiculite, pre-equilibrated at various humidities, R (van Olphen 1965, 1969)

R	x†	$d(001)$‡	Q	H
Na-vermiculite				
0	0	9.8	132.3 >	1.56×10^4
0.082	19.0	11.8	115.8 >	1.58×10^4
0.356	84.1	11.8	58.8 >	0.78×10^4
0.808	220.1	14.8	0	
Mg-vermiculite				
0	0	9.3	232.5 >	13.53×10^4
0.004	1.0	11.6	225.3 >	3.34×10^4
0.018	62.6	11.6	110.9 >	4.25×10^4
0.034	83.8	11.6	90.0 >	1.47×10^4
0.356	188.5	14.3	4.64 >	0.29×10^4
0.808	207.3	14.3	1.65	

† x = means of water sorbed, mg g^{-1}.
‡ Basal spacing, Å.

penetration is achieved, energy is stored in the deformation of the aluminosilicate layers so that water may then diffuse more deeply into the structure. The hysteresis between the adsorption and desorption cycles is much less for Mg-vermiculite.

5.7.3.1 Heats of immersion
When these are measured after pre-equilibration at a range of humidities, the heats of formation at each stage during hydration can readily be calculated. Table 5.5 shows a comparison of these measurements for the various stages in the hydration of Na- and Mg-vermiculite (van Olphen, 1965, 1969). The total heat of immersion for Mg-vermiculite is about twice that for Na-vermiculite, which, allowing for there being twice as many interlayer cations in Na- as in Mg-vermiculite, is in the same proportion as the hydration energies of the cations (Norrish 1954). When the data are examined step by step through the stages of hydration, a very significant feature is the large heat evolution for the initial hydration of the Mg-form, nearly ten times that of the Na-form. The heat of immersion and isotherm data suggest that the integral heats of sorption per mole of water are constant during the formation of each layer, the heat of sorption for the first layer being about twice that for the second layer in both cases.

5.7.3.2 Infrared studies
On the basis of the pleochroism of the OH and OD absorptions for the hydration molecules, Farmer and Russell (1971) found two main orientations for the OH bonds. The higher frequency component corresponds with hydrogen bonds to oxygen of Si—O—Si linkages, and the lower frequency components to water–water bonds and to hydrogen bonds to oxygens of Al—O—Si linkages. The more pleochroic high frequency absorption is associated with OH nearly perpendicular to the layer plane, whereas the OH giving the lower frequency absorption is more nearly parallel to the layer plane.

5.7.3.3 Nuclear magnetic resonance
Because vermiculite has only a small external surface, it is an ideal material for studying the dynamic structure of interlayer water. Early work was done by Graham *et al.* (1964) and by Boss

and Stejskal (1965), and more recently, studies by Hougardy et al. (1976, 1977) have revealed further details. One difficulty with all NMR work is that iron impurity makes a significant contribution to the spin–lattice relaxation time T_1. The approach is therefore to study vermiculite with the lowest iron content (e.g. Llano vermiculite, with 1700 p.p.m. Fe) and to make some estimate of the contribution from the paramagnetic impurities (Fripiat et al. 1982). Deuteron and proton spectra of oriented Na-vermiculite show doublets characteristic of preferred orientation of the water molecules. Analysis of the line shape suggests that in an octahedral arrangement of H_2O about the Na ions, each water molecule rotates about its coordination bond to the cation (C_2 axis) and that the entire coordination shell itself rotates about an axis normal to the layer plane. If these two rotations were independent, Giese and Fripiat (1979) calculated that the activation energy should be about $210\,kJ\,mol^{-1}$, but that coupling of the two rotations could decrease the activation energy below $80\,kJ\,mol^{-1}$, which is in better agreement with the estimate from the T_1 measurement of $40\,kJ\,mol^{-1}$. With Mg-vermiculite, the average occupancy of the interlayer cation sites is low, and Giese and Fripiat (1979) suggest that fast averaging motion allows the Mg^{2+} ion to be shared between the two octahedral water shells. The spectra of the one-sheet complex of Na-vermiculite and the two-sheet complex of Li-vermiculite have not so far been interpreted in detail (Hougardy et al. 1977).

5.7.3.4 Neutron scattering
Diffraction study of a Co(II)-exchanged vermiculite with a basal spacing of 14.25 Å (Adams and Riekel 1980) has shown that the structure of the interlayer water is similar to that proposed by Mathieson and Walker (1954). One hydrogen atom of every water molecule forms a hydrogen bond to another water oxygen in the same sheet, the remaining hydrogen atom forming a hydrogen bond to an oxygen of one of the adjacent silicon tetrahedra.

Quasi-elastic neutron scattering from Ca-vermiculite (and smectites, see later) shows that in general a simple two-dimensional diffusion parallel to the interlayer plane can be eliminated because momentum transfer is not anisotropic parallel and perpendicular to the layer plane (Hall 1981). The observations suggest a spatially bounded jump diffusion process, and the physical interpretation of this is as follows. Water molecules outside the hydration shells undergo rapid jumps within a cage bounded by the silicate surfaces and the hydrated cations; slower- and longer-range motions are interpreted as more occasional jumps between cages. For the Ca^{2+}-two sheet hydrate of vermiculite, both correlation times are about 50% longer than for montmorillonite, and this is accounted for by the greater restrictions to movement caused by the larger density of interlayer cations in the higher charged vermiculite

5.7.4 Montmorillonites and beidellites

5.7.4.1 Sorption isotherms
Probably because of their capacity for taking up large quantities of water and swelling, sorption isotherms for many cation forms of different smectites have been published; the references are summarized in Table 5.6. Although the same general pattern of behaviour has been found, there are some significant differences in detail. Desorption isotherms for Na- and Ca-Wyoming bentonite from three publications are compared in Fig. 5.9 showing that up to 20% difference may be reported in the absolute amount of water sorbed, even for a smectite from a reputedly common origin. Sources of difference may be: (i) preparation of the sample, including the size fraction used; (ii) reference weight of dry clay; (iii) hysteresis in the adsorption–desorption cycle;

TABLE 5.6. Hydration measurements on homoionic dioctahedral smectites

Authors	Measurements	Mineral and cation form
Hendricks et al. (1940)	Isotherms, $d(001)$, DTA	Laurel, Mississippi, Otay, California, Wyoming smectites: Na, Ca, Mg, Sr, Ba, Li, Na, H, K, Cs
Mooney et al. (1952)	Isotherms, $d(001)$	Wyoming montmorillonite: Li, Na, K, Rb, Cs, Mg, Ca, Sr, Ba
Orchiston (1955)	Isotherms	Arizona montmorillonite: H, K, Na, Ca, Mg
Glaeser and Mering (1968)	Isotherms, $d(001)$	Camp Berteau montmorillonite, Unterrupsroth beidellite, hectorite: Na, Ca
Kijne (1969)	Isotherms, heats of wetting	Redhill montmorillonite: Na, Li, NH_4, Cs, Rb, Ca
Roderick et al. (1969)	Isotherms, $d(001)$ bulk swelling	Wyoming montmorillonite: Na, Ca
Tarasevich et al. (1971)	Isotherms, $d(001)$	Pizhevski montmorillonite: Na, Ca, Ba
Keren and Shainberg (1975)	Isotherms, $d(001)$ heats of immersion	Wyoming montmorillonite: Na, Ca

(iv) effect of surface contaminants on swelling, particularly in the higher vapour pressure region; (v) cation form of the smectite.

For the isotherms in Fig. 5.9, sources (iii) and (v) should not have been important, because the appropriate cation forms were compared for the *desorption* cycles, and if the samples were of common origin, (iv) should have been the same for both. The most likely sources are therefore (i) and (ii), the latter being the most troublesome. Clays in general and swelling minerals in particular are notoriously difficult to dehydrate completely, residual water often being trapped in the interlayer and not released under vacuum unless heated. In some instances, quite high temperatures are needed (e.g. 700 °C for vermiculite) and once dehydrated, these structures may be difficult to rehydrate (Walker 1961). Even after vacuum drying for one month over P_2O_5, Ca-Redhill montmorillonite still gave a 12 Å basal reflection (Ormerod and Newman 1983); after long heating at 250 °C halloysite still retained 7% of the original hydration water (Cruz et al. 1978).

It seems, therefore, that the only simple way of establishing the anhydrous weight of clay is by heating to a high temperature, e.g. 950 °C, and making an appropriate correction for the weight loss due to dehydroxylation. This, of course, cannot be done until after the isotherms have been measured, or alternatively, it can be done on a sub-sample of clay pre-equilibrated at a known relative humidity.

As smectites have a much larger external (non-interlamellar) surface than vermiculites, an analysis of the isotherm with the Langmuir equation would not be appropriate, because a significant proportion of water sorption would be on non-interlayer surfaces where multilayers could develop. Likewise, the BET equation would also be inappropriate because the interlayers

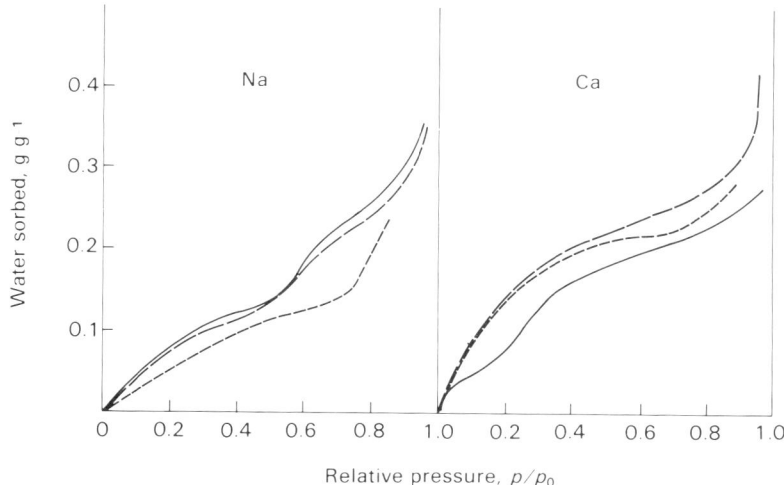

FIG. 5.9. Differences between isotherms reported for desorption of water from Na- and Ca-Wyoming montmorillonite.

sorb water in a step-like manner, and multilayers cannot develop freely. Not only does the BET equation fit isotherms like those in Fig. 5.9 very poorly but also the "monolayer" coverage so calculated is half or less of the total surface (Mooney et al. 1952). This is because over the low vapour pressure region of the isotherm where the BET equation is usually valid, smectite is forming a single sheet complex, effectively a monolayer shared between two surfaces. Only at higher vapour pressures and outside the normal range of the BET equation does the double sheet complex develop. Therefore the simple application of the BET equation for estimating the specific surface of smectites is not recommended.

However, a type of hybrid and empirical analysis may be possible for the cation forms that are stable over a suitable range of humidity, e.g. Mg- or Ca-smectites, in which the double sheet structure remains nearly unchanged from a low to a high humidity. Provided correction is made for such small change as does occur in the interlayer, the observed water sorption should in theory reflect the development of multilayers on the non-lamellar surfaces. This approach was tested by Ormerod and Newman (1983) for Ca-Redhill montmorillonite, by measuring water sorption and basal spacings over the relative pressure range 0.35–0.85. After a correction for change in interlayer volume, the sorption isotherm was similar to isotherms tabulated by Hagymassy et al. (1969) for oxide surfaces, enabling an estimate of the non-interlamellar surface to be made.

5.7.4.2 Heats of immersion

Even more than in isotherm studies, the drying procedure used before measuring heats of immersion is critical to the results obtained. This is because the first addition of water is the most energetically sorbed as was noted in the example of Mg-vermiculite. Thus, Slabaugh (1959) showed that oven-drying above 30 °C doubled the heat of immersion of Ca-Wyoming bentonite to 92 J g^{-1}. Anderson and Sposito (1964) found that when Cheto-montmorillonite was oven-dried the heat of immersion was 96 J g^{-1} but after drying at 100 °C under vacuum, it was increased to 147 J g^{-1}. Keren and Shainberg (1975) note that Ca-Wyoming bentonite dried for a week at 25 °C over P$_2$O$_5$ still contained 3.0% water compared with oven-dried clay, but that degassing at 1.4×10^{-3} Pa (1×10^{-5} mmHg) at 25 °C was as effective as oven-drying. They obtained a heat of wetting for Ca-Wyoming bentonite of 92 J g^{-1}, which agrees with the other results for samples dried at 100 °C and atmospheric pressure. For vacuum-degassed Na-Wyoming and Redhill montmorillonites, and Na-Unterrupsroth beidellite, Greene-Kelly (1962) measured heats of immersion of 68, 73 and 93 J g^{-1} respectively, rather larger than the value of 50 J g^{-1} obtained by Slabaugh (1959) and Keren and Shainberg (1975) for Na-Wyoming bentonite.

In any detailed comparison of heats of immersion, therefore, it is important to take experimental details into consideration before drawing conclusions. Nevertheless, it seems certain that heats of immersion for Na-montmorillonites are approximately half that for Na-vermiculite, 50–73 J g^{-1}, compared with 132 J g^{-1}, and this difference appears to reflect the relative charge densities and numbers of counter-ions at the surface of the minerals: 0.12 C m^{-2} for montmorillonite, compared with 0.25 C m^{-2} for vermiculite, with beidellite intermediate at 0.17 C m^{-2}. An additional difference is that the integral heat of immersion is not constant during the formation of the single and double sheet complexes as in vermiculite (van Olphen 1965) but decreases as the complex develops (Keren and Shainberg 1975).

5.7.4.3 Structure
Compared with vermiculite, the information about the interlamellar structures of smectites obtained by diffraction studies is much more limited. Generally, the layer stacking in montmorillonite is disordered with little evidence of detail in the *hk* bands, so that the amount of structural information obtained from the diffraction pattern is limited. Hawkins and Egelstaff (1980) used neutron diffraction to study deuterated Na-Wyoming montmorillonite at several deuteration stages, and concluded, as did Pezerat and Mering (1967) from X-ray diffraction examination, that the relative lack of structure in the interlayer region showed that interlayer water was not highly ordered into planes of atoms. They showed that the observed pattern of *hk* reflections could be nearly matched by linear combinations of the patterns from dry clay and from D$_2$O, but that small deviations of peak positions indicated that at least some of the atoms in interlayer water molecules are spatially correlated with the aluminosilicate sheets above and below.

By analogy with the concepts of Eisenberg and Kauzmann (1969) for pure water, Sposito and Prost (1982) distinguished between the vibrationally averaged structure (V-structure) and the diffusionally averaged structure (D-structure) of water in smectites. On a time scale that is long relative to the vibration of a hydrogen bond in liquid water, about 10^{-13} s, but short relative to the time for a water molecule to diffuse a distance equal to its own diameter, about 10^{-11} s, the structure perceived by a molecular observer is the V-structure, as detected by the infrared, ESR and incoherent neutron scattering techniques. At the other extreme, observations made on a time scale that is long by comparison with diffusion of water molecules perceives the D-structure, examples being X-ray and neutron diffraction and thermodynamic measurements. Between these time domains, techniques such as NMR and dielectric relaxation probe the motions of rotation and translation whereby the D-structure is formed from the V-structure. The picture, therefore, for water in smectites is a dynamic one, more liquid-like than in the instance of vermiculite, and summarized as follows.

On a short time scale, cations are fixed hydrophilic sites on the surface and in the interlamellar space between the aluminosilicate layers, and by solvation organize the V-structure of the adsorbed water. The water molecules themselves rotate about the coordinating link to the cation and the whole hydration shell rotates around the cation. This picture is very similar to that described above for vermiculite, but there are quantitative differences. Incoherent neutron scattering measurements have been interpreted (Hall 1981) as showing that for the short-range motions in the two-sheet complex of Ca-montmorillonite, the mean jump lengths are 1.7 Å and the residence time 1.0×10^{-11} s, and for the longer range motion, the mean jump length and residence time are 5.0 Å and 1×10^{-10} s. The cage size, bounded by the hydrated cation and the aluminosilicate layers, is about 5–6 Å in diameter, this size being in agreement with the average space between cations. For Ca-vermiculite, both residence times are approximately 50% longer, probably because the denser population of cations in the interlayer space causes greater restriction on movement. In bulk water the mean correlation time is about 10^{-12} s so there is clearly a substantial restriction to movement in the interlayer space.

Infrared and NMR measurements of pleochroism in the absorption spectra show that despite the dynamic general picture, the water molecules maintain some preferred orientation relative to the aluminosilicate layers. If the deficit of cationic charge in the layers originates in the octahedral sheet, this becomes distributed over the whole of the siloxane basal oxygen atoms so that OH orientation is general rather than localized, but substitution of Si by Al in the tetrahedral sheet focuses the charge deficit on particular oxygen atoms and directs hydrogen bonding towards these atoms (Farmer and Russell 1971).

5.7.5 Other smectites

5.7.5.1 Hectorites

As mentioned earlier, NMR results become more difficult to interpret when iron is present and Fripiat *et al.* (1982) have investigated the interaction of water with hectorite and laponite, a synthetic hectorite-like clay (Neumann 1965), both of which contain only a little Fe. Water sorption, heats of immersion and NMR relaxation of ^{1}H and ^{2}H in D_2O and H_2O were measured. The prime objective of the work was to evaluate the range of the surface interaction into the water films on the clay surfaces. The exchange capacity of hectorite is 0.44 meq g^{-1} and its N_2 specific surface 63 m^2 g^{-1}, whereas laponite has an exchange capacity of 0.783 meq g^{-1} and an N_2 surface of 360 m^2 g^{-1} (van Olphen and Fripiat 1979). Laponite consists of very fine needle-shaped particles of 10–100 nm in size along the axis and possibly having one dimension as small as 1 nm. Both water isotherms (Fig. 5.10) fit the BET equation well, giving monolayer capacities of 22 mg g^{-1} for hectorite and 79 mg g^{-1} for laponite; the data suggest that interlamellar volume in laponite is much smaller than in hectorite. Heats of immersion were 100 J g^{-1} for laponite and 75 J g^{-1} hectorite; when three to four statistical layers of water was preadsorbed before immersion, the heat of immersion was not significantly different from the internal energy of water in contact with its vapour. This was taken to indicate that the influence of the surface did not extend beyond the third or fourth layer and was confirmed by the NMR results.

5.7.5.2 Saponites

With the availability of synthetic saponites having a range of compositions (Suquet *et al.* 1977), it has been possible to investigate the effect of substitution and charge on the hydration states and unit cell parameters (Suquet *et al.* 1981). Many saponites show a considerable degree of stacking order (Suquet *et al.* 1975) so that the *b*-parameter can be calculated satisfactorily from the 060, 331 reflection. It was found that in the two-water sheet complexes, the *b*-parameter is related to within-layer interactions between the tetrahedral and octahedral sheets, but that in

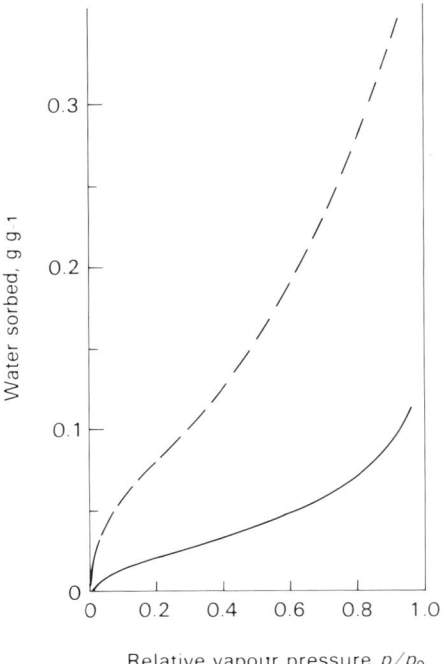

FIG. 5.10. Water sorption isotherms for hectorite and laponite; hectorite: ———; laponite: — — —.

the one-water sheet complexes, the b-parameter also depended on the nature of the interlayer space. It was also shown that there is a continuous increase in basal spacing as the layer charge x decreases from 2 per $O_{20}(OH)_4$ to 0.66; for example, for Ca-saponites the basal spacing increases from 14.68 Å for $x=2$ to 15.30 Å for $x=0.66$. Although this range of minerals has not yet been investigated for their quantitative interaction with water, the earlier study of the natural saponite from Kozakov (Suquet et al. 1975) shows that well-defined hydration domains may be expected, more like beidellite than like montmorillonite.

5.7.5.3 Mixed-layer minerals

Although minerals with smectite layers interstratified with mica, kaolinite and chlorite layers are common in many sedimentary formations and soils, their interaction with water has been little investigated. A recent study (Branson and Newman 1983) included an illite–smectite (85:15) with IMII ordering, and a soil clay containing illite and illite–smectite, both Ca-saturated, and showed that development of multilayers of water was very similar in pattern to that on reference oxides. There was no evidence of microporosity, which is at variance with the X-ray evidence of interstratification with smectite. Further investigation of such minerals is needed.

5.8 THE SEPIOLITE–PALYGORSKITE GROUP

The minerals in this group occur in a variety of macroscopic forms but are fibrous or lath-like on a microscopic scale. In all the minerals, the fibres are grouped together in bundles which are "frayed" or opened up at the ends (Martin-Vivaldi and Robertson 1971). Although the structures of sepiolite and palygorskite differ, they are sufficiently alike for them to be considered together, with particular differences noted. In both, there are structural units of narrow talc-like ribbons that extend along the fibre-axis and these are linked one to the next by inversion of the SiO_4 tetrahedra at basal oxygens along the edges of the ribbons. The minerals differ in that the ribbon in palygorskite is one Si_6O_6 ring wide but in sepiolite the ribbon width is spanned by $2Si_6O_6$ rings (see Fig. 7.5). The coordination of octahedral cations at each edge of the ribbon is completed by two water molecules. Joining ribbons in this way creates channels between adjoining ribbons and at ambient temperatures and humidities these channels contain water molecules not coordinated to cations. By analogy with the channels in zeolites, this water has been called "zeolitic" water, whereas the water coordinated to the octahedral cations is called "bound" or "crystalline" water, the former term being preferred here. In palygorskite the channel dimensions are 3.7×6.4 Å^2 approximately and in sepiolite 3.7×10.6 Å^2; the total theoretical specific surface including that of the channels is about 900 Å^2.

The sorptive abilities of sepiolite and palygorskite depend greatly on the outgassing pretreatment. Sorption of N_2 is enhanced by removal of water from channels achieved by outgassing *in vacuo* at 70–90 °C (Barrer and Mackenzie 1954; Dandy 1968; Fernandez Alvarez 1978) but the inference from these observations that N_2 penetrates the channels when water is removed is not justified. The maximum specific surface to N_2 found after drying is about 380 m^2 g^{-1} for sepiolite (Dandy 1968) and for palygorskite about 150 m^2 g^{-1} (Fernandez Alvarez 1978), both only fractions of the total theoretical surface, whereas the sepiolite surface available to NH_3 under the same conditions is 610 m^2 g^{-1} (Dandy 1971). Repeated flushing and evacuation of sepiolite with ammonia replaces zeolitic water and ultimately all coordinated water as well (Serna and Van Scoyoc 1979) so that clearly NH_3 must be accessible to the whole surface. It seems probable therefore that N_2 cannot penetrate far into the channel structure and is predominantly sorbed on external surfaces.

If sepiolite and palygorskite are further dehydrated by heating to 300 °C *in vacuo* the structures change as about half of the bound water is driven off, alternate ribbons rotating positively and negatively to close the channels, forming what are known as folded structures (Nagata *et al.* 1974; Serna *et al.* 1975; Van Scoyoc *et al.* 1979; Fig. 7.5). Folding is associated with a considerable decrease in surface accessible to N_2 (Barrer and Mackenzie 1954; Fernandez Alvarez 1978) which is explained by the effect of folding closing up the exposed channels or grooves on the superficial surface in which it is believed some N_2 sorption occurs. Sepiolite and palygorskite dehydrated to this stage can be readily rehydrated by exposure to water vapour.

Further heating to 600 °C causes the loss of the second part of the bound water, and also dehydroxylation in palygorskite (Mifsud *et al.* 1978), rehydration, when it can be achieved, may require hydrothermal conditions of pressure and temperature. On the other hand, Nagata *et al.* (1974) reported that sepiolite could not be rehydrated at all after heating for 120 hours at 600 °C, and suggested that once the second part of the bound water was removed, sepiolite could not be rehydrated.

Sorption of water by partially dehydrated palygorskite (Barrer and Mackenzie 1954) apparently involves two processes: sorption on the external surfaces, which is rapid, and penetration into the channels, which is slow and only becomes evident when some zeolitic water is removed. If the initial degassing temperature was low (e.g. 50 °C) equilibration times were quite fast, but as the degassing temperature was raised, the rate of rehydration slowed, many days being required to reach equilibrium after heating at 215 °C. It seems that slow penetration of the channels may be one reason why low specific surfaces have been calculated from water sorption data (e.g. 275 m^2 g^{-1} for sepiolite: Serna and Van Scoyoc 1979). It should, however, be pointed out that even in untreated sepiolites and palygorskites, the total water content is considerably below that required for a monolayer containing 0.28 mg H$_2$O per m^2 of surface covering 900 m^2 g^{-1}, equivalent to about 26% water. The total water content of sepiolites rarely exceeds 20%, including 2.4 to 4% H$_2$O from dehydroxylation, and that of palygorskites 22% including 2 to 6% from dehydroxylation (Weaver and Pollard 1973); Mifsud et al. (1978) concluded that there was usually a deficit of zeolitic water in palygorskites. Turning these observations round, it is clear that the packing of water on the surfaces of palygorskites and sepiolites is very much less dense than that in the "normal monolayers" found for other layer silicates and that this must be taken into account in calculating apparent specific surfaces from water sorption studies. On balance, it is probably safer not to use water sorption measurements in this way. When, in addition, account is taken of the morphology of the bundles of fibres that make up the macroscopic particle, it is clear that the sorption properties of palygorskites and sepiolites present formidable complexities in interpretation.

REFERENCES

ADAMS J.M. & RIEKEL C. (1980) One-dimensional neutron diffraction study of a vermiculite. *Clays Clay Miner.* **28**, 444–445.

ALCOVER J.F. & GATINEAU L. (1980a) Structure de l'espace interlamellaire de la vermiculite Mg bicouche. *Clay Minerals* **15**, 25–35.

ALCOVER J.F. & GATINEAU L. (1980b) Structure de l'espace interlamellaire des vermiculites Ba monocouches. *Clay Minerals* **15**, 193–203.

ALCOVER J.F. & GATINEAU L. (1980c) Facteurs determinant la structure de la couche interlamellaire des vermiculites saturees par des cation divalents. *Clay Minerals* **15**, 239–248.

ALCOVER J.F., GATINEAU L. & MERING J. (1973) Exchangeable cation distribution in Ni- and Mg-vermiculites. *Clays Clay Miner.* **21**, 131–136.

ANDERSON D.M. & SPOSITO G. (1964) Heat of immersion of Arizona bentonite in water. *Soil Sci.* **97**, 214–219.

ANDERSON D.M., RICE A.R. & BANIN A. (1973) The water–ice phase composition of clay–water systems. I: The kaolinite–water system. *Soil Sci. Soc. Am. Proc.* **37**, 819–822.

BAILEY A. & KAY S.M. (1967) A direct measurement of the influence of vapour, of liquid and of oriented monolayers on the interfacial energy of mica. *Proc. Roy. Soc. London* **A301**, 47–56.

BARRER R.M. & MACKENZIE N. (1954) Sorption by attapulgite. Part I: Availability of intracrystalline channels. *J. phys. Chem., Ithaca* **58**, 560–568.

Boss B.D. & STEJSKAL E.O. (1965) Anisotropic diffusion in hydrated vermiculite. *J. chem. Phys.* **43**, 1068–1069.

BOWER C.A. & GOERTZEN J.O. (1959) Surface areas of clays by an equilibrium ethylene glycol method. *Soil Sci.* **87**, 289–292.

BRANSON K. & NEWMAN A.C.D. (1983) Water sorption on Ca-saturated clays. I: Multilayer sorption and microporosity in some illites. *Clay Minerals* **18**, 277–287.

BRINDLEY G.W. (1980) Order–disorder in clay mineral structures. In *Minerals and their X-ray Identification* (G.W. Brindley and G. Brown, eds), pp. 125–195. Mineralogical Society, London.

BRINDLEY G.W. & ERTEM G. (1971) Preparation and solvation properties of some variable charge montmorillonites. *Clays Clay Miner.* **19**, 399–404.

BRINDLEY G.W. & GOODYEAR J. (1948) X-ray studies of halloysite and metahalloysite. II: The transition of halloysite to metahalloysite in relation to relative humidity. *Mineralog. Mag.* **28**, 407–422.

BROWN G. & NEWMAN A.C.D. (1973) The reactions of soluble aluminium with montmorillonite. *J Soil Sci.* **24**, 339–354.

BRYDON J.E. & TURNER R.C. (1972) The nature of Kenya vermiculite and its aluminium hydroxide complexes. *Clays Clay Miner.* **20**, 1–11.

CAIRNS-SMITH A.G. (1971) *The Life Puzzle on Crystals and Organisms and the Possibility of a Crystal as an Ancestor*. Ch. 7, p. 131. Oliver and Boyd, Edinburgh.

CALVET R. (1975) Dielectric properties of montmorillonites saturated by bivalent cations. *Clays Clay Miner.* **23**, 257–265.

CEBULA D.J., THOMAS R.K. & WHITE J.W. (1980) Small angle neutron scattering from dilute aqueous dispersions of clay. *J. Chem. Soc., Faraday Trans. I* **76**, 314–321.

CHURCHMAN G.J. & CARR R.M. (1973) Dehydration of the washed potassium acetate complex of halloysite. *Clays Clay Miner.* **21**, 423–424.

CHURCHMAN G.J., ALDRIDGE L.P. & CARR R.M. (1972) The relationship between the hydrated and the dehydrated states of halloysite. *Clays Clay Miner.* **20**, 241–246.

COLOMBERA P.M., POSNER A.M. & QUIRK J.P. (1971) The adsorption of aluminium from hydroxyaluminium solutions on to Fithian illite. *J. Soil Sci.* **22**, 118–128.

CRUZ M.I., LETELLIER M. & FRIPIAT J.J. (1978) NMR study of adsorbed water. II: Molecular motions in the monolayer hydrate of halloysite. *J. chem. Phys.* **68**, 2018–2027.

DANDY A.J. (1968) Sorption of vapors by sepiolite. *J. phys. Chem., Ithaca* **72**, 334–339.

DANDY A.J. (1971) Zeolitic water content and adsorptive capacity for ammonia of microporous sepiolite. *J. chem. Soc.* (A) 2383–2387.

DEL PENNINO U., MAZZEGA E., VALERI S., ALIETTI A., BRIGATTI M.F. & POPPI L. (1981) Interlayer water and swelling properties of monoionic montmorillonites. *J. colloid interface Sci.* **84**, 301–309.

DIAMOND S. & KINTER E.B. (1958) Surface areas of clay minerals as derived from measurements of glycerol retention. *Clays Clay Miner.* **5**, 334–347.

DYAL R.S. & HENDRICKS S.B. (1950) Total surface of clays in polar liquids as a characteristic index. *Soil Sci.* **62**, 421–432.

EISENBERG D. & KAUZMANN W. (1969) *The Structure and Properties of Water*. Clarendon Press, Oxford.

ELTANTAWY I.M. & ARNOLD P.W. (1974) Ethylene glycol sorption by homoionic montmorillonites. *J. Soil Sci.* **25**, 99–110.

FARMER V.C. (1974) The layer silicates. Ch. 15 in *The Infrared Spectra of Minerals* (V.C. Farmer, ed.), pp. 331–363. Mineralogical Society, London.

FARMER V.C. (1978) Water on particle surfaces. Ch. 6 in *The Chemistry of Soil Constituents* (D.J. Greenland and M.H.B. Hayes, eds), pp. 405–448. John Wiley, Chichester.

FARMER V.C. & RUSSELL J.D. (1971) Interlayer complexes in layer silicates: the structure of water in lamellar ionic solutions. *Trans. Faraday Soc.* **67**, 2737–2749.

FARMER V.C., RUSSELL J.D., McHARDY W.J., NEWMAN A.C.D., AHLRICH J.L. & RIMSAITE J.Y.H. (1971) Evidence for loss of protons and octahedral iron from oxidised biotites and vermiculites. *Mineralog. Mag.* **38**, 121–137.

FERNANDEZ ALVAREZ T. (1978) Effecto de la deshidratacion sobre las propriedades adsorbentes de la palygorskita y sepiolita. I: Adsorcion de nitrogeno. *Clay Minerals* **13**, 325–335.

FORNES V., DE LA CALLE C., SUQUET H. & PEZERAT H. (1980) Etude de la couche interfoliaire des hydrates a deux couches des vermiculites calcique et magnesienne. *Clay Minerals* **15**, 399–411.

FRIPIAT J.J. (1980) The application of NMR to the study of clay minerals. Ch. 5 in *Advanced Chemical Methods for Soil and Clay Minerals Research* (J.W. Stucki and W.L. Banwart, eds), pp. 245–315. D. Reidel Publishing, Dordrecht, Holland.

FRIPIAT J., CASES J., FRANCOIS M. & LETELLIER M. (1982) Thermodynamic and microdynamic behaviour of water in clay suspensions and gels. *J. Colloid Interface Sci.* **89**, 374–377.

GARRETT W.G. & WALKER G.F. (1959) The cation-exchange capacity of hydrated halloysite and the formation of halloysite–salt complexes. *Clay Miner. Bull.* **4**, 75–80.

GIESE R.F. & FRIPIAT J.J. (1979) Water molecule positions, orientations, and motions in the dihydrates of Mg- and Na-vermiculites. *J. Colloid Interface Sci.* **71**, 441–450.

GLAESER R. & MERING J. (1968) Domaines d'hydration homogene des smectites. *C.r. hebd. Séanc. Acad. Sci. Paris*, Serie D, **267**, 463–467.

GRAHAM J. (1964) Adsorbed water on clays. *Rev. Pure and Appl. Chem.* **14**, 81–90.

GRAHAM J., WALKER G.F. & WEST G.W. (1964) Nuclear magnetic resonance study of interlayer water in hydrated layer silicates. *J. chem. Phys.* **40**, 540–550.

GREENE-KELLY R. (1952) A test for montmorillonite. *Nature, Lond.* **170**, 1130.

GREENE-KELLY R. (1955) Lithium adsorption by kaolin minerals. *J. phys. Chem., Ithaca* **59**, 1151–1152.

GREENE-KELLY R. (1962) Charge densities and heats of immersion of some clay minerals. *Clay Miner. Bull.* **5**, 1–8.

GREENE-KELLY R. (1964) The specific surface areas of montmorillonites. *Clay Miner. Bull.*, **5**, 392–400.

GREENLAND D.J. & QUIRK J.P. (1964) Determination of the total specific surface areas of soils by the adsorption of cetyl pyridinium bromide. *J. Soil Sci.* **15**, 178–91.

GREGG S.J. & SING K.S.W. (1982) *Adsorption, Surface Area and Porosity*. Academic Press, London.

HAGYMASSY J., BRUNAUER S. & MIKHAIL R.S. (1969) Pore structure analysis by water vapour adsorption. I: t-curves for water vapour. *J. Colloid Interface Sci.* **29**, 485–491.

HALL P.L. (1981) Neutron scattering techniques for the study of clay minerals. Ch. 3. *Advanced Techniques for Clay Mineral Analysis* (J.J. Fripiat, ed.), pp. 51–75. Elsevier, Amsterdam.

HAWKINS R.H. & EGELSTAFF P.A. (1980) Interfacial water structure in montmorillonite from neutron diffraction experiments. *Clays Clay Miner.* **28**, 19–28.

HAYNES J.M. (1961) The specific surface of clays. *Trans. Brit. Ceram. Soc.* **60**, 691–707.

HEILMAN M.D., CARTER D.L. & GONZALEZ C.L. (1965) The ethylene glycol monoethyl ether (EGME) technique for determining soil-surface area. *Soil Sci.* **100**, 409–413.

HENDRICKS S.B. & JEFFERSON M.E. (1938) Structure of kaolin and talc–pyrophyllite hydrates and their bearing on water sorption of clays. *Am. Miner.* **23**, 863–875.

HENDRICKS S.B., NELSON R.A. & ALEXANDER L.T. (1940) Hydration mechanism of the clay mineral montmorillonite saturated with various cations. *J. Am. chem. Soc.* **62**, 1457–1464.

HOFMANN U. & KLEMEN E. (1950) Loss of exchangeability of lithium ions in bentonite on heating. *Z. anorg. allg. Chem.* **262**, 95–99.

HOMSHAW L.G. (1981) Supercooling and pore size distribution in water saturated porous materials: application to the study of pore form. *J. Colloid Interface Sci.* **84**, 141–148.

HOMSHAW L.G. & CAMBIER P. (1980) Wet and dry pore size distribution in a kaolinitic soil before and after removal of iron and quartz. *J. Soil Sci.* **31**, 415–428.

HOMSHAW L.G. & CHAUSSIDON J. (1978) Pore size distribution in water-saturated calcium montmorillonite using low-temperature heat-flow scanning calorimetery. *Proc. Int. Clay Conf. 1978, Oxford*, pp. 141–151. Elsevier, Amsterdam.

HOMSHAW L.G. & QUIRK J.P. (1980) Sur l'etude par calorimetrie a basse temperature du comportement de l'eau au voisinage de la surface d'une illite. *C.r. hebd. Séanc. Acad. Sci., Paris* **291B**, 77–79.

HOUGARDY J., STONE W.E.E. & FRIPIAT J.J. (1976) NMR study of adsorbed water. I: Molecular orientation and protonic motions in the two-layer hydrate of a Na vermiculite. *J. chem. Phys.* **64**, 3840–3851.

HOUGARDY J., STONE W.E.E. & FRIPIAT J.J. (1977) Complex proton NMR spectra in some ordered hydrates of vermiculites. *J. magn. Reson.* **25**, 563–567.

HOWER J. & MOWATT T.C. (1966) The mineralogy of illites and mixed-layer illite/montmorillonites. *Am. Miner.* **51**, 825–54.

HUANG S-D., PULKRABEK P. & KLIER K. (1978) Surface chemistry of mica–aluminium–phosphate and nucleotide systems. *J. Colloid Interface Sci.* **65**, 583–586.

HUGHES I.R. (1966) Mineral changes of halloysite on drying. *N.Z. Jl Sci.* **9**, 103–113.

JONAS J., BROWN D. & FRIPIAT J.J. (1982) NMR study of kaolinite-water system at high presure. *J. Colloid Interface Sci.* **89**, 374–377.

JONES A.A. (1981) Charges on the surfaces of two chlorites. *Clay Minerals* **16**, 347–359.

JURINAK J.J. (1961) The effect of pretreatment on the adsorption and desorption of water vapour by lithium and calcium kaolinite. *J. phys. Chem., Ithaca* **65**, 62–64.

JURINAK J.J. (1963) Multilayer adsorption of water by kaolinite. *Soil Sci. Soc. Am. Proc.* **27**, 269–272.

KEENAN A.G., MOONEY R.W. & WOOD L.A. (1951) The relation between exchangeable ions and water adsorbed on kaolinite. *J. phys. Chem., Ithaca* **55**, 1462–1474.

KEREN R. & SHAINBERG I. (1975) Water vapor isotherms and heat of immersion of Na/Ca-montmorillonite systems. I: Homoionic clay. *Clays Clay Miner.* **23**, 193–200.

KIDDER G. & REED L.W. (1972) Swelling characteristics of hydroxy-aluminium interlayered clays. *Clays Clay Miner.* **20**, 13–20.

KIJNE J.W. (1969) On the interaction of water molecules and montmorillonite surfaces. *Soil Sci. Soc. Am. Proc.* **33**, 539–543.

KIRKMAN J.H. (1977) Possible structure of halloysite disks and cylinders observed in some New Zealand rhyolitic tephras. *Clay Minerals* **12**, 199–216.

KNUDSON M.I. & MCATEE J.L. (1974) Interlamellar and multilayer nitrogen sorption by homoionic montmorillonites. *Clays Clay Miner.* **22**, 59–65.

LAGALY, G. (1981) Characterization of clays by organic compounds. *Clay Minerals* **16**, 1–21.

LLOYD M.K. & CONLEY R.F. (1970) Adsorption studies on kaolinites. *Clays Clay Miner.* **18**, 37–46.

LOW P.F. (1982) Water in clay–water systems. *Agronomie* **2**, 909–914.

MACEWAN D.M.C. & WILSON M.J. (1980) Interlayer and intercalation complexes of clay minerals. Ch. 3 in *Crystal Structure of Clay Minerals and their X-ray Identification* (G.W. Brindley and G. Brown, eds), pp. 197–248. Mineralogical Society, London.

MACKENZIE R.C. (ed.) (1957) *The Differential Thermal Investigation of Clays*. Mineralogical Society, London.

MACKENZIE R.C. (ed.) (1970) *Differential Thermal Analysis*. Academic Press, London.

MACKENZIE R.C. (1981) Thermoanalytical methods in clay studies. Ch. 1 in *Advanced Techniques for Clay Mineral Analysis* (J.J. Fripiat, ed.), pp. 5–29. Elsevier, Amsterdam.

MACKENZIE R.C. & CAILLERE S. (1975) The thermal characteristics of soil minerals and the use of these characteristics in the qualitative and quantitative determination of clay minerals in soils. Ch. 16 in *Soil Components*. Vol. 2. *Inorganic Components* (J.E. Giesking, ed.), pp. 529–571. Springer-Verlag, New York.

MAMY J. (1973) Relations entre les etats de l'eau adsorbee par les micas alteres et leurs proprietes dielectrique, *Proc. Int. Clay Conf. 1972, Madrid*, pp. 509–517. Division de Ciencias, Madrid.

MARTIN R.T. (1959) Water vapor sorption on kaolinite: hysteresis. *Clays Clay Miner.* **7**, 259–277.

MARTIN R.T. (1955) Ethylene glycol retention by clays. *Soil Sci. Soc. Am. Proc.* **19**, 160–164.

MARTIN R.T. (1958) Water vapor sorption on lithium kaolinite. *Clays Clay Miner.*, **5**, 23–38.

MARTIN R.T. (1962) Adsorbed water on clay: a review. *Clays Clay Miner.* **9**, 28–70.

MARTIN-VIVALDI, J.L. & ROBERTSON R.H.S. (1971) Palygorskite and sepiolite (the hormites). Ch. 8 in *The Electron-Optical Investigation of Clays* (J.A. Gard, ed.), pp. 255–275. Mineralogical Society, London.

MATHIESON A.McL. (1958) Mg-vermiculite: a refinement and re-examination of the structures of the 14.36 Å phase. *Am. Miner.* **43**, 216–217.

MATHIESON A.McL. & WALKER G.F. (1954) Crystal structure of Mg-vermiculite. *Am. Miner.* **39**, 231–255.
MIFSUD A., RAUTUREAU M. & FORNES V. (1978) Etude de l'eau dans la palygorskite a l'aide des analyses thermiques. *Clay Minerals* **13**, 357–374.
MIKHAIL R.S., GUINDY N.M. & HANAFI S. (1979) Vapor adsorption on expanding and non-expanding clay minerals. *J. Colloid Interface Sci.* **70**, 282–292.
MOONEY R.W., KEENAN A.C. & WOOD L.A. (1952) Adsorption of water by montmorillonite. *J. Am. chem. Soc.* **74**, 1367–1374.
MUIR J. (1954) Dielectric loss in water films adsorbed by some silicate clay minerals. *Trans. Faraday Soc.* **50**, 249–254.
NAGASAWA K., BROWN G. & NEWMAN A.C.D. (1974) Artificial alteration of biotite into a 14 Å silicate with hydroxy-aluminium interlayers. *Clays Clay Miner.* **22**, 241–252.
NAGASAWA K. & MIYAZI S. (1976) Mineralogical properties of halloysite as related to its genesis. *Proc. Int. Clay Conf. 1975, Mexico*, pp. 257–265. Applied Publishing Ltd, Wilmette, Illinois, USA.
NAGATA H., SHIMODA S. & SUDO T. (1974) On dehydration of bound water of sepiolite. *Clays Clay Miner.* **22**, 285–293.
NELSON S.M., HUANG H.H. & SUTTON L.E. (1969) Dielectric study of water, ethanol and acetone adsorbed on kaolinite. *Trans. Faraday Soc.* **65**, 225–243.
NEUMANN B. (1965) Behaviour of a synthetic clay in pigment dispersions. *Rheolog. Acta*, **4**, 250–255.
NEWMAN A.C.D. (1955) The electrical properties of adsorbed films. D. Phil. Thesis. Oxford University.
NEWMAN A.C.D. (1967) Changes in phlogopites during their artificial alteration. *Clay Minerals* **7**, 215–227.
NEWMAN A.C.D. & BROWN G. (1966) Chemical changes during the alteration of micas. *Clay Minerals* **6**, 297–309.
NEWMAN A.C.D. & BROWN G. (1969) Delayed exchange of potassium from some edges of mica flakes. *Nature, Lond.* **223**, 175–176.
NORRISH K. (1954) The swelling of montmorillonite. *Disc. Faraday Soc.* **18**, 120–133.
NORRISH K. (1973a) Forces between clay particles. *Proc. Int. Clay Conf. 1972, Madrid*, pp. 375–83. Division de Ciencias, Madrid.
NORRISH K. (1973b) Factors in the weathering of micas to vermiculite. *Proc. Int. Clay Conf. 1972, Madrid*, pp. 417–432. Division de Ciencias, Madrid.
NORRISH K. & RAUSELL-COLOM J.A. (1963) Low-angle X-ray diffraction studies of the swelling of montmorillonite and vermiculite. *Clays Clay Miner.* **10**, 123–149.
OLIVIER J.P. & SENNETT P. (1973) Particle size-shape relationships in Georgia sedimentary kaolins-II. *Clays Clay Miner.* **21**, 403–412.
ORCHISTON H.D. (1954) Adsorption of water vapor. II: Clays at 25°C. *Soil Sci.* **78**, 463–480.
ORCHISTON H.D. (1955) Adsorption of water vapour. III: Homoionic montmorillonites at 25°C. *Soil Sci.* **79**, 71–78.
ORMEROD E.C. & NEWMAN A.C.D. (1983) Water sorption on Ca-saturated clays. II: Internal and external surfaces of montmorillonite. *Clay Minerals* **18**, 289–299.
PASHLEY R.M. (1981a) Hydration forces between mica surfaces in aqueous electrolyte solutions. *J. Colloid Interface Sci.* **80**, 153–162.
PASHLEY R.M. (1981b) DLVO and hydration forces between mica surfaces in Li^+, Na^+, K^+ and Cs^+ electrolyte solutions: a correlation of double-layer and hydration forces with surface cation exchange properties. *J. Colloid Interface Sci.* **83**, 531–546.
PERROTT K.W. (1981) The nature of cationic aluminium species on the cation exchange surface of mica. *J. Colloid Interface Sci.* **82**, 136–140.
PEZERAT H. & MERING J. (1967) Recherches sur la position des cation echangeables et de l'eau dans les montmorillonites. *C.r. hebd. Séanc. Acad. Sci., Paris* **265**, 529–532.
QUIRK J.P. (1955) Significance of surface areas calculated from water vapour isotherms by the use of the B.E.T. equation. *Soil Sci.* **80**, 423–430.
RAUSELL-COLOM J.A., FERNANDEZ M., SERRATOSA J.M., ALCOVER J.F. & GATINEAU L. (1980) Organisation de l'espace interlamellaire dans les vermiculite monocouches et anhydres. *Clay Minerals* **15**, 37–58.
RAWSON R.A.G. (1969) A rapid method for determining the surface areas of aluminosilicates from the adsorption dynamics of ethylene glycol vapour. *J. Soil Sci.*, **20**, 325–335.
RAYNER J.H. (1974) The crystal structure of phlogopite by neutron diffraction. *Mineralog. Mag.* **39**, 850–856.
RICH C.I. (1968) Hydroxy interlayers in expansible layer silicates. *Clays Clay Miner.* **16**, 15–30.
RODERICK G.L., SENICH D. & DEMIREL T. (1969) X-ray diffraction and adsorption isotherm studies of the montmorillonite–water system. *Proc. Int. Clay Conf. Tokyo*, Vol. 1, pp. 659–668. Israel Universities Press, Jerusalem.
ROSS D.K. & HALL P.L. (1980) Neutron scattering methods of investigating clay systems. Ch. 2, in *Advanced Chemical Methods for Soil and Clay Minerals Research* (J.W. Stucki and W.L. Banwart, eds), pp. 93–169. D. Reidel Publishing Co., Dordrecht, Holland.
ROTHBAUER R. (1971) Untersuchung eines $2M_2$-Muscovit mit neutronen strahlung. *Neues Jb. Miner. Monats.* **4**, 143–154.
RUSSELL J.D. (1974) Instrumentation and techniques. Ch. 2 in *The Infrared Spectra of Minerals* (V.C. Farmer, ed.), pp. 11–25. Mineralogical Society, London.
RUSSELL J.D. & FARMER V.C. (1964) Infrared spectroscopic study of the dehydration of montmorillonite and saponite. *Clay Miner. Bull.* **5**, 443–464.

SALOMON M. (1970) The thermodynamics of ion solvation in water and polypropylene carbonate. *J. phy. Chem., Ithaca* **74**, 2519–2524.
SERNA C., AHLRICHS J.L. & SERRATOSA J.M. (1975) Folding in sepiolite crystals. *Clays Clay Miner.* **23**, 452–457.
SERNA C.J. & VAN SCOYOC G.E. (1979) Infrared study of sepiolite and palygorskite surfaces. *Proc. 6th Int. Clay Conf. 1978, Oxford*, pp. 197–206. Elsevier, Oxford.
SHIROZU H. & BAILEY S.W. (1966) Crystal structure of a two-layer Mg-vermiculite. *Am. Miner.* **51**, 1124–1143.
SLABAUGH W.H. (1959) Heats of immersion of preheated homoionic clays. *J. phys. Chem., Ithaca* **63**, 1333–1335.
SPOSITO G. & PROST R. (1982) Structure of water adsorbed on smectites. *Chem. Rev.* **82**, 553–573.
STOCKER P.T. (1969) The structure and swelling of divalent montmorillonites: a review. *Aust. Road Res.* **3**, 49–71.
STONE W.E. (1981) The use of NMR in the study of clay minerals. Ch. 4 in *Advanced Techniques for Clay Mineral Analysis* (J.J. Fripiat, ed.), pp. 77–112. Elsevier, Amsterdam.
SUQUET H., DE LA CALLE C. & PEZERAT H. (1975) Swelling and structural organization of saponite. *Clays Clay Miner.* **23**, 1–9.
SUQUET H., IIYAMA J.T., KODAMA H. & PEZERAT H. (1977) Synthesis and swelling properties of saponites with increasing layer charge. *Clays Clay Miner.* **25**, 231–242.
SUQUET H., MALARD C., COPIN E. & PEZERAT H. (1981) Variation du parametre b et de la distance basal d_{001} dans une serie de saponites a charge croissante. I: Etats hydrates. *Clay Minerals* **16**, 53–67.
TARASEVICH YU.I., ORAZMURADOV D.A. & OVCHARENKO F.D. (1971) On the thermodynamic investigation of the adsorption of water on montmorillonite. *Dokl. Akad. Nauk, SSSR* **196**, 882–884.
VAN OLPHEN H. (1965) Thermodynamics of interlayer adsorption of water in clays. I: Na vermiculite. *J. Colloid Interface Sci.* **20**, 822–837.
VAN OLPHEN H. (1969) Thermodynamics of interlayer adsorption of water in clays. II: Magnesium vermiculite, *Proc. Int. Clay Conf. 1969, Tokyo*, pp. 649–658. Israel Universities Press, Jerusalem.
VAN OLPHEN H. (1970) Determination of surface areas of clays – evaluation of methods. *Surface Area Determination* (D.H. Everett and R.H. Ottewill, eds), pp. 259–269. Butterworths, London.
VAN OLPHEN H. (1975) Water in soils. Ch. 5 in *Soil Components. Vol. 2 Inorganic Components* (J.E. Gieseking, ed.), pp. 497–527. Springer-Verlag, New York.
VAN OLPHEN H. & FRIPIAT J.J. (eds) (1979) *Data Handbook for Clay Minerals and other Non-metallic Minerals*. Pergamon Press, Oxford.
VAN SCOYOC G.E., SERNA C.J. & AHLRICHS J.L. (1979) Structural changes in palygorskite during dehydration and dehydroxylation. *Am. Miner.* **64**, 215–223.
WALKER G.F. (1956) Diffusion of interlayer water in vermiculite. *Nature, Lond.* **177**, 239–240.
WALKER G.F. (1961) Vermiculite minerals. Ch. 7 in *The X-ray Identification and Crystal Structures of Clay Minerals* (G. Brown, ed.). Mineralogical Society, London.
WALKER G.F. & MILNE A. (1950) The hydration of vermiculite saturated with various cations. *Trans. Int. Congr. Soil Sci., Amsterdam* **2**, 62–67.
WEAVER C.E. & POLLARD L.D. (1973) *The Chemistry of Clay Minerals*. Elsevier, London.
WEIR A.H. & GREENE-KELLY R. (1962) Beidellite. *Am. Miner.* **47**, 137–146.
WEIR A.H. & RAYNER J.H. (1974) An interstratified illite–smectite from Denchworth series soil in weathered Oxford clay. *Clay Minerals* **10**, 173–187.
WEISS A., BECKER H.O., ORTH A., MAI G., LECHNER H. & RANGE K.J. (1970) Particle size effects and reaction mechanism of the intercalation into kaolinite, *Proc. Int. Clay Conf. 1969, Tokyo*, Vol. 2, pp. 180–184. Israel Universities Press, Jerusalem.
YARIV S. & SHOVAL S. (1975) The nature of the interaction between water molecules and kaolin-like layers in hydrated halloysite. *Clays Clay Miner.* **23**, 473–474.
ZETTLEMOYER A.C., MICALE F.J. & KLIER K. (1975) Adsorption of water on well-characterized solid surfaces. Ch. 5 in *Water: a Comprehensive Treatise. Vol. 5. Water in Disperse Systems* (F. Franks, ed.), pp. 249–291. Plenum Press, New York.

Chapter 6

Catalytic Properties of Clay Minerals

J. P. RUPERT, W. T. GRANQUIST and T. J. PINNAVAIA

	Page		Page
6.1 INTRODUCTION	275	6.4.1.1 Acid activation	289
6.2 ACIDITY OF CLAY MINERAL SURFACES	276	6.4.1.2 Catalyst properties	291
6.2.1 Brønsted and Lewis Acidity: Brønsted sites and Lewis sites	277	6.4.1.3 Pyrolysis of organo-clay complexes	292
6.2.2 The study of surface acidity	277	6.5 SYNTHETIC LAYER-LATTICE ALUMINOSILICATES AS CATALYSTS	293
6.2.2.1 Titration of aqueous dispersions	277	6.5.1 A synthetic aluminian smectite	293
6.2.2.2 Hammett indicator methods	278	6.5.1.1 Dehydroxylation	294
6.2.2.3 Amine titrations	279	6.5.1.2 Acidity	295
6.2.2.4 Spectroscopic methods	280	6.5.1.3 Catalytic activity	296
6.3 CATALYTIC REACTIONS OTHER THAN CRACKING	283	6.6 CLAY INTERCALATED METAL COMPLEX CATALYSTS	297
6.3.1 Isomerization	284	6.6.1 Hydrogenation–isomerization	298
6.3.1.1 Skeletal rearrangement	284	6.6.2 Asymmetric hydrogenation	305
6.3.1.2 Double bond migration	285	6.6.3 Hydroformylation	307
6.3.2 Polymerization	285	6.6.4 Solvolysis and proton-assisted reactions	308
6.3.3 Desulphurization	287	6.7 PILLARED CLAYS AS CATALYSTS	311
6.3.4 Hydrolysis	287	NOTE	314
6.4 CATALYTIC CRACKING	288	REFERENCES	314
6.4.1 Clay-derived catalysts	289		

6.1 INTRODUCTION

The use of aluminosilicates in heterogeneous catalysis is almost as old as the catalytic concept itself. For example, only 29 years after Berzelius (1836) described the phenomenon, von Liebig (1865)[1] discussed in a textbook the ability of (original in German) "powdered porcelain or ordinary pumice (to) bring about the combination of hydrogen and oxygen to water, and the combination of sulphurous acid with oxygen to sulphuric acid, at temperatures at which these substances would otherwise not combine".

Clays were obvious choices for investigation, and an early favourite was palygorskite, then called attapulgite and referred to variously as floridin, Florida earth, Florida fuller's earth, Attapulgus clay. Montaland (1911) was granted US Patent 999 667 for the isomerization of pinene to camphene over various catalysts including palygorskite. Gurwitsch (1912) used palygorskite to polymerize pentenes and hexenes to di- and trimers and higher polymers. Later, Gurwitsch (1923) described in detail the catalytic activity of the partially dehydrated mineral for isomerization of terpenes and warned of the extreme exothermic nature of the pinene reaction. In the same paper he discussed palygorskite as a catalyst for oxidation of alcohols to aldehydes and for the cleavage of hydrazobenzene to azobenzene and aniline, and pointed out that this clay was not active in the disproportionation of toluene to benzene and xylene. Rideal and Thomas (1922) compared the catalytic activity of Florida, Surrey, and Somerset bleaching earths for the decomposition of hydrogen peroxide and found palygorskite (Florida earth) to be the best catalyst of the three.

The use of clays as cracking catalysts for gasoline production was disclosed in German patents in 1923 and subsequent years (see Ryland *et al.* 1960), but catalytic cracking did not become commercially feasible until Houdry *et al.* (1938) solved the problem of catalyst regeneration. It is particularly satisfying that this application of catalysis, which is of major economic importance and today accounts annually for cracking catalyst sales in excess of \$75 million, first reached the commercial stage in 1936, a century after Berzelius.

The consumption of clay catalyst, prepared for the most part by the acid treatment of montmorillonite and halloysite, (Milliken *et al.* 1955) rose rapidly with the growth of catalytic cracking. Development of silica–alumina gel catalysts led to a decrease in the share of the market available to the clay-based catalysts. According to these authors, clay catalysts accounted for 40% of a USA catalyst market of 470 tons day^{-1} (426 tonne day^{-1}) with an annual (1952) value of approximately \$61 million.

Casual reading of the literature leads to the conclusion that silica–alumina and clay-based catalysts have been supplanted by zeolites (still aluminosilicates, however). It is certainly true that such catalysts are preferred for most cracking operations, with important exceptions. What is not so generally realized is that zeolitic cracking catalysts consist of a minor (10–15% w/w) amount of highly active synthetic faujasite in a less active, and sometimes inert, matrix. Common constituents of this matrix might be silica–alumina gel, kaolinite, halloysite, or mixtures of these minerals. In the authors' opinions, the role of the matrix in these multicomponent catalysts has been neglected.

Development of a synthetic 2:1 layer-lattice silicate (Granquist 1966; Capell and Granquist 1966; Granquist and Pollack 1967) of a high order of catalytic activity may help to re-establish clay-like structures as catalysts of industrial importance. Such a synthetic mineral is an especially suitable matrix for faujasite incorporation.

6.2 ACIDITY OF CLAY MINERAL SURFACES

The carbonium ion mechanisms of the various reactions occurring over aluminosilicate catalysts require acidic surfaces as a condition for such catalysis (see, for example, Thomas 1949 and Greensfelder *et al.* 1949). In a study of propylene polymerization, Gayer (1933) described an active catalyst which contained a small amount of alumina and observed that the surface gave an acidic reaction in an aqueous system when contacted with methyl orange. Tamele (1950) discussed the acidity and activity of silica–alumina cracking catalyst and concluded "the acid strength and number of acid sites are the major factors determining the activity of alumina–silica catalysts".

Walling (1950) defined the acidity of a solid surface as "the ability of the surface to convert an adsorbed neutral base to its conjugate acid". He developed a semi-quantitative method of measuring the acidity by use of indicators with a range of pK_a values, and used iso-octane solutions to eliminate effects due to water. This procedure was extended by Benesi (1956), who concluded among other things that "unused silica–alumina catalysts appear to be at least as strongly acid as 90% sulfuric acid". Mapes and Eischens (1954) used infrared spectroscopy to study surface interactions, a procedure that made possible the study of surface acidity under conditions nearly equivalent to those existing in vapour phase heterogeneous catalysis. In their review of cracking catalysts, Ryland *et al.* (1960) listed over 50 references relating to the acidity of the aluminosilicates. In the past decade, work in this field has continued to expand, stimulated by the advent of the highly active faujasites, the availability of improved spectroscopic techniques, and controversy over the relative importance of Brønsted and Lewis acid sites in cracking analysis.

Techniques applicable to amorphous silica–alumina are also applicable to the clay minerals, and conclusions relative to the former should also apply to the latter. Therefore, these studies of surface acidity are pertinent to the general purpose of this chapter and will be considered further. Whenever possible, emphasis will be placed on measurements made on clay mineral surfaces.

6.2.1 Brønsted and Lewis acidity: Brønsted sites and Lewis sites

The Brønsted–Lowry theory of acidity defines an acid in terms of its ability to act as a proton donor; a base, as a proton acceptor. Neutralization, while not formally defined by the theory, is the transfer of a proton from an acid to a base. The Lewis electronic theory regards an acid as an electron pair acceptor. Bases are electron pair donors, and neutralization is the formation of a coordinate bond between the base and the acid (for a thorough treatment of acid–base concepts, see Moeller 1952). Since the proton is, of course, an ideal electron pair acceptor, the Lewis theory encompasses the Brønsted concept. In work on surface acidity, however, the term "Lewis acid" does not include Brønsted acidity as a special case. It would be better to designate these sites by the terms "protonic" for the Brønsted site and "aprotic" for the Lewis site.

It is clear that the aluminium in aluminosilicates is the source of the acidic nature of the surface (Thomas 1949; Tamele 1950; Ryland et al. 1960; Fripiat et al. 1965). Al^{3+} proxying for Si^{4+} in tetrahedral coordination in an aluminosilicate (tetrahedral Al^{3+} in montmorillonites, vermiculites, micas, zeolites) gives rise to a net negative charge. A charge-balancing H_3O^+ associated with such tetrahedral aluminium corresponds to a Brønsted, or protonic, acid site. An aluminium in three-fold coordination, perhaps occurring at an edge, or arising from a Si—O—Al rupturing dehydroxylation of the Brønsted site, would correspond to the Lewis site. Solomon and Rosser (1965) have proposed that an octahedral Al^{3+}, located at a platelet edge, will function as a Lewis site after thorough dehydration. Such aluminium ions would be electron pair acceptors and would function as aprotic acids in the Lewis sense. Obviously, water (a Lewis base) will convert the Lewis site into a Brønsted site, a fact which limits study of Lewis sites to relatively anhydrous systems. These various configurations of Al^{3+} are summarized in Fig. 6.1.

6.2.2 The study of surface acidity

6.2.2.1 Titration of aqueous dispersions

Bradfield (1923) established the presence, in the clay fraction of soils, of aluminosilicic acids which could be reproducibly titrated with standard base. He observed that the same amount of the colloidal acid was required to neutralize equivalent quantities of two bases, and concluded that the reaction between acid colloidal clays and strong bases seemed to be an ordinary neutralization. Such titrations became standard techniques in soil chemistry, and various soils, clays and clay minerals have been converted to the acid form and neutralized with a variety of bases; see, for example, Marshall (1949).

For the person interested in catalysis, this procedure has several disadvantages. First, H-clays in aqueous systems are unstable and hydrated aluminium ion, removed from the clay structure, spontaneously replaces hydronium ion (Paver and Marshall 1934). Second, the presence of water assures that all acid sites will be protonic. Finally, most applications of

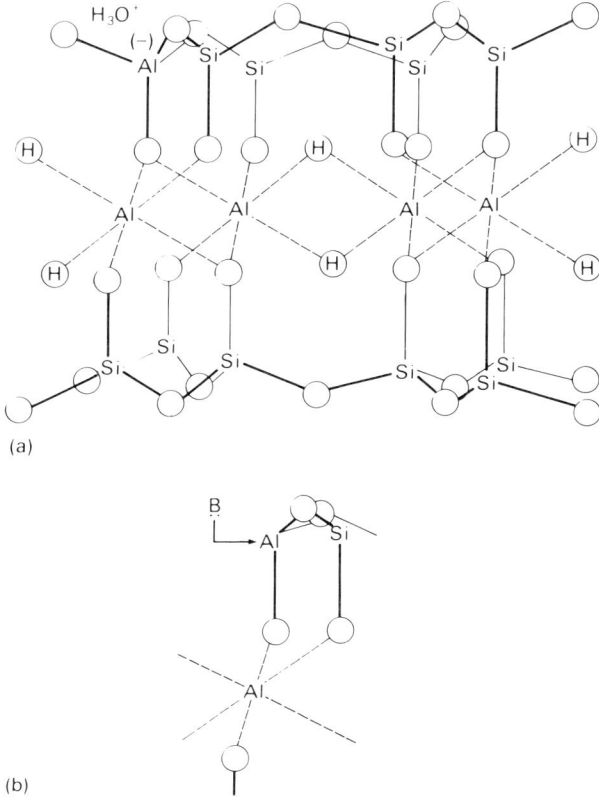

FIG. 6.1. (a) An off-axis projection of a 2:1 dioctahedral clay mineral with a Brønsted acid site. (b) A possible configuration for a Lewis acid site. O = oxygen; Ⓗ = hydroxyl; B = generalized base. Some edge oxygens and bonds are omitted for the sake of clarity.

aluminosilicate catalysts will involve vapour-phase reactions at temperatures where liquid water will not be present (steam is frequently present as a reactant or as a diluent). Therefore, studies of acidity of clay dispersions in water seem ill-suited to the purpose of this writing and will not be discussed further.

6.2.2.2 Hammett indicator methods
Walling (1950) proposed that surface acidity should be measured by observation of the colour of the adsorbed form of selected indicators with a range of pK_a values. He used dried clays and conducted the measurement in iso-octane which contained the indicator in solution. Among his conclusions, he noted that the acid strength was decreased by the presence of weak bases such as acetone and water, and that, while alumina and pure silica had neutral surfaces, silica–alumina was strongly acid. Application of the Hammett and Deyrup (1932) H_0 function

$$H_0 = -\log a_H + f_B/f_{BH^+} \tag{1}$$

to this surface–indicator interaction permitted assignment of an H_0 value to the surface, thus giving a semi-quantitative measure of the acid strength of the surface. Walling (1950) placed the

TABLE 6.1. Indicators used for acid strength measurements[†]

Indicator	Basic colour	Acid colour	pKa	H_2SO_4, wt %
Methyl red	Yellow	Red	+6.8	8×10^{-8}
Phenylazonaphthylamine	Yellow	Red	+4.0	5×10^{-5}
Butter yellow	Yellow	Red	+3.3	3×10^{-4}
Benzeneazodiphenylamine	Yellow	Purple	+1.5	0.02
Dicinnamalacetone	Yellow	Red	−3.0	48
Benzalacetophenone	Colourless	Yellow	−5.6	71
Anthraquinone	Colourless	Yellow	−8.2	90

[†] Reprinted with permission from Benesi (1956). Copyright 1956 American Chemical Society.

TABLE 6.2. Acid strength of various silicate surfaces[†]

Solid	H_0
Sodium kaolinite	−3.0 to −5.6
Ammonium kaolinite	−3.0 to −5.6
Hydrogen kaolinite	−5.6 to −8.2
Sodium montmorillonite	+1.5 to −3.0
Ammonium montmorillonite	+1.5 to −3.0
Hydrogen montmorillonite	−5.6 to −8.2
Silica–Alumina	< -8.2
Silica–Magnesia	+1.5 to −3.0
Clay catalyst (acid-activated)	< -8.2

[†] Reprinted with permission from Benesi (1956). Copyright 1956 American Chemical Society.

following restrictions on the choice of indicators: (1) the basic form must be uncharged and be converted to its conjugate acid by simple proton addition; (2) the indicator must be adsorbed by the surface; and (3) the acid form should be more highly coloured than the basic form.

Benesi (1956) extended this treatment to a wider range of pK_a values. Table 6.1, taken from his paper, lists the indicators, their basic and acid colours, pK_a values, and the sulphuric acid composition corresponding to the mid-point of each indicator transition. He used these indicators to study a variety of surfaces, including several homoionic forms of kaolinite and montmorillonite, as well as silica–alumina, silica–magnesia and acid-activated natural clay catalysts. Table 6.2 contains surface acid strengths selected from his paper.

It is interesting that the hydrogen clays had acid strengths corresponding to sulphuric acid solutions between 71 and 90 wt% H_2SO_4 and that silica–alumina and the acid-activated clay catalyst were acids at least as strong as, and perhaps stronger than, 90% sulphuric acid.

Both authors emphasized that the process observed was the transfer of a proton from the surface to the adsorbed indicator; therefore, the method measures surface acidity arising from protonic sites.

6.2.2.3 Amine titrations

The determination of acid strength of surfaces by Hammett indicator methods does not give information about the total number of acid sites or the distribution of strengths among those

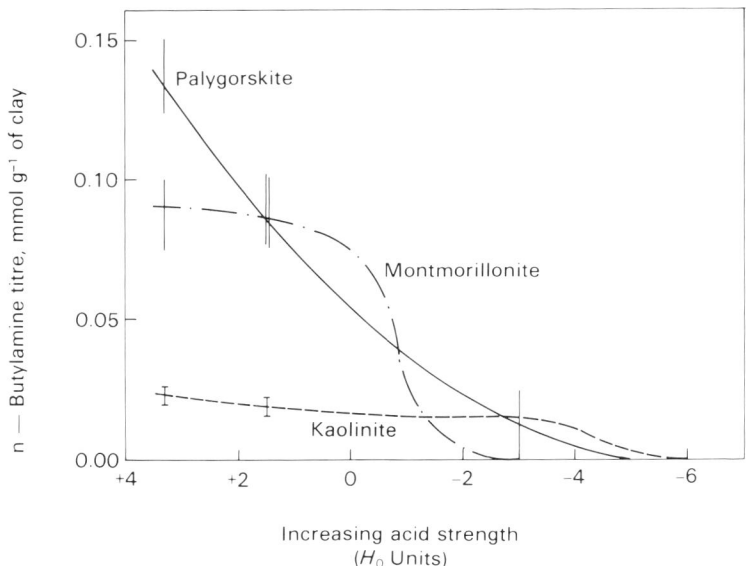

FIG. 6.2. Butylamine titres v. acid strengths for clays dried at 120 °C. Vertical lines denote titre uncertainties. Taken from Benesi (1957). Used by permission: copyright 1957 American Chemical Society.

sites. Benesi (1957) dealt with this problem by titrating benzene suspensions of powdered catalyst with n-butylamine to endpoints determined by Hammett indicators. The catalyst was carefully dried, protected from further contact with moisture, and the n-butylamine titre (mmol g^{-1} of clay) corresponding to a given H_0 range determined by successive approximations. The results were plotted as titre v. H_0, a given titre being a measure of the number of acid centres with acid strengths equal to or lower than the pK_a value of the indicator used. Remember that acid strength increases as H_0 decreases; i.e. becomes more negative.

Kaolinite, palygorskite and Wyoming bentonite were dried at 120 °C and subjected to this procedure. The results are plotted in Fig. 6.2. The curves were interpreted as showing that kaolinite had few acid centres but of high acid strength, bentonite a large total acidity most of which was in a single strength range weaker than that of the kaolinite sites, and palygorskite the highest total acidity with distribution over a wide range of strengths.

Application of the technique to naturally occurring Wyoming bentonite and the sodium and hydrogen forms of the same clay gave the results shown in Table 6.3. Both the natural clay and the sodium form had some surface acidity, with the natural material the more acid. Both the total acidity and the acid strength increased markedly when the clay was converted to the hydrogen form. The titre was less than the exchange capacity (0.98 meq g^{-1}) measured in an aqueous system; this difference was attributed either to the failure of the benzene solution to penetrate all of the interlamellar space or to a partial conversion of the hydrogen clay to a mixed H,Al-form.

6.2.2.4 Spectroscopic methods
Evans (1950) studied the ultraviolet spectrum of Ph$_2$C=CH$_2$(I). The spectrum changed when (I) was dissolved in concentrated H$_2$SO$_4$ and a new band at 430 nm appeared, which was

TABLE 6.3. Acidity of montmorillonite in relation to type of exchangeable cation[†]

	Surface area, m^2/g	Butylamine titre (mmol g^{-1}) in the H_0 range			
		+3.3 to +1.5	+1.5 to −3.0	to < −3.0	Total
Natural form	44	0.00	0.09	0.00	0.09[‡]
Sodium form	41	0.03	0.01	0.00	0.04[‡]
Hydrogen form	52	0.10	0.00	0.55	0.65[§]

[†] Taken from Benesi (1957). Used by permission: copyright 1957 American Chemical Society.
[‡] Titre uncertainty is 0.01 mmol g^{-1}.
[§] Titre uncertainty is 0.05 mmol g^{-1}.

attributed to Ph_2C^+—CH_3 formed by the transfer of a proton from the acid to (I). Furthermore, the freezing point depression of H_2SO_4 caused by (I) gave added evidence of the proton transfer reaction:

$$H_2SO_4 + Ph_2C=CH_2 \rightarrow HSO_4^- + Ph_2C^+-CH_3 \quad (2)$$

Evans found the same ultraviolet absorption peak (430 nm) when (I) was brought into contact with BF_3 in benzene. He also found that if (I) was added to heat-activated palygorskite suspended in a paraffinic solvent, the suspension had a band at 430–440 nm, with the colour located on the surface of the clay particles. Evans concluded that the carbonium ion had been formed by transfer of a proton from the clay surface to the olefin.

Mapes and Eischens (1954) chemisorbed NH_3 on silica–alumina cracking catalyst at 175 °C and studied the infrared spectrum of the ammonia–catalyst surface complex. They found adsorption bands similar to NH_3, and to NH_4^+ in ammonium salts. The bands assigned to NH_3 were more intense than those arising from NH_4^+; addition of H_2O to the system strengthened the NH_4^+ bands at the expense of the NH_3 bands. These results were evidence of the two kinds of acid sites: the NH_3 was chemisorbed by (1) formation of a coordinate bond between the Lewis base (NH_3) and a Lewis acid site; and (2) transfer of a proton from a Brønsted site to the base to form NH_4^+ bonded to the site by coulombic forces. Mapes and Eischens concluded that most catalyst acidity is of the Lewis type.

The power of the spectroscopic approach was immediately evident. Workers interested in catalysis and in aluminosilicates applied the technique to their particular problems, and an abundance of papers appeared and continues to do so. Various authors discussed the interaction of bases with acid surfaces ranging from amorphous silica–alumina to the clay minerals to the synthetic faujasites. A detailed review is not possible in this chapter; the reader is referred to the following representative articles: Mortland et al. (1963); Basila et al. (1964); Hughes and White (1967); Uytterhoeven et al. (1965); Ward and Hansford (1969). The remainder of this section will be devoted to a particular controversy over the relative importance of Lewis and Brønsted sites in a surface–base interaction, and to the application of the spectroscopic technique to the chemisorption of pyridine by a synthetic layer-lattice aluminosilicate. The controversy is reviewed because it affords insight into both the power and the weakness of the technique.

The situation concerned the mechanism of formation of triphenylcarbonium ion from triphenylmethane on the surface of silica–alumina. In brief, Hall and co-workers (1961, 1966a,b) proposed that a Lewis acid site on the silica–alumina surface abstracted, and held, a hydride ion from Ph_3CH, thus forming Ph_3C^+. Hirschler and colleagues (1963, 1964, 1966) argued that the formation of the Ph_3C^+ was through a catalyst proton (Brønsted site) after oxidation of Ph_3CH

to Ph_3COH; i.e.,

$$Ph_3COH + H^+ cat = Ph_3C^+ + H_2O \qquad (3)$$

The arguments and their final resolution were concisely summarized by Wu and Hall (1967), who showed that protonic sites do effect the formation of Ph_3C^+ by the following route:

$$Ph_3CH + Cat\ H^+ \rightarrow PhH + Ph_2C^+H \qquad (4)$$

$$Ph_2C^+H + Ph_3CH \rightarrow Ph_2CH_2 + Ph_3C^+ \qquad (5)$$

That is, Hirschler was correct with respect to the importance of Brønsted sites, but was incorrect as to the mechanism. It is both fascinating and instructive to follow the two lines of reasoning through the various references involved.

Wright et al. (1972) have studied the chemisorption of pyridine, a method introduced by Parry (1963), on acid sites of the synthetic mixed-layer dioctahedral 2:1 layer-lattice aluminosilicate described by Granquist (1966) and Granquist and Pollack (1967). A thin film containing 8.85 mg solids per cm^2 was prepared and mounted in the sample holder of an infrared cell similar to that described by Uytterhoeven et al. (1965). Manipulation of the cell was described in that paper, and by Granquist and Kennedy (1967). Figure 6.3 presents pertinent spectra.

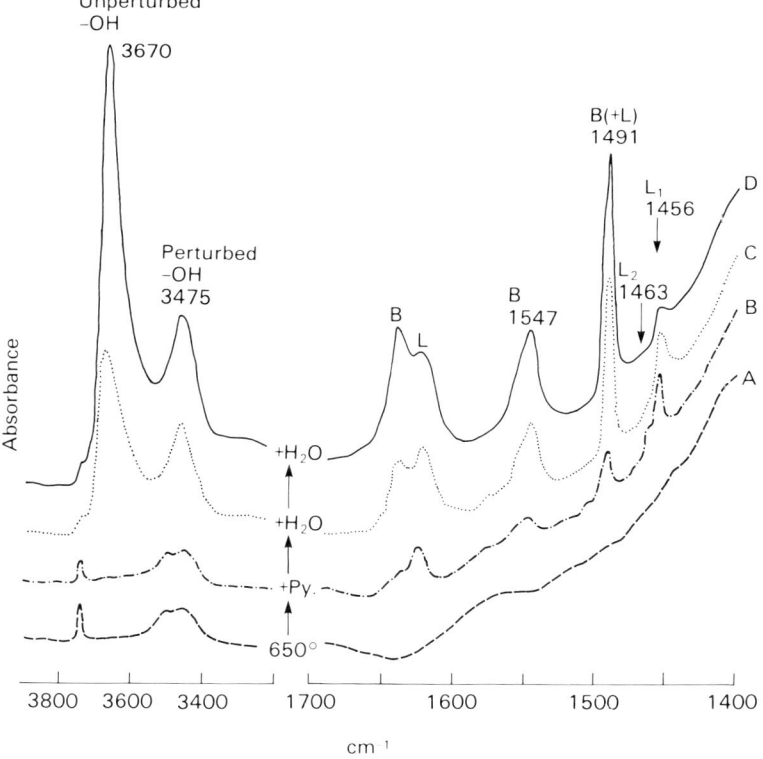

FIG. 6.3. Pyridine chemisorption on a synthetic smectite: transformation of pyridine to pyridinium ion by interaction with water. L = Lewis acid site, B = Brønsted acid site; see text for details.

The film was heated in air at 650 °C for 1 hour, followed by 15 hours at 650 °C and 10^{-6} Torr (10^{-7} kPa). The cell was isolated, cooled to room temperature and curve A obtained (IR-12, absorption mode). Note the lack of any bands in the region 1400–1700 cm^{-1}, the distinct evidence for residual hydroxyl near 3450 cm^{-1}, and the sharp hydroxyl absorbance at 3747 cm^{-1}, attributed to edge silanol groups. The cell was then remounted on the vacuum line and the film exposed to pyridine vapour ($P_{py} = 25$ Torr, 3.3 kPa) for 30 min, followed by removal of free pyridine by evacuation at 150 °C for 2 hours with the temperature increased to 200 °C for an additional 2 hours. The pressure during this stage was again 10^{-6} Torr (10^{-7} kPa). After isolation and cooling, spectrum B resulted.

This spectrum was interpreted as showing evidence of the chemisorption of pyridine at both aprotic and protonic sites. The band at 1456 cm^{-1} was considered diagnostic for the Lewis site; that at 1547 cm^{-1} for the Brønsted site. The reduction in intensity of the 3747 cm^{-1} edge silanol band was considered significant; presumably, the protonic mechanism was operative at such sites. It is apparent from spectrum B that bonding of pyridine at Lewis sites was predominant under these conditions.

The effect of water is demonstrated by curves C and D. Incremental amounts of water vapour were allowed to react with the film at 200 °C. The cell was not evacuated following these additions, so that the amount of pyridine was constant for curves B, C, and D. Water vapour clearly changed much of the chemisorbed pyridine to pyridinium ion; compare the intensities at 1456 cm^{-1} and 1547 cm^{-1}. This shift from pyridine to pyridinium was accompanied by a sharp increase in the structural hydroxyl; note the increase in the —OH stretching frequencies at 3475 cm^{-1} and 3670 cm^{-1}. In part, at least, these changes were rationalized on the basis of water combining with a Lewis site to regenerate a hydroxyl and a pyridinium ion.

Many studies (for example, see Ryland *et al.* 1960) have related acidity and catalytic activity for reactions such as propylene polymerization, and cumene and cetane cracking. However, the relative importance of Lewis and Brønsted sites is still an open question; in the authors' opinion the weight of the evidence favours the latter. In commercial cracking operations, as opposed to more rigidly controlled laboratory experimentation, the usual presence of water vapour at some stage of the process ensures maximization of Brønsted sites at the expense of Lewis sites.

6.3 CATALYTIC REACTIONS OTHER THAN CRACKING

This section considers applications of the clay minerals as diverse as polymerization of olefins, isomerization of terpenes, desulphurization of gasoline, conjugation of double bonds to convert unsaturated oils to drying oils, dimerization of fatty acids to so-called dimer acids, and hydrolysis of esters and disaccharides. The authors can provide no documented evidence for actual commercial use of clay minerals for these purposes in present US practice. However, conversations with various laboratory and production personnel lead to the conclusion that a small amount of palygorskite is used for rosin treatment (probably combining decolorization with some isomerization) and double bond conjugation. It is certain that clays are involved in the preparation of dimer acids. It is also certain that virtually no natural clay catalysts currently find utility in the refining industry, except (previously noted) as inexpensive components of the matrix of synthetic faujasite catalysts.

6.3.1 Isomerization

Isomerization is the process of rearrangement of a molecule to a new molecule with the same empirical formula. Such rearrangements can be skeletal, as, for example, the rearrangement of a normal hydrocarbon to branched-chain structure with the same carbon number; they can involve double bond migration such as the transition of butene-1 to butene-2, or the conjugation of double bonds to produce drying oils from various unsaturated oils; and they can be geometrical (or spatial) as in the configuration about a double bond (for example, *cis*-butene-2 and *trans*-butene-2), the difference between the boat and chair forms of cyclohexane, or the special case of geometrical arrangement of groups to create, in the molecule, one or more asymmetrical carbon atoms, with optical activity as the result. Acidic aluminosilicate surfaces can catalyse various kinds of isomerizations. For detailed discussions of hydrocarbon isomerization, the reader is referred to Egloff *et al.* (1942), Egloff (1943), Condon (1958) and Germain (1969).

6.3.1.1 Skeletal rearrangement
A typical reaction of this kind is typified by the work of Hay *et al.* (1945) on the isomerization of hexene-1 to branched-chain olefins. The reaction proceeds by a carbonium ion mechanism with the following steps (Condon (1958):

1. Proton from catalyst transferred to olefin to yield carbonium ion.
2. Rearrangement of the carbonium ion.
3. Transfer to the catalyst of proton from the rearranged carbonium ion to produce the isomeric olefin.

In this study, 2- and 3-methylpentenes were the primary products; lesser amounts of dimethylbutenes and small amounts of hydrocarbons were formed by secondary cracking and polymerization reactions. The isomerization appeared to be stepwise: hexene-1 → methylpentenes → dimethylbutenes. The catalysts studied included, among others, acid-activated bentonite, palygorskite and silica–alumina. The activity of these catalysts were in the order: acid-activated clay > palygorskite > silica–alumina gel. Of particular interest was the high activity noted when essentially inactive sodium permutit, $Na_2O \cdot Al_2O_3 \cdot 5SiO_2$ was converted to the acid form. The resultant acidic material ranked between the acid-activated clay and palygorskite in activity. It was concluded that the catalyst must have protonic sites to be effective in the isomerization.

The catalytic activity of palygorskite for the isomerization of the terpenes was recognized early; see the discussion in the introduction to this writing. These rearrangements, so catalysed, are violently exothermic, as can be demonstrated readily by cautiously adding lightly calcined palygorskite to gum terpentine in an *open* vessel. Again, the reaction probably proceeds through a carbonium ion intermediate, obtained by transfer of a proton from the catalyst to the terpene, rearrangement of the carbonium ion and subsequent loss of the proton to the catalyst to give the isomer.

The isomerization of alpha-pinene to camphene can be summarized as follows:

(6)

6.3.1.2 Double bond migration

Acidic aluminosilicates can also cause migration of the double bond in olefins, and conjugation of double bonds in molecules containing multiple unsaturations. In particular, Slobodin (see Egloff 1943, pages 6–8, and Condon 1958, pages 111–112 for discussion and references) studied the isomerization by palygorskite of diolefins with allenic and isolated alkene linkages. He observed conjugation of the double bonds and also, in the case of the allenic linkages, formation of alkynes. Once again, the reaction was assumed to proceed by a carbonium ion intermediate which arose by transfer of a proton from the catalyst surface to the reactant molecule.

An application of this effect to a system of commercial importance was described by Turk and Feldman (1943). They studied the shift of double bonds in dienic or polyenic fatty oils from isolated to conjugated positions. By this means, fast drying oils could be obtained from oils (for example, soybean) containing isolated double bonds predominantly. The procedure involved heating the oil/catalyst mixture, with stirring, in an inert atmosphere to temperatures in the range of 200–300 °C, and observing the increase in the refractive index of the recovered oil. Palygorskite was the most active of the catalysts studied.

6.3.2 Polymerization

As noted in the introduction, Gurwitsch (1912) observed polymerization by palygorskite of pentenes and hexenes to di- and trimers and higher polymers. Also, Gayer (1933) found that heat-activated palygorskite catalysed the polymerization of propylene at 350 °C and atmospheric pressure. The liquid polymers consisted mostly of low boiling olefins with a small amount of saturated hydrocarbons. Mild acid treatment improved the activity of the palygorskite. Activity was also improved if the propylene was saturated with water at room temperature; larger amounts of water decreased the activity. The catalyst was poisoned by small amounts of alkali.

These observations are again consistent with a carbonium ion mechanism; for example (see Germain 1969):

$$C-C=C + H^+\text{-clay} \rightarrow C-\overset{\oplus}{C}-C \tag{7}$$

$$C-\underset{\underset{C}{|}}{C^{\oplus}} + C=C-C \rightarrow C-\underset{\underset{C}{|}}{C}-C-\underset{\underset{C}{|}}{C^{\oplus}} \rightarrow C-\underset{\underset{C}{|}}{C}-C=C-C + H^+ \tag{8}$$

An interesting variation of this reaction is the depolymerization of C_8 homopolymers of isobutylene and copolymers of isobutylene and n-butylene. Ciapetta *et al.* (1948) studied this process with preparation of pure isobutylene their objective. Palygorskite was found to be an excellent catalyst for the depolymerization, with product better than 99% isobutylene from homopolymers and 90–98% isobutylene from various codimers. Mass spectrometer analyses provided evidence that skeletal isomerization of the C_8 olefins accompanied depolymerization; for example, the transformation of 2,4,4-trimethylpentene-1 or -2 into 2,3,4-trimethylpentene-1 or -2 of 3,4,4-trimethylpentene-2 into 2,3,3-trimethylpentene-2, and trimethylpentenes into dimethylhexenes. Pilot scale work at throughputs as high as 1145 barrels (bbl) ton^{-1} clay (200 m^3 kg^{-1} clay) at 350 °C gave no indicated loss in activity, which confirmed laboratory indications of long catalyst life. The authors suggested a carbonium ion mechanism with the first step being the transfer of a proton from the catalyst to the C_8 olefin.

Clays catalyse the dimerization of unsaturated fatty acids to dicarboxylic acids, which find use in the preparation of polyamide resins. Fischer (1964) described the state of the art and compared various clays, including acid-activated bentonite, Wyoming bentonite, and an acidic bentonite from Angelina County, Texas, for activity in the dimerization of tall oil fatty acids which contain unsaturated carbon–carbon bonds and are derived from the digestion of pine woods. In general, he found conversions of the order of 60% with dimer/trimer ratios near 5.0; the acidic Texas bentonite was slightly superior to the other clays tested.

den Otter (1970) has described in detail the dimerization of oleic acid with a montmorillonite catalyst. His paper includes a list of references that can serve as a convenient literature guide for this subject. Also, Weiss (1981) has investigated the dependence of this reaction on clay layer charge; the results of this work are discussed later.

Solomon (1968) discussed the behaviour of various monomers in the presence of clay mineral surfaces. It was remarkable, for example, that kaolin catalysed the polymerization of styrene but inhibited the same reaction for methyl methacrylate. Furthermore, for styrene polymerization, various mineral surfaces exhibited markedly different effects. This latter situation is summarized in Table 6.4 taken from the earlier paper of Solomon and Rosser (1965).

The masking of the crystallite edges by polyphosphate reduced the activity of the mineral for the styrene polymerization, and also lessened the inhibition of the methyl methacrylate system (i.e. the system now behaved in the manner of the normal thermally induced polymerization). It was concluded, therefore, that both the catalysis and the inhibition were associated with the crystallite edges. Also of importance was that water or polar organic solvents reduced the catalytic activity for the styrene reaction.

In the case of hydroxy monomers such as hydroxyethyl methacrylate, polymerization was inhibited by silicates that did not form interlayer complexes with the monomer (example, kaolinite). If interlayer complexes did form and the crystal lattice contained a transition metal in a lower valence state (usually Fe^{2+}), then the polymerization was catalysed. The activity of the mineral was not diminished by masking the crystallite edges, nor by the presence of water.

TABLE 6.4. Comparison of mineral structure and catalytic activity[†]

Mineral	Idealized unit cell formula	Layer charge, meq/100 g	Efficiency in polymerizing styrene[‡]
Pyrophyllite	$Al_4Si_8O_{20}(OH)_4$	0	1
Talc	$Mg_6Si_8O_{20}(OH)_4$	0	0
Montmorillonite	$(Al_{4-x}Mg_x)Si_8O_{20}(OH)_4$	90–100	3
Hectorite	$(Mg_{6-x}Li_x)Si_8O_{20}(OH)_4$	80	1
Vermiculite	$(Mg_{6-x}Fe_x)Si_6Al_2O_{20}(OH)_4$	120–200	0
Muscovite	$Al_4Si_6Al_2O_{20}(OH)_4$	250	1
Illite	$Al_4Si_6Al_2O_{20}(OH)_4$	200	3
Kaolinite	$Al_8Si_8O_{20}(OH)_{16}$	3–15	5
Gibbsite	$Al(OH)_3$	0	0
Alumina	Al_2O_3	0	0
Palygorskite	$Mg_5Si_8O_{20}(OH)_2(OH_2)_4 4H_2O$	20–30	5

[†] Adapted from Solomon and Rosser (1965). Used by permission: copyright 1965 John Wiley & Sons, Interscience Publishers.
[‡] The minerals have been given a rating based on yield of polystyrene: 0 represents no polymerization while 5 is quantitative conversion.

Solomon rationalized these observations by proposing that the styrene polymerization was catalysed by a Lewis acid site at an edge octahedral aluminium. Presence of water or polar solvent converted the Lewis site to a protonic site and reduced the activity. The mechanism proceeded through the formation of a cation-radical by transfer of an electron from the styrene to the Lewis acid. He then suggested that the equivalent edge aprotic site would inhibit the methyl methacrylate polymerization by preferential sorption of the propagating free radical.

The behaviour of the hydroxy monomer/clay system was explained by formation of an interlayer complex between the monomer and mineral, with production of a reactive intermediate as a result of electron transfer from the mineral (i.e. oxidation of the transition metal). This reactive intermediate would then propagate between the mineral layers and grow into the bulk monomer. The mechanism was considered analogous to catalysis of olefin polymerization by transition metal catalysts.

6.3.3 Desulphurization

Although no longer practised in the US, clay can be used for the catalytic desulphurization of straightrun and cracked gasolines to improve lead susceptibility. In the process, the gasoline vapours were contacted with a clay catalyst at atmospheric pressure and a temperature of 400 °C to convert mercaptans, sulphides and disulphides to H_2S. The treatment decreased the amount of tetraethyllead (TEL) required to give the desired octane number increase in the final product.

Amero and Wood (1947) compared palygorskite with activated bauxite (γ-Al_2O_3), a Texas bentonite, and acid-activated bentonite for desulphurization of a cracked gasoline at 400 °C and a liquid hourly space velocity of 1.0. With all catalysts, the treatment removed $c.$ 35% of the total sulphur and essentially all of the mercaptan sulphur. The ASTM octane number of the unleaded gasoline increased 2 points, and the addition of 3 cm^3 TEL to the desulphurized product increased octane number 4 points above the value obtained for the same amount of TEL added to the untreated gasoline. The TEL savings in ml bbl^{-1} ($cm^3 m^{-3}$) for a product of constant octane number was greatest for the acid-activated bentonite.

6.3.4 Hydrolysis

The ability of acid clays to catalyse the hydrolysis of esters and disaccharides has been studied by many workers: see, for example, Rice and Osugi (1918) Wiegner (1931), Kayser and Bloch (1952), and McAuliffe and Coleman (1955). The latter authors made a detailed study of the hydrolysis of ethyl acetate to ethanol and acetic acid, and sucrose to D-glucose and D-fructose. This latter reaction is commonly referred to as the "inversion" of sucrose, because the positive specific rotation of sucrose is inverted to a negative value for the mixture of products, a change which arises from the strong laevorotation of D-fructose.

McAuliffe and Coleman (1955) obtained first-order constants for these reactions (pseudo-first order, assuming large excess of water) in the presence of hydrogen-clay and dilute HCl. The clays studied were a Utah bentonite, a Wyoming bentonite and a halloysite. The ratio of the rate constant to the amount of hydrogen ion available (exchangeable H^+ for the clays) was reasonably constant for various concentrations of a particular clay or HCl. It was observed that the acid forms of both the Utah and Wyoming bentonites were more active catalysts than HCl for the ethyl acetate hydrolysis, and that H-halloysite and H-Utah bentonite were slightly more

288 CATALYTIC PROPERTIES OF CLAY MINERALS

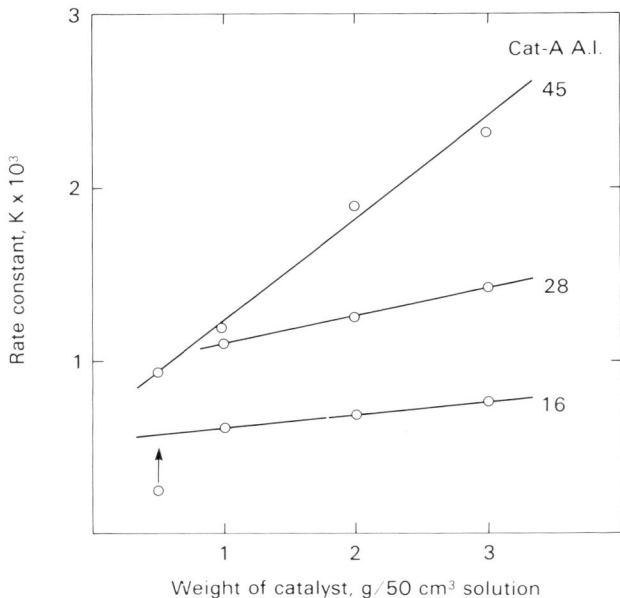

FIG. 6.4. Inversion of sucrose by SiO_2–Al_2O_3 catalysts. Each line corresponds to a catalyst with the cracking activity specified. Taken from Oblad *et al.* (1951). Used by permission of Academic Press Inc.

active than HCl for the inversion reaction. The authors recognized the danger of the spontaneous conversion of the acid to the aluminium form, and used their H-clays promptly, before the transition had time to occur. They prepared the aluminium forms and found such clays to exhibit a manyfold decrease in activity.

Among other things, they concluded that "aged electrodialysed clays and clays leached with 0.5M HCl were found by catalytic activity measurements to be almost completely Al-saturated, while clays leached with cold HCl were largely saturated with H-ions".

The inversion reaction has been used as a measure of acidity and related to the cracking activity of silica–alumina. Oblad *et al.* (1951) published a graph, reprinted in Fig. 6.4, which showed plots of the rate constant for the inversion reaction as a function of catalyst concentration for three catalysts covering a wide range of cracking activity. It is apparent that, for these silica–aluminas at least, catalytic activity for reactions such as hydrolysis and cracking varied in the same order.

6.4 CATALYTIC CRACKING

Catalytic cracking of gas oils is a fundamental refining operation of maximum economic importance, since gasoline and furnace oils are the primary products, and the by-products serve as raw material for a host of satellite units. A modern fluid catalytic cracking (FCC) installation may process as much as 60 000 bbl (9600 m^3) of fresh feed a day, involve catalyst transfer rates as high as 50–60 tons min^{-1} (45–54 Mg min^{-1}) at high linear velocities, feature catalyst/oil contact times measured in seconds, and produce enough heat during regeneration to make the unit self-sustaining despite the endothermic nature of the cracking reaction.

Because of these features, the catalyst used must have in addition to its activity, many characteristics: size and shape (microspheroidal, 20–80 μm range) to fluidize properly in the reactant and regenerate streams; attrition resistance to prevent breakdown (and subsequent loss from the unit) under the turbulent, high velocity flows; and resistance to sintering and steam deactivation (i.e. maintenance of surface area) under the conditions existing in the regenerator. The catalyst must do more than crack (rupture C—C bonds); it must also promote isomerization, cyclization, but should minimize polymerization and dehydrogenation. That is, a catalyst completely selective with respect to cracking probably would not be acceptable for commercial gas oil cracking because of poor gasoline quality and by-products of lower economic value (e.g. n-C_4 rather than iso-C_4).

The requirement for catalyst acidity and the differences in product distributions obtained in catalytic and thermal cracking are best explained by a mechanism dependent on carbonium ion formation. This formation is easily imagined for alkene molecules (proton transfer from catalyst Brønsted sites) but is less easily established for alkanes. One possibility is hydride abstraction from the alkane by Lewis sites on the catalyst; another is a small degree of thermal cracking or possibly even dehydrogenation to give sufficient olefin to generate the required carbonium ion. The possible modes of carbonium ion formation and the probable course of subsequent chain reaction(s) were summarized by Germain (1969, pp. 207–213).

6.4.1 Clay-derived catalysts

The preparation of cracking catalysts from clays was an important commercial operation from the origin of the catalytic cracking process until the advent, in the mid-1960s, of the zeolitic catalysts, many of which continue to use clays as matrices. In general, preparation of a suitable catalyst was accomplished by partial degradation, by acid, of montmorillonite, halloysite, or kaolinite. The chemistry of the acid-activation process and the properties and chemistry of the resultant catalysts have been discussed by (among others): Milliken *et al.* (1955); Ryland *et al.* (1960); Mills *et al.* (1950); Thomas *et al.* (1950); and Hansford (1952).

6.4.1.1 Acid activation
Natural halloysite and montmorillonite have limited cracking activity, with the former clay the more active. Acid treatment increases this activity (*c.* eight-fold for montmorillonite and two- to three-fold for halloysite) to a maximum, but the activity then declines gradually as the dissolution continues. The primary accomplishment of this acid leading procedure is the removal of metallic ions, mostly aluminium and magnesium, from the octahedral sheet of the mineral. Thomas *et al.* (1950) postulated that removal of one of a pair of octahedral aluminium ions from montmorillonite, for example, removes two hydroxyl groups and leaves the other aluminium of the pair in four-fold coordination. This tetrahedral aluminium, with its charge balancing proton, forms a Brønsted site and is the source of the catalytic activity. On the basis of this proposal, all montmorillonites should become active catalysts upon such treatment.

However, as Mills *et al.* (1950) have shown, the cracking activities of acid-treated bentonites from various deposits cover the range from inactive to a level of activity comparable to silica–alumina. This range exists despite almost identical physical properties, after acid treatment, for both activable and non-activable bentonites. A possible basis for differentiating between these two classes may be found in the observation by Milliken *et al.* (1955) that acid activation of raw kaolinite and halloysite, followed by calcination, led to catalysts of inferior quality. However, if calcination at 550 °C precedes the acid leaching step, the activity of the catalyst so obtained from either halloysite or kaolinite was greatly improved. Calcination of the kaolin at this temperature should have caused dehydroxylation with its accompanying transition to metakaolin. This change is believed to be characterized by a shift in the coordination of the aluminium from octahedral to tetrahedral (Fripiat 1963). However, in this configuration the tetrahedra do not share corners as might be expected; rather, they share edges. This hypothesis is supported by the Al K_α energies seen by X-ray fluorescence. A. C. Wright (private communication) has obtained Al K_α shifts relative to Al metal (see Fripiat 1963) (in $\Delta 2\theta$, EDDT analysing crystal) as follows: -0.110 for Al^{VI}, -0.055 for corner-sharing AlO_2 tetrahedra in silicates, and -0.075 for metakaolin. It is significant to the later discussion that these edge-sharing alumina tetrahedra are uncharged and therefore do not contribute to the CEC or the acidity.

No published information is available concerning the detailed structural composition of these activable and non-activable clays, particularly with respect to the extent and nature of isomorphous substitution. Nevertheless, if the bentonite deposits described by Mills et al. (1950) are compared with the similar samples, from nearly the same localities in some cases, studied by Osthaus (1956), it becomes apparent that the distinguishing characteristic of an activable clay may well be the presence of considerable aluminium in tetrahedral coordination.

To explain the pseudo-first-order kinetics (large excess of acid) of the dissolution of bentonite observed by Osthaus (1956), Granquist and Sumner (1959), and others, it is necessary that the area of the clay surface attacked by the acid decreases with time. The surface area measured by nitrogen adsorption increased with time: therefore, the reactive surface must have been either a decreasing part of the measured area or not part of it. Furthermore, the X-ray diffraction and analytical data presented by Granquist and Sumner (1959), although interpreted somewhat differently by them, show that the "00l order", that is, the largest value of l observed for the 00l series of reflections, of the stacking domains decreased more rapidly than the transport of aluminium from the structure to the solution. However, if the acid-leached product had had a structure consisting of unaltered silica sheets separated by an Al^{IV} layer obtained by removal of about half the ions in the original octahedral layer, the 00l order would have decreased less rapidly than the aluminium dissolution. These various observations were best explained in terms of a rather uniform edge attack by the acid on the octahedral layer of the montmorillonite. In such a situation, the reactive surface (octahedral edge area) would decrease with time as demanded by the kinetics, but the total area would increase as observed.

Acid activation thus promotes catalytic activity of bentonite by making more surface accessible for the heterogeneous catalytic reaction and by increasing the number of Brønsted and potential Lewis acid sites (arising from four-coordinated Al^{3+}) in that surface. It is suggested that these sites are not created by the acid treatment, but rather are exposed by that dissolution process. This suggestion requires that the conclusion reached by Osthaus (1956) be correct; i.e. in montmorillonite, the aluminium ions in six-fold coordination go into solution faster than do those in four-fold coordination. It is recognized that the work of Ross (1969) establishes that this conclusion is not generally applicable; specifically, not to the chlorite studied by him. As a result, the first stage of the treatment exposes such Al^{IV} sites and increases their concentration per unit of surface area. More drastic treatment removes Al^{IV} from these exposed sites, and thus decrease the Al^{IV} from these exposed sites, and thus decreases the Al^{IV}/area ratio and causes a corresponding drop in the catalytic activity.

For kaolinite and halloysite the mechanism must be somewhat different since appreciable substitution of Al^{3+} for Si^{4+} does not occur. Also, the "amorphous" metakaolin phase is more readily attacked by acid than is the crystalline mineral. This difference may be due more to the disorder accompanying the transition kaolin → metakaolin than to the relative solubilities of Al^{IV} and Al^{VI}. At any rate, the development of catalytic activity in this case can be attributed to the presence of four-coordinated Al^{3+} in an edge-sharing configuration, brought about by the calcination step, followed by acid-leaching to remove part of the tetrahedra with rearrangement of those remaining to some configuration with a negative charge per tetrahedron. The development of accessible surface of sufficient extent is also a necessary result of the acid treatment.

Admittedly, this hypothesis does not explain why calcination should preferably precede the acid degradation, since the $Al^{VI} \rightarrow Al^{IV}$ transition should lead to the same active site whether it comes before or after the acid treatment. The reason may lie in the surface area/pore volume relationships; i.e. there is no reason to suppose that the same degree of aluminium removal from crystalline kaolin and disordered metakaolin should lead to the same pore geometry (the term "pore" is used in its broadest sense). Even if the Al^{IV}/area ratio were the same in the two cases, which is doubtful, the area may exist as pores (cracks, crevices) too small to be accessible to the reactant molecules.

6.4.1.2 Catalyst properties

Bench-scale methods of catalytic evaluation may involve quantities of catalysts ranging from a few grams to several hundred grams and use a fixed-bed with granular catalyst or a fluidized fixed-bed with microspheroidal (20–80 μm) particles. Feedstocks range from 2,3-dimethylbutane and isopropylbenzene (few and predictable compounds in the cracked product), to *n*-hexadecane, in the boiling range normally encountered in commercial practice and useful for a study of product distribution (branched-chain, cyclic, aromatic compounds must arise by secondary reactions), to commercial gas oils representative of typical refinery practice. The test conditions may employ weight-hourly-space-velocities (WHSV, wt of feed/wt of catalyst/hour) from less than one to values near twenty. The choice of the method will vary with the reason for the evaluation; i.e. whether one is trying to determine commercial acceptability of a particular catalyst, or studying the cracking reaction itself and striving for simplicity in the reaction system, or trying to relate changes in catalyst composition (added ions, for example) to changes in product distribution, and so forth.

Non-cracking parameters related to catalyst performance are characteristics such as surface area (BET method, N_2 adsorption), pore volume (extension of N_2 isotherm to saturation pressure, or mercury intrusion, or oil absorption), pore volume distribution (interpretation of the isotherm or the porosimeter data), particle size distribution and bulk and true density, composition, and trace metal analysis. These properties relate to the accessibility of the surface to the hydrocarbon molecule which is being cracked; the regenerability of the catalyst, a process which is likely diffusion-controlled; the fluidization properties; and the probable behaviour of the catalyst in the commercial operation (stability in the presence of high-sulphur feedstocks, control of secondary reactions leading to excessive coke formation and high gas yields).

For these reasons, it is difficult to select from the literature a set of interrelated, self-consistent data that presents the properties of the clay-based cracking catalysts and places these properties in the correct position relative to those of the gel catalysts. Table 6.5, based on Milliken *et al.* (1955), gives a partial set of results on catalysts (i.e. calcined at 550 °C, but not steam deactivated or previously used).

Upon steam deactivation, the activity index declined 25–35%, with a somewhat greater decrease in the surface area. The halloysite-based catalyst is more stable in this respect than its montmorillonite counterpart. Increase in the alumina content of the synthetic silica–alumina to

TABLE 6.5. Comparative properties of clay catalysts[†]

	SiO_2/Al_2O_3 mole ratio	Surface area, $m^2 g^{-1}$	Catalytic activity index
Synthetic SiO_2/Al_2O_3	11.4	300–600	45–50
Montmorillonite			
Raw	4.3		5
Acid-activated	5.3	300	40
Halloysite			
Raw	1.6		12–20
Acid-activated	2.3	160	34–40

[†] Taken from Milliken et al. (1955). Used by permission of the California Division of Mines and Geology.

give ratios near those shown for the clay catalysts improves the activity and stability of the gel systems and makes them substantially superior to the clay-derived products.

6.4.1.3 Pyrolysis of organo-clay complexes

It is well-known to workers in colloid chemistry of clay minerals that surface areas (BET method, N_2 adsorbate) of montmorillonite do not encompass the large interlamellar area. That is, such an area determination for montmorillonite will give a result of the order of 15–30 $m^2 g^{-1}$, while the interlamellar area (counting all face area) can be calculated as approximately 800 $m^2 g^{-1}$ (Grim 1968). Surface areas determined by water sorption approach this value; in this laboratory, values in excess of 600 $m^2 g^{-1}$ have been measured. Cracking activity of a clay catalyst correlates well with the area measured with nitrogen sorption and one is forced to the conclusion that the interlamellar space is not involved. Rather, the activity is derived from the edges and exposed faces of the collection of stacking domains that make up the gross sample. The reactivity of organic material held in the space between the layers remains of interest, however.

Pertinent to this question, but peripheral to the main theme of this chapter, is the work of Chou and McAtee (1969). They studied the pyrolysis of clay–organic complexes, prepared by ion exchange (Jordan 1949). The product contained as the charge-balancing cation an organic ammonium ion possessing one or more hydrocarbon chains, the longest being C_{18}. Jordan (1949) has shown that in the dried product the chains lie in the interlamellar space along the surface of the crystallites. Chou and McAtee observe the changes occurring when these compounds are heated in air and in an inert atmosphere. As heating proceeded, the first reaction was an extensive dehydrogenation of the chain occurring between 180 and 350 °C; this reaction was followed by some cracking and limited loss of lower molecular weight organic molecules. Dehydroxylation of the clay substrate at about 650 °C caused a partial loss of carbon by reaction with the evolved water to give carbon monoxide and dioxide. The reaction was thus quite different from the catalytic cracking observed when a molecule of about the same size as the C_{18} chain is contacted in the vapour phase with the silicate surface.

One very important objection exists to accepting this work as evidence of the limited reactivity of the interlamellar space. That is, the organic compounds used were all nitrogen bases with the positive charge concentrated near the nitrogen. In the complex, the nitrogens would be seated as closely as possible to the sites of negative charge which probably are also the catalytically active sites on the crystallite faces. Therefore, the method of preparing the complexes may well have poisoned most of the expected catalytic activity (Voge 1958). A similar study, but based on the use of non-ionic complexes chosen to minimize the possibility of site poisoning, would be a most interesting extension of these authors' work.

6.5 SYNTHETIC LAYER-LATTICE ALUMINOSILICATES AS CATALYSTS

Mineral synthesis offers a possible route to layer-lattice aluminosilicates with properties optimized for particular catalytic reactions. Proper control of crystal growth can maximize the edge/face area ratio, for example. By regulating the nature and degree of isomorphous substitution, one can conceivably obtain products of varying surface acidity (both strength and concentration of sites) and catalytic function. That is, by variation of Si/Al within the stability limits of the desired phase the former property is, in principle, controllable. Also, by introduction during synthesis of replacements such as Ni^{2+}/Al^{3+} or Cu^{2+}/Al^{3+}, for example, the catalytic function of the product can be altered, say, from cracking to dehydrogeneration.

6.5.1 A synthetic aluminian smectite

Granquist (1966), Capell and Granquist (1966), Granquist and Pollack (1967) and Wright et al. (1972) have discussed in detail the synthesis, structure and properties of such a mineral. The product was a 2:1 dioctahedral layer-lattice silicate, which contained only aluminium in the octahedral layer, some Al^{3+}/Si^{4+} tetrahedral substitution, and about one F^-/OH^- replacement per unit cell. The unit cell formula could be written as:

$$[(Al_4)^{VI}(Al_xSi^{IV}_{8-x})O_{20}(OH,F)_4]^{x-} \cdot xNH_4^+ \tag{9}$$

For $x = 1.5$, a typical case, X-ray examination of an oriented specimen provided evidence of a

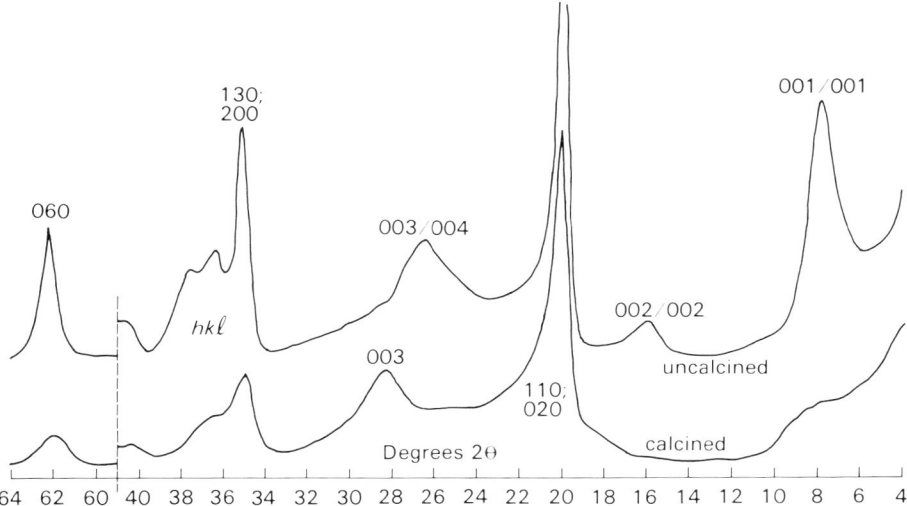

FIG. 6.5. X-ray diffractometer traces of non-oriented synthetic smectite, uncalcined and calcined to equilibrium at 700°C.

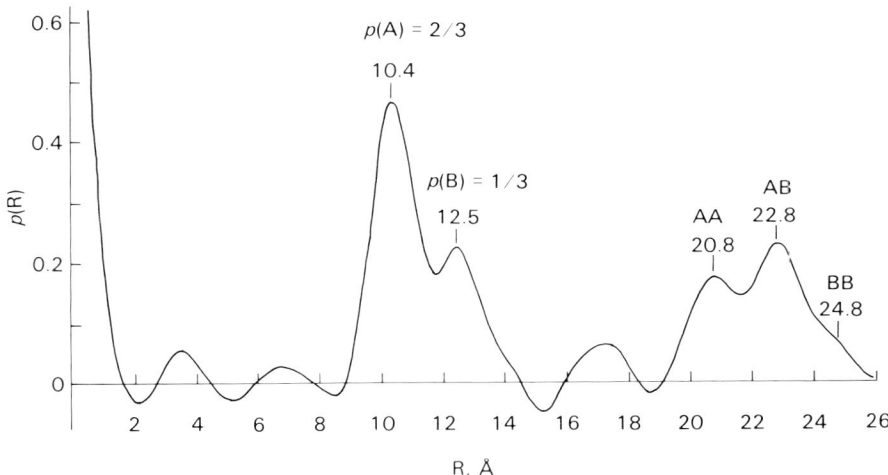

FIG. 6.6. Direct Fourier transform of the diffracted intensity of an oriented sample of synthetic smectite. Interstratification of two spacings, 10.4 Å and 12.5 Å.

non-integral 00l sequence. A direct transform (MacEwan 1958) of the diffracted intensity demonstrated that: the sequence was composed of two layers, one with $d(001)$ (A) = 10.4 Å; the other, $d(001)$ (B) = 12.5 Å; the number density, p, of each of the two kinds of layers was $p(A) = 2/3$, $p(B) = 1/3$; and the stacking sequences approached random interstratification. One possible reason advanced for the existence of two spacings was compositional; i.e. for species A, $1 < x < 2$ and for species B, $0 < x < 1$. If so, A would be mica-like and B, montmorillonite (or beidellite)-like. Since $p(A) > 0.5$, even with random interstratification AAA... sequences must occur frequently in the stacking domains. The presence of weak hkl reflections in the X-ray pattern was attributed to such occurrences.

The X-ray diffractometer traces for a randomly oriented sample, before and after calcination at 700 °C, are shown in Fig. 6.5. The direct transform of the intensity data obtained from the uncalcined oriented specimen appears as Fig. 6.6.

6.5.1.1 Dehydroxylation

Upon calcination, the following sequence of reactions was thought to occur (for simplicity, F is disregarded and $x = 1$):

$$[(Al_4)(AlSi_7)O_{20}(OH)_4]^- \cdot NH_4^+$$
$$\downarrow \begin{array}{c} T \approx 350\,°C \\ -NH_3 \end{array}$$
$$[(Al_4)(AlSi_7)O_{20}(OH)_4]^- \cdot H^+$$
$$\downarrow \begin{array}{c} T \approx 550\text{--}650\,°C \\ -H_2O \end{array}$$
$$[(Al_4)(AlSi_7)O_{22}]^- \cdot H^+$$

While written as sequential reactions, deamination and dehydroxylation probably overlapped considerably. In fact, deamination may well catalyse the dehydroxylation process.

Differential thermal analysis showed the anhydride to be stable to at least 900 °C. For most of the reactions studied in the laboratory, the stable anhydride is the catalytically active species.

The position in the structure and the fate, upon prolonged heating, of the charge-balancing protons was of interest. The perturbed —OH stretch at 3475 cm^{-1} (see Fig. 6.3) provided evidence that the proton may drop into the tetrahedral vacancy adjacent to the hydroxyls in the octahedral layer. Dehydroxylation probably proceeded by interaction of adjacent pairs of hydroxyls and could conceivably have gone to completion without a net loss of charge-balancing protons. Exchange between the two kinds of protons, hydroxyl and those arising from the NH_4^+ decomposition, undoubtedly occurred to a large extent. In fact, Fripiat et al. (1965) have proposed proton delocalization as a necessary preliminary step in the dehydroxylation of mica, and a similar mechanism may have been operative here.

For each molecule of water evolved, the crystallite lost two protons and one site (i.e. hydroxyl oxygen), which in turn decreased the probability of the occurrence of the next H····O····H lattice interaction. Such a process should be first order with respect to OH^-. The presence of fluoride in the structure complicated the picture because it decreased the initial —OH concentration and effectively separated the hydroxyls, thus inhibiting dehydroxylation.

No matter what the mechanism, the process must have ended with charge-balancing protons present in the structure, or lost as water by interaction with oxygens normally in an Al—O—Si, or in an Al—O—Al which originated from the pairwise —OH loss. In these latter two cases, the unit cell charge could have been compensated internally by a mechanism such as that proposed for synthetic faujasite by Uytterhoeven et al. (1965). Even after 16 hours at 650 °C and a vacuum of 10^{-6} Torr (10^{-7} kPa), evidence was still seen in Fig. 6.3 for hydroxyl, both in the perturbed-OH stretch at 3475 cm^{-1}, and in the sharp-OH band at 3750 cm^{-1} attributed to edge silanol groups. As demonstrated in this figure, and discussed in detail by Granquist and Kennedy (1967), the anhydride chemisorbed water to regenerate hydroxyl. This chemisorption was considerable even at 500 °C (c. 2.0% by weight) and was enhanced by the presence of fluoride.

6.5.1.2 Acidity

The acidity of the anhydride was discussed earlier in Section 6.2.2.4. The number of acid sites, corresponding to the Lewis- and Brønsted-bound pyridine of Fig. 6.3, was measured quantitatively and found to be 3 meq/100 g. If it is assumed that the same area is accessible to nitrogen at -165 °C and to pyridine at 200 °C, this value is equivalent to 1.8×10^{17} sites/m^2, for an area of 100 m^2 g^{-1}. Measured areas varied between 100 and 135 m^2 g^{-1}, dependent on the calcination and outgassing temperatures. Since the charge deficiency, and thus the acidity, arose from Al^{3+}/Si^{4+} replacement, the expected acidity would be of the order of the cation exchange capacity. This latter value, calculated from the number of tetrahedral Al^{3+} per unit cell, was near 200 meq/100 g. Measured values, obtained by Kjeldahl distillation of NH_3 from the ammonia clay as synthesized, usually amounted to 40–50 meq/100 g less than this expected value. It was assumed that this discrepancy was caused by the presence in the interlamellar space of some hydroxyaluminium cation. In either case, the CEC was two orders of magnitude higher than the measured value of the acidity. It is instructive to consider reasons for this difference.

From a consideration of X-ray line broadening, nitrogen sorption data, and electron micrographs, Wright et al. (1972) concluded that the stacking domains consisted of N parallel platelets normal to the c axis, with N following a Poisson distribution centred at about $N = 4.5$. The gross, non-oriented, sample was a collection of these domains such that the c axis for a given domain could assume any orientation in space. Nitrogen sorption–desorption isotherms were sigmoid (type II) and showed little or no hysteresis; i.e. the process seemed to be equivalent to multilayered sorption on a planar surface. A logical conclusion was that nitrogen was sorbed on the two faces and the edges of each stacking domain, and that the interlamellar space was not accessible (see, for example, Grim 1968). If it is assumed that the same situation obtains for pyridine chemisorption at 200 °C, this inaccessibility of the interlayer space effectively reduces the measured acidity to $1/N$ of the CEC.

If the stacking domains are considered cylindrical with $d(001) = 10$ Å, the face area contributed by each domain will be $2\pi r^2$; the edge area, $20\pi rN$. The edge/face area ratio will then be $10N/r$. If it is additionally assumed that *only* the *edge* area contributes to the measured acidity, the observed value ought to be the product of these two factors, $(1/N)(10N/r)$, or $10/r$ of the CEC. A more rigorous derivation follows: let the domains be cylindrical of radius r and contain N plates per domain. Then the total face area per domain will be $2\pi r^2 N$, the total edge area, $2\pi r \cdot 10N$. The ratio of edge area to total area is then $(20\pi rN)/(20\pi rN + 2\pi r^2 N) = 10/(10+r) \approx 10/r$. For $N = 5$ and $r = 500$ Å (supported by electron micrographs), the factor becomes $10/510 \approx 1/50$. A measured CEC of 160 meq would then correspond to 3.2 meq of acidity. The agreement between this calculation and the value, 3 meq/100 g, measured by pyridine chemisorption at 200 °C lends credence to the hypothesis that only the crystallite edges are involved in the catalytic process.

6.5.1.3 Catalytic activity
The changes in catalytic activity and other catalyst parameters, which occurred upon crystallization of the smectite from the amorphous silica–alumina reaction mixture, are summarized in Table 6.6. The decrease in exchange capacity and surface area, expected as a result of the crystallization, was accompanied by an unexpected increase in the catalytic activity. In the experiment which resulted in the tabulated data, the silica–alumina was prepared by hydrolysis of $SiCl_4$ and $Al(OCH(CH_3)_2)_3$. These observations, made early in the study of smectite synthesis by Granquist, led to the work summarized in this section of the chapter.

Capell and Granquist (1966) provided a detailed comparison of the properties and cracking activities of this synthetic smectite, so-called low-alumina (13 wt %) silica–alumina gel catalyst, and an acid-activated halloysite catalyst. Table 6.7 presents a summary of their data. It is apparent from the tabulation that the synthetic smectite possessed superior resistance to steam

TABLE 6.6. Changes occurring upon crystallization

Structure:	300 C Reactants ⟶ Product 1240 psi (6.9 MPa) H_2O	
	X-ray amorphous	Smectite
SiO_2/Al_2O_3 mole ratio	3.32	3.00
Predominant exchange cation before calcination	NH_4^+	NH_4^+
Exchange capacity, meq/100 g	285	148
Surface area, BET, $m^2 g^{-1}$	240	98
Relative catalytic activity[†]	1	2.6
Relative unit area activity	1	5

[†] For cracking 2,3-dimethylbutane.

TABLE 6.7. Comparison of synthetic smectite, SiO_2/Al_2O_3, and halloysite catalysts

	Synthetic smectite	Silica–alumina	Halloysite
SiO_2/Al_2O_3 mole ratio:	2.42	1.11	3.19
Surface area, $m^2 g^{-1}$			
Fresh[†]	115.4	579	160
Steam-deactivated[‡]	104	285	79
Saturation volume, $cm^3 g^{-1}$			
Fresh	0.338	0.66	0.30
Steam-deactivated	0.319	0.56	0.26
Downflow activity[§]			
60 min conversion, wt %	45.1	32	23
20 min conversion, wt %	59.4	39.8	32.1

† Calcined 3 h at 565 °C.
‡ Above, steam-treated at 593 °C for 8 h, and 15 psig (104 kPa), followed by recalcination at 565 °C.
§ See Whitaker and Kinzer (1955).

deactivation and that the high order of catalytic activity demonstrated in the 2,3-dimethylbutane experiments (Table 6.6) extended to cracking tests on a commercially realistic feedstock.

Patents by Jaffe (1970), Jaffe and Kittrell (1970), Kittrell (1970a,b), Kittrell and Sullivan (1970), and Csicsery et al. (1970), which described hydrotreating catalysts based on this synthetic smectite, suggest that the range of utility may go well beyond the cracking reactions studied by Granquist.

6.6 CLAY INTERCALATED METAL COMPLEX CATALYSTS[2]

Recent advances in the intercalation chemistry of smectite clays have rekindled interest in these minerals as catalyst supports for metal complex catalysts. Although the immobilization of complex catalysts in clay structures makes it possible to conduct solution-like reactions in the solid state and to minimize many of the technical and economic barriers associated with the use of homogeneous solution catalysts, the advantages of catalyst intercalation go beyond mere immobilization. By modifying the chemical and physical forces acting on interlayer reactants, one often can improve catalytic specificity relative to homogeneous solution.

Metal complex catalysts intercalated in smectite clays are accessible for reaction under ambient conditions, provided that the interlayers are swollen to permit rapid diffusion of reagents. A solution-like environment exists when the interlayers are highly expanded by a large number of molecular layers of solvent. However, the gel-like nature of such highly swollen phases makes them highly resistant to fluid flow and therefore unsuitable as heterogeneous catalysts. For an intercalated clay molecular catalyst to be useful, the interlayers must be mobile under intermediate degrees of swelling where the intercalate retains the mechanical properties of a genuine solid.

Complementary ESR and NMR experiments (Pinnavaia 1980, 1981; Fripiat 1980; Stone 1981), along with quasi-elastic neutron scattering studies (Ross and Hall 1980; Hall 1981), have provided incisive information on the interlayer dynamics of clay intercalates. A general picture of the interlayer environment has emerged from these studies. At low degrees of interlayer solvation, the solvated exchange cations adopt oriented positions on the interlamellar surfaces. Though oriented, the solvated cations are in a dynamic state and undergo anisotropic rotations about specific molecular axes. Uncoordinated water molecules between the solvated cations are capable of translational diffusion within and between the "cages" defined by the solvated cations and the silicate layers (Hall 1981). Restricted motions and preferred orientations also have been observed for intercalated organic species (McBride 1980; Stone 1981; Resing *et al.* 1980).

Clementz *et al.* (1973) and McBride *et al.* (1975a) found that hydrated Cu^{2+}, ions in clay interlayers are highly oriented when solvated by one to three molecular layers of water. However, swelling the interlayers to ~ 12 Å with more water results in rapid tumbling and dynamic Jahn–Teller distortions of the hydrated Cu^{2+} ion. Quantitative estimates of the tumbling motion of Mn^{2+} and VO^{2+} ions under comparable swelling conditions show their correlation times to be only 30–50% larger than those observed for the same ions in dilute aqueous solution (McBride *et al.* 1975b; McBride 1979). It is this labile nature of solvated smectite clay interlayers at intermediate degrees of swelling which makes possible their use of intercalation catalysts under mild reaction conditions.

Although the importance of exchangeable metal cations on the acid catalysed reactions of smectites was revealed more than 40 years ago (Broughton 1940), it has been only recently that authentic metal complex catalysed reactions have been carried out in the interlamellar spaces of these minerals. A number of different types of reactions have been catalysed by metal complexes in clays, including the hydrogenation and isomerization of olefins and acetylenes and olefin hydroformylation (Pinnavaia 1982).

6.6.1 Hydrogenation–Isomerization

Pinnavaia *et al.* (1976) and Pinnavaia and Welty (1975) found that di-rhodium acetate complex cations intercalated in hectorite react with triphenylphosphine from methanol solution to form intercalated $Rh(PPh_3)_n^+$ species:

$$\overline{Rh_2(OAc)_{4-x}^{x+} \xrightarrow[\text{(MeOH)}]{PPh_3} 2Rh(PPh_3)_n^+} \quad (10)$$

where $x = 1$ or 2, $n = 2$ or 3, and the horizontal lines represent the silicate sheets. The $Rh_2(OAc)_{4-x}^{x+}$ ions, which are derived from $Rh_2(OAc)_4$ by reaction with protonic acids, were originally formulated as Rh_2^{4+} ions based on earlier work by Legzdins *et al.* (1970) in which the use of rhodium-bridged carboxylates as homogeneous hydrogenation catalysts was first reported. Later studies of the solution protonation chemistry of $Rh_2(OAc)_4$ (Pinnavaia *et al.* 1979; Wilson and Taube 1975) showed that two or three bridging acetate ligands remain coordinated to the di-rhodium(II) ion in the presence of strong Brønsted acids. Nevertheless, the replacement of coordinated acetate by triphenylphosphine according to equation (10) occurs readily both in the intercalated state and in homogeneous solution.

$Rh(PPh_3)_n^+$ cations are catalyst precursors for the hydrogenation of alkenes and alkynes (Legzdins *et al.* 1970; Schrock and Osborn 1976a,b; Pinnavaia *et al.* 1979). Table 6.8 compares the results for the hydrogenation of 1-hexene in methanol with the intercalated and homogeneous catalyst systems. Under the reaction conditions employed, the hydrogen uptake rate is lower for the intercalated catalyst than for the homogeneous catalyst, but, remarkably, the intercalated catalyst greatly reduces the extent of 1-hexene to 2-hexene isomerization which

CLAY INTERCALATED METAL COMPLEX CATALYSTS 299

TABLE 6.8. Hydrogenation of 1.0M 1-hexene in methanol with $Rh(PPh_3)_n^+$ as the catalyst precursor under hectorite-intercalated and homogenous condition[†]

System	PPh_3/Rh	% Conver.	Rate[‡]	Product distribution %	
				Hexane	2-Hexene
Intercalated[§]	4.0[♦]	5	16	100	
		50	16	100	
Homogeneous	4.0[♦]	5	200	100	
		38	190	63	37
		66	130	65	35

[†] Initial substrate to rhodium ratio is 2000:1; temperature is 25 °C.
[‡] Hydrogen uptake rate, ml min^{-1} mmol^{-1} Rh.
[§] Rhodium loading on hectorite is 0.72 wt%.
[♦] Total PPh_3/Rh ratio.

accompanied hydrogenation. The ability of the intercalated catalyst to inhibit substrate isomerization is related to the presence of a surface equilibrium between rhodium monohydride and dihydride complexes as active catalyst (equation (11)). The dihydride complex is known to

$$Rh(PPh_3)_n^+ \xrightarrow{H_2} RhH_2(PPh_3)_n^+ \leftrightarrows RhH(PPh_3)_n + H^+ \qquad (11)$$

be a good hydrogenation catalyst but a poor isomerization catalyst, whereas the monohydride is both a good hydrogenation catalyst and a good isomerization catalyst (Schrock and Osborn 1976a). Relative to homogeneous solution the clay interlayers shift this equilibrium in favour of the dihydride complex and isomerization thereby is inhibited.

Recent studies using $Rh(NBD)(PPh_3)_2^+$ (NBD = norbornadiene) as a precursor to the dihydride complex (equation (12)), indicate that the inhibition of 1-hexene isomerization under

$$Rh(NBD)(PPh_3)_2^+ \xrightarrow{H_2} RhH_2(PPh_3)_2^+ + \text{norbornane} \qquad (12)$$

intercalated conditions in hectorite very much depends on the composition of the solvating medium (Raythatha and Pinnavaia 1983). With 0.1 wt% water in methanol no isomerization of 0.80M 1-hexene is observed at 40% conversion, but with 0.5 wt% water in the reaction medium about 50% of the product is 2-hexene at 50% conversion. The isomerization reaction with the clay intercalated catalyst is also highly dependent on the initial concentration of 1-hexene. At initial concentration ≲0.6M isomerization accompanies hydrogenation over the entire range of conversion (as it does with homogeneous catalyst), but at initial concentration ≳0.7M isomerization is inhibited until 40–60% of the substrate has been hydrogenated, and then isomerization is initiated. The product distributions for the hydrogenation of 0.1M and 0.8M 1-hexene are illustrated in Fig. 6.7. In marked contrast to the intercalated catalyst, the homogeneous catalyst affords hexane and 2-hexene in ~30:70 relative abundance, independent of conversion, water content of the solvent, or initial concentration.

The dependence of 1-hexene isomerization on water content of the solvent and initial substrate concentration appears to be related to changes in the intrinsic Brønsted acidity of the clay interlayers. In the intercalated catalysts described above, ~90% of the exchange sites are occupied by Na^+. Although Na^+ is not an acidic ion under ordinary hydration conditions in

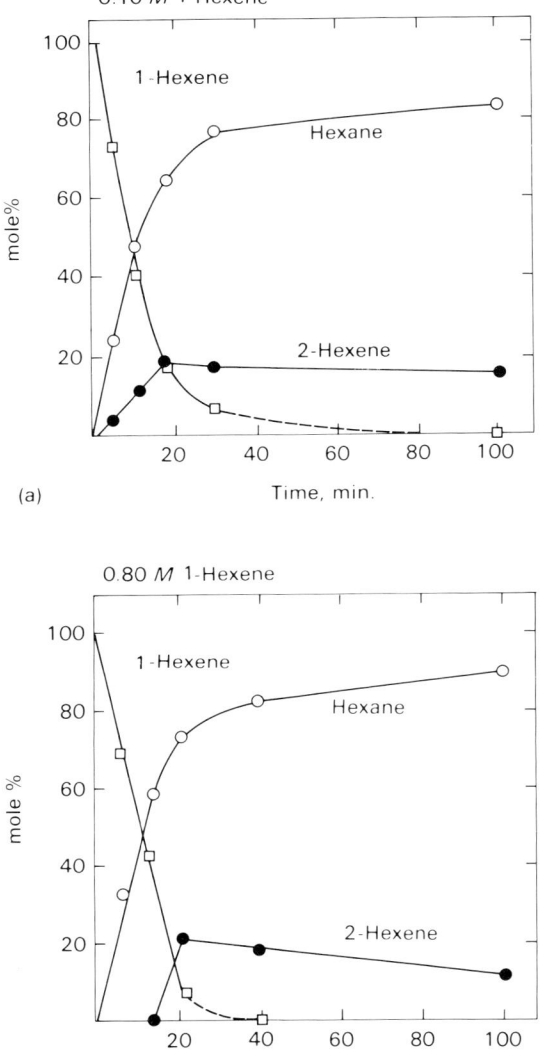

FIG. 6.7. Hydrogenation in methanol (0.20 wt % H_2O) of (a) 0.10M 1-hexene and (b) 0.80M 1-hexene at 25 °C with $Rh(NBD)(PPh_3)_2^+$-hectorite. The amount of intercalated rhodium used was 0.015 mol in 15 ml of solution.

solution, the *partially* hydrated ion on clay interlayers is sufficiently acidic to protonate bases as weak as tetraphenylporphyrin in detectable amounts (Cady and Pinnavaia 1978). Thus the hydrolysis of partially hydrated Na^+ ions (equation (13)) probably plays a major role in

$$\overline{Na^+\text{---}OH_2} \rightarrow \overline{[NaOH]} + H^+ \qquad (13)$$

determining the position of the equilibrium described by equation (11). As the number of water molecules solvating the interlayer Na^+ ions is decreased, the charge on Na^+ should become progressively less shielded, and the extent of water dissociation should increase. Therefore, a low water content should increase the surface proton activity and favour $RhH_2(PPh_3)_2^+$ over $RhH(PPh_3)_2$. Similarly, the interlayer acidity is expected to increase with increasing substrate concentration. As the amount of adsorbed olefin is increased at the expense of the interlayer methanol, the interlayers should become less hydrophilic. A decrease in the hydrophilic character of the interlayers will lower the surface concentration of water and increase the extent of Na^+ hydrolysis. Thus the inhibition of olefin isomerization at low water content and high substrate concentration is qualitatively consistent with an increase in interlayer Brønsted acidity which tends to shift the equilibrium in equation (11) in favour of the dihydride complex.

Although the deviations from solution behaviour for the hydrogenation of 1-hexene with the intercalated catalyst arise from shifts in the position of catalytically important protonic equilibria, substrate size has been found to play a crucial role in the selectivity of intercalation catalysts in alkyne hydrogenation (Pinnavaia *et al.* 1979). Both $RhH_2(PPh_3)_n^+$ and $RhH(PPh_3)_n$ are active for the hydrogenation of alkynes to the corresponding *cis* olefins (Schrock and Osborn 1976b). Table 6.9 shows the dependence of the hydrogenation rates for a series of

TABLE 6.9. Hydrogenation of alkynes in methanol at 25 °C with hectorite-intercalated and homogeneous $Rh(PPh_3)_n^+$ catalyst precursors[†]

Substrate	Initial rates[‡]		R_I/R_H [•]
	Intercalated[§]	Homogeneous	
1-hexyne	2100	2100	1.0
2-hexyne	2200	2400	0.92
2-decyne	1200	2500	0.48
3-hexyne	360	1800	0.20
PhC≡CPh	<1	100	<0.01

[†] Substrate concentration is 1.0M; substrate: Rh = 2000:1.
[‡] Rates are in ml H_2 min^{-1} mmol^{-1} Rh.
[§] Catalyst intercalated in Na^+-hectorite at a loading of 0.72 wt% Rh.
[•] Ratio of intercalated to homogeneous rates.

alkynes in methanol in the presence of the intercalated and homogeneous $Rh(PPh_3)_n^+$ catalyst precursors. With methanol as the solvating medium the interlayer regions of the intercalated hectorite are about 7.7 Å thick. As the steric bulk on either side of the C≡C bond increases, the ratios of reaction rates for the intercalated and homogeneous catalyst (R_I/R_H) decrease. The results of complementary alkyne hydrogenation experiments in which the nature of the swelling solvent is varied and the alkyne size is held constant (using 2-decyne) are shown in Table 6.10. As the average thickness of the interlayers ($\Delta d(001)$) is decreased from 10.0 Å in CH_2Cl_2 to 5.6 Å in benzene (where there is essentially no interlayer swelling), the R_I/R_H values decrease dramatically.

Although the results for alkyne hydrogenation show that the selectivity of the intercalated catalyst depends on substrate size and interlayer swelling, the idea of a molecular sieving effect is untenable based on the data in Table 6.9. Each substrate in this table is sufficiently small in at

TABLE 6.10. Hydrogenation of 2-decyne with hectorite-intercalated and homogeneous $Rh(PPh_3)_n^+$ in solvents of different swelling power[†]

Solvent	Initial rates[‡]		R_I/R_H[•]	$\Delta d(001)(\text{Å})$[∥]
	Intercalated[§]	Homogeneous		
CH_2Cl_2	2800	3300	0.85	10.0
MeOH	1200	2800	0.43	7.7
EtO/MeOH (3:1 v/v)	660	2800	0.24	6.7
C_6H_6	20	1000	0.02	5.7

[†‡§¶] See same footnotes in Table 6.9.
[∥] Average interlayer thickness of the solvated intercalate as determined from 001 X-ray reflections.

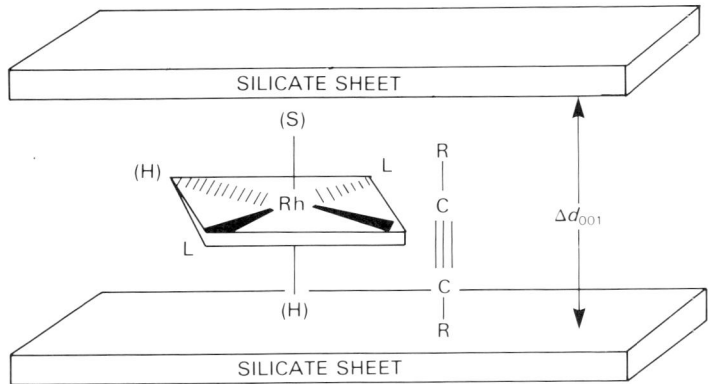

FIG. 6.8. A proposed orientation for the reactive $RhH_2L_2^+$-alkyne (L = triphenylphosphine) intermediate formed in the interlayers of hectorite. Note that the C≡C axis of the alkyne is perpendicular to both the L–Rh–L axis and the plane of the silicate sheets.

least one dimension to penetrate the methanol solvated interlayers. The apparent substrate size selectivity most likely arises instead from the spatial requirements of partially oriented $RhH_2(PPh_3)_2^+$-alkyne intermediates in the restricted interlayer region. One possible orientation is illustrated in Fig. 6.8 where the *trans* Ph_3P—Rh—PPh_3 axis is parallel to the clay layers and the C≡C axis is perpendicular to the layers. Destabilization of the oriented intermediate due to unfavourable repulsions between the silicate layers and R groups on the substrate will be especially severe when the substrate critical dimension for orienting the C—C bond axis perpendicular to the silicate layers is larger than the interlayer swelling. However, a precise relationship between substrate critical dimension, interlayer swelling, and reactivity cannot be expected. In addition to being turbostratic, clays interlayered with metal complex catalysts tend to be interstratified with some interlayers somewhat larger or smaller than the average value indicated by 00*l* X-ray reflections. Thus, efficient catalytic discrimination among substrates on the basis of size or shape can be expected only when the differences in critical dimensions are larger than the spread in interlayer spacings.

1-Hexene in methanol also has been selectively hydrogenated to 1-hexene over a Rh(III)–hectorite catalyst in the presence of PPh_3 as a ligand (Pinnavaia *et al.* 1976). In this case the clay catalyst was prepared by air oxidation of $Rh(OAc)_{4-x}^{x+}$ (originally formulated as Rh_2^{4+}) in the presence of a small amount of HF which acts as a catalyst for the metal oxidation (equation (14)):

$$M^+\text{-hectorite} \xrightarrow[\text{methanol}]{Rh_2(OAc)_{4-x}^{+},O_2,HF} Rh(III)\text{-hectorite} \quad (14)$$

$$M^+ = Na^+, N(CH_3)_4^+ \qquad (40\text{--}45 \text{ mmol Rh/100 g})$$

The clay bound Rh(III) species is believed to be an polymeric oxycation with an average charge per Rh(III) of ~0.8. The intercalate exhibits a single $00l$ X-ray reflection corresponding to $d(001) = 14.7$ Å. In the absence of PPh$_3$, the Rh(III) is reduced rapidly under hydrogenation conditions to rhodium metal. Rhodium metal on hectorite also has been obtained by the reduction of Rh(COD)S$_2^+$–hectorite, where COD = 1,5 cyclooctadiene and S = acetone (Pinnavaia *et al.* (1976). Little is known regarding the nature of the clay-supported metal.

It should be noted that a small but detectable amount of rhodium desorption has been observed for Rh(PPh)$_3^+$ catalyst precursors intercalated in smectite (Pinnavaia *et al.* 1979). The proposed mechanism for desorption is based on the equilibrium shown in equation (11) wherein the neutral monohydride complex is free to desorb from the clay interlayers. The extent of rhodium desorption is increased by the addition of N(C$_2$H$_5$)$_3$ as a base, which tends to support the desorption mechanism. The loss of rhodium can be eliminated, however, by replacing the neutral phosphine ligands with positively charged ligands such as $[(C_6H_5)_2P(CH_2)_2-P(C_6H_5)_2(CH_2C_6H_4)]^+$, abbreviated P–P$^+$. This phosphonium–ligand has been used successfully to form smectite intercalated mixed ligand hydrogenation catalysts of the type RhCl(PPh$_3$)$_x$(P–P$^+$)$_{3-x}$, which are positively charged analogues of Wilkinson's complex (Quayle and Pinnavaia 1979). The ligand has also been useful in forming clay-intercalated olefin hydroformylation catalysts, which are discussed in greater detail below.

Another important feature of clay-intercalated rhodium–triphenylphosphine hydrogenation catalysts should be noted at this point. Because of the relatively large size of the complexes, an upper limit of only ~30% of the Na$^+$ ions initially present in the interlayers can be replaced by metal complex. A study of the effects of Rh(PPh$_3$)$_x^+$ catalyst loading in hectorite on the rate of reduction of 1.0M 1-hexene in methanol showed that the turnover frequency (rate per mole of Rh) remained constant over the range 0.06–0.7 wt% Rh, but at loadings above 0.7 wt% (>10% exchange, >25% surface coverage), the turnover frequency decreased with increased loading (Pinnavaia *et al.* 1979). Simple geometric consideration of particle size and optimum loading suggest that the readily accessible (non-diffusion limiting) metal centres are localized in the clay interlayers within the first 100 Å or so from the platelet edges.

1,3-Butadienes represent another class of inorganic molecules which have been hydrogenated in the presence of an intercalated clay catalyst (Raythatha and Pinnavaia 1981). In this case terminal monoene and internal monoene products are observed as a result of overall 1,2 and 1,4-hydrogen addition, respectively (equation 15)). The cationic metal complexes which

$$\underset{\text{}}{\overset{R\quad R'}{\diagup\!\!\!\diagdown}} \xrightarrow{H_2} \underset{1,2 \text{ addn}}{\overset{R\quad R'}{\diagup\!\!\!\diagdown}} + \underset{1,2 \text{ addn}}{\overset{R\quad R'}{\diagup\!\!\!\diagdown}} + \underset{1,4 \text{ addn}}{\overset{R\quad R'}{\diagup\!\!\!\diagdown}} \quad (15)$$

best catalyse this reaction are of the type Rh(diene)L$_2^+$, where diene is norbornadiene (NBD), for example, and L$_2$ is a bidenate ligand such as (C$_6$H$_5$)$_2$P(CH$_2$)$_2$P(C$_6$H$_5$)$_2$ (dppe) (Schrock and Osborne 1976c). The reaction is believed to occur by an unsaturated pathway in which the 1,3-diene first adds to the metal centre and then hydrogen rapidly adds to eliminate products

TABLE 6.11. Hydrogenation of 1,3-butadienes at 25 °C with Rh(NBD)(dppe)$^+$ as the catalyst precursor under intercalated and homogeneous conditions

Diene	Solvent	Yield of 1,2 additional products (%)		Relative selectivity‡
		Intercalated catalyst	Homogeneous catalyst	
(structure)	Acetone	45	30	1.5
		60	33	1.8
(structure)	Acetone	34	19	1.8
	MeOH	44	20	2.2
(structure)	Acetone	32	17	1.9
	MeOH	39	20	2.0

† Product analysis after >98% conversion of substrate to 1,2 and 1,4 addition products.
‡ Ratio of 1,2 addition products obtained with clay intercalated catalyst to 1,2 addition products obtained with the homogeneous catalyst.

reductively (equation (16)):

$$\text{Rh(NBD)(dppe)}^+ \xrightarrow[-\text{norbornane}]{H_2} \text{Rh(dppe)}^+ \xrightarrow{1,3\text{-diene}} \text{Rh(1,3-diene)(dppe)}^+ \quad (16)$$

$$\downarrow H_2$$

monoenes

Thus the distribution of terminal and internal monoenes most likely is kinetically regulated by the reaction pathways of a common RhH(R)(dppe)$^+$ intermediate formed in the hydrogen addition step.

Table 6.11 compares, for a series of butadienes, the yields of 1,2 and 1,4 hydrogen addition products obtained with Rh(NBD)(dppe)$^+$ intercalated in hectorite and in homogeneous solution. Included in the table is the selectivity of the intercalated catalyst toward 1,2 hydrogen addition products, relative to the homogeneous catalyst. No hydrogenation is observed for the intercalated catalyst in benzene, which fails to swell the interlayer regions of the intercalate. However, in acetone and methanol, which swell the clay interlayer to thicknesses of 12 Å and 15 Å, respectively, reaction is observed. Significantly, the yields of the more valuable 1,2 addition products are 1.3 to 2.3 times larger for the intercalated catalyst than for the homogeneous catalyst.

Relatively little is known regarding the structures of RhH(R)(dppe)$^+$ intermediates leading to monoenes, but reasonable possibilities are η_3- and η_1-allyls of the type

HRh- - - - -⟩   HRh—/=\   and   HRh—⟨=

It has been suggested (Raythatha and Pinnavaia 1981) that surface polarization effects may influence the transfer of hydrogen from metal to the η_3-allyl or, in the case of the η_1 allyls, the relative stabilities of the

HRh—/=\   and   HRh—⟨=

linkages which lead to the final monoenes. If this were the case, it may be that only a fraction of the clay interlayers have the proper spacing or charge density to influence product distribution. An intercalated clay catalyst with a more uniform charge distribution than that provided by hectorite might lead to even greater selectivity toward overall 1,2 hydrogen addition.

6.6.2 Asymmetric hydrogenation

Kagan and Dang (1972) first demonstrated that rhodium complexes of the type Rh(diene)(diphos*)$^+$, where diphos* is a chiral bidentate phosphine ligand, are homogeneous catalyst precursors for the asymmetric hydrogenation of certain prochiral olefins which are incapable of double bond isomerization. α-Acylaminoacrylates, for example, can be catalytically hydrogenated to chiral amino acid derivatives (equation (17)) in high optical yields in the presence of such catalyst (Valentine and Scott 1978; Caplar et al. 1981; Merrill 1981). Many of the chiral amino acids that can be synthesized by direct asymmetric hydrogenation have medicinal utility, such as L-DOPA.

$$\underset{R\quad\quad NHCOR}{\overset{H\quad\quad COOR}{\diagup\!\!\!\diagdown}} \xrightarrow[\text{Rh (diene) (diphos)*}]{H_2} RCH_2C\underset{NHCOR}{\overset{COOR}{\diagup\!\!\!\diagdown}} H \qquad (17)$$

Mozzei et al. (1980) investigated the asymmetric hydrogenation of acetamidoacrylic acids with smectite-intercalated and homogeneous Rh(COD)(PNNP)$^+$, where COD = 1,5-cyclooctadiene and PNNP is the ligand defined below. Several smectite clays were utilized in forming the intercalated catalysts, including hectorite, montmorillonite and nontronite. Also, Chang (1982) has recently completed a systematic study of several acrylic acid derivatives in the presence of homogeneous and hectorite-intercalated Rh(NBD)(diphos*)$^+$ complexes in which diphos* included such dissymmetric ligands such as DIOP(+) and (R)-4-Me-Prophos (see below)

PNNP (+)   DIOP (+)   (R)-4-Me-Prophos

Chang's results indicated that the optical yields (enantiomeric excess of chiral products) obtained with the intercalated catalyst generally compared favourably with those obtained with the homogeneous catalyst, though large differences in the optical yields of the homogeneous and intercalated catalysts were sometimes observed in certain reaction systems. The differences between the intercalated and homogeneous catalyst were dependent on the nature of the substrate and the diphos* ligand. 4-Me-Prophos, which gave the highest optical yields, also gave the highest $d(001)$ spacings when solvated by the solvents used as a reaction medium.

TABLE 6.12. Asymmetric hydrogenation of selected prochiral olefins with Rh(NBD)(4-Me-(R)-Prophos)$^+$ under intercalated and homogeneous reaction conditions[†]

Substrate	Optical yield (%)	
	Intercalated catalyst	Homogeneous catalyst
phenylalanine precursor (furan, C=C with COOH and NHCOCH$_3$)	89.6	92.6
tyrosine precursor (phenol, C=C with COOH and NHCOCH$_3$)	78.5	72.0
L-DOPA precursor (AcO, OMe benzene, C=C with COOH and NHCOCH$_3$)	95.1	95.3

[†] Reactions were carried out at 25 °C, 1 atm pressure, in 95% ethanol. The chemical yields were >98% in each case.

Apparently, as the interlayers become more highly swollen and solution-like, the more closely the intercalated catalyst yields approach the homogeneous catalyst yields. Table 6.12 illustrates some of the optical yields obtained by Chang (1982) with Rh(NBD)(4-Me-Prophos)$^+$ under intercalated and homogeneous conditions. It can be seen that the intercalated and homogeneous optical yields are very similar for the phenylamine, tyrosine and L-DOPA precursors shown.

Mozzei et al. (1980) found that with Rh(COD)(PNNP)$^+$ intercalated in smectite, the optical yields for a given substrate depended on the type of smectite utilized. With a phenylalanine precursor, for example, the optical yield was 49% when the complex was intercalated in hectorite and only 8.5% when intercalated in bentonite (montmorillonite). Also, prehydrogenation of the intercalated catalyst at elevated pressure (20 atm) in the absence of substrate was needed before hydrogenation could be observed. Significantly, the optical yields decreased with increasing number of cycles of the intercalated catalyst. In marked contrast to the results of Mozzei et al., Chang found that facile hydrogenation occurs with hectorite-

TABLE 6.13. Optical yields obtained with Rh(NBD)(DIOP)(+))⁺ and Rh(COD)(PNNP(+))⁺ complexes intercalated in hectorite

Substrate	Rh(NBD)(DIOP(+))⁺†		RH(COD)(PNNP(+))⁺‡	
	Homogeneous	Intercalated	Homogeneous	Intercalated
CH₂=C(COOH)(NHCOCH₃)	76	74	72–75	70
(phenoxy)–C(COOH)=CH₂ with NHCOCH₃	85	85 84 (2nd cycle) 81 (3rd cycle)	80–85	49 38 (2nd cycle) 24 (3rd cycle)
(phenoxy, OAc, OMe)–C(COOH)=CH₂ with NHCOCH₃	88	86	87§	72 58 (2nd cycle) 47 (3rd cycle)

† Data from Chang (1982). Hydrogenations were carried out in 95% ethanol at 25 °C and ambient pressure. The amount of substrate and rhodium used was 4–8 mmol and 0.040 mmol, respectively, in 30 ml of solvent. All products had S-configurations.
‡ Data from Mozzei et al. (1980). Reaction was carried out at 25 °C, 20 atm H_2; substrate concentration is 1.5–1.6×10^{-1}M and substrate/Rh = 100.
§ This result was taken from Fiorini et al. (1978).

intercalated Rh(NBD)(diphos*) catalysts at ambient pressure in presence of substrate without the need for a prehydrogenation step. Also, Chang observed only a small reduction in optical yields upon recycling the catalyst. A comparison of the results of Chang and of Mozzei et al. is provided in Table 6.13. It is possible that these differences in catalytic behaviour for the two types of intercalated complexes are related to solvolytic or oxidation reaction of the PNNP ligand but not the DIOP ligands in the clay interlayers. The loss of PNNP through solvolysis of the P—N bonds or through oxidation to phosphine oxides could lead to the formation of rhodium metal and dramatic reduction of optical yields. With the more stable DIOP and 4-Me-Prophos ligands, however, there is considerable promise that the asymmetric catalytic hydrogenation of certain amino acid derivatives such as L-DOPA, for example, could be accomplished better with a clay-intercalated catalyst than with a homogeneous catalyst, because of the greater efficiency in recovering and recycling the catalyst complex without significant loss of product purity.

6.6.3 Hydroformylation

Although cationic complexes such as Rh(diene)(PPh₃)₂⁺ are active for olefin hydroformylation, they are not suitable for intercalation in layered silicates, because the active species formed under hydroformylation conditions are electrically neutral (Crabtree and Felkin 1979). Since neutral complexes have little or no affinity for the negatively charged silicate sheets, extensive

TABLE 6.14. Hydroformylation of 1-hexene in acetone with homogeneous and intercalated rhodium complex precursors[†]

Rh precursor	P-P/Rh$^+$	Product distribution (%)		
		n-Heptanal	2-Me-Hexanol	2-Hexene
A. Homogeneous catalyst				
[RhCl(COD)]$_2$	4	55	22	23
[Rh(CO)$_2$Cl]$_2$	3	54	26	20
[Rh(COD)]$^+$	2	60	30	10
B. Intercalated catalyst				
[RhCl(COD)]$_2$	4	63	23	8
[Rh(CO)$_2$Cl]$_2$	3	71	23	6
[Rh(COD)]$^+$[‡]	2	70	23	0

[†] 100 °C, 600 psi CO/H$_2$ (1/1).
[‡] This system gave 7% of an unidentified reaction product.

desorption of rhodium occurs during the reaction and most of the observed catalytic activity occurs in the solution phase.

Despite the paucity of cationic hydroformylation catalysts, rhodium desorption from clay interlayers can be effectively eliminated by replacing the neutral phosphine ligands on rhodium with positively charged phosphine ligands such as the P–P$^+$ species described earlier. Table 6.14 compares the results for the hydroformylation of 1-hexene (equation (18)) in acetone with three

$$C_4H_9CH{=}CH_2 + CO + H_2 \xrightarrow{\text{Rh complex}} \underset{\text{n-heptanal}}{C_4H_9CH_2CH_2CHO} + \underset{\text{2-methyl-hexanol}}{C_4H_9CH(CHO)CH_3} \qquad (18)$$

different catalyst precursor systems containing P–P$^+$ as a ligand (Farzaneh and Pinnavaia 1983). For each of the intercalated catalysts, all of the activity occurred in the solid phase; no catalytic activity was observed for the clear filtrates. The clay-intercalated catalysts provide a small but distinct chemical advantage over the homogeneous catalysts. The yields of the synthetically more valuable normal chain aldehyde are consistently higher for the intercalated catalysts. Also, the extent of the 1-hexene isomerization to 2-hexene is lower for the intercalated catalyst than for the homogeneous catalyst. Apparently, the restricted interlayers of the intercalated catalyst favours the formation of the sterically less demanding α-alkyl intermediate. Similar steric factors may also be affecting the isomerization pathway, though the magnitude of such effects is relatively small.

6.6.4 Solvolysis and proton-assisted reactions

Earlier sections of this chapter have described several Brønsted catalysed reactions of synthetic, natural and acid modified smectite clays. Often the Brønsted acid catalytic activity of the clay is influenced greatly by the nature of the interlamellar cation complex, as evidenced by early work of Broughton (1940). It is becoming increasingly more evident that the nature of the exchange cation not only influences the acid strength of the clay but also that the *selectivity* of solvolytic and of proton catalysed reactions in clay interlayers can be greatly influenced by the exchange cation. One example of a selective cation effect in solvolysis was provided by Mortland and Raman (1967) for the hydropyrolysis of thiophosphate esters (equation (19) at acidic pH values in the

$$\text{Cl}-\underset{\text{Cl}}{\underset{|}{\bigcirc}}_{N}-O-\underset{OC_2H_5}{\overset{S}{\underset{|}{P}}}-OC_2H_5 \xrightarrow[pH=5-6]{H_2O} \text{Cl}-\underset{\text{Cl}}{\underset{|}{\bigcirc}}_{N}-OH + PO(OH)(C_2H_5)_2 \quad (19)$$

presence of various metal ion exchange forms of smectite and vermiculite. Cu^{2+}-montmorillonite was as effective as Cu^{2+} in homogeneous solution in catalysing the hydrolysis reaction, but more acidic interlayer cations such as Al^{3+} failed to catalyse the solvolysis reaction. Moreover, Cu^{2+} in nontronite, beidellite and vermiculite showed little or no ability to catalyst solvolysis. It is apparent that the solution-like complexation of the thiophosphate ester by Cu^{2+} played an important role in the solvolysis reaction in montmorillonite, but not in the other clays investigated. Other samples of exchange cation effects on solvolytic and proton-catalysed reactions on clay surfaces can be found in a review by Fripiat and Cruz-Cumplido (1974).

More recently, Thomas and his co-workers (1977, 1982) have investigated a variety of organic reactions in the presence of smectites interlayered with hydrated metal cations. Many of the reactions are stoichiometric chemical conversions (Thomas et al. 1977) whereas others are authentic proton-catalysed processes (Thomas 1982). At the temperatures utilized for proton-catalysed conversions (100–200 °C) by clays interlayered with acidic exchange cations (e.g. Al^{3+}, Fe^{3+}) only a few molecular layers of water occupy the interlamellar regions of the clay. Although the interlayers are highly restricted, organic molecules apparently are capable of penetrating the interlayer regions at the monolayer level.

Direct evidence for interlamellar reaction has been demonstrated (Adams et al. 1979) for the quantitative conversions of C_6–C_8 terminal alkenes to 2,2'-dialkyl ethers (equation (20)) in the one- and two-water layer hydrates of Cu^{2+}-montmorillonite.

$$2C_4H_5CH=CH_2 + H_2O \xrightarrow{(Cu^{2+})} \underset{\underset{CH_3}{|}}{C_4H_5CH}-O-\underset{\underset{CH_3}{|}}{CHC_4H_5} \quad (20)$$

Although this reaction is not self-sustaining and therefore not truly catalytic, it illustrates the high specificity that can be achieved by reaction of molecules in close proximity in clay interlayers. Recent studies by Adams et al. (1982) suggest that ether production results from the reaction of a secondary alcohol formed in a water-rich region of the interlayer (e.g. near hydrated Cu^{2+} ions) and a nearby carbocation formed in a local environment low in water. Although other metal ions in clays catalyse the reaction shown in equation (20), they are not as selective as Cu^{2+} (Adams et al. 1979).

The dehydration of primary alcohols over Al^{3+}-montmorillonite at 200 °C affords 1,1'-dialkyl ethers. In this case reaction is truly catalytic and the clay can be reused. As shown by the data in Table 6.15 (Ballantine et al. 1981a) little or no intramolecular alcohol dehydration to alkenes is observed. Apparently, the dehydration proceeds by nucleophilic displacement of water from adjacent reactant pairs in the clay interlayers (equation (21)). The absence of

$$RCH_2CH_2OH \xrightarrow{(H^+)} RCH_2CH_2OH_2^+ \xrightarrow[-H_2O, -H^+]{(RCH_2CH_2OH)} RCH_2CH_2-O-CH_2CH_2R \quad (21)$$

branched Markownikoff products precludes the intermediacy of carbocations or alkenes. Extensive alkene formation is observed, however, for the dehydration of secondary and tertiary

TABLE 6.15. Product distributions (wt %) for reactions of alcohols over Al^{3+}-montmorillonite at 200 °C for 4 h[†]

Alcohol	Unreacted alcohol	Ethers			Alkenes	Alkene dimers
		1,1'-	1,2'-	2,2'		
Ethanol	35	64				1
1-Propanol	22	63	8		3	4
1-Butanol	44	45	4		6	1
1-Pentanol	46	32	5		12	5
1-Hexanol	38	41	5		14	2
1-Heptanol	50	39	1.5		9	0.5
1-Octanol	59	30			11	1
2-Propanol	29			30	40	1
2-Butanol	10			9	78	2
2-Pentanol	9			7	83	1
2-Hexanol	7			5	82	6
cyclohexanol	6			4	88	1
2-Me-2-propanol	3				18	79[‡]
2-Me-2-butanol	2				93	4

[†] Data from Ballantine et al. 1981a. Reaction was carried out using 0.5 g of clay and 5 g of reactant.
[‡] 50% dimer, 29% tetramer.

alcohols on montmorillonite (cf. Table 6.15). Selective catalytic elimination reactions analogous to those observed for primary alcohols have been reported also for primary amines (Ballantine et al. 1981d) and primary and secondary thiols (Ballantine et al. 1981c).

The catalytic addition of carboxylic acids to alkenes on Al^{3+}-montmorillonite at 200 °C also has been found to be highly specific (Ballantine et al. 1981b). Ethyl acetate and propyl acetate are the sole products formed from the addition of acetic acid to ethylene and propylene, respectively. In this case the reaction proceeds through equilibrated carbocation intermediates, as evidenced by the distribution of Markownikoff addition products obtained in the reaction of acetic acid and 1-hexene: 70% 2-hexylacetate, 30% 3-hexylacetate.

In an earlier section of this chapter we noted the use of smectite clays as catalysts for dimerization of unsaturated fatty acids (Fischer 1964; den Otter 1970). More recently Weiss (1981) has investigated the dependence of oleic acid dimerization on the charge density of the smectite catalyst. His results demonstrate incisively the importance of reactant pair proximity in regulating the selectivity of reactions in clay interlayers. Interlayer reaction pair proximity undoubtedly is also important in many of the highly specific proton-catalysed chemical conversions described above.

Figure 6.9 shows the distribution obtained for reaction of oleic acid in $N(CH_3)_4^+$-montmorillonites of different charge density. The yield of desired dimeric acid is found to be a maximum (66%) when the average layer charge is $\sim 0.3 e/Si_4O_{10}$. At this value the average distance between $N(CH_3)_4^+$ ions in the interlayers permits optimum pairing of oleic acid molecules. As the distance between exchange cations is decreased with increasing layer charge, the number of reacting pairs is decreased by the intervening cations. Consequently, cis–trans isomerization and H-transfer reactions of the monomer compete more favourably with the dimerization reaction. Decreasing the layer charge density below $0.3 e/Si_4O_{10}$ increases the number of three- and four-monomer groupings in the space between exchange cations, and

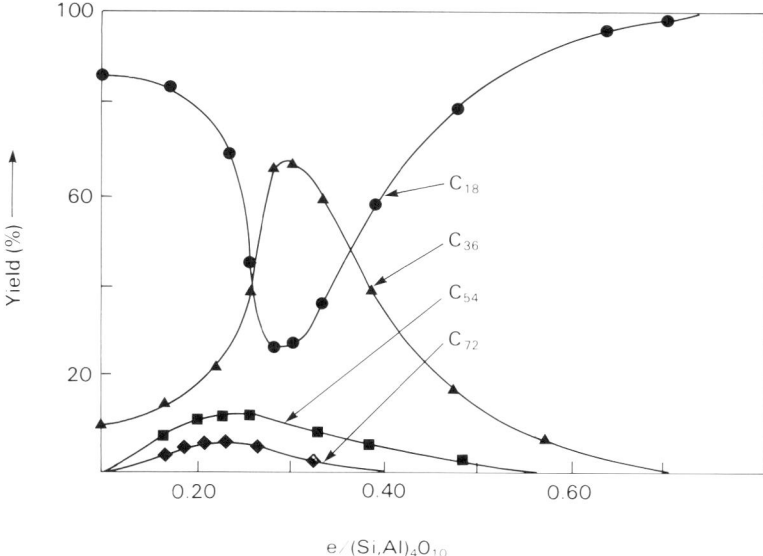

FIG. 6.9. Oligomerization of oleic acid to di-, tri-, and oligocarboxylic acids with $(CH_3)_4$N-montmorillonite as catalyst as a function of the charge density. Starting material: oleic acid 98%; 0.5 g catalyst per 100 ml oleic acid. ●: Oleic acid + stearic acid (C_{18}); ▲: Dicarboxylic acids (C_{36}); ■: Tricarboxylic acids (C_{54}); ◆: Oligocarboxylic acids ($>C_{72}$). (From Weiss 1981. Reprinted by permission of Verlag Chemie GmbH.)

higher oligomers are formed at the expense of dimers. Significantly, no oligomerization is observed for the corresponding C_{18} alcohol and nitrile, which adopt vertical or inclined positions in the interlayer rather than the parallel orientation assumed by the carboxylic acid.

6.7 PILLARED CLAYS AS CATALYSTS

Interlamellar reactions in ordinary metal ion exchange forms of smectite clays at high temperatures (>200 °C) are precluded by the dehydration and collapse of the interlayer region. The limitations imposed by interlayer collapse recently have been circumvented by the intercalation of thermally stable, robust cations which act as molecular props or pillars in keeping the silicate layers separated in the absence of a swelling solvent. As illustrated in Fig. 6.10, various types of cations have been used as pillaring agents, including alkylammonium ions (Barrer 1978), bicyclic amine cations (Mortland and Berkheiser 1976; Shabtai et al. 1976), metal chelate complexes (Knudson and McAtee 1973; Traynor et al. 1978; Loeppert et al. 1979) and polynuclear hydroxy metal cations (Brindley and Sempels 1977; Yamanaka and Brindley 1978).

The concept of pillaring smectite clays was demonstrated first by Barrer and MacLeod (1955) when they utilized tetraalkylammonium ions to induce interlayer porosity in montmorillonite. Though Barrer developed further the chemistry of alkylammonium clays, and even demonstrated their selective adsorption properties (Barrer 1978), the idea of pillaring clays to achieve porous networks seems to have been overshadowed by the rapid advances being made in the synthesis and catalytic applications of zeolites. However, there is renewed interest in pillared clays, because it is now realized that the pore sizes can be designed to be larger than those of faujasitic zeolites. By varying the size of the pillar and/or the spacing between pillars,

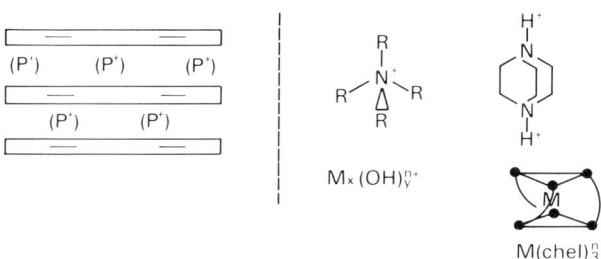

FIG. 6.10. A schematic representation of pillared clay. The pillaring agent (P^+) may be an alkylammonium ion, a bicyclic amine cation, a *tris* metal chelate (e.g. $M = Fe^{2+}$, chel = 1,10-phenanthroline), or a polynuclear hydroxy metal ion (e.g. $Al_{13}O_4(OH)_{28}^{3+}$). The regions between pillaring cations define pores for the adsorption and possible catalysis of organic molecules.

one has the potential of adjusting the pore size to suit a particular application. Thus pillared clays offer new possibilities for catalysis of larger molecules such as those found in residual crude oils.

Among the pillaring agents illustrated in Fig. 6.10 the alkylammonium and bicyclic amine cations decompose below 250 °C and the metal chelates degrade below 450 °C. However, at least two polynuclear hydroxy cations afford pillared phases which are stable above 500 °C. Brindley and Sempels (1977) and Yamanaka and Brindley (1978) were first to report that hydroxy-aluminium and -zirconium cations formed by hydrolysis over a specific range of OH^-/M^{n+} values yielded thermally stable clays with surface areas of 200–500 m² g⁻¹ and interlayer free spacings near 9 Å. Using similar methods of preparation, Lahav *et al.* (1978) and Vaughan and Lussier (1980) independently prepared analogous pillared phases containing hydroxy aluminium ions.

The hydroxy-zirconium pillars are Zr_4 oligomer of the type $Zr_4(OH)_{16-x}^{x+}$ (Yamanaka and Brindley 1978). Although Al_6 oligocations have been proposed as the intercalated species in the hydroxy-aluminium system (Brindley and Sempels 1977), the structure of the pillaring cations is almost certainly an Al_{13} oligomer related to the known cation $Al_{13}O_4(OH)_{24}^{7-}$, as suggested by Vaughan and Lussier (1980). A recent evaluation of ^{27}Al NMR and potentiometric titration data by Bottero *et al.* (1980) for hydrolysed solutions of Al^{3+} indicate $Al_{13}O_4(OH)_{28}^{3+}$ to be the dominant species present in solution at the OH^-/Al^{3+} ratios used to prepare hydroxy-aluminium pillared clays. Also, the interlayer free spacing (~9.5 Å) of smectites interlayered with hydroxy-aluminium is consistent with the expected size of an Al_{13} oligomer.

The remarkable thermal stability of the Zr_4 and Al_{13} pillared smectites has been attributed to the formation of metal oxide clusters upon dehydroxylation of the hydroxy cations at elevated temperature. In the case of Al_{13}-montmorillonite the overall interlayer reaction may be written as

$$Al_{13}O_4(OH)_{28+n}^{(3-n)+} \xrightarrow{-H_2O} 6.5Al_2O_3 + (3-n)H^+ \qquad (22)$$

where the alumina is in the form of small clusters. Little is known at present regarding the structure of the clusters or the location of the protons in the intercalate.

Shabtai (1979, 1980), Shabtai and Lahav (1980), Lussier *et al.* (1980) and Vaughan *et al.*

TABLE 6.16. FCC pilot unit data for pillared clay, cogel and zeolite catalysts[†]

		Pillared clay	Cogel	Zeolite
Conversion:	V%	71.0	55.5	73.5
Hydrogen:	W% FF	0.062	0.041	0.030
$C_1 + C_2$:	W% FF	2.25	2.81	1.56
Total C_3's:	V% FF	10.8	10.7	9.5
$C_3^=$:	V% FF	8.9	9.5	7.5
Total C_4's:	V% FF	16.2	14.2	15.8
$C_4^=$:	V% FF	8.3	9.8	6.6
iC_4:	V% FF	6.8	3.8	7.6
C_5^+ Gasoline:	V% FF	51.5	38.0	58.0
C_5^+ Gaso./Conv.		0.73	0.68	0.79
Octane no.				
RON+O:		90.5	94.0	89.0
MON+O:		80.2	80.7	80.6
Light cycle oil:	V% FF	23.3	30.0	19.4
Gravity:	°API	25.3	28.2	19.9
Aniline pt:	°F	101	129	77
640°F$^+$ Residue:	V% FF	5.7	14.5	7.1
Gravity:	°API	8.6	19.8	3.5
Aniline pt:	°F	113	176	88
Coke:	W% FF	7.1	3.6	4.3

[†] Pilot unit conditions: 4 c/o, 950°F, Sohio (3/79), 40 WHSV; data are from Lussier *et al.* (1980).

(1979, 1981a,b) independently disclosed the molecular sieving and acid catalytic properties of clays pillared by hydroxyl metal ions. For example, Al_{13}- and Zr_4-montmorillonite when utilized as petroleum cracking catalysts give gasoline octane ratings comparable to zeolite catalysts, along with enhanced light cycle oil yields (Lussier *et al.* 1980). Table 6.16 compares the cracking performance of Al_{13}-montmorillonite with an amorphous co-gel and zeolite catalyst.

Al_{13}-montmorillonite is reported to adsorb molecules as large as mesitylene, but not methyl mesitylene, which have kinetic diameters of 7.6 and 8.0 Å, respectively (Vaughan and Lussier 1980). The highly selective molecular sieving properties of clays pillared by hydroxy metal cation requires a regular distribution of pillars and pores in the interlayer region. However, it is known that the layer charge distribution in smectite clays is highly irregular (Stul and Mortier 1974; Peigneur *et al.* 1975; Lagaly and Weiss 1976), with the layer charge varying by as much as a factor of two from interlayer to interlayer. Thus one should expect pillared clays to exhibit a range of pore sizes, particularly for those intercalates containing hydrolytically stable pillaring agents such as $M(chel)_3^{n+}$, where ion exchange is the sole driving force for intercalation. With polynuclear hydroxy metal pillars, however, the ions fill each interlayer region to essentially the same population density, regardless of the layer charge. In this latter case, the pillar spacing is decided by the radius of the *hydrated* cations, and the charge on the cations is regulated by the extent of hydrolysis, which is dependent on layer charge.

Although polynuclear hydroxy metal ions formed by hydrolysis in aqueous solution can yield pillared clays with interlayer free spacings in the range 5–20 Å, the number of metals which form suitable oligomeric species is limited. New approaches to the pillaring of smectite clays are being developed which promise to extend the number of pillaring species. It has been possible, for example, to interlayer montmorillonite with silicic acid by hydrolysis of silicon acetylacetonate in the interlayer region (Endo *et al.* 1980, 1981):

$$\overline{Si(acac)_3^+} \xrightarrow[-H(acac)]{H_2O} \overline{Si(OH)_4} + H^+ \qquad (23)$$

The silicic acid–clay complex exhibits an interlayer free spacing of 3 Å and a nitrogen surface area of $\sim 200\ m^2\ g^{-1}$. Also, a related approach involving hydrolysis and oxidation of metal cluster cations such as $Nb_6Cl_{12}^{2+}$ and $Ta_6Cl_{12}^{2+}$ (equation (24)) affords clays pillared by small clusters of metal oxide approximately 10 Å in diameter and stable to $\sim 400\ °C$ (Christiano *et al.* 1982).

$$\overline{Nb_6Cl_{12}^{2+}} \xrightarrow[-H_2,\ -HCl]{H_2O} \overline{3Nb_2O_5} + 2H^+ \qquad (24)$$

The study of pillar clays is in its infancy. Future developments undoubtedly will lead to new pillared systems and new catalytic applications.

NOTES

1. References were selected from the viewpoint of the clay mineralogist. The subject is vast and a comprehensive review has not been attempted, nor would such be possible within the space limitations.
2. *Added in proof.* Several reviews on catalytic properties of intercalated clays and related derivatives by Ballantine, Poncelet and Schutz, Cornelius and Laszlo, Roudier and Foucaud, and Pinnavaia have recently appeared in *Chemical Reactions in Organic and Inorganic Constrained Systems*, R. Setton ed. (1986). Reidel, Boston.

REFERENCES

ADAMS J.M., BALLANTINE J.A., GRAHAM S.H., LAUB R.J., PURNELL J.H., REID P.I., SHAMAN W.Y.M. & THOMAS J.M. (1979) Selective chemical conversions using sheet silicate intercalates. *J. Catal.* **58**, 238–252.

ADAMS J.M., BYLINA A. & GRAHAM S.H. (1982) Conversion of 1-hexene to di-2-hexyl ether using a Cu^{2+}-smectite catalyst. *J. Catal.* **75**, 190–195.

AMERO R.C. & WOOD W.H. (1947) Catalytic desulfurization of cracked and straightrun gasoline. *Oil Gas J.* **46**, 82–85, 99.

BALLANTINE J.A., DAVIES M., PURNELL H., RAYANAKORN M., THOMAS J.M. & WILLIAMS K.J. (1981a) Chemical conversions using sheet silicates: novel intermolecular dehydration of alcohols to ethers and polymers. *J. Chem. Soc. Chem. Comm.* 427–428.

BALLANTINE J.A., DAVIES M., PURNELL H., RAYANAKORN M., THOMAS J.M. & WILLIAMS K.J. (1981b) Chemical conversions using sheet silicates: facile ester synthesis by direct addition of acids to alkenes. *J. Chem. Soc. Chem. Comm.* 8–9.

BALLANTINE J.A., GALVIN R.P., O'NEIL R.M., PURNELL H., RAYANAKORN M. & THOMAS J.M. (1981c) Chemical conversions using sheet silicates: novel intermolecular eliminations of hydrogen sulphide from thiols. *J. Chem. Soc. Chem. Comm.* 695–696.

BALLANTINE J.A., PURNELL H., RAYANAKORN M., THOMAS J.M. & WILLIAMS K.J. (1981d) Chemical conversions using sheet silicates: novel intermolecular eliminations: novel intermolecular elimination of ammonia from amines. *J. Chem. Soc. Chem. Comm.* 9–10.

BARRER R.M. (1978) *Zeolites and Clay Minerals as Sorbents and Molecular Sieves.* Academic Press, New York, pp. 407–483.

BARRER R.M. & MACLEOD D.N. (1955) Activation of montmorillonite by ion exchange and sorption complexes of tetra-alkylammonium montmorillonites. *Trans. Faraday Soc.* **51**, 1290–1300.

BASILA M.R., KANTNER T.R. & RHEE K.H. (1964) The nature of acidic sites on a silica-alumina. *J. Phys. Chem., Ithaca* **68**, 3197–3207.

BENESI H.A. (1956) Acidity of catalyst surfaces. I. Acid strength from colors of adsorbed indicators. *J. Am. Chem. Soc.* **78**, 5490–5494.

BENESI H.A. (1957) Acidity of catalyst surfaces. II. Amine titration using Hammett indicators. *J. Phys. Chem., Ithaca* **61**, 970–973.
BERZELIUS J.J. (1836) Quelques Idees sur une nouvelle Force agissant dans les Combinaisons des Corps Organiques. *Annls Chim.* **1xi**, 146–151.
BOTTERO J.Y., CASES J.M., FLESSINGER F. & POIRER J.E. (1980) Studies of hydrolyzed aluminum chloride solutions. 1. Nature of aluminum species and composition of aqueous solutions. *J. Phys. Chem., Ithaca* **84**, 2933–2939.
BRADFIELD R. (1923) The nature of the acidity of the colloidal clay of acid soils. *J. Am. Chem. Soc.* **45**, 2669–2678.
BRINDLEY G.W. & SEMPELS R.E. (1977) Preparation and properties of some hydroxyaluminum beidellites. *Clays Clay Miner.* **12**, 229–236.
BROUGHTON G. (1940) Catalysis by metallized bentonites. *J. Phys. Chem., Ithaca* **44**, 180–184.
CADY S.S. & PINNAVAIA T.J. (1978) Porphyrin intercalation in mica-type silicates. *Inorg. Chem.* **17**, 1501–1507.
CAPELL R.G. & GRANQUIST W.T. (1966) Cracking catalyst and process of cracking. US Pat. 3,252,889.
CAPLAR V., COMISSO G. & SUNJIC V. (1981) Homogeneous asymmetric hydrogenation. *Synthesis* 85–116.
CHANG H.M. (1982) Asymmetric hydrogenation of prochiral olefins with layered silicate intercalation catalysts. Ph.D. Thesis. Michigan State University, East Lansing, Michigan, USA.
CHOU C.C. & MCATEE JR., J.L. (1969) Thermal decomposition of organo-ammonium compounds exchanged onto montmorillonite and hectorite. *Clays Clay Miner.* **17**, 339–346.
CHRISTIANO S.P., WANG J. & PINNAVAIA T.J. (1982) Pillaring of montmorillonite by cluster cations of niobium, tantalum, and molybdenum. *Abstracts, Nat. Clay Miner. Soc. Meeting, Hilo, Hawaii*.
CIAPETTA F.G., MACUGA S.J. & LEUM L.N. (1948) Depolymerization of butylene polymers. *Ind. Engng Chem.* **40**, 2091–2099.
CLEMENTZ D.M., MORTLAND M.M. & PINNAVAIA T.J. (1973) Stereochemistry of hydrated copper(II) ions on the interlamellar surfaces of layered silicates: an electron spin resonance study. *J. Phys. Chem., Ithaca* **77**, 196–200.
CONDON F.E. (1958) Catalytic isomerization of hydrocarbons. In *Catalysis* (P.H. Emmett, ed.), Vol. IV, pp. 2091–2099. Reinhold Publishing Co., New York.
CRABTREE R.H. & FELKIN H. (1979) Homogeneous hydroformylation with cationic rhodium diolefin complexes. *J. Mol. Catal.* **5**, 75–79.
CSICSERY S.M., HUGHES T.R. & JAFFE J. (1970) Hydrocracking catalyst and process. US Pat. 3,535,228.
DEN OTTER M.J.A.M. (1970) The dimerization of oleic acid with a montmorillonite catalyst I. *Fette Seifen Anstrichmittel* **72**, 667–673.
EGLOFF G. (1943) The reactions of aliphatic hydrocarbons. In *Organic Chemistry* (H. Gilman, ed.), Vol. I, pp. 2–64. John Wiley and Sons, New York.
EGLOFF G., HULLA G. & KOMAREWSKY V.I. (1942) Isomerization of pure hydrocarbons. Reinhold Publishing Co., New York.
ENDO T., MORTLAND M.M. & PINNAVAIA T.J. (1980) Intercalation of silica in smectite. *Clays Clay Miner.* **28**, 105–110.
ENDO T., MORTLAND M.M. & PINNAVAIA T.J. (1981) Properties of silica-intercalated hectorite. *Clays Clay Miner.* **29**, 153–156.
EVANS A.G. (1950) Contribution to the discussion. *Disc. Faraday Soc.* **8**, 302–303.
FARZANEH F. & PINNAVAIA T.J. (1983) Metal complex catalysts interlayered in smectite clay. Hydroformylation of 1-hexene with rhodium complexes ion-exchanged into hectorite. *Inorg. Chem.* **22**, 2216–2220.
FIORINI M., MARCATI M. & GIONGO G.M. (1978) Chiral aminophosphine–rhodium complexes as catalysts for hydrogenation of olefins. *J. Mol. Catal.* **4**, 125–134.
FISCHER E.M. (1964) Polymeric fat acids. US Pat. 3,157,681.
FRIPIAT J.J. (1963) Surface properties of aluminosilicates. *Clays Clay Miner.* **12**, 327–358.
FRIPIAT J.J. (1980) The application of nmr to the study of clay minerals. Ch 5 in *Advanced Chemical Methods for Soil and Clay Minerals Research* (J.W. Stucki and W.L. Banwart, eds). Riedel, Boston, USA.
FRIPIAT J.J. & CRUZ-CUMPLIDO M.I. (1974) Clays as catalysts for natural processes. *Ann. Rev. Earth Planetary Sci.* **2**, 239–255.
FRIPIAT J.J., LEONARD A. & UYTTERHOEVEN J.B. (1965) Structure and properties of silicoaluminas. II. Lewis and Brønsted acid sites. *J. Phys. Chem., Ithaca* **69**, 3274–3279.
FRIPIAT J.J., ROUXHET P. & JACOBS H. (1965) Proton delocalization in micas. *Am. Miner.* **50**, 1937–1958.
GAYER F.H. (1933) The catalytic polymerization of propylene. *Ind. Engng Chem.* **25**, 1122–1127.
GERMAIN J.E. (1969) *Catalytic Conversion of Hydrocarbons*. Academic Press, New York.
GRANQUIST W.T. (1966) Synthetic silicate minerals. US Pat. 3,252,757.
GRANQUIST W.T. & KENNEDY J.V. (1967) Sorption of water at high temperatures on certain clay mineral surfaces. Correlation with lattice fluoride. *Clays Clay Miner.* **15**, 103–117.
GRANQUIST W.T. & POLLACK S.S. (1967) Clay mineral synthesis II. *Am. Miner.* **52**, 212–226.
GRANQUIST W.T. & SUMNER G.G. (1959) Acid dissolution of a Texas bentonite. *Clays Clay Miner.* **6**, 292–307.
GREENSFELDER B.S., VOGE H.H. & GOOD G.M. (1949) Catalytic and thermal cracking of pure hydrocarbons. *Ind. Engng Chem.* **41**, 2573–2584.
GRIM R.E. (1968) *Clay Mineralogy*, 2nd edn. McGraw-Hill, New York.
GURWITSCH L. (1912) Adsorption. *Z. Chemie Ind. Kolloide* **11**, 17–19.
GURWITSCH L. (1923) Zur Kenntnis der heterogenen Katalyse. *Z. Phys. Chem., Frankfurt* **107**, 235–248.

HALL P.L. (1981) Neutron scattering techniques for the study of clay minerals. Ch. 3 in *Advanced Techniques for Clay Mineral Analysis* (J.J. Fripiat, ed.). Elsevier, New York.

HALL W.K. & LEFTIN H.P. (1961) A spectrophotometric study of the mechanism of chemisorption of paraffin molecules on silica–alumina catalysts. *Actes Congr. Intern. Catalyses, 2ᵉ, Paris, 1960* **1**, 1353–1372.

HALL W.K. & PORTER R.P. (1966a) A comparative spectrophotometric study of the adsorption of triphenylmethane and perylene on silica–alumina. *J. Catal.* **5**, 366–386.

HALL W.K. & PORTER R.P. (1966b) In answer to Hirschler. *J. Catal.* **5**, 544–547.

HAMMETT L.P. & DEYRUP A.J. (1923) A series of simple basic indicators. I. The acidity functions of mixtures of sulfuric and perchloric acids with water. *J. Am. Chem. Soc.* **54**, 2721–2739.

HANSFORD R.C. (1952) Chemical concepts of catalytic cracking. *Adv. Catal.* **4**, 1–29.

HAY R.G., MONTGOMERY C.W. & COULL J. (1945) Catalytic isomerization of 1-hexene. *Ind. Engng Chem.* **37**, 335–339.

HIRSCHLER A.E. (1963) The measurement of catalyst acidity using indicators forming stable surface carbonium ions. *J. Catal.* **2**, 428–439.

HIRSCHLER A.E. (1966) On the generation of carbonium ions by electrophilic sites as proposed by Porter and Hall. *J. Catal.* **5**, 390–392.

HIRSCHLER A.E. & HUDSON J.O. (1964) The reaction of triphenylmethanes and perylene with silica–alumina. The nature of the acid sites. *J. Catal.* **3**, 239–251.

HOUDRY E., BURT W.F., PEW JR. A.E. & PETERS JR. W.A. (1938) Catalytic processing by the Houdry process. *Nat. Pet. News* **48**, R570–R580.

HUGHES T.R. & WHITE H.M. (1967) A study of the surface structure of decationized Y zeolite by quantitative infrared spectroscopy. *J. Phys. Chem., Ithaca* **71**, 2192–2201.

JAFFE J. (1970) Catalyst comprising Rhenium and layered synthetic crystalline aluminosilicate and process using said catalyst. US Pat. 3,535,233.

JAFFE J. & KITTRELL J.R. (1970) Hydrocarbon conversion catalyst comprising a substantially catalytic metals-free layered crystalline clay-type aluminosilicate component in a matrix of other catalyst components, and process using said catalyst. US Pat. 3,535,229.

JORDAN J.W. (1949) Reaction of bentonite with amines. *Miner. Mag.* **28**, 598–605.

KAGAN H.B. & DANG T.P. (1972) Asymmetric catalytic reduction with transition metal complexes. I. A catalytic system of rhodium(I) with (−)-2,3-O-isopropylidene-2,3-dihydroxy-1,4-bis(diphenylphosphine)butane, a new chiral diphosphine. *J. Am. Chem. Soc.* **94**, 6429–6433.

KAYSER F. & BLOCH J.M. (1952) Some catalytic properties of montmorillonite. **234**, 1885–1887. *C.r. hebd. Séanc. Acad. Sci., Paris* **234**, 1885–1887.

KITTRELL J.R. (1970a) Hydrocracking catalyst comprising a layered clay type crystalline aluminosilicate component, a Group VIII component and Rhenium and process of using said catalyst. US Pat. 3,535,230.

KITTRELL J.R. (1970b) Catalyst body consisting of physical mixture of different catalysts, one of which comprises Rhenium. US Pat. 3,535,231.

KITTRELL J.R. & SULLIVAN R.F. (1970) Method of reducing hydrocracking activity decline of catalyst comprising layered crystalline clay-type aluminosilicate component. US Pat. 3,535,272.

KNUDSON M.I. & MCATEE J.L. (1973) The effect of cation exchange of tris-(ethylenediamine)cobalt(III) for sodium on nitrogen sorption by montmorillonite. *Clays Clay Miner.* **21**, 19–26.

LAGALY G. & WEISS A. (1976) The layer charge of smectites layer silicates. *Proc. Int. Clay Conf. 1975, Mexico*, pp. 157–172. Applied Publishing Ltd, Wilmette, Illinois, USA.

LAHAV N., SHANI U. & SHABTAI J. (1978) Cross-linked smectites. 1. Synthesis and properties of hydroxy-aluminum-montmorillonite. *Clays Clay Miner.* **26**, 107–115.

LEGZDINS P., MITCHELL R.W., PEMPEL G.L., RUDDICK J.D. & WILKINSON G. (1970) The protonation of ruthenium- and rhodium-bridged carboxylates and their use as homogeneous hydrogenation catalysts for unsaturated substances. *J. Chem. Soc. (A)* 3322–3326.

LOEPPERT JR., R.H., MORTLAND M.M. & PINNAVAIA T.J. (1979) Synthesis and properties of heat-stable expanded smectite and vermiculite. *Clays Clay Miner.* **27**, 201–208.

LUSSIER R.J., MAGEE J.S. & VAUGHAN D.E.W. (1980) Pillared interlayered clay cracking catalyst—preparation and properties. Preprints, 7th Can. Sym. Catal., Edmonton, Alberta, 88–95.

MCAULIFFE C. & COLEMAN N.T. (1955) Hydrogen-ion catalysis by acid clays and exchange resins. *Soil Sci.* **19**, 156–160.

MCBRIDE M.B. (1979) Mobility and reactions of VO^{2+} on hydrated smectite surfaces. *Clays Clay Miner.* **27**, 91–96.

MCBRIDE M.B. (1980) Application of spin probes to esr studies of organic–clay systems. Ch. 9 in *Advanced Chemical Methods for Soils and Clay Minerals Research* (J.W. Stucki and W.L. Banwart, eds). Reidel, Boston.

MCBRIDE M.B., PINNAVAIA T.M. & MORTLAND M.M. (1975a) Electron spin resonance studies of cation orientation in restricted water layers on phyllosilicate (smectite surfaces). *J. Phys. Chem., Ithaca* **79**, 2430–2435.

MCBRIDE M.B., PINNAVAIA T.J. & MORTLAND M.M. (1975b) Electron spin relaxation and the mobility of manganese(II) exchange ions in smectite. *Am. Miner.* **60**, 66–72.

MACEWAN D.M.C. (1958) Fourier transform methods. II. Calculation of diffraction effects for different types of interstratification. *Kolloidzeitschrift* **156**, 61–67.

MAPES J.E. & EISCHENS R.P. (1954) The infrared spectra of ammonia chemisorbed on cracking catalysts. *J. Phys. Chem., Ithaca* **58**, 1059–1062.

MARSHALL C.E. (1949) *The Colloid Chemistry of the Silicate Minerals*. Ch. 10. Academic Press, New York.
MERRILL R.E. (1981) A chemists tool kit. *Chemtech*, 118–127.
MILLIKEN T.H., OBLAD A.G. & MILLS G.A. (1955) Use of clays as petroleum cracking catalysts. *Clays Clay Miner*. **1**, 314–326.
MILLS G.A., HOLMES J. & CORNELIUS E.B. (1950) Acid activation of some bentonite clays. *J. Phys. Colloid Chem*. **54**, 1170–1185.
MOELLER T. (1952) *Inorganic Chemistry*. Ch. 9. John Wiley and Sons, New York.
MONTALAND L. (1911) Process for converting pinene into camphene. US Pat. 999,667.
MORTLAND M.M. & BERKHEISER V.E. (1976) Triethylenediamine–clay complexes as matrices for adsorption and catalytic reactions. *Clays Clay Miner*. **24**, 60–63.
MORTLAND M.M., FRIPIAT J.J., CHAUSSIDON J. & UYTTERHOEVEN J.B. (1963) Interaction between ammonia and the expanding lattices of montmorillonite and vermiculite. *J. Phys. Chem., Ithaca*, **67**, 248–258.
MORTLAND M.M. & RAMAN K.V. (1967) Catalytic hydrolysis of some organic phosphate pesticides by copper(II). *Agric. Food Chem*. **15**, 163–167.
MOZZEI M., MARCONI W. & RIOCCI M. (1980) Asymmetric hydrogenation of substituted acrylic acids by Rh-aminophosphine chiral complex supported on mineral clays. *J. Mol. Catal*. **9**, 381–387.
OBLAD A.G., MILLIKEN JR., T.H. & MILLS G.A. (1951) Chemical characteristics and structure of cracking catalysts. *Adv. Catalysis* **3**, 199–245.
OSTHAUS B. (1956) Kinetic studies on montmorillonites and nontronite by the acid dissolution technique. *Clays Clay Miner*. **4**, 301–321.
PARRY E.P. (1963). An infrared study of pyridine adsorbed on acidic solids. Characterization of surface acidity. *J. Catal*. **2**, 371–379.
PAVER H. & MARSHALL C.E. (1934) The role of aluminium in the reactions of clays. *J. Soc. Chem. Ind., Lond*. **53**, 750–760.
PEIGNEUR P., MAES A. & CREMERS A. (1975) Heterogeneity of charge density distribution in montmorillonite as inferred from cobalt adsorption. *Clays Clay Miner*. **23**, 71–75.
PINNAVAIA T.J. (1980) Applications of esr spectroscopy to inorganic clay systems. Ch. 8 in *Advanced Chemical Methods for Soil and Clay Minerals Research* (J.W. Stucki and W.L. Banwart, eds). Reidel, Boston.
PINNAVAIA T.J. (1981) Electron spin resonance studies of clay minerals. Ch. 6 in *Advanced Techniques for Clay Mineral Analysis*. (J.J. Fripiat, ed.). Elsevier, New York.
PINNAVAIA T.J. (1982) Intercalation of molecular catalysts in layered silicates. *ACS Symp. Ser*. **192**, 242–253.
PINNAVAIA T.J., RAYTHATHA R., LEE J.G.S., HALLORAN L.J. & HOFFMAN J.F. (1979) Intercalation of catalytically active metal complexes in mica-type silicates. Rhodium hydrogenation catalysts. *J. Am. Chem. Soc*. **101**, 6891–6897.
PINNAVAIA T.J. & WELTY P.K. (1975) Catalytic hydrogenation of 1-hexene by rhodium complexes in the intracrystal space of a swelling layer lattice silicate. *J. Am. Chem. Soc*. **97**, 3819–3820.
PINNAVAIA T.J., WELTY P.K. & HOFFMAN J.F. (1976) Catalytic hydrogenation of unsaturated hydrocarbons by cationic rhodium complexes and rhodium metal intercalated in smectite. *Proc. Int. Clay Conf. 1975, Mexico*, pp. 373–381. Applied Publishing Ltd, Wilmette, Illinois.
QUAYLE W.H. & PINNAVAIA T.J. (1979) Utilization of a cationic ligand for the intercalation of catalytically active rhodium complexes in swelling layer-lattice silicates. *Inorg. Chem*. **18**, 2840–2847.
RAYTHATHA R. & PINNAVAIA T.J. (1981) Hydrogenation of 1,3-butadienes with a rhodium complex–layered silicate intercalation catalyst. *J. Organomet. Chem*. **218**, 115–122.
RAYTHATHA R.H. & PINNAVAIA T.J. (1983) Clay intercalation catalysts interlayered with rhodium phosphine complexes. Surface effects on the hydrogenation and isomerization of 1-hexene. *J. Catal*. **80**, 47–55.
RESING H.A., SLOTFELDT-ELLINGSEN D., GARROWAY A.N., WEBER D.C., PINNAVAIA T.J. & UNGER K. (1980) ^{13}C chemical shifts in adsorption systems: molecular motions, molecular orientations, qualitative and quantitative analysis. In *Magnetic Resonance in Colloid and Interface Science* (J.P. Fraissard and H.A. Resing, eds), pp. 239–258. Reidel, Boston.
RICE F.E. & OSUGI S. (1918) The inversion of cane sugars by soils and allied substances and the nature of soil acidity. *Soil Sci*. **5**, 333–358.
RIDEAL E.K. & THOMAS W. (1922) Adsorption and catalysis in Fuller's earth. *Chem. Soc. J. Trans* **121**, 2119–2123.
ROSS D.K. & HALL P.L. (1980) Neutron scattering methods of investigating clay systems. Ch. 2 in *Advanced Chemical Methods for Soil and Clay Minerals Research* (J.W. Stucki and W.L. Banwart eds). Reidel, Boston.
ROSS G.J. (1969) Acid dissolution of chlorites: release of magnesium, iron, and aluminum and mode of acid attack. *Clays Clay Miner*. **17**, 347–354.
RYLAND L.B., TAMELE M.W. & WILSON J.N. (1960). Cracking catalysts. In *Catalysis* (P.H. Emmett, ed.), Vol. VII, pp. 1–93. Reinhold Publishing Co., New York.
SCHROCK R.R. & OSBORN J.A. (1976a) Catalytic hydrogenation using cationic rhodium complexes. I. Evolution of the catalyst system and the hydrogenation of olefins. *J. Am. Chem. Soc*. **98**, 2134–2143.
SCHROCK R.R. & OSBORN J.A. (1976b) Catalytic hydrogenation using cationic rhodium complexes. II. The selective hydrogenation of alkynes to *cis* olefins. *J. Am. Chem. Soc*. **98**, 2143–2147.
SCHROCK R.R. & OSBORN J.A. (1976c) Catalytic hydrogenation using cationic rhodium complexes. II. The selective hydrogenation of dienes to monoenes. *J. Am. Chem. Soc*. **98**, 4450–4455.
SHABTAI J. (1979) Zeolites and cross-linked silicates as media for selective catalysis. *Chem. L'Indust*. **61**, 734–741.

SHABTAI J. (1980) Class of cracking catalysts acidic forms of cross-linked smectites. US Patent 4,238,364.
SHABTAI J., FRYDMAN N. & LAZAR R. (1976) Synthesis and catalytic properties of 1,4-diazabicyclo(2,2,2)octane–montmorillonite system – a novel type of molecular sieve. *Proc. 6th Inter. Congr. Catal.* **B5**, 1–7.
SHABTAI J. & LAHAV N. (1980) Cross-linked montomorillonite molecular sieves. US Patent, 4,216,188.
SOLOMON D.H. (1968) Clay minerals as electron acceptors and/or electron donors in organic reactions. *Clays Clay Miner.* **16**, 31–39.
SOLOMON D.H. & ROSSER M.J. (1965) Reactions catalyzed by minerals. I. Polymerization of styrene. *J. Appl. Polym. Sci.* **9**, 1261–1271.
STONE W.E.E. (1981) The use of nmr in the study of clay minerals. Ch. 4 in *Advanced Techniques for Clay Mineral Analysis* (J.J. Fripiat, ed.). Elsevier, New York.
STUL M.S. & MORTIER W.J. (1974) The heterogeneity of the charge density in montmorillonite. *Clays Clay Miner.* **22**, 391–396.
TAMELE M.W. (1950) Chemistry of the surface and the activity of silica–alumina cracking catalyst. *Disc. Faraday Soc.* **8**, 270–279.
THOMAS C.L. (1949) Chemistry of cracking catalysts. *Ind. Engng Chem.* **41**, 2564–2573.
THOMAS C.L., HICKEY J. & STECKER G. (1950) Chemistry of clay cracking catalysts. *Ind. Engng Chem.* **42**, 866–871.
THOMAS J.M. (1982) Sheet silicate intercalates: new agents for unusual chemical conversions. Ch. 3 in *Intercalation Chemistry* (M.S. Whittingham and A.J. Jacobson, eds). Academic Press, New York.
THOMAS J.M., ADAMS J.M., GRAHAM S.H. & TENNAKOON D.T.B. (1977) Chemical conversion using sheet silicates. *Adv. Chem. Ser.* **163**, 298–315.
TRAYNOR M.F., MORTLAND M.M. & PINNAVAIA T.J. (1978) Ion exchange and intercalation reactions of hectorite with tris-bipyridyl metal complexes. *Clays Clay Miner.* **26**, 319–326.
TURK A. & FELDMAN J. (1943) Catalytic isomerization of fatty polyenes. *Paint, Oil, and Chem. Rev.* **106**, 10–11.
UYTTERHOEVEN J.B., CHRISTNER L.G. & HALL W.K. (1965) Studies of the hydrogen held by solids. VIII. The decationated zeolites. *J. Phys. Chem., Ithaca* **69**, 2117–2126.
VALENTINE, JR., D. & SCOTT J.W. (1978) Asymmetric synthesis. *Synthesis* 329–356.
VAUGHAN D.E.W. & LUSSIER R.J. (1980) Preparation of molecular sieves based on pillared interlayered clays. *Proc. 5th Int. Conf. Zeolites, Naples, Italy*, pp. 94–101.
VAUGHAN D.E.W., LUSSIER R.J. & MAGEE J.S. (1979) Pillared interlayered clay materials useful as catalysts and sorbents. US Patent, 4,176,090.
VAUGHAN D.E.W., LUSSIER R.J. & MAGEE J.S. (1981a) Stabilized pillared interlayered clays. US Patent, 4,248,739.
VAUGHAN D.E.W., LUSSIER R.J. & MAGEE J.S. (1981b) Pillared interlayered clay products. US Patent, 4,271,043.
VOGE H.H. (1958) *Catalytic Cracking in Catalysis* (P.H. Emmett, ed.). Vol. VI, pp. 407–493. Reinold Publishing Co., New York.
VON LIEBIG J. (1865) Chemische briefe, C.F. Winter'sche Verlagshandung, Leipzig, 96.
WALLING C. (1950) The acid strength of surfaces. *J. Am. Chem. Soc.* **72**, 1164–1168.
WARD J.W. & HANSFORD R.C. (1969) The detection of acidity on silica–alumina catalysts by infrared spectroscopy-pyridine chemisorption. *J. Catal.* **13**, 154–160.
WEISS A. (1981) Replication and evolution in inorganic systems. *Angew. Chem. Int. Ed. Engl.* **20**, 850–860.
WHITAKER A.C. & KINZER A.D. (1955) Effect of residual coke behavior of cracking catalyst. *Ind. Engng Chem.* **47**, 2153–2157.
WIEGNER G. (1931) Some physicochemical properties of clay. II. Hydrogen clay. *J. Soc. Chem. Ind.* **50**, 103–112.
WILSON C.R. & TAUBE H. (1975) Acetate complexes of dirhodium and diruthenium. Aquation and reduction-oxidation. *Inorg. Chem.* **14**, 2276–2279.
WRIGHT A.C., GRANQUIST W.T. & KENNEDY J.V. (1972) Catalysis by layer-lattice silicates. I. The structure and thermal modification of a synthetic ammonium dioctahedral clay. *J. Catal.* **25**, 65–80.
WU C.Y. & HALL W.K. (1967). Mechanism of triphenylcarbonium ion formation on the silica–alumina surface. *J. Catal.* **8**, 394–396.
YAMANAKA S. & BRINDLEY G.W. (1978) High surface area solids obtained by reaction of montmorillonite with zirconyl chloride. *Clays Clay Miner.* **26**, 119–124.

Chapter 7

Thermal, Oxidation and Reduction Reactions of Clay Minerals

GEORGE W. BRINDLEY and JACQUES LEMAITRE

		Page			Page
7.1	SURVEY OF REACTIONS	319		7.6.1.3 Chrysotile reactions	346
7.2	LOW-TEMPERATURE REACTIONS	322		7.6.1.4 Chlorite–olivine reaction	346
	7.2.1 Relocation of small interlayer cations	322		7.6.1.5 Talc–enstatite reaction	347
	7.2.2 Reorganization of layer stacking	323		7.6.1.6 Reactions of Ni-containing talcs	347
7.3	DEHYDROXYLATION REACTIONS	324		7.6.1.7 Saponite–enstatite reaction	348
	7.3.1 Kinetics of dehydroxylation	324		7.6.1.8 Sepiolite and palygorskite reactions	348
	7.3.1.1 Temperature dependence	327		7.6.1.9 Vermiculite reactions	348
	7.3.1.2 Vapour pressure dependence	327	7.6.2	Dioctahedral mineral reactions	349
	7.3.2 Reaction mechanisms	329		7.6.2.1 Pyrophyllite–mullite reaction	349
	7.3.3 Pre-dehydroxylation effects: proton delocalization	331		7.6.2.2 Kaolinite–mullite reactions	349
7.4	DEHYDROXYLATED PHASES	331		7.6.2.3 The spinel phase	350
	7.4.1 Kaolinite	332		7.6.2.4 High-temperature phases from muscovite	354
	7.4.2 Other kaolinite group minerals	333		7.6.2.5 High-temperature phases from montmorillonites	355
	7.4.2.1 Halloysite	334	7.7	OXIDATION REACTIONS	356
	7.4.3 Chlorite	334		7.7.1 Oxidation of berthierine	357
	7.4.3.1 Superlattice effects	335		7.7.2 Oxidation of micas	357
	7.4.4 Serpentine	335		7.7.3 Oxidation of chlorites	359
	7.4.4.1 Nickel-containing serpentine minerals	336		7.7.4 Oxidation in relation to biotite–vermiculite relations	360
	7.4.5 Pyrophyllite and talc	336		7.7.5 The possible role of OH orientations on chemical behaviour	360
	7.4.6 Muscovite, illite	337	7.8	REDUCTION REACTIONS	360
	7.4.6.1 Illites	338		7.8.1 Reduction reactions of nickel-containing silicates	361
	7.4.7 Montmorillonite	339		7.8.1.1 Serpentines	361
	7.4.8 Vermiculite	340		7.8.1.2 Talc	362
	7.4.9 Other dehydroxylated phases	340		7.8.2 Reduction reactions of nickel-hydroxy montmorillonites	362
	7.4.9.1 Saponite	340	7.9	SUMMARY AND CONCLUSIONS	362
	7.4.9.2 Sepiolite and palygorskite (attapulgite)	341	NOTE		363
7.5	REHYDROXYLATION	341	REFERENCES		364
	7.5.1 Hydroxylation reactions	342			
7.6	HIGH-TEMPERATURE PHASES: TOPOTACTIC REACTIONS	344			
	7.6.1 Trioctahedral mineral reactions	344			
	7.6.1.1 Serpentine to forsterite reactions	344			
	7.6.1.2 Reactions of Ni-containing serpentines	346			

7.1 SURVEY OF REACTIONS

Clay minerals are hydrous phyllosilicates principally of Al, Mg, Fe(II) and Fe(III), and less frequently other ions such as Ni. Larger cations such as K, Na and Ca also are involved in certain mineral groups. The main anions involved are O^{2-} and $(OH)^-$ but F^- ions are important in some cases. Almost by definition, clay minerals have particle sizes less than $1 \,\mu m = 10^{-6} \,m = 10^{-4} \,cm$; these small sizes are very important in relation to their thermal reactions.

For some purposes it is advantageous to study the corresponding macrocrystalline minerals but the applicability of the results to the microcrystalline materials must be given careful consideration. It will often be evident from the context where macro- and micro-crystalline data are under consideration.

Frequently it is convenient to treat the chemical behaviour of clay minerals in terms of idealized formulae such as $Al_2Si_2O_5(OH)_4$ for kaolinite, $KAl_2(Si_3Al)O_{10}(OH)_2$ for muscovite and similar simplifications for other "pure" clay minerals, but it should not be forgotten that substitutions such as Fe(III) for Al, Fe(II) or Ni for Mg, or even $2Al\square$ for 3Mg may modify physico-chemical behaviour to a greater extent than might be expected from the small chemical changes involved. In other words, one must be cautious in treating even the purest clay mineral in the same way as "pure" chemical compounds. The large specific surface of clays, ranging from a few $m^2 g^{-1}$ to values approaching $1000 m^2 g^{-1}$, also raise questions concerning bulk versus surface reactions which ought not to be forgotten. To all these considerations can be added the further problem that clay minerals generally exist as mixtures of minerals so that the thermal reactions are those of mixtures, often greatly assisted by their small particle sizes. In view of these many complications, it is reasonable to consider first the optimum conditions for mainly monomineralic materials with "reasonably well known" chemical compositions and crystal-structural characteristics. Most of this chapter will be oriented towards these considerations and the more complex behaviour (or at least, the different behaviour of mixed systems) must be left for others to consider.

The principal thermal reactions and the approximate temperature ranges in which they occur are conveniently considered in the following categories, although in practice they may be less sharply differentiated.

1. *Low-temperature reactions below about 400 °C.* These involve loss of molecular water from between layers of expanding minerals and from channels in sepiolite and palygorskite. Much of this water is related to hydration of cations, although in some minerals, e.g. halloysite, sheets of water molecules exist, probably hydrogen-bonded to surfaces. Removal of hydration water is treated in Chapter 5.

   Also in this category comes a particular reaction in which small interlayer cations, notably Li^+ and Mg^{2+}, migrate into vacant octahedral cation sites of montmorillonite.

2. *Intermediate-temperature reactions, mainly 400–750 °C.* These involve dehydroxylation and the formation of quasi-stable dehydroxylated phases. Prior to dehydroxylation proper, in some minerals delocalization of protons takes place, and as this is a forerunner of dehydroxylation, it is most conveniently considered in this category. Oxidation processes also may occur in this temperature range and often are closely related to dehydroxylation; these reactions will be treated separately.

3. *High-temperature reactions, above 750 °C.* These are recrystallization processes involving the formation of new phases. Reactions in categories (2) and (3) may take place almost concurrently.

4. *Oxidation reactions.* These are particularly important for minerals with variable valence cations.

Table 7.1(a) and (b) summarizes the principal mineral types and reactions to be considered.

The effect of water vapour atmospheres, less than 1 atm pressure, will be considered in relation to the kinetics of dehydroxylation but reactions in hydrothermal atmospheres, pressure $\geqslant 1$ atm, and generally elevated temperatures are not considered here.

The chemistry of clay mineral reactions involves considerations of the compositions of the

TABLE 7.1(a). Summary of thermal reactions – dioctahedral minerals

Mineral					
Kaolinite $Al_2Si_2O_5(OH)_4$ 1:1 type	Predehydroxylation state $\longrightarrow$ 450–550 °C	Metakaolin $Al_2Si_2O_7$	$\longrightarrow$ 900 °C Spinel[†] type phase	$\longrightarrow$ 1000–1100 °C	Mullite[†] $Al_6Si_2O_{13}$
Pyrophyllite $Al_2Si_4O_{10}(OH)_2$ 2:1 type		$\longrightarrow$ 650 °C Pyrophyllite dehydroxylate $Al_2Si_4O_{11}$	$\longrightarrow$ 950–1000 °C		Mullite[†] $Al_6Si_2O_{13}$
Montmorillonite[‡] 2:1 type	$\rightarrow$ 150–250 °C	$\rightarrow$ Montmorillonite anhydride	$\rightarrow$ 700 °C Montmorillonite dehydroxylate	1000 °C $\rightarrow$ Spinel-type phase	1150 °C Mullite[†]
Muscovite 2:1 type		$\longrightarrow$ 750 °C Muscovite dehydroxylate	$\longrightarrow$ 1050 °C	$\longrightarrow$ Spinel-type phase Sanidine	$\longrightarrow$ 1250 °C Corundum α-Al_2O_3 Mullite[†] Other Phases

[†] With separation of silica and/or other phases.
[‡] Ideal formula: $M^+_{0.33}nH_2O \cdot (Al_{1.67}Mg_{0.33})Si_4O_{10}(OH)_2$.
Temperatures variable according to chemical composition, crystal size, conditions of heating.

TABLE 7.1(b) Summary of thermal reactions – trioctahedral minerals

Mineral			
Serpentine $Mg_3Si_2O_5(OH)_4$ 1:1 type	$\longrightarrow$ 550 °C Serpentine dehydroxylate	$\longrightarrow$ 600 °C Forsterite (olivine)[†] Mg_2SiO_4	$\longrightarrow$ 1100 °C Forsterite + enstatite $MgSiO_3$
Talc $Mg_3Si_4O_{10}(OH)_2$ 2:1 type		$\longrightarrow$ 900–1000 °C Enstatite[†]	
Saponite[‡] 2:1	$\longrightarrow$ 550 °C Saponite dehydroxylate	$\longrightarrow$ 750 °C Enstatite[†]	
Vermiculite[§] 2:1	$\longrightarrow$ 500–800 °C Vermiculite dehydroxylate	$\longrightarrow$ 850 °C Enstatite[†]	
Chlorite $(Mg,Al)_6(Si,Al)_4O_{10}(OH)_8$ 2:1 type	$\longrightarrow$ 450–550 °C partial or complete dehydroxylation	$\longrightarrow$ 800 °C Olivine[†]	

[†] With separation of silica and/or other phases.
[‡] Ideal formula: $M^+_{0.33}nH_2O \cdot Mg_3(Si_{3.67}Al_{0.33})O_{10}(OH)_2$.
[§] Typical formula: $Mg^{2+}_{0.38}nH_2O \cdot (Mg_{2.00}Fe^{III}_{0.46}Al_{0.22}Ti_{0.11})(Si_{2.72}Al_{1.28})O_{10}(OH)_2$.
Temperatures variable according to chemical composition, crystal size, condition of heating.

initial minerals and of the products formed at different temperatures, the kinetics and mechanisms of the reactions, and the crystal–chemical relations between the phases. Much of this information has been summarized by Grim (1968, particularly pp. 278–352). Much related information on the thermal investigation of clays has been compiled by Mackenzie (1957), and useful reviews have been given by Taylor (1962), Dent et al. (1962), Brindley (1963), and Brett et al. (1970).

7.2 LOW-TEMPERATURE REACTIONS

Low-temperature reactions are concerned principally with the removal of surface and interlayer water and these aspects are treated in Chapter 5.

With removal of interlayer water, 2:1 type minerals, particularly smectites and vermiculites with hydrated inter-layers, collapse and the basal spacing approaches 10 Å; the exchangeable cations will be situated more or less within the pseudo-hexagonal holes of the Si—O network. The precise value of the dehydrated spacing depends on the particular interlayer cations, the humidity of the ambient conditions, and the particular temperature (such as 100 °C, 110 °C, 150 °C, etc) employed. Rehydration is generally possible provided the heat treatment has not taken place at a too high temperature, but is dependent on the particular cations involved.

7.2.1 Relocation of small interlayer cations

In the case of montmorillonites containing small interlayer cations, notably Li^+, Mg^{2+} and Ni^{2+}, these ions appear to be capable of moving irreversibly into the layer structure at temperatures around 200–300 °C and losing their capacity for exchange reactions and for solvation (Hofmann and Klemen 1950; Gonzáles García 1950; Greene-Kelly 1953, 1955; Calvet and Prost 1971). Hofmann and Klemen attributed these results to a migration of the small cations into vacant octahedral sites and Greene-Kelly supported this by showing that hectorite, which is trioctahedral without vacant sites, and beidellite, which has principally tetrahedral substitutions, do not show similar results. The effect, therefore, may be diagnostic for montmorillonite. However, other points of view have been expressed. Tettenhorst (1962), on the basis of infrared data for montmorillonites, saturated with many different cations and heated for 8 hours at 300 °C, considered that all cations penetrate the Si—O network to an extent determined by their size, but none passes to the octahedral positions. Heller (1965), principally on the basis of glycol and glycerol solvation of montmorillonites heated to 500–600 °C, considered that Li^+ and Mg^{2+} ions "do not differ fundamentally from other cations" in their behaviour. Glaeser and Mering (1967) strongly concur with Greene-Kelly, and note that for a dioctahedral mineral in which the layer charge arises partly in the octahedral and partly in the tetrahedral sheets of the structure, only the octahedral charge is neutralized by Li migration. Schultz (1969) has used this principle for distinguishing octahedral and tetrahedral contributions to the total layer charge of montmorillonites. Further observations by Russell and Farmer (1964) and Farmer and Russell (1967) showed that Li-, Mg-, and NH_4-montmorillonites heated to 350 °C gave acid reactions with indicators in ethylene glycol (used, presumably, to expand the heated clays). They considered that in the case of Li-montmorillonites, part of the Li becomes fixed in the octahedral sites, but some may approach the hydroxyl units causing the release of protons. Thus it seems possible that a complex process takes place in which the small cations penetrate the layer structure in more than one way.

The possible relocation of Li$^+$ ions from interlayer positions into octahedral positions makes it feasible to consider a controlled change of the layer charge and thereby to correlate swelling properties with the resultant layer charge. Brindley and Ertem (1971) prepared Li, Na-, Li,K- and Li,Ca-montmorillonites with various proportions of the pairs of the interlayer cations which were then heated for periods of about 24 hours at 220 C. Due to the fact that Li-, Na- and K-montmorillonites were fully dispersed in water, it was possible to prepare Li,Na- and Li,K-clays by merely mixing in the required proportions of suspensions of Li- and Na- or Li- and K-saturated clays. After centrifugation and washing the exchangeable ions were determined by displacement with 1M ammonium chloride solutions.

Li,Ca-clays were prepared by treating Ca-clay suspensions with LiCl solutions of various concentrations for 2 hour periods and then were subsequently analysed to determine the proportions of Li and Ca exchangeable ions. In accordance with the earlier work of Tettenhorst (1962) it was found that the numbers of exchangeable cations after heat treatment are not entirely in accordance with migration of Li$^+$ ions into vacant octahedral sites up to the limit of the octahedral charge deficiency. The main interest of the study, however, was concerned with the expandability of montmorillonite in various liquids in relation to the numbers of cations remaining in the interlayer space (Table 7.2).

With water, acetone, and 3-pentanone, the montmorillonite swells broadly in accordance with the number and field strength of the exchangeable cations and the dipole moments of the molecules, and is generally consistent with cation–dipole interactions. With ethanol, ethylene glycol and morpholine, swelling is largely independent of the particular cation ratios within the range of exchange capacities used (88–20 meq/100 g) and appears slightly greater when the number of exchangeable cations is small.

In further studies, Ertem (1972) concluded that Li ions occupying exchange sites of montmorillonite can undergo two different reactions upon heating: (i) some migrate into vacant octahedral sites in accordance with the classical Hofmann–Klemen concept; but (ii) some may react with structural OH groups or with residual H$_2$O to form protons as noted by Russell and Farmer (1964) and Farmer and Russell (1967).

The use of Li ions to determine quantitatively the octahedral and tetrahedral charges in dioctahedral smectites evidently must be applied cautiously.

7.2.2 Reorganization of layer stacking

The turbostratic structure of montmorillonite normally gives an X-ray powder pattern of 00l basal reflections, often non-integral, and hk two-dimensional diffraction bands. After Li migration, the diffraction bands with k a multiple of 3 becomes strongly modulated and appear to be resolving into separate hkl diffractions (Mering and Glaeser 1967). The results indicate that a layer stacking order develops in which the pseudo-hexagonal cavities of adjacent layers tend to have a face-to-face arrangement similar to that found in pyrophyllite. They concluded that the initial turbostratic stacking provides a better local neutralization of layer charges than an ordered stacking, but ordering becomes possible when the small monovalent ions have moved from their interlayer positions into the octahedral vacant sites adjacent to the Mg-for-Al substitutions.

324 THERMAL, OXIDATION AND REDUCTION REACTIONS OF CLAY MINERALS

TABLE 7.2. Basal spacings of heat-treated clays after solvation with the liquids indicated. Stippled areas indicate changes in state of solvation

Solvating liquid	Cation other than Li	Basal spacings, in Å, of 220 °C/24 h treated clays							
		Fraction of exchangeable Li ions in pre-heated clays								
		0		0.2		0.4	0.6	0.8		1.0
Water	Na	∞	∞	∞	9.8$_5$ vb	9.6 vb	9.5$_5$ sh	9.5		⎫
	K	∞	10.5 vb	9.9$_0$ b	9.8$_5$	9.8$_5$ sh				⎬ 9.6$_0$
	Ca	19.1$_0$ sh		19.1$_0$		19.2$_0$ 9.5$_5$		9.6$_0$		⎭
Acetone	Na			18.5$_0$	18.5$_0$	13.0$_0$ sh	13 m-l	9.3$_5$ m-l		⎫
	K	13.1 sh	13.0 sh	13.0	13.0	12.9$_5$				⎬ 9.4$_5^{\ddagger}$
	Ca	17.4$_5$ sh		17.4$_5$ sh		17.5$_0$ sh 13.1$_5$ irreg		13.1 irreg		⎭
3-Pentanone	Na	13.2$_0$ v sh		13.1$_5$ v sh	13.2$_5$ v sh	13.1$_0$ sh	13.2$_0$ sh	9.4$_5$ (12.8)†		⎫ 9.6 (13.2)
	K	13.4	13.2$_5$			13.2$_5$ sh				⎭
Ethanol	Na	16.8$_0$		17.0$_5$		17.1$_0$	17.1$_0$	17.0		17 (18.5) m-l
Ethylene glycol	Na	17.0$_0$ v sh		17.2$_0$ v sh		17.0$_5$ v sh	17.1$_0$ v sh	17.1$_5$ sh		⎫ 17.1
	Ca			16.8$_5$ v sh		16.9$_5$ v sh	17.0$_5$ sh	17.0$_5$ v sh		⎭ irreg
Morpholine	Na	14.8$_0$ (9.6)		14.8$_0$ (9.6)			14.8 (9.6)			⎫ 15.1$_5$ (9.5)
	K	9.9								⎭

v sh – very sharp; sh – sharp; b – broad; vb – very broad; irreg – irregular; m-l – mixed layered; ( ) – also present; ∞ – no basal spacing observed; † – trace of 22Å spacing 9.45 + 12.8 = 22.25 Å; ‡ – trace of long spacing.
After Brindley and Ertem (1971); reprinted by permission. Copyright 1971 The Clay Minerals Society, USA.

7.3 DEHYDROXYLATION REACTIONS

Under normal atmosphere conditions, these reactions occur in a temperature range approximately 450–750 °C, as summarized in Table 7.1(a) and (b). Quite low vapour pressures have considerable influence on these reactions, not because of any marked reversibility which requires hydrothermal conditions but mainly because of surface effects. Consequently the results depend in some respects on the sample size, and especially on the form and packing of the microcrystallinte particles, since these factors influence the water vapour atmosphere within the powder sample. The reactions of clay grade materials therefore involve considerations which are not present in single crystal studies.

7.3.1 Kinetics of dehydroxylation

Many experiments on kinetics have been undertaken with a view to determining the mechanism of dehydroxylation reactions but this objective goes beyond what can be obtained from a kinetic study of a solid state reaction. The results are relevant only to the rate-controlling process, which may not involve the primary mechanism.

Kinetic measurements hopefully lead to the determination of rate constants in relation to temperature, vapour pressure and possibly other variables. The rate constants are obtained by fitting observed and theoretical α, t curves, where α is the fraction of material reacted and t is the time. To distinguish different rate-controlling processes, such as diffusion, phase boundary reactions, nucleation, etc., the α, t curve should fit over a wide range of α, if possible approaching 100%.

Isothermal gravimetric methods are well suited to these kinetic studies but require much repetitive work. Dynamic methods, such as dynamic gravimetric analysis and differential thermal analysis, offer great convenience by covering both time and temperature variations in a single run (see Freeman and Carroll 1958; Coats and Redfern 1963; Achar et al. 1966; Sestak et al. 1973), but the difficulties of interpretation are greater than for multiple isothermal runs. Infrared studies have great potential for elucidating mechanisms as distinct from kinetics because they reveal what is happening to the OH radicals.

Many of the earlier dehydroxylation experiments, for example the pioneering experiments of Murray and White (1949, 1955), were interpreted in terms of first-order kinetics, i.e. an equation of the type $-dc/dt = k \cdot c^n$ with $n=1$, where c is concentration of reactant at time t. Many experiments were carried out before important aspects of the process were appreciated, such as the role of vapour pressure, the effect of containers, the use of very thin and loosely compacted samples, and (if possible) powders of uniform particle size and shape. For example, Brindley and Nakahira (1957) studied the dehydroxylation of discs of kaolinite under constant temperature conditions, 497 °C, pressed to various thicknesses ranging from 0.38 to 2.68 mm. They found greatly different rates of reactions. Furthermore, discs of 3 mm thickness were found to be almost 95% reacted in the outer surfaces, but no more than about 40% at their centres when the overall dehydroxylation was near 50%. Evidently the conditions of kinetic studies are of paramount importance in understanding the nature of the processes involved.

Later experiments using micro-balances and thin layers of material tended to confirm diffusion-controlled kinetics, although there is still not complete agreement on this question. However, on this basis the following equations can be given:

$$\alpha^2 = (K/x^2)t$$

(one-dimensional diffusion; $2x$ = particle thickness)

$$(1-\alpha)\ln(1-\alpha) + \alpha = (K/r^2)t$$

(two-dimensional diffusion: r = particle radius)

$$(1 - 2\alpha/3) - (1-\alpha)^{2/3} = (K/r^2)t$$

(three-dimensional diffusion; r = radius of spherical particle)

Numerical evaluation of these and other kinetic equations was tabulated by Sharp et al. (1966), who gave curves of α v. a reduced time (t/t_{50}) where t_{50} = time for 50% of complete reaction. Both α and t/t_{50} are non-dimensional variables. For each type of reaction kinetics (see Fig. 7.1(a)) the reduced time plot gives a single curve which facilitates the matching of experimental data and theoretical curves. As an example of the use of reduced time plots, Fig. 7.1(b) shows very clearly the results for kaolinite with eight different isothermal conditions, with water vapour pressures ranging from 10^{-5} to 175 mm, temperatures from 416 to 537 °C, with one set of data for ambient vapour pressure and 486 °C. The curves agree closely up to about 75% of total reaction. The

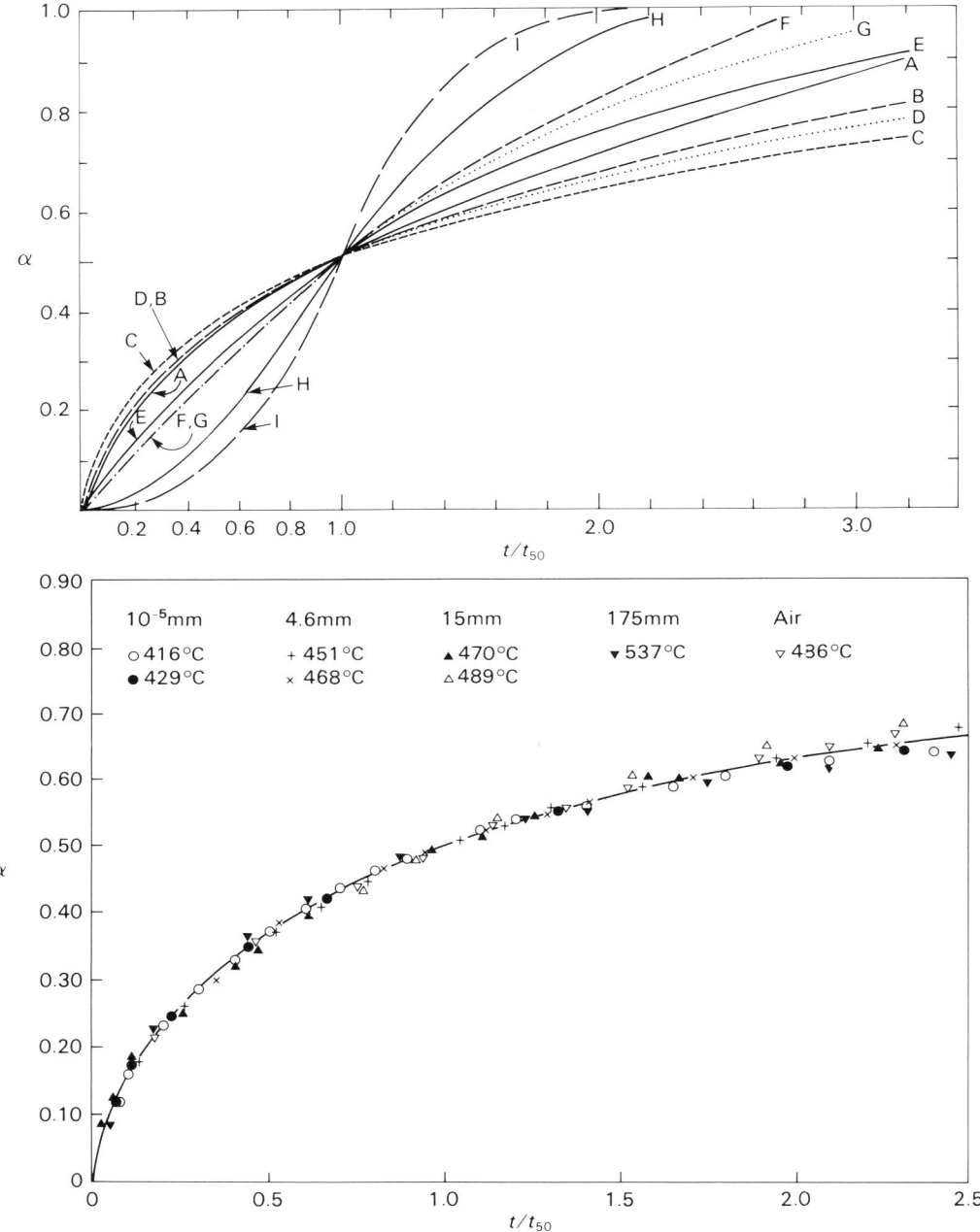

FIG. 7.1. (a) Theoretical curves of α, fraction of material reacted, versus a reduced time t/t_{50}, where t_{50} is the time for 50% of complete reaction. Curves have the following significance: A, B, C, D: Diffusion controlled reactions for particles in the form of plates (A), circular discs (B), sphere (C, D). C is based on the approximate Jander equation. E: First-order kinetic equation. F, G: Phase-boundary controlled reactions in circular discs (F), spheres (G). H, I: Reactions following the Avrami–Erofe'ev equations in two and three dimensions. After Sharp et al. (1966). (b) Fraction reacted, α, versus t/t_{50}, reaction time for 50% reaction, for Florida kaolinite dehydroxylated at temperatures and vapour pressures indicated. From Brindley et al. (1967a). Copyright 1967 by the MSA.

differentiation of first-order kinetics and diffusion-controlled kinetics calls for especially careful considerations because the usual test for first-order kinetics, a linear relation between $\ln(1-\alpha)$ and t, is insufficient (see Brindley *et al.* 1967a, pp. 209–210); additionally the linear plot must pass through $\alpha = 0$ when $t = 0$.

The particle size factor, represented by x or r in the previous equations, also presents a major difficulty, whether one attempts experimentally to obtain particles of uniform size and shape, or to calculate theoretically the effect of a particle size distribution (Gallagher 1965; Johnson and Kessler 1969). The last-named authors concluded that the formation and vaporization of water from the surfaces of kaolinite is the limiting step in the reaction rather than a diffusion control; their analyses involved taking into account the particle size distribution.

7.3.1.1 Temperature dependence
At constant vapour pressure, the reaction constant K varies with the absolute temperature T according to an Arrhenius-type expression:

$$K(T) = A \exp(-\Delta G/RT)$$

where ΔG, the free energy of activation, can be written $\Delta H - T \Delta S$, and

$$K(T) = A' \exp(-\Delta H/RT)$$

The activation energy ΔH, obtained from experimentally determined values of $K(T)$ by plotting $\ln K(T)$ v. $1/T$ is usually of the order of 80–240 kJ mol^{-1}, but no precise significance has been attached to these values.

7.3.1.2 Vapour pressure dependence
Reaction rates vary greatly with the ambient vapour pressure even at low pressures in the range 0.1–10 Torr. At higher pressures, the reaction proceeds only at considerably increased temperatures until eventually the character of the reaction itself changes, as shown by Roy and Weber (1964) for serpentine minerals and by Weber and Roy (1965) for kaolinite. These results, however, belong to the hydrothermal range.

The mechanisms involved at low vapour pressures are still not clear but almost certainly involve surface considerations. A simple model which fits experimental results considers that a fraction, θ, of the surface is covered or blocked by chemisorbed molecules (Toussaint *et al.* 1963; Brindley *et al.* 1967a,b) and that the reaction rate is proportional to the free surface, $(1-\theta)$. With the empirical relation $\theta = mP^n$, the reaction rate $K(P)$ at pressure P, is given by

$$K(P) = K_0(1 - mP^n)$$

where K_0 is the reaction rate *in vacuo*. Experimental results for kaolinite and serpentine, Fig. 7.2, for a range of temperatures and vapour pressures fit this relationship (Brindley *et al.* 1967a,b).

When the vapour pressure is raised to a value such that $mP^n \geq 1$, the reaction rate is reduced to zero. This is illustrated by the results shown in Fig. 7.3 (Brindley and Millhollen 1966) where a kaolinite sample dehydroxylating *in vacuo* at 425 °C ceased to react when water vapour at 47 Torr was admitted. The thermobalance showed a gain of weight equivalent to about a monolayer of water molecules. The dehydroxylation was resumed immediately the vacuum was restored and the monolayer removed.

Water vapour produced by dehydroxylation, entrapped between particles in a compact powder, as noted earlier, exerts a marked retarding influence and gives rise to high apparent

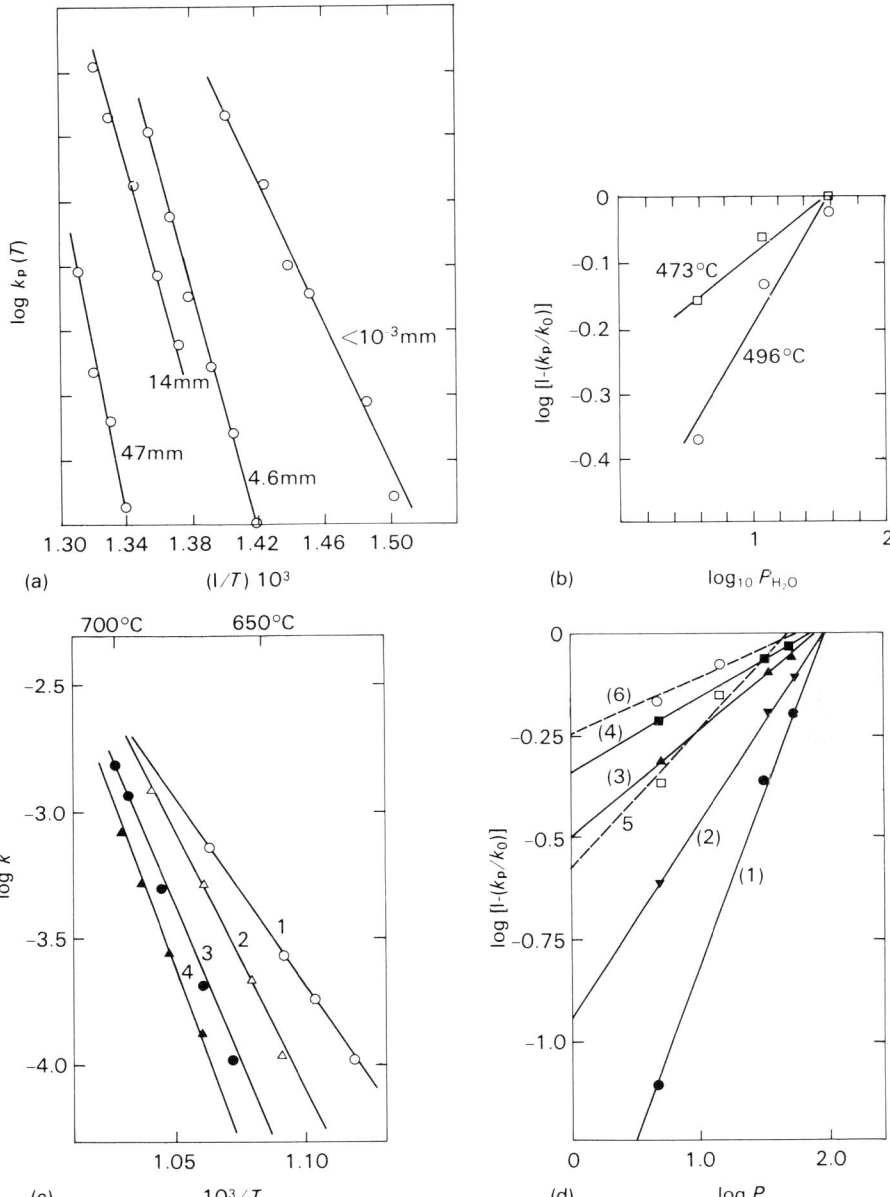

FIG. 7.2. Dehydroxylation rate constants for kaolinite (a) and (b) and serpentine (c) and (d) with respect to temperature at constant vapour pressure, (a) and (c) and to water vapour pressure at constant temperature (b) and (d). In (a) and (c), reciprocal of temperature in Kelvin is plotted, and in (b) and (d) log P in mmHg (Torr) is used. In (a), water vapour pressures are indicated. In (c), lines 1, 2, 3 and 4 correspond to vapour pressures $< 10^{-3}$, 4.6, 30, 0 and 47 mmHg (Torr). In (b), constant temperatures in °C are shown. In (d), lines 1, 2, 3 and 4 correspond to 698, 679, 657 and 636 °C. Figure 7.2 (a) and (b), after Brindley et al. (1967a); Fig. 7.2 (c) and (d) after Brindley et al. (1967b). Copyright 1967 by the MSA.

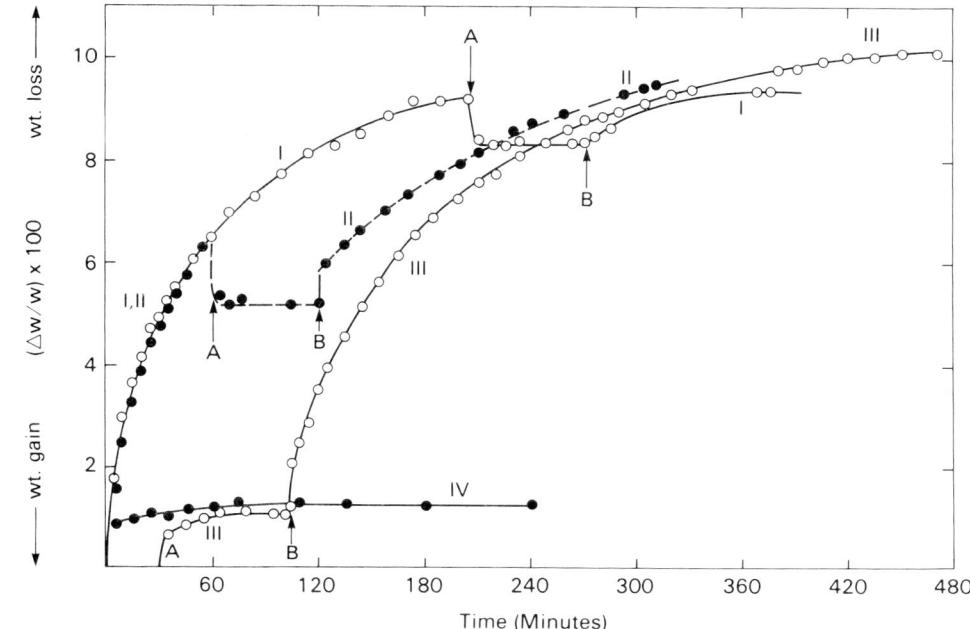

FIG. 7.3. Dehydroxylation of kaolinite; effect of introducing and removing a water vapour atmosphere. Water vapour pressure, 47 mmHg (Torr). Temperature, 425 °C. Curves I, II: Dehydroxylation commences *in vacuo*. At (A), water vapour is introduced, dehydroxylation is halted, chemisorption of water occurs. At (B) vacuum is restored and dehydroxylation is resumed. Curves III, IV: Dehydroxylation commences in water vapour atmosphere. At (B) vacuum is applied. Origin of curve III is displaced to $t = 30$ min. (After Brindley and Millhollen 1966.) Copyright 1966 by the AAAS.

activation energies. Some of the variations reported in the literature have probably arisen from this cause. Techniques using packed powders in cavities, as in many DTA techniques, will be prone to these effects.

The concepts used by Johnson and Kessler (1969) concerning the evaporation of water from clay mineral surfaces as being the controlling factor in dehydroxylation reactions could usefully be extended to provide a better model for the effect of external water vapour pressure on dehydroxylation kinetics. In an experiment not hitherto reported, Brindley et al. examined dehydroxylation behaviour of kaolinite at vapour pressures of 250–300 Torr, i.e. considerably higher than those shown in Fig. 7.1(b). Correspondingly higher temperatures were required to initiate dehydroxylation. Although the temperature was raised gradually, the first indication of any appreciable reaction was always *explosive*, with the thin layer of sample blown entirely from the shallow sample support. This explosive behaviour was checked on several samples.

7.3.2 Reaction mechanism

The simplest concept of the reaction is a proton transfer between two hydroxyl groups:

$$OH^- + OH^- \rightarrow O^{-2} + H_2O \nearrow \tag{1}$$

This reaction may be considered as the result of two separate steps (Pampuch 1971):

$$OH^- \rightleftharpoons H^+ + O^{2-} \qquad (2)$$

$$H^+ + OH^- \rightleftharpoons H_2O \nearrow \qquad (3)$$

The first step is the dissociation of one hydroxyl group to a free proton and an oxygen ion, while the second is the combination of the free proton with a second hydroxyl ion to form a water molecule. The dehydration process can be said to be homogeneous if reactions (2) and (3) involve in succession two adjacent hydroxyl groups. Water molecules will then be produced throughout the structure and must migrate towards a gas–solid interface in order to achieve the dehydration process. It is tempting to correlate this concept with the diffusion control found in many kinetic experiments. However, it is necessary to remember that the basic mechanism need not be the rate-controlling process. Indeed, the dependence of the rate of reaction on quite small vapour pressures (Brindley and Nakahira 1957) and the possibility that small vapour pressures will build up at points where water molecules emerge from the structure suggest that the control of the reaction lies in the removal of water, which is close to the conclusion of Johnson and Kessler (1969); but this leaves wide open the question relating to the *formation* of water.

Since protons can diffuse more easily than water molecules, reaction (3) may occur at favourable sites and be maintained by a migration of protons to a reaction zone. This mechanism, called a *heterogeneous*, or *inhomogeneous* mechanism, was suggested by Ball and Taylor (1961) for the dehydroxylation of brucite and subsequently applied to silicate reactions (Taylor 1962; Brindley 1961, 1963; see later). According to the heterogeneous mechanism, free protons produced in some regions of the solid (called acceptor regions) by reaction (2), migrate through the structure to regions where water is formed by reaction (3) (donor regions). A counter-migration of cations (Mg^{2+}, Fe^{2+}, Fe^{3+}, Al^{3+}, etc) from donor towards acceptor regions will maintain electrical neutrality. Diffusion-controlled kinetics may then be related to the escape of water molecules via pores or other defects, or to the counter-migration of cations, or to both processes.

The concepts that proton migration is closely connected with the basic mechanism is supported by the prehydroxylation effects discussed in the next section. Depending on the system of interest, each of the above mechanisms may be an adequate description of the overall dehydration reaction. According to Pampuch (1971), the homogeneous process is more likely to occur in structures where pairs of hydroxyl groups, exhibiting different acidities, are found; the proton of the more acidic hydroxyl will react with the less acidic one to form water. Water molecules are thus formed with essentially the same probability throughout the volume of the crystal. The simultaneous formation of water in the whole volume of the reacting crystallites, by disturbing the structure, would promote the formation of poorly crystallized products. This could be the case in kaolinite and other layer silicates known to produce poorly crystallized dehydroxylation products. The existence of different adjacent OH groups in kaolinite has been evidenced by IR spectroscopy (Pampuch *et al.* 1971) and many subsequent studies (e.g. Rouxhet *et al.* 1977; Mestdagh *et al.* 1982).

On the other hand, structures in which all the OH groups have the same probability to dissociate according to reaction (2) will most probably dehydrate following the inhomogeneous mechanism, provided cation migration through the oxygen structure is possible. A typical case is the dehydration of brucite, as already mentioned. As the heterogeneous dehydration proceeds through a process of cation migration without major disturbance of the oxygen arrangement, the solid products formed according to this mechanism, as noted by Pampuch (1971), will most likely be well crystallized and exhibit a high degree of crystallographic orientation with respect to the parent phases.

7.3.3 Pre-dehydroxylation effects: proton delocalization

Infrared and electrical conductivity measurements on hydrous layer silicates at temperatures below those where dehydroxylation freely occurs point to a degree of proton delocalization which eventually leads to loss of hydroxyls as water molecules (Fripiat and Toussaint 1960).

The electrical conductivity of kaolinite increases to a maximum at around 350 °C, then diminishes but increases again above 420 °C when dehydroxylation occurs (Fripiat and Toussaint 1963; Nakahira 1964; Weiss and Hartl 1966). Parallel infrared measurements show related changes in the OH stretching frequencies. An interchange of protons and deuterons occurs readily in kaolinite at 300 °C, which is clear evidence for proton delocalization at temperatures below the dehydroxylation range (White et al. 1970). More informative results have been obtained with micas which present more suitable materials for detailed study, namely single crystals which can be oriented within the infrared beam. Vedder (1964) first observed a marked decrease in the OH absorption bands of phlogopite, and subsequently Fripiat et al. (1965) showed that the decrease was common to other micas, muscovites, phlogopite and biotite. The decrease cannot be a consequence of a reorientation of the vibrational moment, which is excluded by measurements made at different crystal orientations, but is attributable to delocalization of protons. Cant and Hall (1968) pointed out a possible source of error in these experiments due to thermal emission from the heated samples, but after correction for this effect, a real diminution of absorption still remains (Fripiat et al. 1968, page 90). For muscovite and phlogopite heated to 600 °C, approximately 20% of the oscillator populations are shifted to energy levels which do not contribute to the discrete OH stretching bands (Fripiat et al. 1969, p. 231).

Delocalization of protons from OH radicals is not confined to the few minerals so far mentioned, but occurs also in zeolites, boehmite, montmorillonite and silica gel; it is not found, however, in gibbsite, bayerite and brucite and therefore cannot be assumed to occur in *all* solids containing hydroxyl ions.

A model for proton delocalization has been suggested in which a proton, localized in a potential well corresponding to a particular oxygen atom, may be excited to a level about 18 kJ mol^{-1} higher from which it can pass to corresponding levels by a tunnelling process through energy barriers (Rouxhet et al. 1969; Freund 1969). At normal temperatures they will be increasingly at higher levels which permit their break away from the local conditions defining the sharp hydroxyl stretching frequency. With further increase of temperature, greater mobility of the protons leads to the formation of water molecules and their loss from the crystalline material. The concept of proton delocalization makes it no longer necessary to envisage only local interactions and the diffusion of molecular units. The model has similarities to that of electrons in metals and perhaps the analogy can be extended to an increased resistance to proton movement when the structure becomes disordered by dehydroxylation or other causes.

7.4 DEHYDROXYLATED PHASES

The dioctahedral minerals kaolinite (and its polymorphic varieties), pyrophyllite, montmorillonite and muscovite give quasi-stable dehydroxylated phases, but the trioctahedral minerals tend to dehydroxylate and recrystallize more or less contemporaneously. Kaolinite especially, and to a lesser extent its polymorphic varieties, have been the subject of many studies and it is appropriate to consider these first and in some detail.

7.4.1 Kaolinite

Dehydroxylation of most kaolinites occurs in normal atmospheres at temperatures above about 400 °C and is accompanied by a weight loss of 14.0% which corresponds with the formula $Al_2Si_2O_5(OH)_4$. Extensive studies by differential thermal analysis (see Holdridge and Vaughan 1957, p. 105) have shown nearly symmetrical endothermic peaks with maximum values near 550–600 °C and with a width which diminishes with increased rate of heating. The exceptionally well-crystallized Keokuk geode kaolinite, however, gives a markedly different endothermic peak (Keller *et al.* 1966) which combines a broad peak near 600 °C and a sharper, higher peak near 700 °C which are not easily resolved into separate reactions. Keller *et al.* consider that the "higher dehydroxylation temperature is the true and characteristic dehydroxylation peak of kaolinite" and that the lower dehydroxylation temperatures shown by most "typical" kaolinites may be attributed "to inferior ordering and crystallinity, plus some effects from the small crystal size".

X-ray powder diffraction studies normally indicate an almost amorphous product after dehydroxylation but single crystal analysis indicates that some structural order is maintained. Retention of structural order evidently is related to the particular polymorphic form of kaolin minerals, to the degree of crystalline order, and perhaps also to crystal size and other variables. Brindley and Hunter (1955) studied the better crystalline polymorph nacrite and Roy *et al.* (1955) applied electron diffraction to single crystals of micron size and in both studies some retention of crystalline order was observed. Structural modifications of dickite, more particularly of the better structurally ordered forms, have been observed by Brindley and Wan (1978), but they observed no similar changes with the Keokuk geode kaolinite although there were marked similarities in their DTA curves. Evidently a complex relationship exists between the polymorphic form of kaolin mineral being considered, its state of crystalline order, the particle size and possibly other variables such as water vapour pressure (i.e. without the higher pressures used in hydrothermal studies).

The work of Brindley and Nakahira (1959) using single crystals of kaolinite shows that order persists mainly within the individual structural layers but not between the layers. The progressive disordering of the basal planes was analysed by Mitra and Bhattacherjee (1969). The *hk* diffractions persist after dehydroxylation but 00*l* and *hkl* diffractions usually are not obtained. However, Radczewski and Schädel (1962) by electron diffraction from dehydroxylated crystals standing on edge succeeded in measuring basal spacings about 6.3 Å which agrees with estimates made by Brindley and Nakahira (1959) from density considerations. If metakaolin is only slightly less dense than kaolinite, as various authors have found (Rieke and Mauve 1942; Brown and Gregg 1952) despite the 14% weight loss due to dehydroxylation, then the layers must collapse to about the same extent, although in a very irregular manner. This argument was not accepted by Freund (1967), who considered that spaces left behind after release of water amounted to 20% by volume. Surface area measurements by BET gas adsorption methods, however, show little change in the available surface of metakaolin (Brown and Gregg 1952; Gastuche *et al.* 1963) which may even decrease slightly.

The structure of metakaolin can be inferred only indirectly for the most part. Maintenance of the a and b parameters means that the Si—O sheets of the kaolinite structure are largely conserved. Electron diffraction measurements (Brindley and Gibbon 1968) show a small but significant increase in b, from 8.95 Å (kaolinite) to 9.145 Å (metakaolin), which corresponds to a relaxation of the twisted Si—O network when the octahedral aluminium sheet is dehydroxylated. Pampuch (1966) came to a similar conclusion from infrared data which he interpreted as showing an increase in symmetry of the Si—O sheets from ditrigonal to hexagonal. Various lines of evidence indicate that Al in metakaolin is in four-fold coordination as had previously been inferred from general considerations, in particular the Al K_α fluorescence X-ray wavelength (Brindley and McKinstry 1961; Gastuche et al. 1963) and infrared spectroscopic data (Stubican and Roy 1961; Pampuch 1966). The easy removal of Al from metakaolin by 1M HCl also may be a consequence of the four-fold coordination (Gastuche et al. 1963).

Metakaolin is usually considered to be the fully dehydroxylated product of kaolinite. It seems, however (Nakahira 1954; Stubican 1959) that the last stages of dehydroxylation may take place with greater difficulty for which various hypotheses can be advanced.

Interactions between pairs of OH ions will become more difficult when few are left but if the model of freely moving protons is accepted then some other explanation is required. Disordering of the structure may increase the resistance to proton mobility. Pampuch (1966) suggested that the last 12% of hydroxyl ions are incorporated in a chain- or ribbon-like arrangement of AlO,OH attached to the Si—O hexagonal network, with the Al ions in four-fold coordination. This structure is considered to become unstable with the final removal of hydroxyls at higher temperatures. Most previous authors (see Brindley and Nakahira 1959) assumed that some form of alumina sheet structure persists. The question whether the retardation of the final removal of hydroxyls is of kinetic or structural origin still poses problems. Others have suggested a chain-like arrangement replacing the alumina sheet (cf. Johns 1953) but often with octahedrally coordinated Al, which is now no longer tenable.

7.4.2 Other kaolinite group minerals

These have received less attention than kaolinite. The polymorphic forms dickite and nacrite probably give similar dehydroxylated phases to metakaolin and work on these minerals has already been mentioned briefly.

A wide range of dickite samples from Jamaica (Brindley and Porter 1978) showed a sequence of endothermic peaks ranging from a broad peak centred at about 600 °C to a sharp peak at 700 °C, with various intermediate forms which combined the broad and sharp peaks. The endothermic behaviour of Keokuk kaolinite was shown to be very similar to the intermediate patterns obtained from the dickites.

The thermal behaviour of single crystals of dickite has been studied by Iwai et al. (1971) and Iwai and Shimamune (1975a,b). Crystals dehydroxylated to the extent of 75% still retain a three-dimensionally ordered structure in which the Al ions are in four-fold coordination. Chains of AlO_4 tetrahedra are developed parallel to [110]. At higher temperatures, Iwai and Shimamune (1975b), using the method of radial distribution analysis, concluded that the metaphase from dickite "has not fixed structure, but has a variable one with heating temperature".

Dickite shows one extraordinary result not shown by either kaolinite or nacrite, namely the development of a broad, strong diffraction peak corresponding to a spacing of about 14 Å or slightly less when dehydroxylation occurs (Hill 1955; Roy and Brindley 1956). Subsequent studies (Brindley and Wan 1978) have shown that, among dickites, well-ordered structure and relatively high dehydroxylation temperatures favour maximum development of the 14 Å phase. Dickites with well-ordered structures and with broad, almost double endotherms, showed a smaller development of the 14 Å phase, which may be related to the higher temperature component of the double endotherm, i.e. to a sharp peak at 700 °C. Less well-ordered dickites show little or no development of this phase. A doubling of the DTA endothermic peak of dickite observed previously by Stoch (1964) was attributed to the OH groups on the exterior of the structural layers dehydroxylating more easily (at lower temperature) than the internal hydroxyls. He also attributed the double peak endotherm to a phase with composition near $Al_4Si_4O_{13}(OH)_2$ which retained a quarter of the original hydroxyls. The 14 Å phase was first thought to be correlated with a composition similar to that suggested by Stoch: indeed, a retention of a quarter of the original hydroxyls would agree exactly with a chlorite after dehydroxylation of the "brucite" layer. It was previously suggested (Brindley 1975) that the phase developed by dickites might resemble a partially dehydroxylated chlorite structure (see below). The most recent studies (Brindley and Wan 1978), however, do not suggest the retention of any appreciable proportion of OH groups in the 14 Å phase. With well-ordered dickites, the 14 Å phase develops as dehydroxylation proceeds and reaches a maximum when dehydroxylation is almost complete. It is possible that, when the first part of a double endotherm has occurred, the average composition of the sample corresponds more or less with that given by Stoch, but this does not prove that each individual crystal is dehydroxylated to this average value. It is therefore thought that the double endothermic peak exhibited by some dickite samples may originate from mixtures of well- and less well-ordered dickite crystals. The description of the 14 Å phase developed by well-ordered dickites as "chlorite-like" seems thus questionable. However, the existence of this modified form seems well established and an extension of the single crystal studies into this domain may resolve the questions involved.

Al K_α fluorescence wavelength measurements indicate that the Al ions are in four-fold coordination in the 14 Å phase. Unfortunately, absence of extended diffraction data makes a structural study difficult. So far, no further information on its nature has been gained by infrared spectroscopic data.

7.4.2.1 Halloysite
Its dehydroxylation behaviour is similar to that of kaolinite but occurs generally at lower temperatures. The following average peak temperatures in DTA runs have been given by Holdridge and Vaughan (1957, p. 123): halloysite, 573 °C; metahalloysite, 583 °C; kaolinite, 607 °C; dickite, 686 °C; nacrite, 666 °C. The sequence corresponds generally with increasing crystal size and degree of crystal-structural order. It is difficult (or impossible) to assess the extent to which each variable is responsible, especially when the influence of entrapped water vapour on dehydroxylation kinetics is considered.

7.4.3 Chlorite

Among trioctahedral layer silicates, chlorites show particularly interesting thermal reactions and were among the first layer silicates to be studied by single crystal methods in relation to topotactic behaviour (Brindley and Ali 1950). The dehydroxylation of magnesian chlorites takes place in two well-defined processes. A rapid reaction near 550 °C corresponds to the dehydroxylation of the interlayer hydroxide sheets with little or no change of the 2:1 layers and usually little change of basal spacing; about three-quarters of the total water is lost in this reaction. A slower reaction, corresponding to the dehydroxylation of the 2:1 layers, occurs from around 600 to 850 C; forsterite or olivine develops more or less simultaneously with the second reaction.

7.4.3.1 Superlattice effects
Re-examination of these reactions (Brindley and Chang 1974), using both naturally occurring and synthetic magnesian chlorites, led to the unexpected discovery of the formation of a 27–28 Å reflection and some of the higher odd-order reflections when the first stage of dehydroxylation occurs. The intensities of these superlattice reflections, particularly of the strong first-order reflection, cannot be explained simply on the basis of a different reorganization of interlayers in an alternating sequence of the type ... A-B-A-B-A .... The 2:1 layers themselves must be drawn together in pairs so that the separation of the 2:1 layers is alternately somewhat smaller and somewhat larger than an average value. The atoms remaining in the dehydroxylated interlayers are not sufficient (whatever state of order or disorder is assumed) to account for the observed long-spacing intensities, and one must suppose that the 2:1 layers are no longer equally spaced as they were in the original chlorite structures.

This surprising result adds one more example of the tendency of layer silicates to develop alternating sequences, i.e. superlattices. As Norrish (1973, see especially pp. 422–423) indicated, there must be some mechanism by which a change of one interlayer is communicated to adjacent interlayers so that the same change is less likely to occur in the next interlayer but may occur in the next-but-one. He suggested that the communication mechanism may be an interaction of the OH dipoles within the 2:1 layer structure.

7.4.4 Serpentine

Serpentine is the trioctahedral analogue of kaolinite with octahedral cation sites filled principally with Mg, but some replacement of Mg and Si by Al is possible. The temperature interval betwen dehydroxylation and recrystallization is usually much smaller for these minerals than for kaolinite and attention has been directed more towards the combined operation than to a pure dehydrogenation process. Dehydroxylation appears to be diffusion-controlled and its dependence on temperature and water vapour pressure has been mentioned already in the section dealing with reaction kinetics (Brindley *et al.* 1967b).

As in the case of kaolinite, dehydroxylation of massive samples is dependent on depth within the sample because of the retarded release of water vapour (Brindley and Hayami 1963a,b). As will be seen later, the subsequent recrystallization reactions are related to the rate of dehydroxylation.

Studies of heat-treated chrysotiles, the fibrous and tubular form of serpentine, show that materials with open textures, as determined by nitrogen adsorption methods, dehydroxylate at lower temperatures than more massive, or "solid", materials (Naumann and Dresher 1966; Murphy and Ross 1977). The open-textured materials form an X-ray amorphous phase analogous to metakaolin, whereas the "solid" materials pass more or less directly into the recrystallized phase, forsterite. The results are clearly related to whether or not a water atmosphere develops which retards the dehydroxylation.

More recent data obtained by Ross and Vishwanathan (1981) show that the dehydroxylation of chrysotile follows kinetic expressions similar to those established for transport-controlled reactions, based on radial diffusion from a cylindrical structure. The authors therefore conclude that the controlling step may be the diffusion of water molecules or related species towards the surface, water being formed by the homogeneous mechanism described in Section 7.3.2.

7.4.4.1 Nickel-containing serpentine minerals

Nickel may partially or totally replace magnesium in serpentine minerals. The thermal transformations of the complete isomorphic series have been studied by Brindley and Wan (1975) in the particular case of the lizardite–nepouite series. Lizardite is the massive form of Mg-serpentine exhibiting a planar layer arrangement. Nepouite is the isomorph of lizardite containing more than 1.5 Ni atoms per formula unit $(Mg,Ni)_3Si_2O_5(OH)_4$, while isomorphs containing fewer Ni atoms are called Ni-lizardites. Dehydroxylation of Ni-lizardites with less than 0.05 Ni per formula unit starts at 450 °C and is about 80% completed by 600–650 °C. The remaining dehydroxylation takes place quite slowly and approaches completion at 850 °C. In contrast, dehydroxylation of nepouites is clearly a two-stage process with about 75% of the process completed at about 700 °C. The remaining 25% hydroxyls are removed at about 1000 °C. All the samples in the series studied by Brindley and Wan (1975) showed the development of a long-spacing phase in the range 600–800 °C (LS-phase), characterized by a broad diffraction peak corresponding to spacings around 11.5–14.5 Å. This behaviour is reminiscent of that of well-ordered dickites (see Section 7.4.2) but, in contrast with this latter case, OH groups seem to be present in the LS-phase developed by serpentine minerals. The LS-phase is formed in the temperature range where dehydroxylation is retarded. It is therefore concluded that this phase is a modified form of the original layer structure retaining a significant proportion of structural "water" and may resemble the chlorite arrangement after its first stage of dehydroxylation.

7.4.5 Pyrophyllite and talc

These are the simplest di- and trioctahedral minerals of the 2:1 structure type. Pyrophyllite gives a well-ordered dehydroxylated phase from around 650 to 900 °C, but talc recrystallizes almost contemporaneously with dehydroxylation. Thilo and Schünemann (1937) described the X-ray powder pattern of the pyrophyllite dehydroxylate as being like that of talc, implying an expansion of the lattice parameters. Lindqvist (1962) confirmed this result and supposed that the Al ions were randomly disposed over the octahedral cation sites. Hydrous mica, montmorillonite and paragonite were considered to behave similarly. However this hypothesis is difficult to accept without more detailed analysis and appears to take no account of the loss of hydroxyls by the dehydroxylation. Brindley and Wardle (1970) showed that two-layer monoclinic and one-layer triclinic forms of pyrophyllite expand with dehydroxylation (Table 7.3) but to a lesser extent than that observed for kaolinite, and only partial relaxation of the initially twisted Si—O tetrahedral network takes place.

In pyrophyllite and possibly in other dioctahedral 2:1 minerals it is generally supposed that dehydroxylation occurs by 2 $(OH)^-$ ions becoming H_2O and O^{2-} with the residual oxygens moving into the same structural level as the Al ions, and this view has received confirmation several times by one-dimensional Fourier synthesis of 00*l* diffraction intensities. Although it has been shown (see Section 7.4.6) that six-fold coordination of Al may still be maintained, a five-

TABLE 7.3. Lattice parameter changes accompanying dehydroxylation

Mineral	Å	Initial	After dehydroxylation	Per cent change
Kaolinite[†]	b	8.95	9.14_5	+2.18
Pyrophyllite, 1Tr[‡]	a	5.173	5.140	−0.64
	b	8.960	9.116	+1.74
	$d(001)$	9.204	9.352	+1.61
Pyrophyllite, 2M[‡]	a	5.172	5.173	0
	b	8.958	9.114	+1.74
	$d(001)$	18.408	18.707	$+1.62_5$
Muscovite[§]	a	5.20	5.24	+0.77
	b	9.05	9.21	+1.36
	$d(001)$	19.95	20.16	+1.05

[†] Brindley and Gibbon (1968).
[‡] Brindley and Wardle (1970).
[§] Eberhart (1963). Similar data given by Nicol (1964), Vedder and Wilkins (1969).

fold coordination is the simplest arrangement which can be adopted. This has been mentioned as a possibility by Heller et al. (1962) and Brett et al. (1970) and a detailed structure analysis by Wardle and Brindley (1972) provides considerable support for such an arrangement.

The dehydroxylation of talc takes place together with recrystallization around 900 °C and the combined process is considered with other high-temperature reactions.

7.4.6 Muscovite, illite

Because mica provides excellent crystalline material for single crystal studies, it is surprising that no intensive crystallographic study of a dehydroxylated mica has yet been undertaken; the need for such a study was emphasized by Brindley (1975, see p. 129). Eberhart (1963), Nicol (1964) and Vedder and Wilkins (1969) agree that the structural parameters of muscovite expand with dehydroxylation (see Table 7.3); the changes are comparable with those for pyrophyllite and may indicate a partial relaxation of the (Si,Al)—O tetrahedral sheets. One-dimensional Fourier analysis (Eberhart 1963) indicates that the "residual" oxygens move to the level of the Al ions. Eberhart, following Bradley and Grim (1951), proposed a modified sheet structure with Al still in six-fold coordination which appeared to be consistent with the intensities of a few $(hk0)$ reflections, but whether this arrangement or the five-fold coordinated arrangement proposed by Wardle and Brindley (1972) is correct must await further study.

Implicit in these structural considerations is that one O^{2-} ion replaces an adjacent pair of 2 $(OH)^-$ so that oxygens as well as proton must migrate when dehydroxylation occurs, and this suggests a homogeneous type of reaction. Those who favour a heterogenous process of proton migration to reaction zones have then to provide an alternative explanation for the one-dimensional Fourier results. Nicol (1964) suggested an averaging of the dehydroxylate proper and the material left in reaction zones, but the concept is difficult to visualize. Vedder and Wilkins (1969) studied by infrared analysis the rehydration of a dehydroxylated muscovite and showed that the original structure was restored: they considered that a homogeneous reaction and H_2O migration provided the simplest interpretation of their results.

The kinetics of dehydroxylation of micas and illites have been studied by several authors.

Isothermal gravimetric measurements on micron-sized particles *in vacuo* (Kodama and Brydon 1968) gave results agreeing with a two-dimensional diffusion model and an activation energy of 225 kJ mol^{-1}. Holt *et al.* (1958, 1964) found the dehydroxylation kinetics of mica flakes difficult to interpret but since they used a temperature range 738–817 °C their results may have been influenced by the development of a mosaic texture such as Sabatier (1955), Eberhart (1963) and Gaines and Vedder (1964) have considered. The infrared studies of Vedder and Wilkins (1969) and of Rouxhet (1970) on the dehydroxylation and rehydroxylation, and also the deuteration of several micas at temperatures sufficiently low to avoid the problems arising from textural changes, mainly around 600 °C, have added considerably to basic knowledge of the processes involved. Dehydroxylation and OH—OD exchange appear to be diffusion-controlled processes taking place normal to (001). For muscovite, the rates of these processes are of the same order with an activation energy of 196 kJ mol^{-1}. Evidently both processes depend on a migration of identical or at least closely related chemical species, and the rate-limiting step appears to be "the movement of a transport complex H_2O or OH^-". However, Rouxhet (1970, pp. 851–852) goes on to say: "this does not require any net oxygen transfer and, between two slow steps occurring at some spots in the crystal, the proton might well jump from one oxygen to another one". "The dehydroxylation process may not be depicted as a diffusion of stable water molecules formed by definitive condensation of two hydroxyls. It is only suggested that the slowest step of the reaction is the jump of the proton included in the transport complex H_2O considered, whatever the life time of the complex."

It appears to the writers that a communication gap exists between specialists working on these elusive problems. Cautious interpretation of kinetic data, precise structure analyses of the dehydroxylated phases of different micas and parallel infrared studies are important requirements for future work.

7.4.6.1 Illites
The dehydration and dehydroxylation of naturally occurring clay-grade micas are of considerable interest but are difficult to evaluate because generally there is no clear break between the loss of surface adsorbed molecular water and the dehydroxylation proper. This behaviour is evidently related to the very small particle size because similar results are obtained with fine fractions of ground muscovite, as shown for example by the thermogravimetric data of Grim *et al.* (1937), and of Wentworth (1967) (Fig. 7.4).

Many naturally occurring clay-grade micas differ compositionally from well-crystallized micas by having proportionally more silica and water, and less K_2O. Brown and Norrish (1952) suggested that the high H_2O+ content of illites may arise from a partial replacement of K^+ by H_3O^+ ions in the interlayered positions, and on this basis they re-evaluated the formulae of several illites. The results came closer to four octahedral cations and two interlayer cations per unit cell than the results obtained by the usual methods of calculation. Muñoz Taboadela and Aleixandre Ferrandis (1957) applied the Brown–Norrish method to other illites and also found rather good results, but Gaudette *et al.* (1966) obtained data which did not support the Brown–Norrish concept. As Wentworth has pointed out, many authors neglect to state the conditions under which H_2O+ is measured and this is of crucial importance. At the present time there seems little hope that interlayer H_2O or H_3O^+ in illites can be unequivocally determined from dehydration–dehydroxylation measurements and other methods, possibly infrared analysis, should be undertaken.

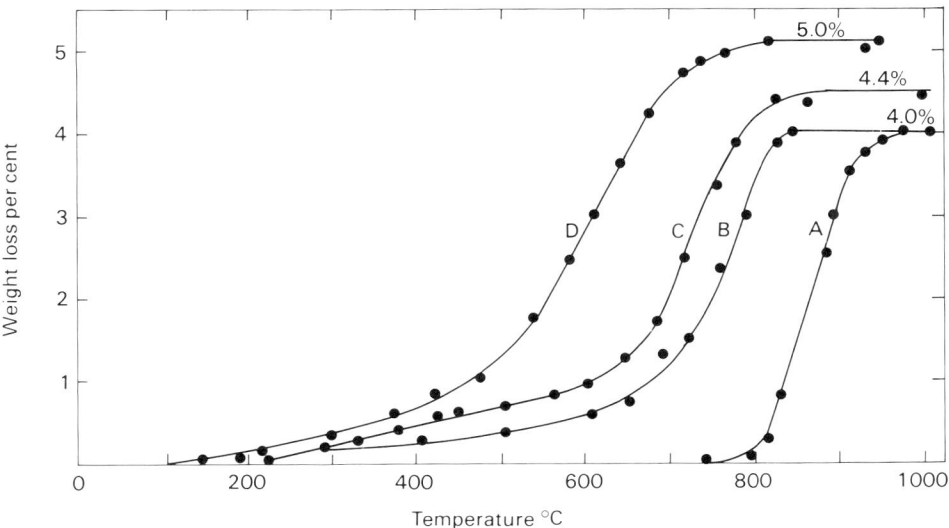

FIG. 7.4. Weight loss curves for muscovite in air at 3 °C min^{-1}: (A) single flake; (B) 5–10 micron particle size, area 2 m^2 g^{-1}; (C) 2–5 micron particle size, area 5 m^2 g^{-1}; (D) <2 micron particle size, area 23 m^2 g^{-1}. (After S. A. Wentworth 1967.)

7.4.7 Montmorillonite

Like other fine-grained minerals, montmorillonites dehydroxylate over a wide temperature range, although the main loss of water generally occurs between about 500 and 700 °C (Earley et al. 1953a). In some cases there appears to be a small, continued loss of water at higher temperatures which is reminiscent of the behaviour of kaolinite. The explanation is not obvious, but prior structural changes may retard the last stages of water removal. Differential thermal analyses indicate that the main endothermic reaction occurs around 600–700 °C but some montmorillonites show a double endotherm between 450 and 700 °C for which no certain explanation has been given apart from the possibility of small amounts of hydrous impurities (Earley et al. 1953b; Grim and Kulbicki 1961). The variable compositions of montmorillonites, of major importance at higher temperatures (see Section 7.6.2.5), also may influence the dehydroxylation process (Grim and Kulbicki 1961).

Horváth and Gáliková (1979) investigated the influence of exchangeable cations on the kinetics of montmorillonite dehydroxylation, and found that monovalent cations, except Na$^+$, promoted first-order kinetics of dehydroxylation while Na$^+$ and divalent cations promoted a diffusion-controlled mechanism. High substitution of Mg^{2+} by Fe^{2+} in the octahedral layer lowers the dehydroxylation temperature of montmorillonites. The process is then preceded by the oxidation of any divalent iron present (Heller-Kallai and Rozenson 1980).

Bradley and Grim (1951) observed modifications in the X-ray powder diagrams of montmorillonites by dehydroxylation around 700 °C, and suggested a reorganization of the oxygens around the octahedral cations which retained six-fold coordination. Although the idealized model involved unrealistically large Al—O distances, it was shown to be broadly consistent with the X-ray powder diffraction intensities and it has served as a model for subsequent studies by Eberhart and others. The validity for montmorillonite of the model suggested by Wardle and Brindley (1972) for pyrophyllite dehydroxylate with five-fold coordinated Al has been evidenced recently by Heller-Kallai and Rozenson (1980) using Mössbauer spectroscopic data. They showed that the Fe(III) ions located in octahedral sites with two OH groups in *cis* arrangement, became five-fold coordinated upon dehydroxylation while iron in octahedral sites with OH groups in *trans* configuration remained in six-fold coordination, but the sites were highly distorted. They also concluded from X-ray diffraction, Mössbauer and infrared spectroscopic data that dehydroxylation of montmorillonite was a three-stage process: proton delocalization was first observed, followed by localized dehydroxylation without significant change in the overall configuration of the octahedral layer; finally, most of the OH groups were lost with changes in the cell dimension. It was also found that no cation migration occurred in the course of heating the specimen for 1–3 hours at 600–700 °C, so that dehydroxylation most probably takes place according to the homogeneous mechanism.

7.4.8 Vermiculite

Vermiculites, like montmorillonites, exhibit intracrystalline hydration and considerable attention has been given to the low-temperature loss of interlayer molecular water and to the structure of the various hydration states (see Chapter 5, Section 5.7.3). Vermiculite is principally a trioctahedral mineral with Mg ions in the octahedral positions and in interlayer positions, though the latter can be exchanged for other cations. The cation exchange capacity is about twice that of smectite so that the exchangeable cations in vermiculites are likely to play a more important role in thermal reactions. Vermiculites exist in macrocrystalline sheet form as well as in clay-grade form. Also, particularly in soils, dioctahedral vermiculites may occur but they are probably much less common than the trioctahedral varieties. The fact that relatively well-crystallized single crystals of vermiculite are obtainable should facilitate the detailed study of the thermal reactions, but such work seems not to have been undertaken.

Two useful reviews (Walker and Cole 1957; Walker 1961) summarized the data available at that time. Dehydroxylation appears to be gradual from about 550 to 850 °C and recrystallization to enstatite begins around 800 °C. The two processes appear to overlap. Thermal analyses show two endotherms separated by an exotherm corresponding to recrystallization. Dehydroxylation seems to start before recrystallization, but to continue after recrystallization has begun. The details depend partly on the vermiculite used and on the exchangeable cations, and possibly also on the dynamic character of thermal analysis. Elucidation of the details needs further study.

7.4.9 Other dehydroxylation phases

7.4.9.1 Saponite

This magnesian trioctahedral smectite, with ideal composition $M^+_{0.67}Mg_6(Si_{7.33}Al_{0.67})O_{20}(OH)_4$, has a small temperature range between loss of water by dehydroxylation, around 500–600°C, and the first appearance of the high-temperature phase, enstatite, around 740 °C (Weiss *et al.* 1955). Within the range 650–750 °C, the dehydroxylated phase gives sharp X-ray diffraction reflections but no study appears to have been made of the structural reorganization accompanying dehydroxylation. Enstatite seems to develop less readily from saponite than from sepiolite, for which there may be a structural explanation (Kulbicki 1959).

7.4.9.2 Sepiolite and palygorskite (attapulgite)
These minerals differ from other phyllosilicates in that the octahedral sheets are in the form of ribbons attached alternately to opposite sides of the continuous sheets of SiO_4 tetrahedra. The tetrahedra point in opposite directions in order to coordinate with the octahedral ribbons. These minerals are essentially hydrous magnesian silicates and the ribbons have a talc-like structure, but various substitutions of Si and Mg by Al and Fe are possible; palygorskites generally have a higher proportion of substitutions than sepiolites. The structures and compositions are described by Caillère and Hénin (1961), by Grim (1968), and more recently by Nagata *et al.* (1974).

The channels formed by the alternating arrangement of the ribbon-like units contain exchangeable cations and hydration water (also referred to as zeolitic water) and also water molecules firmly bonded to the edges of the ribbons and sometimes called "crystal water". The talc-like structure of the ribbons includes hydroxyl ions.

The loss of hydration water around 100 °C occurs with little or no structural change. Sepiolite heated in air loses its bounded "crystal water" in two stages, according to Nagata *et al.* (1974): half of the hydration water is lost between 200 and 380 °C and the remaining half between 380 and 680 °C. The exact temperature ranges will depend on the experimental conditions: they are markedly lowered when heating the specimen *in vacuo* (Rautureau and Mifsud 1975). The two dehydration steps are well evidenced by TGA and correspond to slightly different folded structures, as shown by changes in X-ray powder reflections. Structural changes of sepiolite on heating are sketched in Fig. 7.5.

Palygorskite under vacuum loses water in two stages (Rautureau and Mifsud 1975). The first step is reversible, occurs below 350 °C and involves about half of the bound water, as shown by X-ray diffraction and infrared spectroscopy (Van Scoyoc *et al.* 1979). The remaining bound water, together with the hydroxyls, is lost between 300 and 540 °C and no subsequent rehydration is possible.

For both sepiolite and palygorskite dehydroxylation and recrystallization to enstatite appear to take place more or less concurrently (Preisinger 1959, 1963; Kulbicki 1959). In some instances enstatite forms more easily from palygorskite than from sepiolite which is difficult to understand since Al substitutions for Mg tend to be larger in palygorskite (Kulbicki 1959).

7.5 REHYDROXYLATION

Following the work of Grim and Bradley (1948), various experiments have shown that some degree of rehydroxylation of dehydroxylated clay minerals may occur under relatively mild conditions of hydration, such as exposure to atmospheric moisture, or immersion in water over long periods, or by passing steam through differential thermal equipment when the sample cools from a temperature above the dehydroxylation range. Roy and Brindley (1956) studied the reconstitution of kaolinite group minerals (kaolinite, dickite, nacrite, halloysite) under hydrothermal treatment and showed that in all cases some form of kaolinite was developed. The evidence for rehydroxylation under mild conditions, particularly of montmorillonite, in early experiments was mainly the reappearance of endothermic peaks which correspond partially with those of the initial mineral. A broadly based study of rehydroxylation by Heller *et al.* (1962) of pyrophyllite, montmorillonite and illite utilized chemical, thermal, X-ray and infrared methods. They passed moist nitrogen through the sample after dehydroxylation, either as it cooled slowly or when held at fixed temperatures. They concluded that minerals with high Al-

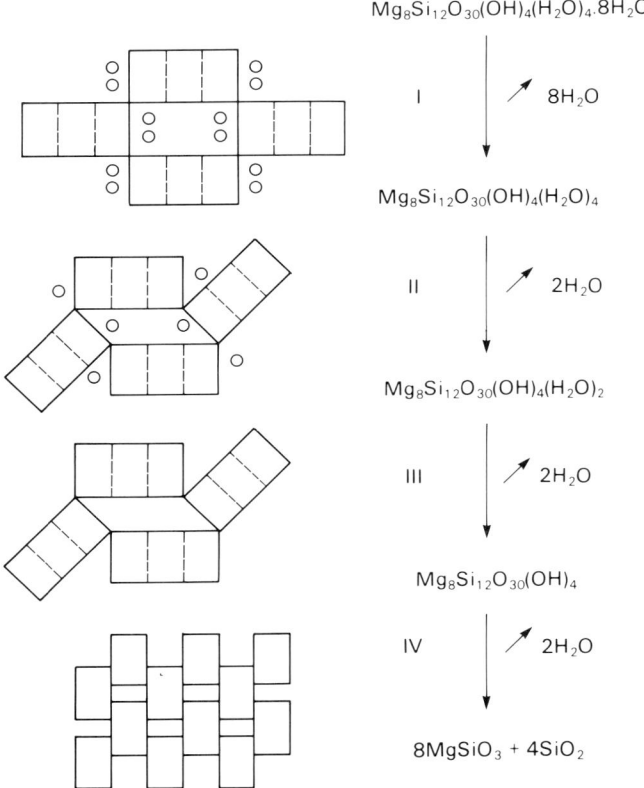

FIG. 7.5. Schematic figures of structural changes of sepiolite on heating. Circles: bound water. (After Nagata et al. 1974.)

for-Si substitution do not rehydroxylate readily under the conditions used, but for pyrophyllite and montmorillonite, rehydroxylation occurs rapidly at about 400–500 °C, and amounts to about 40%. The infrared data indicate that the rehydroxylated montmorillonite show OH frequencies similar to those of pyrophyllite which suggests some reorganization of the Al and Mg ions such that pyrophyllite-like areas develop. Illite also rehydroxylated readily. X-ray analyses gave mainly inconclusive results though changes of some undetermined kind were produced by the hydration treatments. That the infrared evidence appeared clearer than the X-ray evidence suggests that short-range rather than long-range ordering processes were involved.

7.5.1 Hydroxylation reactions

Hydroxyl reactions, as well as hydration reactions, of expandable clays can occur at normal temperatures under appropriate chemical conditions. Their discussion in the present chapter is relevant only in so far as the thermal behaviour of the products is used for their characterization. Under conditions where hydroxylation as well as hydration of interlayer cations occurs, the resulting products, whether formed under natural conditions or in the laboratory, are intermediate between smectites or vermiculites (with hydrated interlayer cations) and chlorites (with hydroxylated cations). They are frequently described as "intergrade" minerals; Barnhisel (1977, p. 335) has listed 15 different terms used for them. Extensive literature reviews have been given by Rich (1968) and by Barnhisel (1977) and some additional references are given by Brindley and Kao (1980).

The tests usually applied to characterize these minerals involve the use of ethylene glycol to test smectite or vermiculite swelling behaviour, and X-ray diffraction after heating to temperatures in the range 100–500°C to test the thermal stability of the layer structures and possibly to observe changes in intensities of basal reflections such as are often observed with chlorites.

A major difficulty arises in attempting to obtain structural formulae for intergrade minerals, even if one assumes that all layers have essentially the same chemical composition. The usual procedure to convert a chemical analysis, expressed in percentages of oxides, to cations and anions in a mineral formula is to normalize with respect to a total cation valence of $+22$ for micas, smectites and vermiculites, and of $+28$ for chlorites. Intergrade minerals fall between these limits. A detailed study by Brindley and Kao (1980) of hydroxy-Mg and hydroxy-Al intergrades was based on the carefully determined composition of the initial Na-montmorillonite, namely:

$$(Na_{0.30}K_{0.01}Ca_{0.04}H_2O_{0.35})(Al_{1.54}Mg_{0.20}Fe_{0.20})(Si_{3.93}Al_{0.07})O_{10}(OH)_2$$

Normalization of the intergrade products was then based on $Si_{3.93}$ in tetrahedral sites, together with $Al_{0.07}$ to fill the tetrahedral positions. Octahedral and interlayer compositions were then determined.

When converted to Mg- and Al-montmorillonite by treatment with the respective nitrate solutions, the interlayer compositions showed respectively $Mg_{0.31}Na_{0.04}$ and $Al_{0.35}K_{0.01}$. In fact the total chemical analyses showed $[Mg_{0.31}(OH)_{0.32}(H_2O)_{0.78}]$ and $[Al_{0.35}(OH)_{0.76}(H_2O)_{0.84}]$ in the interlayers. Evidently the Na^+ ions were replaced by either $[Mg(OH)(H_2O)_2]^+$ or $[Al(OH)_2(H_2O)_{2.5}]^+$.

The explanation for these results, as discussed earlier by Fripiat et al. (1971, p. 206), lies in the partial dissociation of H_2O in near proximity to the highly charged cations:

$$Al^{3+}(H_2O)_6 \rightarrow [Al(OH)(H_2O)_5]^{2+} + H^+$$
$$\rightarrow [Al(OH)_2(H_2O)_4]^+ + 2H^+$$

and similarly for the Mg case.

The H^+ ions or H_3O^+ ions participate in further cation exchange reactions. In the experiments in question, the H^+ ions were titrated with 0.1M NaOH solution so that additional Al and Mg ions in the ambient solutions entered the interlayers. In the final stages, the interlayer composition approached $Mg_3(OH)_6$ and $Al_2(OH)_6$. External precipitation of $Mg(OH)_2$ and $Al(OH)_3$ became appreciable only towards the end of the internal reactions. Unfortunately, this external precipitation limited the determination of the final interlayer compositions.

We can return now to the discussion of the thermal and X-ray data for the intergrade products. For details, the original tabulations of results must be seen. For the Mg-hydroxy montmorillonite, the basal spacing was initially near 15 Å and at 300°C it collapsed to about 9.6 Å with low proportions of hydroxy-Mg between the layers; with high proportions of hydroxy-Mg, a basal spacing of 14.5 Å was maintained to 300°C. For the hydroxy-Al products more variable data were obtained probably because $[Al(OH)_2]^+$ and $[Al(OH)]^{2+}$ ions were involved. Nevertheless, with low proportions of hydroxy-Al ions, the basal spacing collapsed to near 10 Å at 300 °C and to a mixed-layer situation from 100 to 300 °C. With high proportions of hydroxy-Al between the layers, a basal spacing near 14.2–14.5 Å was maintained to 300 °C. Brindley and Kao (1980) concluded that "swelling in ethylene glycol and in water vapor and the stability or instability to heating depend on the OH/Al and OH/Mg ratios in the interlayers". Furthermore, they considered that the "results confirm most of the published data on these systems, but show more clearly the progressive changes of composition and properties with the amount of NaOH added".

7.6 HIGH-TEMPERATURE PHASES: TOPOTACTIC REACTIONS

High-temperature phases develop mainly at temperatures exceeding about 800 °C and many of the reactions are strongly topotactic, i.e. the main products have important and usually simple crystallographic relations to the parent mineral. The recognition of the topotactic behaviour gave rise initially to descriptions of these reactions as taking place homogeneously by a reorganization of cations within a largely constant framework of oxygen anions. This simplistic point of view, however, provided no satisfactory explanation for the liberation of silica which occurs in many of these reactions and which may appear as cristobalite or as quartz, or (when not directly observable) as a phase, crystalline or amorphous, to balance a chemical equation.

The topotactic behaviour shown in these reactions is consistent with the classical concepts of nucleation and phase growth. The strain energy due to misfits between the new and old phases is likely to be minimized by the particular orientations and consequently these orientations are preferentially developed. The activation energy for phase growth also may be minimized when structural reorganization is small. The phases formed are not always those expected from equilibrium diagrams and metastable situations can develop. Non-observable "silica" may incorporate atoms other than SiO_2. Quantitative analyses are difficult to make and are seldom attempted, and consequently chemical equations may be oversimplified.

The topotactic behaviour of trioctahedral magnesian minerals is usually simpler than that of dioctahedral minerals, and the most likely reason is that the main products, forsterite (or olivine) and enstatite have crystal structures which are related rather simply to those of the parent minerals. The corresponding relations are less simple for dioctahedral minerals where mullite is the main product.

7.6.1 Trioctahedral mineral reactions

7.6.1.1 Serpentine to forsterite reactions
Many investigations have been made of the dehydroxylation of serpentine minerals and of the resulting products under dry conditions and under hydrothermal conditions. More attention has been given to the fibrous form of serpentine, chrysotile, than to the platy forms, lizardite and antigorite, but both the platy and the fibrous forms clearly show topotactic behaviour. As stated earlier, the dehydroxylation reaction may take place prior to recrystallization, but often they occur more or less simultaneously. The topotactic relations will be considered, therefore, between the initial serpentine and the resulting forsterite. For single crystals of lizardite, the following relations were established (Brindley and Zussman 1957; Brindley 1963) where suffix S

denotes the serpentine mineral, and suffix F forsterite:

$$2a_S \ (10.62 \ \text{Å}) \qquad b_F \ (10.20 \ \text{Å}) \qquad (4.0\% \ \text{contraction})$$

$$2b_S \ (18.40 \ \text{Å}) \qquad 3c_F \ (17.94 \ \text{Å}) \qquad (2.5\% \ \text{contraction})$$

$$2d(001)_S \ (14.64 \ \text{Å}) \qquad 3a_F \ (14.27 \ \text{Å}) \qquad (2.5\% \ \text{contraction})$$

$$8V_S \qquad 9V_F \qquad (9.0\% \ \text{volume contraction})$$

These are vector relations, signifying that $2a_S$ is approximately equal in magnitude and parallel to b_F, and likewise for the other relations. V_S and V_F are unit cell volumes, and $8V_S$ develops into $9V_F$ with about a 9.0% volume contraction. The compositions are as follows:

$$8V_S: 48 \ \text{Mg}, \ 32 \ \text{Si}, \ 144 \ \text{O}, \ 64 \ \text{H}$$

$$9V_F: 72 \ \text{Mg}, \ 36 \ \text{Si}, \ 144 \ \text{O}$$

The constancy of oxygen content is obvious and the overall reaction is an exchange of 64H by 24Mg + 4Si. The H is probably lost as water from reaction zones and a compensating migration of Mg and Si ions maintains the electrical neutrality.

The overall chemical reaction can be expressed as follows:

$$Mg_6Si_4O_{10}(OH)_8 \rightarrow xMg_2SiO_4 + yH_2O + \text{X-ray amorphous material}$$

If the amorphous phase is SiO_2, a simple equation can be written:

$$Mg_6Si_4O_{10}(OH)_8 \rightarrow 3Mg_2SiO_4 + 4H_2O + SiO_2$$

This reaction has been studied quantitatively. The H_2O product is measured gravimetrically as a weight loss. The SiO_2 is not directly observed. The forsterite has been measured quantitatively by X-ray diffraction (Brindley and Hayami 1965). The composition of the dehydroxylated phase in terms of oxides is $3MgO \cdot 2SiO_2$ and according to the $MgO-SiO_2$ equilibrium diagram this should exsolve into forsterite and enstatite, as follows:

$$3MgO \cdot 2SiO_2 \rightarrow 2MgO \cdot SiO_2 + MgO \cdot SiO_2$$
$$\text{forsterite} \qquad \text{enstatite}$$
$$58.3\% \qquad \quad 41.7\%$$

A non-equilibrium reaction yielding maximum forsterite and residual silica is:

$$3MgO \cdot 2SiO_2 \rightarrow 1.5(2MgO \cdot SiO_2) + 0.5SiO_2$$
$$\qquad \qquad 87.5\% \qquad \quad 12.5\%$$

Serpentine fired at 800 °C for period up to 100 hours yielded 75–80% of forsterite in the final product, a result which is very close to the second reaction. The forsterite X-ray reflections were broadened as compared with those from high-temperature produced forsterite and an estimated crystal domain size of 400 Å was obtained which may correspond to the distance over which proton migration and (Mg + Si) counter-migration take place.

At temperatures exceeding 1000 °C, enstatite makes its appearance corresponding to a reaction between the excess forsterite and SiO_2. The fact that the measured forsterite was somewhat less than the theoretical maximum could be taken as an indication that possibly some of the magnesia component was contained in the X-ray amorphous "silica" phase.

7.6.1.2 Reactions of Ni-containing serpentines

The high-temperature transformations of Ni-serpentines have been investigated by Pham Thi Hang and Brindley (1973); a detailed study of the lizardite–nepouite series has been made by Brindley and Wan (1975). These authors found that the high-Ni members of the series formed a transitional face-centred cubic phase (FCC-phase) upon heating between about 800 and 1000 °C, which may be a very finely divided form of nickel oxide. The high-temperature phases appearing above 800 °C are olivine and enstatite from low-Ni materials and olivine and cristobalite from high-Ni ones. Higher nickel substitution in the mineral tends to increase the temperature at which olivine is formed. No enstatite formation was found in nepouite samples, even after heating for about 4 hours at 1150 °C.

7.6.1.3 Chrysotile reactions

The thermal reactions of this fibrous form of serpentine have been studied by many authors. Hey and Bannister (1948) showed a beautiful rotation-type X-ray photograph of forsterite obtained by heating a chrysotile fibre at 1000 °C for 12 hours. Ball and Taylor (1963) affirmed that dry heating below 1000 °C gave results agreeing with those for lizardite. They discussed at length an inhomogeneous mechanism involving cation migrations and supposed that early in the reaction Mg and Si ions migrate in opposite directions so that Mg-rich regions develop from which forsterite and enstatite respectively recrystallize, with the forsterite appearing below 1000 °C and enstatite mainly above 1000 °C. This description is consistent with forsterite and enstatite exsolving from the dehydroxylated phase, $3MgO \cdot 2SiO_2$. The concept of an amorphous phase of enstatite composition, given also by Koltermann and Rasch (1964), was recently confirmed from electron microscopic evidence by Martin (1977). This author observed the first change in the chrysotile structure at 580 °C, where cavities started to develop between (001) layers as the fibres were dehydrated, giving rise to strong low-angle diffraction. Although no evidence was found of any structure in the remaining material, some degree of the original arrangement was probably preserved because subsequently forsterite and enstatite developed in certain preferred orientations. In samples held between 600 and 800 °C, forsterite developed slowly with little further disruption of the fibre, while above 800 °C, the remaining amorphous areas rapidly recrystallized to a mixture of forsterite and enstatite. Similar observations by Helena de Souza Santos and Yada (1979), using high resolution electron microscopy, showed the formation at 600 °C of a new fringe system of 10–15 Å spacing, appearing sporadically parallel to the 7.3 Å spacing of the chrysotile. Areas of these extra fringes seemed to constitute favourable sites for subsequent nucleation of forsterite.

An interesting relation between dehydroxylation and recrystallization was found in experiments using small blocks of microcrystalline serpentine (Brindley and Hayami 1963a,b). Dehydroxylation proceeds most readily near the surface but recrystallization proceeded slowly, while within the block dehydroxylation proceeds more slowly and recrystallization more rapidly. This inverse relationship suggests that when water vapour escapes quickly, the dehydroxylated phase is formed quickly, is highly disordered, and recrystallization is thereby slowed down. Conversely, slow dehydroxylation within the block causes less disorder and facilitates recrystallization. Corresponding to this variable behaviour a "spectrum" of activation energies was derived for the recrystallization reaction ranging from about 320 to 400 kJ mol^{-1}.

7.6.1.4 Chlorite–olivine reaction

This reaction is similar to the serpentine–forsterite reaction and was the earliest layer silicate reaction to be studied in detail by single crystal methods (Brindley and Ali, 1950; Bradley and Grim 1951). The topotactic relations are similar to those for serpentine, with $d(001)_{Ch}(14.37$ Å) written in place of $2d(001)_S(14.64$ Å). Since chlorites have partial replacement of Mg^{2+} and Si^{4+} by Al^{3+}, and since Al ions are not accepted by the olivine structure, the chlorite reactions are dependent on the amount of Al present. As noted earlier, dehydroxylation of chlorites often takes place in two stages, the first around 550 °C corresponding to decomposition of the hydroxide or brucite-like layer, and the second around 800 °C when the 2:1, talc-like layer dehydroxylates. The intensities of the basal reflections are changed markedly by the first stage of reaction, but the spacing is modified only slightly; the intensity changes may be a useful diagnostic feature for chlorites in mixtures. However, many fine-grained chlorites show little or no separation of the two reactions and the entire reaction may take place around 450 °C.

7.6.1.5 Talc–enstatite reaction

This reaction takes place around 900–1000 °C. The topotactic relations have been studied by Nakahira and Kato (1964) who gave the following results:

$$a_T(5.29 \text{ Å}) \simeq c_E(5.18) \text{ Å} \quad (2.1\% \text{ contraction})$$
$$b_T(9.16 \text{ Å}) \simeq b_E(8.81 \text{ Å}) \quad (3.8\% \text{ contraction})$$
$$d(001)_T(18.70 \text{ Å}) \simeq a_E(18.23 \text{ Å}) \quad (2.5\% \text{ contraction})$$
$$V_T \simeq V_E \quad (8.4\% \text{ contraction})$$

The unit cell compositions are:

$$V_T: 12 \text{ Mg, } 16 \text{ Si, } 48 \text{ O, } 8 \text{ H}$$
$$V_E: 16 \text{ Mg, } 16 \text{ Si, } 48 \text{ O}$$

The inhomogeneous mechanism can be hypothesized as a migration of protons to reaction zones to liberate water molecules, with a counter-migration of 4Mg for each 8H. The "unwanted" silica is left in the reaction zones. The overall chemical equation can be written:

$$Mg_3Si_4O_{10}(OH)_2 \rightarrow 3MgSiO_3 + SiO_2 + H_2O$$

The dehydroxylation kinetics of talc, studied by Ward (1975) in the temperature interval 825–885 °C, showed first-order kinetics and the activation parameters measured were consistent with a mechanism consisting of Mg—OH bond breaking and subsequent migration of Mg. The silica, in the form of cristobalite, was identified by electron diffraction in samples heated to 1200 °C, with the [111] direction normal to (001) of talc, which corresponds to the (111) oxygen planes of cristobalite lying parallel to the oxygen planes of talc. As remarked earlier, these relations must not be extended to give an oversimplified view of the actual reaction. Enstatite has a chain structure which necessitates a structural reorganization. von Gehlen (1962a), studying the reactions of fine-grained talc, considered that proto-enstatite forms at 1300–1400 °C. Above 1400 °C, clino-enstatite appears to form (Koltermann 1964).

7.6.1.6 Reactions of Ni-containing talcs

High-temperature transformations of Ni-containing talcs have been investigated by Pham Thi Hang and Brindley (1973) and subsequently, in a more systematic way, for the kerolite–pimelite series by Brindley *et al.* (1979). As with the talc–enstatite reaction, few changes occur until dehydroxylation begins around 700–800 °C. Enstatite is the first product in all cases, irrespective of the relative proportion of Ni and Mg, reflecting the topotactic character of the talc–enstatite transformation upon dehydroxylation. A poorly defined face-centred cubic phase (FCC-phase) is formed together with enstatite in all the samples. This FCC-phase is very similar to that observed in the study of the lizardite–nepouite series (see Section 7.6.1). In the temperature range 800–1000 °C where the FCC-phase is formed, the initial green colour of the mineral changes to black and the higher the nickel content, the deeper is the black colour. The green colour, however, is restored beyond 1100 °C due to olivine formation. It is thought that the black colour and the FCC-structure indicate the formation of a defective NiO. That forsterite does not develop from low-Ni kerolite is due to the considerable structural reorganization required for the formation of the forsterite phase. The appearance of this phase in high-Ni kerolites and pimelites is due presumably to the instability of the nickel analogue of enstatite.

7.6.1.7 Saponite–enstatite reaction

Saponite is a magnesian smectite with ideal composition $M^+_{0.33}Mg_3(Si_{3.67}Al_{0.33})O_{10}(OH)_2$ which differs from talc in the Al-for-Si replacement and the presence of interlayer cations denoted here by M^+. The structurally ordered dehydroxylated phase reacts to give enstatite around 740 °C (Weiss et al. 1955). The process is probably similar to that involved in the talc reactions, but is complicated by the necessity of eliminating the Al ions, possibly as Al_2O_3, from the tetrahedral sheets. Weiss et al. mention a diffuse X-ray diffraction ring which may arise from this "alumina". The high-temperature reactions of chlorites also involved a consideration of the "unwanted" Al_2O_3.

7.6.1.8 Sepiolite and palygorskite reactions

These essentially magnesian silicates with talc-like structures, have been described briefly in relation to the dehydration and dehydroxylation behaviour. They recrystallize to enstatite and cristoballite around 800 °C more or less concurrently with dehydroxylation (Preisinger 1959, 1963; Kulbicki 1959). Their ideal formulae, after dehydroxylation are similar to that of talc; in the following list, all formulae are scaled to $Si_{12}O_{30}$:

Sepiolite anhydride	$Mg_8Si_{12}O_{30}(OH)_4$ (Brauner–Preisinger formula)
	$Mg_9Si_{12}O_{30}(OH)_6$ (Nagy–Bradley formula)
Palygorskite anhydride ($\times 1.5$)	$Mg_{7.5}Si_{12}O_{30}(OH)_3$ (Bradley)
Talc ($\times 3$)	$Mg_9Si_{12}O_{30}(OH)_6$

(For derivation of these formulae, see Caillère and Hénin 1961; Grim 1968.)

All these minerals give rise to enstatite and cristobalite at high temperatures but, as Kulbicki points out, enstatite forms more readily from sepiolite and palygorskite than from talc, possibly because the ribbon-like structures in sepiolite and palygorskite resemble more closely the chain structure of enstatite than does the layer structure of talc. On the other hand, the ionic replacements of Mg by Al, particularly in palygorskite, probably have an adverse effect on the development of enstatite; sillimanite has been mentioned (Caillère and Hénin 1961, p. 351) as a product which may contain the alumina.

7.6.1.9 Vermiculite reactions

The high-temperature reactions of this hydrous magnesian silicate have been given little attention compared with the numerous detailed studies of the low-temperature hydration states. Walker and Cole (1957) reported that after heating to around 800–900 °C, the exothermic reaction shown by differential thermal analysis corresponds to the formation of enstatite, more particularly when the exchangeable cations are Mg, but when the mineral is saturated with other cations an olivine-like phase appears which persists along with enstatite up to 1050 °C. The thermal analysis curves at high temperatures vary with the nature of exchangeable cations, but there seems to be no agreed interpretation of the results. Walker and Cole remarked "that much more experimental evidence is required before a full understanding ... can be achieved", and this still seems to be true.

7.6.2 Dioctahedral mineral reactions

As stated earlier, dehydroxylated phases exist over wide temperature ranges and from these the high-temperature phases develop at temperatures exceeding about 900 °C. Topotactic development of the dehydroxylates occurs to varying degrees and suffices to cause topotactic reactions also at higher temperatures. From kaolinite and pyrophyllites, an essentially pure aluminosilicate, mullite, $3Al_2O_3 \cdot 2SiO_2$, is the main reaction product, but the chemical equations require considerable liberation of silica:

$$3(Al_2O_3 \cdot 2SiO_2) \rightarrow (3Al_2O_3 \cdot 2SiO_2) + 4SiO_2$$
kaolinite dehydroxylate   63.9%   36.1%

$$3(Al_2O_3 \cdot 4SiO_2) \rightarrow (3Al_2O_3 \cdot 2SiO_2) + 10SiO_2$$
pyrophyllite dehydroxylate   41.5%   58.5%

These percentages of silica are considerably higher than for the corresponding trioctahedral minerals, namely 12.5% from serpentine and 16.6% from talc. Evidently greater structural changes are involved in the dioctahedral mineral reactions.

7.6.2.1 *Pyrophyllite–mullite reaction*

Topotactic development of mullite from pyrophyllite was studied by Bradley and Grim (1951) by X-ray diffraction and by Nakahira and Kato (1964) by electron diffraction. The chain-type structure of mullite develops with the c-axis, the chain axis, parallel to b, or an equivalent direction making 120° with b of the original pyrophyllite. The electron diffraction data additionally gave evidence for the topotactic development of cristobalite, with the [111] axis of the cubic form normal to (001) of pyrophyllite. This orientation corresponds to the oxygen planes in the cristobalite lying parallel to the oxygen planes of the pyrophyllite.

7.6.2.2 *Kaolinite–mullite reactions*

The sequence of topotactic reactions leading to the formation of mullite and cristobalite has been very extensively studied. It was first clearly shown by Brindley and Hunter (1955) using the kaolinite polymorph, nacrite, and by Brindley and Nakahira (1959) using macrocrystalline kaolinite. Others, including Comeforo et al. (1948), Roy et al. (1955), Comer (1960, 1961), also recognized some features of this topotactic sequence. The main experimental difficulty in these studies lies in obtaining suitable single crystals, which are rarely available for X-ray studies of kaolinite but are more easily available for single crystal electron diffraction studies. However, the latter can seldom be studied in different orientations such as are possible in single crystal X-ray diffraction work. The difficulties of studying the first stage, the formation of metakaolin, have been described already. The second stage, involving the formation of a spinel-like phase around 900 °C, has been equally difficult to elucidate. Hyslop and Rooksby (1928) made a big advance by their recognition of a spinel-like structure which they identified as γ-Al_2O_3, undoubtedly the best description at the time, but the exact nature of this phase has been much disputed.

7.6.2.3 The spinel phase
The nature of this phase has been strongly debated from many standpoints. Is it γ-Al_2O_3, or an Al,Si-spinel with Si ions occupying tetrahedral sites, or some combination of these forms?

The topotactic formation of the spinel phase, clearly shown by X-ray diffraction (Fig. 7.6), was thought by Brindley and Nakahira (1959) to indicate a minimum separation of silica from metakaolin in order to facilitate its orderly development. They suggested the reaction

$$2Al_2O_3 \cdot 4SiO_2 \rightarrow 2Al_2O_3 \cdot 3SiO_2 + SiO_2$$
$$\text{metakaolin} \qquad \text{spinel phase}$$

In terms of 32 oxygens per unit cell, the proposed spinel has the formula $Si_8[Al_{10\frac{2}{3}}\square_{5\frac{1}{3}}]O_{32}$, which can be compared with the formula $[Al_{21\frac{1}{3}}\square_{2\frac{2}{3}}]O_{32}$ which corresponds to γ-Al_2O_3. Evidence for and against Si ions entering the spinel phase can be summarized as follows.

Direct evidence based on lattice parameter measurements is difficult to obtain. Yamada and Kimura (1962) observed that γ-aluminas prepared in the presence of colloidal silica have a variable unit cell size, and when the cell parameter is plotted against the ratio $Al_2O_3/(Al_2O_3 + SiO_2)$, the resulting straight line runs parallel to that through the parameters measured by Brindley and Nakahira for γ-Al_2O_3 and the spinel phase from kaolinite.

Weiss et al. (1970) claimed that pure Al,Si-spinel was obtained by leaching heat-treated kaolinite with pyrocatechol which reacts with amorphous silica. They write: "Exhaustive leaching ... finally yields the Al,Si-spinel, the formula of which is in excellent agreement with the theoretical formula derived by Brindley and Nakahira (1959)." It is very unfortunate that they did not report the lattice parameter of the purified spinel phase.

Much evidence has accumulated pointing to γ-Al_2O_3 rather than a Si-containing spinel phase. Infrared data given by Pampuch (1966) and by Percival et al. (1974) support the formation of γ-Al_2O_3, though both studies concede the possibility that some Al,Si-spinel may be formed though in lesser amounts.

Application of Warren's method of radial electron density analysis to the kaolinite–mullite sequence (Lemaitre et al. 1975b; Léonard 1977) pointed to the formation of γ-Al_2O_3 rather than a Si-containing spinel. A variety of other evidence has been obtained indicating an extensive segregation of silica when the spinel phase is formed. Careful absolute density measurements (Lemaitre et al. 1975b) and also data on the shift of the Al K_α X-ray fluorescence peak of kaolinite samples after various heat treatments (Léonard 1977; Bulens et al. 1978) support a complete separation of alumina and silica before mullite formation. Kinetic studies on pure and mineralized samples (Lemaitre et al. 1975a), including the effect of the crystallinity of the starting kaolinite (Bulens and Delmon 1977a) and of artificial nucleation of mullite (Bulens and Delmon 1977b), have led them to propose the following scheme for the metakaolin–mullite sequence in the 900–1000 °C range:

$$Al_2O_3 \cdot 2SiO_2 \xrightarrow{A} \tfrac{1}{3}(3Al_2O_3 \cdot 2SiO_2) + \tfrac{4}{3}SiO_2$$
$$\text{metakaolin} \qquad\qquad\qquad \text{mullite}$$
$$\searrow_{B_1} \quad \gamma\text{-}Al_2O_3 + 2SiO_2 \quad \nearrow_{B_2}$$

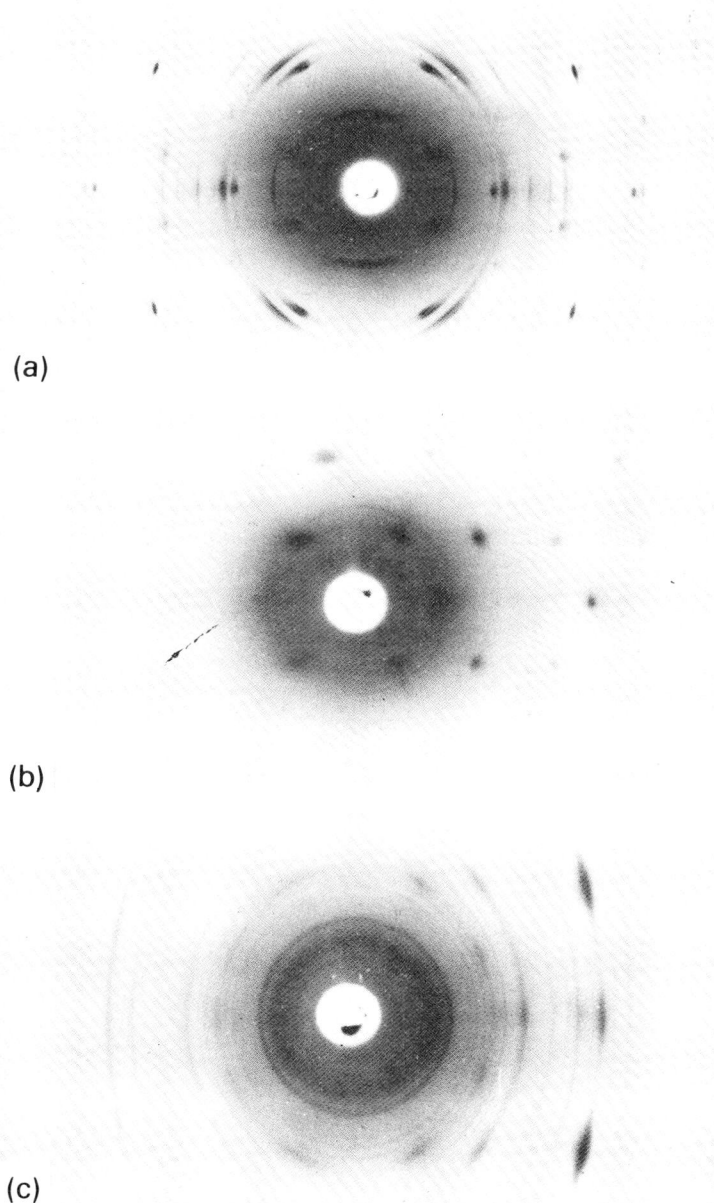

FIG. 7.6. X-ray diffraction patterns of: (a) single crystal of kaolinite rotated about b or pseudo-b axis; (b) the same crystal after heating at 850 °C; (c) the same crystal after heating at 950 °C. (After Brindley and Nakahira 1959.)

Path A, corresponding to direct mullite formation (primary mullite), is favoured by the presence, in well-crystallized material, of well-dispersed CaO. Path B involves a two-step formation of mullite. In a first step metakaolin transforms to γ-Al_2O_3 with an extensive segregation of amorphous silica. In a second step, free alumina and silica recombine at higher temperatures to

TABLE 7.4. Amounts of free silica produced by the transformation of metakaolinite, according to various hypotheses

Reaction products	Free SiO_2 (wt%)
Si,Al-spinel	11.6
γ-Al_2O_3	46.6
3:2 Mullite	31.0

form further mullite. Path B is favoured by a low degree of crystallinity of the starting kaolinite and, for well-crystallized samples, by the presence of well-dispersed MgO.

Evidence in favour of the Si,Al-spinel hypothesis, proposed recently by Chakraborty and Ghosh (1978), is based on a selective alkaline extraction of free silica from samples submitted to various heat treatments and on the thermal behaviour of etched samples. These authors found that a maximum of 38–42 wt% SiO_2 could be extracted from kaolinite samples after a short heating up to 900 °C. The residue of extraction showed the characteristic diffraction pattern of a cubic phase and, after heating as 1200 °C, showed diffraction peaks typical of mullite. Upon subsequent alkaline extraction, no supplementary SiO_2 was removed. These results were interpreted as supporting the transformation of metakaolin to a Si,Al-spinel having an Al/Si ratio very close to that of normal mullite (Al/Si = 3). However, the amount of silica extracted from the sample heated at 980 °C (42 wt%) is closer to the amount expected if complete segreation to alumina and silica is assumed (see Table 7.4). Moreover it has been shown that mullite-like structures without silica can develop from alumina containing small amounts of sodium (2.4 mol% Na_2O) (Duvigneaud 1974; Perrotta and Young 1974). It is therefore not unlikely that the residue studied by Chakraborty and Ghosh could develop a mullite-like structure upon heat treatment as a consequence of sodium contamination brought about by alkaline extraction.

On the basis of available evidence, it seems that the spinel phase is usually close to γ-Al_2O_3, but whether or not some Si ions enter tetrahedral sites is difficult to decide. The possibility that the nature of the spinel may depend on the nature of the preceding phases should not be overlooked. In particular, the crystalline order–disorder of the original kaolinite, the retention of hydroxyl ions or their absence in metakaolin and the precise state of the metakaolin, and perhaps also the presence of impurities (such as titanium) in the original kaolinite, may all have some influence on the spinel phase. It has to be recalled that the single crystal kaolinite used by Brindley and Nakahira (1959) may not have been either chemically or structurally equivalent to the less ideal material commonly used in these experiments.

The formation of mullite and cristobalite, which takes place slowly in the range 1000–1100 °C and abundantly above 1150 °C, also has provided problems. The electron-optical work of Comer (1960, 1961) has shown that mullite develops as short needle-like crystals oriented at 120° to each other (Fig. 7.7). Bradley and Grim (1951) deduced a similar arrangement for mullite derived from pyrophyllite. von Gehlen (1962b) showed that the (111) plane of the spinel phase, parallel to (001) of kaolinite, becomes the (310) plane of mullite. Cristobalite appears more or less contemporaneously with the mullite, but it is not possible to say exactly when it forms and probably a slow process of crystallization from a silica-rich disordered phase is involved which is strongly time- and temperature-dependent.

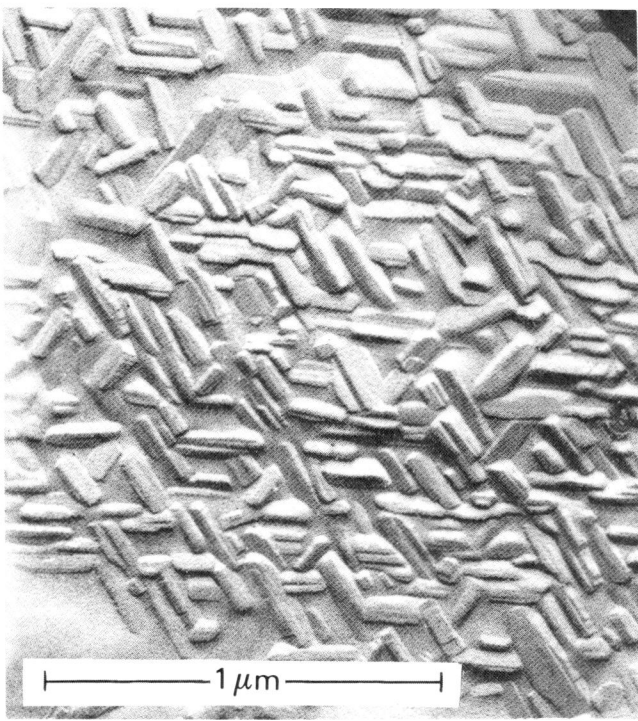

FIG. 7.7. Electron micrograph of a preshadowed carbon replica of a kaolinite crystal heated 20 hours at 1110 C showing development of mullite rod-shaped crystals. (Photograph supplied by J. J. Comer. After Brindley and Nakahira 1959.)

An important reason why Brindley's hypothesis of an intermediate Si,Al-spinel in the kaolinite–mullite sequence was found necessary was the view that the topotactic character of the thermal transformation strongly suggested a structural continuity of the oxygen lattice, and that such a structural continuity could hardly be expected if a major segregation of silica and alumina (as γ-Al_2O_3) occurred in the course of the transformation. The argument based on topotaxy was probably overemphasized and instead, the orientation relations are now seen as arising from the oriented nucleation of new phases preferentially on certain planes, the product phase being developed by the growth of these nuclei (Brindley 1975). A remarkable example of a topotactic transformation for which clearly no structural continuity exists is the thermal decomposition of pyroaurite [$Mg_6Fe_2(OH)_{16}$. CO_3 . $4H_2O$] to MgO and $MgFe_2O_4$ (Rouxhet and Taylor 1969). It is thus reasonable to think that the cubic phase nucleates in preferred orientations relative to the parent metakaolin, in order to minimize interface energy. Crystallites of the cubic phase will then presumably grow in an ordered way (possibly by an inhomogeneous mechanism) by expelling silica. At this stage, the disagreement between the Si,Al-spinel and the γ-Al_2O_3 hypotheses is the extent to which silica is expelled. The oriented crystallization of mullite from segregated silica and alumina may be explained in a similar way: as mullite is formed by solid-state reaction at the interface between free silica and ordered γ-Al_2O_3, it is likely that the nuclei of the new phase will appear with preferred orientations relative to the parent γ-Al_2O_3

crystallites, in order to minimize interface energy, and thus grow in an ordered manner by interdiffusion of silica and alumina.

The order in which the high-temperature phases appear from the various kaolinite group minerals has received much attention, and some kinetic studies also have been made. The results show considerable variation and even contradictions. Grim (1968, pp. 304–307, 311–312), summarizing work up to the early 1960s, shows that many factors may be involved, including the following. (1) The state of crystalline order and disorder in the initial mineral; differences in behaviour of kaolinite, disordered kaolinite, and halloysite have usually been attributed to this factor. (2) The conditions of heating; static heating for long periods, and quenching after heating to a particular temperature in a DTA equipment represent extremes of heat treatments. (3) The presence of impurities; these may be inevitable mineralogical impurities, or the gas atmosphere, or the exchangeable ions associated with the clay.

The order in which the high-temperature phases appear in X-ray powder diffraction patterns appears to depend on the crystalline order of the initial mineral. With well-ordered kaolinite, the spinel phase appears around 950 °C, poorly developed mullite around 1000–1050 °C, and cristobalite at a somewhat higher temperature (Richardson 1951; Glass 1954) but a more or less simultaneous formation of the spinel phase and mullite has also been claimed (Tsuzuki 1961; Tsuzuki and Nagasawa 1969). With disordered kaolinite and halloysite, the first appearance of mullite seems to be retarded, sometimes considerably (Richardson 1951; Glass 1954; Tsuzuki 1961; Bulens and Delmon 1977a); better crystalline mullite and easily recognized cristobalite develop around 1250 °C.

Kinetic studies of mullite formation (Duncan et al. 1969; Mackenzie 1969a) from well-crystallized kaolinite and from halloysite, by X-ray and infrared methods at temperatures in the range 1100–1220 °C have shown a very similar time and temperature dependence, and when the amount of mullite is plotted against a reduced time parameter, the data for both minerals fall on a common curve, showing that the rate-controlling process is the same for both. The results agree with first-order kinetics which is interpreted as a dependence of the reaction on the formation of nuclei, in turn dependent on the volume of unreacted material. The rate of growth is considered to be rapid compared with the rate of nucleation, so that the latter is the rate-controlling process. Direct evidence for nucleation-controlled mullite formation has been found by Bulens and Delmon (1977b).

In studies of the effect of impurities on mullite formation from halloysite (Mackenzie 1969b), exchangeable cations are shown to affect the nucleation rate but not the mechanism of formation, probably "by providing lattice sites of suitable energy for nucleation". Anion exchange is shown to reduce the rate of mullite formation, sometimes to the point of inhibiting the reaction (studied at 1100 °C). The firing atmosphere also is shown to be very important. Water vapour and a vacuum accelerate the reaction considerably; nitrogen and hydrogen have less effect but oxygen and carbon dioxide retard the reaction.

7.6.2.4 High-temperature phases from muscovite
Commenting on the variety of results found by different investigators, Yoder and Eugster (1955, see p. 229) emphasized that the products obtained are not equilibrium assemblages. On the basis of the K_2O–Al_2O_3–SiO_2 equilibrium diagram, the anhydrous muscovite composition should give leucite + K-feldspar + mullite below 1140 °C, leucite + mullite + liquid from 1140 to 1315 °C, and leucite + corundum + liquid above 1315 °C. There is considerable evidence for the formation of a spinel-type phase among the first products formed (Roy 1949; Sundius and Bystrom 1953; Eberhart 1963; Nicol 1964) which has been called γ-Al_2O_3 but may well be

similar to the phase developed from kaolinite. Eberhart observed the topotactic development of the spinel phase with $(111)_{sp}$ parallel to (001) of muscovite and $(110)_{sp}$ parallel to (010) of muscovite. By using very thin lamellae of muscovite and a 100 kV electron microscope, very small and uniformly oriented crystals of the spinel phase could be seen directly. Lattice parameter measurements showed that the composition of the spinel varied between that of γ-Al_2O_3 and $MgAl_2O_4$ according to the chemical composition of the initial mica. Nicol confirmed the topotactic behaviour and showed that water may play a catalytic role. Corundum also has been reported by many investigators. Brindley and Maroney (1960) found the lattice parameter was 0.7% greater than that of pure corundum and suggest possible substitutions of some Fe^{3+} for Al ions. Mullite has been recorded by Sundius and Bystrom (1953), Eberhart (1963), Nicol (1964), leucite by Eberhart, sanidine (K-feldspar) by Yoder and Eugster (1955) and kalsilite, also formed topotactically, by Nicol.

The products are mainly those to be expected from the equilibrium diagrams, but the amounts and the associations differ probably because of the precise chemical compositions of the micas and the heating treatments employed. The spinel-type phase would not be expected and its topotactic formation suggests that it is a transitional stage.

7.6.2.5 High-temperature phases from montmorillonites

The high-temperature phases of montmorillonites vary considerably and the main reason appears to be the variable composition of this mineral. Grim and Kulbicki (1961) studied about 40 samples heated to 1400 °C and found two principal types of behaviour, which they called "Cheto-type" and "Wyoming-type" corresponding to two localities. These differ principally in the octahedral cation compositions, with the Cheto-type containing an average of 0.61% Mg^{2+} per formula unit and the Wyoming-type 0.16% Mg^{2+} per formula unit. The overall layer charges are about -0.60 and -0.34 per half-formula unit for the two clay types respectively.[1]

The Cheto-type clays form β-quartz between 900 and 1000 °C and cordierite, $Al_3(Mg,Fe)_2(Si_5Al)O_{18}$, between 1200 and 1300 °C; the quartz converts to cristobalite around 1100 °C. The Wyoming-type clays form cristobalite and mullite, $Al_6Si_2O_{13}$, around 1150–1200 °C. The X-ray diffractions from the initial minerals disappear around 850–900 °C for the Cheto-type clays and around 925 °C for the Wyoming-type clays. The formation of cordierite in the one case and mullite in the other, as Bradley and Grim (1951) noted much earlier, appears to be determined primarily by the chemical compositions of the systems, with the higher proportion of Mg in the Cheto-type clays favouring formation of cordierite. However, this may be an oversimplification because addition of magnesium to the Wyoming-type clays did not produce high-temperature phases characteristic of the Cheto-type clays, and conversely leaching of the clays with HCl to remove octahedral cations did not change the high-temperature reactions. Consequently, Grim and Kulbicki sought a structural explanation and suggested some inverted tetrahedra in the layer structure of the Cheto-type clays. This generally disfavoured concept, however, need more justification before it can be accepted as an explanation. Bradley and Grim (1951) contrasted the formation of the spinel-like phases from some montmorillonites and of quartz from others and pointed out that these correspond with octahedrally directed crystallization in the former case, and tetrahedrally directed in the latter. Kulbicki (1958) also emphasized the possibly important role of exchangeable cations on the high-temperature reactions. Evidently a complete understanding of the high-temperature behaviour of montmorillonites has not yet been achieved.

7.7 OXIDATION REACTIONS

Oxidation of minerals is concerned principally with the conversion of ferrous to ferric iron and concomitant processes required to maintain electrical neutrality. The subject is one with a long history and much of the early work can be found in a publication by Chamberlin (1908). The present account is focused on more recent, quantitative work. Oxidation of minerals occurs under normal atmospheres and is part of the weathering process. It may be simulated in the laboratory by use of oxidizing agents such as H_2O_2 and Br_2 (Farmer *et al.* 1971). Saturated bromine water has been shown to bring very reproducible results and to produce oxidized minerals identical with those appearing in soils upon weathering (Gilkes *et al.* 1972; Ross 1975). Oxidation occurs also when minerals are heated in normal atmospheres and often takes place in about the same temperature range as dehydroxylation. Oxidation at elevated temperatures may occur in the absence of external oxygen by an internal oxidation process. A thermomagnetic study by Escoubes and Karchoud (1977) showed that in most iron-containing clay minerals, except biotite (see below), Fe^{2+} ions migrate easily during the dehydroxylation process and precipitate as magnetite (Fe_3O_4) or hematite (α-Fe_2O_3) depending on the atmosphere.

These various reactions can be described formally as follows:

1. Precipitation of iron(II) oxide

$$Fe^{2+} + 2OH^- \rightarrow FeO + H_2O \tag{4}$$

2. Formation of magnetite (oxygen absent)

$$3FeO + H_2O \rightarrow Fe_3O_4 + H_2 \tag{5}$$

3. Formation of hematite (oxygen present)

$$2FeO + 1/2 O_2 \rightarrow Fe_2O_3 \tag{6}$$

The continued loss of hydroxyls and oxidation may appear as a single reaction. The reactions are not restricted to layer silicates and have in fact been much studied in relation to chain silicates (see, for example, a series of papers by Addison and co-workers (1962a,b), Addison and Sharp (1962a,b, 1968), Hodgson *et al.* (1965a,b).

In the case of iron-rich, trioctahedral micas (see also Section 7.7.2), Fe^{2+} ions are not precipitated upon dehydroxylation. Under a vacuum, they can be oxidized *in situ* with hydrogen evolution. In the presence of oxygen, some Fe^{2+} ions can be extracted from the outmost layers of the mineral to form α-Fe_2O_3 (Escoubes and Karchoud 1977).

Reactions of several kinds are possible which can be represented as follows:

$$4Fe^{2+} + 4OH^- + O_2 \rightarrow 4Fe^{3+} + 4O^{2-} + 2H_2O \tag{7}$$

$$4Fe^{2+} + 4OH^- \rightarrow 4Fe^{3+} + 4O^{2-} + 2H_2 \tag{8}$$

$$4Fe^{2+} + O_2 \rightarrow 4Fe^{3+} + 2O^{2-} \tag{9}$$

$$4OH^- \rightarrow 2O^{2-} + 2H_2O \tag{10}$$

Reactions (7) and (8) change the charges but not the numbers of structural anions and cations; they are dehydrogenation processes with hydrogen lost molecularly in (8) and by aerial oxidation as water in (7). Reactions (9) and (10) together give reaction (7) and are respectively oxygenation and dehydroxylation processes. Reactions (7) and (9) require the presence of oxygen in the atmosphere. Reactions (7) and (8) occur with a small weight loss of one H per Fe(II) ion; reaction (9) occurs with a weight gain of one O per two Fe(II); and reaction (10) involves a large weight loss. Besides these reactions involving OH^- ions, other reactions may involve F^- ions when these form part of the anionic content of the silicate. Therefore it is important to emphasize that weight losses cannot always be equated with the formation of H_2O.

7.7.1 Oxidation of berthierine

The oxidation of this 1:1 trioctahedral mineral (berthierine was previously called chamosite), which approximates to a ferrous analogue of serpentine (Brindley and Youell 1953; Brindley 1982), takes place by reactions (4) and (7). One sample studied in detail had the approximate formula $(Fe(II)_{1.8}Mg_{0.2}Al_{0.8})(Si_{1.3}Al_{0.7})O_5(OH)_4$ and after oxidation by heating in air at 400 °C, this became $(Fe(III)_{1.8}Mg_{0.2}Al_{0.8})(Si_{1.3}Al_{0.7})O_5O_{2.3}(OH)$. The oxidation process took place only in the presence of atmospheric oxygen which suggests that reaction (7) is involved. The compositional change can be represented as a sum of reactions (7) and (10), as follows:

$$1.8Fe^{2+} + 1.8OH^- + \underset{\text{(atmospheric)}}{0.9O} \rightarrow 1.8Fe^{3+} + 1.8O^{2-} + 0.9H_2O \tag{7}$$

$$1.2OH^- \rightarrow 0.6O^{2-} + 0.6H_2O \tag{10}$$

The total reaction is:

$$1.8Fe^{2+} + 3OH^- + 0.9O \rightarrow 1.8Fe^{3+} + 2.4O^{2-} + 1.5H_2O$$

This last reaction agrees closely with the change in mineral composition. The loss of hydroxyls is seen to be 60% due to oxidation of iron, and 40% due to "normal" dehydroxylation.

Escoubes and Karchoud (1977) showed by thermomagnetic analysis that when berthierine was heated *in vacuo*, iron was precipitated as magnetite (Fe_3O_4) from the very beginning of the dehydroxylation process.

7.7.2 Oxidation of micas

The compositional changes of micas, particularly biotites, when heated in air at 0.1 Torr pressure, and in argon, at various temperatures ranging from 450 to 1200 °C, have been studied in detail by Rimsaite (1970). Measurements were made of fluorine, FeO, Fe_2O_3, and all the oxides normally included in full analyses. The calculation of structural formulae presents the biggest problem in these studies. This is a problem which already has been mentioned (Section 7.5.1) in another connection. The customary idealizations which assume certain fixed charges (or valencies) or numbers of ions per unit cell were not followed; instead the compositions were normalized with respect to a total valence of $(44+Z)$ per unit cell as compared with $+44$ for the ideal mica composition. Various procedures were used to estimate Z, but these are too detailed to present here and the original paper must be consulted. The highest Z number was found in a natural oxidized biotite (2.56) and in a high-iron biotite heated in air (2.93). In iron-poor phlogopite, muscovite and lepidolite, Z is close to zero and the structural formulae can be derived on the basis of 44 valencies/unit cell provided the normal amount of potassium is retained. In altered micas with low potassium, Z may be positive or negative depending on the amount of oxidized iron in relation to the lost potassium (or other cations).

Prior to heating, the biotite samples already had positive values of Z due to partial oxidation under natural conditions. There were also deficiencies of hydroxyls which could be attributed in some cases to dehydroxylation by reaction (10) but in other cases to a replacement of OH^- by O^{2-} to balance oxidation, or a combination of these processes. Further complications arise when F^- partly replaces OH^-.

When biotites are heated, hydroxyl decomposition can occur in three ways depending on the temperature, the amount of ferrous iron and the presence or absence of oxygen. At low temperatures, 450–650 °C, in an oxygen-free atmosphere, oxidation by reaction (8) may prevail over dehydroxylation by reaction (10). Dehydroxylation increases with temperature to about 900 °C, when oxygen ions may replace fluorine ions. When atmospheric oxygen is available, oxidation by reaction (7) occurs. Fluorine is given off at lower temperatures in air than in any oxygen-free system.

In contrast with other iron-containing clay minerals, biotite was found to oxidize without appreciable precipitation of iron oxides (Escoubes and Karchoud 1977; Tripathi et al. 1978). Iron is oxidized *in situ* according to reaction (8) *in vacuo*, and the reaction (7) in air, without formation of free iron oxides. Biotite also shows a distinct behaviour with respect to the influence of the atmosphere on its rate of dehydroxylation. Escoubes and Karchoud (1977) found that dehydroxylation–oxidation of iron-rich biotites started at about 600 °C in air while no weight loss occurred *in vacuo* below 800 °C. In contrast, other iron-containing minerals underwent dehydroxylation *in vacuo* at 100–200 °C lower temperatures than in air.

Thermal decomposition of biotites in air has been studied by Hogg and Meads (1975), and of phlogopite and vermiculite by Tripathi et al. (1978), using Mössbauer spectroscopy. According to these authors, free α-Fe_2O_3 is not found below 900 °C, at which temperature structural breakdown of biotite commences. The progressive changes in the Mössbauer spectra were correlated with the following processes: (1) Oxidation of Fe^{2+} in isolated octahedral sites to give $Fe^{3+}(O_5OH)$. (2) Oxidation of Fe^{2+} in adjacent a-octahedral (*cis* OH^-) sites to give $Fe^{3+}(O_6)$. (3) Oxidation of Fe^{2+} in adjacent a- and b-octahedral (*trans* OH^-) sites to give either $Fe^{3+}(O_5OH)$ or $Fe^{3+}(O_6)$ depending on whether a shared OH group remains intact or not. These processes are observed upon heating for 24 hours in the range 300–500 °C. At higher temperatures, $Fe^{3+}(O_5OH)$ converts to $Fe^{3+}(O_6)$. During all these transformations the iron remains in the biotite structure. Some degree of reversibility of these reactions is possible since ferric ions formed previously can be reduced under vacuum to ferrous ions, at temperatures above which the loss of hydroxyl groups is complete (Rouxhet et al. 1972).

Iron in biotite can be largely oxidized to the ferric state upon weathering by three possible mechanisms (Farmer et al. 1971), formulated as follows for the ferrous end-members of the phlogopite–biotite series:

1. Loss of interlayer cation

$$(Si_3AlO_{10})Fe_3^{2+}(OH)_2 \cdot K + \tfrac{1}{4}O_2 + \tfrac{1}{2}H_2O \rightarrow$$
$$(Si_3AlO_{10})Fe_2^{2+}Fe^{3+}(OH)_2 + KOH$$

2. Loss of hydroxyl proton

$$(Si_3AlO_{10})Fe_3^{2+}(OH)_2 \cdot K + \tfrac{1}{2}O_2 \rightarrow$$
$$(Si_3AlO_{10})Fe^{2+}Fe_2^{3+}O_2 \cdot K + H_2O$$

3. Loss of octahedral iron

$$(Si_3AlO_{10})Fe_3^{2+}(OH)_2 \cdot K + \tfrac{3}{4}O_2 + \tfrac{1}{2}H_2O \rightarrow$$
$$(Si_3AlO_{10})Fe_2^{3+}(OH)_2 \cdot K + FeO \cdot OH$$

The first mechanism, associated with vermiculitization, plays a minor role in high-iron biotites, as the amount of iron oxidized during vermiculitization is very much greater than the decrease in layer charge (Newman and Brown 1966). The second mechanism has been well established for thermal oxidation of biotites (see above) but is less certain in oxidative weathering as the excess water always present is difficult to distinguish from constitutional hydroxyls (Rimsaite 1970). The third mechanism involving the formation of iron oxide has long been discounted on the ground that no discrete iron oxide phase was detectable. Farmer et al. (1971) have given infrared evidence for ejection of octahedral iron upon oxidation of vermiculitized biotite in aqueous solutions of hydrogen peroxide or bromine. Electron microscopic examination showed the formation of free iron oxide in the form of β-FeOOH (alkaganeite) in samples treated with bromine while those treated with hydrogen peroxide indicated the formation of amorphous iron oxide. These results have been confirmed subsequently for biotites oxidized in saturated bromine water by Gilkes et al. (1972) who showed that the loss of octahedral cations is accompanied by a corresponding loss of interlayer cations in order to maintain electrical neutrality. This results in a decrease of the b-axis dimension of the mineral and its increased stability upon subsequent weathering.

7.7.3 Oxidation of chlorites

An important experiment relating to the evolution of hydrogen when ferrous iron-containing minerals are heated in an oxygen-free system was carried out by Orcel and Renaud (1941), who compared the thermal behaviour of two chlorites, a ripidolite with 18.73% FeO and a white chlorite with 1.24% FeO, when heated in an evacuated system. Evolved water was trapped by a solid CO_2-cooled U-tube, and evolution of hydrogen, being very much more for the ripidolite than for the white chlorite, was demonstrated spectroscopically. The interpretation placed on the results was not the single reaction (8), but a two-stage operation in which water, lost by dehydroxylation, is decomposed to furnish oxygen for the oxidation process and to liberate hydrogen. In a static system, an equilibrium is established between water, hydrogen, ferrous and ferric iron.

The importance of oxidation in relation to chlorite compositions has been recognized for many years. The general formula of ferromagnesian chlorites can be written $(Mg_{6-x-y}Fe(II)_y Al_x)(Si_{4-x}Al_x)O_{10}(OH)_8$. This formula corresponds to the orthochlorites of Tschermak (1890, 1891). Other chlorites, called leptochlorites by Tschermak, usually containing considerable ferric iron, show big deviations from this type of formula. Winchell (1936), Holzner (1938) and others have shown that when leptochlorite compositions are recalculated with Fe(II) replacing Fe(III) and with corresponding adjustments of O^{2-} to OH^-, the results in many cases come close to orthochlorite compositions. In other words, many leptochlorites probably owe their "abnormal" compositions to oxidation *after* initial formation as ferrous orthochlorites. However, not *all* ferric iron in a chlorite has been produced by subsequent oxidation.

It is known that chlorites, like micas, may weather in soils to vermiculites (Makumbi and Herbillon 1972). Ross (1975) has shown that pure, unoxidized chlorites containing an appreciable amount of ferrous iron may be converted to vermiculite in the laboratory by reacting them in saturated bromine water on a steambath. He showed that oxidation of the structural iron was necessary for converting chlorite to vermiculite, since no transformation was observable in the absence of an oxidizing agent.

7.7.4 Oxidation in relation to biotite–vermiculite relations

Field evidence shows that biotites frequently weather to form vermiculites and in the process oxidation of ferrous to ferric iron takes place. It has seemed reasonable, therefore, to consider that the lower layer charge of most vermiculites is the result of this oxidation and that expansion of the layer structure has thereby become easier and facilitated an exchange of the interlayer K ions of biotites for the Mg ions usually found in vermiculite. However, a converse argument also has been put forward, namely that leaching of K ions from biotites by magnesian solutions has opened up the layer structure by hydration of the Mg ions, and that oxidation with some loss of interlayer cations has been a concomitant process. These questions have been discussed at length by Barshad (1948), Roy and Romo (1957), Foster (1963), Robert and Pedro (1965), Wey *et al.* (1966), Wey and LeDred (1968) and Newman and Brown (1966). Opinion seems to favour cation exchange as the primary alteration process leading to vermiculite formation. Phlogopites and low-iron biotites can be converted to vermiculite, but commonly the amount of iron is such that oxidation is an important aspect of the total change.

7.7.5 The possible role of OH orientations on chemical behaviour

Bassett (1960) firstly clearly showed the possible role of hydroxyl orientation in mica structures in relation to chemical alteration. In trioctahedral micas, such as phlogopites and ferrous iron biotites, the O—H orientation is normal or nearly normal to 001, but in dioctahedral micas, O—H is oriented more nearly parallel to 001. These orientations are determined mainly by the repulsion of the proton by the associated octahedral cations. Since the hydroxyl ion is directly opposite the interlayer K ion, the orientation of the dipole strongly influences the bonding of the K ion, and conversely the interlayer cation influences the hydroxyl stretching frequency. Barshad and Kishk (1968) have shown that oxidation of ferrous iron in soil vermiculites and biotites increases the "potassium fixation capability", defined as the percentage of K not replaceable by NH_4^+. When Fe^{2+} becomes Fe^{3+}, the deviation of the O—H orientation away from the normal to 001 places the K ion in a "more electrically negative environment". An experimental confirmation of the reorientation of the OH dipole when ferrous biotites are oxidized has been obtained by Juo and White (1969). Ismail (1969, 1970) studied the oxidation of finely ground biotites by solutions of hydroxylamine, and under basic conditions found a close correspondence between Fe(II) oxidized, the amounts of Fe(II) and Mg released to the solution, and the interlayer (K + Na) released to the solution. The following illustrates his results. In terms of meq/100 g, the oxidation of Fe(II) to Fe(III) increased the layer charge by 70.9 units, but loss of Fe(II) to solution diminished the layer charge by 26.2 units. The net increase, 44.7 units, is balanced by a corresponding release of (K + Na) ions to the solution. The release of octahedral cations to the solution is an additional change which takes place. Under acid conditions, the decomposition of the octahedral layer proceeds much further and a simple balance, as described above, is not possible.

7.8 REDUCTION REACTIONS

Reduction studies on clay minerals have been concerned almost exclusively with nickel-containing hydrous silicates of the serpentine and talc families (garnierites). The interest towards these minerals has increased recently owing to their increasing importance as alternative nickel ores, and also because of their possible use as precursors of nickel-on-silica catalysts, either in their natural or synthetic form (Swift 1977). Until now only a few studies have been available, all using hydrogen as a reducing agent. In earlier work, reduction experiments have been conducted in static conditions, conversion being monitored gravimetrically (Martin *et al.* 1970; Scholten and Kiel 1975). It is usual now to perform reduction studies under non-isothermal conditions. The method is referred to as "Temperature Programmed Reduction" (TPR for short) and was proposed first by Robertson *et al.* (1975). From a TPR experiment, a thermogram is obtained in which the rate of reduction is plotted versus temperature. With well-defined experimental conditions, a sample yields a characteristic thermogram exhibiting reduction peaks or bands and so carrying information on its mineralogical nature. The area under the TPR bands is proportional to the amount of reducible species (e.g. Ni^{2+}) and so can be used for analytical purposes (Lemaitre and Gérard 1981a).

7.8.1 Reduction reactions of nickel-containing silicates

7.8.1.1 Ni-serpentines

Both natural and synthetic nepouites (the Ni-rich end-members of the serpentine-like minerals) exhibit a double TPR band with maxima around 500–580°C and 610–700°C (see Fig. 7.8) (Lemaitre and Gérard 1981b). The first component is completely suppressed in favour of the second when samples have been submitted to a preliminary heat treatment above their dehydroxylation temperature. The first component of the TPR band of fresh nepouite can therefore be ascribed to the reduction of the hydrated phase; reduction and dehydroxylation reactions occurring in the same temperature range, a more or less important fraction (depending on the exact experimental conditions) of the hydrated mineral will be converted to the LS-phase (see Section 7.4.4) before having the opportunity to react with hydrogen. The dehydroxylated fraction will react at higher temperature, thereby producing the second TPR peak which persists when dehydroxylated samples are used and therefore is characteristic of the reactivity of the LS-

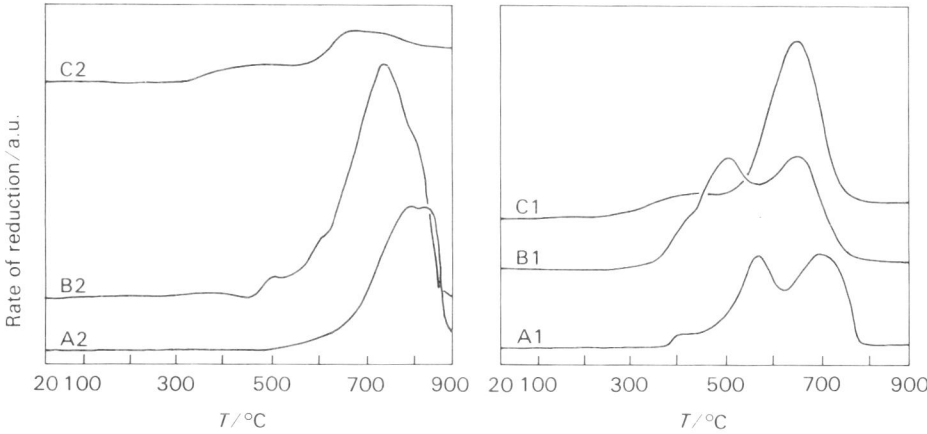

FIG. 7.8. Temperature-programmed reduction patterns of garnierites. A1: fresh synthetic nepouite; B1: fresh natural nepouite; C1: natural nepouite after a short heating up to 900°C (10°C min^{-1}). A2: fresh synthetic pimelite; B2: fresh natural pimelite; C2: natural pimelite after a short heating up to 900°C (10°C min^{-1}). (After Lemaitre and Gérard 1981b.)

phase. A third reduction peak around 700 °C also has been attributed to the reduction of the FCC-phase which starts to form at this temperature at the expense of the LS-phase (see Section 7.6.1).

7.8.1.2 Ni-talcs
Natural and synthetic pimelites both exhibit a rather sharp TPR band in the range 730–830 °C (Lemaitre and Gérard 1981b) (see Fig. 7.8). This band exhibits a shoulder at high temperature, which has been assigned to the reduction of the transitory FCC-phase produced by the thermal decomposition of the unreduced fraction of the mineral. The amount of reducible nickel is considerably decreased after heat treating pimelites above 900 °C. Some reduction at about 700 °C is still observed in natural pimelites heat treated for a short time at 900 °C, and has been consistently assigned to minor amounts of dehydroxylated Ni-serpentine present in the sample. By such a heat treatment the respective contributions to reduction of serpentine- and talc-like fractions of a mixed sample can be separated, and an estimate of these fractions can be derived from TPR results. (Lemaitre and Gérard 1981a).

The above results can be summarized in a tentative classification of the ease of reduction of various nickel-containing phases found in fresh or heat-treated Ni-hydrous silicates: FCC-phase $\geqslant$ pecoraite > LS-phase > pimelite > high-temperature phases (enstatite or olivine). In fact, synthetic, well-crystallized pecoraite has been found more reactive than the product FCC-phase (Lemaitre unpublished result). The above classification is thus only approximate in that it does not take fully into account such factors as particle size, crystallinity and degree of substitution of Ni by Mg.

7.8.2 Reduction reactions of nickel-hydroxy montmorillonites

Nickel-hydroxy montmorillonites are usually obtained by progressive alkaline titration of a soluble Ni-salt in an aqueous suspension of montmorillonites (Yamanaka and Brindley 1978). A similar procedure has been described earlier in Section 7.5.1 for the development of hydroxy-Mg and hydroxy-Al montmorillonites. Similar experiments also have been made on hydroxy-Cr montmorillonites (Brindley and Yamanaka 1979) where diagrams have been given showing how the hydroxy-Cr polymeric forms may develop in the interlayers of montmorillonite. This procedure has been applied particularly in attempts to prepare synthetic chlorite-like Ni-containing phases (Gupta and Malik 1969).

Three kinds of Ni-species can be found in those materials, depending on the degreee of saturation of the ion exchange capacity and of the interlayer space by nickel (Yamanaka and Brindley 1978; Ghesquière *et al.* 1982): (i) Exchangeable hydroxy-Ni ions approximated by the formula $[Ni(OH)_{0.5}]^{1.5+}$; (ii) interlayered hydroxy-Ni species, very similar to the brucite layer in the structure of chlorite; and (iii) amorphous nickel hydroxide precipitated outside the montmorillonite crystals.

Ghesquière *et al.* (1982) have shown, using TPR analysis (see above) that these various Ni-species behave differently upon reduction by hydrogen. The most reactive species is free nickel hydroxide which produces a reduction peak in the range 300–400 °C. Interlayered hydroxy-Ni is less reactive (TPR band between 400 and 550 °C), possibly due to a shielding effect of the montmorillonite layers. Samples containing only exchangeable Ni exhibit only one TPR peak at about 700–800 °C.

That exchangeable nickel reacts with hydrogen in a temperature range characteristic of Ni-talc seemed at first amazing. It has been shown, however (Calvet and Prost 1971) that exchangeable Ni ions, just like Li and Mg (see Section 7.2.1) can migrate into the vacant octahedral sites of the montmorillonite sheet upon mild thermal treatment (below 200 °C). This effect presumably takes place during the temperature increase of a TPR experiment and so explains why exchangeable nickel exhibits the same reactivity as nickel located in the octahedral layer of a 2:1 mineral.

7.9 SUMMARY AND CONCLUSIONS

The present chapter shows the constantly broadening scope of the thermal, oxidation and reduction reactions of the principal layer silicate minerals. Although attention has been directed principally towards the fine-grained minerals of this group, i.e. the clay minerals, it has also been necessary to give some attention to the corresponding macrocrystalline minerals, particularly the micas and chlorites, from which much of the detailed knowledge of their chemistry and structure has been obtained.

The treatment of the subject has been simplified in two respects, namely (i) by restricting the discussion to pure or mainly pure representatives of the various mineral groups, and (ii) by omitting mixed-layer systems. It has to be recognized that "pure" clay minerals are seldom pure in a strictly chemical sense. Not only do they commonly contain a variety of isomorphous chemical replacements, but they also may contain small but not insignificant admixtures of other components. Furthermore, clay minerals nominally of the same group may vary appreciably in their degree of structural order or disorder. Thus clay minerals are materials showing variable structural and chemical characteristics such that their thermal and chemical behaviour is likely to vary from one sample to another. In making these general statements, we exclude the mixed-layer systems which have additional problems arising from the statistics of the layer sequence. Furthermore we have not considered the behaviour of grossly mixed components which form the basis of much of the ceramic industry.

The chapter has distinguished five main categories of reaction. (1) Dehydration reactions which occur mainly at low temperatures in a range roughly up to about 200 °C. (2) Dehydroxylation reactions which occur mainly in the temperature range 400–750 °C. (3) Recrystallization reactions which occur mainly above 900 or 1000 °C. Sharp lines of distinction cannot be drawn between these main types of thermal reactions. Whether they are largely distinct or overlap depends on the particular minerals considered, not only the broad mineral group but also the detailed chemical and physical characteristics of the samples studied.

Two further categories are considered. (4) Oxidation reactions and (5) Reduction reactions. Attention is directed principally to the role of iron in mineral structures, but is not restricted to this element. It seems certain that in the future increasing attention will be given to these reactions as they occur not only in the laboratory under well-chosen conditions, but also in nature under more complex and often less well-defined conditions.

Although much factual information has been condensed in this chapter, it has been necessary to be selective in dealing with the history of the subject. It can be anticipated that many further developments will occur in the future.

NOTE

1. The classification of "Cheto-type" montmorillonite is based on their behaviour when

heated. Grim and Kulbicki (1961) included an Otay and a Tatatilla montmorillonite in the "Cheto" group. From the classification proposed by Schultz (1969), summarized in Chapter 1, the layer charge of Otay and Tatatilla types varies between 0.4 and 0.55 per O_{11}.

REFERENCES

ACHAR B.N.N., BRINDLEY G.W. & SHARP J.H. (1966) Kinetics and mechanisms of dehydroxylation processes, III. Applications and limitations of dynamic methods. *Proc. Int. Clay Conf. 1966*, Jerusalem, Vol. 1, pp. 67–73. Israel Program for Scientific Translations, Jerusalem.

ADDISON C.C., ADDISON W.E., NEAL G.H. & SHARP J.H. (1962a) Amphiboles, Part I. Oxidation of crocidolite. *J. Chem. Soc.* 1468–1471.

ADDISON W.E., NEAL G.H. & SHARP J.H. (1962b) Amphiboles, Part II. Kinetics of oxidation of crocidolite. *J. Chem. Soc.* 1472–1475.

ADDISON W.E. & SHARP J.H. (1962a) Amphiboles, Part III. Reduction of crocidolite. *J. Chem. Soc.* 3693–3698.

ADDISON W.E. & SHARP J.H. (1962b) Mechanism for the oxidation of ferrous iron in hydroxylated silicates. *Clay Miner. Bull.* **5**, 73–79.

ADDISON W.E. & SHARP J.H. (1968) Redox behavior of amosite. *Mineralog. Soc. I.M.A. Volume*, I.M.A. Meeting 1966, 305–311.

BALL M.C. & TAYLOR H.F.W. (1961) Dehydration of brucite. *Mineralog. Mag.* **32**, 754–766.

BALL M.C. & TAYLOR H.F.W. (1963) Dehydration of chrysotile in air and under hydrothermal conditions. *Mineralog. Mag.* **33**, 467–482.

BARNHISEL R.I. (1977) Chlorites and hydroxy interlayered vermiculite and smectites. Ch. 10 in *Minerals in Soil Environments* (J.B. Dixon and S.B. Weed, eds), pp. 331–356. Soil Science Society of America, Madison, Wisconsin, USA.

BARSHAD I. (1948) Vermiculite and its relation to biotite. *Am. Miner.* **33**, 655–678.

BARSHAD I. & KISHK F. (1968) Oxidation of ferrous iron in vermiculite and biotite after fixation and replaceability of potassium. *Science* **162**, 1401–1402.

BASSETT, W.A. (1960) Role of hydroxyl orientation in mica alteration. *Bull Geol. Soc. Am.* **71**, 449–456.

BRADLEY W.F. & GRIM R.E. (1951) High temperature thermal effects of clay and related materials. *Am. Miner.* **36**, 182–201.

BRETT N.H., MACKENZIE K.J.P. & SHARP J.H. (1970) Thermal decomposition of hydrous layer silicates and their related hydroxides. *Quart. Rev.* **24**, 185–207.

BRINDLEY G.W. (1961) Role of crystal structure in the dehydration reactions of some layer-type minerals. *J. Miner. Soc., Japan.* **5**, 217–237.

BRINDLEY G.W. (1963) Crystallographic aspects of some decomposition and recrystallization reactions. *Prog. Ceram. Sci.* **3**, 3–55.

BRINDLEY G.W. (1975) Thermal transformations of clays and layer silicates. *Proc. Int. Clay Conf. 1975, Mexico City*, pp. 119–129. Applied Publishing Ltd, Wilmette, Illinois, USA.

BRINDLEY G.W. (1982) Chemical compositions of berthierines – A review. *Clays Clay Miner.* **30**, 153–155.

BRINDLEY G.W., ACHAR B.N.N. & SHARP J.H. (1967b) Kinetics and Mechanisms of dehydroxylation processes. II. Temperature and vapor pressure dependence of dehydroxylation of serpentine. *Am. Miner.* **52**, 1697–1705.

BRINDLEY G.W. & ALI S.Z. (1950) X-ray study of thermal transformations in some magnesian chlorite minerals. *Acta Crystallogr.* **3**, 25–30.

BRINDLEY G.W., BISH D.L. & HSIEN-MING WAN (1979) Composition, structures and properties of nickel-containing minerals in the kerolite–pimelite series. *Am. Miner.* **64**, 615–625.

BRINDLEY G.W. & CHANG T.S. (1974) Development of long basal spacing in chlorites by thermal treatments. *Am. Miner.* **59**, 152–158.

BRINDLEY G.W. & ERTEM G. (1971) Preparation and solvation properties of some variable charge montmorillonites. *Clays Clay Miner.* **19**, 399–404.

BRINDLEY G.W. & GIBBON D.L. (1968) Kaolinite layer structure: Relaxation by dehydroxylation. *Science* **162**, 1390–1391.

BRINDLEY G.W. & HAYAMI R. (1963a,b) Kinetics and mechanism of dehydration and recrystallization of serpentine. Part I, II. *Clays Clay Miner.* **12**, 35–54.

BRINDLEY G.W. & HAYAMI R. (1965) Mechanism of formation of forsterite and enstatite from serpentine. *Mineralog. Mag.* **35**, 189–195.

BRINDLEY G.W. & HUNTER K. (1955) Thermal reactions of nacrite and the formation of metakaolin, γ-alumina and mullite. *Mineralog. Mag.* **30**, 574–584.

BRINDLEY G.W. & CHUH-CHIN KAO (1980) Formation, compositions and properties of hydroxy-Al- and hydroxy-Mg-montmorillonite. *Clays Clay Miner.* **28**, 435–443.

BRINDLEY G.W. & MARONEY D.M. (1960) High-temperature reactions of clay mineral mixtures and their ceramic properties. II. *J. Am. Ceram. Soc.* **43**, 511–516.
BRINDLEY G.W. & MCKINSTRY H.A. (1961) The kaolinite–mullite reaction series: IV. The coordination of aluminum. *J. Am. Ceram. Soc.* **44**, 506–507.
BRINDLEY G.W. & MILLHOLLEN G.L. (1966) Chemisorption of water at high temperatures on kaolinite; effect on dehydroxylation. *Science* **152**, 1385–1386.
BRINDLEY G.W. & NAKAHIRA M. (1957) Role of water vapor in the dehydroxylation of clay minerals. *Clay Miner. Bull.* **3**, 114–119.
BRINDLEY G.W. & NAKAHIRA M. (1959) The kaolinite–mullite reaction series: I, Survey of outstanding problems. II, Metakaolin. III, The high-temperature phases. *J. Am. Ceram. Soc.* **42**, 311–324.
BRINDLEY G.W. & PORTER A.R.D. (1978) Occurrence of dickite in Jamaica – Ordered and disordered varieties. *Am. Miner.* **63**, 554–562.
BRINDLEY G.W., SHARP J.H., PATTERSON J.H. & ACHAR B.N.N. (1967a) Kinetics and mechanism of dehydroxylation processes, I. Temperature and vapor pressure dependence of dehydroxylation of kaolinite. *Am. Miner.* **52**, 201–211.
BRINDLEY G.W. & HSIEN-MING WAN (1975) Composition, structures and thermal behavior of nickel-containing minerals in the lizardite–nepouite series. *Am. Miner.* **60**, 863–871.
BRINDLEY G.W. & HSIEN-MING WAN (1978) The 14 Å phase developed in heated dickites. *Clay Minerals* **13**, 17–24.
BRINDLEY G.W. & WARDLE R. (1970) Monoclinic and triclinic forms of pyrophyllite and pyrophyllite anhydride. *Am. Miner.* **55**, 1259–1272.
BRINDLEY G.W. & YAMANAKA S. (1979) A study of hydroxy-chromium montmorillonites and the form of the hydroxy-chromium polymers. *Am. Miner.* **64**, 830–835.
BRINDLEY G.W. & YOUELL R.F. (1953) Ferrous chamosite and ferric chamosite. *Mineralog. Mag.* **30**, 57–70.
BRINDLEY G.W. & ZUSSMAN J. (1957) Thermal transformation of serpentine minerals to forsterite. *Am. Miner.* **42**, 461–474.
BROWN G. & NORRISH K. (1952) Hydrous micas. *Mineralog. Mag.* **29**, 929–932.
BROWN M.J. & GREGG S.J. (1952) A study of the effect of heat on kaolinite by adsorption methods. *Clay Miner. Bull.* **2**, 228–235.
BULENS M. & DELMON B. (1977a) The exothermic reaction of metakaolinite in the presence of mineralizers. Influence of crystallinity. *Clays Clay Miner.* **25**, 271–277.
BULENS M. & DELMON B. (1977b) Kinetic control of the formation of high-temperature phase in the kaolinite–mullite sequence. *Bull. Soc. Chim., Belges* **86**, 405–411.
BULENS M., LEONARD A.J. & DELMON B. (1978) Spectroscopic investigation of the kaolinite–mullite reaction sequence. *J. Am. Ceram. Soc.* **61**, 81–84.
CAILLÈRE S. & HÉNIN (1961) "Sepiolite", Ch. VIII, "Palygorskite", Ch. IX in *X-ray Identification and Crystal Structures of Clay Minerals* (G. Brown, ed.). Mineralogical Society, London.
CALVET R. & PROST R. (1971) Cation migration into empty octahedral sites and surface properties of clays. *Clays Clay Miner.* **19**, 175–186.
CANT N.W. & HALL W.K. (1968) Studies of the hydrogen held by solids. Part 13. Effect of temperature on hydroxyl band intensities of some solid oxides. *Trans. Faraday Soc.* **64**, 1093–1101.
CHAKRABORTY A.K. & GHOSH D.K. (1978) Re-examination of the kaolinite-to-mullite reaction series. *J. Am. Ceram. Soc.* **61**, 170–173.
CHAMBERLIN R.T. (1908) The gases in rocks. Publn. No. 106, Carnegie Inst. of Washington.
COATS A.W. & REDFERN J.P. (1963) Thermogravimetric analysis (with 322 references). *Analyst* **88**, 906–924. See also *Nature*, **201**, 68–69 (1964).
COMEFORO J.E., FISCHER R.B. & BRADLEY W.F. (1948) Mullitization of kaolinite. *J. Am. Ceram. Soc.* **31**, 254–259.
COMER J.J. (1960) Electron microscope studies of mullite development in fired kaolinites. *J. Am. Ceram. Soc.* **43**, 378–384.
COMER J.J. (1961) New electron optical data on the kaolin–mullite transformation. *J. Am. Ceram. Soc. Soc.* **44**, 561–563.
DENT L.S., GLASSER F.P. & TAYLOR H.F.W. (1962) Topotactic reactions in inorganic oxycompounds. *Quart. Rev.* **16**, 343–360.
DE SOUZA SANTOS H. & YADA K. (1979) Thermal transformation of chrysotile studied by high resolution electron microscopy. *Clays Clay Miner.* **27**, 161–174.
DUNCAN J.F., MACKENZIE K.J.D. & FOSTER P.K. (1969) Kinetics and mechanism of high-temperature reactions of kaolinite minerals. *J. Am. Ceram. Soc.* **52**, 74–77.
DUVIGNEAUD P.H. (1974) Existence of mullite without silica. *J. Am. Ceram. Soc.* **57**, 224.
EARLEY J.W., MILNE I.H. & MCVEAGH W.J. (1953a) Thermal dehydration and X-ray studies on montmorillonites. *Am. Miner.* **38**, 770–783.
EARLEY, J.W., OSTHAUS B.B. & MILNE I.H. (1953b) Purification and properties of montmorillonite. *Am. Miner.* **38**, 707–724.
EBERHART J.P. (1963) Transformation du mica en muscovite par chauffage entre 700 et 1200 C. *Bull. Soc. franç. Minér. Cristallogr.* **86**, 213–251.
ERTEM G. (1972) Irreversible collapse of montmorillonite. *Clays Clay Miner.* **20**, 199–205.
ESCOUBES M. & KARCHOUD M.M. (1977) Contribution à l'étude du comportement des ions fer au cours de la déshydroxylation des minéraux argileux. *Bull. Soc. Fr. Céram.* **114**, 43–55.

FARMER V.C. & RUSSELL J.D. (1967) Infrared absorption spectrometry in clay studies. *Clays Clay Miner.* **15**, 121–142.
FARMER V.C., RUSSELL J.D., MCHARDY W.J., NEWMAN A.C.D., AHLRICHS J.L. & RIMSAITE J.Y.H. (1971) Evidence for loss of protons and octahedral iron from oxidized biotites and vermiculites. *Mineralog. Mag.* **38**, 121–137.
FOSTER M.D. (1963) Interpretation of the composition of vermiculites and hydrobiotites. *Clays Clay Miner.* **10**, 70–89.
FREEMAN E.S. & CARROLL B. (1958) Application of thermoanalytical techniques to reaction kinetics. *J. Phys. Chem., Ithaca* **62**, 394–397.
FREUND F. (1967) Kaolinit-Metakaolinit, Modellfall eins Festkörpers mit extrem hohen Störstellenkonzentrationen. *Ber. deut. Keram. Ges.* **44**, 5–13.
FREUND F. (1969) Dehydroxylation mechanism of clay minerals. The initial conversion of hydroxyl groups into water molecules by proton tunnelling. *Proc. Int. Clay Conf. 1969, Tokyo*, Vol. 1, pp. 121–128. Israel Universities Press, Jerusalem.
FRIPIAT J.J., CHAUSSIDON J. & JELLI A. (1971) *Chimie-physique des phénomènes de surface. Applications aux oxides et aux silicates.* Masson et Cie. Ed. Paris.
FRIPIAT J.J., ROUXHET P.G. & JACOBS H. (1965) Proton delocalization in micas. *Am. Miner.* **50**, 1937–1958.
FRIPIAT J.J., ROUXHET P.G., JACOBS H. & JELLI A. (1968) La délocalisation des protons dans les solides inorganiques. *Bull. Grpe franç. Argiles* **19**, 87–95.
FRIPIAT J.J., ROUXHET P.G., JACOBS H. & JELLI A. (1969) Proton delocalization in inorganic solids. In *Molecular Dynamics and Structure of Solids* (R.S. Carter and J.J. Rusk, eds). Special Publn. 301, National Bureau of Standards, pp. 227–231. Washington.
FRIPIAT J.J. & TOUSSAINT F. (1960) Predehydroxylation state of kaolinite. *Nature, Lond.* **186**, 627–628.
FRIPIAT J.J. & TOUSSAINT F. (1963) Dehydroxylation of kaolinite. II. Conductometric and infrared spectroscopy. *J. Phys. Chem., Ithaca* **67**, 30–36.
GAINES G.L. & VEDDER W. (1964) Dehydroxylation of muscovite. *Nature, Lond.* **201**, 495.
GALLAGHER K.J. (1965) Effect of particle size distribution on the kinetics of diffusion reactions in powders. In *Reactivity of Solids* (5th Int. Symp., Munich, 1964), pp. 192–203. Elsevier Publishing Co., Amsterdam.
GASTUCHE M.C., TOUSSAINT F., FRIPIAT J.J., TOUILLEAUX R. & VAN MEERSCHE M. (1963) Study of intermediate stages in the kaolin–metakaolin transformation. *Clay Miner. Bull.* **5**, 227–236.
GAUDETTE H.E., EADES J.L. & GRIM R.E. (1966) The nature of illite. *Clays Clay Miner.* **13**, 33–48.
VON GEHLEN K. (1962a) Orientierte Umwandlung von Talk in Protoenstatit in gebrannten Presskörpern von Speckstein. *Ber. deut. Keram. Ges.* **39**, 155–161.
VON GEHLEN K. (1962b) Die orientierte Bildung von Mullit ans Al-Si-Spinell in der Umwandlungsreihe Kaolinit-Mullit. *Ber. deut. Keram. Ges.* **39**, 315–320.
GHESQUIÈRE C., LEMAITRE J. & HERBILLON A.J. (1982) An investigation of the nature and reducibility of Ni-hydroxymontmorillonites using various methods including temperature-programmed reduction (TPR). *Clay Minerals* **17**, 217–230.
GILKES R.J., YOUNG R.C. & QUIRK J.P. (1972) The oxidation of octahedral iron in biotite. *Clays Clay Miner.* **20**, 303–315.
GLAESER R. & MERING J. (1967) Effet de chauffage sur les montmorillonites saturées de cations de petit rayon. *C.r. hebd. Séanc. Acad. Sci., Paris* **265**, 833–835.
GLASS H.D. (1954) High-temperature phases from kaolinite and halloysite. *Am. Miner.* **39**, 193–207.
GONZÁLEZ GARCÍA (1950) Silicates del grupo de la montmorillonita. *Anal. Eda. Fis. Veg., Madrid* **9**, 149–185.
GREENE-KELLY R. (1953) Irreversible dehydration in montmorillonite. *Clay Miner. Bull.* **2**, 52–56.
GREENE-KELLY R. (1955) Dehydration of the montmorillonite minerals. *Mineralog. Mag.* **30**, 604–615.
GRIM R.E. (1968) *Clay Mineralogy* (2nd edn). McGraw-Hill, New York.
GRIM R.E. & BRADLEY W.F. (1948) Rehydration and dehydration of the clay minerals. *Am. Miner.* **33**, 50–59.
GRIM R.E., BRAY R.H. & BRADLEY W.F. (1937) The mica in argillaceous sediments. *Am. Miner.* **22**, 813–829.
GRIM R.E. & KULBICKI G. (1961) Montmorillonite: High temperature reactions and classification. *Am. Miner.* **46**, 1329–1369.
GUPTA G.C. & MALIK W.V. (1969) Transformation of montmorillonite to nickel-chlorite. *Clays Clay Miner.* **17**, 233–239.
HELLER L. (1965) Sorption of glycol and glycerol by preheated monoionic montmorillonite. *Proc. Int. Clay Conf. 1963, Stokholm*, Vol. 2, pp. 105–113. Pergamon Press, London.
HELLER L., FARMER V.C., MACKENZIE R.C., MITCHELL B.D. & TAYLOR H.F.W. (1962) Dehydroxylation and rehydroxylation of trimorphic dioctahedral clay minerals. *Clay Miner Bull.* **5**, 56–72.
HELLER-KALLAI L. & ROZENSON I. (1980) Dehydroxylation of dioctahedral phyllosilicates. *Clays Clay Miner.* **28**, 355–368.
HEY M.H. & BANNISTER F.A. (1948) Thermal decomposition of chrysotile. *Mineralog Mag.* **28**, 333–337.
HILL R.D. (1955) 14 Å spacings in kaolin minerals. *Acta Crystallogr.* **8**, 120.
HODGSON A.A., FREEMAN A.G. & TAYLOR H.F.W. (1965a) Thermal decomposition of crocidolite from Koegas, South Africa. *Mineralog. Mag.* **35**, 5–30.
HODGSON A.A., FREEMAN A.G. & TAYLOR H.F.W. (1965b) Thermal decomposition of amosite. *Mineralog. Mag.* **35**, 445–463.
HOFMANN U. & KLEMEN R. (1950) Verlust der Austauschfähigkeit von Lithiumionen an Bentonit durch Erhitzung. *Z. anorg. Chemie* **262**, 95–99.

HOGG C.S. & MEADS R.E. (1975) A Mössbauer study of thermal decomposition of biotites. *Mineralog. Mag.* **40**, 79–88.
HOLDRIDGE D.A. & VAUGHAN F. (1957) The kaolin minerals, Ch. 4 in *The Differential Thermal Investigation of Clays* (R.C. Mackenzie, ed.). The Mineralogical Society, London.
HOLT J.B., CUTLER I.B. & WADSWORTH M.E. (1958) Rate of thermal dehydration of muscovite. *J. Am. Ceram. Soc.* **41**, 242–246.
HOLT J.B., CUTLER I.B. & WADSWORTH M.E. (1964) Kinetics of the thermal dehydration of hydrous silicates. *Clays Clay Miner.* **12**, 55–67.
HOLZNER J. (1938) Eisenchlorit aus dem Lahngebiet; chemische Formel und Valenzausgleich bei den Eisenchloriten. *Neues Jb. Min., Abt. A, Beil.-Bd.* **73**, 389–418.
HORVÁTH I. & GÁLIKOVÁ L. (1979) Mechanism of the $H_2O(g)$ release during a dehydroxylation of montmorillonite. *Chem. Zvesti* **33**, 604–611.
HYSLOP J.F. & ROOKSBY H.P. (1928) Further notes on crystalline break-up of kaolin. *Trans. Ceram. Soc., England* **27**, 299–302.
ISMAIL F.T. (1969) Role of ferrous iron oxidation in micaceous minerals and the type of clay minerals formed in soils of arid and humid regions. *Am. Miner.* **54**, 1460–1466.
ISMAIL F.T. (1970) Oxidation–reduction mechanism of octahedral iron in mica-type structures. *Soil Sci.* **110**, 167–171.
IWAI S. & SHIMAMUNE T. (1975a) Thermal dehydroxylation process of dickite. In *Contribution to Clay Mineralogy dedicated to Professor Toshio Sudo* (K. Henmi, ed.), pp. 26–29. Tokyo University of Education, Bunkyo-Ku, Tokyo.
IWAI S. & SHIMAMUNE T. (1975b) X-ray study of the metakaolin state of dickite. In *Contribution to Clay Mineralogy dedicated to Professor Toshio Sudo* (K. Henmi, ed.), pp. 30–33. Tokyo University of Education, Bunkyo-Ku, Tokyo.
IWAI S., TAGAI H. & SHIMAMUNE T. (1971) Untersuchung des Vorgangs der Structürveränderung des Dickits beim Entwässerung. *Acta Crystallogr. (B)* **27**, 248–250.
JOHNS W.D. (1953) High-temperature phase changes in kaolinites. *Mineralog. Mag.* **30**, 186–198.
JOHNSON H.B. & KESSLER F. (1969) Kaolin dehydroxylation kinetics. *J. Am. Ceram. Soc.* **52**, 199–204.
JUO A.S.R. & WHITE J.L. (1969) Orientation of the dipole moment of hydroxyl groups in oxidized and unoxidized biotites. *Science* **165**, 804–805.
KELLER W.D., PICKETT E.E. & REEMAN A.L. (1966) Elevated hydroxyl temperature of the Keokuk geode kaolinite – a possible reference material. *Proc. Int. Clay Conf. 1966, Jerusalem*, Vol. 1, pp. 75–85. Israel Program for Scientific Translations, Jerusalem.
KODAMA H. & BRYDON J.E. (1968) Dehydroxylation of microcrystalline muscovite. *Trans. Faraday Soc.* **64**, 3112–3119.
KOLTERMANN M. (1964) Der thermische Zerfall von Talk. *N. Jb. Miner. Mh.* 97–106.
KOLTERMANN M. & RASCH H. (1964) Die thermische Umwandlung der Serpentinminerale. *Schweiz. Min. Petr. Mitt.* **44**, 499–516.
KULBICKI G. (1958) High-temperature phases in montmorillonites. *Clays Clay Miner.* **5**, 144–158.
KULBICKI G. (1959) High-temperature phases in sepiolite, attapulgite and saponite. *Am. Miner.* **44**, 752–764.
LEMAITRE J., BULENS M. & DELMON B. (1975a) Influence of mineralizers on the 950°C exothermic reaction of metakaolinite. *Proc. Int. Clay Conf. 1975, Mexico City*, pp. 539–544. Applied Publishing Ltd, Wilmette, Illinois.
LEMAITRE J. & GÉRARD P. (1981a) Thermoprogrammed reduction: A new way of characterizing nickel minerals. *Proc. 2nd Eur. Symp. Thermal Anal., Aberdeen*, 525–528.
LEMAITRE J. & GÉRARD P. (1981b) Characterization of hydrous nickel-containing silicates by temperature-programmed reduction. *Bull. Minéralog.* **104**, 655–660.
LEMAITRE J., LÉONARD A.J. & DELMON B. (1975b) The sequence of phases in the 900–1050°C transformation of metakaolinite. *Proc. Int. Clay Conf. 1975, Mexico City*, pp. 545–552. Applied Publishing Ltd, Wilmette, Illinois.
LÉONARD A.J. (1977) Structural analysis of the transition phases in the kaolinite–mullite thermal sequence. *J. Am. Ceram. Soc.* **60**, 37–43.
LINDQVIST B. (1962) Polymorphic phase changes during heating of dioctahedral layer silicates. *Geol. Fören. Stokholm Förhand.* **84**, 224–229.
MACKENZIE R.C. (ed.) (1957) *The Differential Thermal Investigation of Clays.* Mineralogical Society, London.
MACKENZIE K.J.D. (1969a) Infrared kinetic study of high-temperature reactions of synthetic kaolinite. *J. Am. Ceram. Soc.* **52**, 635–637.
MACKENZIE K.J.D. (1969b) The effect of impurities on the formation of mullite from kaolinite-type minerals. I. Effect of exchangeable cations. II. Effect of exchangeable anions. III. Effect of the firing atmosphere. *Trans. Brit. Ceram. Soc.* **68**, 97–109.
MAKUMBI L. & HERBILLON A.J. (1972) Vermiculitisation expérimentale d'une chlorite. *Bull. Grpe franç. Argiles* **24**, 153–164.
MARTIN C.J. (1977) The thermal decomposition of chrysotile. *Mineralog. Mag.* **41**, 453–459.
MARTIN G.A., RENOUPREZ A., DALMAI-IMELIK G. & IMELIK B. (1970) Synthèse du talc et de l'antigorite de nickel, étude de leur décomposition thermique et de leur réduction en vue d'obtenir des catalyseurs de nickel sur silice. *J. Chim. Phys. Physicochim. Biol.* **67**, 1149–1160.
MERING J. & GLAESER R. (1967) Réarrangement structural de la montmorillonite-Li sous l'effet du chauffage. *C.r. hebd. Séanc. Acad. Sci., Paris* **265**, 1153–1156.

MESTDAGH M.M., HERBILLON A.J., RODRIQUE L. & ROUXHET P.G. (1982) Evaluation du rôle du fer structural sur la cristallinité des kaolinites. *Bull. Minéral.* **105**, 457–466.

MITRA G.B. & BHATTACHERJEE S. (1969) X-ray diffraction studies on the transformation of kaolinite into metakaolin. I. Variability of interlayer spacings. *Am. Miner.* **54**, 1409–1418.

MUÑOZ TABOADELA M. & ALEIXANDRE FERRANDIS V. (1957) The mica minerals. Ch. VI in *Differential Thermal Investigation of Clays* (R.C. Mackenzie, ed.). Mineralogical Society, London.

MURPHY W.J. & ROSS R.A. (1977) A comparative study of thermal effects on surface and structural parameters of natural Californian and Quebec chrysotile asbestos up to 700 °C. *Clays Clay Miner.* **25**, 78–89.

MURRAY P. & WHITE J. (1949) Kinetics of the thermal dehydration of clays. *Trans. Brit. Ceram. Soc.*, **48**, 187–200.

MURRAY P. & WHITE J. (1955) Kinetics of the thermal dehydration of clays. *Trans. Brit. Ceram. Soc.* **54**, 137–238.

NAGATA H., SHIMODA S. & SUDO T. (1974) On dehydration of bound water of sepiolite. *Clays Clay Miner.* **22**, 285–293.

NAKAHIRA M. (1954) The thermal transformation of kaolinite and halloysite. *Mineral J. Japan* **1**, 129–139.

NAKAHIRA M. (1964) Electrical resistance measurements of kaolinite and serpentine powders during dehydroxylation. *Clays Clay Miner.* **12**, 29–33.

NAKAHIRA M. & KATO T. (1964) Thermal transformations of pyrophyllite and talc as revealed by X-ray and electron diffraction studies. *Clays Clay Miner.* **12**, 21–27.

NAUMANN A.W. & DRESHER W.H. (1966) Influence of sample texture on chrysotile dehydroxylation. *Am. Miner.* **51**, 1200–1211.

NEWMAN A.C.D. & BROWN G. (1966) Chemical changes during the alteration of micas. *Clay Minerals* **6**, 297–310.

NICOL A.W. (1964) Topotactic transformation of muscovite under mild hydrothermal conditions. *Clays Clay Miner.* **12**, 11–19.

NORRISH K. (1973) Factors in the weathering of mica to vermiculite. *Proc. Int. Clay Conf. 1972, Madrid*, pp. 417–432. Division de Ciencias, Madrid.

ORCEL J. & RENAUD P. (1941) Dégagement d'hydrogène associé au départ de l'eau de constitution des chlorites ferromagnésiennes. *C.r. Acad. Sci., Paris* **212**, 918–921.

PAMPUCH R. (1966) Infrared study of thermal transformations of kaolinite and the structure of metakaolin. *Polka Akad. Nauk, Prace Mineralogiczne* **6**, 53–72.

PAMPUCH R. (1971) Le mécanisme de la déshydroxylation des hydroxydes et des silicates phylliteux. *Bull. Grpe franç. Argiles* **23**, 107–118.

PAMPUCH R., KAWALSKA M. & PTAK W. (1971) Thermal dissociation of hydroxyl groups in 1:1 layer lattice silicates. *Pol. Acad. Nauk, Oddzial Kradowie, Pr. Kom. Ceram., Ceram.* **17**, 63–75.

PERCIVAL H.J., DUNCAN J.F. & FOSTER P.K. (1974) Interpretation of the kaolinite–mullite reaction sequence from infrared absorption spectra. *J. Am. Ceram. Soc.* **57**, 57–61.

PERROTTA A.J. & YOUNG JR. J.E. (1974) Silica-free phases with mullite-type structures. *J. Am. Ceram. Soc.* **57**, 405–407.

PHAM THI HANG & BRINDLEY G.W. (1973) The nature of garnierite. III. Thermal transformations. *Clays Clay Miner.* **21**, 51–57.

PREISINGER A. (1959) X-ray study of the structure of sepiolite. *Clays Clay Miner.* **6**, 61–67.

PREISINGER A. (1963) Sepiolite and related compounds: its stability and application. *Clays Clay Miner.* **10**, 365–371.

RADCZEWSKI O.E. & SCHÄDEL J. (1962) Ultramicrotomschnitte von Kaolin. Ein Betrag zum Metakaolinit-Problem. *Ber. deut. Keram. Ges.* **39**, 48–51.

RAUTUREAU M. & MIFSUD A. (1975). Précisions apportées par les analyses thermiques de la sépiolite et de la palygorskite, sous vide et dans des conditions normales. *C.r. hebd. Séanc. Acad. Sci., Paris* **281(D)**, 1071–1074.

RICH C.I. (1968) Hydroxy interlayers in expansible layer silicates. *Clays Clay Miner.* **16**, 15–30.

RICHARDSON H.M. (1951) Phase changes which occur on heating kaolin clays. Ch. III in *X-ray Identification and Crystal Structures of Clay Minerals* (G.W. Brindley, ed.), pp. 76–85. Mineralogical Society, London.

RIEKE R. & MAUVE L. (1942) Zur Frage des Nachweises der mineralischen Bestandteile der Kaolin. *Ber. deut. Keram. Ges.* **23**, 119–151.

RIMSAITE J. (1970) Structural formulae of oxidized and hydroxyl-deficient micas and decomposition of the hydroxyl group. *Contr. Miner. Petrol.* **25**, 225–240.

ROBERT M. & PEDRO G. (1965) La vermiculitisation expérimentale de la phlogopite. *C.r. hebd. Séanc. Acad. Sci., Paris* **261**, 4147–4150. See also *Bull. Grpe franç. Argiles* **17**, 3–17 (1966).

ROBERTSON S.D., MCNICOL B.D., DE BAAS J.H., KLOET S.C. & JENKINS J.W. (1975) Determination of reducibility and identification of alloying in copper–nickel-on-silica catalysts by temperature-programmed reduction. *J. Catal.* **37**, 424–431.

ROSS G.J. (1975) Experimental alteration of chlorites into vermiculites by chemical oxidation. *Nature, Lond.* **255**, 133–134.

ROSS R.A. & VISHWANATHAN V. (1981) Dehydration reactions of chrysotile asbestos below 500 °C. *Surf. Techn.* **14**, 233–240.

ROUXHET P.G. (1970) Kinetics of dehydroxylation and of OH–OD exchange in macrocrystalline micas. *Am. Miner.* **55**, 841–853.

ROUXHET P.G., GILLARD J.L. & FRIPIAT J.J. (1972) Thermal decomposition of amosite, crocidolite and biotite. *Mineralog. Mag.* **38**, 583–592.

Rouxhet P.G., Samudacheata N., Jacobs H. & Anton O. (1977) Attribution of the OH stretching bands of kaolinite *Clay Minerals* **12**, 171–179.
Rouxhet P.G. & Taylor H.F.W. (1969) Thermal decomposition of sjögrenite and pyroaurite. *Chimia* **23**, 480–485.
Rouxhet P.G., Touillaux R., Mestdagh M. & Fripiat J.J. (1969) New considerations about the dehydroxylation processes of minerals. *Proc. Int. Clay Conf. 1969, Tokyo*, Vol. 1, pp. 109–119. Israel Universities Press, Jerusalem.
Roy R. (1949) Decomposition and resynthesis of the micas. *J. Am. Ceram. Soc.* **32**, 202–209.
Roy R. & Brindley G.W. (1956) Hydrothermal reconstruction of the kaolin minerals. *Clays Clay Miner.* **4**, 125–132.
Roy R. & Romo L.A. (1957) Weathering studies, I. New data on vermiculite. *J. Geol.* **65**, 603–610.
Roy R., Roy D.M. & Francis E.E. (1955) Thermal decomposition of kaolinite and halloysite. *J. Am. Ceram. Soc.* **38**, 198–205.
Roy R. & Weber J.N. (1964) Pressure–temperature relations for the dehydration of metastable serpentine at pressures from 15–20,000 psi. *22nd Sess. Internat. Geol. Congress, New Delhi, India*. See also J.N. Weber and R.T. Greer, Dehydration of serpentine: Heat of reaction and reaction kinetics at $P_{H_2O} = 1$ atm. *Am. Miner.* **50**, 450–464 (1965).
Russell J.D. & Farmer V.C. (1964) Infra-red spectroscopic study of the dehydration of montmorillonite and saponite. *Clay Miner. Bull.* **5**, 443–464.
Sabatier G. (1955) Les transformations du mica muscovite aux environs de 700°C. *Bull. Grpe franç. Argiles* **6**, 35–39.
Scholten J.J.F. & Kiel A.M. (1975) Dehydration and reduction of synthetic garnierite. *J. Mater. Sci.* **10**, 1182–1187.
Schultz L.G. (1969) Lithium and potassium absorption, dehydroxylation temperature, and structural water content of aluminous smectites. *Clays Clay Miner.* **17**, 115–149.
Sestak J., Savata V. & Wendlandt W.W. (1973) The study of heterogeneous processes by thermal analysis. *Thermochim. Acta* **7**, 333–556.
Sharp J.H., Brindley G.W. & Achar B.N.N. (1966) Numerical data for some commonly used solid state reaction equations. *J. Am. Ceram. Soc.* **49**, 379–382.
Stoch L. (1964) Thermal dehydroxylation of minerals of the kaolinite group. *Bull. Acad. Polonaise Sci.* **12**, 173–180.
Stubican V. (1959) Residual hydroxyl groups in the metakaolin range. *Mineralog. Mag.* **32**, 38–52.
Stubican V. & Roy R. (1961) Proton retention in heated 1:1 clays studied by infrared spectroscopy, weight loss and deuterium uptake. *J. Phys. Chem., Ithaca* **65**, 1348–1351.
Sundius N. & Bystrom A.M. (1953) Decomposition products of muscovite at temperatures between 1000°C and 1260°C *Trans. Brit. Ceram. Soc.* **52**, 632–642.
Swift H.E. (1977) Catalytic properties of synthetic layered silicates and aluminosilicates. In *Advanced Materials in Catalysis* (J.J. Burton and R.L. Garten, eds), pp. 203–233. Academic Press, New York.
Taylor H.F.W. (1962) Homogeneous and inhomogeneous mechanisms in the dehydroxylation of minerals. *Clay Miner. Bull.* **5**, 44–55.
Tettenhorst R. (1962) Cation migration in montmorillonites. *Am. Miner.* **47**, 769–773.
Thilo E. & Schünemann H. (1937) Über das Verhalten des Pyrophyllits beim Erhitzen und die Existenz eines "wasserfreien Pyrophillits". *Z. anorg. allg. Chemie* **230**, 321–335.
Toussaint F., Fripiat J.J. & Gastuche M.C. (1963) Dehydroxylation of kaolinite, I. Kinetics. *J. Phys. Chem., Ithaca*, **67**, 26–30.
Tripathi R.P., Chandra Usha, Chandra Ramesh & Lokanathan S. (1978). A Mössbauer study of the effect of heating biotite, phlogopite and vermiculite. *J. Inorg. Nucl. Chem.* **40**, 1293–1298.
Tschermak G. (1890) Die Chloritgruppe. *S. B. Akad. Wiss., Wien, Abt. I* **99**, 174–266.
Tschermak G. (1891) Die Chloritgruppe. *S. B. Akad. Wiss., Wien, Abt. I* **100**, 29–107.
Tsuzuki K. (1961) Mechanism of the 980°C exotherm of kaolin minerals. *J. Earth Sciences, Nagoya University* **9**, 305–344.
Tsuzuki Y. & Nagasawa K. (1969) Transitional stage to the 980°C exotherm of kaolin minerals. *Clay Science, Japan* **3**, 87–102.
Van Scoyoc G.E., Serna C.J. & Ahlrichs J.L. (1979) Structural changes in palygorskite during dehydration and dehydroxylation. *Am. Miner.* **64**, 215–223.
Vedder W. (1964) Correlation between infrared spectrum and chemical composition of mica. *Am. Miner.* **49**, 736–768.
Vedder W. & Wilkins R.W.T. (1969) Dehydroxylation and rehydroxylation, oxidation and reduction of micas. *Am. Miner.* **54**, 482–509.
Walker G.F. (1961) Vermiculite minerals. Ch. VII in *X-ray Identification and Crystal Structures of Clay Minerals* (G. Brown, ed.), 2nd edn, pp. 297–324. Mineralogical Society, London.
Walker G.F. & Cole W.F. (1957) Vermiculite minerals. Ch. VII in *Differential Thermal Investigation of Clays* (R.C. Mackenzie, ed.), pp. 191–206. Mineralogical Society, London.
Ward R. (1975) Kinetics of talc dehydroxylation. *Thermochim. Acta*, **13**, 7–14.
Wardle R. & Brindley G.W. (1972) The crystal structures of pyrophyllite, 1Tc, and of its dehydroxylate. *Am. Miner.* **57**, 732–750.
Weber J.N. & Roy R. (1965) Dehydroxylation of kaolinite, dickite and halloysite: D.T.A. curves under $p(H_2O) = 15$ to 10,000 psi. *J. Am. Ceram. Soc.* **48**, 309–311.
Weiss A. & Hartl (1966) Das elektrische Leitvermögen von Kaolin-Einkristallen während des thermischen Abbaues. *Proc. Int. Clay Conf. 1966, Jerusalem*, Vol. 1, pp. 87–91. Israel Program for Scientific Translations, Jerusalem.

Weiss A., Koch G. & Hofmann U. (1955) Zur Kenntnis von Saponit. *Ber. deut. Keram. Ges.* **22**, 12–17.
Weiss A., Range K.-J. & Russow J. (1970) The Al, Si-spinel phase from kaolinite. Isolation, chemical analysis, orientation and relations to its low-temperature precursors. *Proc. Int. Clay Conf. 1969, Tokyo*, Vol. 2, pp. 34–37. Israel Universities Press, Jerusalem.
Wentworth S.A. (1967) Investigation of fine-grained micas with emphasis on their hydrous character. Ph.D. Thesis. The Pennsylvania State University. See also S.A. Wentworth (1970) Illite. *Clay Science. Japan* **3**, 140–155.
Wey R. & LeDred R. (1968) Influence des ions échangeables sur la transformation biotite–vermiculite. *Bull. Grpe franç. Argiles* **20**, 55–67.
Wey R., LeDred R. & Schoenfelder J. (1966) Transformation d'un mica partiellement vermiculitisé en vermiculite par oxidation du fer(II). *Bull. Grpe franç. Argiles* **17**, 107–114.
White J.L., Laycock A. & Cruz M. (1970) Infrared studies of proton delocalization in kaolinite. *Bull. Grpe franç. Argiles* **22**, 157–165.
Winchell A.N. (1936) Third study of chlorite. *Am. Miner.* **21**, 642–651.
Yamada H. & Kimura S. (1962) Studies on co-precipitates of alumina and silica gels and their transformation at higher temperatures. *J. Ceram. Assoc. Japan (Yogyo Kyokai Shi)* **70**, 65–71.
Yamanaka S. & Brindley G.W. (1978) Hydroxy-nickel interlayering in montmorillonite by titration method. *Clays Clay Miner.* **26**, 21–24.
Yoder H.S. & Eugster H.P. (1955) Synthetic and natural muscovites. *Geochim. Cosmochim. Acta* **8**, 225–280.

Chapter 8

Reactions of Clays with Organic Substances

J. A. RAUSSELL-COLOM and J. M. SERRATOSA

	Page		Page
8.1 INTRODUCTION	371	8.4.4 Hydrogen bonding to surface oxygen and between molecules themselves	395
8.2 COMPLEXES FORMED BY CATION EXCHANGE	373	8.5 INTERCALATION PROCESS IN KAOLINITE	396
8.2.1 Reaction mechanism	375	8.5.1 Interlayer forces in kaolinite	397
8.2.2 Steric hindrance	376	8.5.2 Reaction mechanisms	398
8.2.3 Adsorption excess over CEC	377	8.5.3 Factors affecting the rate	399
8.2.4 Bonding mechanisms	378	8.5.4 Displacement reactions	401
8.2.5 Competitive adsorption	381	8.5.5 Intercalation capacity of kaolins	402
8.3 ADSORPTION ON CLAYS SATURATED WITH ORGANIC CATIONS	381	8.5 CLAY MINERALS AS TEMPLATES FOR ORIENTED REACTIONS	406
8.3.1 Swelling with organic solvents	382	8.7 ADSORPTION ON PALYGORSKITE AND SEPIOLITE	407
8.3.2 Swelling with water	384	8.7.1 Sorption active centres	408
8.4 ADSORPTION ON CLAYS SATURATED WITH INORGANIC CATIONS	385	8.7.2 Clay organic interactions	409
8.4.1 Direct coordination to cations	386	8.7.2.1 Sorption of non-polar compounds	409
8.4.1.1 Non-transition metal ions	388	8.7.2.2 Sorption of polar organic compounds	409
8.4.1.2 Transition metal ions	388		
8.4.2 Coordination via water bridges	391	8.7.3 Selectivity of fibrous clay minerals for adsorption of organic compounds	411
8.4.3 Protonation of organic bases	392		
8.4.3.1 Protonation on H^+- or NH_4^+-saturated clays	392	8.8 ORGANO-MINERAL DERIVATIVES	412
8.4.3.2 Proton transfer from coordinated water molecules	392	NOTE	415
		REFERENCES	415

8.1 INTRODUCTION

The study of complexes formed by clay and organic substances was initiated in the decade 1930–1940 from areas of research of apparently unrelated disciplines, such as colloid science, soil science, sedimentary geology and petroleum technology.

Experiments involving simple organic chemicals and pure bentonite clay soon made clear that the exchangeable inorganic cations could be replaced by organic cations (Gieseking 1939; Hendricks 1941; Grim *et al.* 1947), and also that uncharged polar compounds could enter the interlayer space without cations being released (Bradley 1945; MacEwan 1948; Greene-Kelly 1955).

Developments coincided with the rapidly increasing use of X-ray diffraction techniques in clay minerals research. The most obvious manifestation of the introduction of organic molecules into the interlayer space of clays is a modification of the basal spacing of the mineral; therefore, formation of a complex could be ascertained from X-ray diffraction diagrams, and clues on the probable packing and orientation arrangement of the molecules could be obtained by combining spacing measurements with considerations of molecular geometry.

For some complexes, quantitative measurements of diffraction intensities were made, from which monodimensional electron density maps were derived, thus providing more rigorous information on the orientation adopted by molecules in the interlayer space (Brindley 1956; Bradley et al. 1963; Johns and Sen Gupta 1967).

As a result of these chemical and X-ray studies, mechanisms of interaction between adsorbed molecules and the clay surfaces were postulated. The existence of H-bonds of the type $NH \cdots O$, $OH \cdots O$, and also $CH \cdots O$ was proposed, and the importance of van der Waals' forces in giving stability to the organic complexes was recognized. A distinction was made between complexes with monomolecular or with bimolecular layers, and also between complexes in which molecules lie flat or stand upright on the silicate surfaces (α- and β-type complexes: Aragón et al. 1959). Other complexes were prepared with long chain alkylammonium ions, which found technical applications because of their organophilic properties (Jordan 1949a,b). Complexes with long chain alkylammonium ions show long basal spacings related to the length of the carbon chain, and quantitative relationships were found to exist between the layer charge of the silicate, the shape and cross-sectional area of the molecules and the resulting basal spacings of the complexes formed (Weiss 1963). Incorporation of organic molecules into the interlayers space of uncharged layer silicates was first reported for halloysite (MacEwan 1946), and later for kaolinite (Wada 1961). Evidence of lattice expansion in the 001 direction was conclusive proof of interlayer penetration.

By means of quantitative determinations of the amount adsorbed at a fixed temperature (adsorption isotherms) the ease of replacement of inorganic by organic cations has been estimated, and the number of neutral molecules associated with each exchangeable cation has been determined. The nature of the inorganic interlayer cations, their field strengths and solvation energies, as well as their coordination properties all were found to have a strong influence on the adsorption process (Glaeser 1948, 1954; Gutiérrez-Rios and Rodríguez 1961). When combined with calorimetric, DTA and gas chromatographic techniques, adsorption studies have provided data from which heats of adsorption and other thermodynamic constants have been calculated, giving the energetics of the process and elucidating the nature of the adsorption mechanisms involved (Bissada and Johns 1969; Fenoll and Martin-Vivaldi 1970; Van Assche et al. 1973).

Spectroscopic studies using special techniques of sample preparation have led to major breakthroughs in the investigation of clay organic complexes (Fripiat et al. 1962; Serratosa 1965; Farmer and Mortland 1965; Ledoux and White 1966). Infrared (IR) spectra provide valuable information on molecular orientation in the interlayer space, and on the mechanisms of interaction with the silicate surface, with the interlayer cations, with residual water, and also between the molecules themselves. The method can detect changes in chemical composition resulting from reactions taking place in the interlayer space under the effect of heat, evacuation or changing humidity conditions. In contrast with X-ray diffraction methods, IR spectroscopy deals directly with molecular interactions, and does not require the existence of long-range ordering; when present, the combinations of the two techniques has provided structural information on aspects such as conformational changes of adsorbed species which, otherwise, could not have been accessible (Martin-Rubi et al. 1974). IR studies have contributed to the understanding of the nature of clay–organic association by demonstrating that molecules are protonated in the interlayer space of clays, that hemisalts are formed when neutral organic bases are adsorbed on H-clays in excess over the number of protons available, that molecules may be coordinated to the inorganic saturating cations either directly or through water bridges, that coordination through π electrons results when benzene or other aromatic compounds are adsorbed on clay saturated with Cu^{2+} or Ag^+ ions, and that H-bonds are formed between molecules and the silicate surface in intercalation complexes of kaolinite.

Ultraviolet and visible, nuclear magnetic resonance, electron spin resonance and other spectroscopic techniques are now being used to provide complementary information on types of bonding, on kinetics of the adsorption and on the nature of the reactions taking place at clay surfaces. Recent reviews on the use of these techniques for clay research have been published in specialized monographs by several authors (Stucki and Banwart 1980; Fripiat 1981).

Even techniques such as differential refractometry and optical microscopy have proved to be valuable tools for quantitative determinations of adsorption from liquid media and for measurements of rates of exchange with organic cations from aqueous solutions (Walker 1959; Brindley 1971).

Over the years the study of clay–organic interaction has been extensive, and important generalizations have emerged in the understanding of the reaction mechanisms. Brindley (1971) has written "Perhaps the most important feature common to much of the current work is the realization that clay organic interactions are multivariable reactions involving the silicate layers, the inorganic cations, water and the organic molecules...." Studies continue to be actively pursued, with new incentives arising from the use and misuse of organic herbicides and pesticides, from the property of clay minerals of acting as templates for reactions of oriented polymerization, from the possibility of incorporating clay–organic derivatives into plastic materials for the improvement of their mechanical characteristics, and from the probable role played by clay suspensions in prebiotic times on the synthesis, concentration and storage of organic molecules of biological interest. Excellent reviews have been published by Greenland (1965), Weiss (1969), Mortland (1970), Brindley (1970), Bailey and White (1970), Cloos (1972), Calvet and Chassin (1973), Theng (1974, 1979, 1982), van Olphen (1977) and MacEwan and Wilson (1980).

The present chapter deals mainly with the mechanisms of interaction and with the manner in which organic reactants are arranged or retained on the mineral substrate. Properties resulting from the electric field distribution near clay surfaces and the catalytic influence on the reactivity of adsorbed species are discussed in Chapter 6 and will not be treated here again.

8.2   COMPLEXES FORMED BY CATION EXCHANGE

Organic cations may be adsorbed on clay minerals by replacement of the inorganic metal ions saturating the structural negative charge on the silicate layers. The reaction may be expressed as

$$RH^+ + M^+\text{-clay} \rightleftharpoons RH^+\text{-clay} + M^+$$

R being any organic base capable of protonation.

Organic cations are furnished from aqueous solutions of organic-base salts, such as the hydrochlorides, or else from solutions of the free base in water with subsequent pH adjustment to a suitable value. At a given pH the concentration of cations in solution relative to the concentration of uncharged molecules is dependent on the pK value of the base.[1] If the pH is adjusted to be equal to the pK value, then the ratio of cation to free base is equal to unity, and cation exchange may be accompanied by adsorption of neutral molecules of the same organic species. For cations to be the dominant species in solution, the pH should be at least one or two units lower than the pK. If too acidic, adsorption may be hindered due to competition with H^+ ions or with metal cations released from the silicate lattice by acid attack. Adsorption will

374　REACTIONS OF CLAYS WITH ORGANIC SUBSTANCES

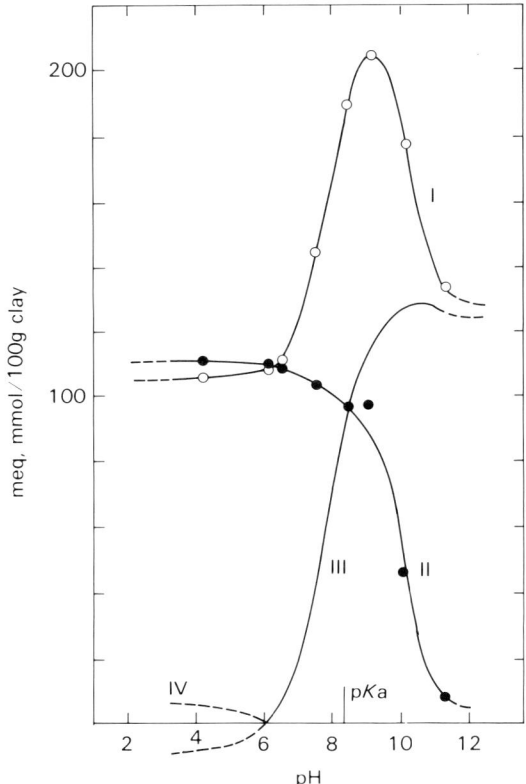

FIG. 8.1. Sorption of morpholine by Ca-montmorillonite, in relation to pH. (After Brindley and Tsunashima 1972.) Curve I = total sorption. Curve II = inorganic cation desorbed = organic cation sorbed. Curve III = Curve I − Curve II = neutral organic molecules sorbed. Curve IV = hydrogen ion sorption. Arrow marks pK_a of morpholine.

depend, too, on the solubility of the base in water which, in turn, may be pH dependent. Figure 8.1 and Table 8.1 illustrate the dependence of adsorption on pK and solubility.

The accessibility of the exchange surface sites varies with the mineral system. Water suspensions of Na-montmorillonite and of certain Li-vermiculites contain the mineral as fully dissociated layers, 10 Å thick, at large distances from each other (> 20–30 Å), this being so even when gelling or aggregation occurs for clay concentrations exceeding certain levels. When organic cations are added to these suspensions, all the clay surface area is accessible to the solution phase, and exchange occurs readily. Most generally, however, clay particles in suspension are domains consisting of stacks of elementary layers with $d(001)$ spacings of less than 20 Å, depending on the hydration properties of the interlayer cation. Exchange then requires that organic cations penetrate first into the interlayer region to displace the inorganic metal ions attached to the internal surfaces. Virtually any organic cation having less than 6 Å as its smallest dimension can be introduced by ion exchange. The larger the size of the cation, the wider is the resulting spacing between adjacent layers in the complex formed.

TABLE 8.1. Physical and chemical properties of organic herbicides and per cent of CEC satisfied by adsorbate, for Na- and H-montmorillonite, at maximum extent of exchange (data from Bailey et al. 1968)

Adsorbate[†]	Chemical name	Water solubility p.p.m. (°C)		pK	(°C)	Maximum amount adsorbed, μmol g^{-1}		Per cent CEC	
						Na-mont.	H-mont.	Na-mont.	H-mont.
Simetone	2,4-bis(ethylamino)-6-methoxy-s-triazine	3200	(20–22)	4.17	(20)	113	511	27.2	139
Atrazine	2-chloro-4-ethylamino-6-isopropylamino-s-triazine	70	(27)	1.68	(22)	3.85	22.7	0.93	6.2
Trietazine	2-chloro-4-diethylamino-6-ethylamino-s-triazine	20	(20–22)	1.88		2.24	6.2	0.54	1.7
Propazine	2-chloro-4,6-bis(isopropyl-amino)-s-triazine	8.6	(20–22)			0.34	2.7	0.08	0.74
Diuron	3-(3,4-dichlorophenyl)-1,1-dimethylurea	42		−1 to −2		1.51	6.8	0.18	0.92
Dicryl	3′,4′-dichloro-2-methyl-acrylanilide	8–9				0.22	0.84	0.03	0.12

[†] Initial concentration of solution approximating maximum solubility.

8.2.1 Reaction mechanism

The exchange reaction of clays in aqueous salt solutions has been conveniently studied using millimetre size single crystals of vermiculite as models. It has been demonstrated that the reaction starts at the edges of the particle, and proceeds towards the centre in a highly regular fashion. When conditions are favourable, that part of the crystal which has reacted is separated from the unaltered part by a well-defined boundary, the movement of which may be easily observed by optical microscopy (Walker 1959). Rate measurements have shown that, to a first approximation, penetration is linearly related to the square root of time of immersion, indicating a diffusion-controlled process.

The kinetics of the process has been studied by several authors (Mifsud et al. 1970; Mackintosh et al. 1971). Both the extent of displacement and the rate of the reaction are affected by a number of variables. The general characteristics of the adsorption process and its variables are discussed below.

Rates of penetration increase as the temperature of the reaction increases. Reaction temperatures of 60–70°C are recommended if they do not interfere with the stability of the organic material.

Chemical and structural characteristics of the clay mineral have a marked influence on the rates of exchange. Total charge on the silicate layers and its location (octahedral or tetrahedral) as well as the size and valence of the saturating inorganic cations determine the strength of the electrostatic attraction between layers and influence the mobility of cations in interlayer space. Attraction between layers is greatest for the higher charged silicates saturated with the larger, less hydrated, monovalent cations, K^+, Rb^+, NH_4^+, Cs^+, leading to contraction of the basal spacing to approximately 9.8 to 10.8 Å. Contrarily, with Na^+, Li^+, Ca^{2+} or Mg^{2+} ions, because of their larger hydration energies, attraction is overcome and expansion to between 15 and 20 Å or even to complete dissociation occurs, so that cations are more mobile and easier to replace. Exchange can be completed after reaction times of the order of *minutes* for the montmorillonites or *hours* for the vermiculites, but it may take several months for completely exchanging the potassium from micas. Of the latter, the dioctahedral muscovite requires considerably longer reaction times than the trioctahedral biotites and phlogopites, because the potassium ions in the latter are less efficiently held by the lattice due to the perturbing effect of the structural OH directed at right angles to the layer (with the proton located immediately above and below the potassium in the structure).

As exchangeable proceeds, the concentration in solution of the extracted metal cations may build up to critical equilibrium levels which prevent further displacement. Critical equilibrium levels depend on the selectivity of the clay for the different metal cations. Montmorillonites are rather insensitive to their presence in the organic solution and can adsorb organic cations even in the presence of brines (Doehler and Young 1962), but, for biotite, 300 p.p.m. of potassium ions in solution prevent the exchange with dodecylammonium ions, and concentrations as low as 3 p.p.m. become critical for lepidolite and muscovite (Mackintosh *et al.* 1971). For vermiculite, rates of exchange with L-ornithine cations are significantly reduced by the presence in solution of 24 p.p.m. of Mg^{2+} or 90 p.p.m. of Sr^{2+}, and exchange ceases at concentrations of 60 p.p.m. of Mg^{2+} or 500 p.p.m. of Sr^{2+} (Mifsud 1975).

The concentration of organic cations in solution may be varied over a wide range with no appreciable effect on the rate. At very high and at very low concentrations, however, the rate is reduced. If too low, diffusion of cations in the external solution phase towards the clay particles may become rate determining. At very high concentrations osmotic forces may reduce the interlayer separation in the crystallites, thus making diffusion more difficult. In addition, considerable amounts of anions enter the interlayers, with a corresponding excess of cations over the exchange capacity.

For crystallites of a given size and shape, completion times are related to the square of the minimum lateral dimension of the particles. The effect is illustrated in Fig. 8.2, showing the influence of particle size on the rate and extent of substitution of potassium by dodecylammonium ions in biotite (Mackintosh *et al.* 1971). From the data of Fig. 8.2, it is also apparent that, for the smallest particles, the exchange reaction, although initially more rapid, does not proceed to total potassium depletion. It is an established experimental fact that the amount of interlayer potassium that can be replaced from micas decreases as the particle size is reduced to the submicron level. Only the mica minerals are known to exhibit this behaviour. The effect has been related to distortions taking place in the mica layers as potassium is replaced from near the edges of the particles (Bassett 1959; Scott, 1968; Reichenbach and Rich 1969), but the precise nature of the changes of configuration of the electrical field that impart stability to the central core is still unknown (Sawhney 1972; Norrish 1973).

8.2.2 Steric hindrance

Restrictions on quantitative exchange can arise when the size of the cation to be introduced is too large in relation to the surface available per exchange position in the silicate layers, i.e. when the projected area of the cation is larger than the equivalent charge on the surface. For energetic reasons cations in interlayer space are mostly arranged in *single layers*; however, if they have asymmetric structures with the charge at one end, the cations may also form *double layers*, each one in contact with one silicate surface (Weiss 1969), or they may stand at high angles to the

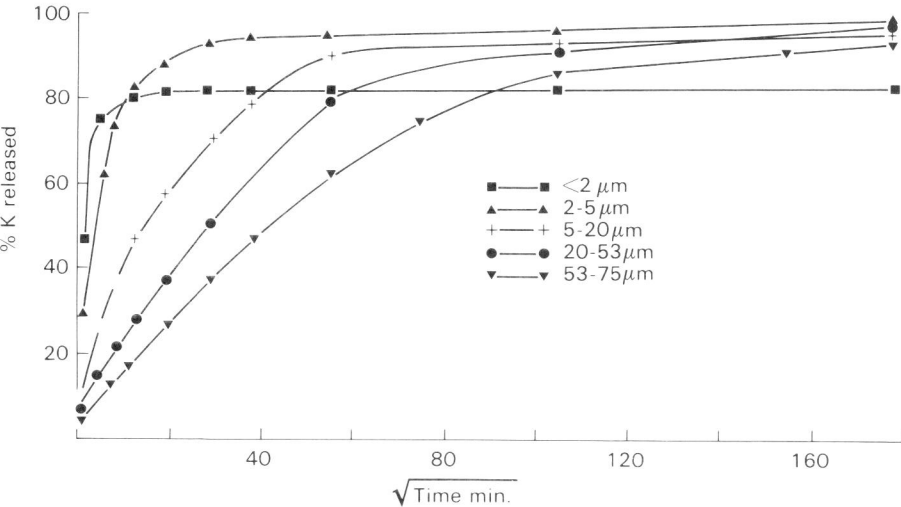

FIG. 8.2. Effect of particle size on the displacement of potassium from Ontario biotite. Samples of 20 mg shaken at 70 °C in 200 ml of 0.025M dodecyl ammonium chloride. (After Mackintosh et al. 1971.)

basal plane. In either case exchange will not proceed beyond complete surface coverage, so that, if cations are too large, the area available may become insufficient to accommodate all the cations (Weiss and Kantner 1960; Weiss 1963). The relationship existing between surface charge and extent of replacement with an organic cation of a given size is illustrated in Fig. 8.3 (Weiss 1963). There, amounts of exchangeable cations (meq/100 g clay) are plotted against the corresponding equivalent surface areas, i.e. area available on the silicate layer per monovalent cation (Å^2/unit charge). In the absence of steric hindrance the amount of organic cations fixed should increase hyperbolically as the equivalent area decreases (dashed line). If the organic cation occupies an area of 50 Å^2, replacement should drop below the exchange capacity for minerals with an equivalent surface area smaller than that value, unless an arrangement in a bimolecular layer is possible. Experimental points represent the extent of exchange for the alkaloid codeine (46–50 Å^2 per molecule) on clay minerals of increasing charge, and the agreement between the calculated line and the experimental data is quite satisfactory.

8.2.3 Adsorption in excess over CEC

Research on adsorption of alkylammonium compounds by clays has shown that these may be taken up in amounts exceeding the exchange capacity of the mineral (Cowan and White 1958; Weiss 1963; Theng et al. 1967; Johns and Sen Gupta 1967). Electroneutrality is maintained by simultaneous adsorption of anions. Large cations containing ten C atoms or more in the aliphatic chain are adsorbed from fairly concentrated aqueous solutions in great excess over the exchange capacity, irrespective of whether or not all of the saturating inorganic cations are replaced. Table 8.2 illustrates the effect of molecular size and concentration on cation and anion uptake and on Na$^+$ replacement for the adsorption of n-alkylpyridinium bromides by Na-montmorillonite (Greenland and Quirk 1962). For the smaller cations, the excess salt accumulated in interlayer space may be easily desorbed by washing with water or alcohol–water

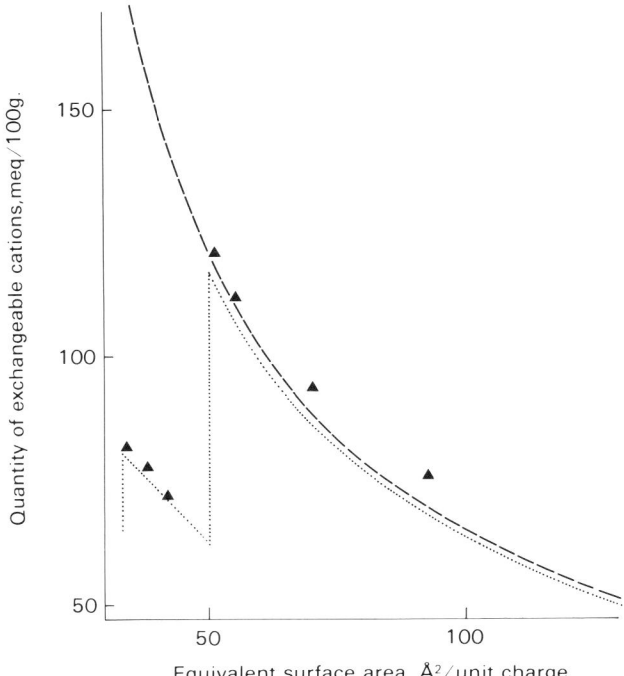

FIG. 8.3. Relationship between cation exchange capacity and the magnitude of the equivalent surface area in montmorillonite and vermiculites. (After Weiss 1963.) ······ Calculated exchange capacity for a cation with an area of 50 Å². ---- Calculated exchange capacity without steric hindrance. ▲ Exchange capacity determined for codeine cations.

mixtures (Garrett and Walker 1962), but for large cations the excess adsorbed will resist washing (Furukawa and Brindley 1973).

8.2.4 Bonding mechanisms

The linkage between organic cations and the charged silicate surface is fundamentally electrostatic but physical, non-coulombic forces, contribute to adsorption. van der Waals' attraction between the aliphatic residues and the surface, as well as between adjacent molecules, adds to the adsorption forces and becomes progressively significant as the molecular weight increases. The study of adsorption isotherms shows that, for large cations, van der Waals' forces dominate the adsorption process, the principal interaction being between the adsorbed molecules themselves rather than between molecules and the clay surface (Greenland and Quirk 1962; McAtee and Hackman 1964; Theng 1964). Cetylpyridinium cations become so closely packed that from the knowledge of the amount adsorbed and of the surface covered per cation, specific surface area of clay minerals may be determined (Greenland and Quirk 1964).

Within the sequence methyl- to butyl-ammonium ions, Na-montmorillonite shows also greater affinity for the larger cations, but here the study of the corresponding adsorption isotherms has demonstrated that the exchange reaction is ruled mainly by entropy effects, and that van der Waals' interaction is irrelevant. Consequently, it has been concluded that the

TABLE 8.2. Adsorption of 1-n-alkyl-pyridinium bromides by Na-montmorillonite (Greenland and Quirk 1962)

Pyridinium compound	Equilibrium concentration		Pyridinium adsorbed (meq/100 g of air-dry clay)[†]	Bromide adsorbed (meq/100 g of air-dry clay)	Na$^+$ replaced from clay (meq/100 g of air-dry clay)	Final pH of suspension
	(g/100 ml)	(mmol l^{-1})				
Ethyl	0.004	0.213	26.3			
	0.112	5.90	47.8	3.3 ± 1.5	55.0	7.0
	0.165	8.76	75.3	4.9 ± 2.5	59.0	7.0
	0.248	13.1	75.7	11.5 ± 3.0	59.0	6.9
Butyl	0.028	1.25	65.5	2.0 ± 1.0	55.0	6.7
	0.158	7.30	94.9		59.0	5.2
	0.320	14.8	102.0	9.1 ± 4.0	61.0	4.8
	0.400	18.5	83.8	8.3 ± 4.5	61.0	4.65
Octyl	0.050	1.84	78.0			
	0.290	10.3	74.4	5.8 ± 3.0	57.0	6.9
	0.345	12.7	76.9	9.4 ± 4.0	58.0	7.0
	0.415	15.2	65.8	8.1 ± 4.5	58.0	7.0
	0.510	18.7	59.9	8.1 ± 5.0	60.0	6.75
Dodecyl	0.006	0.183	82.6	1.0 ± 1.0	89.0	5.7
	0.055	1.68	106.0	12.5 ± 1.5	87.0	4.3
	0.124	3.78	121.0	25.5 ± 2.0	87.0	3.9
	0.190	5.79	140.0	42.0 ± 2.5	87.0	3.9
	0.220	6.80	151.0	58.0 ± 2.5	91.0	3.7
	0.330	10.1	171.0	75.0 ± 3.0	92.0	3.65
	0.400	12.2	163.0	89.0 ± 3.5		3.6
Cetyl	0.005	0.12	70.0	0.0 ± 0.0	49.0	6.45
	0.006	0.14	107.0	22.0 ± 1.0	61.0	6.25
	0.009	0.22	125.0	37.0 ± 1.5	49.0	6.05
	0.016	0.385	210.0	120.0 ± 2.0	67.0	5.30
	0.145	3.50	264.0		65.0	

[†] Water content of the air-dry clay determined by drying over P$_2$O$_5$ *in vacuo* at 100 C was 12%.

increased affinity for the larger cations is due to their greater disruptive effect on the structure of the hydration shell around the interlayer Na$^+$ ions (Vansant and Uytterhoeven 1972). Earlier reports, indicating that Na$^+$ is more efficiently replaced by organic cations than Ca^{2+} or Mg^{2+}, are consistent with the above interpretation (Garrett and Walker 1962; Doehler and Young 1962; Theng *et al.* 1967).

Fixation is favoured if cations contain radicals capable of interacting by H-bonding with the surface oxygens of the silicate. Since these are arranged in ditrigonal six-membered rings, the presence of groups with trigonal symmetry, i.e. —NH$_3^+$ groups in alkylammonium or guanidinium ions, favours reactions (Weiss *et al.* 1958). When the charge on the silicate is tetrahedrally located the combined effect of electrostatic and H-bond interactions determines the "keying" of the —NH$_3^+$ groups of alkylammonium ions into the surface cavities (Walker 1963, 1967; Johns and Sen Gupta 1967; Serratosa *et al.* 1970; Martin-Rubi *et al.* 1974). The possibility of —CH$_2$— or terminal —CH$_3$ groups being H-bonded to the surface oxygens has also been investigated: IR spectra of montmorillonite complexes with amines (Fripiat *et al.* 1962) and with amino acids and peptides (Laby, 1962) show no conclusive evidence in favour of the existence of such bonds. However, in the case of alkylammonium complexes of vermiculite

IR spectra show a splitting of the symmetric deformation vibration band of —CH_3 at 1380 cm^{-1}, with a new, perturbed, dichroic component appearing at 1395 cm^{-1} (Gonzalez-Carreño et al. 1977): this could be interpreted as indicative of terminal —CH_3 groups having their C_3 axes perpendicular to the surface, and of a weak interaction between these groups and the silicate oxygen atoms.

Hydrogen bonding of —NH_3^+ groups to surface oxygens may be prevented if the base itself contains functional groups acting as electron donors. If they also carry a negative charge, then coulombic and/or H-bond interaction may be preferentially directed to those groups rather than to the surface. This is the case for L-ornithine cations adsorbed on vermiculite, where the presence of a charged carboxyl group causes both —NH_3^+ groups to be located away from the surface and directed towards the carboxyls of neighbouring cations (Rausell-Colom and Fornés 1974). Non-protonated —NH_2 groups from molecules of the same species, adsorbed from solution as neutral molecules or as cations of lower valency, may also act as electron donors. Thus, benzidine is adsorbed from aqueous solutions on Ca- and Na-montmorillonite as a mixture of monovalent and divalent cations at a pH below 3, and even as neutral molecules when the pH is higher. Aniline is also adsorbed as a mixture of monovalent cations and neutral molecules at pH = 3.2. Amounts adsorbed are in excess of the exchange capacity of the clay. In both cases intermolecular association by hydrogen bonds of the type

$$^+H_3N—R—NH_2\cdots\cdots\,^+H_3N—R—NH_3^+$$

and

$$Ph—NH_3^+\cdots\cdots H_2N—Ph \qquad (Ph = phenyl)$$

is assumed to exist between cations and molecules in interlayer space (Furukawa and Brindley 1973). Similarly, ethylenediamine adsorbed from solution on montmorillonite is found to exist in interlayer space as trimers of the form

$$^+H_3N—R—NH_2\ldots\,^+H_3N—R—NH_3^+\ldots H_2N—R—NH_3^+$$

(Cloos and Laura 1972). The role played by intermolecular H-bonding is more strikingly illustrated by the specific co-adsorption of purines and pyrimidines on montmorillonite: thymine and uracil, which are not adsorbed from solution alone, are appreciably adsorbed if solutions contain also adenine or 2,6-diamino purine, which are themselves adsorbed. Co-adsorption is attributed to H-bonding between molecules in solution, rather than to adsorption of the purine cations and subsequent uptake of pyrimidine molecules by the complex (Lailach and Brindley 1969).

Weak organic bases are adsorbed as uncharged molecules from aqueous solution by H-montmorillonite, where they become protonated and form the corresponding cationic complexes. The amount of base adsorbed may exceed the number of protons available, and, if the molecular ratio is approximately 2:1, then two molecules compete for the proton and a cation is formed in interlayer space of the type $(B_2-H)^+$. Such complexes are called hemisalts, and are characterized by the fact that the proton is shared on an equal basis by the two molecules, forming a strong symmetrical hydrogen between them. Urea and various amides are known to form hemisalts with H-montmorillonite (Mortland 1966; Tahoun and Mortland 1966). Spectroscopy evidence indicates that protonation occurs at the C=O groups, and that participation of hydrated protons $(H_3O)^+$ is required in these systems.

Each one of the forces discussed above cooperates to a greater or lesser extent in the fixation of organic bases by ion exchange. The stability of the complex formed will depend on the combined strength of all the operating bonding mechanisms. An assessment of the affinity of the clay for the different organic cations can be obtained by determining the ease with which interlayer cations are displaced by equilibration with solutions containing increasing amounts of organic bases. The binding strength of montmorillonite for alkylammonium compounds has been studied in this manner (Weiss 1963; Tahoun and Mortland 1966; Theng *et al.* 1967) and was found to decrease in the sequence

$$R_3NH^+ > R_2NH_2^+ > R\text{—}NH_3^+$$

The replacement of ethylammonium ions from EA^+-montmorillonite and EA^+-vermiculite by quaternary ammonium ions (tetramethyl- to tetrapropylammonium) has been studied in relation to ion size and layer charge on the silicate clay (McBride and Mortland 1973). Cation preference in exchange was found to be crucially related to interlayer spacing of clay in suspension, this being specially true for the more contracted, high layer-charge clay, where energy is required to expand the lattice so that exchange by large cations of small hydration energies can occur. Thus, on EA^+-vermiculite effectiveness of replacement with TMA, TEA and TPA cations is inversely related to molecular weight, because the larger TPA^+ ions are unable to expand the lattice and replace the small EA^+ ions ($d(001) = 12.7$ Å for EA^+-vermiculite; 14.5 Å being the $d(001)$ spacing required for TPA^+-vermiculite). In contrast, with EA^+-montmorillonite, effectiveness is directly related to molecular weight at low replacement levels (absence of steric hindrance), but higher replacement levels are achieved with smaller ions.

8.2.5 Competitive adsorption

A relationship has been found to exist between surface charge density of the adsorbent clay (montmorillonite or vermiculite) and the effectiveness of exchange with two bipyridylium salts, diquat (1,1'-ethylene-2,2'-bipyridylium dibromide) and paraquat (1,1-dimethyl-4,4'-bipyridylium dichloride). Competitive adsorption of the two cations on montmorillonite and on vermiculite was studied by Weed and Weber (1968, 1969). Both compounds are adsorbed from solution as divalent cations. They adopt a planar disposition in interlayer space, the distance between layers, 3.5 Å, being slightly less than the van der Waals' thickness of the pyridyl nucleus. The separation between positive charges is 7 Å for paraquat, and 3.5 Å for diquat. Thus, diquat can be fitted more efficiently than paraquat on the highly charged surfaces of vermiculite, while for the lower charged surfaces of montmorillonite, closer fitting between charges favours paraquat adsorption. Based on this property, a method for estimating the charge density of clays has been proposed by Philen *et al.* (1970, 1971). Calorimetric measurements by Hayes *et al.* (1973) demonstrate that paraquat is more firmly adsorbed than diquat on montmorillonite, while diquat is more strongly held on vermiculite than paraquat.

8.3 ADSORPTION OF CLAYS SATURATED WITH ORGANIC CATIONS

Once the saturating cation on the clay is an organic cation, neutral molecules of other organic species may be introduced in interlayer space by adsorption from the vapour phase, by immersion of the complex into the pure liquid or by mixing and heating above the melting point if the reactant is a solid. Introduction of the new compound results in interactions with the clay or with the organic cations already present, or with both.

When the molecules introduced are of the same species as the existing cations then hemisalt complexes are generally formed. Thus, pyridine and ethylamine are adsorbed into pyridinium-montmorillonite or EA⁺-montmorillonite, respectively, and the corresponding hemisalts are obtained in interlayer space (Farmer and Mortland 1965, 1966).

On adsorption of molecules of a different species, proton transfer reactions occur in interlayer space if the molecules introduced are bases capable of competing with the cations for the proton. The reaction is of the type

$$AH^+ + B \rightleftharpoons BH^+ + A$$

where AH^+ represents the organic cation already present on the clay exchange sites and B are neutral molecules of the base introduced. Systems in which proton transfer reactions are known to occur are montmorillonite and nontronite complexes with pyridinium⁺, ethylammonium⁺, methylammonium⁺ and urea⁺ equilibrated with pyridine, methylamine and 3-aminotriazole pure liquids (Raman and Mortland 1969). Two factors were found to affect the extent to which the adsorbed molecules were protonated: first, the relative basicities of the two interacting compounds; and second, the relative abundance of reactants and products in interlayer space, so that if there is a large excess of B adsorbed, protonation still occurs even if the cations are stronger bases than the molecules. The nature of the clay substrate had little or no influence on the process.

Asymmetric hydrogen bonding to the organic cations occurs if the adsorbed molecules contain functional groups capable of acting as electron donors. This has been demonstrated for the adsorption on trimethylammonium montmorillonite of dialkylamides and other carbonyl compounds such as aldehydes, phenyl-alkyl-diamides and the herbicide ethyl-N,N-di-n-propyl thiocarbamate (EPTC). Spectroscopic evidence of hydrogen bonding was obtained from the shifts of C—O and N—H stretching frequencies of the amides and the alkylammonium ion. The greater the electronegativity of the amides and of the adsorbed compound, the greater were the frequency shifts observed, i.e. the stronger the H-bonds formed (Doner and Mortland, 1969a). Adsorption of EPTC on pyridinium-montmorillonite was found to result from an identical mechanism (Mortland 1968a). Introduction of guest molecules resulted in variations of the interlayer spacing of no more than about 1 Å, and this only for the larger molecules.

8.3.1 Swelling with organic solvents

When the interlayer organic cations are capable of solvation, van der Waals' interactions are established between the organophilic residues of these and the adsorbed molecules. Accommodation of the adsorbed molecules in interlayer space may affect the orientation of the organic cation, so that swelling of the complex is generally observed. Thus, adsorption of benzene or chlorobenzene in pyridinium-montmorillonite changes the disposition of the pyridinium cation from parallel to normal to the silicate layers, and an increase of $d(001)$ from 12.5 to 15 Å is observed (Farmer and Mortland 1966; Serratosa 1968a).

If interlayer cations are long n-alkylammonium ions and the adsorbed species are polar n-alkyl compounds, then considerably more swelling takes place. Clay organic derivatives of this type are plentiful, and are extensively documented in the literature. Reviews of the subject have been published by Weiss (1963, 1969) and may be summarized as follows. Strong van der Waals' interactions between the alkyl chains of molecules and cations cause them to be densely packed in bimolecular layers between the silicate plates. When the adsorbent clay is montmorillonite then molecules and cations are arranged with their longitudinal axes perpendicular to the plates, whereas in vermiculite they are inclined 54° to the plates. The charged and polar groups are in contact with the surface, probably keyed into the hexagonal cavities of the tetrahedral sheets. In certain vermiculite-n-alkylammonium-n-alkanol complexes, however, IR evidence would suggest that a fraction of the adsorbed alkanol molecules have their —OH ends directed towards the central plane of the structure (Gonzalez-Carreño et al. 1977). When the number of carbon atoms is the same in both alkyl chains, then the interlayer volume becomes fully occupied, with a stoichiometry of two alkyl chains per Si_3AlO_{10} unit and an interlayer separation dependent on chain length but independent of layer charge and the chemical nature of the polar compounds. If the chain length of the polar molecule exceeds that of the organic cation, then lattice expansion is proportional to the total number of —CH_2— contained in the interlayer volume, and is, therefore, inversely related to layer charge. A method for charge determination based on this property has been proposed by Lagaly and Weiss (1969), consisting of measuring the $d(001)$ spacing of the clay derivative with n-dodecylammonium plus n-tetradecanol.

For short alkylammonium complexes, adsorption with lattice expansion is possible only if the alkyl chain of the cation exceeds a critical length that is inversely related to layer charge. Five to six carbon atoms are required for vermiculites, but at least eleven atoms are required for lower charged montmorillonite. Adsorption of polar molecules with branched alkyl chains, or with aromatic or cyclic rings on n-alkylammonium derivatives, also produces significant lattice expansion, but no geometrical or stoichiometric relationship of the kinds discussed above has been established. Expansion will also occur with the di- and tri-alkylammonium derivatives. In either case the molecules cause first a readjustment of the orientation of the alkylammonium chains, and then they adapt themselves to the voids created in interlayer space in the most space-saving manner, so that maximum adsorption takes place.

The structural ordering of bimolecular films in complexes of montmorillonite and vermiculite has been extensively studied (Pfirrmann et al. 1973; Lagaly 1976), and it is now known that in many cases the molecules forming the film are not fully stretched as outlined above but that, because of rotations around C—C bonds, they undergo conformational changes in which the stretched nature of the alkyl chains is essentially retained although slightly modified. Such conformational changes are called kinks and were first proposed to occur as defects and structural elements in crystals of the paraffins and polymers. Phase transitions observed under heat treatment have been explained by the nucleation of kink-block defects in the structure of the interlayer biomolecular film. Such defects may take place cooperatively so that, if a kink is formed on a molecule in a position favourable with respect to a neighbouring molecule, steric effects cause a kink to be formed on its chain, and the reaction continues to the next chain in an ordered fashion. Figure 8.4 illustrates nucleation and growth during kink-block formation.

Polar organic liquids, and also aromatic hydrocarbons, are adsorbed by N-(n-alkyl)-pyridinium beidellite (Lagaly et al. 1973). Molecules become close packed in interlayer space, forming quasi-crystalline structures. The pyridine nuclei of the organic cations remain in

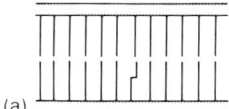
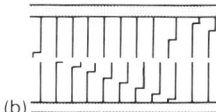

FIG. 8.4. Conformation of alkyl chains in a bimolecular film. (a) All-*trans* block with a kink as an isolated defect. (b) Kink block. (From Lagaly 1976.)

contact with the clay surfaces, and the alkyl residues adapt themselves to the molecules of the liquid by conformational changes around C—C bonds. Basal spacings of the complexes range from 30 to 50 Å, depending on the length of the alkyl chains on the organic cation.

Long chain alkylammonium derivatives of montmomorillonite have the property of dispersing in polar organic liquids, forming thixotropic gel structures with very high liquid contents. The organophilic character of these derivatives was first investigated by Jordan (1949a,b), and was subsequently summarized by Weiss (1963) and by Grim (1968). Development of good organophilic properties requires that more than half of the available clay surface be covered by the hydrocarbon chains of the alkylammonium cations, and that an interlayer separation of about 8 Å, which is the thickness of a flat bimolecular layer, be initiated. For the complexes with single chain primary amines, alkyl chains of 12 carbon atoms or more are adequate to achieve that separation. Maximum swelling occurs, then, with organic liquids such as nitrobenzene or benzonitrile, combining a high polarity and high organophilic characteristics. The complex sorbs first an amount of liquid sufficient to cover the uncoated part of the silicate surface, and then gelation occurs with the remainder of the liquid. With other less polar liquids, or with the non-polar hydrocarbons, swelling is enhanced if small amounts of polar additives, such as alcohols, esters or aldehydes, are dissolved in them; once adsorbed in the complex they will render the surface completely organophilic. Complexes with unsymmetrical di-, tri- or tetra-alkylammonium cations having two long aliphatic chains are more organophilic because a larger proportion of the silicate surface is covered by the alkyl chains. They swell with the less polar liquids and even with pure unsaturated hydrocarbons, but not with the saturated hydrocarbons. It has been claimed that intensive gel formation is always associated with the presence of traces of water (Weiss 1963), as the water molecules favour interlinkage of the montmorillonite particles by hydrogen bridges to form wide-mesh frameworks (edge-to-edge association).

8.3.2 Swelling with water

In inverse relation to the development of organophilic properties, the water-adsorbing properties of montmorillonite become gradually reduced as the mineral surfaces are covered by alkylammonium ions. The water adsorption properties of alkylammonium-substituted montmorillonites have been investigated by Gieseking (1939), Hendricks (1941), Grim et al. (1947) and Jordan (1949a,b), and their data show that the larger the size of the aliphatic chain on the organic ion, the smaller is the water content of the complex in equilibrium with atmospheres of a given relative humidity. One-dimensional structure analyses have been made of complexes of n-alkylammonium ions with single crystals of vermiculite (Johns and Sen Gupta 1967; Martin-Rubi et al. 1974) showing the existence of hydrated phases; both the position found for the water molecules (at 6 Å from the octahedral Mg plane) and the water content of the hydrates (two molecules per structural $Si_6Al_2O_{20}$ unit) indicate the association of the water molecules with the uncovered part of the silicate surface.

Considerably more hydration takes place in complexes with organic cations containing functional groups capable of interacting with water molecules by H-bonding. Mifsud et al. (1970) have reported the existence of hydrated phases in complexes of L-ornithine with single crystals of vermiculite, having crystalline $d(001)$ spacings as high as 42 Å. Between 20.2 and 29.6 Å intermediate crystalline phases were observed at discrete $d(001)$ intervals of 3.1–3.2 Å, all being present simultaneously in the same vermiculite crystal. One-dimensional structure analyses of the most stable hydrates indicate that part of the total water content is associated with the silicate surface, the remainder being in association with the functional groups of the amino acid cations (Rausell-Colom and Fornés 1974).

Complexes of vermiculite single crystals with short chain alkylammonium ions swell in water or in dilute solutions of the alkylammonium salt, forming coherent gel-like structures twenty to thirty times thicker than the original crystal, the water content of these being 10 g H_2O per gram of mineral, or higher (Garrett and Walker 1962). In these, the silicate layers are separated at distances of several hundreds of angstroms (Rausell-Colom 1964). Vermiculite saturated with certain amino acids in cationic form swells in water in the same manner. Swelling occurs, too, when natural Mg-vermiculite crystals are immersed in concentrated amino acid solutions at or near the isoelectric point (Garrett and Walker 1961): in this case, the interlayer Mg^{2+} ions are first replaced by amino acid cations present in solution in equilibrium with the dipolar ions, so that swelling takes place after formation of the corresponding vermiculite–amino acid$^+$ complex (Rausell-Colom and Salvador 1971a). Subsequent research, including swelling pressure measurements, supports the assumption that diffuse ionic "double layers" are formed around the silicate layers, and that repulsion forces originating from "double layer" interaction cause the silicate plates to move apart from each other to the separation found in the gel structure (Rausell-Colom and Salvador 1971b).

8.4 ADSORPTION ON CLAYS SATURATED WITH INORGANIC CATIONS

Since the early work of Bradley (1945) and MacEwan (1948) the adsorption of neutral organic molecules on phyllosilicates has been studied extensively. Neutral molecules penetrate into the interlayer space of clays when the energy released in the adsorption process is sufficient to overcome the attraction between layers.

Possible adsorption sites in the clay structure are, firstly, the exchangeable metal ions with which sorbate molecules may form coordination compounds and, secondly, the surface oxygens of the tetrahedral sheets which may act as proton acceptors for the formation of H-bonds with molecules containing —OH or —NH groups. Both adsorption mechanisms may act simultaneously, but their relative contribution to the adsorption process will depend on the nature of the sorbate molecules and on the kind of exchangeable cations present in the clay.

Earlier studies had attributed a predominant role to the interaction of the sorbate molecules with the silicate surfaces, but after the use of IR spectroscopic techniques became widespread, the relevance of the interactions with the exchangeable cations has been recognized. van der Waals' attraction between molecules and the mineral substrate contributes to the adsorption forces, but its significance is secondary except for organic compounds of large molecular weight.

Organic molecules may be adsorbed from the vapour phase, from the pure liquid, or from solutions in water or in other solvents. In either case, adsorption is influenced by the state of hydration of the clay. Sometimes clay samples are thoroughly dehydrated prior to adsorption, but more frequently they are simply air-dried. When interlayer water is present, the cohesion forces of the clay are greatly reduced and, consequently, penetration of the sorbate molecules is facilitated. Molecules, then, compete with water for coordination sites around the cations and, depending on the relative values of the hydration and solvation energies of these, they will:

(a) replace water and become coordinated to the cations; or
(b) occupy sites in a second sphere of coordination around the cations, being bonded to them through bridging water molecules; or

(c) accept a proton from the water of coordination around the cations, or from the cations themselves if the clay is saturated with H^+ or NH_4^+.

8.4.1 Direct coordination to cations

Metal ions, particularly those of the transition elements, have the property of forming coordination compounds with anions or with neutral molecules capable of donating electrons. The coordination number, i.e. the total number of ligands, depends on the electronic configuration of the cation as well as on the nature of the ligand molecule and on the environment in which the complex is formed.

When molecules that have functional groups with free electron pairs have penetrated in the interlayer space of charged phyllosilicates, the possibility exists that they will form coordination compounds with the exchangeable cations. It should be expected, however, that the structure and properties of the complexes formed will be influenced by the combined effect of the electric field of the silicate layers and of steric restrictions in interlayer space. The interlayer space provides a special environment in which a particular chemistry occurs, and it is possible that complexes not known in solution chemistry may be obtained in this environment.

Direct coordination to the exchangeable cations has been recognized in complexes of montmorillonite with a wide variety of organic compounds, such as alcohols and ketones (Glaeser 1954; Gutiérrez-Ríos and Rodriguez 1961; Dowdy and Mortland 1967, 1968; Parfitt and Mortland 1968; Bissada et al. 1967; Tarasevich et al. 1970; Annabi-Bergaya et al. 1980, 1981), aliphatic and aromatic amines (Farmer and Mortland 1965; Heller and Yariv 1969; Van Assche et al. 1973), nitriles (Gutiérrez Ríos et al. 1962; Rodriguez et al. 1967; Serratosa 1968b; Yamanaka et al. 1971; Tarasevich and Ovcharenko 1973), alkylphosphates (González-García et al. 1970), urea and amides (Mortland 1966; Tahoun and Mortland 1966; Farmer and Ahlrichs 1969), pyridine (Farmer and Mortland 1966; Serratosa 1966), nitrobenzene (Yariv et al. 1966), and with many other compounds used as pesticides (Bailey and White 1970; Cloos 1972). Evidence for coordination is obtained from IR spectra showing the perturbation of characteristic vibrations of the sorbate molecules and the effect of the polarizing power of the cations on the displacement of the characteristic adsorption frequencies (Fig. 8.5).

Interest has recently been shown in the complexes formed by homoionic smectites and vermiculites with macrocyclic organic ligands of the type crown ethers and criptands (Ruiz-Hitzky and Casal 1978; Guinard and Pezerat 1979; Casal 1983) known to be very efficient host molecules for trapping inorganic metal cations, with which they form very stable polydentate coordination compounds.

Crown ethers and cryptands are adsorbed by smectites and vermiculite from methanol solution. Interlayer water is excluded, and the compounds are coordinated directly to the interlayer cations in monolayer or bilayer complexes of the types illustrated in Fig. 8.6. Both the stoichiometry and the interlayer arrangement are governed by the ratio r_c/r_i where r_c is the radius of the macrocyclic cavity and r_i is the radius of the saturating cation. If the ratio is greater than 1, then the cation is occluded into the cavity of the cyclic ligand. Consequently, the number of adsorbed ligand molecules per cation is equal to 1, and one layer ($\Delta d_L = 4$ Å (Fig. 8.6(a))) or two-layer ($\Delta d_L = 8$ Å (Fig. 8.6(c))) complexes result depending on the area of equivalent charge on the silicate surface relative to the projected area of the molecule. When the ratio r_c/r_i is smaller than 1 then each cation is generally sandwiched between two cyclic ligands, and two-layered ($\Delta d_L = 8$ Å (Fig. 8.6(d))) complexes result. However, in the case of K^+- or

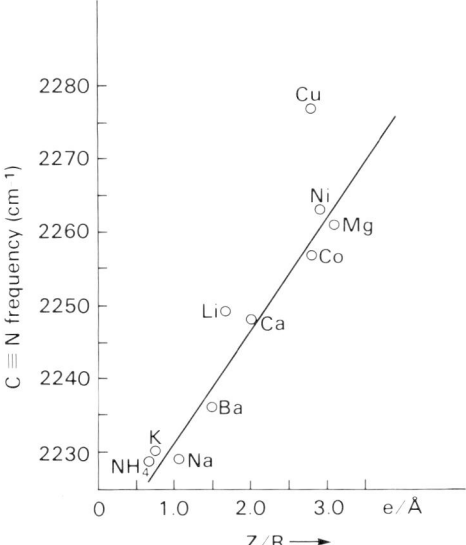

FIG. 8.5. CN stretching frequencies of acrylonitrile adsorbed on montmorillonite as a function of the polarizing power of interlayer cations (after Yamanaka et al. 1971).

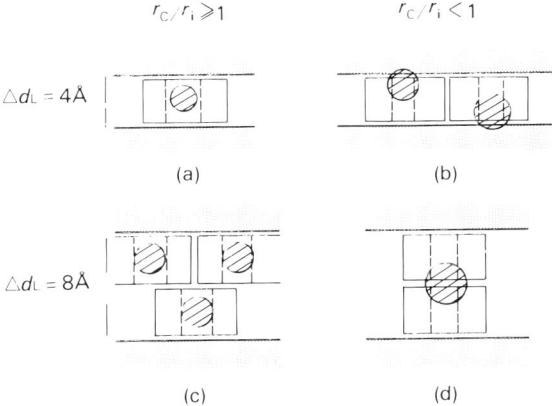

FIG. 8.6. Proposed interlayer arrangement of exchange cations and associated cyclic ligands in complexes with montmorillonite and vermiculite. Ligand molecules parallel to silicate layers. r_c = radius of macrocyclic cavity; r_i = radius of saturating cation; Δd_L = interlayer separation (see text). (After Casal 1983.)

NH_4^+-saturated clays, one-layer complexes may be formed in which the cations are coordinated to the oxygens of the cyclic ligand from one side and to the oxygens of the ditrigonal cavities on the silicate surface from the other (Fig. 8.6(b)). IR spectra from NH_4^+-montmorillonite with 18C6 crown ether indicate a change of the symmetry of NH_4^+ from T_d to C_{3v}, suggesting hydrogen bonding of NH_4^+ to the crown oxygens.

8.4.1.1 Non-transition metal ions

The nature of the complexes by ethanol and acetone with cations of the most highly electropositive, non-transition elements in the interlayer space of montmorillonite has been investigated in detail by Bissada et al. (1967) and by Bissada and Johns (1969). Their results have shown that the interlayer separations are conditioned by the number of molecules adsorbed per exchangeable cation, which, in turn, depend on the interlayer cation present. Thus two molecules exist per each interlayer K^+ and three molecules per each interlayer Na^+, the resulting complexes having, in both cases, basal spacings of 13–14 Å (one-layer complexes). For the divalent cations Ca^{2+} and Ba^{2+}, molecule/cation ratios are 8–10 for complexes with 17 Å basal spacing (two-layer complexes), and 4–5 when the more stable 13–14 Å complexes are formed. Gas–solid chromatographic determinations of heats of adsorption gave bond energies of 53 kJ mol^{-1} for K-montmorillonite complexes, and 125 kJ mol^{-1} for the Ca-montmorillonite complexes. A good agreement was found to exist between these energies and theoretical bond energies calculated from equations based on a simplified electrostatic model where point charges and dipoles were assumed; this led the above authors to the conclusion that the association between molecules and cations is governed mainly by electrostatic cation–dipole interactions. Bruque et al. (1982) have studied the complexes of lanthanide-montmorillonite with amines and have also concluded that true coordination compounds of the ion–dipole type are formed in the interlayer space of the silicate.

8.4.1.2 Transition metal ions

Coordination properties of transition metal ions, with partially filled d or f orbitals, are not sufficiently explained by metal–ligand interaction models as simplified as the above model, where spherical symmetry for the electron density distribution on the cation is assumed. More adequate theories on the coordination bond have been evolved in theoretical chemistry, based on models in which the necessary refinements are incorporated. Thus, crystal field theory considers the splitting of d or f orbitals by the electrostatic field of the ligand molecules, and the resulting gain in bond energy by preferential filling of low-lying orbitals. At the opposite extreme, the molecular orbital theory considers the overlap of electron orbitals of the cation and the ligand molecules, and the formation of the corresponding molecular orbitals in which electrons are delocalized.

These theories account for the stronger bond energies in coordination complexes with the transition metal ions, and for the fact that no straightforward relationship exists between bond energy and ionic charge and radius. They also afford an explanation for ligand exchange reactions where water ligands are replaced by organic molecules.

As referred to above, smectites saturated with transition metal ions form coordination complexes with many organic substances adsorbed from the vapour phase, from the pure liquid or from solutions in inert solvents. At pH values where cation exchange reactions do not occur complexes are also readily formed from aqueous solutions, for example with purines and pyrimidines (Lailach et al. 1968), with amino acids (Bodenheimer and Heller 1967; Sung Do Jang and Condrate 1972; Heller-Kalai et al. 1973), with ethylendiamine (Laura and Cloos 1972; Schoonheydt et al. 1979a,b), with 3-aminotriazole (Russell et al. 1968) and also with fulvic acids (Schnitzer and Kodama 1972). Most of these complexes are described as chelate-type complexes.

The preparation and properties of smectite complexes with coordination compounds of transition metal cations and 2,2'-bipyridine (bp) have been extensively studied in recent years (Schoonheydt et al. 1977; Traynor et al. 1978; Krenske et al. 1980; Abdo et al. 1981; Cruz et al.

1982). The interest resides in the fact that such compounds are used as sensitizers for the photodecomposition of water into H_2 and O_2. When these compounds are adsorbed in the interlayer space of smectites, both their reactivity and photophysical properties are different from the same properties in homogeneous solution; also, these properties depend on the water content of the clay. Thus, the emission quantum yields of $Ru(bp)_3^{2+}$ adsorbed on hectorite can reach values between 2 and 100 times higher than those observed in aqueous solution (Krenske et al. 1980). Similarly, $Ru(bp)_3^{2+}$ adsorbed in hectorite is not oxidized readily by PbO_2 or Ce^{4+}, even though these oxidations do occur in solution (Traynor et al. 1978).

The majority of the studies referred herein deal with the identification, conditions of formation and properties of the respective complexes. Of special relevance are the studies of Pleysier and Cremers (1975), Maes et al. (1976, 1978) and Peigneur et al. (1979) in which free energies of formation have been determined for the complexes formed by polyamines with Cu-, Ni-, Zn-, Cd-, and Hg-montmorillonite and by thiourea with Ag-montmorillonite. Their results show that thermodynamic stability constants for the complexes in interlayer space are two to three orders of magnitude higher than those for the same complexes in solution. Yariv et al. (1964) showed that complexes not formed in solution media, for instance the Fe-pyrocatechol complex cation, can be stabilized in the interlayer space of montmorillonite. Additional examples of complex stabilization will be treated in following paragraphs.

Research in this field is rather scarce, and studies including determination of bond energies, stability constants and other thermodynamic parameters relative to metal ion–ligand interactions in complexes formed in the interlayer space of clays should, in the opinion of the writers, be pursued for they will afford important contributions to the advancement of the chemistry of coordination compounds.

The complexes discussed so far are formed with organic compounds that have basic or polar substituents capable of acting as electron donors for coordination with the exchangeable metal ions. Other complexes have been obtained in recent years with Cu^{2+}- or Ag^+-saturated smectites and aromatic compounds that do not contain those functional groups, e.g. benzene and its aliphatic derivatives. Such complexes are highly coloured, and their IR spectra show upward shifts of C—H out-of-plane vibrations that are indicative of a perturbation of the electronic system of the organic molecules.

Coordination compounds via donation of π electrons are well known in inorganic chemistry. The prototype compound, dibenzene chromium, $(C_6H_6)_2Cr$, is formed by direct interaction of the aromatic hydrocarbon with the metal halide in the presence of a proper reducing agent and halogen acceptor, and of a Friedel–Crafts activator; it is transformed to the cationic form $(C_6H_6)_2Cr^+$ by hydrolysis of the reaction mixture with diluted acid. In view of this, it is to be expected that π coordination compounds may also be formed in the interlayer space of smectites, but the fact remains that some of the compounds so formed have no counterpart in reactions taking place in solution media. More precisely, Cu^{2+} ions form π coordination compounds with arenes in montmorillonite (Doner and Mortland 1969b; Mortland and Pinnavaia 1971), but it has not been shown that they can do so in homogeneous solution. In addition, the clay itself must be a stabilizing factor in preserving the complexes, since aromatic compound–metal complexes commonly decompose when exposed to the air.

The water of hydration must be removed from the Cu^{2+}-montmorillonite in order that the arene–Cu^{2+} complex be formed in the interlayer space. Two kinds of complexes are obtained with benzene, depending on the extent of dehydration. Type I complex (green) is formed simply by exposing the clay to benzene vapour in a desiccator with P_2O_5; benzene is truly coordinated to the cation through the π electron system, and the aromaticity of the molecule is preserved.

Type II complex (red) is formed by further dehydration of the former; for that complex, evidence by IR and ESR spectroscopy indicates that the benzene ring no longer preserves its D_{6h} symmetry, but is part of a radical cation represented tentatively as

$$Cu^+ \langle + \rangle$$

A brown complex is formed from the type I (green) complex after long exposure to atmospheric conditions (Vandepoel et al. 1973). The structure suggested is of type II and conforms also to that of a radical cation,

$$Cu(OH) \langle + \rangle$$

Type II complexes are formed with Cu^{2+}-montmorillonite and thiophene (Cloos et al. 1973) and with biphenyl, naphthalene and anthracene (Rupert 1973). With alkyl substituted benzene only type I complexes are obtained (Pinnavaia and Mortland 1971; Fenn and Mortland 1973). Silver montmorillonite also forms complexes with arenes, but only type I complexes have been identified (Clementz and Mortland 1972).

Arene complexes are also formed with other Cu^{2+}- or Ag^+-saturated smectites. Charge location (tetrahedral or octahedral) has an influence on the formation of these compounds, complexation being easier in octahedrally charged smectites. This results from the fact that water is more firmly bound and more difficult to eliminate in tetrahedrally charged smectites; thus, aromatic molecules will find more difficulty for entering into direct coordination with the cations (Clementz and Mortland 1972). Also, arene complexes are formed more easily with Ag-nontronite than with Cu-nontronite, due to the lower hydration energy of Ag^+ ions.

Arene complexes can also be formed with smectites saturated with cations such as VO^{2+} and Fe^{3+}, since the role of these ions is to function as oxidizing agents for radical cation formation (Pinnavaia et al. 1974).

Formation of π coordinated complexes and the resulting possible disruption of the bond resonance in the benzene ring suggest important new pathways for organic synthesis via the complex (Pinnavaia 1977). The reactions of methyl-phenyl-ether (anisole) and Cu^{2+}-hectorite afford an interesting example of these possibilities. The process has been described in the following manner (Fenn et al. 1973); first, anisole is physically adsorbed on the clay; next, anisole is oxidized and polymerized to 4,4'-dimethoxybiphenyl after ageing for some time at room temperature, polymerization resulting from the formation either of carbonium ion or a radical cation as an intermediate step; finally the oxidation product is stabilized as complexes of types I and II in interlayer space. The proposed scheme is illustrated as follows:

$$Cu^{2+}\text{-hectorite} \rightarrow \text{phys. ads. anisole} \overset{-H_2O}{\rightleftarrows} \text{Type II anisole complex}$$
(blue)

$$H_2O \updownarrow -H_2O$$

Type II 4,4'- ← Type I 4,4'- ← Type I anisole complex
dimethoxybiphenyl   dimethoxybiphenyl   (tan)
complex   complex
(green)   (tan)

8.4.2 Coordination via water bridges

The existence of association via water bridges was first demonstrated for pyridine adsorbed on montmorillonite (Farmer and Mortland 1966), and has been shown to occur in many other systems since then. Evidence for this type of association is obtained through the displacements from their normal position of the IR adsorption bands of both the water and the organic molecules, as a consequence of H-bond formation.

Complexes have been identified in which one or both protons of each water molecule participate in H-bonding so that, if R represents the organic species, water bridges are of the types

$$\text{clay} -- M^{n+} -- O \begin{smallmatrix} H \\ \\ H--R \end{smallmatrix} \quad \text{or} \quad \text{clay} -- M^{n+} -- O \begin{smallmatrix} H--R \\ \\ H--R \end{smallmatrix}$$

The first type of complex is represented by the montmorillonite–benzonitrile system

$$\text{clay} -- M^{n+} --- O -\!\!\!-\!\!\!- H_a --- N \equiv C - \langle \bigcirc \rangle$$
$$\hspace{4.5cm}|$$
$$\hspace{4.3cm}H_b$$

studied in detail by Serratosa (1968b). When M^{n+} is an alkaline earth cation the resulting basal spacing of the complex is approximately 15 Å. The organic molecules are disposed with the plane of the benzene ring at a high angle to the silicate layers and with the principal axis (N≡C bond) parallel to the layers. The linking water molecules are asymmetrically perturbed; their C_{2v} symmetry no longer is preserved, and the two OH groups vibrate almost independently. The IR adsorption frequency of the OH_a bond is shifted to positions depending on the polarizing power of the interlayer cation present. H-bonding displaces also the position of the C≡N band, but this frequency is not affected by the cation. The strength of the H-bonds formed depends on the basicity of the adsorbate (Table 8.3).

TABLE 8.3. OH_a stretching frequencies for organic complexes of smectites

$$M^{n+} \cdots O - H_a \cdots R$$
$$\hspace{1.5cm}|$$
$$\hspace{1.3cm}H_b$$

Complexes	Interlayer cation	OH_a stretching frequency cm^{-1}
Benzonitrile–montmorillonite[†]	Ba^{2+}	3405
	Sr^{2+}	3390
	Mg^{2+}	3348
Nitrobenzene–montmorillonite[‡]	Mg^{2+}	3500
Pyridine–saponite[§]	Mg^{2+}	2780

[†] Serratosa (1968b).
[‡] Yariv et al. (1966).
[§] Farmer and Mortland (1966).

For complexes of the second type, in which both protons of each water molecule participate in H-bonding, the Mg-saponite–pyridine system, phase of 23 Å basal spacing, is the most representative. That complex is obtained by exposing the phase of 14.8 Å to pyridine vapour. Here, IR spectra show one position for OH stretching adsorption band of water at 3049 cm^{-1} (Farmer and Mortland 1966).

Association of organic molecules to the interlayer cations via water bridges has also been demonstrated for complexes of montmorillonite with ketones (Parfitt and Mortland 1968), with nitrobenzene and benzoic acid (Yariv et al. 1966), with pyridine-N-oxide (Olejnik et al. 1973), with aniline and its derivatives (Yariv et al. 1969), with nitriles (Sánchez et al. 1973), and with phenol (Fenn and Mortland 1973).

Molecules capable of forming multiple bonds of this type, such as polyethylene glycol, are strongly adsorbed by montmorillonite even from aqueous solutions (Parfitt and Greenland 1970).

8.4.3 Protonation of organic bases

Many organic bases become protonated when they are adsorbed as neutral molecules into layer silicates. Sources of protons are: (a) hydrogen ions in H^+- or NH_4^+-saturated clays, and (b) water molecules coordinated to interlayer metal ions.

8.4.3.1 Protonation on H^+- or NH_4^+-saturated clays

The protonation of molecules by H^+-clays is a neutralization reaction that is represented by the equation:

$$H^+\text{-clay} + B \rightleftarrows BH^+\text{-clay}$$

When bases are adsorbed in excess over the number of available protons, hemisalts may be formed by mechanisms described in former sections. The possibility of hemisalt formation must be taken into consideration when using infrared spectra as evidence for protonation.

When interlayer ammonium cations are present as saturating cations, they are capable of donating a proton to an adsorbed base. The proton transfer reaction may be represented as follows:

$$NH_4^+\text{-clay} + B \rightleftarrows BH^+\text{-clay} + NH_3$$

The extent of the reaction will depend on the relative strength of the two bases (NH_3 and B) and on the concentrations or activities of the reactants in the interlayer space. The reaction is similar to that described previously for cationic organic complexes. Molecules known to accept a proton from NH_4^+ are pyridine, methylamine, ethylendiamine and 3-aminotriazole (Russell et al. 1968; Mortland and Raman 1968; Cloos et al. 1975).

8.4.3.2 Proton transfer from coordinated water molecules

The most common mechanism by which adsorbed organic molecules may be protonated is by proton transfer from interlayer water molecules. Metal ions containing water ligands are acids of measurable strength. Table 8.4 gives the acid ionization constants determined in water for a number of these complex ions (Hunt 1963). The values refer to the reaction:

$$[M(H_2O)_x]^{n+} + H_2O \rightleftarrows [MOH(H_2O)_{x-1}]^{(n-1)+} + H_3O^+$$

Therefore, it is to be expected that water coordinated to interlayer cations will show acid

TABLE 8.4. Hydrolysis data for some hydrated metal ions (Hunt 1963)

Acid	Conjugate base	pK (25 °C)
$[Fe(H_2O)_x]^{3+}$	$[FeOH(H_2O)_{x-1}]^{2+}$	2.17
$[Al(H_2O)_x]^{3+}$	$[AlOH(H_2O)_{x-1}]^{2+}$	5.02
$[Mg(H_2O)_x]^{2+}$	$[MgOH(H_2O)_{x-1}]^{+}$	11.4
$[Ca(H_2O)_x]^{2+}$	$[CaOH(H_2O)_{x-1}]^{+}$	12.7
$[Na(H_2O)_x]^{+}$	$[NaOH(H_2O)_{x-1}]$	14.6

characteristics that will depend on the nature of the metal ions. The pK values of interlayer water, however, should not necessarily be the same as in solution, and in fact it has been found that the acidity of interlayer water is greater than that deduced from Table 8.4, this reflecting the influence of the electric field due to the silicate layers.

Also, hydrates with a number of molecules less than that corresponding to normal coordination exist in the interlayer space of phyllosilicates. As the water content decreases, the polarizing effect of the cation on the remaining water will be stronger, and this will result in an increased acidity. The higher acidity of interlayer water has been established by various techniques, such as conductivity measurements (Fripiat et al. 1965), nuclear magnetic resonance (Ducros and Dupont 1962; Hecht et al. 1966; Touillaux et al. 1968), adsorption of inorganic and organic bases (Mortland et al. 1963) and by the use of Hammett indicators (Benesi 1956; Fowkes et al. 1960; Solomon et al. 1971; Frenkel 1974).

Quantitative data on the ionization constants of interlayer water are difficult to obtain. Pulsed NMR studies indicate that the degree of dissociation of adsorbed water on montmorillonite is 7 orders of magnitude greater than that of bulk water. Mortland and Raman (1968) have studied the reaction:

$$[M(H_2O)_x]^{n+} + NH_3 \rightleftharpoons [M(OH)(H_2O)_{x-1}]^{(n-1)+} + NH_4^+$$

and were able to estimate the equilibrium constants:

$$K_e = \frac{[OH^-] \cdot [NH_4^+]}{[H_2O] \cdot [NH_3]}$$

for different cations at various water contents. Their results show that the equilibrium constants decrease in the order Al^{3+}, Mg^{2+}, Ca^{2+}, Li^+, K^+ and that for a given cation, they are larger in systems with lower water contents.

The acidity of coordinated water has a major influence on the adsorption of organic molecules because, when molecules capable of protonation are adsorbed by clays, they are retained as cations. The reaction can be represented by the equation

$$[M(H_2O)_x]^{n+} + B \rightleftharpoons BH^+ + [M(OH)(H_2O)_{x-1}]^{(n-1)+}$$

where B is a neutral organic molecule.

Evidence for the formation of protonated organic cations is obtained by infrared spectroscopy, as the adsorption bands for neutral molecules and protonated species are separated in the spectrum. The phenomenon was observed for the first time by Fripiat et al. (1962), for the adsorption of short chain aliphatic amines by montmorillonite. In the case of

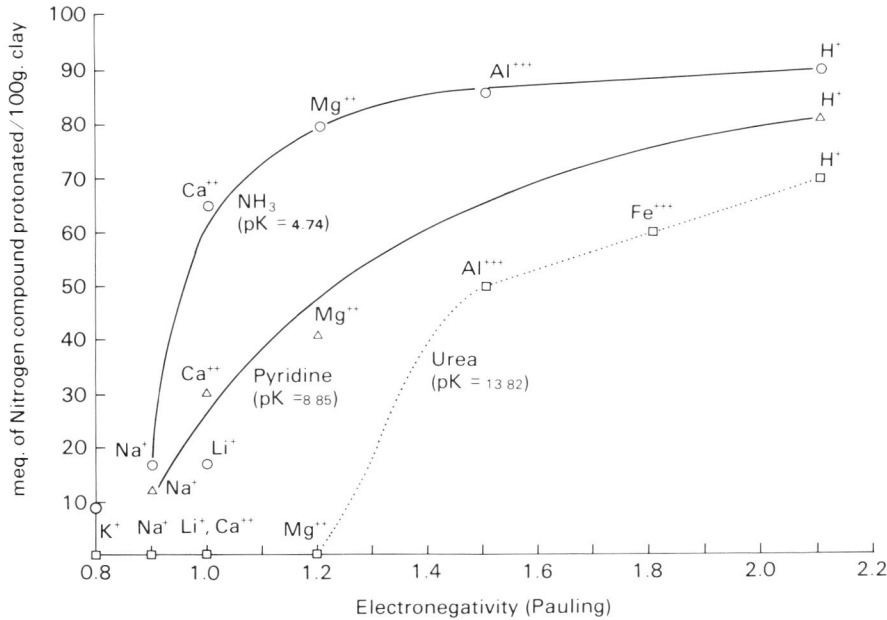

FIG. 8.7. Extent of proton donation by residual water in montmorillonite to three different bases (ammonia, pyridine, urea) as related to the electronegativity (Pauling) of the exchangeable cation. (After Mortland 1968b.) Ammonia data from Russell (1965), pyridine data from Farmer and Mortland (1966), urea data from Mortland (1966).

molecules containing a carbonyl and an amino group, such as urea and the amides, the proton is taken by the CO rather than by the NH group (Mortland 1966; Tahoun and Mortland 1966). For protonation of 3-aminonitrazole, 5-triazines and other molecules used as herbicides the review of Bailey and White (1970) must be consulted.

The amount of the base which is protonated depends upon: (1) the field strength of the inorganic cation; (2) the basicity of the adsorbed molecules; (3) the water content of the system; and (4) the site of the layer charge (tetrahedral or octahedral).

The influence of the first two factors is shown in Fig. 8.7 where the degree of protonation of organic bases of different strengths is plotted as a function of the electronegativity of the interlayer cations (Mortland 1968b). Ammonia and pyridine are protonated at least partially even by monovalent cations of relatively small field strength, whereas urea, a weaker base, is protonated only by Al^{3+}, Fe^{3+} and H^+ ions.

The influence of the water content on the clay is illustrated by the fact that pyridine adsorbed on hydrated Mg-montmorillonite is not protonated, but that pyridinium ions are formed upon partial dehydration (Farmer and Mortland 1966). At nearly zero water content, residual interlayer water has the properties of a strong, concentrated, mineral acid. This has been demonstrated by the protonation and demetallation in the interlayer space of hectorite of Sn(IV)-tetrapyridyl-porphyrin when the corresponding complex is dehydrated (Abdo et al. 1980). For the free Sn(IV)-porphyrin, an acidic condition approaching that of 100% H_2SO_4 is required for the reaction to take place.

After removal of the last traces of adsorbed water, the proton-donating ability of the clay is sharply reduced: Brønsted acid sites are still present in anhydrous clays, but it has been shown that the acidity is due to hydroxyl groups on edge surfaces (Davidtz 1976).

The acidity of adsorbed water can also activate the transformation or decomposition of organic molecules adsorbed on clays. One clear example is the conversion of nitriles into amides in the interlayer space of montmorillonite (Sánchez et al. 1973):

$$CH_2Cl-CN + H_2O \xrightarrow{H^+} CH_2Cl-CONH_2$$

Other reactions involve the formation of carbonium ions as intermediate species, for example the reversible transformation of triphenyl carbinol into triphenylmethylcarbonium ion by montmorillonites (Fripiat et al. 1964):

$$Ph_3C-OH \underset{H_2O}{\overset{H^+}{\rightleftarrows}} Ph_3C^+$$

The site of layer charge (tetrahedral or octahedral) seems also to affect the proton transfer reaction between coordinated water and adsorbed base, but the mechanism is not well understood and results are sometimes conflicting. According to Mortland and Raman (1968), the effect may be attributed to the different interaction between interlayer cations and negative charges on the silicate layer, which indirectly influences the polarizability of the associated water and consequently affects its acidity.

8.4.4 Hydrogen bonding to surface oxygens and between molecules themselves

The study of the interaction between clay surfaces and adsorbed organic molecules by infrared spectroscopy has demonstrated that contrary to what was formerly believed, surface oxygens are weak electron donors, and that formation of hydrogen bonds with functional NH or OH groups does not contribute significantly to the adsorption forces. The subject has been treated in detail by Mortland (1970) and by Farmer (1971), and it is now accepted that the attraction between surfaces and groups capable of hydrogen bonding is considerably weaker than interactions with the exchangeable cations.

Based on evidence from infrared absorption and other data, the above authors recognize that hydrogen bonding to the surface may be significantly stronger when the charge on the silicate layers is tetrahedrally located, so that, for exchangeable cations of low solvation energy and organic molecules with multiple OH or NH groups, the contribution to the absorption process of hydrogen bonding with surface oxygen atoms may become relevant.

Farmer and Russell (1971) consider that, for each positive charge deficiency on the tetrahedral sheet, the resulting negative charge will be distributed over the three surface oxygens coordinated to the substituting Al^{3+} ion, and that the same charge will be distributed over at least the ten surface oxygens of the four silica tetrahedra linked to a site of octahedral substitution. By this criterion, surface oxygens of saponite and vermiculite are more electronegative and stronger electron donors than those of montmorillonite and hectorite. Consistent with that consideration are the different OH stretching frequencies found for water adsorbed on clays. Water hydroxyls bonded to oxygens coordinated to Al^{3+} absorb at 3330–

$3350\ cm^{-1}$ while water hydroxyls interacting with oxygens of Si—O—Si linkages absorb at higher frequencies (3480–$3630\ cm^{-1}$). Thus, the strength of the hydrogen bonds formed with the surface is smaller than, or at best equal to that of the bonds formed between the interlayer water molecules themselves (3350–$3480\ cm^{-1}$).

The different capabilities for hydrogen bond formation are more apparent when alkylammonium cations rather than neutral molecules are the species present in the interlayer space. As previously stated, alkylammonium ions adsorbed in montmorillonite have the —NH_3^+ groups disposed with their C_3 axes parallel to the layer, and no hydrogen bonding to the surface can be detected by infrared spectroscopy. Contrarily, when they are adsorbed on vermiculite, the three protons of each —NH_3^+ group participate in hydrogen bonding with surface oxygens, the bonds are relatively strong, and C_3 axes are maintained perpendicular to the layers (Martín-Rubí et al. 1974).

Hydrogen bond interactions between clay surface and neutral organic species are difficult to study by infrared techniques, as the evidence provided by the spectra is always inconclusive due to the combined effects of the various adsorption mechanisms involved. Reduction of the CEC of montmorillonite by the Hofmann–Klemen (1950) effect (see Chapter 5, Section 5.3.3) provides a means for studying those interactions under conditions where the overriding effects due to the exchangeable cations are minimized. This has been done by Brindley and Gozen Ertem (1971) who found a minimum value for the montmorillonite charge below which ketones are no longer adsorbed in the interlayer space, while alcohols are still adsorbed. The energies for the cation–dipole interaction are nearly identical for the two compounds (Bissada and Johns 1969), but alcohols can form hydrogen bonds with the clay surfaces whereas ketones cannot; the authors concluded, therefore, that, for alcohols, hydrogen bonding to surface oxygens becomes important as an additional mechanism in determining adsorption.

Recent studies by Annabi-Bergaya et al. (1980, 1981) on the adsorption of methanol vapours by homoionic montmorillonites has led to the recognition that hydrogen bonding between the alcohol molecules themselves has also a strong influence in determining adsorption. In fact, at 35 °C and for relative alcohol vapour pressures between 0.1 and 0.9, only between 10 and 25% of the total amount of molecules adsorbed by the clay are retained in the cation coordination shell in the interlayer space, i.e. one molecule per cation is influenced directly by Na^+, two molecules are influenced by Ba^{2+}, 1.3–1.9 molecules by Li^+, and four molecules by Ca^{2+}. The remaining molecules are H-bonded to each other, either in the interlamellar space or in the micropore volume between crystallites.

Thus, in sorption processes by smectites, interactions between adsorbed molecules are important for species with high association properties (methanol or water), where the tendency of the interlayer cations to build their own solvation shell is complementary to a tendency of the sorbate to grow a continuous network of molecules in the interlamellar space.

8.5 INTERCALATION PROCESSES IN KAOLINITE

The minerals of the kaolin group have been classified traditionally as non-expandable clays, halloysite being the only member of the group that, besides occurring in nature as an interlayer hydrate, was also known to have a capacity for interlayer adsorption of certain organic compounds (MacEwan 1946). For the remaining members of the group common experience in clay minerals laboratories has shown that water or organic reagents capable of penetrating between the unit layers of smectites left the structures of kaolinites unaffected.

An entirely new field of study of clay–organic interactions was opened up when expansion of kaolinite clays was first achieved by treatment with K-acetate and with other salts of organic acids of low molecular weight (Wada 1961). The resulting complexes were called "intersalation" compounds, a term that was later replaced by "intercalation" compounds when it was found that urea, formamide, dimethylsulphoxide and many other non-saline organic substances penetrate also between the kaolinite layers (Weiss 1961; González-García and Sánchez Camazano 1965).

Intercalation compounds have been classified in two main groups according to the method of preparation (Weiss *et al.* 1966):

1. Compounds which can be prepared by direct reaction of the mineral with the organic substance.
2. Compounds which can be prepared only by replacement of a substance previously intercalated by direct reaction of the above type.

Experiments have shown that many organic substances that form intercalation compounds of type 2 either do not react directly with kaolinite or else their intercalation rates are negligibly small.

Molecules which form readily type 1 complexes are of various natures:

(a) Salts of the organic acids of low molecular weight (acetates, cynoacetates and propionates) with large monovalent cations of low hydration energy.
(b) Compounds with a strong tendency for hydrogen bond formation, such as urea, formamide, acetamide and imidazole.
(c) Molecules having a high dipole moment or with mesomeric structures, such as dimethylsulphoxide and pyridine-N-oxide.
(d) Molecules combining two or more of the above mentioned characteristics, such as ammonium acetate, N-methylacetamide and the potassium salt of picolinic acid-N-oxide.

These compounds intercalate directly either from the liquid (hydrazine), from the melt (acetamide), or from concentrated aqueous solutions (10M solution of urea).

8.5.1 Interlayer forces in kaolinite

According to Cruz *et al.* (1973) interlayer bonding in the kaolinite microcrystals arises from van der Waals' attraction between layers, from hydrogen bonding between octahedral OH groups on one layer and tetrahedral O atoms on the adjacent layer, and from electrostatic interactions arising from the fact that each layer, although electrically neutral, bears a net fractional charge of opposite sign on each basal surface. Calculations based on simplifying assumptions have led the above authors to estimate the total cohesion energy as 210 kJ per formula unit, the main part of which arises from the electrostatic energy term.

For penetration of organic molecules, sufficient energy must be provided to overcome the cohesion between layers. The intercalation process, then, may be envisaged as resulting from the tendency of the dipolar kaolinite layers to become solvated with molecules.

Infrared spectra of intercalated kaolinite complexes afford evidence on the type of interaction between the organic molecules and the mineral substrate. When an organic substance is intercalated, the bands corresponding to OH groups on the surface of the kaolinite

TABLE 8.5. Stretching frequencies of structural OH in some intercalation complexes of kaolinite (Cruz et al. 1973)

Complexes	OH frequency cm^{-1}
Formamide[†]	3590
Urea[‡]	3585–3560 (shoulder)
Methyl urea[‡]	3595 3540 3515
Dimethyl urea[‡]	3565 3515
Methyl formamide[†]	3550
Dimethylsulphoxide[§]	3546 3507
Dimethyl formamide[†]	3415

[†] Cruz et al. (1969).
[‡] Cruz et al. (1970).
[§] Jacobs and Sterckx (1970).

layers are displaced to lower frequencies (Table 8.5), clearly demonstrating the existence of hydrogen bonds between the surface hydroxyls and the intercalated molecules. From the frequency shifts observed, the strength of the hydrogen bonds may be estimated. By this procedure, Cruz et al. (1973) have calculated the contribution of hydrogen bonding to the net enthalpy change for the intercalation of dimethylsulphoxide into kaolinite as 46 kJ per formula unit, which is less than the 210 kJ necessary to overcome the cohesion of the layers. They conclude that other factors should contribute to the energy balance, and postulate an additional contribution, either due to a decrease of the electrostatic attraction caused by a higher dielectric constant in the interlayer volume after intercalation, or due to a compensation of the internal dipole moment of the kaolinite layers by the dipole moment of the intercalated phase.

8.5.2 Reaction mechanisms

The intercalation reaction starts at the edges of the kaolinite microcrystals and molecules penetrate towards the interior with the same rate between each pair of layers. That part of the crystal where molecules have penetrated has a basal spacing about 50% greater than the original 7.2 Å phase. Molecules act as wedges on the kaolinite crystal and cause the layers to be elastically deformed at the interphase. Because of their unsymmetrical structure, all layers in the crystal are bent in only one direction.

According to Weiss et al. (1969) the kinetics of the intercalation reaction cannot be described by a simple diffusion process, the relationship between the fraction of reacted kaolinite and the time being best expressed by the logarithmic equation

$$\sqrt{-\ln(1-\alpha)} = kt$$

that is, by an expression describing reactions which are typically cooperative. The length of the elastically deformed zone in the interior of the crystal is taken as the "cooperative action length", a_0, and depends on the net increase in basal spacing as well as on the elastic properties of the layers, which are themselves related to the crystallinity of the kaolinite used: the better the crystallinity, the longer a_0 and the faster the reaction, since the weaker cohesion on that zone of the crystal favours further penetration.

The ratio between a_0 and crystal diameter becomes crucial in determining the mechanism of reaction. Thus, for large crystals, intercalation starts from all crystal edges but, for crystals which are too small, the elastic deformations induced by molecules penetrating from one side are transmitted across the entire particle causing a contraction of the layers on the opposite side. Intercalation from that side is hindered, so that for small particles the reaction proceeds at a slow rate via a "one side mechanism", while for large particles, where nucleation can start from all edges, the reaction proceeds faster via a "ring mechanism".

8.5.3 Factors affecting the rate

The effect of particle size distribution and of degree of crystallinity on the intercalation properties of different kaolinites has been investigated by several authors, but their conclusions do not always agree (Bodenheimer *et al.* 1967; Range *et al.* 1969; Wiewiora and Brindley 1969). According to the model of Weiss *et al.* (1969) both characteristics should have analogous effects on the reaction rate but, since they are not always directly correlated, the combined effect of the two may be difficult to assess by measuring only the fraction of material that has reacted after a selected reaction time. More definite patterns are observed when the complete kinetics of the reaction is studied and the intercalation properties are expressed as a function of the reaction rate. This is clearly illustrated from the data of Figs 8.8 and 8.9 (Weiss *et al.* 1969) showing that, for a given kaolinite, intercalation rates first increase with particle size and then decrease, a trend also observed for the variation of the index of crystallinity of the different size fractions.

Reaction rates are also affected by factors that bear no relation to the characteristics of the mineral substrate. These include the molecular size of the intercalating material, the concentration of the aqueous or alcoholic solutions if solutions are used in lieu of the pure substance, the solution pH, and also the reaction temperature.

The effects of adddition of water to the intercalating agent have been studied for the reaction of kaolinite with hydrazine (Weiss *et al.* 1963), with formamide and with N-methylformamide (Olejnik *et al.* 1970). It was observed that, in all cases, rates were significantly increased when water was added to the liquid, becoming maximum at a certain water content that varies widely with the system involved. Thus, the maximum rate was attained at 10% water content for the formamide–water and for the N-methylformamide–water systems, while for the hydrazine–water system the maximum rate was attained at 10M concentration (approximately 68% water in the mixture).

As pure liquids, those compounds are either hydrogen-bonded or dipole-associated polymeric networks, with very few free or unbonded molecules coexisting at normal temperature. It has been suggested (Olejnik *et al.* 1970) that the addition of water or other polar molecules favours intercalation because the associated structure of the pure liquid is broken up, thus increasing the proportion of unbonded molecules. However, if excess water is added, then the solute becomes solvated by the solvent, and molecules free to intercalate are fewer. The rate thus passes through a maximum because of the competition for unsolvated molecules between the kaolinite crystal and the liquid phase.

400 REACTIONS OF CLAYS WITH ORGANIC SUBSTANCES

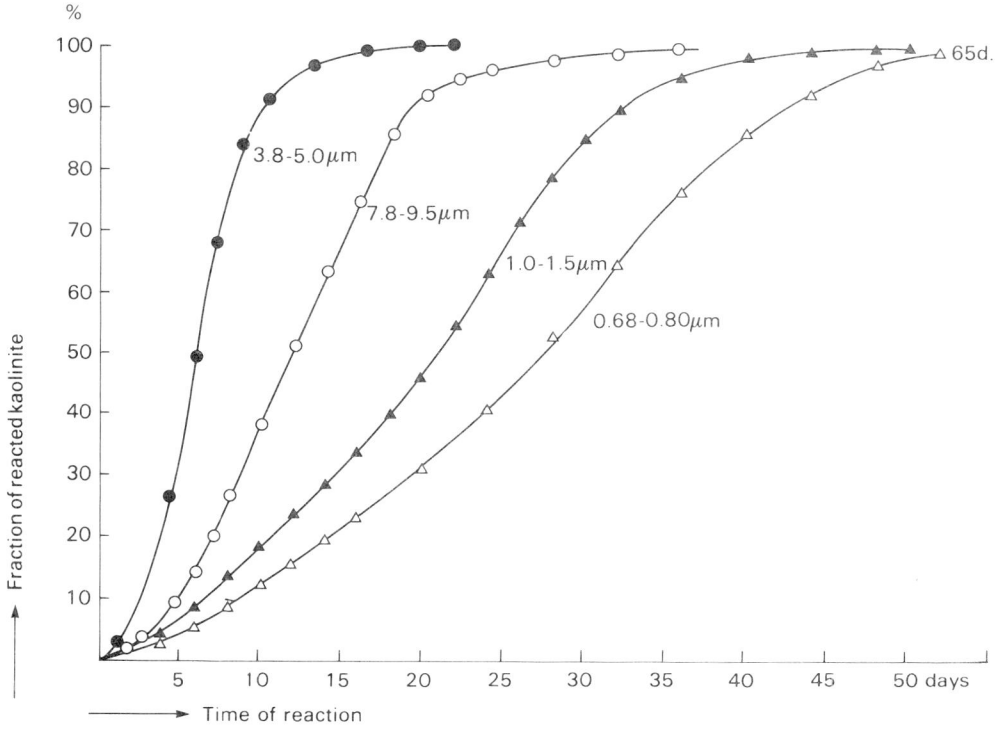

FIG. 8.8. Effect of particle size on the intercalation of urea into kaolinite from a saturated aqueous solution. (After Weiss *et al.* 1969.)

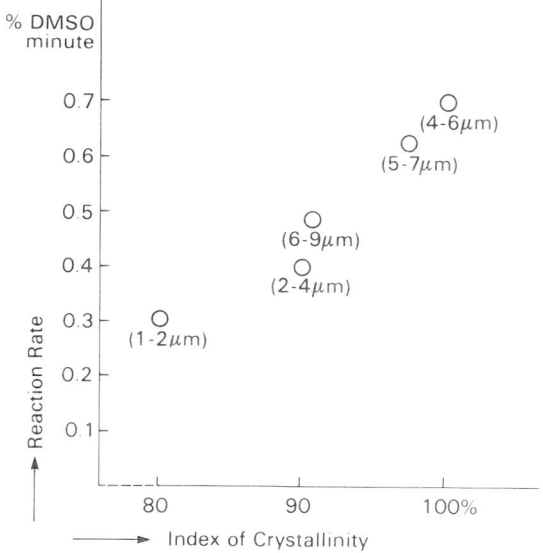

FIG. 8.9. Reaction rate versus index of crystallinity. Intercalation of dimethyl sulphoxide into fractions of kaolinite KI S, Schnaittenbach. (After Weiss *et al.* 1969.)

Higher temperature should also favour intercalation because of its disruptive effect on the liquid structure. This is in accordance with the increased rates reported for the intercalation of urea (Weiss et al. 1963) and dimethylsulphoxide (Weiss et al. 1966; Mata et al. 1970) into kaolinite.

No definite relationship exists between molecular size and intercalation ability. Considering that the work needed for lattice expansion should be the greater the larger the resulting $d(001)$ spacing, it might be expected that rates would increase with decreasing molecular weight of the intercalating agent. However, rate determinations for compounds from homologous series such as formamide and acetamide and their monomethyl and dimethyl derivatives do not indicate such a tendency (Olejnik et al. 1970). Undoubtedly, any possible influence of molecular size is overshadowed by differences in the extent of molecular association in the liquids or melts.

The intercalation of organic salts (the NH_4, K, Rb and Cs acetates) is strongly influenced by the pH value of the saturated solution used. It has been observed (Weiss et al. 1963, 1966) that the reaction takes place in an alkaline medium only, the maximum rate being attained in the pH range of 8 to 11. At pH values higher than 13 the reaction is also inhibited. No explanation has been offered for that behaviour.

8.5.4 Displacement reactions

With few exceptions, intercalation complexes are stable only in the presence of an excess of the organic medium. Incorporated molecules are easily extracted either by leaching with water or even by air drying. Washing collapses the lattice to the 7.2 Å spacing characteristic of kaolinite, but its original crystallinity is never restored, possibly because traces of the organic substance are retained between the layers (Wiewiora and Brindley 1969).

Certain kaolinites yield hydrates ($d(001) = 10$ Å) when their complexes with hydrazine are leached with water. That water may subsequently be replaced by ethylene glycol (but not by glycerol) to give a complex with 10.8 Å spacing. In contrast, water from hydrated halloysite may be replaced by glycerol ($d(001) = 11.1$ Å). These properties have been proposed as a means for differentiating kaolinite types, as well as for unequivocal identification of halloysite (Range et al., 1969).

Once the kaolinite layers are separated in a complex of type 1, then the intercalated molecules may be displaced by repeated washing with other organic compounds that do not react directly with the mineral. Type 2 complexes have been prepared by displacement reactions with a large variety of substances. These include:

1. Polar compounds: acetone, glycols, acetonitrile and nitrobenzene (Sánchez-Camazano and González García 1966).
2. Organic bases: alkyl- and aromatic amines, alkylene diamines, pyridine, purines and pyrimidines, morpholine (Weiss 1969).
3. Amino acids, peptides and their salts: K-glycocolate, alaninate and lysinate (Weiss et al. 1963).
4. Other organic salts: Na-acetate and K-oxalate and lactate (Weiss et al. 1963).

The complex with acid ammonium acetate CH_3—$COONH_4$. CH_3—$COOH$ has been used as starting material for displacement reactions because of the large separation it produces between the kaolinite layers ($d(001) = 17$ Å), but other type 1 complexes with smaller $d(001)$

spacings, e.g. the dimethylsulphoxide, the hydrazine or the dimethylformamide complexes are also effective.

8.5.5 Intercalation capacity of kaolins

Rates of intercalation with K-acetate, hydrazine or DMSO have been measured for kaolins from various origins and types, residual or sedimentary, ball-clays, fire-clays, fint- and semiflint-clays, and also for kaolins extracted from different soils, in the hope that they could constitute a convenient index for characterizing varieties for the industrial applications of these minerals (Weiss *et al.* 1963; Alietti 1966; Smith *et al.* 1966; Gomes 1967; Wiewiora and Brindley 1969; Range *et al.* 1969).

Besides differences in rate, results have also shown that certain kaolins do not undergo intercalation, and that for others, the reaction does not proceed to completion. Thus, the terms intercalation capacity, or intercalation grade, have been introduced to indicate the weight fraction of the mineral that can, or cannot, be expanded after a certain reaction time, generally 1 hour at 70 °C.

The relationship existing between rates and intercalation capacity on one side and the morphological, granulometric, structural and genetic characteristics of the kaolin on the other can be summarized as follows (Gomes 1982).

Intercalation capacities depend on the cohesion energies of the kaolinite stackings which, in turn, are strongly correlated to structural disorder. Generally, residual kaolins are more ordered and have higher intercalation capacities than sedimentary kaolins. As for the intercalation capacities, they depend on particle size in as much as crystal size is related to crystal imperfection. Translational or rotational stacking faults (Plançon 1976; Plançon and Tchoubar 1977a,b) do not affect appreciably the intercalation capacity. Contrarily, both the capacity and the rate are strongly dependent on the presence of defects arising from isomorphous substitutions or from a random occupancy of octahedral cation sites, which would disturb the electrostatic valence balance within the layers and will add to the electrostatic component of the cohesive energy of the layer stackings. Defects affecting the stacking sequence, i.e. the presence of randomly interstratified single layers of smectite or illite, are also strongly correlated to the intercalation properties. Such defects are more abundant in the smaller particle size fractions, and are generally responsible for the asymmetry of 001 X-ray reflections. Figure 8.10 shows the relationship between the abundance of defects of this kind and intercalation behaviour for several kaolins (Gomes 1982).

8.6 CLAY MINERALS AS TEMPLATES FOR ORIENTED REACTIONS

The reactivity of molecules adsorbed on phyllosilicates may be influenced, both qualitatively and quantitatively, by limitations that the interlayer space imposes on the orientation and packing arrangement of the molecules. The effect is essentially of a steric nature and differs from the catalytic influences resulting from the electric field distribution in the interlayer volume. In general, however, both influences operate simultaneously and, therefore, it will be difficult to evaluate their separate contributions to the result of a particular reaction.

The specificity of a clay with respect to reactions taking place in its interlayer space should be considered as a particular aspect of the chemistry in preoriented media, a subject that has attracted considerably attention in the last few years. The aim of studies in this field is directed to

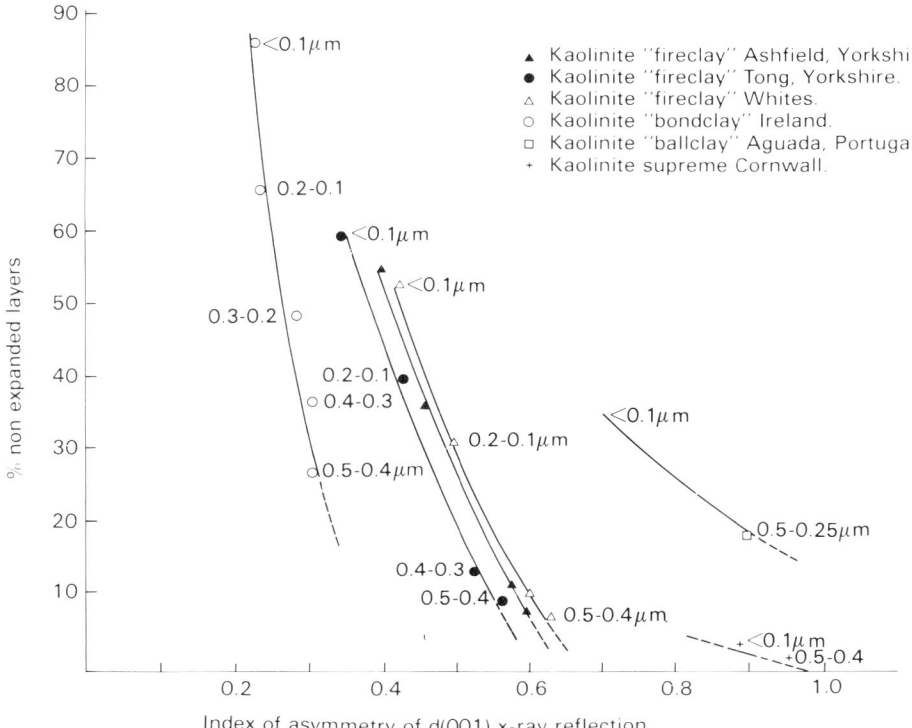

FIG. 8.10. Relationship between index of asymmetry of $d(001)$ X-ray reflection and percentage of non-expanded layers of kaolinite by hydrazine (after Gomes, 1982).

the search for organic reactions of high specificity and stereoselectivity, differing from those taking place in solution, where randomness prevails.

Reactions taking place in oriented media can be influenced by the substrate in any of the following ways:

1. Blocking of certain functional groups by structural discontinuities or active surface sites on the substrate, that will render such groups non-reactive.
2. Imposing a particular packing arrangement that will position the functional groups in neighbouring molecules so that a certain type of molecular approach is favoured and others prevented.
3. Inducing changes on the molecular configuration of the reactants or products. For example, only one of several possible configurations might be adopted by the molecules when they are adsorbed on an orienting substrate. It is even possible that the configuration adopted is one not normally adopted by that molecular species when it is in solution.

The particular case of the reactivity of preoriented organic–clay mineral systems has not been treated abundantly, and no systematic work has been conducted on the subject. There are examples, however, that illustrate the specificity of the interlayer space of phyllosilicates as a reaction medium. One of these is the condensation of L-ornithine cations

$$^+H_3N-(CH_2)_3-CH\begin{matrix}COO^-\\ \\NH_3^+\end{matrix}$$

in the interlayer space of vermiculite (Fornés et al. 1973), with the formation of the cyclic peptide

$$^+H_3N-(CH_2)_3-CH\begin{matrix}CO-NH\\ \\NH-CO\end{matrix}CH-(CH_2)_3-NH_3^+$$

The starting L-ornithine complex is a hydrate, and condensation takes place upon dehydration by heating at 180 °C. In the hydrated complex, L-ornithine cations lie flat on the vermiculite surfaces, with the COO^- and NH_3^+ groups directed towards the central plane of the interlayer volume (Rausell-Colom and Fornés 1974). Each position on the surface occupied by a cation alternates with a site of identical shape and area which is covered by water molecules (Fig. 8.11), and the stacking of the vermiculite layers is such that the ornithine cations on one surface are positioned against the water molecules of the adjacent surface. Furthermore, both the COO^- and the NH_3^+ groups from each ornithine cation on one layer are facing the same groups of each cation on the adjacent layer. Such disposition favours condensation through the $-NH_3^+$ and $-COO^-$ groups and, since identical groups are facing each other, formation of chain peptides would be expected, but this would require some disruption of the existing packing arrangement.

Cyclic peptides are obtained instead, and this suggests that the existing order is preserved and that condensation occurs after a catalytic racemization of the amino acid has taken place. Such possibility seems likely in view of the environmental conditions prevailing, i.e. the proximity of electronegative surfaces and the high temperature.

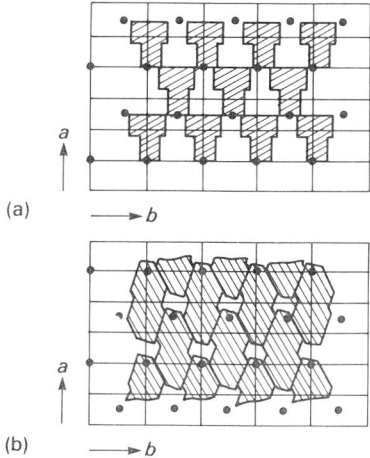

FIG. 8.11. Surface coverage and interlayer packing of adsorbed species. (a) Vermiculite-L-ornithine complexes; (b) Vermiculite–peptide complex (after Rausell-Colom and Fornés 1974).

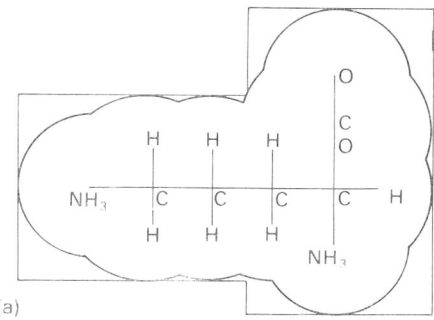

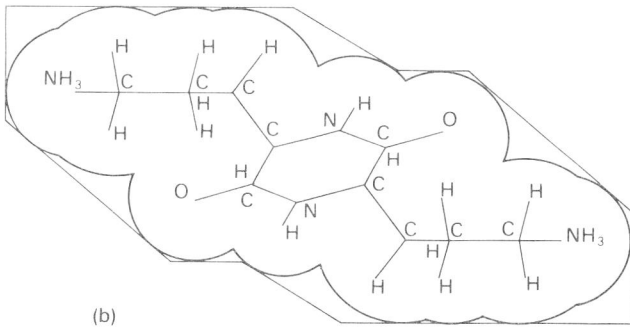

FIG. 8.12. Projection on the *ab* plane of vermiculite of adsorbed species. (a) L-Ornithine cation. (b) Peptide cation.

The structural order is best preserved if the peptides formed have a *trans* configuration with respect to the plane of the cyclic ring. This is, indeed, the configuration obtained in the resulting complex, although in order to reduce surface coverage to the extent required, both lateral chains are rotated by 60° around the corresponding CH—CH$_2$ bond axis (Fig. 8.12). Peptide molecules with the *cis* configuration would be formed with equal probability by condensation of the racemic amino acid in a solution medium, but they are not formed when condensation takes place in the interlayer space of vermiculite.

The selective polymerization of oleic acid in the interlayer volume of montmorillonite is also a process being governed by the particular packing arrangement of the reactants, which is in turn imposed by the charge density on the silicate layers and by the type and distribution of interlayer saturating cations (Weiss 1981). When adsorbed on montmorillonite oleic acid monomers lie flat on the mineral surfaces, and polymers are formed by cross-linking via transposition of one H atom between two unsaturated C atoms of adjacent molecules, the reaction being activated by the proximity of the double bond to the silicate surface. The montmorillonite is saturated initially with $(CH_3)_4$—N^+ cations which fill partially the interlayer volume distributed uniformly over the mineral surface. The remaining volume is available for adsorption of the oleic acid molecules, and steric effects are imposed on the packing of the monomers by the existing organic cations. At high cation density, neighbouring acid molecules are separated from each other by the cations, and dimerization cannot occur even if double bonds are activated. At medium density, most acid molecules are

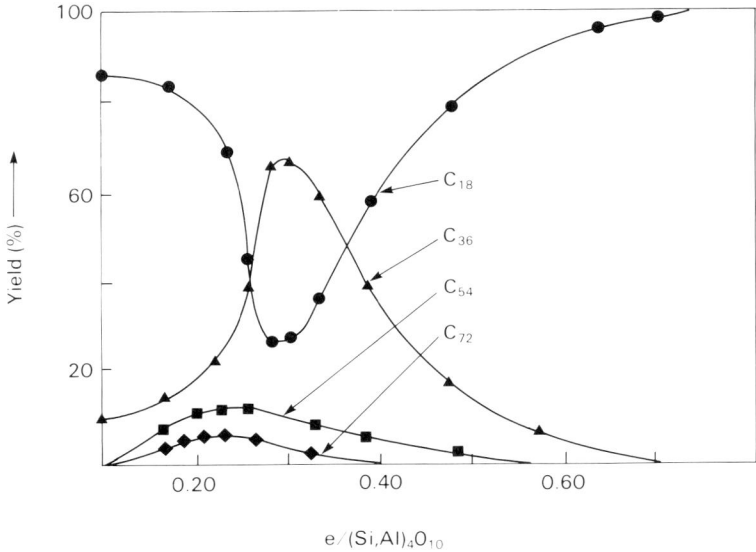

FIG. 8.13. Oligomerization of oleic acid to di-, tri- and oligocarboxylic acids with $(CH_3)_4$N-montmorillonite as catalyst as a function of the charge density. ●: Oleic acid + stearic acid (C_{18}); ▲: Dicarboxylic acids (C_{36}); ■: Tricarboxylic acids (C_{54}); ♦: Oligocarboxylic acids ($\geqslant C_{72}$) (from Weiss 1981).

grouped in pairs, and activation leads to the dimeric acid. At still lower density, triplets or adruplets are more abundant, and trimers or tetramers are formed. Yields for the different polymerization reactions are given in Fig. 8.13 as a function of the charge density of montmorillonite.

Other cases of specific chemical conversions involving montmorillonite intercalates have been reviewed by Thomas et al. (1977); a particularly efficient one is the thermal decomposition of the diprotonated 4,4'-diamino-trans-stilbene montmorillonite complex ($d(001) = 14.5$ Å), for the production of aniline as the sole gaseous product. The reaction proceeds at a fast rate at 270 °C with a yield of 45% aniline from the parent diamine (Tennakoon et al. 1974).

In recent years an increasing emphasis has been placed on bio-organic clay mineral studies. According to Degens (1971) many structural and functional characteristics of clays resemble remarkably those of biological macromolecules. Both are polymers that are composed of one or more types of monomeric blocks and exhibit degrees of polymerization. Monomers are linked together in a specific manner, and give rise to a structural order where differently charged ionic groups coexist within the same molecular framework. Because of their structure and functionality, both can act as templates for de novo synthesis or act in the capacity of catalysts. Experiments have shown that urea, formamide, amino acids and polypeptides may be synthesized from simple inorganic molecules, such as CO_2, NH_3, HCN and H_2O, in the presence of the clays (Harvey et al. 1971; Cruz et al. 1974); from aqueous solutions containing paraformaldehyde, polysaccharides, particularly hexose, are formed and esterification of the fatty acids with formation of glycerides is induced. All these reactions take place at a fast rate and in a well-coordinated fashion. In view of this it seems highly possible that clay minerals may have played an important role (a) in the abiotic synthesis of complex organic molecules, and (b) on their final incorporation into coacervates or cellular systems. It is well known that ionizing radiation is capable of producing in certain atmospheric systems organic molecules considered as the basic building blocks of life, and many authors have concluded that such compounds were derived in this fashion in prebiotic times. In contrast, it may well be that processes operating in sediments, and which rely on energies stored on clay mineral surfaces, represent a more plausible and natural mechanism for the synthesis, storage and replication of biological materials (Degens et al. 1970; Paecht-Horowitz et al. 1970; Jackson 1971; Paecht-Horowitz and Katchalsky 1973; Degens and Mopper 1975).

The formation of "insertion polymers" in clays is also a line of interest in research on reactions in preoriented media. Those compounds have been studied in some detail, and reaction mechanisms have been proposed to explain the role played by the clay substrate and the nature of the products formed (Blumstein 1961, 1965, 1970; Blumstein et al. 1974). The effect of the clay substrate is to impose a two-dimensional organization to the monomer molecules, so that planar polymers with extensive branching are formed if small amounts of a cross-linking agent are added to the system. Also, when monomers susceptible to cationic initiation are adsorbed in the interlayer space, the exchange cations themselves may act as initiators. Thus, styrene polymerizes spontaneously in H-montmorillonite, but not in Na-montmorillonite (Pezerat and Vallet 1973). In such cases polymerization involves protonation of the monomer with formation of carbonium ions that subsequently polymerize with monomer molecules, but it is the existing two-dimensional order that leads to the particular stereospecificity of the polymer chains. Polymerization of formaldehyde with other organic molecules in the interlayer space of montmorillonite has been reported by Aragon and Evole (1974).

Apart from inducing modifications on the reactivity of adsorbed substances, the orientating influence of the interlayer space might result in an enhanced stability for molecules whose steric configuration fits the silicate pattern. Thus, the thermal stability of porphyrine, which as a free salt decomposes at 220 C, is maintained up to 280 C when it forms the porphyrine–montmorillonite complex (Weiss and Roloff 1964).

8.7 ADSORPTION ON PALYGORSKITE AND SEPIOLITE

Palygorskite and sepiolite are porous, fibrous clay minerals that have a high adsorptive capacity and good suspending or gelling characteristics. They have applications as fillers for gas chromatography columns (Manara and Taramasso 1972), and as catalysts or catalyst carriers (Whittam 1977; Inooka et al. 1978a,b; Ioka et al. 1978; Dandy and Nadiye-Tabbiruka 1982). They are also used to stabilize suspension fertilizer mixes, to make granular pesticide formulations and to suspend or dilute medicinal compounds or drugs. Most of their properties are attributed to their extensively convoluted external surface of $200-400 \text{ m}^2 \text{ g}^{-1}$, the even larger internal pore surface (structural channels), their fibrous or needle-like morphology and their rather low base exchange capacity. A review of surface properties of sepiolite and palygorskite has been published by Serratosa (1979) and may be summarized as follows:

Palygorskite and sepiolite crystals are submicroscopic fibres whose dimensions are quite variable depending on their origin; widths range between 100 and 300 Å, thicknesses between 50 and 100 Å and lengths between 0.2 and 2 μm (Martin-Vivaldi and Robertson 1971). Such morphology confers a high external surface to the crystallites. Exceptionally, sepiolite from Ampandandrava, Madagascar, grows fibres several centimetres long. The extent to which fibres of this particular sepiolite expose external surfaces is illustrated (Fig. 8.14) by high resolution electron micrographs of sections cut perpendicular to the fibre axis (Tchoubar et al. 1973). It appears from Fig. 8.14 that external surfaces consist predominantly of (110) crystal faces,

FIG. 8.14. High resolution electron micrograph of sepiolite. Section normal to the fibre axis (Tchoubar et al. 1973).

although crystal habits with (100) external faces are also abundant, as demonstrated by Guinier–Wolf X-ray photographs from oriented aggregates (Kubler 1962) as well as by dichroic effects observed on IR spectra (Serna et al. 1977). Taking into account average fibre sizes and relative abundances of (110) and (100) planes as observed in electron micrographs (Rautureau and Tchoubar 1972), average values for external and internal surfaces have been estimated as 400 $m^2 g^{-1}$ external and 500 $m^2 g^{-1}$ internal for Vallecas sepiolite, and 300 $m^2 g^{-1}$ external and 600 $m^2 g^{-1}$ internal for Georgia palygorskite (Serna and Van Scoyoc 1979).

These minerals have structures consisting of talc-like ribbons parallel to the fibre axis, assembled in such a way that tetrahedral sheets are continuous throughout but invert apical directions in adjacent ribbons, each ribbon alternating with channels along the fibre axis (Bradley 1940; Brauner and Preisinger 1956).

The width of the ribbons and, consequently, the width of the channels is different in the two minerals. In palygorskite the width corresponds to five octahedral cations sites, normally occupied by two Mg^{+2}, two Al^{+3} and a vacant site, in a dioctahedral pattern with the sequence Mg–Al–Vac–Al–Mg (Drits and Aleksandrova 1966; Serna et al. 1977). The channel cross-section is of 6.4×3.7 $Å^2$. In sepiolite there are eight octahedral cation sites, all occupied by Mg^{2+} ions in a trioctahedral pattern, and the channel cross-section is 10.6×3.7 $Å^2$.

8.7.1 Sorption active centres

Three kinds of sorption centres may be distinguished in sepiolite and palygorskite:

1. *Oxygen ions* on the tetrahedral sheet of the ribbons. On account of the few isomorphous substitutions present in the tetrahedral sheet of these minerals, these oxygen ions behave as weak electron donors and the interactions with sorbed species should also be weak.
2. *Octahedral Mg^{2+} ions* at the edges of the structural ribbons. Adsorption can occur by replacement of coordinated water (two molecules per cation) or by hydrogen bonding to the water. Mechanisms are similar to those for the smectites, with the difference that Mg^{2+} is not readily exchangeable and that coordinated water does not show the acidic properties of interlayer water in smectites. This is best demonstrated by the absence of

protonation after NH_3 adsorption on sepiolite and palygorskite (Van Scoyoc et al. 1979; Serna and Van Scoyoc 1979).

3. *SiOH groups.* The structure and morphology of palygorskite and sepiolite crystals are such that a large number of terminal silica tetrahedra on the ribbons are present at external surfaces. Broken Si—O—Si bonds compensate their residual charge by accepting a proton or a hydroxyl and become Si—OH groups. These groups interact with molecules adsorbed on external surfaces, as demonstrated by a shift to lower frequencies of the v_{OH} band at $3720\,cm^{-1}$. Even non-polar compounds such as fluorolube and nujol will perturb this vibration (Ahlrichs et al. 1975).

Si—OH groups occur at intervals of approximately 5 Å along the fibre axis, and their abundance can be related to the dimension of the fibres and also to crystal imperfections. Large compact crystals have fewer exposed edges and less Si—OH groups. Short, thin fibres with rough surface topography have a greater number of exposed Si—OH. Relative abundances of Si—OH groups have been determined by IR spectroscopic techniques (Ahlrichs et al. 1975; Serna et al. 1975). In palygorskite they are less abundant than in sepiolite, which is consistent with the smaller external surface area of the former mineral.

The abundance of Si—OH groups becomes relevant for the capability of these minerals to form true covalent bonds with certain organic reagents, as will be discussed in the following section.

8.7.2 Clay organic interactions

In sepiolite and palygorskite, molecules of organic sorbates mainly interact with the edges of the silicate ribbons, either in intracrystalline channels or in "gutters" at external surfaces. Exposed Mg^{2+} ions play a role similar to that of interlayer, exchangeable cations, in smectites. The clay, however, is non-expansible and steric hindrance severely limits intracrystalline adsorption, so that only compounds of a small molecular size and a high polarity can penetrate into the structural channels. For larger molecules adsorption is generally confined to external surfaces. Adsorption by ion exchange is of secondary importance, since only minor quantities of adsorbate are involved.

8.7.2.1 Sorption of non-polar compounds
Sorption of non-polar organic molecules on sepiolite and palygorskite appears to be restricted to external surfaces (Barrer and Mackenzie, 1954; Dandy 1968; Serna and Fernández Alvárez, 1975). Interactions are mainly of van der Waals' type, although some interaction with the Si—OH groups probably adds to the adsorption forces. This interaction accounts for the observed shift towards lower frequencies of the $3720\,cm^{-1}$ v_{OH} absorption band of sepiolite.

8.7.2.2 Sorption of polar organic compounds
Mechanisms of interaction of primary alcohol molecules with sepiolite and palygorskite have been studied in detail by IR techniques by Serna and Van Scoyoc (1979). When alcohol vapour is adsorbed on the clay just after zeolitic water has been outgassed, molecules interact with the coordinated water through H-bonding, as demonstrated by the perturbation of the IR adsorption bands of the water (Fig. 8.15). Next, after repeated adsorption–desorption cycles, a

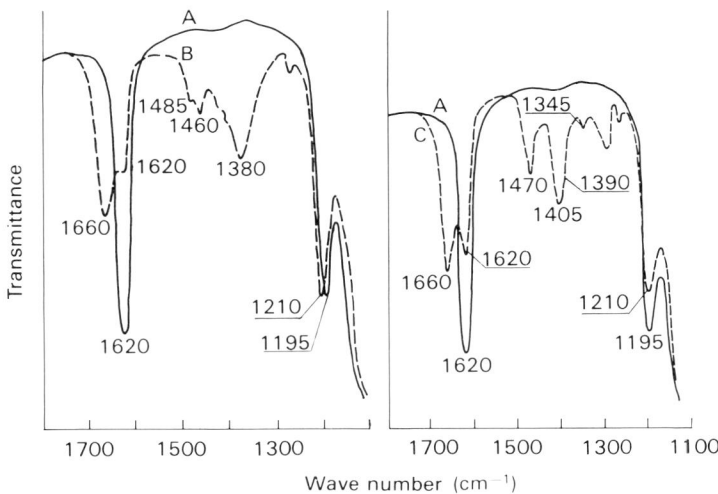

FIG. 8.15. Adsorption of alcohols on sepiolite evacuated at room temperature A: initial surface; B: exposed to an atmosphere of methanol ($p/p_0 = 1$); C: exposed to an atmosphere of n-butanol ($p/p_0 = 1$). (After Serna and Van Scoyoc 1979.)

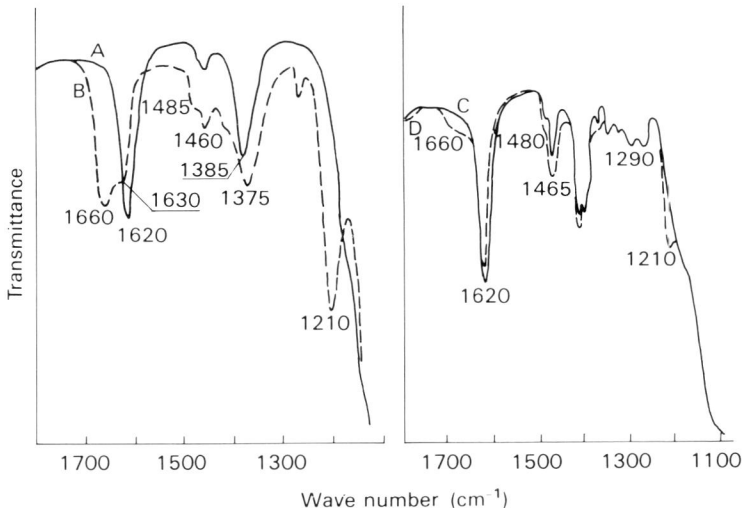

FIG. 8.16. Infrared spectra of sepiolite under vacuum after several adsorption–desorption of alcohols. A: methanol; C: n-butanol; B: as in A but re-exposed to an atmosphere of methanol; D: as in C but re-exposed to an atmosphere of n-butanol. (After Serna and VanScoyoc 1979.)

portion of the coordinated water is replaced by the alcohol, the amount depending on the alcohol as well as on the mineral. Molecules are now coordinated directly to exposed Mg^{2+} cations, and the IR spectra (Fig. 8.16) show the δ_{H_2O} band at 1620 cm^{-1} less intense and, also, the δ_{OH} band from the alcohol shifted 40 cm^{-1} to higher frequencies from its unperturbed value in vapour (1340 cm^{-1}). Regarding intracrystalline penetration the authors conclude that only short chain alcohols (methanol and ethanol) can penetrate into the internal channels. This is based on the observation that, upon further adsorption of alcohol, the remaining, non-substituted, water can form H-bonds with methanol and ethanol, but not with propanol or butanol.

Studies by Serna (1973) and Fernandez-Alvarez and Fernandez-Hernandez (1983) indicate that the adsorption of acetone and primary amines obeys the same interaction mechanisms. Thus, on adsorption of these molecules, hydrogen bonds are first formed with the coordination water, and subsequently part of that water is replaced, but, according to these authors, adsorption seems to be restricted to open channels at external surfaces. However, dimethylsulphoxide (DMSO) molecules do penetrate in the internal channels as clearly evidenced by infrared spectroscopy (Van Scoyoc personal communication).

8.7.3 Selectivity of fibrous clay minerals for adsorption of organic compounds

The surface characteristics of sepiolite and palygorskite particles confer a specificity to these minerals for organic sorbates having molecular sizes and shapes that fit closely to grooves or surface corrugations of nearly analogous dimensions, which are undoubtedly the open channels at exposed crystal faces.

This fact was first demonstrated (Nederbragt and De Jong 1946) by percolating binary mixtures of saturated hydrocarbons, i.e. long chain n-paraffins and naphthenes dissolved in pentane, through palygorskite columns. Extremely good separations of the two components were obtained, the long chain n-paraffins being the most retained by the clay. Barrer *et al.* (1954) have also found that free energy, heat and entropy changes for the n-paraffins (n-pentane) sorbed on palygorskite are greater than for the adsorption of branched-chain paraffins (isopentane or neo-pentane), and that the former are adsorbed in larger amounts than the latter. Sepiolite shows a similar behaviour but its selectivity is different from that of palygorskite; sepiolite sorbs n- and iso-pentane equally strongly but neo-pentene is less strongly adsorbed. In sepiolite the channel width is larger than in palygorskite, and it seems that the molecules of the former sorbates can be accommodated into the channels at the external surface, but not so the molecules of neo-pentane.

In both minerals, selectivity is lost after outgassing at temperatures above 100 °C, when the structure is folded and the open channels at the external surfaces collapse.

Highly specific directional binding of dyes such as acridine orange into sepiolite has been recently reported by Ridler and Jennings (1980). The dye molecules were shown to adsorb on the clay surface with their long axes parallel to the sepiolite fibre axis. Acridine orange is a triple ring, planar, cationic dye molecule whose shape approximates that of a parallelepipedic block 12 Å long, 7.2 Å wide and 3.6 Å thick, that fits remarkably well the cross-sectional dimensions of exposed channels or ribbons in the clay surface. Other dyes with planar molecular configurations of the same thickness, i.e. methylene blue or natural indigo, are also firmly adsorbed on sepiolite or palygorskite. In particular, it has been suggested that Maya blue, an exceptionally stable pigment used in Indian pottery and murals during the pre-Spanish period, is an indigo–palygorskite complex (van Olphen 1966). The colour cannot be extracted with acetone, resists hot concentrated mineral acids and persists upon heating to about 250 °C. An analogous stable pigment can be prepared from sepiolite and indigo. In both cases stability is achieved by heating the freshly prepared complex for several days to ~ 100 °C. No stable pigment can be prepared from clays with plate-like structures. Considering that, once freed of water, open channels on (010) crystal faces have the width and depth just appropriate for holding tightly the dye molecule from both sides, it appears that the constraint imposed on the molecules might indeed be the key to the high specific adsorption at the clay surface.

8.8 ORGANO-MINERAL DERIVATIVES

Under the name of organo-mineral derivatives are included those compounds in which true covalent bonds are formed between the mineral substrate and the organic molecules.

Reactions giving rise to those compounds always take place through the Si—OH groups present at the edges of the mineral particles. Appropriate organic substances react with the exposed OH groups to produce covalent bonds.

The reactivity of silanol groups towards organic molecules has been extensively studied in silica and in porous glass surfaces. As a classical example, the alkylation of silica surfaces by treatment with alcohol at relatively high temperature and pressure proceeds according to the reaction (Iler 1953)

$$\equiv Si-OH + R-OH \rightarrow Si-OR + H_2O$$

Numerous other reactions have been studied on silica surfaces (Little 1966; Hair 1967; Iler 1979), some of which have also been investigated with clay minerals. Clay surfaces have Si—OH groups only at the edges of individual particles, their content being generally low. Fibrous clays, sepiolite in particular, have contents of reactive Si—OH groups considerably larger than plate-like clays and, thus, they are adequate mineral substrates for direct grafting reactions. The following types of grafting reactions have been experimented successfully with sepiolite:

1. Grafting through Si—O—Si bonds, when the mineral reacts with organochlorosilanes (Ruiz-Hitzky and Fripiat 1976a)

$$\equiv Si-OH + Cl-\underset{R_3}{\overset{R_1}{Si}}-R_2 \rightarrow Si-O-\underset{R_3}{\overset{R_1}{Si}}-R_2 + HCl$$

The linkage of the organic molecules to the mineral substrate through Si—O—Si bonds is resistant to hydrolysis.

2. Grafting through Si—O—C bonds
   (a) by addition of alkyl- or phenyl-isocyanates (Fermández-Hernández and Ruiz-Hitzky 1979)

   $$\equiv Si-OH + O=C=N-R \rightarrow Si-O-CO-NH-R$$

   (b) by reaction with 1,2-epoxides (Casal and Ruiz-Hitzky 1977)

   $$\equiv Si-OH + R-\underset{O}{\underset{\diagdown\diagup}{CH-CH_2}} \rightarrow Si-O-\underset{CH_2OH}{\overset{}{CH}}-R$$

or (c) by reaction with diazomethane (Hermosin et al. 1982)

$$\equiv Si-OH + CH_2N_2 \rightarrow Si-O-CH_3 + N_2$$

Grafting through Si—O—C bonds leads to products which are easily hydrolysed, since those bonds are rather unstable in the presence of water.

Direct grafting reactions using montmorillonite as the mineral substrate were also investigated some years ago by several authors. This was done in an attempt to provide evidence in support of the existence of abundant Si—OH groups at the mineral surfaces, consistent with one of the two structural models proposed for the mineral at the time (Edelman and Favejee 1940). It was claimed that montmorillonite could be readily methylated with diazomethane (Berger 1941), alkylated by 1,2-epoxides (Deuel et al. 1950) or grafted directly with the anhydrides of the carboxylic acids or their halogen derivatives (Gieseking 1949), and that these reactions proceeded with the incorporation of substantial amounts of organic material into the interlayer space. However, neither the true mechanism of the above reactions nor the exact nature of the products formed has even been conclusively established, as the interpretation of the results is always obscured by the presence of residual water on the clay, which hydrolyses the majority of the organic reactants. It is now well known that montmorillonite layers expose only oxygen ions at their basal surfaces, and that Si—OH groups are only present in small amounts at the edges of particles. Consequently, only limited amounts of the organic substance can be linked by the methods described above, and when amounts in excess are found present in interlayer space, they probably are products of hydrolysis of the organic reactants and are merely retained on the clay by adsorption forces.

It seems clear, therefore, that for most clays a large number of Si—OH groups has to be created in the mineral substrate prior to the reaction if organic derivatives with substantial amounts of organic material incorporated are to be obtained. This can be easily accomplished by elimination of octahedral cations by acid attack but, in order to preserve the mineral framework, it is necessary that the removal of octahedral cations is followed by immediate fixation of the organic molecules. This is essentially the method devised by Lentz (1964) for the preparation of organic derivatives of olivine and hemimorphite: the minerals are suspended in isopropyl alcohol while HCl and hexamethyldisiloxane are added simultaneously.

Fripiat and Mendelovici (1968) have applied this method to the preparation of organic derivatives of chrysotile. According to these authors the process occurs in the following steps:

1. Creation of silanol groups by acid solution of octahedral cations

$$\equiv Si-O-Mg-O-Si\equiv + 2H_3O^+ \rightarrow 2\equiv Si-OH + Mg^{2+} + 2H_2O$$

2. Hydrolysis of hexamethyldisiloxane with formation of protonated silanol

$$(CH_3)_3Si-O-Si(CH_3)_3 + H_3O^+ \rightarrow (CH_3)_3Si-OH_2^+ + (CH_3)_3-SiOH$$

3. Condensation of the protonated silanol with the surface Si—OH groups

$$\equiv Si-OH + (CH_3)_3Si-OH_2^+ \rightarrow \equiv Si-O-Si(CH_3)_3 + H_3O^+$$

The same final product is obtained if trimethylchlorosilane $(CH_3)_3SiCl$ is used instead of hexamethyldisiloxane. That product becomes hydrolysed in the acid medium, and protonated silanol groups are formed which can be condensed with the mineral SiOH groups created by acid attack.

The trimethylsilyl derivative of chrysotile, obtained by the above method, has been well characterized by several techniques. Its infrared spectrum shows bands corresponding to the organic groups, together with bands characteristic of the silica substrate. Electron diffraction diagrams indicate that the crystalline habit of the untreated mineral is preserved after the reaction: the b-axis dimension of the derivative is 9.02 ± 0.06 Å, which corresponds well with that of a bidimensional silica sheet.

Vinyl derivatives of chrysotile, vermiculite and sepiolite have been obtained by Zapata *et al.* (1972) and by Ruiz-Hitzky and Fripiat (1976b) using methylvinyldichlorosilane. By changing the experimental conditions (temperature and concentration of chlorosilane) they were able to prepare a series of organo-mineral compounds with different contents of organic reagent attached.

The process is similar to that described above but, due to the bifunctional character of the dichlorosilanes, short polysiloxane chains are formed which may be linked either by one end or by both ends to the mineral substrates. The reactions can be written as follows:

$$\equiv Si-O-\underset{R_2}{\underset{|}{Si}}-OH + nR_1R_2Si(OH)_2 \xrightarrow{H_3O^+} \equiv Si-O-\left[\underset{R_2}{\underset{|}{Si}}-O\right]_n \underset{R_2}{\underset{|}{Si}}-OH + nH_2O$$

and

$$\equiv Si-O-\left[\underset{R_2}{\underset{|}{Si}}-O\right]_{n+1}-H + H-\left[O-\underset{R_2}{\underset{|}{Si}}\right]_m-O-Si\equiv$$

$$\xrightarrow{H_3O^+} \equiv Si-O-\left[\underset{R_2}{\underset{|}{Si}}\right]_{n+m}-\underset{R_2}{\underset{|}{Si}}-O-Si\equiv + H_2O$$

At the same time, competitive reactions take place within the liquid phase with formation of cyclosiloxanes of four to six silicon atoms, as the major subproduct (Ruiz-Hitzky and Van Meerbeek 1978). These products are soluble in benzene, and can be eliminated easily from the organo-mineral compound.

Determination of the amounts of octahedral cations released and of the C contents of the synthetic organo-mineral product at different reaction times has shown that the kinetics of the reaction may be expressed by equations of the type (Zapata *et al.* 1972):

$$\%C = a\sqrt{t} - b(1 - e^{-kt^3})$$

where %C is the percentage of carbon in the organo-mineral derivative, t is time and a, b and k are constants depending on the experimental conditions. The rate-limiting step of the reaction is the release of cations from the octahedral sheet, which is a diffusion-controlled process.

Several organo-mineral derivatives of palygorskite have been obtained by Mendelovici and Carroz-Portillo (1976), and the trimethylsilyl derivative of halloysite by Kuroda and Kato (1979). In both cases, the method of co-hydrolysis was used to increase the content of surface silanol groups. The corresponding organic derivative of kaolinite could not be obtained by the same procedure.

Interlamellar grafting producing expansion of $d(001)$ X-ray spacings has been achieved with lamellar compounds other than clay minerals, such as layered silicic acids obtained from natural magadiite or from synthetic alkaline layer silicates (Ruiz-Hitzky and Rojo 1980). Pretreatment with polar organic substances such as dimethylsulphoxide, *N*-methylformamide and *N,N*-

dimethylformamide is required. The swelling produced by the polar molecules makes the basal Si—OH groups accessible to the grafting reagent so that subsequent grafting can occur. Reagents used were trimethylchlorosilane and hexamethyldisilazane. X-ray basal spacings correspond to the presence of two monomolecular layers of grafted material in the interlayer space. These materials can be thought of as planar silicones.

The organic derivatives of clay minerals are organo-mineral macromolecules whose organic content can be controlled. They are materials of interest because they have the surface properties and the reactivity corresponding to the grafted organic molecules, while they preserve the mechanical and thermal properties of the mineral framework. When the molecules attached contain unsaturated groups it becomes possible to copolymerize the organo-mineral compound with several monomers such as methylmethacrylate, methyl-, ethyl-, n-butyl, n-dodecyl acrylates, vinyl acetate, styrene, methyl-vinyl-cetone and diacetone-acrylamide. The addition of vinyl derivatives of chrysotile or sepiolite to natural or synthetic rubber and subsequent polymerization produces elastomers with special physical properties (Zapata *et al.* 1973; Ruiz-Hitzky 1974). Finally, double bonds of vinyl- or allyl-derivatives can fix chemically compounds of transition metal elements as osmium tetroxide (Barrios *et al.* 1974, 1981); this possibility opens ways for the preparation of new supported catalysts.

NOTE

1. For definition of pK convention see D.D. Perrin, *Dissociation Constants of Organic Bases in Aqueous Solution*, Butterworths, London. 1965.

REFERENCES

ABDO S., CANESSON P., CRUZ M.I., FRIPIAT J.J. & VAN DAMME H. (1981) Photochemical and photocatalytic properties of adsorbed organometallic compounds. II. Structure and photoreactivity of tris (2,2'-bipyridine) Ruthenium (II) and Chromium (III) at the solid–gas interface on hectorite. *J. Phys. Chem., Ithaca* **85**, 797–809.

ABDO S., CRUZ M.I. & FRIPIAT J.J. (1980) Metallation–demetallation reaction of tin tetra (4-pyridyl) porphyrin in Na-hectorite. *Clays Clay Miner.* **28**, 125–129.

AHLRICHS J.L., SERNA J.C. & SERRATOSA J.M. (1975) Structural hydroxyls in sepiolites. *Clays Clay Miner.* **23**, 119–124.

ALIETTI A. (1966) Identification of disordered kaolinites. *Clay Minerals* **6**, 229–231.

ANNABI-Bergaya F., CRUZ M.I., GATINEAU L. & FRIPIAT J.J. (1980) Adsorption of alcohols by smectites: III Nature of bond. *Clay Minerals*, **15**, 225–238.

ANNABI-BERGAYA F., CRUZ M.I., GATINEAU L. & FRIPIAT J.J. (1981) Adsorption of alcohols by smectites: IV Models. *Clay Minerals* **16**, 115–122.

ARAGON F., CANO-RUIZ J. & MACEWAN D.M.C. (1959) β-type interlamellar sorption complexes. *Nature* **183**, 740–741.

ARAGON F. & EVOLE N. (1974) Polymerization in the interlamellar space of montmorillonite, 2nd Meeting European Clay Groups, Strasbourg, Abstracts, p. 35.

BAILEY G.W. & WHITE J.L. (1970) Adsorption of pesticides in soils, *Residue Rev.* **32**, 29–92.

BAILEY G.W., WHITE J.L. & ROTHBERG T. (1968) Adsorption of organic herbicides by montmorillonite. *Soil Sci. Soc. Am. Proc.* **32**, 222–234.

BARRER R.M. & MACKENZIE N. (1954) Sorption by attapulgite. Part I. Availability of intracrystalline channels. *J. Phys. Chem., Ithaca* **58**, 560–568.

BARRER R.M., MACKENZIE N. & MACLEOD D.M. (1954) Sorption by attapulgite. Part II. Selectivity shown by attapulgite, sepiolite and montmorillonite for n-paraffins. *J. Phys. Chem., Ithaca*, **58**, 568–572.

BARRIOS J., PONCELET G. & FRIPIAT J.J. (1981) Nitrogen retention by an osmium complex supported on sepiolite. *J. Catal.* **68**, 362–370.

BARRIOS J., RODRIQUE L. & RUIZ-HITZKY E. (1974) Mise en évidence de groupements organiques insaturés greffés sur la sepiolite. *J. Microscopie* **20**, 295–298.

BASSETT W.A. (1959) Vermiculite deposit at Libby, Montana. *Am. Miner.* **44**, 282–299.

BENESI H.A. (1956) Acidity of catalyst surfaces. I. Acid strength from colors of adsorbed indicators. *J. Am. Chem. Soc.* **78**, 5490–5494.

BERGER G. (1941) Struktur van montmorilloniet. *Chem. Weekblat* **38**, 42–43.

BISSADA K.K. & JOHNS W.D. (1969) Montmorillonite–organic complexes. Energies of interactions, *Clays Clay Miner.* **17**, 197–204.

BISSADA K.K., JOHNS W.D. & CHENG F.S. (1967) Cation–dipole interactions on clay organic complexes. *Clay Minerals* **7**, 155–166.

BLUMSTEIN A. (1961) Etude des polymerisations en couche adsorbée. *Bull. Soc. Chim., France* 899–914.

BLUMSTEIN A. (1965) Polymerization of adsorbed monolayers. *J. Polym. Sci.* **A3**, 2653–2644.

BLUMSTEIN A. (1970) Polymerization in preoriented media. *Adv. Macrom. Chem.* **2**, 123–148.

BLUMSTEIN R., BLUMSTEIN A. & PARIKH K.K. (1974) Polymerization of monomolecular layers adsorbed on montmorillonite: cyclization in polyacrylonitrile and polymethacrylonitrile. *Appl. Polym. Symp.* **25**, 81–88.

BODENHEIMER W. & HELLER L. (1967) Sorption of α-amino-acids by Cu-montmorillonite. *Clay Minerals* **7**, 167–176.

BODENHEIMER W., HELLER L. & BARTURA J. (1967) Intersalation of K-acetate in flint clay. *Clay Minerals* **7**, 237–239.

BRADLEY W.F. (1940) Structure of attapulgite. *Am. Miner.* **25**, 405–410.

BRADLEY W.F. (1945) Molecular associations between montmorillonite and organic liquids. *J. Am. Chem. Soc.* **67**, 975–981.

BRADLEY W.F., WEISS E.J. & ROWLAND R.A. (1963) A glycol–sodium vermiculite complex. *Clays Clay Miner.* **10**, 117–122.

BRAUNER K. & PREISINGER A. (1956) Struktur und Entstehung des Sepioliths. *Tschermaks Miner.-Petrogr. Mitt.* **6**, 120–140.

BRINDLEY G.W. (1956) Allevardite, a swelling double layer mica mineral. *Am. Miner.* **41**, 91–103.

BRINDLEY G.W. (1970) Organic complexes of silicates. *An. Reun. Hisp. Belga Min. Arci., Madrid* 55–56.

BRINDLEY G.W. (1971) Clay organic studies in the USA. A review. US–Japan Seminar on Clay–Organic Complexes. *Abstracts*, 1–4.

BRINDLEY G.W. & GOZEN ERTEM (1971) Preparation and solvation properties of some variable charge montmorillonite. *Clays Clay Miner.* **19**, 399–404.

BRINDLEY G.W. & TSUNASHIMA A. (1972) Montmorillonite complexes with dioxane, morpholine and piperidine. Mechanisms of formation. *Clays Clay Miner.* **20**, 233–240.

BRUQUE S., MORENO-REAL L., MOZAS T. & RODRIGUEZ-GARCIA A. (1982) Interlayer complexes of lanthanide–montmorillonites with amines. *Clay Minerals* **17**, 201–208.

CALVET R. & CHASSIN P. (1973) Complexes organiques des argiles. *Bull. Grpe Franc. Argiles* **25**, 87–112.

CASAL B. (1983) Estudio de la interacción de compuestos macrocíclicos (éteres-corona y criptandos) con filosilicatos. Ph.D. Thesis. Univ. Complutense. Madrid.

CASAL B. & RUIZ-HITZKY E. (1977) Reaction of epoxides on mineral surfaces. Organic derivatives of sepiolite. *Proc. 3rd Europ. Clay Conf.*, Oslo, pp. 35–37.

CLEMENTZ D.M. & MORTLAND M.M. (1972) Interlamellar metal complexes in layer silicates. III. Silver(I)–arene complexes in smectites. *Clays Clay Miner.* **20**, 181–186.

CLOOS P. (1972) Interaction entre pesticides et la fraction minérale du sol. *Pédologie* **22**, 148–173.

CLOOS P. & LAURA R.D. (1972) Adsorption of ethylenediamine (EDA) on montmorillonite saturated with different cations. *Clays Clay Miner.* **20**, 259–275.

CLOOS P., LAURA R.D. & BADOT C. (1975) Adsorption of ethylenediamine (EDA) on montmorillonite saturated with different cations – V. Ammonium- and methylammonium-montmorillonite: ion exchange, protonation and hydrogen bonding. *Clays Clay Miner.* **23**, 417–423.

CLOOS P., VANDEPOEL D., & CAMERLYNCK J.P. (1973) Thiophene complexes on montmorillonite. *Nature (Phys. Sci)* **243**, 54–55.

COWAN C.T. & WHITE D. (1958) Exchange reactions of montmorillonite with alkylammonium salts. *Trans. Faraday Soc.* **54**, 691–697.

CRUZ M., JACOBS H. & FRIPIAT J.J. (1973) Interlayer bonding in kaolin minerals. *Proc. Int. Clay Conf. 1972, Madrid*, pp. 35–46. Division de Ciencias, C.S.I.C., Madrid.

CRUZ M., KAISER P.G., ROUXHET P. & FRIPIAT J.J. (1974) Adsorption and transformation of HCN on the surface of copper and calcium montmorillonite. *Clays Clay Miner.* **22**, 417–425.

CRUZ M., LAYCOCK A. & WHITE J.L. (1969) Perturbation of OH groups in intercalated kaolinite donor–accepted complexes. *Proc. Int. Clay Conf. 1969, Tokyo*, Vol. 1, pp. 775–789. Israel Universities Press, Jerusalem.

CRUZ M., LAYCOCK A. & WHITE J.L. (1970) Complexes de kaolinite avec urée et ses derivés. *Proc. Reunion Hispano-Belga de Minerales de la Arcilla*, pp. 148–153. Consejo Superior de Investigaciones Cientificas, Madrid.

CRUZ M.I., NIJS H., FRIPIAT J.J. & VAN DAMME H. (1982) Photochemical and photocatalystic properties of adsorbed coordination compounds. 3. Cis $\rightleftharpoons$ trans isomerization of Ru $(bpy)_2(H_2O)_2^{2+}$ on layer lattice silicates. *J. Chim. Phys.* **79**, 753–757.

DANDY A.J. (1968) Sorption of vapors by sepiolite. *J. Phys. Chem., Ithaca* **72**, 334–339.

DANDY A.J. & NADIYE-TABBIRUKA M.S. (1982) Surface properties of sepiolite from Amboseli, Tanzania, and its catalytic activity for ethanol decomposition. *Clays Clay Miner.* **30**, 347–352.

DAVIDTZ J.C. (1976) The acid activity of 2:1 layer silicates. *J. Catal.* **43**, 260–263.
DEGENS E.T. (1971) Cellular processes at work in sediments. US–Japan Seminar on Clay–Organic Complexes. *Abstracts* 85–89.
DEGENS E.T., MATHEJAR J. & JACKSON T.A. (1970) Template catalysis: asymmetric polymerization of aminoacids on clay minerals. *Nature, Lond.* **227**, 492–493.
DEGENS E.T. & MOPPER K. (1975) Early diagenesis of organic matter in marine soils. *Soil Sci.* **119**, 65–72.
DEUEL H., HUBER G. & IBERG R. (1950) Organische Derivate von Tonmineralien. *Helv. Chim. Acta* **33**, 1229–1232.
DOEHLER R.W. & YOUNG W.A. (1962) Adsorption of quinoline by clay minerals. *Clays Clay Miner.* **9**, 468–483.
DONER H.E. & MORTLAND M.M. (1969a) Intermolecular interactions in montmorillonite: NH–CO systems. *Clays Clay Miner.* **17**, 265–271.
DONER H.E. & MORTLAND M.M. (1969b) Benzene complexes with Cu-montmorillonite. *Science* **166**, 1406–1407.
DOWDY R.H. & MORTLAND M.M. (1967) Alcohol–water interactions on montmorillonite surfaces. I. Ethanol. *Clays Clay Miner.* **15**, 259–271.
DOWDY R.H. & MORTLAND M.M. (1968) Alcohol–water interactions on montmorillonite surfaces. II. Ethylene glycol. *Soil Sci.* **105**, 36–45.
DRITS V.A. & ALEKSANDROVA V.A. (1966) The crystallochemical nature of palygorskite. *Zap. Vses. Miner. Obshch.* **95**, 551–560.
DUCROS P. & DUPONT M. (1962) Etude par RMN des protons dans les argiles. *Bull. Grpe Franc. Argiles* **13**, 59–63.
EDELMAN C.H. & FAVEJEE S.C.L. (1940) On the crystal structure of montmorillonite and halloysite. *Z. Kristallogr.* **102**, 417–431.
FARMER V.C. (1971) Adsorption bonds in clays. *Soil Sci.* **112**, 62–68.
FARMER W.J. & AHLRICHS J.L. (1969) Adsorption of urea and derivatives by montmorillonite. *Soil Sci. Soc. Am. Proc.* **33**, 254–258.
FARMER V.C. & MORTLAND M.M. (1965) Infrared study of complexes of ethylamine with ethylammonium and copper ions in montmorillonite. *J. Phys. Chem., Ithaca* **69**, 683–686.
FARMER V.C. & MORTLAND M.M. (1966) Infrared study of the coordination of pyridine and water to exchangeable cations in montmorillonite and saponite. *J. Chem. Soc.* (A) 344–351.
FARMER V.C. & RUSSELL J.D. (1971) Interlayer complexes in layer silicates. The structure of water in lamellar ionic solutions. *Trans. Faraday Soc.* **67**, 2737–2749.
FENN D.B. & MORTLAND M.M. (1973) Interlamellar metal complexes in layer silicates: II. Phenol complexes in smectites. *Proc. Int. Clay Conf. 1972, Madrid*, pp. 591–603. Division de Ciencias, C.S.I.C., Madrid.
FENN D.B., MORTLAND M.M. & PINNAVAIA T.J. (1973) The chemisorption of anisole on Cu(II) hectorite. *Clays Clay Miner.* **21**, 315–322.
FENOLL P. & MARTIN-VIVALDI J.L. (1970) Complejos de montmorillonita con alcoholes alifáticos. *An. Real Soc. Esp. Fis. Quim.* **66B**, 133–140.
FERNANDEZ-ALVAREZ T. & FERNANDEZ-HERNANDEZ M.N. (1983) Efecto de la deshidratación sobre las propiedades adsorbentes de sepiolita y palygorskita. II. Adsorción de sustancias orgánicas. *An Quim.* **79B**, 342–347.
FERNANDEZ-HERNANDEZ M.N. & RUIZ-HITZKY E. (1979) Interracción de isocianatos con sepiolita. *Clay Minerals* **14**, 295–305.
FORNES V., RAUSELL-COLOM J.A., HIDALGO A. & SERRATOSA J.M. (1973) Une réaction de condensation dans l'espace interlamellaire de la vermiculite. *C.r. hebd. Séanc. Acad. Sci., Paris* **227B**, 635–637.
FOWKES F.M., BENESI H.A., RYLAND R.B., SAWYER W.M., DETLING K.D., LOEFFLER E.S., FOLCKEMER F.B., JOHNSON M.R. & SUN Y.P. (1960) Clay-catalyzed decomposition of insecticides. *J. Agric. Food Chem.* **8**, 203–210.
FRENKEL M. (1974) Surface acidity of montmorillonites. *Clays Clay Miner.* **22**, 435–441.
FRIPIAT J.J. (ed.) (1981) *Advanced Techniques for Clay Mineral Analysis*. Elsevier, Amsterdam.
FRIPIAT J.J., HELSEN J. & VIELVOYE L. (1964) Formation des radicaux libres sur la surface des montmorillonites. *Bull. Grpe Franc. Argiles* **15**, 3–10.
FRIPIAT J.J., JELLI A.N., PONCELET G. & ANDRE J. (1965) Thermodynamic properties of adsorbed water and electrical conduction in montmorillonite. *J. Phys. Chem., Ithaca* **69**, 2185–2197.
FRIPIAT J.J. & MENDELOVICI E. (1968) Dérivés organiques des silicates. *Bull. Soc. Chim., France* 483–492.
FRIPIAT J.J., SERVIAS A. & LEONARD A. (1962) Adsorption des amines par montmorillonites. *Bull. Soc. Chim., France* 635–644.
FURUKAWA T. & BRINDLEY G.W. (1973) Adsorption and oxidation of benzidine and aniline by montmorillonite and hectorite. *Clays Clay Miner.* **21**, 279–288.
GARRETT W.G. & WALKER G.F. (1961) Complexes of vermiculite with aminoacids. *Nature, Lond.* **191**, 1389–1390.
GARRETT W.G. & WALKER G.F. (1962) Swelling of some vermiculite organic complexes in water. *Clays Clay Miner.* **9**, 557–567.
GIESEKING J.E. (1939) Cation exchange in smectites. *Soil Sci.* **47**, 1–14.
GIESEKING J.E. (1949) The clay minerals in soils. *Adv. Agron.* **1**, 59–204.
GLAESER R. (1948) Montmorillonite–acetone complexes. *C.r. hebd. Séanc. Acad. Sci., Paris* **226**, 935–937.
GLAESER R. (1954) Complexes organo-argileux. Thése (Paris).
GOMES C.S.F. (1967) Kaolinite displaying a particular behaviour on KCH_3COO intercalation. *Mem. Nat. Mus. Lab. Min. Geol. Univ. Coimbra* **62**, 47–52.

GOMES C.S.F. (1982) Relacao entre capacidade de intercalacao en caolinites e defeitos estructurais. Bol. Soc. Geol. Portugal **23**, 55–64.

GONZALEZ CARREÑO T., RAUSELL-COLOM J.A. & SERRATOSA J.M. (1977) Complexes vermiculite–alkylammonium. Evidence of interaction of terminal —CH_3 groups with the silicate surface. *Proc. 3rd Europ. Clay Conf. Oslo*, pp. 73–74.

GONZALEZ-GARCIA S. & SANCHEZ-CAMAZANO M. (1965) Complejos minerales de la arcilla cor. dimetilsufóxido. *An. Edafol. y Agrobiol.* **24**, 495–590.

GONZALEZ-GARCIA S., SANCHEZ-CAMAZANO M. & GONZALEZ-ZAPATERO M. (1970) Complejos de fosfatos de alquilo con vermiculita. *Proc. Reunión Hispano-Belga de Minerales de la Arcilla*, pp. 103–114. Consejo Superiores de Investigaciones Scientificas, Madrid.

GREENE-KELLY R. (1955) Sorption of organic compounds by montmorillonite. *Trans. Faraday Soc.* **51**, 412–430.

GREENLAND D.J. (1965) Interaction between clays and organic compounds in soils Parts I and II. *Soils and Fertilizers* **28**, 415–425 and 521–532.

GREENLAND D.J. & QUIRK J.P. (1962) Adsorption of 1-n alkylpyridinium bromides by montmorillonite. *Clays Clay Miner.* **9**, 484–499.

GREENLAND D.J. & QUIRK J.P. (1964) Surface area determination on soils. *J. Soil Sci.* **15**, 178–191.

GRIM R.E. (1968) *Clay Mineralogy*, pp. 392–401. McGraw-Hill, New York.

GRIM R.E., ALLAWAY W.H. & CUTHBERT F.L. (1947) Reactions of clays with organic cations. *J. Am. Ceram. Soc.* **30**, 137–142.

GUINARD J. & PEZERAT H. (1979) Etude des interactions entre montmorillonites et agents organiques complexants. *Clay Minerals* **14**, 259–265.

GUTIÉRREZ-RIOS E. & RODRÍGUEZ A. (1961) Complejos montmorillonita-acetona. *An. Real Soc. Esp. Fis. Quim.* **57B**, 117–130.

GUTIÉRREZ-RIOS E., RODRÍGUEZ, A. & GALACHE M.I. (1962) Complejos organicos de montmorillonita. *An. Real. Soc. Esp. Fis. Quim.* **58B**, 53–68.

HAIR M.L. (1967) *Infrared Spectroscopy in Surface Chemistry*. Marcel Dekker, New York.

HARVEY G.R., DEGENS E.T. & MOPPER K. (1971) Synthesis of nitrogen heterocycles on kaolinite from CO_2 and NH_3. *Naturwissenschaften*, **58**(12), 624–625.

HAYES M.H.B., PICK M.E., STACEY M. & TOMS B.A. (1973) Interactions between clay minerals and bipyridilium salts. *Proc. Int. Clay Conf. 1972, Madrid*, pp. 675–682. Division de Ciencias, C.S.I.C., Madrid.

HECHT A.M., DUPONT M. & DUCROS P. (1966) Etude par RMN de l'eau adsorbée dans les argiles. *Bull. Soc. Franc. Miner. Cristallogr.* **89**, 6–13.

HELLER L. & YARIV S. (1969) Sorption of some anilines by montmorillonite. *Proc. Int. Clay Conf. 1969, Tokyo*, Vol. 1, pp. 741–755.

HELLER-KALLAI L., YARIV S. & RIEMER M. (1973) Effect of the acidity on the sorption of histidine by montmorillonite. *Proc. Int. Clay Conf. 1972, Madrid*, pp. 651–662. Division de Ciencias, C.S.I.C., Madrid.

HENDRICKS S.B. (1941) Base exchange of montmorillonite for organic cations and its dependence upon adsorption due to van der Waals forces. *J. Phys. Chem., Ithaca* **45**, 65–81.

HERMOSIN M.C., CORNEJO J. & PEREZ-RODRIGUEZ J.L. (1982) Reacción de sepiolita y palygorskita con diazometano I. Reunión Iberoamericana de Arcillas (Abstracts). Torremolinos, Spain.

HOFMANN U. & KLEMEN R. (1950) Verlust der Austauschfahigkeit von Lithiumionen an Bentonit durch Erhitzung. *Z. Anorg. Allg. Chem.*, **262**, 95–99.

HUNT J.P. (1963) *Metal Ions in Aqueous Solution*. Benjamin Inc., New York.

ILER R.K. (1953) Esterifying the reactive siliceous surfaces of supercolloidal substrates. US Patent, 2.657.149 (*Chem. Abstr.* **48**, 2336c, 1954).

ILER R.K. (1979) *The Chemistry of Silica*. John Wiley, New York.

INOOKA M., KASUYA M. & MATSUDA M. (1978a) Three-stage hydrocracking of fuel oils. Japan Kokai 78.101,004. (*Chem. Abstr.* **90**, 57732, 1979).

INOOKA M., NAKAMURA M. & MORIMOTO T. (1978b) Two-stage hydrodesulfurization of heavy fuel oils. Japan Kokai 78.098,308 (*Chem. Abstr.* **90**, 74251, 1979).

IOKA M., WAKBAYASHI M. & OGUCHI Y. (1978) Hydrogenation catalysts for hydrocarbon oil. Japan Kokai 78.034,691 (*Chem. Abstr.* **89**, 8724, 1978).

JACKSON T.A. (1971) Preferential polymerization and adsorption of L-optical isomers of amino acids relative to D-optical isomers on kaolinite templates. *Chem. Geol.* **7**, 295–306.

JACOBS H. & STERCKX M. (1970) Contribution a l'étude de l'intercalation du dimethylsulfoxide dans le reseau de la kaolinite. *Proc. Reunion Hispano Belga de Minerales de la Arcilla*, pp. 154–160. Consejo Superiores de Investigaciones Cientificas, Madrid.

JOHNS W.D. & SEN GUPTA, P.K. (1967) Vermiculite alkylammonium complexes. *Am. Miner.* **52**, 1706–1724.

JORDAN J.W. (1949a) Reaction of bentonite with amines. *Mineralog. Mag.* **28**, 598–605.

JORDAN J.W. (1949b) Organophilic bentonites. *J. phys. colloid. Chem., Ithaca* **53**, 294–306.

KRENSKE D., ABDO S., VAN DAMME H., CRUZ M. & FRIPIAT J.J. (1980) Photochemical and photocatalytic properties of adsorbed organometallic compounds. I. Luminescence quenching of tris (2,2′-bipyridine) Ruthenium(II) and Chromium(III) in clay membranes. *J. Phys. Chem., Ithaca* **84**, 2447–2457.

KUBLER B. (1962) L'Ochningien (Tortonien) du Locle (Suisse occidentale). *Beitr. Miner. Petrogr.* **8**, 267–314.
KURODA K. & KATO C. (1979) Synthesis of the trimethylsilylation derivative of halloysite. *Clays Clay Miner.*, **27**, 53–56.
LABY R.H. (1962) Adsorption of amino acids and peptides by montmorillonite. Ph.D. Thesis. Adelaide.
LAGALY G. (1976) Kink-block and gauche-block structures of bimolecular films. *Angew. Chem. Int. Ed. Eng.* **15**, 575–586.
LAGALY G. & WEISS A. (1969) Determination of layer charge in mica-type layer silicates. *Proc. Int. Clay Conf. 1969, Tokyo,* Vol. 1, pp. 61–68. Israel Universities Press, Jerusalem.
LAGALY G., STANGE H. & WEISS A. (1973) Adaptation of long chain molecules onto aromatic liquids in mica type layer silicates. *Proc. Int. Clay Conf. 1972, Madrid,* pp. 693–704. Division de Ciencias, C.S.I.C., Madrid.
LAILACH G.E. & BRINDLEY G.W. (1969) Specific coadsorption of purines and pyrimidines by montmorillonite. *Clays Clay Miner.* **17**, 95–100.
LAILACH G.E., TOMPSON T.D. & BRINDLEY G.W. (1968) Adsorption of pyrimidines, purines and nucleosides by Co-, Ni-, Cu-, and Fe(III)-montmorillonite. *Clays Clay Miner.* **16**, 295–301.
LAURA R.D. & CLOOS P. (1972) Adsorption of ethylendiamine on Cu-montmorillonite. *Proc. Reunión Hispano-Belga de Minerales de la Arcilla,* pp. 76–86. Consejo Superiores de Investigaciones Scientificas, Madrid.
LEDOUX R.L. & WHITE J.L. (1966) Infrared studies of hydrogen bonding between kaolinite surfaces and intercalated potassium acetate, hydrazine, formamide and urea. *J. Colloid Interface Sci.* **21**, 127–152.
LENTZ C.W. (1964) Trimethylsilyl derivatives of silicate minerals. *Inorg. Chem.* **3**, 574–579.
LITTLE L.H. (1966) *Infrared Spectra of Adsorbed Species.* Academic Press, London.
MACEWAN D.M.C. (1946) Halloysite–organic complexes. *Nature* **157**, 159–160.
MACEWAN D.M.C. (1948) Complexes of clays with organic compounds I. Complex formation between montmorillonite and halloysite and certain organic liquids. *Trans. Faraday Soc.,* **44**, 349–368.
MACEWAN D.M.C. & WILSON M.J. (1980) Interlayer and intercalation complexes of clay minerals. Ch. 3 in *Crystal Structures of Clay Minerals and their X-ray Identification* (G.W. Brindley and G. Brown, eds). Mineralogical Society, London.
MACKINTOSH E.E., LEWIS D.G. & GREENLAND D.J. (1971) Dodecylammonium mica complexes. I. Factors affecting the exchange reaction. *Clays Clay Miner.* **19**, 209–219.
MAES A., PEIGNEUR P. & CREMERS A. (1976) Thermodynamics of transition metal ion exchange in montmorillonite. *Proc. Int. Clay Conf. 1975, México,* pp. 319–329. Applied Publishing Ltd, Wilmette, Illinois.
MAES A., PEIGNEUR P. & CREMERS A. (1978) Stability of metal–uncharged ligand complexes in ion exchangers. II. The copper ethylendiamine complex in montmorillonite and sulphonic resin. *J. chem. Soc. Faraday I* **74**, 182–189.
MANARA G. & TARAMASSO M. (1972) Sieving effects of hormites in gas-adsorption chromatography. *J. Chromatog.* **65**, 349–353.
MARTÍN-RUBI J.A., RAUSELL-COLOM, J.A. & SERRATOSA J.M. (1974) Infrared absorption and X-ray diffraction study of butylammonium complexes of phyllosilicates. *Clays Clay Miner.* **22**, 87–90.
MARTIN-VIVALDI J.L. & ROBERTSON R.H.S. (1971) Palygorskite and sepiolite (the hormites). In *Electron-optical Investigation of Clays* (J.A. Gard, ed.), pp. 255–276. Mineralogical Society, London.
MATA A., RUIZ-AMIL A. & INARAJA E. (1970) Cinética del proceso de sorción del DMSO en caolinita: estudio por difracción de rayos X. *Proc. Reunión Hispano-Belga de Minerales de la Arcilla,* pp. 115–120. C.S.I.C., Madrid.
MCATEE J.L. & HACKMAN J.R. (1964) Exchange equilibria on montmorillonite, *Am. Miner.* **49**, 1569–1577.
MCBRIDE M.B. & MORTLAND M.M. (1973) Segregation and exchange properties of alkylammonium ions in a smectite and vermiculite. *Clays Clay Miner.* **21**, 323–329.
MENDELOVICI E. & CARROZ-PORTILLO D. (1976) Organic derivatives of attapulgite. I. Infrared spectroscopy and x-ray diffraction studies. *Clays Clay Miner.* **24**, 177–182.
MIFSUD A. (1975) Cinética de la reacción de cambio iónico de vermiculita con monocloruro de L-ornitina. Ph.D. Thesis. Univ. Complutense, Madrid.
MIFSUD A., FORNES V. & RAUSELL-COLOM J.A. (1970) Cationic complexes of vermiculite with L-ornithine. *Proc. Reunión Hispano-Belga de Minerals de la Arcilla,* pp. 121–127. Consejo Superior de Investigaciones Cientificas, Madrid.
MORTLAND M.M. (1966) Urea complexes with montmorillonites. *Clay Minerals,* **6**, 143–156.
MORTLAND M.M. (1968a) Pyridinium montmorillonite complexes with EPTC. *Agric. Food Chem.* **16**, 706–707.
MORTLAND M.M. (1968b) Protonation of compounds at clay surfaces. *Trans 9th Int. Congress Soil Sci., Adelaide* **1**, 691–699.
MORTLAND M.M. (1970) Clay organic complexes and interactions. *Adv. Agron.* **22**, 75–117.
MORTLAND M.M., FRIPIAT J.J., CHAUSSIDON J. & UYTTERHOEVEN J.B. (1963) Interaction between ammonia and montmorillonite and vermiculite. *J. phys. Chem., Ithaca* **67**, 248–258.
MORTLAND M.M. & PINNAVAIA T.J. (1971) Formation of copper(II) arene complexes on the interlamellar surfaces of montmorillonite. *Nature (Phys. Sci.)* **229**, 75–77.
MORTLAND M.M. & RAMAN K.V. (1968) Surface acidity of smectites in relation to hydration, exchangeable cation and structure. *Clays Clay Miner.,* **16**, 393–398.
NEDERBRAGT G.W. & DE JONG J.J. (1946) The separation of long-chain and compact molecules by adsorption. *Rec. Trav. Chim.* 831–834.
NORRISH K. (1973) Weathering of mica to vermiculite. *Proc. Int. Clay Conf. 1972, Madrid,* pp. 417–432. Division de Ciencias, C.S.I.C., Madrid.

OLEJNIK S., POSNER A.M. & QUIRK J.P. (1970) Intercalation of polar organic compounds into kaolinite. *Clay Minerals*, **8**, 421–434.

OLEJNIK S., POSNER A.M. & QUIRK J.P. (1973) Adsorption of pyridine N-oxide into montmorillonite. *Clays Clay Miner.* **21**, 191–198.

PAECHT-HOROWITZ M., BERGER J. & KATCHALSKY A. (1970) Prebiotic synthesis of polypeptide by heterogeneous polycondensation of aminoacid adenylates. *Nature, Lond.* **228**, 630–639.

PAECHT-HOROWITZ M. & KATCHALSKY A. (1973) Synthesis of amino acyl-adenylates under prebiotic conditions. *J. Molec. Evolution* **2**, 91–98.

PARFITT R.L. & GREENLAND D.J. (1970) The adsorption of poly(ethylene glycols) on clay mineral. *Clay Minerals*, **8**, 305–316.

PARFITT R.L. & MORTLAND M.M. (1968) Ketone adsorption on montmorillonite. *Soil Sci. Soc. Am. Proc.* **32**, 355–363.

PEIGNEUR P., MAES A. & CREMERS A. (1979) Ion exchange of the polyamine complexes of some transition metal ions in montmorillonite. *Proc. Int. Clay Conf. 1978, Oxford*, pp. 207–216. Elsevier Sci. Publ. Co., Amsterdam.

PEZERAT H. & VALLET M. (1973) Formation de polymere inseré dans les couches interlamellaires de phyllites gonflantes. *Proc. Int. Clay Conf. 1972, Madrid*, pp. 683–691. Division de Ciencias, C.S.I.C., Madrid.

PFIRRMANN G., LAGALY G. & WEISS A. (1973) Phase transitions in complexes of nontronite with n-alkanols. *Clays Clay Miner.* **21**, 239–247.

PHILEN O.D., WEED S.B. & WEBER J.B. (1970) Estimation of surface charge density of clay minerals. *Soil Sci. Soc. Am. Proc.* **34**, 527–531.

PHILEN O.D., WEED S.B. & WEBER J.B. (1971) Surface charge characterization of layer silicates. *Clays Clay Miner.* **19**, 295–302.

PINNAVAIA T.J. (1977) Metal catalyzed reactions in the intracrystal space of layer lattice silicates. In *Catalysis in Organic Synthesis* (G.V. Smith, ed.), pp. 131–138. Academic Press, New York.

PINNAVAIA T.J., HALL P.L., CADY C.S. & MORTLAND M.M. (1974) Aromatic radical cation formation on the intracrystal surfaces of transition metal layer lattice silicates. *J. Phys. Chem., Ithaca* **78**, 994–999.

PINNAVAIA T.J. & MORTLAND M.M. (1971) Interlamellar metal complexes on layer silicates. I. Copper(II) arene complexes on montmorillonite. *J. phys. Chem., Ithaca* **75**, 3957–3962.

PLANCON A. (1976) Phenomene de diffraction produit par les systemes stratifiés comportant simultanement des feuillets de nature differente et des fautes d'empilement. Application a l'étude qualitative des défauts dans les kaolinites partiellement desordonnées. Ph.D. Thesis. Univ. Orleans.

PLANCON A. & TCHOUBAR C. (1977a) Determination of structural defects in phyllosilicates by X-ray powder diffraction. I. Principle of calculation of the diffraction phenomenon. *Clays Clay Miner.* **25**, 430–435

PLANCON A. & TCHOUBAR C. (1977b) Determination of structural defects in phyllosilicates by X-ray powder diffraction. II. Nature and proportion of defects in natural kaolinites. *Clays Clay Miner.* **25**, 436–450.

PLEYSIER J. & CREMERS A. (1975) Stability of silver–thiourea complexes in montmorillonite clay. *J. Chem. Soc., Faraday I* **71**, 256–264.

RAMAN R.V. & MORTLAND M.M. (1969) Proton transfer reactions at clay mineral surfaces. *Soil Sci. Soc. Am. Proc.* **33**, 313–317.

RANGE K.J., RANGE A. & WEISS A. (1969) Fire-clay type kaolinite or fire-clay mineral? Experimental classification of kaolinite-halloysite minerals. *Proc. Int. Clay Conf. 1969, Tokyo*, Vol. 1, pp. 3–13. Israel Universities Press, Jerusalem.

RAUSELL-COLOM J.A. (1964) Small angle X-ray diffraction study of the swelling of butyl-ammonium vermiculite. *Trans. Faraday Soc.* **60**, 190–201.

RAUSELL-COLOM J.A. & FORNES V. (1974) Monodimensional Fourier analysis of some vermiculite-L-ornithine complexes. *Am. Miner.*, **59**, 790–798.

RAUSELL-COLOM J.A. & SALVADOR P. (1971a) Complexes vermiculite–aminoacides. *Clay Minerals* **9**, 139–149.

RAUSELL-COLOM J.A. & SALVADOR P. (1971b) Gelification de vermiculite dans des solutions d'acide γ-aminobutyrique. *Clay Minerals* **9**, 193–208.

RAUTUREAU M. & TCHOUBAR C. (1972) Etude morphologique de la sepiolite par microscopie electronique. *J. Microscopie* **14**, 139–146.

REICHENBACH H. Graf von & RICH C.I. (1969) Potassium release from muscovite as influenced by particle size. *Clays Clay Miner.* **17**, 23–29.

RIDLER P.S. & JENNINGS B.R. (1980) Transient fluorescence studies on sepiolite suspensions. *Clay Minerals* **15**, 121–133.

RODRIGUEZ A., SANTOS A. & GUTIERREZ-RIOS E. (1967) Propiedades magnéticas y constitución del complejo de montmorillonite-Ni con acetonitrillo. *An. Real Soc. Esp. Fis. Quim.* **63B**, 303–306.

RUIZ-HITZKY E. (1974) Contribution a l'etude des reactions de greffage de groupements organiques sur les surfaces minerales. Greffage de la sepiolite. Ph.D. Thesis. Université de Louvain.

RUIZ-HITZKY E. & CASAL B. (1978) Crown ether intercalations with phyllosilicates. *Nature, Lond.* **276**, 596–597.

RUIZ-HITZKY E. & FRIPIAT J.J. (1976a) Organomineral derivatives obtained by reacting organochlorosilanes with the surface of silicates in organic solvents. *Clays Clay Miner.* **24**, 25–30.

RUIZ-HITZKY E. & FRIPIAT J.J. (1976b) Derivés organiques des silicates. III Le derivé vinilique de la sepiolite. *Bull. Soc. Chim.* 1341–1348.

RUIZ-HITZKY E. & ROJO J.M. (1980) Intracrystalline grafting on layer silicic acids. *Nature, Lond.* **287**, 28–30.
RUIZ-HITZKY E. & VAN MEERBEEK A. (1978) Mechanism of the grafting of organosilanes on mineral surfaces. I. Nature and role of the hydrolysis products of the methylvinyldichlorosilanes in the grafting of silicates in hydrochloric acid and isopropanol. *Colloid Polym. Sci.* **256**, 135–139.
RUPERT J.P. (1973) Electron spin resonance spectra of interlamellar copper(II)–arene complexes on montmorillonite. *J. Phys. Chem., Ithaca* **77**, 784–790.
RUSSELL J.D. (1965) Infrared study of the reaction of ammonia with montmorillonite and saponite. *Trans. Faraday Soc.* **61**, 2284–2294.
RUSSELL J.D., CRUZ M.I. & WHITE J.L. (1968) The adsorption of 3-aminotriazole by montmorillonite. *J. Agric. Food Chem.* **16**, 21–24.
SANCHEZ A., HIDALGO A. & SERRATOSA J.M. (1973) Adsorption des nitriles dans la montmorillonite. *Proc. Int. Clay Conf. 1972, Madrid*, pp. 617–626. Division de Ciencias, C.S.I.C., Madrid.
SANCHEZ-CAMAZANO M. & GONZALEZ-GARCIA S. (1966) Complejos interlaminares de caolinita y haoloisita con liquidos polares. *An. Edafol. y Agrobiol.* **25**, 9–25.
SAWHNEY B.L. (1972) Selective sorption and fixation of cations by clay minerals. A review. *Clays Clay Miner.* **20**, 93–100.
SCHNITZER M. & KODAMA H. (1972) Reaction between fulvic acid and Cu-montmorillonite. *Clays Clay Miner.* **20**, 359–367.
SCHOONHEYDT R.A., PELGRIMS J., HEROES V. & UYTTERHOEVEN J.B. (1977) Characterization of tris (2,2'-bipyridyl) ruthenium(II) on hectorite. *Clay Minerals* **13**, 435–438.
SCHOONHEYDT R.A., VELGHE F., BAERTS R. & UYTTERHOEVEN J.B. (1979a) Complexes of diethylenetriamine (dien) and tetraethylenepentamine (tetren) with Cu(II) and Ni(II) on hectorite. *Clays Clay Miner.* **27**, 269–278.
SCHOONHEYDT R.A., VELGHE F. & UYTTERHOEVEN J.B. (1979b) Characterization of $[Ni(en)_x]^{2+}$ ($x=1,2,3$, en = ethylenediamine) on the surface of montmorillonites. *Inorg. Chem.* **18**, 1842–1847.
SCOTT A.D. (1968) Effect of particle size on K-exchange in micas. *Trans. 9th Int. Congr. Soil Sci., Adelaide* **II**, 649–660.
SERNA J.C. (1973) Naturaleza y propiedades de la superficie de la sepiolita. Ph. Thesis Universidad Complutense, Madrid.
SERNA J.C., AHLRICHS J.L. & SERRATOSA J.M. (1975) Folding in sepiolite crystals. *Clays Clay Miner.* **23**, 452–457.
SERNA J.C. & FERNANDEZ-ALVAREZ T. (1975) Adsorción de hidrocarburos en sepiolita. II. Propiedades de superficie. *An. Quim.* **71**, 371–376.
SERNA J.C. & VAN SCOYOC G.E. (1979) Infrared study of sepiolite and palygorskite surfaces. *Proc. Int. Clay Conf. 1978, Oxford*, pp. 197–206. Elsevier Sci. Publ. Co., Amsterdam.
SERNA J.C., VAN SCOYOC G.E. & AHLRICHS J.L. (1977) Hydroxyl groups and water in palygorskite. *Am. Miner.* **62**, 784–792.
SERRATOSA J.M. (1965) Use of infrared spectroscopy to determine orientation of pyridine sorbed on montmorillonite. *Nature, Lond.* **208**, 679–681.
SERRATOSA J.M. (1966) Infrared analyses of the orientation of pyridine molecules in clay complexes. *Clays Clay Miner.* **14**, 385–391.
SERRATOSA J.M. (1968a) Infrared study of the orientation of chlorobenzene sorbed on pyridinium montmorillonite. *Clays Clay Miner.*, **16**, 93–97.
SERRATOSA J.M. (1968b) Infrared study of benzonitrile–montmorillonite complexes. *Am. Miner.* **53**, 1244–1251.
SERRATOSA J.M. (1979) Surface properties of fibrous clay minerals (palygorskite and sepiolite) *Proc. Int. Clay Conf. 1978, Oxford*, pp. 99–109. Elsevier Sci. Publ. Co., Amsterdam.
SERRATOSA J.M., JOHNS W.D. & SHIMOYAMA A. (1970) IR study of alkylammonium vermiculite complexes. *Clays Clay Miner.* **18**, 107–113.
SMITH D.L., MILFORD M.H. & ZUCKERMAN J.J. (1966) Mechanism for intercalation of kaolinite by alkali acetates. *Science*, **153**, 741–743.
SOLOMON D.M., SWIFT J.D. & MURPHY A.J. (1971) The acidity of clay minerals in polymerization and related reactions. *J. Macromol. Sci. Chem.* **A5**, 587–601.
STUCKI J.W. & BANWART W.L. (eds) (1980) *Advanced Chemical Methods for Soil and Clay Minerals Research.* Reidel Publ. Co., New York.
SUNG DO JANG & CONDRATE R.A. (1972) The IR spectra of lysine adsorbed on several cation-substituted montmorillonites. *Clays Clay Miner.* **20**, 79–82.
TAHOUN S.A. & MORTLAND M.M. (1966) Complexes of montmorillonite with primary, secondary and tertiary amides. II Coordination of amides on the surface of montmorillonite. *Soil Sci.* **102**, 248–254.
TARASEVICH YU.I. & OVCHARENKO F.D. (1973) Interaction between nitrogenous organic substances and montmorillonite. *Proc. Int. Clay Conf. 1972, Madrid*, pp. 627–636. Division de Ciencias, C.S.I.C., Madrid.
TARASEVICH YU.I., RUDENKO J.M., SHARKINA E.V. & OVCHARENKO F.D. (1970) Adsorption of alcohol on montmorillonite and vermiculite. *Kolloidn. Zh.* **32**, 266–271.
TCHOUBAR C., RAUTUREAU M., CLINARD CH. & RAGOT J.P. (1973) Technique d'inclusion appliquée a l'étude des silicates lamellaires et fibreux. *J. Microscopie* **18**, 147–154.
TENNAKOON D.T.B., THOMAS J.M., TRICKER M.J. & GRAHAM S.H. (1974) Selective organic reactions in sheet-silicate intercalates. Conversion of 4,4'-diaminotrans-stilbene into aniline. *J. Chem. Soc. Chem. Commun.* 124–125.

THENG B.K.G. (1964) Organic complexes of montmorillonite. Ph.D. Thesis. Adelaide.
THENG B.K.G. (1974) *The Chemistry of Clay Organic Reactions.* Adam Hilger, London.
THENG B.K.G. (1979) *Formation and Properties of Clay–Polymer Complexes.* Elsevier, Amsterdam.
THENG B.K.G. (1982) Clay activated organic reactions. *Proc. Int. Clay Conf. 1981, Bologna, Pavia,* pp. 197–238. Elsevier, Amsterdam.
THENG B.K.G., GREENLAND D.J. & QUIRK J.P. (1967) Adsorption of alkylammonium cations by montmorillonite. *Clay Minerals* 7, 1–17.
THOMAS J.M., ADAMS J.M., GRAHAM S.H. & TENNAKOON D.T.B. (1977) Chemical conversion using sheet-silicates intercalates. In *Solid State Chemistry of Energy Conversion and Storage* (J.B. Goudenough and M.S. Whittingham, eds), pp. 298–315. *Adv. Chem. Ser.* **163**. American Chemical Society, Washington D.C.
TOUILLAUX R., SALVADOR P., VANDERMEESCHE C. & FRIPIAT J.J. (1968) Study of water layers adsorbed in Na- and Ca-montmorillonite by the pulsed NMR technique. *Israel J. Chem.* **6**, 337–348.
TRAYNOR M.F., MORTLAND M.M. & PINNAVAIA T.J. (1978) Ion exchange and intersalation reactions of hectorite with tris-bipyridyl metal complexes. *Clays Clay Miner.* **26**, 318–326.
VAN ASSCHE J.B., VAN CAUWELAERT F.H. & UYTTERHOEVEN J.B. (1973) Sorption of organic polar gases on montmorillonite. *Proc. Int. Clay Conf. 1972, Madrid,* pp. 605–615. Division de Ciencias, C.S.I.C., Madrid.
VANDEPOEL D., CLOOS P., HELSEN J. & JANNINI E. (1973) Adsorption du benzene sur la montmorillonite cuivrique. *Bull. Grpe Franc. Argiles,* **25**, 115–126.
VAN OLPHEN H. (1966) Maya blue: a clay-organic pigment? *Science* **154**, 645–646.
VAN OLPHEN H. (1977) *An Introduction to Clay Colloid Chemistry.* 2nd edn. Ch. 11. John Wiley and Sons, New York.
VANSANT E.F. & UYTTERHOEVEN J.B. (1972) Thermodynamic of the exchange of n-alkylammonium ions in Na-montmorillonite. *Clays Clay Miner.* **20**, 47–54.
VAN SCOYOC G.E., SERNA J.C. & AHLRICHS J.L. (1979) Structural changes in palygorskite during dehydration and dehydroxylation. *Am. Miner.* **64**, 215–223.
WADA K. (1961) Lattice expansion of kaolin minerals. *Am. Miner.* **46**, 79–91.
WALKER G.F. (1959) Diffusion of exchangeable cations in vermiculite. *Nature,* **184**, 1392–1394.
WALKER G.F. (1963) Ion exchange in clay minerals. Introductory speech. *Proc. Int. Clay Conf. 1963, Stockholm,* Vol. 2, pp. 259–261. Pergamon Press, Oxford.
WALKER G.F. (1967) Interactions of n-alkylammonium with mica-type layer lattices. *Clay Minerals* **7**, 129–143.
WEED S.B. & WEBER J.B. (1968) Competitive adsorption of divalent organic cations by layer silicates. *Am. Miner.* **53**, 478–490.
WEED S.B. & WEBER J.B. (1969) Retention of diquat and paraquat by clay minerals. *Soil Sci. Soc. Am. Proc.* **33**, 379–382.
WEISS A. (1961) Eine schichteinschlussverbindung von kaolinit mit Harnstoff. *Angew. Chem.* **73**, 736.
WEISS A. (1963) Organic derivatives of mica-type layer-silicates. *Angew. Chem. Int. Ed. Engl.* **2**, 134–143.
WEISS A. (1969) *Organic Geochemistry.* Ch. 31, pp. 737–781. (Eglinton–Murphy, eds). Springer-Verlag, Berlin.
WEISS A. (1981) Replication and evolution in inorganic systems. *Angew. Chem. Int. Ed. Engl.* **20**, 850–860.
WEISS A., BECKER H.O., ORTH H., MAI G., LECHNER H. & RANGE K.J. (1969) Particle size effects and reaction mechanism of the intercalation into kaolinite. *Proc. Int. Clay Conf. 1969, Tokyo,* Vol. 2, pp. 180–184. Israel Universities Press, Jerusalem.
WEISS A. & KANTNER I. (1960) Uber eine einfache Möglichkeit zur Abschätzung der Schichtladung glimmerartiger Schichtsilikate. *Z. Naturforsch* **16b**, 804–807.
WEISS A., MICHEL E. & WEISS AL. (1958) Wasserstoffbrukenbindungen. Ein-und zweidimensionale innerkristalline Quellungsvorgänge. Hydrogen bonding (a symposium) (D. Hadzi, ed.), pp. 495–508. Pergamon Press, London.
WEISS A. & ROLOFF G. (1964) Hamin montmorillonite. *Z. Naturforsch* **19b**, 533–534.
WEISS A., THIELEPAPE W., GORING G., RITTER W. & SCHAFFER H. (1963) Kaolinit-Einlagerungsverbindungen. *Proc. Int. Clay Conf. 1963, Stockholm,* Vol. 1, pp. 287–305. Pergamon Press, Oxford.
WEISS A., THIELEPAPE W. & ORTH H. (1966) Neue Kaolinit-Einlagerungsverbindungen. *Proc. Int. Clay Conf. 1966, Jerusalem.* Vol. 1, pp. 277–293. Israel Program for Scientific Translations, Jerusalem.
WHITTAM T.V. (1977) Catalyst mass for removal of hetero atoms from hydrocarbons. Ger. Offen. 2.655.879 (*Chem. Abstr.* **87**, 91415, 1977).
WIEWIORA A. & BRINDLEY G.W. (1969) Potassium acetate intercalation in kaolinite and its removal; effect of material characteristics. *Proc. Inter. Clay Conf. 1969, Tokyo,* Vol. 1, pp. 723–733. Israel Universities Press, Jerusalem.
YAMANAKA S., KANAMARU F. & KOIZUMI M. (1971) Reaction of montmorillonite with acrylonitrile. US Japan Seminar on Clay Organic Complexes. *Abstracts,* 31–35.
YARIV S., BODENHEIMER W. & HELLER L. (1964) Fe-pyrocatechol clay complex. *Israel J. Chem.* **2**, 201–208.
YARIV S., RUSSELL J.D. & FARMER V.C. (1966) Adsorption of benzoic acid and nitrobenzene in montmorillonite. *Israel J. Chem.* **4**, 201–213.
YARIV S., HELLER L. & KAUFHER N. (1969) Sorption of aniline derivatives in montmorillonite. *Clays Clay Miner.* **17**, 301–308.
ZAPATA L., CASTELEIN J., MERCIER J.P. & FRIPIAT J.J. (1972) Derivés organiques des silicates. II. Les derivés vinyliques et allyliques du chrysotile et de la vermiculite. *Bull. Soc. Chim.* 54–63.
ZAPATA L., VAN MEERBEEK A., FRIPIAT J.J., DELLA FAILLE M., VAN RUSSELT & MERCIER J.P. (1973) Synthesis and physical properties of graft copolymers derived from phyllosilicates: I. The chrysotile and vermiculite derivatives. *J. Polymer Sci., Symp.* **42**, 257–272.

Chapter 9

Petrologic Phase Equilibria in Natural Clay Systems

B. VELDE and A. MEUNIER

	Page		Page
9.1 INTRODUCTION	423	9.3.2 Systems closed to chemical migrations – example: shale diagenesis	437
9.2 INTERPRETATIONAL NOTE	424		
9.2.1 Definition of thermodynamic terms	424	9.3.3 Open microsystems – example: glauconite pellets	440
9.9.2 Chemiographic analysis	424		
9.2.3 Mineral compositions and system compositions	426	9.3.4 The effect of bulk composition through oxidation–reeduction – example: burial diagenesis	441
9.2.3.1 Components as extensive variables	426	9.4 CLAY GENESIS IN WEATHERING AREAS	443
9.2.3.2 Components as intensive variables	430	9.4.1 Types of microsystems	443
9.3 DEFINITION OF TYPES OF CHEMICAL PROCESSES	433	9.4.2 Weathering of granitic rocks	444
		9.4.3 Weathering of basic rocks	448
9.3.1 System of chemical concentration	433	9.4.4 Weathering of ultrabasic rocks	451
9.3.1.1 Sepiolite–palygorskite sedimentation in saline lakes and basins	433	9.5 CLAY MINERAL GENESIS IN HYDROTHERMAL ALTERATION	453
		9.5.1 Example of potassic alteration	453
9.3.1.2 Calcrete–caliche soil formation	434	9.5.2 Example of phyllic alteration	454
		9.6 CONCLUSIONS	455
9.3.1.3 Deep-sea sediments	436	NOTE	456
		REFERENCES	456

9.1 INTRODUCTION

This chapter is designed to illustrate an analytical method which can be used with great profit to interpret the occurrence of clay mineral assemblages in nature. The ideas are not new in that they were set out at the end of the nineteenth and beginning of the twentieth century. The systematic application of these ideas was made in metamorphic petrology in the mid-twentieth century. This method can profitably be used in clay mineralogy. The object of this chapter is to introduce the use of ideas which can be easily applied to clays by choosing the correct variables which were active in nature chemistry, temperature and chemical potential or its derivative activity. The representation, graphically, is important also for the simple reason that it often allows us to see what should be evident and to communicate this in a compact and simple form to our fellow clay mineralogists.

The choice of examples which was made here is based upon ease of representation, completeness of data for a given system and the space available for this discussion. The result is that much effort is spent on examples of rock weathering for which, at present, most detail is available through the use of modern methods such as electron microprobe analyses and diffractometer step-scanning of small samples. Other systems can certainly be used and one would hope will be in the future.

9.2 INTERPRETATIONAL NOTE

9.2.1 Definition of thermodynamic terms

We shall be using several simple terms commonly used in verbal thermodynamic descriptions – stable, metastable and unstable, as well as inert and mobile chemical components associated with intensive and extensive variables. There is no need to go through the mathematical definitions of these terms, for it is easily found in most college texts. However, it might be worth while to be reminded how these terms can be used in interpreting clay mineral occurrences in nature.

First, stable means a phase or assemblage which will never be replaced by another at given P–T–x (pressure–temperature–composition) conditions. Metastable is used to describe a phase or assemblage which should be replaced by another but which remains present for geologists to see during geological time spans, say a million years or so. Unstable means a phase or assemblage which will be replaced within the million years or so, at least partially. If a stable phase cannot crystallize because of an energy barrier of some sort, a metastable form will replace it and this form will control chemical activity of ions in solution and eventually other phases which will be present if their rate of crystallization is rapid enough. One can observe reactions between metastable phases; these reactions can be reversed in nature as well as in the laboratory. One will not see reversible reactions between unstable phases. Our job is then to identify stable and metastable phases on the one hand and unstable phases on the other. The basic assumption made here is that phases which crystallize commonly in nature are stable or metastable. The difference between the two will be measured by their relative abundance as a function of time under the same P–T conditions which must be verified. Then one must see whether or not this metastable–stable transition influences the other phases present.

The second sequence of terms is "inert" and "mobile" components used in the context of intensive or extensive variables. An "inert" component is one where the mass of the elements or their relative atomic abundance decides the phases which will form. It is an extensive variable. One can consider this as representing the instances where a group of solids imposes the chemical activity of their components on the solution with which they are in contact. However, if this solution moves out of the system, some of the components of the solids will be transported also, even if in small quantities. This being the case, the solids buffer the chemical potential of the elements in solution, yet these so-called "inert" components do move out of (or possibly into) the system.

Now when we consider a so-called "mobile" component, an intensive variable, the solution controls the content or chemical potential of the element in the solids. If the chemical potential is low, the solids will give up this element to the solution. If the chemical potential of an element is high, the solids might well incorporate more of it into their bulk composition. The chemical potential originates outside the system in question. It is convenient to think of these elements as mobile, as defined by Korzhinskii (1959). However, both "inert" and "mobile" chemical components can be transported into and out of a given system. The important relationship is whether the solution buffers the solid phases or vice versa. Correct use of these simple and old concepts allows us to interpret phase assemblages found in weathering and hydrothermal clay mineral environments.

9.2.2 Chemiographic analysis

The first task we must undertake in a phase analysis of clay minerals is to place the different mineral types in an appropriate chemical context. It is best to start with the basic ionic species in each clay. This initial step is one of simplification. One must determine which elements play homologous roles in the chemistry of clay minerals. The divalent "metallic" cations such as Mg^{2+}, Fe^{2+}, Mn^{2+} can be grouped together as soon as they occur in octahedrally coordinated sites and seem, for the most part, to be entirely interchangeable in any proportion. This is certainly true for the Fe^{2+}–Mg^{2+} identity but there is less information for manganese. However, manganoan species are quite rare and thus their occurrence will not substantially alter our general analysis. The next simplification is the assimilation of Al^{3+} and Fe^{3+}. This is less likely to be rigorously adhered to in nature. Certainly there appears to be a certain lack of continuity between aluminous and ferric iron phases such as illite and glauconite (Velde and Odin 1976). This is possibly true also for the beidellite–nontronite series. However, in many analyses it is useful to group Al and Fe^{3+} together, especially when the mineral system is predominantly aluminous or ferric. The above simplifications give us two components which we can designate as R^2 and R^3. The R^2 component is almost uniquely found in the octahedral site while the R^3, especially alumina, is found in both octahedral and tetrahedrally coordinated sites of clays. One difference between Al^{3+} and Fe^{3+} is that ferric iron is most often an octahedral ion while aluminium can be found in both sites.

A second type of ion which we must consider is the alkali–alkaline earth group: Na, K and Ca. These ions are the most mobile in the clay mineral systems, i.e. they appear to be among the most rapidly transported into and out of a given environment. Further, the concentration or chemical potential of these ions in a given system will determine the type of phase present. High concentrations give framework silicate minerals such as feldspars and zeolites. Intermediate values produce neutral lattice phases such as kaolinite and chlorites.

As we shall see, certain situations will produce high alkali and neutral lattice phases at the same time but only when the intermediate phases – micas and expanding minerals – are no longer stable. The alkali and alkaline earth ions will be designated as M^+ ions. They are principally found in interlayer sites in phyllosilicates.

Next, we must consider the relative importance of silica concentration or chemical potential. In most clay mineral environments, silica is present in excess, i.e. a free silica phase is present, notably quartz. However, it is well known (Krauskopf 1956; Harder and Fleming 1970; Mackenzie and Gees 1971) that most low temperature environments will not precipitate quartz and as a result silica concentrations in solution most often exceed those which should represent quartz equilibrium. In strict thermodynamic terms, all phases which crystallize under silica activities above those of quartz stability are metastable phases. These observations in fact relegate clay mineralogy to a shadowy, "unreal" state of metastable equilibrium.

There is in fact nothing wrong with this, especially since clay minerals are an everyday fact, occurring with great regularity in given chemical geologic environments. One can, and often does, reverse metastable reactions in the laboratory. Thus our knowledge of this state of affairs is quite advanced. The only problem is to know when metastable equilibrium will pertain and when another set of stable phases will replace older, metastable ones. If one finds the transformation metastable → stable occurring, the rate at which this will proceed becomes a vital factor for a geologist. If we observe clays to form at surface conditions at reasonably rapid rates, thousands of years for rock weathering for example, this might indicate that the rate of formation of the metastable assemblage is rapid while that of the stable form is much slower, being inhibited by the configuration of the material present. We can cite the example of silica.

Saturation of aqueous solutions at 25 °C with silica occurs with the precipitation of an amorphous, non-crystalline phase which transforms, as far as we can see by X-ray diffraction, into a form of cristobalite which is more or less hydrated. Either as a function of time or temperature, and possibly pH or alkali content of the solutions, the cristobalite gradually transforms into quartz. Ocean bottom sediments show that the transformation occurs in Cretaceous sediments. This indicates that the chemical potential of silica in silicate assemblages at low temperatures will be governed by amorphous silica for long periods of time and that quartz equilibrium is not important. However, experiments and observations on natural assemblages show that at 100 °C, the conversion to quartz is rapid and in a geological sense, instantaneous. Here, we can expect that new phases will represent a different set of equilibria from those at low temperature. The point which is most difficult to resolve is how rapidly the change from metastable to stable assemblage occurs as a function of temperature. If this change occurs in a systematic and regular manner as a function of temperature, one can use it to establish certain temperature conditions as palaeo-environments. If it is subject to differences in the initial configuration, the problem is too complex to be useful as a geologic indicator. Further, it perturbs a "phase diagram" approach to clay mineralogy. However, as we will see, it seems that such problems are encountered in the temperature range 50–100 °C and the clay assemblages above and below appear to be time–temperature stable as far as geological observation can discern them.

We can say that quartz is not active in the chemistry of phase crystallization at surface conditions. Amorphous silica – which becomes chert – is the phase which limits the quantity of silica in solution. Above 100 °C, it seems that quartz is the silica phase present. Thus, strictly speaking, one should have either phase present before one can disregard silica as a possible component in clay mineral phase equilibria. However, as we shall see, only a few instances of undersaturation in silica in clay mineral environments have been demonstrated.

As a result, we have the following components as element groups: R^2, R^3, M^+ and Si. These will be our basic elements of reference which can be questioned as to their appropriateness as the occasion rises.

9.2.3 Mineral compositions and system compositions

9.2.3.1 Components as extensive variables
M^+R^3–$2R^3$–$3R^2$ *component phases.* If we wish to establish relations between clay mineral species, we must establish which chemical components will represent a mineralogical assemblage frequently encountered in nature. First, we can place the common clay minerals in a very general framework which represents most clays although owing to its general nature, it must contain generalizations which show definite weaknesses. Such a system is defined by the M^+R^3–$2R^3$–$3R^2$ coordinates, which combine some components and use multiples of others in order to describe common mineral chemistry in a convenient, expanded graphical analysis which does not distort chemical space too much.

The first chemical unit is M^+R^3. This represents the chemical relation of one alkali ion and one R^3 replacing one Si in tetrahedral substitution. Such a relation exists in tectosilicates (alkali feldspar and zeolites) and micas (muscovite and phlogopite). It seems that this alkali–tetrahedral ion relation can be used to describe a fundamental type of substitution in low temperature silicates. If one considers calcium as two alkali ion charges then, $Ca = 2M^+$ and the M^+–R^3 relation, such as found in plagioclase, still holds.

The second chemical unit concerns trivalent ions, principally Al^{3+} and Fe^{3+}. Here we assume that a pure R^3-bearing phase will be dioctahedral in character. Thus one unit is considered to be $2R^3$. Likewise, the third chemical pole will be R^2 where trioctahedral minerals will occur and the unit "cell" will contain $3R^2$ ions per octahedral unit.

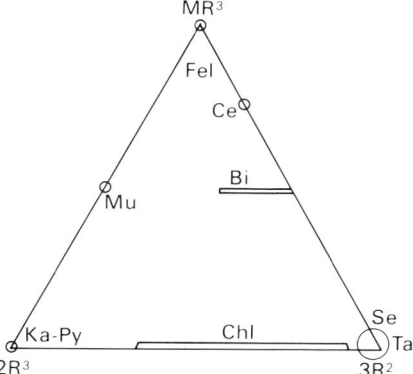

FIG. 9.1. Representation of the ideal compositions of some major phyllosilicate phases in the MR^3–$2R^3$–$3R^2$ coordinates. Mu = muscovite; Fel = feldspar; Ce = celadonite; Bi = biotite; Ka = kaolinite; Py = pyrophyllite; Chl = chlorite; Se = serpentine; Ta = talc.

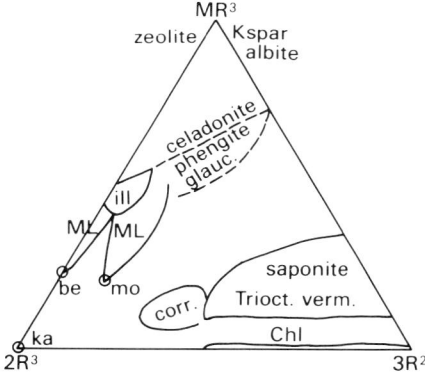

FIG. 9.2. Representation of the compositional fields of some clay minerals. ill = illite; glauc = glauconite; ML = mixed-layer minerals; mo = montmorillonite; be = beidellite; corr = corrensite; Trioct. verm = trioctahedral vermiculite; Chl = chlorite; ka = kaolinite.

We can then place micas between the feldspar–zeolite pole and the others. Muscovite contains one M^+R^3 element ($KAl^{(IV)}$) and one $2R^3$ element ($2Al^{(VI)}$) with the result that its composition plots midway between the M^+R^3 and $2R^3$ corners. Phlogopite then falls midway between M^+R^3 and $3R^2$. Figure 9.1 indicates the position of some important phases to be considered in clay mineral analyses: feldspar, muscovite, pyrophyllite, celadonite, kaolinite, talc, serpentine and the chlorites. As one can see, these phases occur along the edges of our system of chemical coordinates. The common clay mineral families will be found in the interior, a general chemiographic disposition of these phases is shown in Fig. 9.2.

It can be seen that there are overlaps of some of the groups in this projection. One task of further analysis is to decide whether or not this is maintained in other chemical projections.

The M^+R^3–$2R^3$–$3R^2$ projection does not consider the relative quantities of silica present or active in the mineral chemistry of the phases. It is initially assumed that a variation of the silica chemical potential in our clay mineral system will not provoke the crystallization of a new phase not will it suppress another. This is true in many natural clay mineral environments but not in all by any means. We can consider our system to be valid for pelitic sedimentation, most burial diagenesis and hydrothermal alteration systems. It will not be applicable to certain magnesian sedimentation environments nor in certain weathering situations.

The greatest utility of the M^+R^3–$2R^3$–$3R^2$ projection is the representation of both trioctahedral and dioctahedral phases in our alkali environment. One can show most phase relations found in the pelitic rock composition groups. These include sedimentary sandstones and shales as well as their diagenetic equivalents. One can show the weathering and hydrothermal alteration mineral assemblages of acidic and intermediate eruptive rocks as well. The limitations are of course that most of the elements present in clays are shown in fixed coordinates of mass and thus one does not show gradual variation in chemical potential of an element.

Si–R^2–R^3 component phases. If we consider clay minerals as alkali-bearing and non-alkali-bearing, thus grouping together mica and smectites, and serpentine with chlorite, sepiolite and palygorskites, it is possible to treat the latter group in two major representations. If the relative silica content is included, then we have Si–R^2–R^3 as variables, while the other representation concentrates on non-silica variables giving Mg–Al–Fe. This latter representation leaves ambiguous the oxidation state of iron.

The first triad allows one to classify the trioctahedral–dioctahedral minerals according to their silica content. Figure 9.3(a) shows these relations where we find the minerals sepiolite, palygorskite, talc, serpentine, saponite, the 7 Å serpentines and the 14 Å chlorites, kaolinite and the hydrous oxides brucite and gibbsite. For the moment we will ignore the possibility of magnesian dioctahedral smectites, i.e. expanding phases with interlayer ion sites occupied by magnesium exchange ions. This representation shows the tendency for sepiolite–palygorskite minerals and serpentine (7 Å and chlorites (14 Å) to cover a large range of R^2/R^3 compositions. Also, we can see the importance of silica. It is obvious that chlorites will not occur in silica-rich environments (where silica is precipitated from solution) whereas sepiolite–palygorskites are stable.

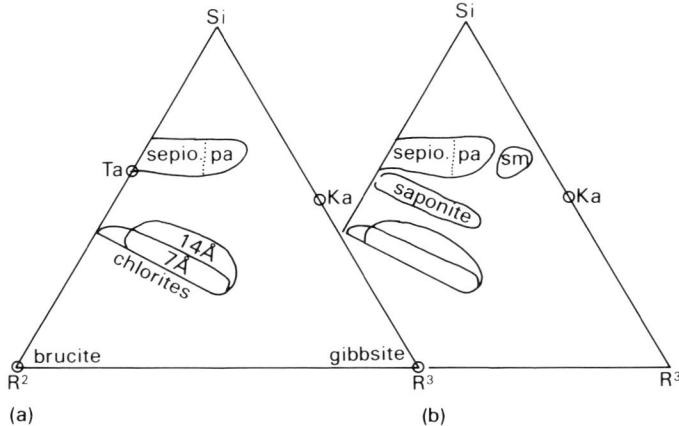

FIG. 9.3. Compositions of natural minerals as a function of the variables Si, R^2 and R^3. (a) sepio. = sepiolite; pa = palygorskite; Ta = talc; Ka = kaolinite. (b) Sm = dioctahedral smectite.

Now, if we assume that a continuous compositional series exists between alkali, alkaline earth and magnesian interlayer-ion smectite species, the relative positions of these minerals are shown in Fig. 9.3(b).

The outlined areas suggest the range of saponites (trioctahedral smectites) and beidellite–montmorillonites (dioctahedral smectites). These phases can be produced in the laboratory at temperatures near 300 C or below 1–2 kbar pressure in a magnesian system. Again, the importance of R^2–R^3 substitution is evident as it creates a third sequence of mineral compositions which separates siliceous and less siliceous minerals from one another. Silica chemical potential in solutions should then be quite important in determining which clays will persist or precipitate from solution. These effects will be most important in silica-poor environments such as those of weathering in basic rocks.

Fe–Mg–Al representation. Chlorites are composed of four major cation types: Fe^{2+}, Mg, Al and Si. Since the system is essentially constrained by a small range of stoichiometric variation, one can represent compositional variation by using Fe^{2+}, Mg, Al coordinates. This allows us to distinguish between the different chlorite types as they are produced in various geological environments. If we wish to use such a representation to establish a *P–T* vs. composition relation, we must be sure that all accompanying phases, that is, the global chemical system, are equivalent. It must not be possible to demonstrate that a chlorite group forms through special chemical conditions (Fig. 9.4).

Comparison of microprobe analyses of 14 Å chlorites from various broad geological environments gives two groups – those of almost constant alumina content, found in metamorphic and high grade diagenetic rocks and more aluminous forms found in diagenetic shales and sandstone clay mineral facies assemblages. In these two groups there seems to be a fairly wide range in Mg–Fe content even in the same rock sample. Apparently, the local chemistry determines the chemistry of the phases. This is especially true for diagenetic clay mineral assemblages. However, samples from black shales in the upper ranges of clay mineral stability (pyrophyllite zone) show a very restricted range of compositions (Paradis 1981). This is undoubtedly due to the bulk composition of the rock which produces much ferrous iron and aluminium that is available to silicates. The common assemblages are chlorite–pyrophyllite or chlorite–pyrophyllite–chloritoid, the latter phase being highly iron-rich. It is evident that the

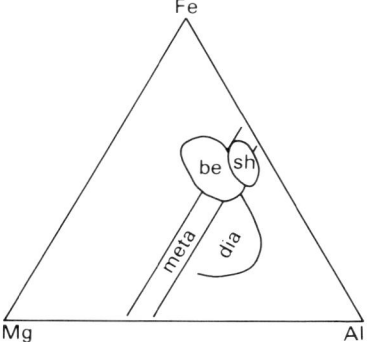

FIG. 9.4. Field compositions in the Fe–Mg–Al system of berthierines (be) and chlorites of different origins: metamorphic (meta), diagenetic (dia); black shales (sh).

compositions of these chlorites will be set aside from those in most pelitic sedimentary rocks. However, within the black shale facies one can probably distinguish various P,T-dependent changes in chlorite composition due, in fact, to the restricted or special chemical composition of these rocks.

In the chemiographic representation above, all of the chemical components are assumed to be extensive variables, whose mass in the system determines their chemical participation in the phases produced. In a general way, these elements do not change their quantity during the course of the geological process involved in the production of a clay mineral assemblage. They are often called immobile components. The chemical composition of the assemblage found now tells the whole story concerning the phases present. In theory, one should be able to find the relative proportion of the phases present knowing their compositions and those of the whole rock. Concerning their number, they can be at a maximum possible, given slightly variable P–T conditions. This number will be equal to that of the major participating chemical components as represented by the poles of the diagrams.

9.2.3.2 Components as intensive variables

If the chemical potential and possibly the quantity of a given element is controlled from outside the system or sample with which we are concerned, this element is considered as an intensive variable; it can be called mobile and will influence the number and kind of phases by its chemical potential. Representations of such systems must combine the aspect of chemical potential (often indicated as the activity of an element in the current literature) and that of inert components (extensive variables). One fundamental aspect of such diagrams is that as chemical elements become mobile (where their chemical potential is the variable) one cannot distinguish various species within a range of compositions (solid solution) of this element in the solid phase. We can take the example of Si in alkali zeolites. If one considers zeolites as alkali–Si–Al component minerals, there is an almost continuous series of compositions between natrolite and clinoptilolite. In making the calculation of phase boundaries (and hence phase reactions) in systems where chemical potential of silica (approximated by activity) is a variable, one generally assumes that the reference state for the element concerned is that of the solid which has an activity coefficient of one. However, if the silica content in that phase can be variable, it is not possible to assign a simple corresponding silica activity to its coexisting aqueous solution because obviously something else must dictate which silica content the zeolite will have. Solid solutions imply variable chemical potential of the element in the solid as well as the solution phase. The result is that phase boundaries will be bands in chemical potential space and phase relations become difficult to represent. It is then often more convenient to present phase relations in chemical potential–composition space (μ–x). The element of variable chemical potential should be present in the solids in nearly fixed or non-variable quantities.

The first series of chemiographic representations show the chemical variation of clay minerals in composition space. When one uses μ–x representations, the chemical definition is less precise, but one sees the phase relations in a much clearer manner. This is particularly important in clay mineralogy, for we know how important transport of dissolved species is in most clay mineral geological processes. The examples of weathering and hydrothermal alteration are particularly striking.

As an example of the relative effect which μ versus x has on the appearance of a phase diagram representation, we can use the system (K,Na), Al, Si which can represent granite or non-R^2 containing pelitic assemblages at low temperatures (< 50 °C). The first representation (Fig. 9.5(a)) shows the phases which we shall consider and their observed natural associations as

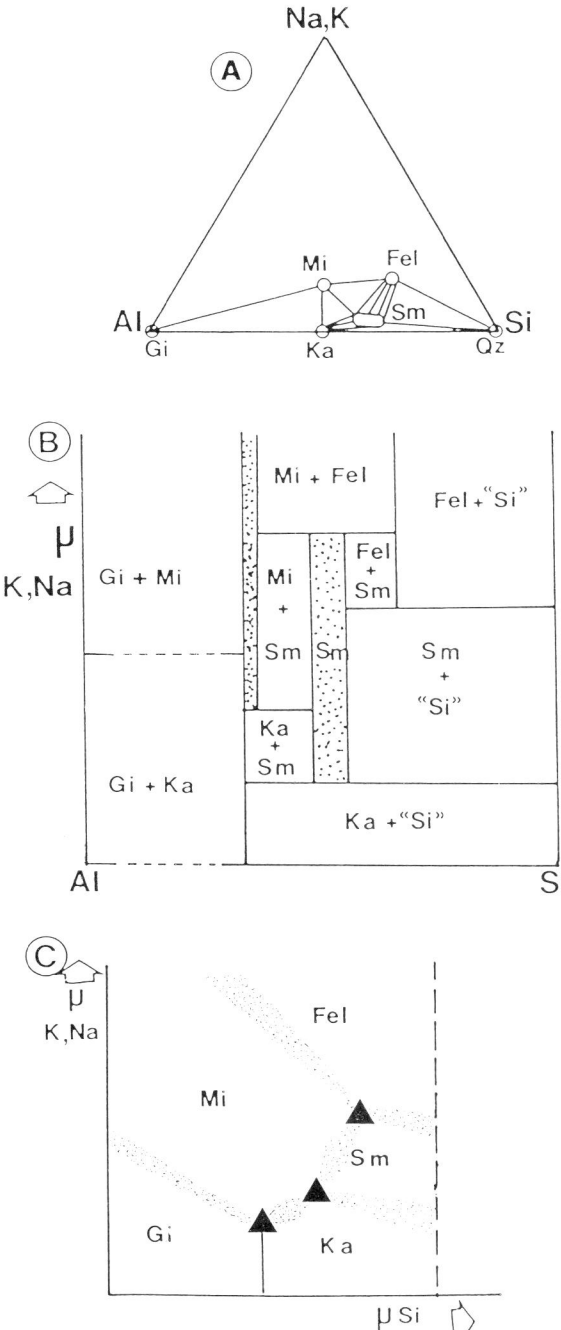

FIG. 9.5. Chemiographic representations for granite weathering. (a) System with three inert components: Si, Al (K,Na). Mi = dioctahedral mica; Fel = feldspars (orthoclase, albite); Gi = gibbsite; Ka = kaolinite; Sm = dioctahedral smectite; "Si" = solid silica phase, not quartz; Qz = quartz. (b) System with two inert components (Si,Al) and one intensive variable (μK, Na). (c) System with one inert component (Al) and two intensive variables (μSi, μK, Na).

projected first into the (K,Na)–Al–Si coordinates. As it turns out, most natural systems contain more silica than alumina in the presence of alkalis and this produces assemblages devoid of hydrous alumina (gibbsite or other forms).

The compositional range of zeolites includes that of microcline, the only alkali feldspar which forms at low temperatures where alkali zeolites are found. In order to simplify the system, we shall consider only feldspar instead of the whole range of tectosilicate phases. Alkali zeolites are found in sediments where solutions were highly alkaline. The effect of pH might well govern the occurrence of zeolites in that high pH favours the increased solubility of silica in solution. Lower pH, and lower silica chemical potential in solution could well favour feldspar. We shall assume for simplicity that we only have feldspar in our system. The smectites are shown by a range of Si and alkali components which indicates that a considerable range of charge is possible, about 0.2 to 0.5 charges per $O_{10}(OH)_2$ layer structure unit. It is important to note that there is a series of mica–feldspar and smectite–feldspar tie lines which represent the different silica contents of the phyllosilicates.

If one wishes to describe conditions of varying alkali chemical potential such as the weathering of a granite, one should use μ(alkali)–Al–Si coordinates. This is shown in Fig. 9.5(b). Solid solution in Al–Si coordinates is shown by the wide zone where one simple phase (e.g. smectite or mica) is found. If the phases had a fixed stoichiometry, their composition would be represented by a vertical line such as is the case for kaolinite. In this type of a diagram there is already a loss of definition concerning the alkali content of the phases; however, we can see the sequence of phase assemblages which will occur if equilibrium is attained throughout the sample as alkali chemical potential changes. This is of course not easily done on a strict composition plot. The major difference between the two plots is that, as we change from three to two inert or extensive chemical variables, the number of phases coexisting over a small range of space in the diagram changes from a maximum of three to two. We find two-phase instead of three-phase zones in the diagram.

The problem of definition is increased as one moves towards more chemical components behaving as intensive variables (mobile components). One can take the example of alkali and silica being mobile components where their chemical potential defines their role in the system. This leaves us with alumina as the only inert or extensive chemical variable. The diagram thus produced gives one-phase fields but the phase boundaries are indefinite due to solid solution (continuous variation of both K and Si contents). Nevertheless one can see the approximate relations where one phase is succeeded by another in chemical potential "space" (Fig. 9.5(c)).

This series of phase diagrams gives an example of the potential use of phase diagrams which could be used to describe the relations between clay mineral phases as a function of the type of chemical components and their specific function in the physico-chemical setting of the geological environment. The following section will deal with specific examples which have been described concerning their clay mineral associations. Using the chemistry of the phases, the chemistry of the general system in which they form and the specific textural relations between the phases which crystallize, we shall attempt to demonstrate in more detail how one can effectively describe the system which formed the clays and therefore explain why certain phases are present together. In fitting the phases into a convenient or correct chemiographic description, we can better understand the chemical forces which were active in forming the clay and then one can hope to predict new assemblages which are likely to occur. It should be kept in mind that we intend to demonstrate only several examples. There are certainly other systems which can be used to look at clay mineral assemblages.

9.3 DEFINITION OF TYPES OF CHEMICAL PROCESSES

9.3.1 System of chemical concentration

9.3.1.1 Sepiolite–palygorskite sedimentation in saline lakes and basins

As an example of a simple system which can be used to describe a series of clay mineral facies, we can examine the mineral parageneses of sepiolite and palygorskite. Chemical analyses have been compiled by Velde (1977). The basic elements which occur in these minerals are MgO, Al_2O_3 and SiO_2. Minor K_2O, CaO and FeO seem not to influence the solid solutions nor to influence the other phases present.

Millot (1964) has gathered the pertinent clay mineral data involving the parageneses of sepiolite and palygorskite in peripheral Tertiary sedimentary basins of north and northwestern Africa. The accumulated information seems to demonstrate the possibility of a direct application of phase analysis to geological processes involving clay minerals. The mineralogical observations all pertain to equilibria in closed sedimentary basins fed by uplands subjected to tropical weathering producing laterites. The detrital material was primarily composed of kaolinite, some smectite, illite and quartz, materials which are impoverished in Fe, Ca and alkalis. The streams were therefore considered by Millot to have carried high concentrations of dissolved Mg, Ca, Si and alkalis.

A cross-section of sedimentation in one of these basins would show a decrease in aluminous materials and an increase in magnesian silicates in the sediments as one progresses from the edge to the centre of the basin (Fig. 9.6(A)). As the quantity of detrital sediments decreases, the following mineralogical sequence is observed: kaolinite, smectite with minor illite and chlorite; smectite and palygorskite; palygorskite and sepiolite; sepiolite. Chert is a common accompanying phase, forming at times discrete layers in the sedimentary sequence. Of course, this schematic interpretation is not without the local contradictions found in natural assemblages but seems, however, to be generally true.

Comparison of these assemblages with the phase diagram deduced for the system of inert components R^2–R^3–Si shows the evolution of bulk chemical composition which successive zones have made in the sequence of sedimentation. The initial sedimentation of detrital material is dominated by the aluminous phases kaolinite and smectite (dioctahedral in this case). As the detritus decreases, the phases containing more Mg and Si begin to appear. This indicates an approach to equilibrium between solids and liquid (sea or saline lake water). Eventually the sedimentary assemblage is dominated by phases precipitated from solution in zones where the detritus is absent. By referring to the phase diagram for closed systems at low temperatures (Fig. 9.6(B), pathline I), this sequence of assemblages outlined by Millot can be described as a passage through the multiphase areas 1 to 5. Zone 1 (kaolinite–dioctahedral smectite–silica) is assumed by Millot to be purely detrital but it could be also due in part to a reaction between the detrital material and the lake or sea water solution. The presence of 14 Å chlorite indicates a non-equilibrium detrital component. The second phase-field is one containing smectite (dioctahedral) and amorphous silica. The third field in a traverse from right to left (R^3 to R^2) contains palygorskite in addition to smectite (dioctahedral) and silica. The fourth field contains silica–palygorskite and then silica–sepiolite (5 on Fig. 9.6(b)). Finally, Mg-sepiolite is present where only Mg and Si are available in the solution and a detrital component is almost non-existent. This is the pure chemical precipitate; Al^{3+} is in low concentration in the solution.

The use of the "closed" system to describe the assemblages in these closed basis seems justified in that frequently, almost always in fact, the number of clay minerals present in the

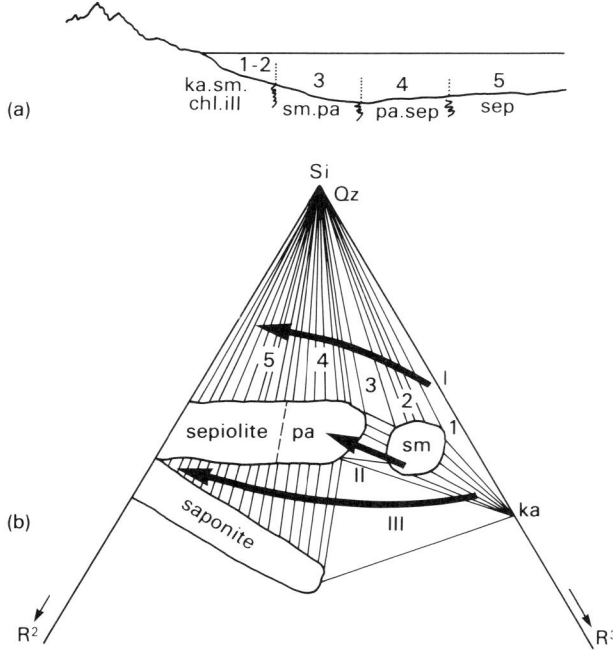

FIG. 9.6. (a) Schematic cross-section of a typical northwest Africa marginal sedimentary basin containing sepiolite (sep), palygorskite (pa): sm = dioctahedral smectites; ka = kaolinite; chl = chlorite; ill = illite. (After the summary presented by Millot 1964.) (b) Phase relations at low temperature (<80 °C) in the system Si–R^2–R^3. I = trend of the mineral reactions in sedimentary basin (numbers relate to Fig. 9.6(a)); II = mineral reactions in calcrete–caliche soil formation; III = mineral transformation in deep sea sediments.

sediments discussed above is two or more. The presence of amorphous silica or chert raises the total number of phases to three in many instances. In an essentially three-component system, Mg–Si–Al or possibly four if H^+ is considered, this indicates that the chemical components of the minerals are present in relatively fixed quantities in the chemical system which produces the mineral assemblages. None of the first three components is "mobile", i.e. its activity is independent of its relative mass in the solids or crystals present. However, there are sediments which present a monophase assemblage where only one variable need be fixed. Under these conditions sepiolite can be precipitated.

Data from Trauth (1977) on Tertiary sedimentary basins allows us to fix the compositions of the smectites, both di- and trioctahedral, which can coexist with sepiolite and palygorskite. Figure 9.6(b) shows the compositions of aluminous dioctahedral smectites and low alumina, magnesian saponites which were found in palygorskite and sepiolite assemblages respectively. The possible compositional limits to the low silica side of the saponite field are based upon compositions deduced from experimental studies at 2 kbar pressure and temperatures above 300 °C (Velde 1973). It is possible that the compositional range is not as extensive as indicated but for lack of other data it has been left as shown.

9.3.1.2 Calcrete–caliche soil formation
In most instances where sepiolite or palygorskite is present in bed-rock material, they are seen to disappear rapidly during weathering (Millot 1964; Paquet 1970; Bigham *et al.* 1980). This is

equally true of talc or serpentine, the other magnesian silicates with clay mineral-type structures. Since both sepiolite and palygorskite are known to form from highly concentrated aqueous solutions, it is not surprising to find that they are destroyed by weathering. Obviously the low activity of Mg and Si in common weathering solutions will be unfavourable to their persistence.

There are, however, certain soil profiles which indicate that both minerals can be stable in soils and, in fact, are formed in the weathering environment. These examples (Vanden Heuvel 1966; Paquet 1970) are found in caliche or the carbonate-precipitating soils of arid climates. In all cases the presence of sepiolite and palygorskite is closely related to the formation of the carbonate. This latter mineral is considered to be formed through the evaporation of subsurface waters which move upwards or downwards by capillary action under highly evaporative conditions. During the early stage water dissolves salts (and silicates) which are later precipitated when the water content is decreased by evaporation. The pH of these solutions should be 7.8 in order to precipitate calcite (see Garrels and Christ 1965). We find that sepiolite is found in the carbonate zone, and beyond these zones smectite is present. In some profiles, palygorskite can comprise 100% of the clay-size fraction silicate material (Paquet 1970).

Water infiltrates rapidly and easily through desiccation cracks to depths below the B and C horizons into the altered rocks. With general evaporation, the water in the lower parts of the profile is brought back up by capillary action and as it evaporates during the movement it becomes more concentrated in dissolved salts which were brought down from the A and B regions of the profile. In the C horizons these elements become active in the sulphate–carbonate–silicate system. The details of the soil profile studied by Vanden Heuvel (1966) show how the magnesio-silicates are associated with the caliche (calcrete) layer that occurs in the soil profile itself (in the instance studied it covers the upper C horizons).

It is evident here that sepiolite is associated with calcite deposition. However, palygorskite, although probably related to this process, appears to be less well correlated with carbonate content. If the top and bottom of the profile are assumed to contain equivalent assemblages, it is possible to follow the evolution of the bulk soil silicate chemistry traced as before on the $Si-R^3-R^2$ triangular diagram. The silicate phases sepiolite–palygorskite, kaolinite and smectite (Fig. 9.6(b), pathline II) are found in the soil sequence. The presence of quartz in the soils is not necessarily restrictive because amorphous silica and not quartz will control silica equilibria in low temperature environments, quartz remaining an inert phase. This soil mineral sequence is then distinct from that found in saline lakes where free silica is present and where the assemblage kaolinte–palygorskite is not found. It is probable that the smectite reported by Vanden Heuvel was at certain points in the profile in fact a saponite.

Since a soil profile is obviously not a closed system, in the sense that material does enter and leave a given horizon, we cannot consider the bulk chemistry as being totally fixed in its molecular proportions of the constituent elements. On the other hand, the mineralogy of the soils is seen to be always multiphase, usually containing three newly formed phases which can be described by (Mg,Ca)–Al–Si coordinates. It is thus unlikely that the chemical potential of Mg^{2+} or Si^{4+} is totally independent of the silicate present at a given point in a horizon. Mg^{2+} and Si^{4+} should not be considered to be perfectly mobile components. Equilibrium (or a reasonable approach to it) can be assumed to have been established in the different horizons of the soil profile between elements in solution and elements in the solid state. This equilibrium is related to the relative concentrations of the elements, and thus the system is considered as being "closed", i.e. containing mainly inert chemical components. However, some elements do migrate to or from a given soil horizon. Therefore, the major control in the formation of a given mineral assemblage at a given depth depends upon the relative proportions of the elements in the solution and in the solids present. The overall trend in the profile is one of chemical migration (an open system). Thus the assemblage of horizons can be considered "open", but each single horizon produces phase assemblages typical of "closed" systems where local equilibrium is attained.

9.3.1.3 Deep-sea sediments
Couture (1977) has made a general survey of palygorskite occurrence reported in the Pacific deep-sea sediments as well as those from the Indian Ocean. These two areas present samples from the abyssal depths where terrigenous air-borne sepiolite–palygorskite material can be excluded as a major component and where platform basin or continental margin basins can be excluded. These palygorskites are strictly deep-sea phases. The silicate material is of two origins: wind-borne continental clays which are illite plus chlorite for the most part, and volcanic ash which contains large quantities of glassy material. A certain amount of silica of diatomaceous origin can be found also. Mixed with these sediments one finds layers of basalt.

The conclusion which Couture makes is that one finds the association palygorskite–smectite (dioctahedral)–alkali zeolite (strictly clinoptilolite according to Stonecipher 1976) to be very common. Varying proportions of the components can be distinguished: frequently almost monomineral palygorskite layers tens of centimetres thick can be found. Since the samples are from drill cores, it is possible to know the lateral extent of these layers which would give us a better idea of the type of event which produced them. We can say that there are several elements which are mobile, being transported to form the monomineralic layer. In general, several clay minerals are present together which were produced at ocean bottom conditions. Both the magnesian and alkali minerals are the most silica-rich in their system solid-solutions.

What then can be the origin of these magnesium-rich aluminosilicates? It is evident that the detrital material is not of a special character since it is found in most ocean sediments and palygorskite is not particularly common. There is no particular evidence of immediate high temperature hydrothermal activity in the sediments such as that which produces sulphide-Fe,Mn oxide deposits (Lonsdale *et al*. 1980). Bonatti and Hoenesu (1968) propose reaction of sediments with hydrothermal solutions. Church and Velde (1979) using isotope and trace element measurements show that some of the material in the palygorskite is derived from terrigenous sources yet the clay is not in isotopic equilibrium with sea water. This suggests a flux of material interacting with the sediments.

It is proposed here that alumina-rich, but most importantly magnesian, pore solutions carry material towards the sediment–sea water interface. These solutions have their origin in basalt–hydrothermal fluid interaction. The content of the dissolved ions is controlled by sediments–pore water equilibria which allow concentrations above those in the more dilute sea water solution. At the pelitic sediment–sea water interface the dissolved species concentrations become unstable, or at least become more reactive than when in the pore water state. Here they react with the aluminous material to form palygorskite or palygorskite–zeolite assemblages. A similar process is proposed by Couture (1977) (Fig. 9.6(b), pathline III). The high alkali (Na principally) and calcium activities form zeolites when the silica activity is very great. One can suppose that the presence of biogenic silica or volcanic glass could supply silica in the necessary form. Magnesium will be incorporated in palygorskite. Both phases are very silica-rich. In this way bi- or mono-mineral layers will be produced. Alumina will be the inert or low solubility component while Mg, Na and Si are the mobile components. Figure 9.7 shows the approximate phase relations in such a system.

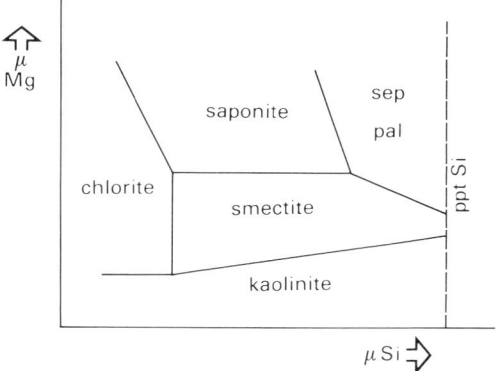

FIG. 9.7. Approximate phase relations in an aluminium inert, magnesium and silica mobile system. The phase field boundaries are only tentative in that solid solution occurs in each phase (kaolinite excepted) and thus precise reaction relations cannot be written. Sep = sepiolite, pal = palygorskite.

9.3.2 Systems closed to chemical migrations – example: shale diagenesis

It has been known for some time (Burst 1959) that argillaceous rocks frequently present a continuous sequence of clay mineral assemblages as depth of burial increases. Weaver (1959) attempted to simulate these transformations by applying high pressure to natural montmorillonites at room temperature. Since these studies were made, further investigations of deeply buried sediments have been completed; all show similar parageneses (Dunoyer de Segonzac 1970; Mitsui 1975). The most valuable information gathered in such studies is that of mineralogical variation as a function of both depth and temperature. With such data one can correlate the mineral facies observed in the context of a geothermal gradient and ultimately delimit the existence of critical assemblages as a function of pressure and temperature. Eventually a grid of diagenetic faces can be produced for argillaceous rocks.

Dunoyer de Segonzac (1969), Iijima (1970), Perry and Hower (1970, 1972), van Moort (1971), Weaver and Beck (1971), Schmidt (1973), Hower et al. (1976), Aoyagi and Kazama (1980) and Lahann (1980), have studied the sequences of clay minerals found in deeply buried sediments, both terrigenous and tuffaceous.

In geothermal areas, disordered interlayered minerals are succeeded by ordered ones as physical conditions become more intense. Schoen and White (1965), Steiner (1968), Muffler and White (1969), Browne and Ellis (1970), Eslinger and Savin (1973) and McDowell and Elders (1980) report the sequences of clay minerals in rocks of varying origin – volcanic tuffs to normal terrigenous sediments – which are found in "hydrothermal" areas where the geothermal gradient is high (above 230 °C at 1 km depth).

It is important to note that in all cases where the bulk composition of the argillaceous samples was determined no major systematic bulk compositional variation was observed as a function of depth (Dunoyer de Segonzac 1969; Perry and Hower 1970; Weaver and Beck 1971; van Moort 1971; Hower et al. 1976; Heling, 1978). This is especially true for potassium. Thus the occurrence of illite or mica in shales is apparently not a function of bulk composition but one of pressure–temperature conditions. A second important observation, made by van Moort, is that the sequence of mineral change does not appear to be related to the age of rocks older than Tertiary, i.e. younger rocks do not appear to be richer in montmorillonite than older rocks for

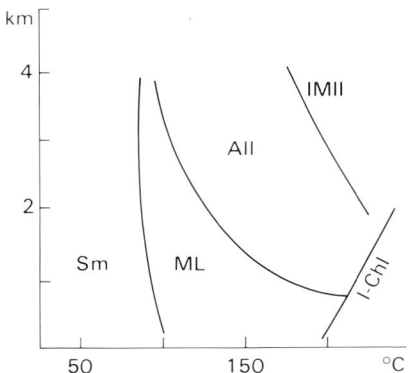

FIG. 9.8. Depth–temperature plot of mixed-layered minerals showing the maximum solid solution of smectite component for a given temperature and pressure. Sm = dioctahedral smectite; ML = 90–30% expanding illite–smectite interlayered mineral; AII = allevardite-type structure (30% smectite mixed-layered mineral with two layer ordering reflection); IMII = four layer ordered mixed-layered phase; I = illite; Chl = chlorite.

given $P–T$ conditions and fully expandable minerals can be found in pre-Cambrian shales. In the deep drill hole studies of more recent sediments (Tertiary) the ages of the rocks vary sufficiently for it to be obvious that physical conditions are the predominant factors in forming the mineral assemblages, but there is also a kinetic factor involved at temperatures below 100 °C.

Figure 9.8 compiles the available information on a temperature–depth plot; the smectites ("fully expandable phase", between 100 and 70% smectite), the mixed-layered minerals with between 70 and 50% smectite, and the ordered (allevardite-like)[1] 30–40% expandable minerals have definite zones of occurrence. However, information is not abundant for the IMII–illite boundary at great depths and low geothermal gradient, i.e. the zone where no partially expandable dioctahedral phase can exist in a pelitic mineral assemblage. It has been suggested by Hower et al. (1976) that this limit approaches 200 °C at greater than 6 km depths. Notable in this diagram is the enlarged field for allevardite at great depths. This suggests that the initial stage of mixed layering, between 70 and 30% expandable, will be greatly reduced at great depths and low temperatures. Such an effect has been observed by Weaver and Beck (1971). There is an obviously influence on pressure on the appearance of the superstructure ordering reflection.

Velde (1977) has performed a number of experiments on natural materials which have various types of interlayering and various non-expandable phases present such as illite, kaolinite and quartz. The starting materials were chosen in order to determine the effect of composition as well as physical conditions upon the transformation of expandable minerals. This expands the information from the simple system described above by adding the chemical variables Mg and Fe to those of Si, Al and K.

Briefly, the samples chose verified the construction of the phase relations proposed in the simple system. The only difference between what can be surmised from the data on natural minerals and the experimental results is the production of a corrensite-like ordered trioctahedral mixed-layered mineral under conditions approaching those which produce ordering in the dioctahedral mixed-layering minerals. We will discuss these results here only as they affect the sequence of ordered minerals found as $P–T$ conditions increase. Figure 9.9 summarizes the experimental results as a type of phase diagram where $P–T$ conditions effect

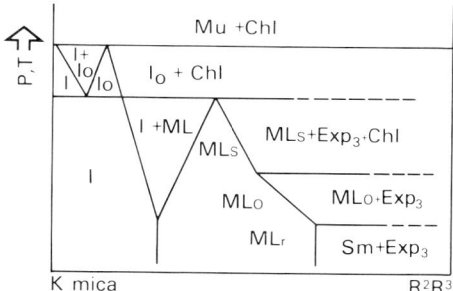

FIG. 9.9. *P–T–x* plot of general phase relations between potassic, aluminous mica (K mica) and non-alkali compositions (R^2, R^3). Exp_3 = trioctahedral expanding phase; Sm = dioctahedral smectite; ML_r = randomly interstratified illite–smectite; ML_0 = ordered mixed-layer illite–smectite; ML_s = allevardite-type ordered phase; I_0 = IMII type interlayered phase; Mu = muscovite; I = illite; Chl = chlorite.

different arrangements in the mica-expanding minerals present between a muscovite and a chlorite composition. We see that the lowest temperatures would give disordered interstratified minerals. The second step with metamorphism is a reduction in the expanding layers present in the structure and an ordering of the units.

Considering the compositions of the mixed-layered minerals found in sedimentary rocks it is obvious that magnesian–iron expandable dioctahedral minerals will be in equilibrium not only with kaolinite but also in many instances with a magnesian–iron phase, either chlorite or an expanding trioctahedral mineral. In such a situation the slope in *P–T* space of the reaction mixed layered mineral → allevardite + phyllosilicate will be controlled not by the aluminium silicate phases, as in the muscovite–pyrophyllite system, but by the production of chlorite or another trioctahedral phase. The reaction most likely to occur in the range of conditions concerned is the transition between a trioctahedral expandable phase and 14 Å chlorite + quartz. In the experimental system $MgO-Al_2O_3-SiO_2-H_2O$ investigated by Velde (1973), this transition has been observed to maintain a negative slope in *P–T* space. Of course, the temperatures at which this phase change occurs in the synthetic system are well above those found in the natural system, due mainly to the presence of iron in nature, but the implication remains that such a reaction "fits" the slope defined by the data for natural assemblages. If the parallel between aluminous and magnesio-aluminium systems can be assumed, kaolinite stability and that of expandable chlorite can be considered analogous in their influence upon the amount of interlayering in the illite–montmorillonite structures present. If we look back to the experimental studies on natural expandable minerals at high pressures, it can be recalled that the production of a chlorite phase occurred when interlayering in the natural dioctahedral mineral had reached about 30% interlayering. It is possible that below this transition only expandable phases are present for most magnesium–iron compositions; one is dioctahedral, the other would be trioctahedral. Thus, at temperatures below the transition to an ordered allevardite-type phase, dioctahedral mixed-layering minerals will coexist with expandable chlorites or metamorphic vermiculites as well as kaolinite. The distinction between these two phases is very difficult because both respond in about the same manner when glycolated. There can also be interlayering in both di- and trioctahedral minerals. The temperature of mineral transition to non-expanding minerals will be a function of the Mg–Fe content in the system as a whole. It follows then that magnesian 14 Å chlorite in sediments will be in disequilibrium until allevardite is stable. It can be suggested then that most of the 14 Å chlorite in sediments is of

440   PETROLOGIC PHASE EQUILIBRIA IN NATURAL CLAY SYSTEMS

detrital origin. The data of Perry and Hower (1970), van Moort (1971) and Hower et al. (1976) suggest such a conclusion in that chlorite content appears to decrease until the allevardite zone is reached where chlorite content begins to increase.

If one considers the four horizontal boundaries in the diagram $P,T-x$ (Fig. 9.9) as they are placed in depth–temperature space, it is evident that the effect of depth is to increase the temperature range of the ordered mixed-layered mineral facies; the one largely eliminated is mixed-layered ordered non-superstructure type. Disordered high expandability is not greatly affected.

9.3.3 Open microsystems – example: glauconite pellets

Electron microprobe analysis of glauconite pellets reveals what has long been suspected by clay mineralogists: glauconite forms in an isolated chemical system at the sea water–sediment interface through a transfer of material into and out of the pellets. A tentative phase diagram using Al–Fe and potassium variation appears to explain the evolution of multiphase clay mineral assemblages into a monomineral pellet. Although many elements are transferred into and out of the system, it remains most convenient to represent the phase assemblages in a composition diagram because the bulk compositions of the solids – clay minerals and oxides – appear to describe a continuous compositional series between glauconites and non-potassic compositions. In a system of varying chemical potential, one should see abrupt changes in the composition of the solids which is the case for glauconites. Each pellet is almost homogeneous in composition. However, when one plots a sequence of individual systems (pellets), they appear to form a continuous compositional series even though no gradual change exists in each system. The theoretical system, solution plus solids, is shown in Fig. 9.10 where μK is the intensive variable and Al–Fe^{3+} are extensive. Different glauconization paths are shown in Fig.

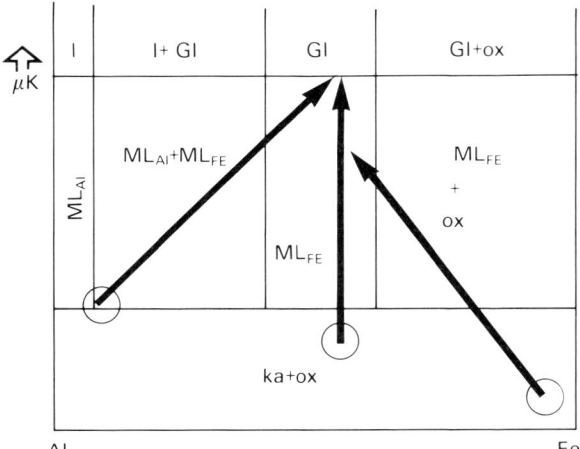

FIG. 9.10. Chemiographic representation of iron–aluminium:inert and potassium:mobile systems. Arrows show the path three different starting materials might take during the process of glauconitizations. Trends may be reversed during glauconite weathering. I = illite; Gl = glauconite; ox = Fe oxide; ML_{Al} = aluminous illite–smectite interlayered mineral; ML_{Fe} = iron-rich glauconite mica–smectite interlayered mineral; ka = kaolinite.

9.10, both for aluminous starting material (smectite–kaolinite) and also for more iron-rich sediments. Ratios of Al–Fe^{3+} change as the process is completed. These two elements are inert (extensive variables) but do not maintain a fixed content throughout the process.

If the process of glauconitization is an equilibrium system, one should be able to reverse the chemical trend given the proper chemically varying system. This means that the reactions should be reversible and instead of forming glauconite gradually from smectite–kaolinite–iron oxides one should be able to destabize glauconite gradually to form less potassium-rich assemblages. Loveland (1981) and Courbe *et al.* (1981) present microprobe data which demonstrate precisely the reverse of the glauconitization process. Weathering profiles of glauconitic sandstones were investigated using microprobe techniques. It was found that new, less potassic mixed-layered minerals (green in thin section) were formed outside the old glauconite pellets which themselves became brown. This new interstitial glauconitic material has a large range of potassium content, 0.5–1.0 atom, indicating that it can be a mixed-phase assemblage such as glauconite plus kaolinite. The general compositional trend during weathering is one of simultaneous potassium and iron loss from the silicates which is the same process but in reverse, as the major trend of glauconitization described by Hower (1961). It crosses the two-phase field of Al and Fe^{3+} mixed-layered series. Few bulk compositions indicate the coexistence of kaolinite and iron oxides which are of relatively low abundance in the profile. In a few secondary "argillan" zones near pores, new phases are potassic and aluminous. This indicates the crystallization of the illite–smectite mineral series within the glauconite profile. Thus the two series, Al and Fe^{3+}, are present in the same rock, produced under different local chemical conditions.

Most notable in this weathering profile is the loss of iron from the system. It seems evident that the key to glauconitization reversal is that iron must be mobile and easily ejected. Loss of potassium is, of course, to be expected during weathering. If iron remains in the local environment, oxides will form, immediately generating the oxide–kaolinite assemblages described by Parron and Nahon (1980). These authors show the rapid reaction of glauconite → iron hydroxide + kaolinite + isolated concentrations of silica which occurs during lateritic weathering of glauconitic sands. The accumulation of iron forms oolites in the weathered rock.

The role of iron mobility in the glauconite system is then the key to its origin and stability. Iron must enter or leave the silicate phase without blocking the system by forming hydroxides. We do not know how the iron is transferred from oxide to silicate in the process of glauconitization but it is certain that iron oxide must be less stable than the potassic iron-rich silicate which is present. The whole system – silicate, oxide and solution – must be considered when treating the stability of glauconite.

9.3.4 The effect of bulk composition through oxidation–reduction – example: burial diagenesis

A possible effect of bulk composition on clay mineral stability can be seen in the data presented by Hower *et al.* (1976). In their exhaustive study of the mineralogical changes occurring during burial metamorphism in the Gulf Coast sediments of the United States, they indicate great changes in mineralogy over a short range of temperature. The major reaction invoked by the authors is: montmorillonite + K^+ → illite + chlorite. This is effected gradually via a decrease in the proportion of the expanding layers in the mixed-layered mineral by loss of mica as an individual phase and feldspar. However, they also note that there is a great loss of kaolinite

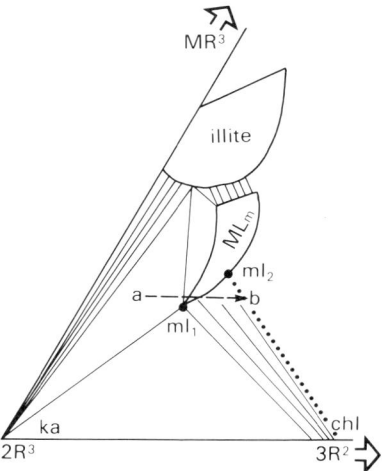

FIG. 9.11. Portion of MR^3–$2R^3$–$3R^2$ plot concerning the phases illite–kaolinite–mixed layered mineral and chlorite. Point "a" shows initial rock composition with a kaolinite (ka)–illite–mixed-layered illite–montmorillonite mineral (ML_m) of composition ml_1. Change in oxidation state of iron, R^2 to R^3, shifts the bulk composition of the system such that "b" represents the new phase assemblage of chlorite (chl) and another mixed-layered phase composition, ml_2, which has fewer expandable layers than that of the assemblage at composition "a".

when the increase in illite layers occurs. We also know that the character of the organic matter changes with a concurrent decrease in the Fe^{3+} content of the rock.

If we consider the phase diagrams in systems using synthetic minerals or those using natural minerals it is evident that an increase in temperature increases the amount of illite present. This is observed as a general fact in the depth–temperature plot. However, if we look at the details of the M^+R^3–$2R^3$–$3R^2$ plot concerning illite and mixed-layered minerals (Fig. 9.11) we can see what happens as we increase the Fe^{2+} content ($3R^2$) through the reduction of iron oxide. The Fe^{2+} (R^2) species will enter into the silicate system and the bulk composition will shift from point "a" to point "b" in Fig. 9.11. Initially we are in three-phase field where illite, kaolinite and a mixed-layered illite–montmorillonite mineral of composition a (ml_1) occur together. If we shift to composition b through reduction of Fe^{3+} we are in a two-phase field where a mixed layered mineral and chlorite coexist. However, the composition of the mixed-layered mineral in assemblage b is one with a greater proportion of illite layers than that in the assemblage a (ml_2). The reaction observed is illite + kaolinite + mixed-layered mineral (1) + Fe^{2+} → chlorite + mixed-layered mineral (2). At constant temperature we can thus observe an increase in the illite component of mixed-layering through a change in oxidation state of iron.

It is highly probable that the reaction as proposed by Hower et al. (1976) does take place as temperature increases. The loss of potassium feldspar from the coarse-grained fraction of the rock most probably enters into the reaction described by these authors. However, the loss of kaolin from the average mineral assemblages in the rocks studied and the corresponding increase in chlorite suggest that a more complex reaction occurs where the mixed-layered mineral has changed composition due in part to a displacement in rock bulk composition or at least in that of the silicates. We can then propose two parallel reactions:

mixed-layered mineral + K^+ (from feldspar) → illite (in mixed layers) and

kaolinite + illite + mixed-layered mineral ml_1 + Fe^{2+}

$\rightarrow$ chlorite + mixed-layered mineral ml_2

It is possible that this type of coupled reaction can account for some of the scatter on the depth–temperature plot concerning the appearance of allevardite-type ordering as mentioned previously.

9.4 CLAY GENESIS IN WEATHERING AREAS

Weathering is an important natural process of clay genesis. The circulation of low temperature fluids ($\leqslant 25\,°C$ on continental countries; $\leqslant 4\,°C$ in oceanic bottom) through crystalline or sedimentary rocks provokes the following mineral reactions:

1. Dissolution of primary minerals.
2. Crystallization of clay and associated minerals.

The overall result is a loss of solid material to the solution. Weathered rocks present a complex petrography because of the small size of the destabilization–recrystallization sites. As a general rule several different clay parageneses coexist in a sample the size of a single thin-section. Consequently, it is necessary to characterize the secondary minerals in their site of crystallization by micropicking and X-ray diffraction on small quantites of matter (Meunier and Velde 1982) in order to identify their crystalline structures. Chemical compositions may be determined by electron microprobe with good accuracy. In this way, the argillaceous phases may be described in terms of mineralogical assemblages or parageneses which characterize precise geochemical microsystems.

9.4.1 Types of microsystems

The chemical properties of these microsystems depend on two important variables:

1. The quantities of fluid flowing in the site, which is a direct function of local porosity.
2. The crystallochemical characteristics of the pre-existing minerals in which the clays appear.

The first factor depends upon the overall chemistry of the host rock which "precedes" that observed. This water reacts with the system observed according to the different types of microsystems present: contact microsystems along intergranular joints, primary plasmic microsystems inside destabilized parental minerals, secondary plasmic microsystems in the argillized saprolite zones, fissural microsystems inside the big fractures of the rock. The predominance of one of these different types of microsystems in the sample determines the macroscopic aspect of the weathered rock. Each profile can be divided into few horizons which are:

4—fresh rock
3—weathered coherent rock (contact microsystem – low porosity)
3—friable rock of saprock in which the initial texture is preserved (primary plasmic microsystem)
1—neostructured horizon or saprolite (secondary plasmic microsystems).

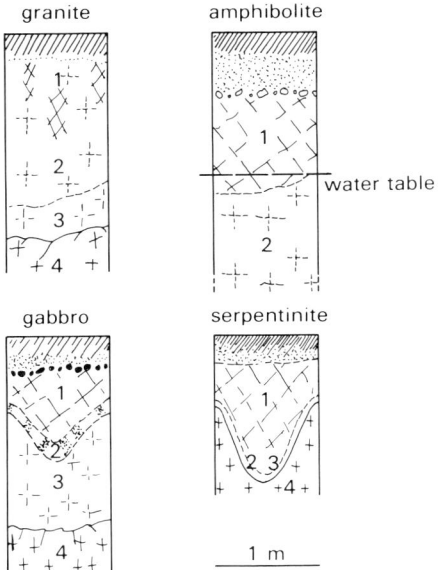

FIG. 9.12. Schematic representation of weathered profiles on acid, basic and ultrabasic rocks under temperate conditions. 1 = saprolite or saprolitic zones; 2 = saprock; 3 = altered coherent rock; 4 = unaltered rock.

All of these horizons are crossed by big fractures in which fluids circulate in large quantities (fissural microsystem). Each fissure can be subdivided into several specific subsystems.

The volume of the different horizons in a given profile depends on the second variable cited above, that is to say, on the crystallochemical characteristic of the primary minerals in the rock. The weathered coherent rock and the saprock are well developed on granites while saprolite is predominant on ultrabasic rocks (Ildefonse *et al.* 1979), Figure 9.12 shows a schematic aspect of weathered profiles in temperate climate conditions. We shall describe the different clay assemblages in the weathering profiles of acid, basic and ultrabasic rocks.

9.4.2 Weathering of granitic rocks

Two weathering profiles described by Meunier (1980) may be used as examples of granitic transformation under temperate conditions. According to the general model described above, it is possible to summarize the mineral reaction processes (Table 9.1).

Figure 9.13 shows the petrographic reaction of clay minerals in the different horizons of the profile under an acid brown soil.

In the weathered coherent rock, solutions flow in very narrow fissures (intergranular joints, microcracks, mineral cleavages, etc.) and provoke the crystallization of an illite mica along muscovite–orthoclase boundaries (Meunier and Velde 1976) (Fig. 9.14). Biotites are slightly oxidized and plagioclases are destabilized along cleavages and microcracks producing kaolinite.

In the saprock where initial rock texture is preserved, the primary minerals are partly transformed into clay assemblages whose nature depends on the chemical properties of the host crystal as well as upon the composition of the flowing solutions. The presence of Ca^{2+} ions in the fluid greatly influences the nature of the crystallizing clays (Hoda and Hood 1972).

CLAY GENESIS IN WEATHERING AREAS 445

TABLE 9.1. Mineral transformations in rock weathering

Horizons		Altered profile under acid brown soil	Altered profile under liassic marl
Weathered coherent rock		Illite at grain contacts muscovite–orthoclase	
Saprock (internal mineral transformation)	Biotite	Trioctahedral vermiculite + kaolinite + Fe oxides + residual biotite	Regular 1/1 mixed layer biotite–vermiculite
	Muscovite	Dioctahedral vermiculite + kaolinite	Kaolinite
	Orthoclase	Dioctahedral vermiculite + kaolinite	Beidellite + kaolinite
	Plagioclase	Kaolinite	Kaolinite
Fissural deposits		Kaolinite + Fe oxides	
Saprolite areas		Beidellitic smectite	

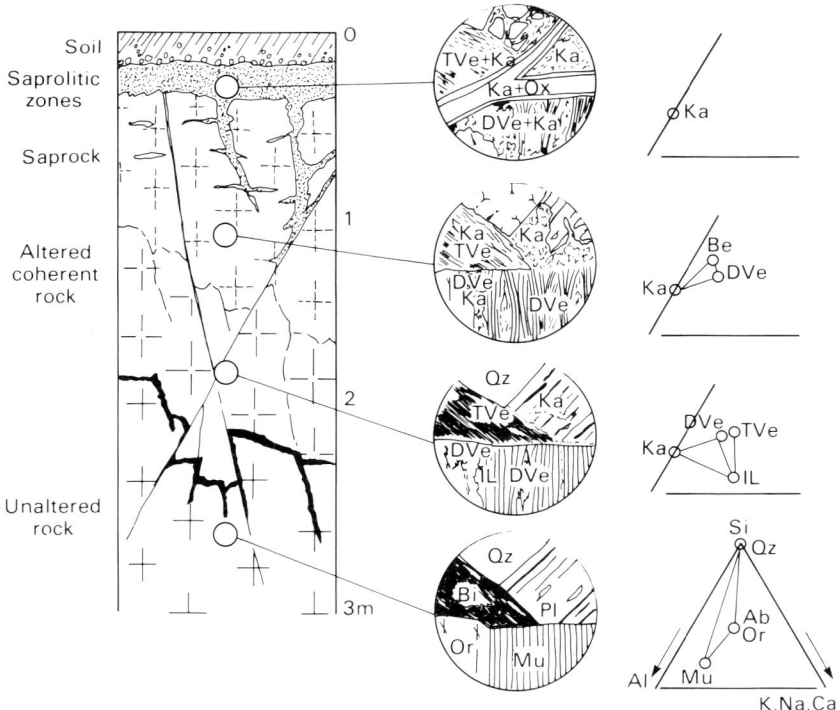

FIG. 9.13. Weathering profile on granitic rock under temperate conditions. Petrographic inserts show the weathering assemblages. They are interpreted in the Si–Al–K,Na,Ca system. Or = orthoclase; Ab = albite; Bi = biotite; Mu = muscovite; Pl = plagioclase; Qz = quartz; IL = illite; DVe = dioctahedral vermiculite; TVe = trioctahedral vermiculite; Ox = Fe oxide; Ka = kaolinite; Be = beidellite.

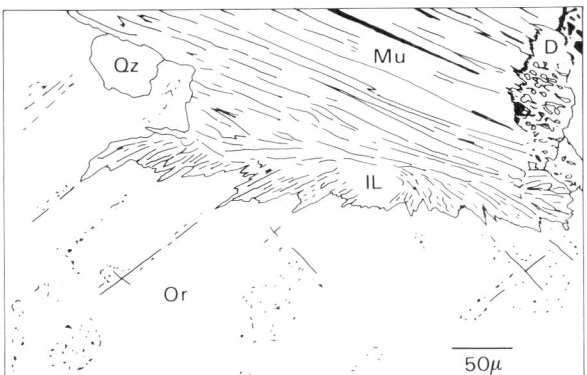

FIG. FIG. 9.14. Crystallization of illite (IL) along a muscovite (Mu)–orthoclase (Or) join in the early stages of weathering. Qz = quartz; D = deuteric syncrystallization of muscovite and quartz. (Drawing after a photograph.)

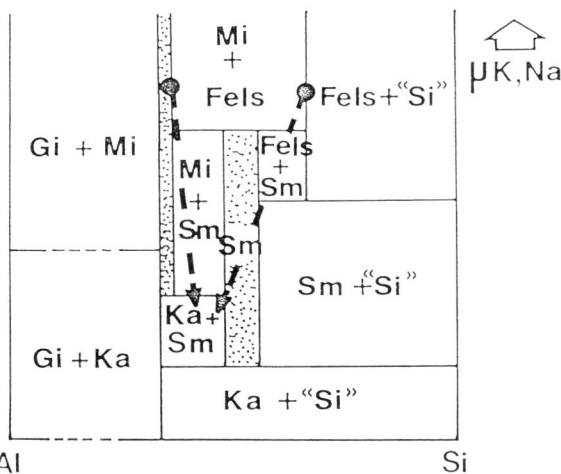

FIG. 9.15. Phase relations in granite weathering process considered as a two inert component system (Si, Al) with an intensive variable (μK, Na). Mi = mica; Fels = feldspar; Sm = dioctahedral expanding phase; "Si" = amorphous or solubilized forms of silica; Ka = kaolinite; Gi = gibbsite.

In the two profiles there is no distinct saprolite horizon but saprolitic zones appear around the major fractures of the rock. In these zones, the initial texture disappears and the vermiculitic clay diminishes while beidellitic smectite becomes more important. The fractures themselves are coated by argillans which are composed exclusively of kaolinite and iron oxides whatever their situation in the profile may be.

If we consider only the acid minerals – feldspars, muscovite and quartz – it is possible to describe most mineral reactions observed which occur during the different stages of weathering. Figure 9.15 is based upon the tie-lines which would be found in the fixed composition system NaK–Al–Si, namely, smectite–quartz and smectite–feldspar. This excludes the possibility of a muscovite–quartz tie-line.

Figure 9.15 allows the construction of a system where two variables are relatively inert, Si and Al, and where alkalis (potassium for the most part) have a varying chemical potential in the system. Thus with two chemical components as extensive variables we can have two-phase fields in a phase diagram. The key reactions with increasing alkali (potassium) potential are kaolinite–smectite, smectite–mica, mica–kaolinite, mica–feldspar. We assume that there is no complete solid solution between smectite or expanding phases and the micaceous phase (illite). None has been observed in the studies yet.

In the fracture systems of granites one finds kaolinite plus oxides. However, in the lowest sectors of the profiles, gibbsite is a common mineral (Calvert *et al.* 1980). The initial stages of feldspar alteration by water were described thermodynamically by Helgeson (1970) as giving gibbsite then kaolinite. Just such a sequence does occur in the lowest portions of fracture systems in granites. The rest of the rock contains no gibbsite and only small quantities of kaolinite in temperate climates. What then are the reactions in the rock which is unaffected by the fracture system?

First, the reactions seen in the lowest zone of the saprock are found at grain contacts between mica and orthoclase. A new, siliceous mica (illite) is crystallized at the edges of the feldspar. This reflects a general lowering of the alkali chemical potential which permits a low alkali, silica-rich mica to form at the expense of the feldspar. Loss of alkali from a siliceous phase occurs in order to form the new mica. These grain boundary effects are small and they do not persist into the overlying saprock zone.

Next, the individual mineral grains are seen to destabilize along minute grain fissures in their interior. The following reactions were noted in two different profiles (a and b):

1. (a) muscovite → kaolinite
   (b) muscovite → kaolinite + vermiculite (aluminous smectite) (arrow 1, Fig. 9.15)
2. (a) orthoclase → smectite + kaolinite
   (b) orthoclose → vermiculite (aluminous smectite) + kaolinite (arrow 2, Fig. 9.15)
3. (a) and (b) plagioclase → kaolinite

If we assume that smectite and vermiculite are functionally the same phase and if we take into account the fact that quartz grains appear to be totally unaffected and thus inert to the chemical reaction, we can place the bulk composition of the system between muscovite (calcic plagioclase also) and orthoclase. The ratio of Al to Si in the magmatic minerals will be within this range. The tendency then will be for the silica-rich phase (orthoclase) to destabilize to a more aluminous assemblage and for the silica-poor phase (muscovite) to become a more siliceous one. The convergence of orthoclase and muscovite assemblages to kaolinite plus smectite (or aluminous vermiculite) then suggest that the altering fluids decrease alkali potential and average the Al–Si relations to those between the initial reacting phases. Plagioclase reacts directly to form kaolinite since all calcium goes into solution and so little alkali is present in the solution that the high Si–Al concentration precipitates the crystallization of kaolinite immediately. The effective composition of plagioclase is at the kaolinite level of alkali chemical potential.

Thus, in the first stages of weathering, there is a tendency to conserve the Si–Al ratio in the system as a whole and to diminish the alkali content. Each magmatic phase reacts to adjust to the new alkali chemical potential of the altering solutions.

The fracture zones present a new chemical environment where the chemical potential of alkalis is still lower. Here we descend to the kaolinite–quartz zone. There appears to be a decided barrier to form quartz at low temperatures, but the solutions are never concentrated enough to allow the precipitation of amorphous silica. The result is a systematic elimination of this element. Thus the fracture zones become monomineralic, silica behaving as a mobile element. The production of kaolinite is a minor effect in the mass of the system; it does not persist much into the soil horizon and thus the appearance there of this phase will represent an inheritance of lower zone chemical reaction.

When the rock material reaches the saprolite stage, argillization, it appears that a dioctahedral smectite is the phase which crystallizes. The overall weathering of the granite in the saprock zone gives a mixture of clay phases which arrive in the saprolite–soil horizon. They are then transformed into a vermiculite phase as described by Jackson *et al.* (1952) in their summary of soil sequences in temperate climate. Kaolinite is formed in the rock-weathering process at many points but this does not mean that the climate was tropical in nature. It means, however, that in the initial stages of rock transformation certain sectors of the system had chemical potentials which dictated the presence of this phase. Kaolinite is often part of a multiphase assemblage and thus must be considered as a stage in the alteration process. It can originate in the fracture or vein system also.

In areas where the rainfall is great and temperatures are sufficiently high to promote rapid chemical reaction, the process of dissolution will go one step further when the alkalis are exhausted. From the kaolinite-plus-quartz zone of Fig. 9.15, the next change is towards an increase in the least soluble element which is aluminium. The change in composition is towards an alumina-rich assemblage by elimination of silica below the concentration which permits kaolinite to be stable. This is the same reaction as that seen in fracture zones in temperate climates. The displacement from the kaolinite composition toward the gibbsite plus kaolinite field (Fig. 9.15) shows the ultimate stage in weathering which is seen commonly in tropical soils (Pédro 1964, 1966). As has been noted by Millot (1964), this stage can be "reversed" by an accumulation of silica in other zones which kaolinitizes gibbsite. In these sequences, there is a significant accumulation of iron oxide along with kaolinite and gibbsite. This is typical of pedologic accumulations of alumina.

9.4.3 Weathering of basic rocks

The description of the weathered metagabbro from Le Pallet (Loire-Atlantique, France) by Ildefonse (1980) is a good example of the weathering of basic rocks under temperate conditions. The weathering profile is considered to have two parts (Fig. 9.15): the base, where the structure of parental metagabbro is preserved (saprock), and the upper part, which is strongly argillized and neostructured (saprolite).

Inside the weathered coherent rock (initial stages of weathering), grain boundary reactions between amphiboles and plagioclases give rise to iron beidellite which crystallizes inward towards the feldspar (Fig. 9.17). In the lower part of the profile where the parental rock texture is preserved, plagioclases are less altered than amphiboles which recrystallize into a polyphased assemblage: nontronite + talc + Fe oxides. In the upper argillized part of the profile both plagioclase and amphibole are altered, the former into a vermiculite phase which is aluminous and dioctahedral and the latter into ferromagnesian and trioctahedral vermiculite. Inside the argillaceous matrix in which is disseminated the debris of parental material, a new trioctahedral vermiculitic phase crystallizes which is more aluminous than the amphiboles. The new assemblage is ferric beidellite + trioctahedral vermiculite. Big fractures in the profile are coated

CLAY GENESIS IN WEATHERING AREAS   449

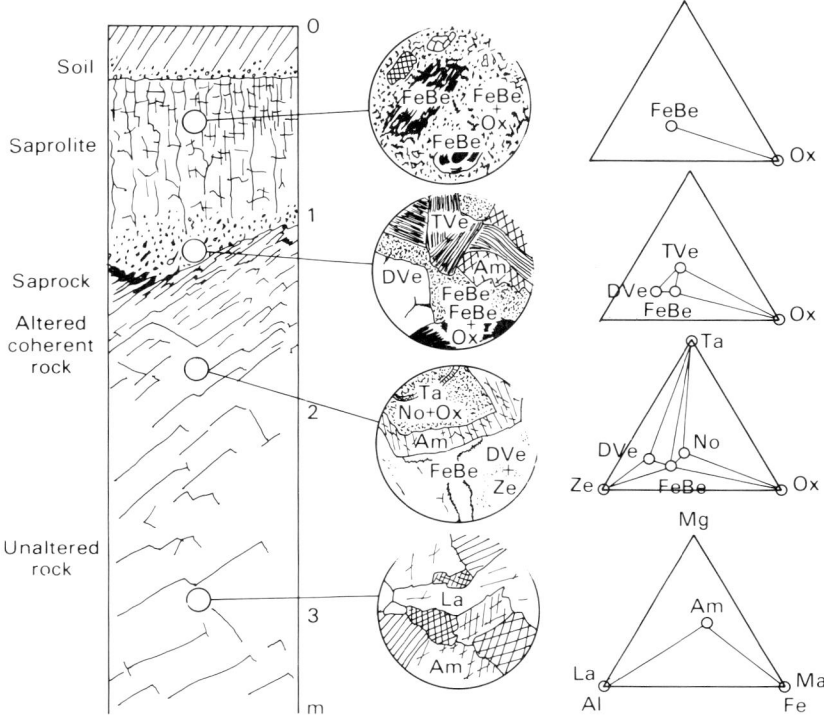

FIG. 9.16. Weathering profile on metagabbroic rock under temperate conditions. Petrographic inserts show the weathering assemblages. They are interpreted in the Mg–Al–Fe system. Am = amphibole; La = labradorite; FeBe = ferric beidellite; DVe = dioctahedral vermiculite; Ze = zeolite; No = nontronite; Ta = talc; Ox = Fe oxide; TVe = trioctahedral vermiculite; Ma = magnetite–ilmenite.

by argillans of very constant composition: iron beidellite + Fe oxides. Petrographic and mineralogic evolutions are summarized in Fig. 9.16.

Considering the inert or less mobile elements, it is possible to represent the sequences of mineralogical assemblages from the altered coherent rock–saprock horizons to the saprolite argilized horizons in Mg–Al–Fe coordinates (Fig. 9.16):

Saprock
1 – Fe beidellite + Fe oxides   = contact microsystems
2 – Dioct. vermiculite + zeolite   = internal destabilization of feldspars
3 – Nontronite + talc + Fe oxides = internal destablization of actinolites

Saprolite
4 – Dioct. vermiculite + trioct.
    vermiculite + Fe beidellite   = argilized matrix in horizon 3
5 – Fe beidellite + Fe oxides   = fissural system
6 – Fe beidellite + Fe oxides   = argillaceous matrix in horizon 4

One can see in Fig. 9.16 that the number of weathering phases decreases from the altered coherent rock horizon to the argilized horizon. This implies that some elements which are inert

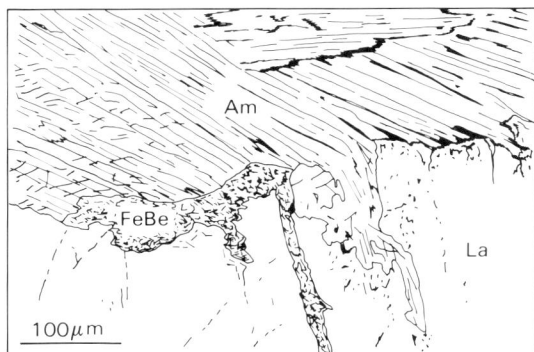

FIG. 9.17. Crystallization of ferric beidellite (FeBe) along an amphibole (Am)–Labradorite (La) join and inside a crack in the labradorite crystal. (Drawing after a photograph.)

in the first stages of the weathering processes become mobile in the argillized levels and in the fissural system. The evolution from the bottom to the top of the profile can be considered as a decrease in the number of independent microsystems. The altered coherent rock is a "mosaic" of active geochemical microsites (amphibole–labradorite contact, internal destabilization of plagioclase and amphibole) while the argillized horizon is more homogeneous: the matrix is composed of ferric beidellite associated with oxides.

The process of homogenization may be considered as the result of the decreasing of μCa and the oxidation of Fe^{2+}. It is possible to interpret the three component phase diagrams of Fig. 9.16 in "two inert – one mobile" component. If we consider that silica content variation is not vitally important, which seems to be the case in the whole rock chemical analyses, it is possible to fix R^2 and R^3 ions as inert components and to designate calcium chemical potential as an intensive variable in the system. Chemical analyses show that Ca-content decreases in the clay phases as weathering is more intense. The large grained vermiculite of the saprolite contains less calcium than the nontronite–beidellite assemblage found below it, in the saprock, for example.

Figure 9.18 shows the phase relations of such a system where two-phase trends occur as the new clay minerals form in the rock. We must treat the rock as two separate systems where certain elements are lost or gained as weathering proceeds. Each mineral, amphibole and plagioclase, maintains its identity during weathering and it thus becomes a separate microsystem. The stages of rock, saprock, saprolite and argillite (soil) are numbered in the figure. Arrows show the chemical "path" taken by each of the reacting systems – amphibole and plagioclase.

In order to produce the aluminous, ferric trioctahedral vermiculite with 2.5 octahedral ions, the composition of the amphibole system must change, losing some of its R^2 component and possibly gaining R^3. Whole rock analyses show there is a small loss of Mg during weathering and a significant oxidation of iron in the saprolite zone. However, magnesium drops to two-thirds and ferrous iron to 40% of its initial value in the argillic soil horizon. It is here that negatively charged ferric beidellite is the dominant phase. In the more open system where the constituent elements are mixed in the solifluxed zone, the chemistry appears to dictate the presence of a single low-charge nontronite. This suggests that most elements are mobile and that the system is homogeneous. The convergence of the arrows in the figure indicates this effect.

An amphibolite in a poorly drained profile reported by Proust and Velde (1976) shows the importance that the configuration of the migrating, altering fluid can have. In the preceding

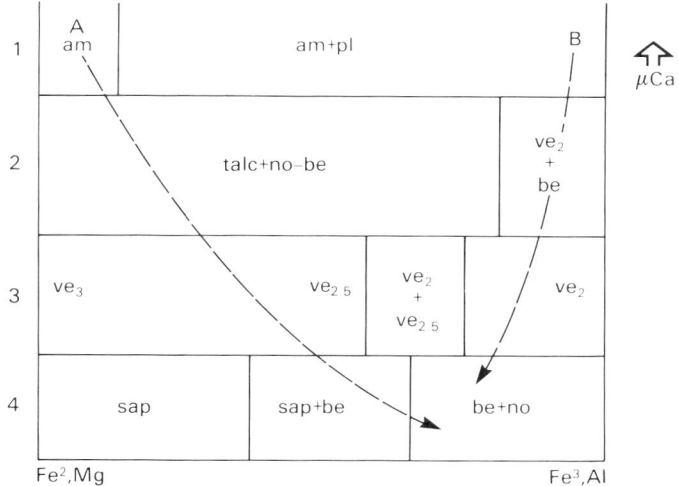

FIG. 9.18. Phase relations in basic rock weathering process considered as a system with two inert components ($Fe^2 + Mg$; $Fe^3 + Al$) and an intensive variable (μCa). Arrows starting from points A and B show respectively mineral reactions of amphibole and plagioclase weathering. am = amphibole; pl = plagioclase; no-be = high charge ferric nontronite–beidellite phase; ve_2 = dioctahedral vermiculite; ve_3–$ve_{2.5}$ = trioctahedral vermiculite solid solution; sap = saponite; be and no = low charge ferric beidellite and nontronite; 1 = rock; 2 = saprock; 3 = saprolite; 4 = soil weathering stage.

example, a well-drained profile, individual grain contacts and internal grain fractures and cracks were the initial site of alteration. It can be assumed that the altering fluids passed through these channels and thus reaction between fluids and unstable silicate minerals occurred here. The first remarkable aspect of the study by Proust and Velde is the lack of grain contact reaction. Secondly, instead of distinct multimineral clay assemblages occurring in each grain system, amphibole and plagioclase, the minerals which form are basically the same. In areas within the grains where new phases are found, one sees a vermiculite layer next to the metamorphic mineral. The clay mineral present in the centres of each grain is a beidellite which has a different composition depending upon the type of mineral from which it is formed. It appears that in a "wet" profile the stage of saprock alteration is more or less skipped over in favour of the later stages of saprolite alteration. There is a jump from high calcium potential to a much lower one because the solutions are much more abundant and the reacting silicates cannot produce a high calcium content in them. In this situation the system has many mobile elements which diminish the number of solid phases present.

9.4.4 Weathering of ultrabasic rocks

The two factors which dominate the weathering of ultrabasic rocks are loss of magnesium from the system and oxidation of iron which might be present. The phyllosilicate mineral phases that are initially present are serpentine, chlorite and occasionally talc. Differences in these profiles are between clay mineral silicates and oxides (essentially Fe_2O_3) present (Wildman et al. 1968; Rimsaite, 1972; Wackerman 1975; Trescases 1979; Eggleton and Boland, 1982.

Observations by Fontanaud and Meunier (1983) allow one to see how much an alteration system can change the composition and mineralogy of a rock (Fig. 9.19). The initial

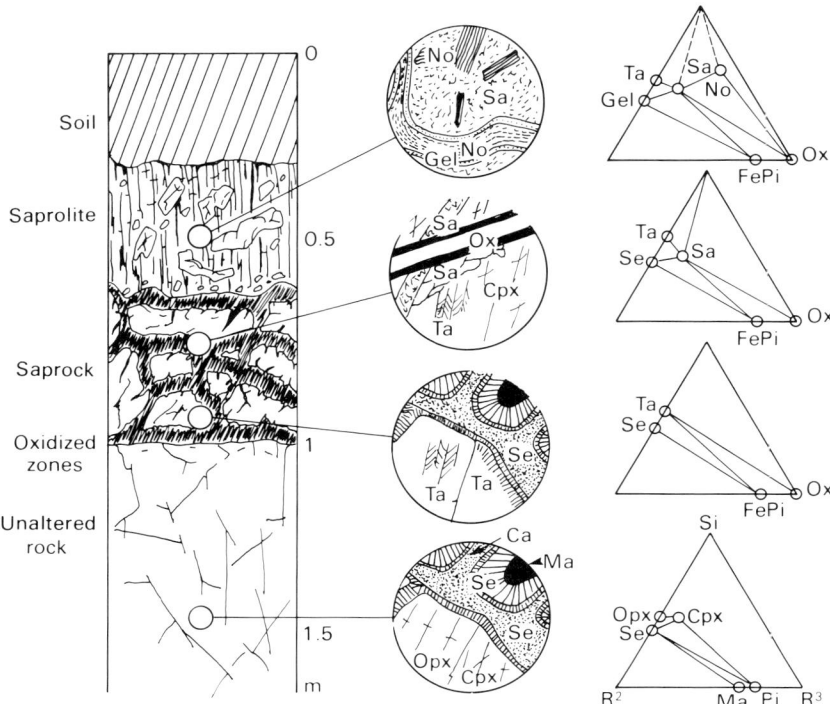

FIG. 9.19. Weathering profile on a lherzolitic rock under temperate conditions, interpreted in the Si–R^2–R^3 system. Opx = orthopyroxene; Cpx = clinopyroxene; Se = serpentine; Ma = magnetite; Ca = calcite; Ta = talc; Sa = saponite; Ox = Fe^{3+} oxides; No = nontronite; Gel = magnesian gel; FePi = ferripicotite; Ma = magnetite–ilmenite.

serpentinized lherzolite minerals were chrysotile, enstatite, diopside, picotite, magnetite and calcite. The first weathering reaction shows the destabilization of clinopyroxene (Cpx) into talc and oxides. Some silica will be lost as well as calcium from the pyroxene and calcite; presumably both are transported in the solutions as we find no apparent trace of them in the clay mineral phases. Talc is a highly siliceous, low temperature phase, as would be expected from the reaction Cpx → talc + Ca^{2+} + Si$_{aq}$. Serpentine (se), forming mostly from olivine (ol), is not siliceous as seen in ol → se + Mg^{2+}. The initial assemblage will be considered as serpentine–clinopyroxene–spinel (Fig. 9.19). The reaction to form talc produces the assemblage serpentine–talc–spinel in the saprock zone. The second weathering reaction (in the saprolite zone) is the formation of saponite at the expense of the other phases. The composition of saponite lies between talc and the oxide (R^3) corner of the system and these minerals contain more aluminium than Fe^{3+}. Saponite is pseudomorphic after both talc and serpentine and the new assemblage shows a bulk composition poorer in R^2 ions than does its predecessor. There is a decided tendency to form a single-phase assemblage.

In fissures of the saprock and saprolite one finds that saponite is unstable and is replaced by nontronite, a ferrisilicate. The shift in bulk composition is towards the Si–R^3 side of the system due to a loss of magnesium and oxidation of iron.

In zones of high water flux (larger fissures and their edges) nontronite does not form and we have an Si–Mg–Fe "gel" plus iron oxides. This must certainly indicate that the recrystallization process saponite → nontronite passes through a solution phase when the Mg–Fe^3–Si ions are in a highly concentrated state, nontronite being formed from this gel. This mechanism has been demonstrated by Harder (1976) for a number of phases believed to crystallize at low temperature. However the gel composition in the ultrabasic rock is more Mg-rich than the nontronite.

The bulk composition of the weathering profile changes: as the above mineral assemblages are transformed, there is an increase in alumina content and alkalis, especially K$_2$O. However both remain low, 6% and 0.4% by weight respectively. Concentrations of these elements can be found in the smectite which is dominantly iron-rich, tending towards a nontronite in composition. Some iron (Fe^{3+}) is found in the tetrahedral site. Calcium is notably absent. The gel when it occurs contains almost no alumina and no alkalis or calcium. Thus the clay mineralogy is dominated by non-aluminous phases and the phase diagram Si–R^2–R^3 could well be Si–Mg–Fe^{3+} at this stage.

9.5 CLAY MINERAL GENESIS IN HYDROTHERMAL ALTERATION

Our treatment of hydrothermal alteration will be brief because there is not yet much detailed petrographic and mineralogical information available concerning the phyllosilicate phases formed in these environments. Hydrothermal systems show the influence of chemical potential as well as temperature variations which will determine the phases formed. If we have μ–x–T variables in a system, it is obvious that very careful mineralogy must be done in order to establish the active factors in the clay mineral genesis. We do know of course that the general sequence of mixed-layered, aluminous minerals occurs as a function of depth and temperature in a hydrothermally active area such as the Salton Sea geothermal field. But this is "regional" in importance. Many authors have observed that clays in vein systems in geothermal fields were not the same as those in country rock (Schoen and White 1965; Steiner 1968 for example). In consequence, there will be regional or country rock assemblages, affected by temperature, as well as local thermal gradient assemblages. Further, it is well known that hydrothermal fluids vary greatly in composition, especially in volatiles, and one can thus expect that the chemical potential of certain elements will be affected by the solutions. Variables such as pH, f_{O_2}, f_{CO_2}, f_{S_2} should be taken into consideration. There is not enough information available at present to generalize in a detailed manner beyond what has been proposed as alteration facies using key minerals or mineral assemblages (Lowell and Guilbert 1970). These facies, potassic, phyllic, argillic and propylitic, are certainly due to a combination of T–x–μ variables.

It should be noted that the types of clay found in hydrothermal alteration of acidic rocks cover the range of those found in sedimentation and diagenesis environments as would be expected. We will treat only one series of clay mineral facies here as an example of the type of analysis which can be performed.

9.5.1 Example of potassic alteration

Beaufort (1981) has studied the alteration in microgranites in a mineral prospect found in France (Sibert prospect, Rhône). The rock presents a generalized potassic alteration facies which can be represented by the system shown in Fig. 9.20. This alteration is pervasive in the rock but is well illustrated near the quartz–orthoclase–chalcopyrite veins which traverse it. The

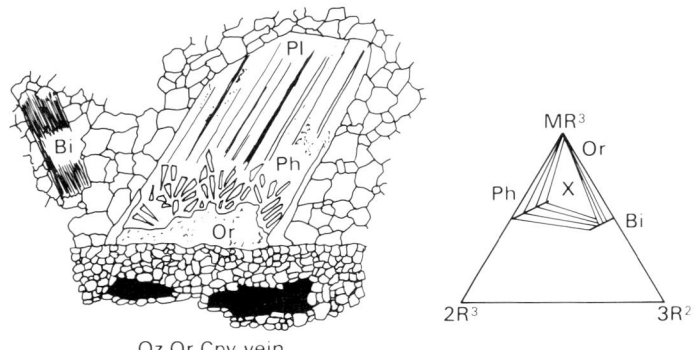

FIG. 9.20. Rock alteration around a potassic vein (drawing after a photograph) and related phase interpretation in the MR^3–$2R^3$–$3R^2$ system. Qz = quartz; Or = orthoclase; Cpy = chalcopyrite; Ph = phengite; Pl = plagioclase; Bi = magnesian biotite. Cross (×) shows bulk composition of the rock.

basic granite mineralogy of two feldspars–biotite–quartz is changed to those of aluminous phengite–phlogopitic mica–orthoclase–albite assemblages. The presence of an aluminous, magnesian phlogopite is typical of potassic alteration facies (Moore and Czamanske 1973; Beane 1974; Jacobs and Parry 1976). In fact, this facies is very close to that of the magmatic rock. The bulk composition of the rock is little changed except for a loss of calcium and a gain in potassium. This potassic alteration facies can then be represented as the same mineral paragenesis as that of magmatic rocks although the temperature conditions under which it formed are certainly lower.

9.5.2 Example of phyllic alteration

We have now "set the scene" for the episode of alteration in which we are interested. Veins cross the potassic altered granite giving it a new mineralogy which can be called phyllic – at least in the vein centres. Phyllic alteration is typified by veins of "sericite mica", quartz and perhaps sulphides (Sales and Meyer 1950; Moolik and Durek 1966; Lowell and Guilbert 1970). In the case described by Beaufort (1981), sericitic veins alter the potassic facies "country rock' giving the following newly formed mineral assemblage sequences outward from the vein; muscovite–phengite–phlogopite, illite–corrensite–orthoclase. These phases are produced at the expense of biotite, plagioclase and to a certain extent orthoclase. Within some biotite grains, the assemblage corrensite–chlorite–illite is formed in the outermost zones which are adjacent to the "unaltered" potassic facies assemblage of orthoclase–phengite–orthoclase. Figure 9.21 shows these relations.

Two observations must be made here. Firstly, the bulk composition of the rock changes from the vein outwards. The crosses in the diagrams show what must be the bulk composition according to the minerals found. It is evident that in going from the vein outwards, the composition of the country rock is attained. Thus we see that the phyllic vein must bring in solutions, material of the muscovite composition which is depositioned in the vein, and exchanged with the enclosing rock. This is the chemical potential effect.

The second observation is that the mineral pair muscovite–magnesian phlogopite is no longer stable in the outer alteration zone. It is replaced by the pair orthoclase–corrensite. This latter clay mineral assemblage is obviously not a magmatic one, and should then represent a

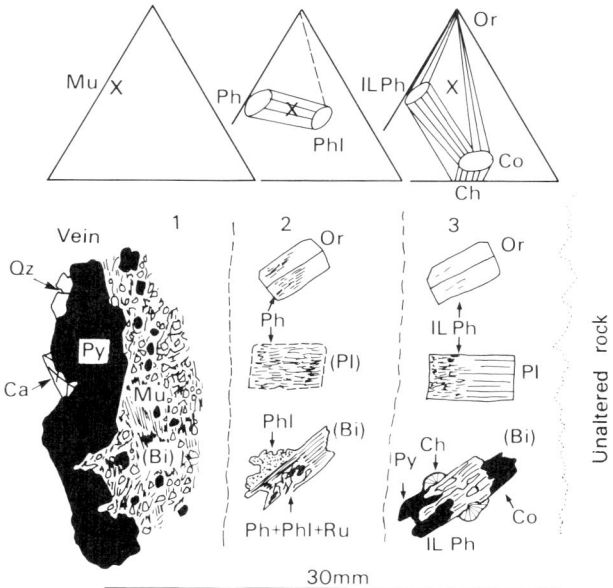

FIG. 9.21. Rock alteration around a phyllic vein and related phase interpretation in the MR^3–$2R^3$–$3R^2$ system. Ca = calcite; Qz = quartz; Py = pyrite; Mu = muscovite; (Bi) = altered biotite; Or = orthoclase; Ph = phengite; Pl = plagioclase; (Pl) = altered plagioclase; Phl = phlogopite; Ru = rutile; ILPh = illite–phengite micaceous phase; Ch = chlorite; Co = corrensite. Crosses (×) show bulk composition of the rock in the different zones.

lower temperature facies. It is then apparent that there is a temperature effect which acts on the rock as well as one of changing chemical potential.

This demonstration is certainly not the only one to be found in hydrothermal alteration zones, but it can possibly be used as a model to explain certain superpositions of alteration facies. It is necessary to look closely at the phases present in each zone, as well as their spatial distribution. This can at times be useful in interpreting the spatial, chemical and temperature relations of hydrothermal alteration.

9.6 CONCLUSIONS

Since this chapter is only one among many, the material dealt with is not exhaustive and thus the conclusions cannot be either. One might say that, in the end, the most important factor is recognition of the active variables in a given clay-bearing system. One must distinguish between chemical components controlled by the solids present and those controlled by the solutions affecting the system. It should be remembered that "inert" is a relative term and that an inert component can move in or out of a chemical system.

The final goal of chemiographic analysis is to simplify the representation of geologic observations; thus one should choose the most simple system possible without losing sight of critical relations.

NOTE

1. Editor's note. According to current definitions, the name rectorite is preferred to allevardite for interstratified minerals with regular alternation of non-expanding (mica-like) and expanding (smectite-like) interlayers. However custom has continued the use of allevardite in expressions such as "allevardite-like ordering" for minerals with some but less regular alternation of non-expanding potassium-containing interlayers with smectite-like interlayers. It is in this latter sense that the name allevardite is used in this chapter (see Section 1.9.3).

REFERENCES

AOYAGI K. & KAZAMA T. (1980) Transformational changes of clay minerals, zeolites and silica minerals during diagenesis. *Sedimentology* **27**, 179–188.

BEANE R.E. (1974) Biotite stability in the porphyry copper environment. *Econ. Geol.* **69**, 241–256.

BEAUFORT D. (1981) Etude pétrographique des altérations hydrothermales superposées dans le porphyre cuprifère de Sibert (Rhône). Influence des microsystèmes géochimiques dans la différenciation des micas blancs et des phases trioctaédriques. Unpub. Thesis. Univ. Poitiers, 147 pp.

BIGHAM J.M., JAYNES W.F. & ALLEN B.L. (1980) Pedogenic degradation of sepiolite and palygorskite on the Texas High Plain. *Soil Sci. Soc. Am. J.* **44**, 159–167.

BONATTI E. & HOENESU O. (1968) Palygorskite from Atlantic deep sea sediments. *Am. Miner.* **53**, 925–983.

BROWNE D.R.L. & ELLIS A.J. (1970) The Oaki–Broadlands hydrothermal area, New-Zealand: mineralogy and related geochemistry. *Am. J. Sci.* **269**, 97–131.

BURST J.F. (1959) Post-diagenetic clay mineral environmental relationships in the Gulf Coast Eocene. *Clays Clay Miner.* **6**, 327–341.

CALVERT C.S., BUOL S.W. & WEED S.B. (1980) Mineralogical characteristics and transformation of a vertical rock–saprolite–soil sequence in the North Carolina Piemont. I – Profile morphology, chemical composition and mineralogy. *Soil Sci. Soc. Am. J.* **44**, 1056–1103.

CHURCH T.M. & VELDE B. (1979) Geochemistry and origin of a deep-sea Pacific palygorskite deposit. *Chem. Geol.* **25**, 31–39.

COURBE C., VELDE B. & MEUNIER A. (1981) Weathering of glauconites: reversal of the glauconitization process in a soil profile in Western France. *Clay Minerals* **16**, 231–243.

COUTURE R. (1977) Composition and origin of palygorskite-rich and montmorillonite-rich zeolite containing sediments from the Pacific Ocean. *Chem. Geol.* **19**, 113–130.

DUNOYER DE SEGONZAC G. (1969) Les minéraux argileux dans la diagenèse. Passage au métamorphisme. *Mém. Serv. Carte Géol. Als.-Lorr.* **29**, 317 pp.

DUNOYER DE SEGONZAC G. (1970) The transformation of clay minerals during diagenesis and low-grade metamorphism: a review. *Sedimentology* **15**, 281–346.

EGGLETON R.A. & BOLAND J.N. (1982) Weathering of enstatite to talc through a sequence of transitional phases. *Clays Clay Miner.* **30**, 11–20.

ESLINGER E.V. & SAVIN S.M. (1973) Mineralogy and oxygen isotope geometry of the hydrothermally altered rocks of the Oaki–Broadlands, New Zealand geothermal area. *Am. J. Sci.* **273**, 240–269.

FONTANAUD A. & MEUNIER A. (1983) Mineralogical facies of a weathered serpentinized lherzolite from Pyrénées (France). *Clay Minerals* **18**, 1–11.

GARRELS R.M. & CHRIST C.L. (1965) *Solutions, Minerals and Equilibrium*. Harper and Row ed., 1 vol., 450 pp.

HARDER H. (1976) Nontronite synthesis at low temperatures. *Chem. Geol.* **18**, 169–180.

HARDER H. & FLEMING W. (1970) Quartz synthese bei tiefen temperaturen. *Geochim. Cosmochim. Acta* **34**, 295–305.

HELGESON H.C. (1970) Description and interpretation of phase relations in geochemical processes involving aqueous solutions. *Am. J. Sci.* **268**, 415–438.

HELING D. (1978) Diagenesis of illite in argillaceous sediments of the Rhine Graben. *Clay Minerals* **13**, 211–220.

HODA S.N. & HOOD W.C. (1972) Laboratory alteration of trioctahedral micas. *Clays Clay Miner.* **20**, 343–358.

HOWER J. (1961) Some factors concerning the nature and origin of glauconite. *Am. Miner.* **47**, 886–896.

HOWER J., ELSINGER E.V., HOWER M.E. & PERRY E.A. (1976) Mechanism of burial metamorphism of argillaceous sediment: 1 – Mineralogical and chemical evidence. *Bull. Geol. Soc. Am.* **87**, 725–737.

IIJIMA A. (1970) Present day zeolitic diagenesis of the Neogene geosyneclined deposits on the Niigata Oil Field, Japan. *2nd Int. Zeolite Conf.* American Chemical Society.

ILDEFONSE P. (1980) Mineral facies developed by weathering in a metagabbro, Loire–Atlantique, France. *Geoderma* **24**, 257–274.

ILDEFONSE P., PROUST D., MEUNIER A. & VELDE B. (1979) Rôle de la structure dans l'altération des roches cristallines au sein des microsystèmes. Mise en évidence de la succession des phénomènes de déstabilisation–recristallisation. *Bull. Assoc. Fr. Et. Sol.* **2–3**, 239–257.

JACKSON M.L., HZEUNG Y., COREY R.B., EVANS E.J. & VANDEN HEUVEL R.C. (1952) Weathering sequence of clay size minerals in soils sediments. II. Chemical weathering of layer silicates. *Soil Sci. Soc. Am. Proc.* **16**, 3–6.

JACOBS D.C. & PARRY W.T. (1976) A comparison of the geochemistry of biotite from some Basin and Range stocks. *Econ. Geol.* **71**, 1029–1035.

KORZHINSKII D.S. (1959) *Physicochemical basis of the analysis of the paragenesis of minerals* (Translation from Russian). New York Consultant Bureau.

KRAUSKOPF K.B. (1956) Dissolution and precipitation of silica at low temperatures. *Geochim. Cosmochim. Acta* **10**, 1–27.

LAHANN R.W. (1980) Smectite diagenesis and sandstone cement: the effect of reaction temperature. *J. Sedim. Petrol.* **50**, 755–760.

LONSDALE P.F., BISCHOFF J.L., BURNS V.M., KASTNER M. & SWEENEY R.E. (1980) High temperature hydrothermal deposit on the sea-bed at a Gulf of California spreading center. *E.P.S.L.* **49**, 8–20.

LOVELAND P.J. (1981) Weathering of a soil glauconite in Southern England. *Geoderma* **29**, 35–54.

LOWELL J.D. & GUILBERT J.M. (1970) Lateral and vertical alteration–mineralization zoning in porphyry ore deposits. *Econ. Geol.* **65**, 373–408.

MCDOWELL S.D. & ELDERS W.A. (1980) Authigenic layer silicate minerals in Borehole Elmore 1, Salton Sea geothermal field, California, USA. *Contr. Miner. Petrol.*, **74**, 293–310.

MACKENZIE F.T. & GEES R. (1971) Quartz: synthesis at earth–surface conditions. *Science* **165**, 171–173.

MEUNIER A. (1980) Les mécanismes de l'altération des granites et le rôle des microsystèmes. Etude des arènes du massif granitique de Parthenay (Deux-Sèvres). *Mém. Soc. Géol. Fr.* **140**, 80 pp.

MEUNIER A. & VELDE B. (1976) Mineral reactions at grain contacts in early stages of granite weathering. *Clay Minerals* **11**, 235–240.

MEUNIER A. & VELDE B. (1982) X-ray diffraction of oriented clays in small quantities (0.1 mg). *Clay Minerals* **17**, 253–256.

MILLOT G. (1964) *Géologie des argiles*. Masson ed.

MITSUI K. (1975) Diagenetic alteration of some minerals in argillaceous sediments in Western Hokkaido, Japan. *Science Rep. Tohoku Univ.*, 3rd ser., **13**, 13–65.

MOOLIK R.C. & DUREK A.D. (1966) The Morenci district. In *Geology of the Porphyry Copper Deposits, South Western North America* (Titley and Hicks, eds), pp. 221–231.

MOORE W.J. & CZAMANSKE G.K. (1973) Compositions of biotites from unaltered monzonitic rocks in the Bingham mining district, Utah. *Econ. Geol.* **68**, 269–274.

MOORT J.C. VAN (1971) A comparative study of the diagenetic alteration of clay minerals in Mesozoïc shales from PaPua New Guinea, and in tertiary shales from Louisiana, USA. *Clays Clay Miner.* **19**, 1–20.

MUFFLER L.J.P. & WHITE D.E. (1969) Active metamorphism of upper cenozoic sediments in the Salton Sea geothermal field and the Salton Trough, Southeastern California. *Bull. Geol. Soc. Am.* **80**, 157–182.

PAQUET H. (1970) Evolution géochimique des minéraux argileux dans les altérations et les sols des climats méditerranéens tropicaux à saisons contrastées. *Bull. Serv. Carte Als. Lorr.* **30**, 212 pp.

PARADIS S. (1981) Le métamorphisme hercynien dans le domaine centre armoricain occidental: essay de caractérisation par l'étude des phyllites des formations gréso-pelitiques. Unpub. Thesis. Univ. Brest, 1 vol., 167 pp.

PARRON C. & NAHON D. (1980) Red bed genesis by lateritic weathering of glauconitic sediments. *J. Geol. Soc.* **137**, 689–693.

PEDRO G. (1964) Contribution à l'étude expérimentale de l'altération géochimique des roches cristallines. *Ann. Agron* **15**, 85–191, 243–333 and 339–456.

PEDRO G. (1966) Essai sur la caractérisation géochimique des roches superficielles (cycle aluminosilicique). *C.r. hebd. Séanc. Acad. Sci., Paris* **262**, 1828–1831.

PERRY E.A. & HOWER J. (1970) Burial diagenesis in Gulf Coast pelitic sediments. *Clays Clay Miner.* **18**, 165–178.

PERRY E.A. & HOWER J. (1972) Late stage dehydration in deeply buried pelitic sediments. *Bull. Am. Assoc. Petr. Geol.* **56**, 2013–2021.

PROUST D. & VELDE B. (1976) Beidellite crystallization from plagioclase and amphibole precursors: local and long-range equilibrium during weathering. *Clay Minerals* **13**, 199–209.

RIMSAITE J. (1972) Genesis of chlorite, vemiculite, serpentine, talc and secondary oxides in ultrabasic rocks. *Proc. Int. Clay Conf.* 1972, Madrid, pp. 291–302. Division de Ciencias, C.S.I.C., Madrid.

SALES R.H. & MEYER C. (1950) Interpretation of wall rock alteration at Butte, Montana. *Quart. Color. Sch. Mines* **45**, 261–274.

SCHMIDT G.W. (1973) Interstitial water composition and geochemistry of deep gulf coast shales and sandstones. *Bull. Am. Assoc. Petr. Geol.* **57**, 321–337.

SCHOEN R. & WHITE D.E. (1965) Hydrothermal alteration in GS-3 and GS-4 drill holes, main Terrace, Steamboat Springs, Nevada. *Econ. Geol.* **60**, 1411–1421.

STEINER A. (1968) Clay minerals in hydrothermally altered rocks at Wairakei, New Zealand. *Clays Clay Miner.* **16**, 193–213.

STONECIPHER S.A. (1976) Origin, distribution and diagenesis of philipsite and clinoptilolite in deep-sea sediments. *Chem. Geol.* **17**, 307–318.

TRAUTH N. (1977) Argiles évaporitiques dans la sédimentation carbonatée continentale et épicontinentale tertiaire. Bassins de Paris, de Mormoiron et de Salinelles (France), Jbel Ghassoul (Maroc). *Mém. Sci. Geol.*, 1 vol., 195 pp.

TRESCASES J.J. (1979) Remplacement progressif des silicates par les hydroxydes de fer et de nickel dans les profils d'altération tropicale des roches ultrabasiques. Accumulation résiduelle. *Sci. Géol. Bull.* **32**, 181–188.

VANDEN HEUVEL R.C. (1966) The occurrence of sepiolite and attapulgite in the calcareous zone of a soil near Las Cruces, New Mexico. *Clays Clay Miner.* **13**, 193–207.

VELDE B. (1973) Phase equilibria studies in the system $MgO–Al_2O_3–SiO_2–H_2O$: chlorites and associated minerals. *Mineralog. Mag.* **39**, 297–312.

VELDE B. (1977) *Clays and Clay Minerals in Natural and Synthetic Systems.* Elsevier, Amsterdam.

VELDE B. & ODIN G.S. (1976) Further information related to the origin of glauconite. *Clays Clay Miner.* **23**, 376–381.

WACKERMAN J.M. (1975) L'altération des massifs cristallins basiques en zone tropicale semi-humide. *Cah Q.R.S.T.M. sér. Géol.* **7**, 167–172.

WEAVER C.E. (1959) The clay petrology of sediments. *Clays Clay Miner.* **6**, 154–187.

WEAVER C.E. & BECK K.C. (1971) Clay–water diagenesis during burial: how mud becomes gneiss. *Geol. Soc. Am., Sp. paper*, 134 pp.

WILDMAN W.E., JACKSON M.L. & WHITTIG L.D. (1968) Iron-rich montmorillonite formation in soils derived from serpentinization. *Soil Sci. Soc. Am. Proc.* **32**, 787–794.

Subject Index

(Numbers followed by T are references to tables)

Adsorption
  free energy of, 136–7
  heats of, 239, 372
Adsorption at oxide surfaces, 135–40
  see also: Sorption isotherm
  phosphate, 139
  pyrophosphate, 138
  silicate, 138, 139
  zinc, 139
Akaganeite, 132T, 150, 155
Albite, 445
Alkylation of silanol groups, 412–15
Allevardite, 99, 438, 439
"Allevardite-like" ordering, 99
Allietite, 94T, 98–9
Allophane, 114–16
  disordered, 114–15
  Si/Al ratios in, 114–15
Alteration of micas, 106
Aluminium-hydroxy interlayers, 91–2
Aluminium oxides, 161–73
  anion adsorption, 171–2
  associated trace elements, 170
  cation adsorption, 170–1
  crystallization of, 163
  dissolution, 173
  mechanisms of accumulation, 162–3
  mineral associations, 172–3
  occurrence and formation, 161–2
  phosphate adsorption, 171
  relative stabilities, 167
Alurgite, 7, 83
Amesite, 27, 29, 31, 33T
Analysis
  chemical methods, 3–6
  "classical" methods, 3, 4
  instrumental methods, 3, 6–8
  rapid analysis methods, 4
Analysis of clays, elemental, 3–9
Anandite, 84
Anatase, 132T, 186, 187
  synthetic, containing Fe(III), 187
Anauxite, 21
Annite, 63
Antigorite, 27, 29, 33, 40, 37T
Arc-spark spectroscopy, 6
Atomic absorption spectrophotometry, analysis by, 7
Atomic emission spectrophotometry, analysis by, 6–7
Attapulgite, *see* Palygorskite

Bauxite, 163, 167, 168, 169, 170
Bayerite, 132T, 161, 165–6
Beidellite, 49, 50, 52T, 427, 445
  surface charge density, 243
  swelling and dispersion, 259
  water sorption isotherm, 258
Berthierine, 27, 29, 31, 34T, 429
  thermal oxidation, 357
BET equation, 246, 247
  non-applicability to swelling minerals, 246, 248
Biotite, 77T, 78, 427, 445
  "sodium biotite", 79
  thermal oxidation, 357–8
  weathering-oxidation, 358–9
Biotite, ferroan, 77T
Biotite, magnesian, 76T, 77T, 454
Birnessite, 132T, 174T, 177–8
  convertion to cryptomelane, 181
  relation to rancieite, 178
  structure, 177
  surface charge on, 139
  synthesis, 178
  trace elements in, 177
Bityite, 84
Boehmite, 132T, 161, 167–8
Brammallite, 69, 71
Brindleyite, 29, 35T
Brittle mica, 69, 75, 84
  unit cell contents, 14–17
Brookite, 132T, 186, 187–8
  Fe(III) content, 187
Brucite, 428
Brunsvigite, 86T; *see also* Clinochlore, Chamosite
Burial diagenesis, reduction of iron in, 441–3
Buserite, 178

Caesium in micas, 81
Caliche formation, 434–6
Caryopilite, 27, 37, 38T
Catalysts, desirable properties, 291–2
Cation exchange capacity, 218–19, 226, 242–3
Cation exchange isotherms, 231
Cation exchange reactions, 225
  definitions, 225–7
  equilibrium dialysate for, 227
  equivalent convention, 226
  equivalent fractions, 228
  molar convention, 226
  with more than two counter ions, 233–5

Cation exchange reactions, *continued*
  with organic cations, 373–6
  selectivity coefficient for, 229
  standard states for, 229
  thermodynamic equilibrium constant for, 229
  thermodynamics of, 227–30
Cation hydration energies, 259–60
Cation hydration numbers, 227, 230
Cation selectivities in clays, 230–1
  effect of cation hydration, 232–3
  effect of surface charge density, 232
Cation solvation by organic molecules, 385–6
Cation surface excess, 225–6
CEC, *see* Cation exchange capacity
Celadonite, 65, 70, 72–5, 74T, 427
Chamosite, 84–5
  aluminian ferrian, 87T
  ferrian, 86T
  magnesian, 85, 86T
Charge heterogeneity in swelling minerals, 59–61
Chemical components, 424–32
  Fe–Mg–Al projections, 429–30
  inert, defined, 424
  M^+, defined, 425
  $M^+R-2R^3-3R^2$ projections, 426–8
  mobile, defined, 424
  R^2, defined, 425
  R^3, defined, 425
  silica, 425–6
  $Si-R^2-R^3$ projections, 428–9
  simplified, 425
Chemical potential-composition representation, 430
Chernykite, 83
Chlorite, 84–92, 427, 428, 429, 434, 437, 438, 439, 442
  dehydroxylation, 334–5
  dioctahedral, 84, 85–9, 90–91T
  di, trioctahedral, 84, 85–9, 88T
  di, trioctahedral, Li contents, 88–9
  octahedral cations per unit formula, 86–9
  oxidation of Fe^{2+} in, 359
  swelling, 90, 92
  trioctahedral, 84–5
  with incomplete hydroxide layers, 89–92, 98
Chromium minerals, *see* Unusual elements in layer silicates
Chrysotile, 27, 29, 32, 36T–37T, 40
  dehydroxylation, 336
  recrystallization to fosterite, 346
  vinyl derivatives, 414
Classification of clay minerals, 11–13
Clay catalysed reactions
  alcohol oxidation, 275
  alkenes to dialkyl ethers, 309

Clay catalysed reactions, *continued*
  asymmetric hydrogenation, 305–7
  carbonium formation, 280–2
  cracking, 288–9, 313
  dehydration of alcohol to ether, 309
  depolymerization of C_8 homopolymers, 285
  disaccharide hydrolysis, 287–8
  double bond migration, 285
  ester hydrolysis, 287
  gasoline disulphurization, 287
  hexene isomerization, 275, 284
  hexene polymerization, 285
  hydrazobenzene cleavage, 275
  hydroformylation, 307–8
  hydrogenation of alkenes, 298–300, 302
  hydrogenation of alkynes, 301–2
  hydrogenation of 1,3-butadiene, 303–5
  hydropyrolysis of thiophosphate esters, 308–9
  oleic acid polymerization, 405–6
  pentene polymerization, 275, 285
  pinene isomerization, 276
  pyrolysis, 292
  synthesis of chiral aminoacids, 305–7
  styrene polymerization, 286–7
  terpene isomerization, 275, 276, 284
  unsaturated fatty acid dimerization, 286, 310–11
Clay mineral catalysts, 286, 402–3, 407
  acid activation, 289–91
  Al coordination in, 290–1
  intercalated metal complex, 297–311
  metal-hydroxy complexes, 311–14
  pillared clays, 311–14
  synthetic smectite, 293–7
Clay mineral, definition of, 2
Clay-organic complexes, 292, 372–4
  α-type complexes, 372
  β-type complexes, 372
  π-complexes, 389–90
  base protonation, 392–5
  cation solvation, 386–90
  kink-block structures, 383–4
  pyrolysis of, 292–3
  structural substitution and complexation, 390, 395
  with transition metals, 388–90
  with water bridges, 392
Clay-organic complexes, specific compounds
  acetone, 388, 396
  acrylonitrile, 387
  adenine, 380
  n-alkanols, 383
  alkylammonium salts, 377–85
  alkylpyridinium bromides, 377
  amino acids, 379

Clay-organic complexes, *continued*
  arene-copper, 390
  arene-silver, 390
  atrazine, 375
  benzene, 382
  benzene-copper, 389–90
  benzonitrile, 391
  cetyl pyridinium bromide, 247, 378
  chlorobenzene, 382
  codeine, 377–8
  crown ethers, 386–7
  cryptands, 386–7
  dialkylamides, 382
  dibenzene-chromium, 389
  dicryl, 375
  diquat, 381
  diuron, 375
  dodecylammonium cations, reactions with mica, 376–7
  ethanol, 388, 396
  ethylamine, 381–2
  ethylene glycol, 246
  ethylene glycol monoethyl ether, 246
  ethyl-N, N-di-n-propyl thiocarbamate (EPTC), 382
  guandinium salts, 379
  methylene blue, 247
  morpholine, 374
  paraquat, 381
  peptides, 379
  propazine, 375
  purines, 380
  pyridine, 381–2, 391
  pyrimidines, 380
  simetone, 375
  thiophene-copper, 390
  triphenylmethylcarbonium, 395
  trietazine, 375
Clay surfaces, 241–2
  interlamellar, 244–5
Clay templates, 402–7
  for biologically-active compounds, 406–7
  for insertion polymers, 407
Clinochlorie, 84–5
  aluminian, 85, 86T
  chromian, 85, 87T
  ferroan, 85, 86T
  manganoan, 85
Clinoptilolite, 436
Clintonite, 75
  surface charge density, 243
  unit cell contents, 14–17
Coagulation, 203, 205
  critical coagulation ratio, 203
Co-ions, 204, 215
Cookeite, 88–9, 89T

Coronadite, 181
Corrensite, 103, 104, 427
Corundun, 132T, 161, 169, 321
Counter-ions, 204, 215, 218
Cronstedtite, 27, 29, 31, 35T
Cryptomelane, 132T, 174T, 181
  surface charge on, 139
Cubic close packing in oxides, 130–3

Deep-sea sediments, 436–7
Dehydroxylation mechanisms, 329–31
Dehydroxylation rates, 324–31, 337–8
  diffusion controlled, 325–7
  effect of powder packing on, 327, 329
  effect of temperature on, 327
  effect of vapour pressure on, 327
Depth-temperature and mineralogy, 438
Deweylite, 47
Diabantite, 86T; *see also* Chamosite
Diaspore, 132T, 161, 166, 168–9
  associated minerals, 169
  Fe substitution in, 168, 169
  manganoan, 169
  occurrence, 168–9
Dickite, 22, 24T, 25T
  Cr-rich, 25T, 26
  dehydroxylation, 333
  14 Å dehydroxylate, 333–4
Dielectric permittivity, 241
Differential heat flow scanning calorimetry, 241
Dissolution of clays for analysis, 8–9
Dissolution of oxides, *see* selective dissolution treatments
Donbassite, 88, 90–91T

Electrical double layers, 204
  experimental test of theory, 222–3
  Gouy–Chapman model, 214, 221
  interacting planar, 219–20
  on clay minerals, 208
  Poisson–Boltzmann equation, 215
  potential and charge distribution, 215–19
  repulsive force between interacting, 220
  Stern model, 215, 221–2
  theory, 214–23
Electron spin resonance, 241, 297–8, 373
Electron microprobe, analysis by, 8
  applications, 23, 35, 38, 41, 47, 73, 79, 145
Enstatite, 321, 347, 348
Ephesite, 84
Equilibrium dialysis, 227

Feldspar, 427, 430
Feroxyhite, 132T, 148
  formation, 148
Ferric hydroxide, *see* Ferrihydrite

Ferrihydrite, 132T, 147–8, 149
　colour variation in, 147
　effect of organic matter on formation of, 148
　formation of, 143, 147, 150, 151
Ferriphlogopite, 75
Ferripyrophyllite, 42
Ferrous iron, determination of, 4
Flocculation, 203, 205
　edge-to-edge and edge-to-face, 208
　flocculation ratio, 205
　of clay suspensions, 209
　of montmorillonite by salts, 209
　sensitizing with polyelectrolytes, 213–14
Fluorine, determination of, 5–6
Forsterite, 321, 335, 344–6
Frenkel–Halsey–Hill equation, 248
Fuchsite, 81–2

Garnierite, 29–30, 47, 360
Gel formation, 207
　by montmorillonite, 209
Gibbs–Duhem equation, application to cation exchange, 227–8
Gibbsite, 132T, 161, 166, 428, 430
　adsorption of silicate, 138
　association with kaolinite, 164
　structure, 164
　surface anions and crystallization, 165
Glauconite, 65, 70, 72–5, 74T, 427, 440
　interlayer hydroxy complexes in, 75
　marine formation, 440–1
Goethite, 131, 132T, 142–6, 149
　adsorption of silicate, 139, 145
　adsorption of pyrophosphate, 138
　adsorption of zinc, 139
　Al substitution in, 143
　crystallite boundaries in, 146
　formation of, 143, 147, 151
　reaction with phosphate, 139, 155–6
　trace and minor elements in, 145
　morphologies of, 144, 145, 146
Greenalite, 27, 34, 38T
Green rust, 150, 151–2, 153
　influence of Al on oxidation, 152
1:1 Group minerals, 11, 12
　definition of, 11
　dioctahedral, 22–6
　trioctahedral, 22, 26–40
2:1 Group minerals, 11
　definition of, 11
　smectite-vermiculite subgroups, 48–63
　talc-pyrophyllite subgroups, 40–8
Groutite, 174T

Halloysite, 22, 25T, 26
　acid-activation, 289–91

Halloysite, *continued*
　catalysis with acid-activated, 276, 287, 291–2
　cation exchange properties, 26, 252
　Cr-rich, 25T, 26
　dehydroxylation, 334
　Fe-rich, 25T, 26
　morphology and composition, 26
Halloysite (10 Å), 252
　NMR measurements on, 253
　water desorption from, 252–3
Halloysite (7 Å), 252
　indirect rehydration, 253
　residual water in, 253
　water sorption by, 252–3
Hausmannite, 174T
Hectorite, 54, 57T
　complex with Ru (bipyridine)$_3^{2+}$, 389
　oxidation of Rh-complex, 302–3
　Rh-complex catalysts, 298–308
　water sorption and N surface, 267
Hematite, 131, 132T, 148–9
　Al in, 149
　formation of, 143, 147, 148–9
　silica in, 149
Hemisalt complexes, 380, 381
Hendricksite, 84
Heterogeneous reaction mechanism, 330
Hexagonal close packing in oxides, 130–3
High temperature transformations
　chlorite to olivine, 346–7
　kaolinite to mullite, 349–50
　metakaolin to spinel, 350–4
　pyrophyllite to mullite, 349
　saponite to enstatite, 348
　sepentine minerals to forsterite, 344–6
　talc to enstatite, 347
Hofmann–Klemen effect, *see* Lithium migration
Hollandite, 132T, 174T, 180, 181
"Hydrobiotite", 106, 108T
Hydrogen bonding to clay surfaces, 385, 395–6, 397–8
Hydrogen bonding in complexes, 372, 379–80, 381–2, 384–6, 391–6
Hydrogen bonding in kaolinite, 397
Hydrogen, determination of, 4–5
Hydromica, 69, 70
Hydromuscovites, 70, 72T
Hydrotalcite, 151
Hydrothermal alteration, 453–5
　phyllic alteration, 454–5
　potassic alteration in microgranites, 453–4
Hydrous micas, 65, 69, 70
Hydroxy interlayers containing Al, Bi, Cr, Cu, Fe, Mg, Nb, Ni, Si and Ta, 17–19, 92, 311–14, 342–4, 362–3

Illite, 65, 70, 71, 72T–73T, 93, 427, 434, 438, 439, 440, 442, 445–7
  dehydroxylation, 338
  diagenesis, 64, 99
  H_3O^+ in, 71, 338
  NH_4^+ in, 71
  "trioctahedral", 78
  water sorption by, 254–6
Ilmenite, 132T, 186, 188
Imogolite, 114, 115–16
Impurity in minerals 65, 74–5, 85, 87; see also Unit cell contents, Structural formula
Intergradient chlorite-vermiculite, 90, 104–5
Interlamellar swelling, 61–3, 80, 255–60
  cation solvation and, 61–2, 255–60
  in chlorites, 90–2
  "equivalent anion radius" and, 62, 62T
  relation to structural substitution, 61
  sorption of water and, 257, 384
Interstratified minerals, 65–7, 92–107
  chlorite-smectite, 28 Å, 105
  component layers in, 92
  description, 93
  glauconite-smectite, 70, 101, 103T
  illite-smectite, 70, 100–1, 102T
  IS ordering in, 99, 102T, 103T
  ISII ordering in, 99, 100, 102T, 103T
  "Kalkberg ordering", 100, 102T
  kaolinite-smectites, 105–7, 106T
  mica-vermiculite, 106–7, 108T
  non-regular chlorite-smectite-vermiculite, 104–5
  ordering in, 93
  partially ordered mica-smectites, 100–1, 102T
  randomly interstratified illite-smectite, 100–1, 102T
  regular dioctahedral chlorite-smectite, 101–4
  regular mica-dioctahedral smectite, 99
  regular talc-chlorite, 94T, 98
  regular talc-saponite, 94T, 98–9
  regular trioctahedral chlorite-saponite, 95T, 104
  structural formulae for, 93–7
  surface charge density of, 243–4
Iron oxidation in minerals, 356–7, 451
Iron oxides, 141–61, 440, 445
  adsorption of NH_3, NO, CO_2, CH_3OH and pyridine, 158–9
  adsorption of transition elements by, 158
  anion adsorption by, 157T
  coatings on aluminosilicates, 160–1
  dissolution, 159–60
  formation in soils, 141–2
  impurities in, 142
  "ligand" exchange at surface of, 155–7
  oxidation of olefins at surface, 158

Iron oxides, *continued*
  synthesis of, 142
Isoelectric point, 135, 140, 155

Kaolinite, 22–6, 24–25T, 427, 430, 437, 440, 442, 445
  atomic ratios to Si in, 23
  catalytic activity, 286
  cation exchange properties, 24
  Cr-rich, 24, 25
  dehydroxylation, 331, 332; see also Metakaolin
  dehydroxylation rate, 327–9
  dielectric absorption in water films, 251
  heats of immersion, 251
  hydration of Li-kaolinite, 250–1
  iron content of, 23
  NMR in water films, 251
  pH-dependent charges, 24
  specific surfaces to N_2, 250–1
  surface acidity of, 280
  surface charge density, 24, 250
  water sorption by, 249–51
Kaolinite intercalation complexes
  amides, 397
  dimethylsulphoxide, 397, 400, 401
  displacement complexes, 401
  hydrazine, 253, 397, 399
  intercalation capacity, 402, 403
  N-methylacetamide, 397
  organic acid salts, 253, 397, 401
  pyridine-N-oxide, 397
  reaction kinetics, 398
  reaction rate and crystallinity, 399–401
  reaction rate and particle size, 399–401, 403
Kaolinite sub-group minerals, 22–6, 24–25T
Kellyite, 29, 31, 33T
Kerolite, 45T, 47–8
Kulkeite, 94T, 98

Langmuir equation, 261
Laponite, 54, 57T
  water sorption and N_2 surface, 267
Lepidocrocite, 132T, 149–51
  Al in, 150
  formation of, 143, 147, 149–51
  occurrence, 149
  transformation to γ-Fe_2O_3, 150
Lepidolite, 81, 82T
Lepidomelane, 77T, 78, 78T
Leucoxene, 186
Lithiophorite, 132T, 174T, 179
  structure, 179
Lithium migration in montmorillonite, 322–4, 396
  structural changes, 323

Lizardite, 27, 29, 32–33T, 40
　dehydroxylation of Ni, 336
Loughlinite, 114
Lyophilic colloids, 203
Lyophobic colloids, 203
　preparation, 206–7

Maghemite, 132T, 152–5
　Fe(II) content, 152, 154
　formation, 143, 151, 153–5
　occurrence in soils, 153
　Ti-rich, 153
Magnesium-hydroxy interlayers, 91–2
Magnesium migration, 322
Magnetite, 132T, 152–5
Manganese oxides, 173–85
　amorphous, 182
　anion adsorption, 183
　cation sorption, 183
　compositions, 174T, 177T
　conditions of formation, 175
　dissolution, 185
　point of zero charge, 182–3
　polymorphism of MnO_2, 174
　synthesis, 176
　trace elements in soil extracts, 184–5
Manganite, 174T
Margarite, 68T, 69
Mariposite, 82
Maya blue, 411
Medmontite, 55
Metakaolin, 332
　transformation to spinel phase, 350–4
　structure, 332–3
Mica minerals, 63–4; *see also* Mica mineral
　　names
　dioctahedral, Cr^{3+}-rich, 81–2
　dioctahedral, tetrahedral Fe in, 65–6
　dioctahedral, tetrasilicic, 64, 430
　dioctahedral, trisilicic, 64
　exchange of potassium from, 376
　fluorine in, 75, 81
　Li-rich, 81
　trioctahedral, 75–80
　trioctahedral, Na-rich, 79–80
　trioctahedral, OH + F deficit in, 78
　unusual elements in, 81
Mica weathering, 360
Microcline, 432
Mineral-alkyl derivatives, 412–15
Minnesotaite, 44T, 46, 79, 80
Mixed-layer minerals, 427, 438, 439, 440, 442;
　see also Interstratified minerals
Monolayer, 246, 247–9
Montmorillonite, 49–53, 50–52T, 427
　acid activation, 289–91

Montmorillonite, *continued*
　Al^{3+}-catalyst, 309–10
　n-alkylammonium complexes, 378–85
　bentonite-organic chemical reactions, 371
　catalysis with acid-activated, 276, 284, 286,
　　287, 291–2, 297
　cation hydration in, 256–60
　Chambers type, 50
　Cheto type, 49
　Cu^{2+}-catalyst, 309
　dehydroxylate, 321, 339
　dehydroxylation mechanism, 339–40
　heats of immersion, 265–6
　high temperature transformations, 355
　homoionic, hydration measurements on, 264
　iron rich, 52T
　"non-ideal", 53
　osmotic swelling, 374, 384
　Otay type, 49
　residual water in, 264
　Rh-intercalated catalyst, 306
　surface acidity of, 280
　surface charge density, 242
　swelling and dispersion, 259
　Tatatila type, 50–3
　transition metal complexes, 388–90
　water sorption isotherms, 257, 258, 265
　Wyoming type, 49
Mössbauer spectroscopy, applications, 4, 54,
　　149, 358
Mullite, 321, 349–54
Muscovite, 64–9, 68T, 427, 439, 445–7
　dehydroxylate, 331, 337–8
　dehydroxylation mechanism, 337–8
　high-temperature transformations, 354–5
　Mn-rich, 83
　surface charge density, 243
　surface hydration, 254
　water sorption by, 254–6

Nacrite, 22, 24T
　dehydroxylation, 332
Nepouite, 27, 29–30
　dehydroxylation, 336
　high temperature reduction, 361
　high temperature transformations, 346
Neutron scattering, 240
Nickel migration, *see* Lithium migration
Nickel minerals, *see* Unusual elements in layer
　　silicates
　reduction of, 360–3
Nimite, 84–5
　magnesian, 85, 87T
Nontronite, 52T, 53
　tetrahedral Fe^{3+} in, 53–4
Norstrandite, 132T, 161, 166

Nsutite, 174T
Nuclear magnetic resonance, 240–1, 297, 373

Olivine, 321, 335, 346–7
Organic cation-organic solvent complexes, 382–4
Organic molecules in surface reactions
  acetone, 323–4
    1,5 cyclooctadiene, 303
    4,4′-diamino-*trans*-stilbene, 406
    ethanol, 323–4
    ethylene glycol, 323–4
    morpholine, 323–4
    norbornadiene, 299–307
    oleic acid, 310–11, 405–6
    L-ornithine, 404–5
    3-pentanone, 323–4
    tetraphenylporphyrin, 300
    triphenylmethylcarbonium, 281–2
    triphenylphosphine, 298–303
Orthoclase, 445–7
Osmotic swelling, 215, 220, 222, 244, 259
Oxide structures, 130–3
  Greek notation for, 133
Oxide surfaces
  adsorption at, 135–40, 155–9, 170–2, 188–9, 207
  aluminosilicate clays, 241–2
  hydroxylation of, 134–5
  "ligand exchange" at, 137, 155
  surface potential at, 136–7
  water sorption isotherms for, 247–9

Palygorskite, 107, 109–12, 110T, 111T, 407–8, 428, 433–7
  catalytic activity, 283, 284, 285
  dehydration, 269, 341
  dehydroxylation, 341
  high-temperature recrystallization, 348
  Mn-rich, 111
  sedimentation in saline basins, 433–4
  sorption of N_2, 269
  sorption of organic compounds, 411–12
  structural formulae for, 109, 110
  surface acidity of, 280
  surfaces, 408–9
  water in, 109, 110, 269, 341, 408
Paragonite, 68T, 69
Particle-particle interaction
  Born repulsion, 213
  entropic repulsion, 212
  edge and face association, 207–8
  effect of particle shape, 207
  lyosphere repulsion, 212–13
  net potential energy of, 205–6

Partridgeite, 174T
Pecoraite, 27, 33, 36T
Pennantite, 84–5, 87T
Penninite, 86T; *see also* Clinochlore
Peptization, 203, 207
Peptizing clays, 210–12
  by edge-charge reversal, 210–11
  by face-charge reversal, 211–212
  by sequestration, 211
  with sodium metaphosphates, 211
  with sodium salts of organic acids, 211
Perovskite, 132T, 186, 188
Phases
  metastable, definition, 424
  stable, definition, 424
  unstable, definition, 424
Phengite, 65–9, 68T
Phlogopite, 63, 75–8, 76T
  Mn-rich, 84
  sodium, 79–80
  Ti-rich, 76T
Pimelite, 45T, 47, 55
  high temperature transformations, 347–8
  high temperature reduction, 362
Plagioclase, 445, 447
Plasma excitation, analysis by, 7
Point of zero charge, 135–7, 204, 207, 208
Poisson–Boltzmann equation, 215
Polylithionite, 81, 82T
Potassium fixation, 360
Potential determining ions, 135, 204, 208
Preiswerkite, 79
Pressure-temperature-composition and illite-smectite formation, 438–40
Protective colloids, 213, 214
Proton delocalization, 331, 340
Pseudoboehmite, 167
Purification of clays, 9–10
  density fractionation, 10
  fractional sedimentation, 9
  magnetic fractionation, 9–10
  selective dissolution, 10, 23, 159–160
Purity of clays, 9
Pyroaurite, 151, 353
Pyrolusite, 132T, 174T, 180, 181–2
Pyrophyllite, 41–4, 42–43T, 427
  dehydroxylate, 331, 336–7
  "extra" OH in, 41
  hydrophobicity of surface, 254
  Si/R^{3+} ratios in, 41
  substitutions in, 41
  transformation to mullite, 349

Quartz, 430, 445
  unit cell contents, 13

Ramsdellite, 132T, 174T
Rancieite, 174T, 178
Rectorite, 93–7, 94T, 99–100
    asymmetric Si occupancy in, 96–7
    interlayer Na, K or Ca, 94T, 99–100
    structural formulae, 93–7
Rehydroxylation, 341
Ripidolite, 86T; see also Chamosite
Romancheite, 180
Roscoelite, 83
Rubidium in micas, 81
Rutile, 132T, 186–7

Sanidine, 321
Saponite, 54, 56–57T, 434, 437
    dehydroxylation, 340
    Fe-rich, 54
    hydration states, 267
    layer charge and basal spacing, 268
    swelling and dispersion, 259
Sauconite, 55
Selective dissolution treatments
    aluminium oxides, 173
    iron oxides, 159–60
    manganese oxides, 185
Sepiolite, 107, 112–14, 112–113T, 407–8, 428, 433–7
    dehydration, 269
    dehydroxylation, 341
    high-temperature recrystallization, 348
    Na-rich, 114
    organic derivatives, 412, 414
    sedimentation in saline basins, 433–4
    sorption of methanol and ethanol, 409–11
    sorption of N_2, 269
    sorption of NH_3, 269, 409
    sorption of organic compounds, 411–12
    structural formula for, 113
    surfaces, 408–9
    water in, 109, 269, 408
Sericite, 69, 70, 72T
Serpentine, 9-layer aluminian, 27, 29
Serpentine sub-group minerals, 26–40, 427
    anomalous structural formulae for, 27–8
    dehydroxylation, 335
    dehydroxylation rate, 327–8
    Fe, Mn and Ni substitution in, 28
    high-temperature transformations, 344–6
    Ni-rich, 27, 29–31
    Ni serpentines, high-temperature reduction, 361–2
    Ni serpentines, high-temperature transformations, 346
    relation to chlorite, 39–40
    structure related to composition, 26–7, 40
    tetrahedral and octahedral R^{3+} in, 30
Shale diagenesis, 437–40

Sheridanite, 86T; see also Clinochlore
Siderophyllite, 77T, 78, 78T
Smectite, 48–56; see also Montmorillonite, Beidellite, Nontronite, Saponite, Volkhonskoite, Sauconite, Pimelite
    charge heterogeneity in, 59–61
    Cr-rich, 55
    dioctahedral, 48–53, 52T, 428, 434, 437, 438, 439, 447
    Fe-rich, dioctahedral, 52T, 448–51
    Fe-rich, trioctahedral, 54, 56T
    Li-rich, 54
    Ni-rich, 55
    trioctahedral, 54–7, 79, 80, 439
    V-rich, 55–6
    Zn-rich, 55
Smectite, synthetic catalyst, 293–7
    catalytic activity, 296–7
    dehydroxylation, 294–5
    structure, 294
    surface acidity, 295–6
Sodium absorption ratio, 234
Sorption isotherms, 238–9, 372
    comparison of, 248–9
    normalized, 247–9
    reduced, 247–9
Specific surface of clays, 242, 245–7
    determination with cetyl pyridinium bromide, 247
    determination by heats of wetting, 247
    determination with N_2, 246
    determination with polar molecules, 246–7
Spinel-type phases, 321, 350–4
Stevensite, 54–5, 57T
Structural formula, calculation of, 19–21; see also Unit cell contents, calculation of
    for clintonite, 20
    effect of impurity on, 21
    errors in, 21
    for interstratified minerals, 21–2
    for montmorillonite with Al-hydroxy interlayer, 20–1
Structure unit contents, see Unit cell contents
Sudoite, 87–8, 88–89T
Surface acidity of clays, 276, 392–5
    amine titrations for determination of, 279–80
    Brønsted sites, 277
    determination of, 277–83
    Hammett indicators for, 278–9
    NH_3 as surface probe, 281, 394
    pyridine as surface probe, 282–3, 391, 394
    spectroscopic study of, 280–3
Surface charge, 204
    by adsorption, 204
    on aluminosilicates, 133

Surface charge, *continued*
  effect of impurities on, 140–1
  on oxides, 133–5, 204
  permanent, 204, 208
  pH-dependent, 134, 135–7
  reversal of, 207, 208, 210–12
Surface charge density, 242–4
  alteration of, 244, 323, 360
  estimation of, 242–44
  relation to cation selectivity, 232
  relation to swelling, 61, 323
Swinefordite, 54

Taeniolite, 81, 83T
Talc, 44–45T, 44–48, 79, 80, 427, 428, 448–51
  aluminian, 44T
  dehydroxylation, 336–7
  hydrophobicity of surface, 254
  Ni-talc, *see* kerolite, pimelite
  sodian aluminian, 45T, 79, 80
  substitutions in, 46–7
  transformations to enstatite, 347
Talc hydrate (10 Å), 48
"Tarasovite", 94T, 100
Temperature programmed reduction, 361
Thermal dehydration, 323
Thermal reactions of clay minerals, 320–2
Thuringite, 87T; *see also* Chamosite
Titanium oxides, 186–9
  in clays, 186
  colloidal TiO_2, 186, 187
  pH dependent charge on, 189
  surface adsorption, 189
  titanium "hydroxide", 186
Todorokite, 132T, 174T, 179–81
  relation to Na buserite, 180
  structure, 180
  transition elements in, 180
Topotactic transformations, 334–5, 347–9, 353
Tosalite, 27, 29, 38, 39T
Tosudite, 94T, 95T, 103, 104
  Cr-rich, 104
  Li-rich, 95T, 104
Trilithionite, 81
Type 2 swelling, *see* Osmotic swelling

Unit cell contents, calculation of, 12–17
  ab initio, 12–15
  for clay minerals, 15–19
  effect of impurity on, 19
  F present, 16
  if density unknown, 15
  on water-free basis, 17
Unusual elements in layer silicates
  Ba, 83, 84
  Be, 84
  Cr, 24, 25T, 26, 81–2, 85, 87T, 104

Unusual elements in layer silicates, *continued*
  Cu, 55
  Ni, 27, 29–30, 33, 35T, 36T, 44T, 45T, 46–8, 84–5, 346, 361–2
  V, 55–6, 83
  Zn, 55, 84

Van der Waals force, 204, 205, 223, 372, 378, 382, 397
Variables, extensive, 424
Variables, intensive, 424
Vermiculite, 48–9, 56–9, 58–59T, 79, 80, 427, 445–7, 448–51
  n-alkylammonium complexes, 379–85
  charge heterogeneity in, 60–1
  dehydroxylation, 340
  heats of immersion, 262
  high-temperature transformation, 348–9
  L-ornithine complex, 376, 380, 384, 404–5
  osmotic swelling, 374, 385
  oxidation of Fe^{2+} in, 57–8
  stability of hydration states, 261
  structure of interlamellar water in, 261, 263
  vinyl derivatives, 414
  water sorption isotherm, 258, 261–2
Vernadite, 132T, 174T, 178–9
  relation to birnessite, 179

Water in clay minerals, 17, 227, 230, 233, 384–5
  acidity of, 308–9, 392–5
  cation polarization of, 245, 392–5
  diffusion of, 261, 263, 266–7
  effect of counter-cation on, 250–1, 255
  infrared absorption by, 239, 253, 262, 267
  interaction with mineral surface, 253, 263, 385
  orientation of, 239, 251, 253, 262–3, 267
  residual, 253, 264
  rotation of, 263, 266–7
  structure of interlamellar, 245, 260–3, 266–7
  "zeolitic", 245
Water, determination of, 4–5
Water sorption isotherms
  beidellite, 251
  effect of counter cation on, 256–7
  effect of particle size on, 257–8
  halloysite, 252
  illite, 254–5
  kaolinite, 250
  montmorillonite, 257, 258, 265
  muscovite, 254, 255
  reference, 248
  vermiculite, 258
Weathering reactions, 443
  in amphibole-plagioclase, 448–51

Weathering reactions, *continued*
   decreasing alkali potential, 446
   decreasing Ca potential, 450–51
   in granite, 431–2, 444–8
   in granitic acid brown soil, 444–5
   in liassic marl, 445
   loss of Mg, 451
   in metagabbro, 448–51
   in mica-feldspar, 446–8
   oxidation of Fe, 451
   in serpentine-clinopyroxene-spinel, 452–3
   in serpentinized lherzolite, 452–3
   weathering profiles, 443–4

Willemseite, 44T, 46, 79
Wonesite, 79

X-ray emission spectrometry, analysis by, 8

Yofortierite, 111

Zeolites, composition range, 432
Zero point of charge, *see* point of zero charge
Zinnwaldite, 81, 83T

Author Index

Abdo, S., 388, 394
Abdulvaliev, R.A., 170
Achar, B.N.N., 325
Adams, J.M., 263, 309
Adams, S.N., 183
Addison, C.C., 356
Addison, W.E., 356
Afanasev, G.D., 69
Ahlrichs, J.L., 386, 409
Ahmed, S.M., 189
Aidinyan, N.Kh., 69
Akitt, J.W., 162, 163
Albee, A.L., 68
Alcover, J.F., 260, 261
Aleixandre Ferrandis, V., 338
Aleksandrova, L.N., 159
Aleksandrova, V.A., 91, 109, 408
Ali, S.Z., 334, 347
Alietti, A., 94, 98, 402
Altschuler, Z.S., 105
Alysheva, E.I., 88
Amero, R.C., 287
Anderson, D.M., 251, 265
Angel, B.R., 23
Annabi-Bergaya, F., 386, 396
Annersten, H., 153, 154
Aomine, S., 115
Aoyagi, K., 437
April, R.H., 105
Aragón, F., 372, 407
Arnold, P.W., 246
Atkinson, R.J., 134, 155, 156
Aylmore, L.A.G., 157

Babcock, K.L., 185, 233, 234
Bailey, A., 254
Bailey, G.W., 373, 375, 386, 394
Bailey, S.W., 2, 12, 22, 29, 32, 33, 37, 41, 46, 49, 62, 63, 69, 73, 75, 79, 83, 84, 88, 93, 99, 100, 104, 107, 260
Bain, D.C., 104
Baird, T., 157
Baker, W.E., 185
Balistrieri, L., 140, 157
Ball, M.C., 330, 346
Ballantine, J.A., 309, 310
Bannister, F.A., 69, 71, 346
Banwart, W.L., 373
Bardossy, G., 165
Barnhisel, R.I., 165, 343

Barrer, R.M., 269, 311, 409, 411
Barrios, J., 415
Barrow, N.J., 135, 139, 183
Barshad, I., 360
Bartlett, R.J., 175
Basila, M.R., 281
Bassett, W.A., 63, 92, 360, 376
Bates, T.F., 166
Bauer, J.F., 41, 48
Bayliss, P., 32, 37, 38, 85, 105
Beane, R.E., 454
Beaufort, D., 453, 454
Beck, K.C., 437, 438
Beckwith, R.S., 157
Bell, H., 143
Belokopytov, Yu.V., 158
Belov, N., 85
Benesi, H.A., 276, 279, 280, 281, 393
Beneslavsky, S.I., 149, 163, 169
Bentor, Y.K., 74, 161, 165
Berg-Madsen, V., 71, 74
Berger, G., 413
Berkheiser, V.E., 311
Bernal, J.D., 142, 150, 151, 153, 155
Bernas, B., 9
Berry, R., 10
Berube, Y.G., 189
Berzelius, J.J., 275
Besson, G., 54
Betekhtin, A.G., 178
Bhattacherjee, S., 332
Biais, R., 168, 169
Bigham, J.M., 434
Bingham, F.T., 157, 171, 172
Bish, D.L., 29, 31, 35, 47, 85, 151
Bissada, K.K., 372, 386, 388, 396
Blackmore, A.V., 160
Blake, R.L., 46
Bloch, J.M., 287
Bloomfield, C., 141
Bloss, F.D., 75
Blount, A.M., 24
Blum, W.E., 158
Blumstein, A., 407
Blumstein, R., 407
Blyholder, G., 134
Boar, P.L., 6, 9
Bodenheimer, W., 388, 399
Boettcher, A.L., 106, 107, 108
Boland, J.N., 451

Bolland, M.D.A., 24
Bolt, G.H., 217
Bonatti, E., 436
Bond, R.D., 175
Borkowska, A., 142
Boss, B.D., 262
Bottero, J.Y., 312
Bourguignon, P., 88
Bowden, J.W., 134–40
Bower, C.A., 246
Bradfield, R., 277
Bradley, W.F., 13, 105, 109, 110, 112, 337, 339, 341, 347, 349, 352, 355, 371, 372, 385, 408
Brannock, W.W., 4
Branson, K., 247, 254, 255, 256, 268
Brauner, K., 112, 113, 408
Breeuwsma, A., 134, 135, 157, 158, 189
Brett, N.H., 322, 337
Bricker, O.P., 173, 178
Bridge, J., 167
Brigatti, M.F., 53, 104
Brindley, G.W., 12, 17, 26, 27, 29, 30, 31, 32, 34, 36, 41, 44, 45, 47, 48, 53, 55, 92, 98, 100, 104, 106, 114, 151, 212, 244, 252, 253, 311, 312, 322–37, 339, 343–53, 357, 362, 372, 373, 374, 378, 380, 396, 399, 401, 402
Bromfield, S.M., 171
Brookins, D.G., 25
Brosset, C., 162
Broughton, G., 298
Brown, G., 17, 18, 41, 44, 69, 71, 92, 93, 95, 99, 162, 244, 254, 260, 338, 359, 360
Brown, M.J., 332
Browne, D.R.L., 437
Bruggenwert, M.G.M., 230
Bruque, S., 388
Brutsch, R., 150
Brydon, J.E., 70, 260, 338
Buckley, H.A., 72, 74
Bulens, M., 350, 354
Burau, R.G., 135, 183, 184
Burns, R.G., 174, 175, 177–81, 183
Burns, V.M., 174, 175, 177–81
Burst, J.F., 437
Buseck, P.R., 174, 180
Buser, W., 178, 181
Busing, W.R., 131
Byrne, P.J.S., 59
Bystrom, A., 181
Bystrom, A.M., 181, 354, 355

Cabrera, F., 139, 189
Cady, S.S., 300
Cahoon, H.P., 56
Caillère, S., 91, 109, 110, 112, 169, 238, 341, 348

Cairns-Smith, A.G., 231
Calvert, C.S., 447
Calvet, R., 241, 322, 363, 373
Cambier, P., 241
Cant, N.W., 331
Capell, R.G., 276, 293–7
Caplar, V., 305
Carlson, L., 148, 177, 178
Carman, J.H., 62, 80
Carr, R.M., 26, 253
Carroll, B., 325
Carroz-Portillo, D., 414
Carstea, D.D., 92
Casal, B., 386, 387, 412
Cashen, G., 24
Cebula, D.J., 240
Cerny, P., 25, 89
Chakraborty, A.K., 352
Chamberlin, R.T., 356
Chang, H.M., 305–7
Chang, T.S., 334
Chapman, D.L., 214
Chappell, B.W., 8
Chassin, P., 373
Chaussidon, J., 217, 241
Chou, C.C., 292
Chourabi, B., 60
Christ, C.L., 435
Christensen, A.N., 149
Christensen, H., 149
Christiano, S.P., 314
Christoph, G.G., 167
Chu, S.Y., 235
Chukhrov, F.V., 40, 42, 55, 141, 143, 147, 148, 149, 155, 174, 177, 178, 179, 180
Church, T.M., 436
Churchman, G.J., 253
Ciapetta, F.G., 285
Cimbalnikova, A., 74, 103
Clements, R.L., 5
Clementz, D.M., 298, 390
Cliff, G., 8, 41
Cline, M.G., 157, 171
Cloos, P., 373, 380, 386, 388, 390, 392
Coats, A.W., 325
Cole, W.F., 340, 349
Coleman, N.T., 106, 287
Collepardi, M., 148
Colombera, P.M., 244
Comeforo, J.E., 349
Comer, J.J., 349, 352, 353
Commonwealth Bureau of Soils, 164, 171
Condon, F.E., 284, 285
Condrate, R.A., 386
Conley, R.F., 250
Cooke, C.J., 8

Coombe, A.D., 76
Cornell, R.M., 160
Correns, C.W., 186
Cotton, F.A., 186
Courbe, C., 441
Couture, R., 436
Cowan, C.T., 377
Cowking, A., 54
Crabtree, R.H., 307
Cradwick, P.D.G., 115, 116
Crawford, D.V., 158
Creer, M.H., 5
Cremers, A., 231, 232, 233, 389
Cross, W., 24
Cruz, M.I., 253, 264, 388, 397, 398, 406
Cruz-Cumplido, M.I., 309
Csicsery, S.M., 297
Cuttler, A.H., 23
Czamanske, G.K., 454

D'yachkova, I.B., 149
Dandy, A.J., 269, 407, 409
Dang, T.P., 305
Davidtz, J.C., 394
Davis, J.A., 139
Day, R.E., 189
De Boer, F.E., 153
de Bruyn, P.L., 189
De Endredy, A.S., 159
De Jong, J.J., 411
De Sitter, J., 154
De Souza Santos, H., 346
de Villiers, J.M., 150
De Waal, S.A., 41, 44, 46, 87
Dean, J.A., 7
Deb, B.C., 159
DeBruyn, P.L., 134, 135
Deelman, J.C., 187
Deer, W.A., 32, 40, 41, 42, 63, 149, 169, 187, 188
Defay, R., 225
Degens, E.T., 406, 407
del Pennino, U., 271
Delmon, B., 350, 354
Den Otter, M.J.A.M., 286, 310
Dent, L.S., 322
Derjaguin, B., 204
Deshpande, T.L., 160, 172
Detournay, J., 142, 150
Deuel, H., 413
Deyrup, A.J., 278
Diamond, S., 246
Dixon, J.B., 9
Doehler, R.W., 376, 379
Dolcater, D.L., 23
Doner, H.E., 382, 389

Dowdy, R.H., 386
Dresher, W.H., 335
Drits, V.A., 89–91, 105, 106, 109, 408
Dromashko, S.G., 110
Drosdoff, M., 175
Dubois, P., 173
Ducros, P., 393
Duncan, J.F., 354
Duncomb, P., 8
Dunn, P.J., 37
Dunoyer de Segonzac, G., 437
Dupont, M., 393
Durek, A.D., 454
Duvigneaud, P.H., 352
Dyal, R.S., 133, 246

Earley, J.W., 50, 51, 105, 339
Eberhart, J.P., 337, 338, 354, 355
Eberl, D., 62, 64, 231, 232
Eckhardt, F.J., 180
Edelman, C.H., 413
Egelstaff, P.A., 240, 266
Eggleton, R.A., 146, 451
Egloff, S., 284, 285
Einaudi, M., 84
Eischens, R.P., 276, 281
Eisenberg, D., 266
El Swaify, S.A., 173
El-Attar, H.A., 5
Elders, W.A., 437
Ellis, A.J., 437
Ellis, J., 155
Elprince, A.M., 233–5
Eltantawy, I.M., 246
Emerson, W.W., 173
Endo, T., 314
Eremeev, A.F., 170
Eriksson, E., 232
Ernst, W.G., 68
Ertem, G., 244, 323, 324, 396
Ervin, G., 168
Escoubes, M., 356, 357, 358
Eslinger, E.V., 437
Essene, E.J., 37
Eswaran, H., 182
Eugster, H.P., 65, 78, 354, 355
Evans, A.G., 280
Evans, R.C., 131, 133, 187
Evans, W.H., 5
Evole, N., 407

Fahey, J.J., 32, 36, 37, 113
Fancher, D., 114
Farmer, V.C., 58, 116, 238, 239, 244, 262, 267, 322, 323, 356, 358, 359, 372, 382, 386, 391, 392, 394, 395

Farmer, W.J., 386
Farthing, A., 162, 163
Farsaneh, F., 308
Fasiska, E.J., 152
Faust, G.T., 29, 32, 33, 36, 37, 55, 57, 177, 178
Favejee, S.C.L., 413
Fawcett, J.J., 41, 46
Feitknecht, W., 141, 150, 151, 153, 173, 178
Feldman, J., 285
Felkin, H., 307
Fenn, 390, 392
Fenoll, P., 372
Fernández Alvárez, T., 269, 409, 411
Fernandez-Hernandez, M.N., 411, 412
Ferris, A.P., 24, 209
Fiorini, M., 307
Fischer, E.M., 286, 310
Fischer, W.R., 141, 147, 148
Fitzpatrick, R.W., 153, 187, 188
Flegmann, A.W., 24
Fleischer, M., 38, 174, 177, 178, 181
Fleming, W., 425
Floran, R.J., 44, 46
Folk, R.L., 59
Follett, E.A.C., 13, 109
Fontanaud, A., 451
Forbes, E.A., 158
Forbes, W.C., 40, 46
Fordham, A.W., 141, 171, 189
Forman, S.A., 14, 75
Fornés, V., 261, 380, 385, 404
Foster, M.D., 48, 50, 51, 52, 53, 58, 63, 68, 72, 75–8, 81–3, 85–7, 360
Fowkes, F.M., 393
Francis, C.W., 10
Frank-Kamenetskii, V.A., 83
Fransolet, A.-M., 88
Freeman, E.S., 325
Frenkel, M., 393
Freund, F., 331, 332
Frey, A., 71
Fripiat, J.J., 5, 57, 60, 109, 160, 240, 241, 251, 263, 267, 277, 290, 295, 297, 309, 331, 343, 372, 373, 393, 395, 412, 413, 414
Frondel, C., 34, 37, 84, 180
Fuchs, M., 205
Fuerstenau, D.M., 189
Furukawa, T., 378, 380

Gaines, G.L., 229, 338
Gáliková, L., 339
Gallagher, K.J., 327
Gallaher, R.N., 177
Gamaleya, Yu.N., 82
Gard, J.A., 13, 109
Gardner, L.R., 162

Garrels, R.M., 435
Garrett, W.G., 25, 26, 252, 378, 379, 385
Gast, R.G., 134, 135, 137
Gastuche, M.C., 142, 149, 160, 163, 165, 332, 333
Gatineau, L., 261
Gattow, G., 175
Gaudette, H.E., 67, 70, 72, 338
Gayer, F.H., 276, 285
Gees, R., 425
Gérard, P., 361, 362
Germain, J.E., 284, 285, 289
Ghesquiere, C., 362
Ghosh, D.K., 352
Gibbon, D.L., 333, 337
Giese, R.F., 61, 263
Gieseking, J.E., 371, 384, 413
Gilbert, H., 149
Gilkes, R.J., 356, 359
Gillery, F.H., 39, 98
Giovanoli, G., 150
Giovanoli, R., 133, 174, 176, 177, 178, 179
Gissel-Nielsen, G., 157
Glaeser, R., 257, 264, 322, 323, 372, 386
Glass, H.D., 354
Goertzen, J.O., 246
Gomes, C.S.F., 402, 403
Gonzáles-García, S., 322, 386, 397, 401
Gonzalez-Carreño, T., 380, 383
Goodman, B.A., 4
Goodyear, J., 26, 253
Gotz, J., 116
Gout, R., 168, 169
Gouy, G., 214
Graham, J., 189, 238, 262
Grahame, D.C., 215
Granquist, W.T., 276, 282, 290, 293–7
Greene-Kelly, R., 243, 244, 246, 251, 266, 322, 371
Greenfield, S., 7
Greenland, D.J., 157, 160, 161, 247, 373, 377–9, 392
Greensfelder, B.S., 276
Gregg, S.J., 245, 332
Grim, R.E., 49, 292, 296, 322, 337, 338, 339, 341, 347, 348, 349, 352, 354, 355, 371, 384
Grimme, H., 135, 158
Groves, A.W., 5
Gruner, J.W., 33, 35, 44, 46
Guggenheim, S., 27, 32, 35, 37, 38, 39, 41, 43, 46
Guidotti, C.V., 69
Guilbert, J.M., 453, 454
Guinard, J., 386
Gupta, G.C., 92, 362
Gur, Y., 232

Gurwitsch, L., 275, 285
Gutiérrez-Ríos, E., 372, 386
Guven, N., 55, 68

Hackman, J.R., 378
Hafner, S.S., 153, 154
Hagymassy, J., 247, 265
Hair, M.L., 412
Hall, A., 77
Hall, P.L., 23, 240, 241, 263, 267, 297
Hall, W.K., 281, 282, 331
Halma, G., 10
Hamaker, H.C., 205
Hamdy, A.A., 157
Hammett, L.P., 278
Handreck, K.A., 171
Hannaker, P., 7, 9
Hansford, R.C., 281, 289
Harder, E.C., 167
Harder, H., 425, 453
Hardy, W.B., 204
Hariya, Y., 174, 178
Hartl (+Weiss), 331
Hartman, P., 61
Harvey, G.R., 406
Harward, M.E., 171
Hatcher, J.T., 172
Hathaway, J.C., 166
Hawkins, R.H., 240, 266
Hay, R.G., 284
Hayami, R., 335, 345, 346
Hayashi, H., 8, 105, 106
Hayden, P.L., 162, 163
Hayes, M.H.B., 381
Haynes, J.M., 247
Hazen, R.M., 81
Heald, W.R., 233
Healy, T.W., 183, 213
Hecht, A.M., 393
Heckroodt, R.O., 24
Heilman, M.D., 244, 246
Heinrich, E.W., 68, 82, 83
Heintze, S.G., 185
Helgeson, H.C., 447
Heling, D., 437
Heller, L., 322, 337, 341, 386, 388
Heller-Kallai, L., 339, 340, 388
Hem, J.D., 165, 175, 176
Hendricks, S.B., 31, 35, 50, 51, 52, 74, 133, 238, 246, 255, 264, 371, 384
Hénin, S., 109, 110, 112, 341, 348
Henley, K.J., 69
Henmi, T., 114–16
Herbillon, A.J., 23, 163, 359
Hermosin, M.C., 412
Herrera, R., 92

Hess, H.H., 37
Hesselink, F.Th., 213
Hewett, D.F., 174
Hewitt, D.A., 62, 80
Hey, M.H., 4, 346
Heystek, H., 52, 56
Hildebrand, E.S., 158
Hill, R.D., 334
Hill, R.J., 168
Hillebrand, W.F., 3
Hingston, F.J., 135–8, 157, 171, 172
Hirschler, A.E., 281
Hoda, S.N., 444
Hodgson, A.A., 356
Hoenesu, O., 436
Hofmann, D.A., 167
Hofmann, U., 244, 322
Hogfeldt, E., 226
Hogg, C.S., 358
Holdridge, D.A., 332, 334
Holmes, A., 76
Holt, J.B., 338
Holzner, J., 359
Homshaw, L.G., 241
Hood, W.C., 444
Hornig, C.A., 143
Horváth, I., 339
Hossner, L.R., 233, 234
Houdry, E., 276
Hougardy, J., 263
Hower, J., 65, 66, 70, 71, 72, 74, 75, 92, 93, 99, 100, 101, 102, 103, 243, 437, 438, 440, 441, 442
Hower, W.F., 55
Howie, R.A., 7
Hsu, Pa Ho, 135, 161, 163, 164, 165, 166, 167, 171, 172
Huang, P.M., 163
Huang, S-D., 244
Huang, W.G., 6
Huang, W.H., 162
Hughes, I.R., 41, 43, 252
Hughes, J.C., 9
Hughes, T.R., 227, 281
Hunt, J.P., 392, 393
Hunter, K., 332, 349
Hunter, R.J., 223
Hutton, C.O., 76
Hutton, J.T., 186–8
Hyslop, J.F., 350

Ianicelli, J., 10
Idzikowski, S., 158
Iijima, A., 437
Ildefonse, P., 444, 448
Iler, R.K., 412

Imai, N., 111, 112
Ingamells, C.O., 5, 9
Ingram, B.L., 6
Ingram, L.K., 6, 9
Inooka, M., 407
Inouye, K., 142
Ioka, M., 407
Ismail, F.T., 360
Ito, J., 84
IUPAC, 12
Iwai, S., 333
Iwao, S., 41, 42, 43

Jackson, M.L., 9, 158, 159, 166, 169, 186, 448
Jackson, T.A., 407
Jacobs, D.C., 454
Jacobs, H., 398
Jacobs, L.W., 171
Jaffe, J., 297
Jahanbagloo, I.C., 29, 33, 40
James, D.P., 105
Janekovic, A., 173
Janot, C., 149
Jefferson, M.E., 238
Jeffery, P.G., 3, 4, 5
Jenkins, H.D.B., 61
Jennings, B.R., 411
Jepson, K., 170
Jepson, W.B., 8, 23, 24, 209
Johns, W.D., 6, 60, 333, 372, 377, 379, 384, 386, 396
Johnson, H.B., 327, 329, 330
Jonas, J., 251
Jonas, K., 143
Jones, A.A., 242
Jones, L.H.P., 171, 177
Jordan, J.W., 212, 292, 372, 384
Jorgensen, P., 10
Juo, A.S.R., 360
Jurinak, J.J., 134, 157, 248, 250, 251

Kacsalova, L., 165
Kagan, H.B., 305
Kalbasi, M., 170
Kamphorst, A., 230
Kantner, I., 377
Kao, Chuh-Chin, 343, 344
Karapetyan, E.T., 160
Karchoud, M.M., 356, 357, 358
Karpova, G.V., 71, 73
Kastner, M., 74
Katchalsky, A., 407
Kato, C., 414
Kato, T., 347, 349
Kauffman, A.J. Jr, 112
Kauzmann, W., 266
Kavanagh, B.V., 172

Kay, S.M., 254
Kayser, F., 287
Kazama, T., 437
Keenan, A.G., 250, 251
Keller, G., 141, 150, 151
Keller, W.D., 24, 74, 162, 168, 332
Kelley, W.P., 75
Kemper, W.D., 160, 232, 233
Kennedy, G.C., 168
Kennedy, J.V., 282, 295
Keren, R., 264, 265, 266
Kessler, F., 327, 329, 330
Keusen, H.R., 62, 79
Khorosheva, D.P., 165, 168, 169
Khoury, H.N., 55
Kidder, G., 244, 260
Kiel, A.M., 361
Kienast, J.R., 69
Kijne, J.W., 264
Kikuchi, T., 178
Kimura, S., 350
Kinniburgh, D.G., 158, 170, 171
Kinter, E.B., 246
Kinzer, A.D., 297
Kirkman, J.H., 26, 253
Kishk, F., 360
Kiskyras, D., 169
Kittrell, J.R., 297
Kittrick, J.A., 162, 164, 167
Kiyama, M., 187
Klemen, R., 322
Klingsberg, C., 81
Klotz, I.M., 227
Knight, R.J., 141, 143
Knudson, M.I., 246, 311
Kodama, H., 70, 93–7, 99, 102, 148, 161, 173, 185, 338, 388
Koelmans, H., 212
Koltermann, M., 346, 347
Komusinski, J., 23
Koons, R.D., 142
Korolev, Ya.M., 94, 100
Korzhinskii, D.S., 424
Kramm, U., 87–9
Krause, A., 142, 147
Krauskopf, K., 175, 425
Kren, E., 141
Krenske, D., 388, 389
Krishna Murti, G.S.R., 157
Kubler, B., 408
Kulbicki, G., 49, 339, 341, 348, 355
Kuroda, K., 414
Kuznetsov, V.A., 158
Kwong, Ng Kee, 163

Laby, R.H., 379

Lafont, R., 149
Lagaly, G., 60, 61, 97, 247, 313, 383
Lahann, R.W., 437
Lahav, N., 92, 312
Lailach, G.E., 380, 388
La Mer, V.K., 213
Landau, L.D., 204
Langmuir, I., 220
Langmyhr, F.J., 9
Langston, R.B., 21
Lapham, D.M., 85, 87
Larsen, E.S., 25
Larson, L.T., 180
Laudelout, H., 227, 229–34
Laura, R.D., 380, 388
Lavrenchuk, V.N., 170
Lazarenko, E.K., 89, 90, 94, 100
Leckie, J.O., 139
Ledoux, R.L., 372
LeDred, R., 360
Lee, G.F., 142
Lee, J.H., 41, 43
Lee, S.Y., 23
Legzdins, P., 298
Lemaitre, J., 350, 361, 362
Lentz, C.W., 413
Léonard, A.J., 350
Le Roux, J., 153, 188
Levinson, A.A., 83
Levy, H.A., 131
Lewis, D.G., 143, 147
Lewis, J.F., 154
Lewis, L.L., 5
Lim, C.H., 25
Lindqvist, B., 336
Lingane, J.J., 6
Lippens, B.C., 167
Little, L.H., 412
Lloyd, M.K., 250
Loeppert Jr, R.H., 311
Loganathan, P., 135, 183, 184
Long, J.V.P., 8
Lonsdale, P.F., 436
Lorimer, G.W., 8
Loughnan, F.C., 186, 187
Loveland, P.J., 441
Low, P.F., 238
Lowell, J.D., 453, 454
Lubner, K.E., 147
Lussier, R.J., 312
Lyklema, J., 135, 157

MacEwan, D.M.C., 61, 104, 238, 245, 253, 257, 294, 371, 372, 373, 385, 396
Mackay, A.L., 155
Mackenzie, F.T., 425

Mackenzie, K.J.D., 354
Mackenzie, N., 409
Mackenzie, R.C., 56, 238, 269, 322
Mackintosh, E.E., 375, 376, 377
Mackor, E.L., 212
MacLeod, D.N., 311
Maeda, T., 114
Maes, A., 232, 233, 389
Mahanty, J., 205
Majumbar, A.J., 47
Maksimov, D., 189
Maksimovic, Z., 25, 26, 29, 31, 35, 48, 55, 104
Makumbi, L., 359
Malati, M.A., 189
Malden, P.J., 23
Malik, W.U., 92, 362
Malone, P.G., 147
Mamy, J., 241
Manara, G., 407
Mann, P.J.G., 185
Mapes, J.E., 276, 281
Marabini, A., 189
Maroney, D.M., 355
Marshall, C.E., 277
Marti, W., 178
Martin, C.J., 346
Martin, G.A., 361
Martin, H., 233
Martin, R.T., 238, 246, 250, 251
Martin-Ramos, J.D., 82
Martin-Rubí, J.A., 372, 379, 384, 396
Martin-Vivaldi, J.L., 369, 372, 407
Masson, C.R., 116
Mata, A., 401
Mathieson, A.McL., 260, 263
Matkjevic, E., 162
Mattigod, S.V., 226, 234
Mauve, L., 332
Mavrodineanu, R., 7
Maxwell, J.A., 3
McAtee, J.L. Jr, 59, 246, 292, 311, 378
McAuliffe, C., 287
McBride, M.B., 232, 298, 381
McCafferty, E., 134
McDowell, S.D., 437
McHardy, W.J., 141, 150
McKeague, J.A., 157, 159, 160, 171, 174
McKenzie, R.M., 135, 139, 142, 143, 150, 151, 155, 173, 174, 176, 178, 180–5
McKie, D., 40, 44, 46
McKinstry, H.A., 333
McLaren, R.G., 158
McLaughlin, 186
Meads, R.E., 23, 358
Medlin, J.H., 6, 9
Mehra, O.P., 159

Mejsner, J., 94, 98
Mendelovici, E., 413, 414
Mering, J., 257, 264, 266, 322, 323
Merrill, R.E., 305
Mestdagh, M.M., 23, 330
Meunier, A., 443, 444, 451
Meyer, C., 454
Michel, A., 141, 153
Mifsud, A., 269, 270, 341, 375, 376, 384
Mikhail, R.S., 254, 255, 256
Millhollen, G.L., 327
Milligan, W.D., 187
Milliken, T.H., 276, 289, 291, 292
Millot, G., 433–4, 448
Mills, G.A., 289
Milne, A., 177, 259, 260
Misawa, T., 153
Mitra, G.B., 332
Mitsui, K., 437
Miyazi, S., 253
Mizota, C., 25, 26
Moeller, T., 277
Moldan, B., 7
Montaland, L., 275
Moolik, R.C., 454
Mooney, R.W., 247, 255–7, 260, 264, 265
Moore, W.J., 454
Mopper, K., 407
Morgan, J.J., 175, 183
Mortier, W.J., 313
Mortland, M.M., 281, 308, 311, 372, 373, 380, 381, 382, 386, 389, 390, 391, 392, 393, 394, 395
Mowatt, T.C., 66, 70, 72, 101, 102, 243
Mozzei, M., 305–7
Muffler, L.J.P., 437
Muir, J., 241, 251
Muljadi, D., 171
Muller, G., 90
Mullins, C.E., 153
Munoz, J.L., 81
Muñoz Taboadela, M., 338
Muradi, Enver, 152
Murata, K.J., 55
Murphy, P.J., 141
Murphy, W.J., 335
Murray, D.J., 184
Murray, J.W., 135, 140, 157, 183, 184
Murray, P., 325
Music, S., 158
Mustoe, G.E., 175

Nadiye-Tabbiruka, M.S., 407
Nagarajah, S., 157
Nagasawa, K., 92, 98, 244, 253, 254
Nagata, H., 109, 113, 269, 341, 342

Nagy, B., 112
Nahon, D., 441
Nakahira, M., 325, 330, 331, 332, 333, 347, 349–53
Nakai, M., 143
Nalovic, L., 142, 143, 149
Nardozzi, M.J., 5
Naumann, A.W., 335
Nederbragt, G.W., 411
Nelson, B.W., 39
Nelson, S.M., 241, 251
Nemecz, E., 99
Neumann, B.S., 57, 267
Newman, A.C.D., 5, 17, 18, 63, 75, 76, 92, 162, 241, 244, 247, 254, 255, 256, 260, 264, 265, 268, 359, 360
Nickiforoff, C.C., 175
Nicol, A.W., 337, 354, 355
Ninham, B.W., 205
Nishiyama, T., 94T, 95T, 99, 100
Norrish, K., 8, 57, 58, 59, 63, 71, 73, 89, 106, 109, 133, 141, 143–5, 156, 160, 169, 171, 184, 189, 237, 244, 258–62, 335, 338, 376

Oades, J.M., 141, 153, 160, 161, 173
Oblad, A.G., 288
Odin, G.S., 425
Oelkrug, D., 189
Okada, K., 114
Olejnik, S., 392, 399, 401
Olivier, J.P., 250
Olsen, E.J., 33
Orcel, J., 359
Orchiston, H.D., 254, 255, 264
Ormerod, E.C., 247, 264, 265
Osborn, E.F., 168
Osborn, J.A., 298–301, 303
Osthaus, B., 290
Osugi, S., 287
Otsuka, R., 112
Ottow, J.C.G., 141, 147
Ovcharenko, F.D., 386
Overbeek, J.Th.G., 204, 212, 220

Paecht-Horowitz, M., 407
Page, N.J., 40
Pampuch, R., 330, 333, 350
Papike, J.J., 44, 46
Paquet, H., 434, 435
Paradis, S., 429
Parfitt, G.D., 189
Parfitt, R.L., 155–7, 171, 172, 386, 392
Parissis, C.M., 9
Parks, G.A., 134, 135, 140
Parron, C., 441
Parry, E.P., 282

Parry, W.T., 454
Pashley, R.M., 254
Pask, J.A., 21
Pattiaratchi, D.B., 84
Pauling, L., 134
Paus, P.E., 9
Paver, H., 277
Pawluk, S., 160
Peacor, D.R., 29, 31, 33, 37
Peck, L.C., 5
Pédro, G., 170, 360, 448
Peech, M., 92
Peigneur, P., 313, 389
Percival, H.J., 350
Perinet, G., 149
Perrault, G., 111
Perrott, K.W., 244
Perrotta, A.J., 352
Perry, E.A., 457, 440
Perseil, E.A., 178
Peters, Tj, 62, 79
Petrova, L.P., 160
Pevear, D.R., 94, 99, 100
Pezerat, H., 266, 386, 407
Pfirrmann, G., 383
Pham, Ti Hang, 30, 47, 212, 346, 347
Philen, D.D., 381
Phillips, T.L., 85, 87
Pickering, J.G., 71, 73, 109
Pinnavaia, T.J., 297–304, 308, 389, 390
Plançon, A., 23, 402
Platonov, B.E., 183
Pleysier, J., 389
Pollack, S.S., 276, 282, 293–7
Pollard, L.D., 26, 114, 270
Poppi, L., 53, 104
Porter, A.R.D., 333
Potter, R.M., 174, 178, 179, 180
Pouillard, E., 141
Preisinger, A., 112, 113, 114, 341, 348, 408
Prider, R.T., 76
Prigogine, I., 225
Prost, R., 238, 239, 266, 322, 363
Proust, D., 450

Quakernaat, J., 57
Quayle, W.H., 303
Quirk, J.P., 241, 247, 377–9

Radczewski, O.E., 332
Radjaipour, M., 189
Radoslovich, E.W., 69
Raghu Moran, N.G., 182
Rains, T.C., 7
Raman, K.V., 308
Raman, R.V., 382, 392, 393, 395

Ramirez-Munoz, J., 7
Range, K.J., 399, 401, 402
Rankama, K., 186, 188
Rasch, H., 346
Raupach, M., 171
Rausell-Colom, J.A., 75, 77, 259, 261, 380, 385, 404
Rautureau, M., 113, 341, 408
Ravina, I., 232
Rawson, R.A.G., 246
Rayner, J.H., 41, 44, 102, 240, 243, 244
Raythatha, R., 299, 303, 304
Razumeenko, M.V., 188
Redfern, J.P., 325
Reed, L.W., 244, 260
Reeve, R., 157
Reisenauer, H.M., 135, 157, 171
Renaud, P., 359
Rengasamy, P., 23, 173
Resing, H.A., 298
Reyes, E.D., 157
Reynolds, R.C., 93, 99, 100, 101
Rhoades, J.D., 106
Rice, F.E., 287
Rich, C.I., 90, 92, 165, 244, 343, 376
Richardson, E.A., 134
Richardson, H.M., 354
Rideal, E.K., 275
Ridler, P.S., 411
Rieder, R., 81–3
Rieke, R., 332
Riekel, C., 263
Rimsaite, J., 357, 359, 451
Rinelli, G., 189
Robbins, C.W., 231
Roberson, C.E., 161, 163–7
Robert, M., 360
Robertson, R.H.S., 24, 269, 407
Robertson, S.D., 361
Rochester, C.H., 158, 159
Roderick, G.L., 264
Rodríguez, A., 372, 386
Rodriguez-Gallego, M., 82
Rogers, L.E.R., 111–13
Rojo, J.M., 414
Roloff, G., 407
Romo, L.A., 360
Rooksby, H.P., 167, 350
Rosell, R.A., 185
Rosenberg, P.E., 40, 41
Rosenfeld, J.L., 68
Ross, C.S., 50, 51, 52, 55, 56, 57, 74
Ross, D.K., 240, 297
Ross, D.S., 175
Ross, G.J., 163, 166, 290, 356, 359
Ross, R.A., 335

Ross Jr, S.J., 173, 174, 177–9, 185
Rosser, H., 156, 189
Rosser, M.J., 277, 286
Rossman, G.R., 174, 178, 179, 180
Rothbauer, R., 240
Rouaix, S., 110
Rousseaux, J.M., 75
Rouxhet, P.G., 330, 331, 338, 353, 358
Rowse, J.B., 8, 24
Roy, D.M., 81
Roy, R., 39, 81, 327, 332, 333, 334, 341, 349, 354, 360
Rozenson, I., 339, 340
Rozhdestvenskaya, I.V., 83
Ruben, A.J., 162, 163
Rubeska, I., 7
Ruehrwein, R.A., 213
Ruiz-Hitzky, E., 386, 412, 414, 415
Rupert, J.P., 390
Russell, J.D., 104, 147, 156, 157, 239, 262, 267, 322, 323, 388, 392, 394, 395
Ryden, J.C., 156
Ryland, L.B., 276, 277, 283, 289

Sabatier, G., 338
Sahama, Th.G., 186, 188
Saini, G.R., 172
Sakharov, B.A., 105, 106
Sales, R.H., 454
Salomon, M., 260
Salvador, P., 385
Samson, H.R., 24, 208
Sánchez, A., 392, 395
Sánchez-Camazano, M., 397, 401
Sato, T., 91
Savin, S.M., 437
Sawhney, B.L., 92, 376
Schadel, J., 332
Schafer, H.N.S., 4
Schahabi, S., 160
Schaller, W.T., 64, 67, 84
Scheffer, F., 147
Schellmann, W., 141, 142, 143, 147, 148, 170, 180
Schlanger, S.O., 166
Schmidt, E.R., 24, 56
Schmidt, G.W., 437
Schnitzer, M., 148, 161, 173, 185, 388
Schoen, R., 161, 163–7, 437, 453
Schofield, R.K., 24, 208, 217
Scholten, J.J.F., 361
Schoonheydt, R.A., 388
Schreyer, W., 41, 45, 47, 79, 94, 98
Schrock, R.R., 298–301, 303
Schultz, L.G., 49, 52, 53, 105, 106, 322
Schulze, D.G., 9, 143

Schünemann, H., 336
Schweisfurth, R., 175
Schwertmann, W., 10, 141–60, 163, 164, 173
Sclar, C.B., 41, 48
Scott, A.D., 376
Scott, J.W., 305
Seager, N.J., 189
Segalen, P., 141, 159, 161, 164, 165, 167
Selwood, P.W., 153, 175
Semples, R.E., 92, 311, 312
Sen Gupta, P.K., 372, 377, 379, 384
Sennett, P., 250
Sergeant, G.A., 5
Serna, C.J., 111, 269, 270, 408, 409, 410, 411
Serratosa, J.M., 372, 379, 382, 386, 391, 407
Sestak, J., 325
Shabtai, J., 92, 311, 312
Shainberg, I., 232, 233, 264, 265, 266
Shapiro, L., 4
Sharp, J.H., 325, 326, 356
Shcheka, S.A., 154
Sherman, G.D., 182, 186, 187
Shimamune, T., 333
Shimoda, S., 55, 94T, 95T, 99, 100
Shimoyama, A., 106, 107, 108
Shirozu, H., 86, 88, 90, 260
Shoval, S., 253
Sidhu, P.S., 155
Sijaric, G., 169
Sims, J.R., 157, 171, 172
Sing, K.S.W., 245
Skinner, S.I.M., 173
Slabaugh, W.H., 265, 266
Slavin, W., 7
Smart, R.St.C., 156, 157
Smith, D.L., 402
Smith, K., 146
Smith, V.C., 5
Smith, W.C., 87
Sokolova, T.A., 175
Solomon, D.H., 277, 286
Solomon, D.M., 393
Solymar, K., 143
Speakman, K., 47
Spear, F.S., 40, 45, 47, 79
Sposito, G., 226, 230, 234, 235, 238, 239, 265, 266
Springer, G., 114
Srodon, J., 105
Stace, H.C.T., 167
Steggerda, J.J., 167
Steiner, A., 437, 453
Steinfink, H., 75
Stejskal, E.O., 263
Stephen, I., 104, 110, 111
Sterckx, M., 398

Stern, O., 215
Sterne, E.J., 71
Stoch, L., 334
Stocker, P.T., 238
Stone, W.E.E., 240, 297, 298
Stonecipher, S.A., 436
Straczek, J.A., 180
Stubican, V., 81, 333
Stucki, J.W., 373
Stul, M.S., 313
Stumm, W., 175, 183
Sudo, T., 54, 57, 72, 91, 95, 99, 105, 106
Suhr, N.H., 9
Sullivan, R.F., 297
Sumner, G.G., 290
Sundius, N., 354, 355
Sung, Do Jang, 388
Suquet, H., 54, 56, 267, 268
Suzuki, T., 100
Svab, E., 141
Sveen, S., 9
Swanson, G.A., 9
Sweatman, T.R., 8
Swift, H.E., 361
Swindale, L.D., 41, 43
Sylva, R.N., 141, 143

Tahoun, S.A., 380, 381, 386, 394
Takada, T., 187
Takahashi, H., 112
Takematsu, N., 158
Talibudeen, O., 217
Tamele, M.W., 276, 277
Tamura, T., 10, 98
Taramasso, M., 407
Tarasevich, Yu.I., 264
Taube, H., 298
Taylor, H.F.W., 322, 330, 346, 353
Taylor, R.M., 133, 141–60, 169, 173–5, 177–183
Tazaki, K., 8, 26
Tchoubar, C., 23, 113, 402, 407, 408
Temple, A.K., 188
Tennakoon, D.T.B., 406
Tettenhorst, R., 59, 167, 322, 323
Teufer, B., 188
Tewari, P.H., 158, 171
Tha, Hla, 77
Thalmann, H., 150
Theng, B.K.G., 373, 377–9, 381
Thiel, R., 143
Thilo, E., 336
Thomas, C.L., 276, 277, 288
Thomas, G.W., 140
Thomas, H.C., 229, 231
Thomas, J.M., 309, 406

Thomas, W., 275
Thompson, G.R., 65, 66, 70, 74, 75, 92, 99, 101, 103
Thompson, M., 7
Thrierr-Sorel, A., 149
Tien, P.-L., 54
Tillmans, E., 186
Topham, S.A., 158, 159
Touillaux, R., 393
Toussaint, F., 327, 331
Townsend, W.N., 153
Trauth, N., 434
Traynor, M.F., 311, 388, 389
Trescases, J.J., 451
Tripathi, R.P., 358
Tschermak, G., 359
Tsunashima, A., 374
Tsuzuki, Y., 354
Turk, A., 285
Turner, R.C., 163, 166, 260
Turner, S., 174, 180

Udagawa, S., 41, 42, 43
Uehara, G., 140
Uyeda, N., 30, 31
Uytterhoeven, J.B., 281, 282, 295, 379

Valentine Jr, D., 305
Vallet, M., 407
Van Assche, J.B., 372, 386
van Bladel, R., 227
Van Loon, J.C., 9
Van Meerbeek, A., 414
Van Moort, J.C., 437, 440
Van Olphen, H., 57, 109, 208–12, 214, 216, 219–21, 238, 255, 259, 261, 262, 266, 267, 373, 411
Van Oosterhout, G.W., 150
van Rooyen, T.H., 150
Van Schuylenborgh, J., 140
Van Scoyoc, G.E., 109, 110, 269, 270, 341, 408, 409, 410, 411
van Silfhout, A., 205
Vanden Heuvel, R.C., 435
Vandepoel, D., 390
Van der Marel, H.W., 98
Vansant, E.F., 379
Vaughan, D.E.M., 312
Vaughan, F., 332, 334
Veblen, D.R., 8, 47, 79, 80
Vedder, W., 331, 337, 338
Velde, B., 66, 68, 69, 425, 433, 434, 436, 438, 439, 443, 444, 450
Veniale, F., 98
Verwey, E.J.W., 204, 220
Vincent, W.E.J., 23

Violante, A., 163, 165, 166
Violante, P., 163, 165, 166
Vishwanathan, V., 335
Voge, H.H., 293
Volborth, A., 3
Vold, M.J., 205
von Gehlen, K., 347, 352
von Knorring, O., 32
Von Liebig, J., 275
Von Reichenbach, Graf H., 376
Von Smoluchowski, M., 205
Voznesenskii, S.A., 170

Wackerman, J.M., 451
Wada, K., 114–16, 372, 397
Wada, S.-I., 25, 26
Wadsley, A.D., 173, 176, 177
Walker, G.F., 25, 26, 245, 252, 259, 260, 261, 263, 264, 340, 349, 373, 375, 378, 379, 385
Walker, J.L., 186
Walkley, A., 173
Wall, J.R.D., 166
Walling, C., 276, 278
Walsh, J.N., 7
Wan, H.-M., 27, 29, 30, 31, 332, 334, 336, 346
Wann, S.S., 140
Ward, D.W., 213
Ward, J.W., 281
Ward, R., 347
Wardle, R., 41, 336, 337, 339
Warkentin, B.P., 217
Washington, H.S., 3
Way, J.T., 225
Weaver, C.E., 8, 23, 26, 105, 114, 270, 437, 438
Weber, J.B., 381
Weber, J.N., 327
Weber, M.D., 160
Weed, S.B., 381
Wefers, K., 168
Weir, A.H., 50, 51, 52, 69, 93, 99, 102, 243, 244
Weiser, H.B., 187
Weiss, A., 56, 60, 61, 261, 286, 310, 311, 313, 331, 340, 348, 350, 372, 373, 376, 377, 378, 379, 381–4, 397–402, 405, 406, 407
Welty, P.K., 298
Wentworth, S.A., 338
Wey, R., 360
Wherry, E.T., 25
Whitaker, A.C., 297
White, D., 377

White, D.E., 457, 453
White, H.M., 28
White, J., 325
White, J.L., 25, 26, 165, 331, 360, 372, 373, 386, 394
White, K.L., 188
Whittaker, E.J.W., 27, 29, 34, 40
Whittam, T.V., 407
Wicks, F.J., 27, 29, 34, 40
Wiedenfeld, R.P., 233, 234
Wiegner, G., 287
Wiewiora, A., 106, 399, 401, 402
Wildman, W.E., 451
Wilke, B.M., 164
Wilkins, R.W.T., 337, 338
Wilkinson, G., 186
Willard, H.H., 5
Wilson, A.D., 4, 5
Wilson, C.R., 298
Wilson, G.W., 234
Wilson, M.A., 116
Wilson, M.J., 61, 238, 245, 253, 257, 373
Winchell, A.N., 359
Winefordner, J.D., 7
Winter, O.B., 5
Wollast, R., 109
Wones, D.R., 62, 77, 78, 80, 81
Wood, W.H., 287
Wright, A.C., 282, 290, 293–7
Wu, C.Y., 282

Yada, K., 346
Yamada, H., 350
Yamanaka, S., 55, 92, 311, 312, 362, 386, 387
Yankovskaya, A.K., 158
Yariv, S., 253, 386, 389, 391, 392
Yoder, H.S., 41, 46, 64, 354, 355
Yoshimura, T., 27, 38
Yoshinaga, N., 115, 116, 143
Youell, R.F., 17, 357
Young, W.A., 376, 379
Young Jr, J.E., 352
Yule, J.W., 9

Zapata, L., 414, 415
Zen, E.-an, 68
Zettlemoyer, A.C., 134, 239
Ziv, E.F., 187
Zoltai, T., 29, 33, 40
Zussman, J., 34, 36, 37, 344